Lecture Notes in Computer Science 815
Edited by G. Goos and J. Hartmanis

Robert Valette (Ed.)

# Application and Theory of Petri Nets 1994

15th International Conference
Zaragoza, Spain, June 20-24, 1994
Proceedings



Springer-Verlag
Berlin Heidelberg New York
London Paris Tokyo
Hong Kong Barcelona
Budapest

Volume Editor

Robert Valette
LAAS-CNRS
7 Av. du Colonel Roche, F-31077 Toulouse Cedex, France



CR Subject Classification (1991): F.1-3, C.1-2, D.4, J.6

ISBN 3-540-58152-9 Springer-Verlag Berlin Heidelberg New York
ISBN 0-387-58152-9 Springer-Verlag New York Berlin Heidelberg

CIP data applied for.


Printed in Germany

Typesetting: Camera-ready by author
SPIN: 10131293 45/3140-543210 - Printed on acid-free paper

## Preface

This volume contains the proceedings of the $15^{th}$ International Conference on Application and Theory of Petri Nets. The aim of the Petri net conferences is to create a forum for discussing progress in the application and theory of Petri nets. Typically the conferences have 150-200 participants - one third of these coming from industry while the rest are from universities and research institutions. The conference always take place in the last week of June.

The previous conferences (1980-1993) were held in Strasbourg (France), Bad Honnef (Germany), Varenna (Italy), Toulouse (France), Aarhus (Denmark), Espoo (Finland), Oxford (United Kingdom), Zaragoza (Spain), Venice (Italy), Bonn (Germany), Paris (France), Aarhus (Denmark), Chicago (USA).

The conferences and a number of other activities are coordinated by a steering committee with the following members: M. Ajmone Marsan (Italy), J. Billington (Australia), H.J. Genrich (Germany), C. Girault (France), K. Jensen (Denmark), G. de Michelis (Italy), T. Murata (USA), C.A. Petri (Germany; honorary member), W. Reisig (Germany), G. Roucairol (France), G. Rozenberg (The Netherlands; chairman), M. Silva (Spain).

This $15^{th}$ conference is organized for the second time in Zaragoza (Spain). In addition to the conference, an exhibition of Petri net tools, and several tutorial lectures (both at the introductory and at a more advanced level), are organized.

We have received 85 submissions from 18 countries of the five continents and 28 have been accepted for presentation. Invited lectures are given by E.M. Clarke (USA), F. DiCesare (USA) and U. Montanari (Italy).

The submitted papers were evaluated by a programme committee with the following members: J. Billington (Australia), S. Christensen (Denmark), J. Desel (Germany), S. Donatelli (Italy), U. Goltz (Germany), R. Gorrieri (Italy), S. Haddad (France), S. Kumagai (Japan), L. Pomello (Italy), B. Rozoy (France), R. Shapiro (USA), M. Silva (Spain), D. Simpson (Great Britain), K. Trivedi (USA), R. Valette (France; chairman), K. Voss (Germany), J. Winkowski (Poland), W. Zuberek (Canada). The programme committee meeting took place at L.A.A.S-C.N.R.S. Toulouse in France.

I should like to express my gratitude to all authors of submitted papers, to the members of the programme committee and to the 178 referees who assisted us. The names of the referees are listed in the following page. I also thank L.A.A.S.-C.N.R.S. and especially E. Dufour for technical help. I am also grateful to K. Jensen for his assistance. For the local organization of the conference in Zaragoza, all of us are thankful to M. Silva and J.M. Colom and to the support received from "el Departamento de Ingeniería Eléctrica e Informática Universidad de Zaragoza" and "la Comision Inter ministerial de Ciencia y Tecnologia de España". Finally, I should like to mention the excellent cooperation with Springer-Verlag in the preparation of these proccedings.

April 1994, Toulouse, France

Robert Valette

## List of Referees

M. Ajmone Marsan
C. Anglano
D. Aquilano
Y. Atamna
P. Azema
M. Balakrishnan
J.A. Banares
K. Barkaoui
D. Barnard
S. Basagni
E. Battiston
N.G. Bean
L. Bernardinello
G. Berthelot
B. Berthomieu
E. Best
J. Billington
A. Bobbio
P. Boldi
D. Bose
O. Botti
J. Campos
L. Capra
S. Caselli
A. Cerone
G. Chiola
A. Choquet-Geniet
P. Chretienne
S. Christensen
P. Chrzastowski-Wachtel
P. Ciancarini
F. De Cindio
H. Clausen
J.M. Colom
C. Constantinescu
G. Conte
M. Courvoisier
J.M. Couvreur
A. Coyle
V. Crespi
G. Cutts
P. Darondeau
R. Davoli
P. Degano
P. Dembinski
H. Demmou
J. Desel
M. Diaz
M. Dickson
S. Donatelli
R. Durchholz
C. Dutheillet
W. Elliott
J. Esparza
J. Ezpeleta
C. Ferigato
A. Ferscha
G. Findlow
A. Finkel
K. Forward
G. Franceschinis
F. Gadducci
R. Gaeta
H. Garavel
S. Garg
F. Garcia
A. Geniet-Choquet
H.J. Genrich
C. Girault
R. Gold
D. Gomm
R. Gorrieri
V. Grassi
I. Guessarian
T.B. Haagh
S. Haddad
N.D. Hansen
D. Hauschildt
W. Henderson
J.C. Hennet
K. Hiraishi
M. Huhn
JM. Ilié
R. Janicki
I. Jelly
K. Jensen
C. Johnen
J.B. Jørgensen
G. Juanole
M. Kaâniche
E. Kindler
S. Kumagai
C. Lakos
J.L. Lambert
L. Lavagno
H.G. Linde-Goers
C. Lyon
A. Maggiolo-Schettini
P. Maigron
J. Mann
O. Marcé
J. Martinez
A. Mazurkiewicz
G. De Michelis
R. Mirandola
D. Moldt
P. Moreaux
K.H. Mortensen
J. Murphy
P.R. Muro-Medrano
Y. Nagao
S. Natkin
K.-P. Neuendorf
R. De Nicola
P. Niebert
M. Nielsen
A. Oberweis
E. Ochmański
K. Onaga
M. Paludetto
K.R. Parker
E. Paviot-Adet
E. Pelz
A. Peron
L. Petrucci-Dauchy
M. Pezzé
J.F. Peyre
V. Pinci
H. Pingaud
G.M. Pinna
D. Poitrenaud
L. Pomello
L. Popova
L. Portinale
A. Rensink
M. Ribaudo
M. Roccetti
B. Rozoy
N. Sabadini
K. Schmidt
P. Schnoebelen
P. Sénac
M. Sereno
A. Sesterhenn
R. Shapiro
H. Shiizuka
C. Sibertin-Blanc
J. Sikakis
G. Siliprandi
M. Silva
C. Simone
D. Simpson
E. Smith
P. Starke
M. Syrjakow
H. Szczerbicka
M. Taghelit
S. Takai
A. Tarlecki
R. Taylor
E. Teruel
T. Thielke
M. Torelli
S. Trainis
K. Trivedi
T. Uchio
R. Valette
I. Vernier
S. Vigna
J.L. Villarroel
W. Vogler
K. Voss
R. Walter
T. Watanabe
H. Wehrheim
J. Winkowski
P. Ziegler
W.M. Zuberek

# Table of Contents

## Invited Papers



## Full Papers

# Automatic Verification of Finite-state Concurrent Systems

Edmund M. Clarke

School of Computer Science
Carnegie Mellon University
Pittsburgh, Pennsylvania 15213-3891, USA
e-mail emc@cs.cmu.edu

## Abstract

Logical errors in finite-state concurrent systems such as sequential circuit designs and communication protocols are an important problem for computer scientists. They can delay getting a new product on the market or cause the failure of some critical device that is already in use. My research group has developed a verification method called *temporal logic model checking* for this class of systems. In this approach specifications are expressed in a propositional temporal logic, while circuits and protocols are modeled as state–transition systems. An efficient search procedure is used to determine automatically if a specification is satisfied by some transition system. The technique has been used in the past to find subtle errors in a number of non-trivial examples.

During the last few years, the size of the state-transition systems that can be verified by model checking techniques has increased dramatically. By representing transition relations implicitly using Binary Decision Diagrams (BDDs), we have been able to check some examples that would have required $10^{20}$ states with the original algorithm. Various refinements of the BDD-based techniques have pushed the state count up to $10^{100}$. By combining model checking with various abstraction techniques, we have been able to handle even larger systems. In one example, we were able to verify a pipelined ALU with more than $10^{1300}$ states (including 64 registers of 64 bits each).

Recently, we have used model checking techniques to verify the cache coherence protocol in the IEEE Futurebus+ Standard. We found several errors that had been previously undetected. Apparently, this is the first time that formal methods have been used to find nontrivial errors in an IEEE standard. The result of the project is a concise, comprehensible, and unambiguous model of the cache coherence protocol that should be useful both to the Futurebus+ Working Group members who are responsible for the protocol and to actual designers of Futurebus+ boards. We believe this experience demonstrates that model checking techniques are already sufficiently powerful to be useful in verifying real industrial designs.

# The Application of Petri Nets to the Modeling, Analysis and Control of Intelligent Urban Traffic Networks

Frank DiCesare, Paul T. Kulp, Michael Gile, and George List

Rensselaer Polytechnic Institute, Troy NY 12180, USA

**Abstract.** This paper focuses on the development of Petri net models and tools for the control and performance analysis of signalized traffic intersections and networks of connected intersections. The potential for Petri nets as a single representation for multiple tools for addressing traffic network problems is discussed. These tools can contribute to the modeling, analysis, performance evaluation, control design and direct control code generation for urban traffic network control. A colored Petri net model of an urban traffic network for the purpose of performance evaluation is presented and discussed. The subnets for the network, the intersections, the external traffic inputs and control are each described. A six intersection urban network example is given to illustrate the subnets and a model step-through is given to show how the subnets work together. This net is used as the input to a simulation using POSES to evaluate the performance of various control strategies. Runs are made using different control strategies and statistics on intersection queue lengths and wait times are given. Conclusions and future research directions are presented.

## 1 Introduction

This paper presents work on the application of Petri nets (PNs) for the analysis and control of vehicular traffic on signalized urban arterial networks. The problem of urban traffic congestion is ubiquitous and well known. Congestion gives rise to costs which include increased time delays, air pollution, and fuel consumption. While urban expressways were considered one solution, the space and expense for such highways presently make it a prohibitive solution. Thus the current emphasis is on finding solutions in better traffic management and control schemes. This, in part, is driven by the development of low cost, high performance technologies such as embedded microprocessors, telecommunications, image processing, sensors and satellite positioning systems. The vision is an intelligent vehicular/highway system. There is a need for innovative systematic modeling, analysis, performance evaluation and control software development methods in order to take full advantage of this hardware based technology. The authors view Petri nets as basis for tools to address the needs of modern traffic systems and their control.

The focus of this paper is the development of PN models and tools for the control and performance analysis of signalized traffic intersections and networks of intersections.

## 1.1 The Traffic Network Control Problem

The nature of a traffic system is one of shared control and resource allocation. The driver controls the motion dynamics of each vehicle, the route of travel and other choices for moving about the urban area. There are thousands if not millions of drivers in each metropolitan area, each competing for system resources to achieve their travel objectives. The resources include road segments, lanes, and intersection rights-of-way. The traffic system controls and allocates these resources by the use of signal lights, signs, road stripes, barriers and protocols for driver and vehicle interaction. Control is needed for safety as well as to provide orderly traffic flow. Among other functions, control attempts to reduce conflict, prevent collisions, reduce delay and prevent gridlock. We concentrate here on the traffic signals (lights) as the principal means of control and the work is aimed at improving traffic flow through networks.

## 1.2 Current Traffic Analysis and Control Methods

Traffic engineering is a well established profession using a variety of methods [13, 15, 14] for analysis and design. Performance evaluation has traditionally been accomplished using analytical methods [2] such as queuing theory, graphical data found in handbooks [7] and simulation [3]. Because of the assumptions required, queuing theory is of limited usefulness in dealing with complex intersections and networks. Charts and graphs found in handbooks generally are derived from current or past traffic systems and may not be very helpful in testing new control strategies. The method which has been most successfully used to analyse and design new control startegies is simulation. Our experience has been that currently available traffic network simulation packages are not flexible enough for to test new sensing capabilities and resulting innovative control signalization schemes. Developing our own simulation [9] using GPSS, a comprehensive simulation language, again proved to be a massive task and involves enormous work in changing network control strategies. The search for a more flexible method lead us to Petri nets. Further, in our communication with the traffic safety group of the New York State Department of Transportation and with Traffic Control Technologies, Inc., a major U.S. producer of traffic control hardware and software, we determined that traffic control software development is also a significant problem. Many current intersection controllers are microprocessor systems with 20,000-25,000 lines of assembly language code. Any changes to this code presents a major challenge. We look to PNs to play a major role in software development and management in traffic control in addition to their role in performance evaluation.

## 2 The Potential Contribution of Petri Net Methods

Traffic and traffic control can be viewed as discrete event systems. Traffic is event driven (asynchronous) as opposed to time dependent. Traffic control may currently be time dependent but the trend is to make control responsive to flow and events. Traffic systems exhibit a high degree of concurrency, are characterized by shared resources and resource conflicts and a tendency to deadlock and overflow, and require synchronization, scheduling and control for optimal or even satisfactory performance.

The idea of applying PNs to traffic control is not new. It at least goes back to the example of a colored Petri net of traffic lights given by Jensen [8] in 1986. Cois, Fanni and Giua [1] recognized the problem of control modeling as a synchronization of mutual exclusion nets for each flow and provided a method for control using the resulting net. Further, Giua [5] demonstrated that this Petri net control with a scheduler could be easily implemented on a microcontroller. Wang et al. [20] illustrated simple intersection and network control using timed nets and performance evaluation through Petri net based simulation using Simnet [16, 17, 18, 19].

The work cited in the preceding paragraph represents a beginning of the application of PNs to the urban traffic problem. The authors believe the remaining potential for contribution is enormous. The research goal is to have a comprehensive set of Petri net based tools for modeling, analysis, performance evaluation, control design and direct control code generation. One major advantage of using nets is that the same model can be the basis for all of these activities related to traffic control.

The fact that nets are defined graphically as well as mathematically gives them an advantage over other modeling paradigms. As will be illustrated later, there are two major types of models, one to specify the control logic and another to represent the traffic flow for simulation. When the nets are merged, the control net synchronizes the simulated traffic flow. The hope is that the control net file may also be down loaded to token players in the traffic control boxes and respond to traffic and control the signals in the network. This aspect will be addressed in the future. The models for control and for traffic flow are relatively straightforward but the issue of traffic engineers using PNs as a specification tool remains open.

The analysis possible because of the mathematical definition of nets is a great advantage. While this paper does not address this directly, the possibility exists of using invariants and other methods to verify that the control logic enforces all required mutual exclusions, that capacities are not exceeded and that gridlock will not occur as a result of the logic.

The use of nets for performance evaluation using simulation is one of the foci of this paper. It would also be interesting to see how well the net based Markov chain performance approaches [10, 11, 12] could address this problem or if any insights could be gained by net based symbolic analysis using moment generating functions [6].

The automatic generation of control code using the Petri net graph as input may be the tool that will ultimately attract practitioners to this approach. With the net specified, the traffic engineer may be able to validate that it is logically correct and to obtain performance statistics. Once satisfied, he or she will be more confident of the control strategies and may appreciate an automatic translation to code to be used in the field.

This paper presents a model for the control of traffic through an urban network. The model is currently used for performance evaluation of coordinated and distributed signalization strategies.

## 3 A Petri Net Model of Network Traffic Control and Flow

This section describes an example of how a moderate size traffic network, with six signalized intersections, as shown in Figure 1, is modeled using PNs. The purpose of this model is to provide the ability to implement traffic signal control strategies and evaluate their impact on traffic flow. Since the model is intended for testing, it allows easy implementation of different signal control strategies and it also represents the behavior of the automobiles moving through the system. Further, the developed model is compact, modular, and adaptable. It permits changes in network topology as well as control over phase sequencing and timing at each intersection. This model is designed to be implemented on a PN simulator, providing the traffic engineer with simulated traffic statistics.

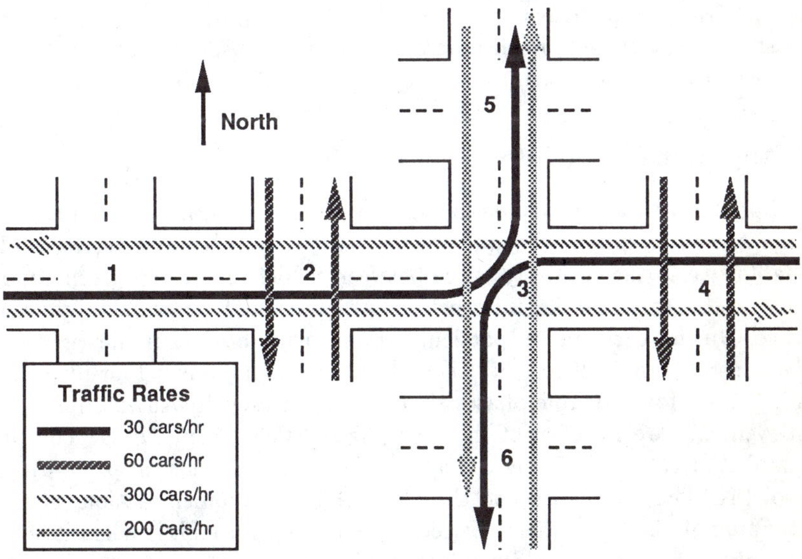


**Fig. 1.** A six intersection network showing mean input traffic flow rates.

### 3.1 Model Description

The model can be viewed as having two major interacting parts: traffic signal control and the network traffic flow. These parts are further broken down into PN modules. The network topology is represented by constructing the Connect PN module. One function of this module is to represents the links for traffic flows between intersections. The Control PN describes the operation of each of the traffic signals. This handles the sequence and logic of the control, i.e., in a given direction, the signal is green for a period of time, followed by a period of yellow, then red. This PN also guarantees properties such mutual exclusions, e.g., no two conflicting directions be given a green signal at the same time. The Approach PN models the paths and dynamics of the cars into and through the intersections. There are direct connections to the Control PN where the motion of the cars is affected by the signals at the intersection. There are also some feedback connections to provide data to the Control PN concerning the presence of cars at various locations in the system. The Sources PN represents the interface between the traffic network being modeled and the "outside world." In this net, traffic input rates to the edges of the system are defined, and where cars exit after having passed through the network.

### 3.2 Connect: The network configuration module

The Connect PN as shown in Fig. 2 allows the user of the system to model different traffic network configurations. This is done by connecting the macroplaces (double circles) representing signalized intersections using transitions and arcs as shown. Travel time delays are modeled with randomly timed (exponential in this instance) transitions. Each macroplace is a subnet, e.g., Approach Subnet 1, which models traffic flow into, through and out of an intersection.

### 3.3 Approaches: The intersection module

The Approach PN, as illustrated in Fig. 3, models the path by which cars move through the intersection, the queuing of vehicles at the approaches, and the vehicle flow time and headway (time between vehicles). At any given time, the possible paths through the intersection is determined by which signal phase is active. During each phase, particular movements are given the right-of-way by the control signals. The traffic signal in this example is a four-phase signal meaning it has four possible phases that constitute a cycle. An example of what the movements are for each of the four phases is shown in Fig. 4. The phases are enabled at transition t2 by the appropriate colored token in place p7 in the Control PN. The connections to the Control PN implement actions such as a car stopping at the light when it is red, and moving through it when it is green. The number of tokens in place p26 represent the number of queued vehicles at each approach to the intersection. Tokens in the input place p17 come from the Connect PN or a Source PN. The Approach PN also models the time it takes a car to move out of the intersection. This time can be approximated

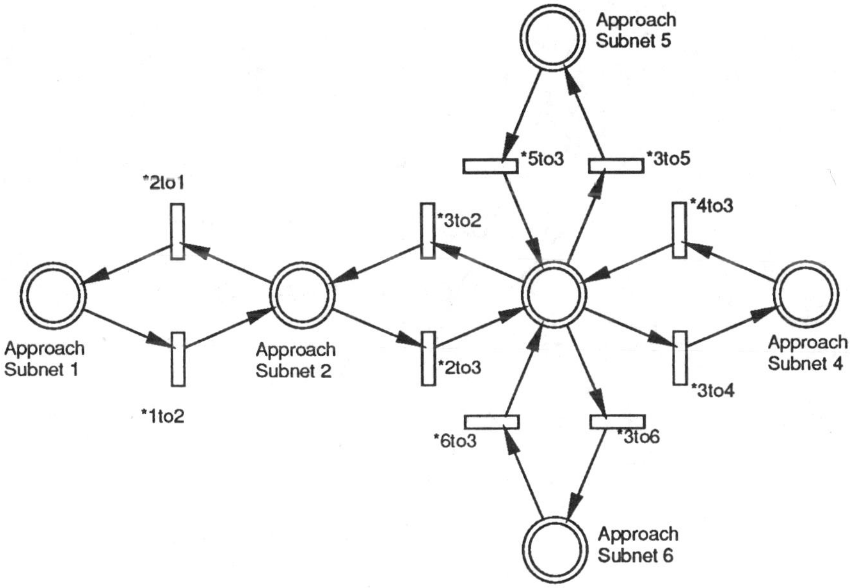


**Fig. 2.** Connect: The network configuration Petri net module.

by a certain amount of fixed time representing minimum travel time through the intersection, plus a random (exponential) time for differences between the starting and movement characteristics of the various cars. There is also feedback to the Control PN indicating when cars pass through the intersection so that the green time of a phase can be extended based on that information if the phase execution logic so requires.

The Source PN as shown in Fig. 5 is where the traffic engineer provides the rates at which cars will be input at various external points of the network. For instance, an engineer might specify that 30 cars/hour will be entering the network at the Eastbound approach of intersection no.1. The tokens which represent the cars in the system are implemented as a tuple which indicates the path of vehicle through the system.

### 3.4 The Control Net

The strategies for the control of an intersection can be divided into two basic types. The first involves strategies for the execution of a phase. In each of these phases, a particular sequence of signals is executed to facilitate this traffic movement. The sequence of signals within a phase is not the primary interest here. The PN that generates that sequence of signals can be altered to model different execution strategies, but the logic used here is representative of a typical policy in use currently. The second type involves the choice of the order in which the various phases are to be executed. This is of primary interest to minimize travel time through the system. For instance, we might want to implement a strategy

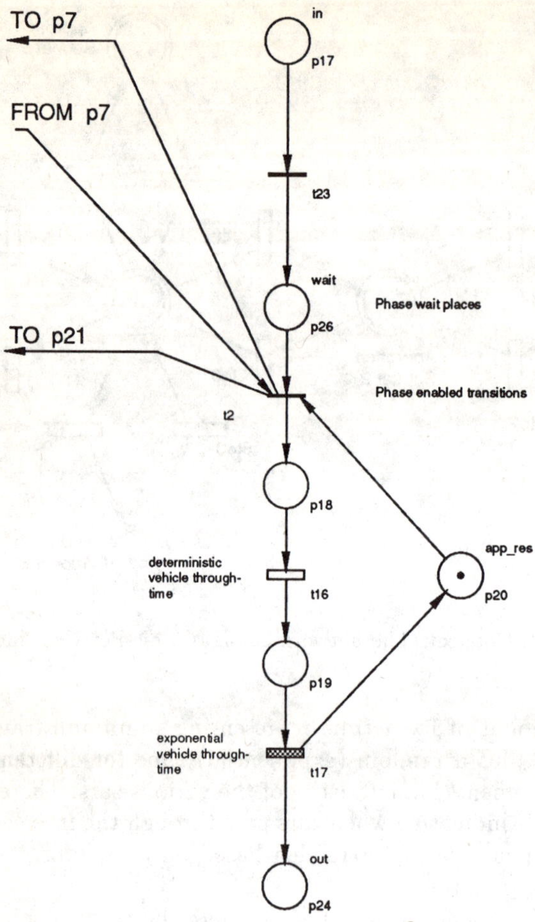


**Fig. 3.** Approach: The intersection Petri net module.

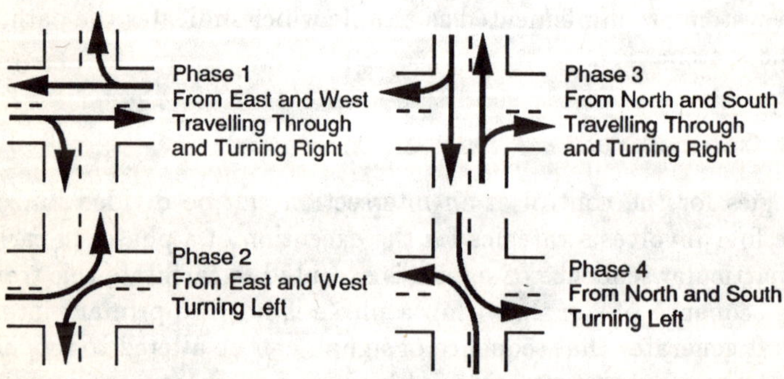


**Fig. 4.** The four phases used for all intersections in this example.

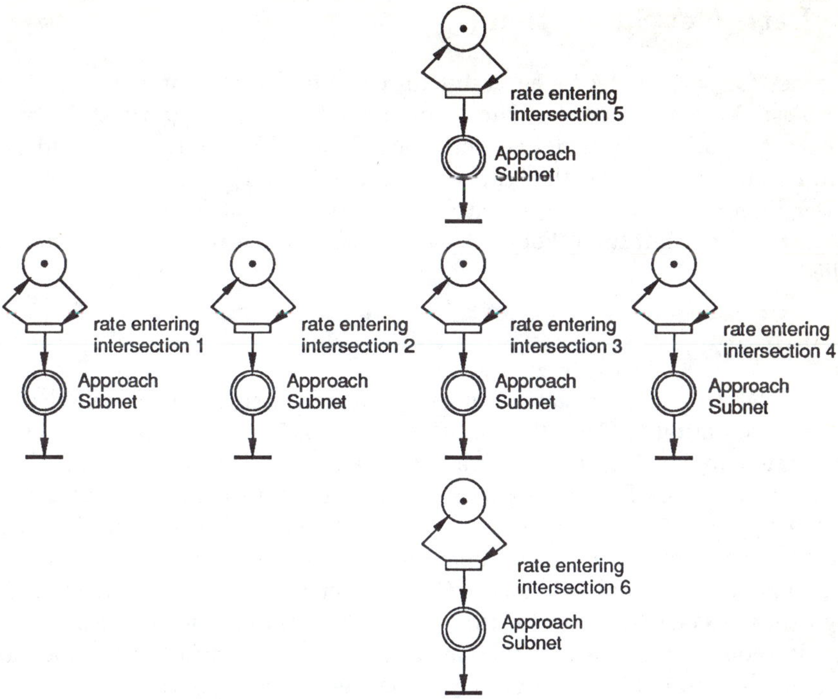


**Fig. 5.** Sources: The external traffic input and output Petri net modules.

that determines which movement has the most cars waiting and execute the phase which permits those cars to pass through the system. Other movements which have fewer cars waiting might be delayed in the interest of moving the bulk of the traffic through the system as efficiently as possible.

The Control PN allows for implementation of different strategies in both of those categories. The phase execution can be demonstrated by playing the "token game". The Control PN shown later in Fig. 6 is a colored Petri net. The color of a token that moves through the Control PN represents the phase that is currently being executed. The basic parts of a phase, green-yellow-red, etc. are common for all phases, so this logic can be contained in a single colored net. The various time parameters required for the execution of a phase are contained in the times of deterministic timed transitions. For this model, these parameters include a maximum and minimum green time, a fixed yellow time, and an increment by which the green time can be extended by the presence of cars moving through the intersection. The phase execution represented here is such that once a phase is "green" for a minimum green time, the green time can be extended an incremental amount by the passage of a car through the intersection, up to a maximum green time as specified.

# 4 Petri Net Simulation

In order for this model to be useful to a traffic engineer, it requires the use of a Petri Net simulator. The simulator is a software package which is capable of executing a Petri Net and keeping statistics on the results. Some additional useful features would be the ability to augment the functionality of the Petri Net with code to make mathematical calculations, and for advanced decision making and optimization. For the rest of this example we will use a simulator called POSES.

## 4.1 POSES

POSES [4] is a commercial Petri Net simulator written by Gesellschaft für Prozeßautomation & Consulting mbH, Chemnitz, Germany. The name POSES is an acronym which stands for Prädikat-Transitions-Netz Orientiertes Simulations und Entwurfs System (Predicate Transition Net Oriented Simulation and Design System). It is capable of modeling colored PNs with up to 250 places and 250 transitions. It can model deterministic and stochastic transitions with several possible random functions. POSES also provides excellent capabilities to augment the Petri Net model with Pascal code to implement complex functions and decision making. Pascal functions can be attached to the firing of a transition and used to monitor and modify the current marking of the net, or perhaps transition delays. POSES creates an executable simulation program which allows the user to modify various parameters in the PN and take statistics without regenerating the model for each change. There is also a debugging capability which permits the user to see exactly what is going on in the net while the simulation is running.

It is necessary to give some background on the transition firing policies used by POSES before explaining the specific function of the PN model. POSES uses a pre-selection firing policy. This means that the token that is going be fired through a transition is pre-selected at random, instead of being chosen based on random-switch probability or probability based on all of the enabled transitions. Also a timed transition in POSES can fire more that one token in parallel. For example, if a transition T (with fixed time d) is enabled by token A at time $t_1$, the token A is preselected (not available for any other transitions). At time $t_1 + d$ the token A will emerge via the output arc(s) of transition T. If another token B arrives at time $t_2$, before time $t_1 + d$ (i.e. $t_1 \leq t_2 \leq t_1 + d$), it also is preselected, and will emerge via the output arc(s) at time $t_2 + d$, unaffected by the fact that the transition is already processing token A. Note that from time $t_1$ through $t_1 + d$, transition T is processing two tokens "in parallel".

## 4.2 Model step-through

To illustrate how the model functions, lets take a look at how the control net executes a phase. The initial marking for the Control net is as show in Fig. 6. The $< 1 >$ marking in P1 represents a colored token with a value of 1, indicating that

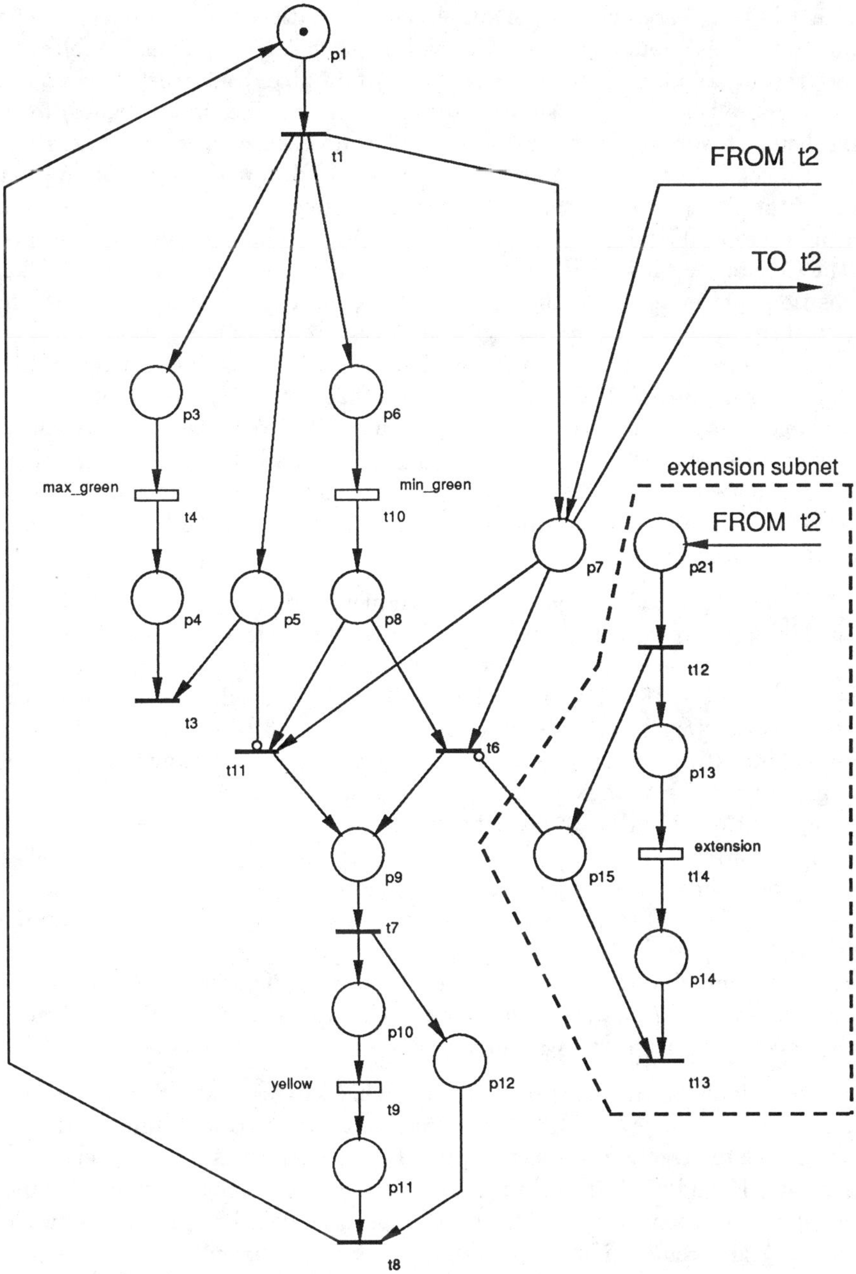


**Fig. 6.** Control.

phase 1 is the phase currently being executed. With a token in p1, immediate transition t1 fires putting $< 1 >$ tokens in p3, p5, p6, p7. This enables fixed time transitions t4 and t10 (max green and min green) and they begin to fire. Let us ignore the extension subnet (set off in dotted lines in the figure) for the time being, assuming it to be inactive. The presence of a $< 1 >$ token in p7 allows tokens which are enabled by phase No.1 to flow through the approach part of the PN (i.e. the lights for phase 1 are green). Note that whenever the token is removed from p7 by t7, it is replaced immediately, so for the purposes of the Control net, it is as if the token remains in p7. At this point no additional transitions are enabled, so the next event to occur is the end of the t10 (min green) firing (min green $\leq$ max green). T10 completes firing after min green time and places a $< 1 >$ token in p8. T7 is disabled by the token in p5. Since we have assumed the extension subnet to be inactive, there are no tokens in p15 to disable t6, so t6 fires and places $< 1 >$ in p9. The removal of the token from p7 when t6 fires marks the end of the green for phase 1. T7 fires next placing tokens in p10 and p12. T9 (yellow) then starts to fire. After the yellow time has finished, t9 places the $< 1 >$ token into p11, enabling t8. The firing of t8 marks the end of phase 1.

T8 is an augmented transition. The augmentation is some functionality contained in Pascal code which runs when t8 fires. This code determines which phase should be executed next, based on information from the net. The color token that comes out of t8 (into p1) is the result of that calculation. For instance, if phase 3 is the next phase that should be executed, t8 will place a $< 3 >$ token into p1, initiating phase 3. The purpose of the extension subnet is to extend the green time (up to max green) as long as there are still cars passing through the intersection. In order to describe the max green function, we will assume a continuous stream of cars. If there is a constant stream of cars, we will always have a token in p15, inhibiting t6. This means the phase will continue to stay green even after t10 (min green) has completed firing. This situation continues until t4 (max green) completes firing. When it does, it places $< 1 >$ in p6. t3 then fires, removing the tokens from both p5 and p6. This enables t11 to fire, since we have a token waiting in p7 and p8 since t6 was disabled. Tokens are then removed from p7 and p8 and the phase goes to yellow as before.

Now a look at the extension subnet. This subnet takes advantage of the "parallel firing" policy of POSES as described earlier. Every time a car passes through the intersection, a token is placed in p21. t12 fires immediately placing tokens in p13 and p15. The token in p13 enables t14 (extension) which begins to fire. After the extension time expires, the token is placed in p14, t13 is enabled, and the token removed from p14 and p15. As mentioned earlier, due to the "parallel firing" of t14, every token that enters this subnet will leave extension time later. p15 will contain at least one token as long as a car has passed through the intersection in the last 'extension' time. This is the criterion for the extension of the phase green time. If no cars arrive for more than 'extension' time, and 'min green' time has already passed, then the t6 will be enabled and the phase will go to yellow. If there is a constant stream of cars as mentioned earlier, the phase will remain in green until t4 (max green) fires after 'max green' time expires.

The Approach PN (Fig. 3) is considerably simpler than the Control PN. P17 is the entry place for this intersection. All cars that have to pass through this intersection come into this place. T2 is enabled only if the direction the 'token' wants to go is enabled by the phase token that is currently in p7. When t2 fires, it takes the token from p7 and replaces it, takes the token from the resource place p20 and moves the car to p18. The deterministic time transition t16 fires, followed by the exponential time t17 as described earlier approximating the time it takes a car to clear the intersection. T17 finishes firing replacing the resource in p20, and placing the car in the "out" place p24. All of the tokens that go through this intersection leave through p24. The Connect and Sources PN's are the mechanism by which the cars are placed in the "in" places of the Approach PN's and removed from the "out" places. Connect PN handles the interconnecting links between the intersections.

### 4.3 Simulation Results

**Table 1.** Simulation statistics: Average queue length at each intersection.

| | | Ave. queue length at intersection no. | | | | | |
|---|---|---|---|---|---|---|---|
| | | 1 | 2 | 3 | 4 | 5 | 6 |
| Control Strategy | Standard sequence | 7.99 | 11.81 | 14.52 | 11.08 | 5.73 | 6.00 |
| | Weighted function 1 | 4.37 | 3.39 | 7.62 | 3.33 | 1.54 | 1.56 |
| | Weighted function 2 | 2.42 | 6.65 | 11.10 | 6.50 | 3.33 | 3.44 |

**Table 2.** Simulation statistics: Average wait at each intersection.

| | | Ave. wait at intersection no. | | | | | |
|---|---|---|---|---|---|---|---|
| | | 1 | 2 | 3 | 4 | 5 | 6 |
| Control Strategy | Standard sequence | 26.53 | 33.69 | 39.54 | 32.94 | 28.44 | 29.50 |
| | Weighted function 1 | 14.34 | 16.84 | 26.69 | 16.48 | 13.30 | 13.37 |
| | Weighted function 2 | 48.30 | 58.85 | 51.54 | 55.87 | 49.47 | 51.59 |

As an example of a potential use of this system, we will examine the 6 intersection traffic network shown in Fig. 1. The traffic rates in the legend indicate rates of traffic at the external input of each of the paths indicated by the arrows. Note that this example does not include all possible flows through the network. We show the evaluation statistics for two traffic control strategies. These statistics are the average number of vehicles waiting and the average delay at each intersection. The first strategy we will consider is a standard cycle where the control at each intersection executes each phase in order, once per cycle. The second strategy determines the next phase to execute based on a weighted function of the number of cars waiting and the time since the intersection was last serviced. Both of these strategies are easily implemented in this model by slight alterations to the code augmented to transition t8 of the Control PN. This is the transition that determines which phase will be fired next. The simulation was run with both control strategies with two different weights for the second one. These runs produced the output data in Tables 1 and 2. The average number of cars waiting at each intersection was determined by examining the expected (average) number of tokens held in p26 (wait) of the Approach PN, which is easily available from POSES. Other data such as the number of firings of transitions, transition throughputs, percent utilization of places, etc are also readily obtained from a POSES simulation. The simulation shows a marked difference between the two strategies. The simulation could be used to further fine tune a strategy to produce even better results.

## 5 Conclusions and Future Research

With the current trend of application of state of the art technologies to the development of intelligent vehicle and highway systems, the need exists for for software tools which can model, analyze, evaluate performance and automatically generate control code. A case is presented for Petri nets to play a major role in the development of these software tools. This paper concentrated on the creation of a colored Petri net model to evaluate performance using simulation. The model achieves modularity through the use of subnets allowing changes to be made to one component without affecting the models for other parts. The model allows easy changes to the network configuration, the traffic control logic, timing and coordination, the model for intersection dynamics and the assumptions on network input flows. The use of the model as input to a simulation is illustrated by an six intersection network example. Results of the simulation for different control strategies are presented.

Future research will concentrate in two areas: developing innovative intelligent distributed traffic control strategies that take full advantage of advanced technologies and the development of tools that allow Petri nets be used to their fullest potential in intelligent vehicle/highway system research and implemention. These include analysis, synthesis, performance evaluation and control. The goal is for Petri nets to become a single representation on which an array of tools are based.

## 6 Acknowledgements

The authors acknowledge the sponsorship for this research by the New York State Energy Research and Development Authority under Grant Number 1936-EEED-POP-93 and the cooperation of Traffic Control Technologies, Inc. and the New

York State Department of Transportation and Gesellschaft für Prozeßautomation & Consulting mbH, Chemnitz for the use of POSES.

## References


1. Cois, A., Fanni, A. and Giua, A.: An Expert System for Designing and Supervising a Discrete Event Model, Proc. IEEE Int. Work. on Intelligent Motion Control (Istanbul, Turkey), pp. 103-107, 1990.
2. Cronje, W.B.: Analysis of existing formulas for delay, overflow and stops, Transportation Research Record 903.
3. Eiger, A, Chin, S. M.: First Generation UTCS Simulation, Transportation Research Record 906, 1983.
4. Gesellschaft für Prozeßautomation & Consulting mbH, Chemnitz: Das Programmsytem POSES, Version 4.31, Germany, 1992.
5. Giua, A.: A Traffic Light Controller Based on Petri Nets, CML Final Project Report, Department of Electrical, Computer, and Systems Engineering (Prof. F. DiCesare), Rensselaer Polytechnic Institute, Troy, New York, 1991.
6. Guo, D., DiCesare, F., Zhou, M.: An Analytical Performance Evaluation Approach for Arbitrary Stochastic Petri Nets, IEEE Trans. Automatic Control, Vol. 38, No.l 2, Feb., 1993, pp. 321-327.
7. Highway Capacity Manual, Special Report 209, Transportation Research Board, Washington, D.C., 1985.
8. Jensen, K.: Coloured Petri Nets, Petri Nets: Central Models and Their Properties, Advances in Petri Nets Part 1 Vol. 254, W. Brauer, W. Reisig and G. Rosenberg, Eds., Springer-Verlag, 1986.
9. Leong, S. M.: Modeling and Analysis of Traffic Adaptive Control Strategies, Doctoral Dissertation, Rensselaer Polytechnic Institute, Troy, N.Y., 1992.
10. Marsan,M.A., Balbo,G., Chiola, G., Conte,G., Donatelli,S. and Franceschinis, G.: An introduction to generalized stochastic Petri nets, Microelectron Reliability, **31** (4) (1991).
11. Marsan,M.A., Chiola,G.and A.Fumagalli: Improving the efficiency of the analysis of DSPN models, Advances in Petri Nets (1989).
12. Marsan,M.A., Chiola,G.: On Petri nets with deterministic and exponentially distributed firing times, Advances on Petri Nets (1987).
13. May, A.D.: Traffic Flow Fundamentals, Prentice-Hall, Inc., 1990.
14. McShane, W.R., Roess, R.P.: Traffic Engineering, Prentice-Hall, Inc., 1990.
15. Pignataro, L.J.: Traffic engineering: theory and practice, Prentice-Hall, 1973.
16. Törn, A.A.: Simulation graphs: a general tool for modeling simulation design, SIMULATION 37:6, 1981.
17. Törn, A.A.:Simulation modeling using Extended Petri Nets Graphs, in Singh, M.(ed.). Encyclopedia of Systems and Control. Pergamon Press, Oxford, England 1988.
18. Törn, A.A.:Decision support by rapid simulation using Simulation Nets, Decision Support Systems, 6, 1990.
19. Törn, A.A.:The simulation net approach to modeling and simulation, SIMULATION 57:3, 1991.
20. Wang, H., List, G. and DiCesare, F.: Modeling and Evaluation of Traffic Signal Control Using Timed Petri Nets, Proc. 1993 IEEE Int. Conf. Systems, Man and Cybernetics, Vol. 2, pp. 180-185, Le Touquet, France, 1993.

# On the Model of Computation of Place/Transition Petri Nets

*José Meseguer** *Ugo Montanari*** *Vladimiro Sassone***

*SRI International, Menlo Park, CA 94025, USA

**Dipartimento di Informatica, Università di Pisa, Italy

ABSTRACT. In the last few years, the sematics of Petri nets has been investigated in several different ways. Apart from the classical "token game", one can model the behaviour of Petri nets via non-sequential processes, via unfolding constructions, which provide formal relationships between nets and domains, and via algebraic models, which view Petri nets as essentially algebraic theories whose models are monoidal categories.

In this paper we show that these three points of view can be reconciled. More precisely, we introduce the new notion of decorated processes of Petri nets and we show that they induce on nets the same semantics as that of unfolding. In addition, we prove that the decorated processes of a net $N$ can be axiomatized as the arrows of a symmetric monoidal category which, therefore, provides the aforesaid unification.

## Introduction

*Petri nets*, introduced by C.A. Petri in [18] (see also [21]), are a widely used model of concurrency. This model is attractive from a theoretical point of view because of its simplicity and because of its intrinsically concurrent nature, and has often been used as a semantic basis on which to interpret concurrent languages (see e.g. [27, 17, 26, 5]). Concerning Petri nets themselves, several different semantics have been proposed in the literature. Most of them can be coarsely classified as *process-oriented* semantics, *unfolding* semantics, or *algebraic* semantics, though the latter is not as clearly delimited and is not as widely known as

*Supported by Office of Naval Research Contract N00014-92-C-0518, National Science Foundation Grant CCR-9224005, and by the Information Technology Promotion Agency, Japan, as a part of the R & D of Basic Technology for Future Industries "New Models for Software Architecture" sponsored by NEDO (New Energy and Industrial Technology Development Organization).

**Partially supported by the EU SCIENCE Programme, Project MASK, and by the Italian National Research Council (CNR), Progetto Finalizzato Sistemi Informatici e Calcolo Parallelo, obiettivo Lambrusco.

the former two classes. Of course, such classes are not at all incomparable, as this paper aims to support. We further discuss these approaches below.

To account for computations involving many different transitions and for the *causal connections* between the "events" which constitute them, the basic notion of computation of Petri nets has been formalized using various notions of *process* [20, 8, 2]. The main criticism raised against process models is that they do not provide a semantics for a net as a whole, but specify only the meaning of single, deterministic computations, while the accurate description of the fine interplay between concurrency and nondeterminism is one of the most valuable features of nets.

Other semantic investigations have capitalized on the *algebraic structure* of PT nets, first noticed by Reisig [21] and later exploited by Winskel [29]. The clear advantage of these approaches resides in the fact that they tend to clarify both the structure of the single PT net, so giving insights about their essential properties, and the global structure of the class of all nets. They provide, for example, useful combinators able to describe operations such as parallel and nondeterministic composition of nets [28, 29, 11, 3, 4, 13].

The formal framework which has proved superior for this kind of investigations is *category theory*. The discovery of categories, occurred in the context of algebraic topology in the early fourties, emphasized the by now well established convinction that mathematical entities are to be studied in terms of their structure, i.e, in terms of the abstract properties that they enjoy, rather than in terms of their actual elements. Indeed, the theory of categories builds on such conceptual guidelines introducing a new idea: the entities we intend to investigate can be equipped with a notion of *morphism* by means of which all their relevant structural properties can be expressed. (Of course, the actual meanings of "morphism" and "structure" depend on the specific nature of the subject one is considering.) This paradigm is clearly well suited for the study of models of computation, where the entities one considers, i.e., system or behaviour descriptions of some kind, come naturally with an associated notion of "morphism", e.g., simulations, bisimulations, or similar behavior-based relationships, which encapsulates their real essence. This is in fact also the case of Petri nets whose very structure suggests a notion of morphism which captures the intuitive idea of simulation and, therefore, the idea of behaviour itself. Then, with this understanding of the role of category theory, founding an algebraic theory of Petri nets on categories simply means considering an abstract framework in which *behaviour* is a "first class citizen". One of the first direct benefits of the use of a categorical framework is that, as a generalization of universal algebra, it provides *universal constructions* which can give fully satisfactory justifications to otherwise ad hoc defined combinators. For example, the parallel and not deterministic compositions of nets discussed above can be understood, respectively, as products and coproducts in the category of nets. An original interpretation of the algebraic structure of PT nets has been proposed in [11], where the theory of *monoidal categories* is exploited to the purpose. Unlike the preceding approaches, [11] yields an algebraic theory of Petri nets in which notions such as firing sequence, case graph, relationships between net descriptions at different levels of abstraction, duality, and invariants find adequate algebraic/categorical (universal) formulations. Alternative interesting categorical approaches are [3, 4].

In addition to that, since from the formal viewpoint categories are simply *algebraic graphs*, and in particular graphs whose arcs are closed under an operation of sequential composition, it is often the case that the computations of a single behavioural entity, say a Petri net, can be modelled themselves as a category, yielding in this way an axiomatization of its *space of computations.* One may call this use of categories "in the small", as opposed to their use "in the large" to study the global properties of the entire class of nets as illustrated above. This idea has been exploited in [6], where it is shown that the *commutative processes* [2] of a net $N$ are isomorphic to the arrows of a symmetric monoidal category $\mathcal{T}[N]$. Moreover, [6] introduced the *concatenable processes* of $N$—a slight variation of Goltz-Reisig processes [8]—and structured them as the arrows of the symmetric monoidal category $\mathcal{P}[N]$. In particular, the distributivity of tensor product and arrow composition in monoidal categories is shown to capture the basic identifications of net computations, thus providing a *model of computation* for Petri nets.

Roughly speaking the *unfolding semantics* consists, as the name indicates, in "unfolding" a net to simple denotational structures such that the identity of every event in their computations is unambiguous. However, *not* every assignment of denotations yields an appropriate semantics for nets. In other words, when defining an unfolding semantics, an integral part of the work is to provide some justification of adequacy of the obtained semantics. Exploiting the categorical framework, it is possible to achieve such a justification implicitly and more satisfactorily than appealing to mere intuition. The idea is to ensure that the denotation assigned to each net enjoys a certain universal property whose role is exactly to guarantee that, for the given target category, the assignment is, informally speaking, "as good as possible". The theory of categories provides the right notion to express this: the notion of *adjunction.* Thus, one would like to identify an *adjoint functor* assigning a denotation to each PT net and preserving certain compositional properties in the assignment. This is exactly what the present authors—building on Winskel's work on safe nets [28]—have done in [12, 13] for PT nets (see [7, 9] for related approaches).

In Winskel's work—which in turn builds on the previous work [15]—the denotation of a *safe* net is a coherent finitary prime algebraic *Scott domain* [25], or *dI-domain* [1]. Winskel shows that there exists a coreflection—a particularly nice form of adjunction—between the category $\underline{\mathsf{Dom}}$ of (coherent) *finitary prime algebraic domains* and the category $\underline{\mathsf{Safe}}$ of *safe Petri nets.* This coreflection factorizes through the chain of coreflections

$$\underline{\mathsf{Safe}} \;\overset{\mathcal{U}[\_]}{\underset{}{\rightleftarrows}}\; \underline{\mathsf{Occ}} \;\overset{\mathcal{E}[\_]}{\underset{\mathcal{N}[\_]}{\rightleftarrows}}\; \underline{\mathsf{PES}} \;\overset{\mathcal{L}[\_]}{\underset{\mathcal{P}r[\_]}{\rightleftarrows}}\; \underline{\mathsf{Dom}}$$

where PES is the category of *prime event structures* (with binary conflict relation), which is equivalent to Dom, Occ is the category of *occurrence nets* [28] and $\hookrightarrow$ is the inclusion functor. In [12, 13], such a chain has been extended to a quite general category PTNets of PT nets by defining the *unfoldings* of PT nets and relating them by means of an *adjunction* to occurrence nets and therefore—exploiting the already existing adjunctions—to prime event structures and finitary prime algebraic domains. Namely, the adjunction between Dom and PTNets is the composition of the chain of adjunctions

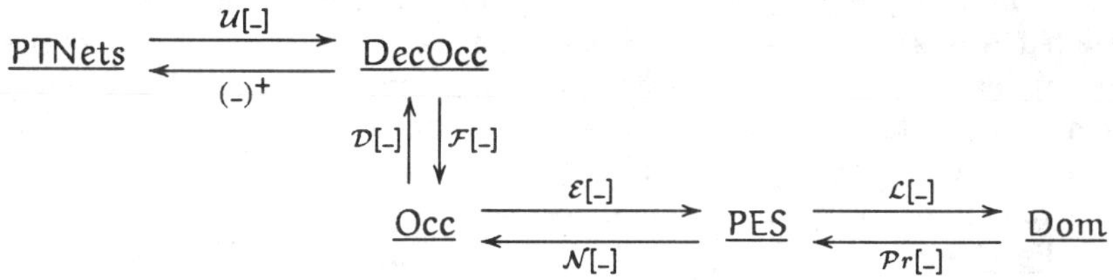


where DecOcc is the *"key"* category of *decorated occurrence nets.* These are occurrence nets in which places belonging to the post-set of the same transition are partitioned into *families.* In this way, since families are used to relate places corresponding in the unfolding to multiple instances of the same place in the original net, they naturally represent the unfoldings of PT nets and can account for the multiplicities of places in transitions.

We have already mentioned that these three views of net semantics are not mutually exclusive and, in fact, we have discussed how [6] provides a unification of the process-oriented and algebraic views via the categories $\mathcal{T}[N]$ and $\mathcal{P}[N]$ modelling, respectively, *commutative* and *concatenable processes.* Concerning the relationships between process and unfolding semantics, in the case of safe nets the question is easily answered by exploiting the existence of a coreflection of Occ into Safe, which directly implies the existence of an isomorphism between the processes of $N$ and the deterministic finite subnets of $\mathcal{U}[N]$, i.e., the finite configurations of $\mathcal{EU}[N]$. Thus, in this case, the process and unfolding semantics coincide, although it should not be forgotten that the latter has the great merit of collecting together all the processes of $N$ as a *whole*, thus accounting at the same time for concurrency and nondeterminism.

In this paper we study the relationships between the algebraic paradigm, the process semantics described above, and the unfolding semantics for PT nets given in [12, 13]. We find that, in the context of general PT nets, the latter two notions do not coincide. In particular, the unfolding of a net $N$ contains information strictly more concrete than the collection of the processes of $N$. However, we show that the difference between the two semantics can be axiomatized quite simply. In particular, we introduce a new notion of process, whose definition is suggested by the idea of families in decorated occurrence nets, and which are therefore called *decorated processes*, and we show that they capture the unfolding semantics, in the precise sense that there is a one-to-one translation between decorated processes of $N$ and finite configurations of $\mathcal{EFU}[N]$.

Then, following the approach of [6], we axiomatize the notion of decorated (concatenable) process in terms of monoidal categories. More precisely, we define an abstract symmetric monoidal category $\mathcal{DP}[N]$ and we show that its arrows represent *decorated concatenable processes.*

The natural environment for the development of a theory of net processes based on monoidal categories is, as illustrated in [6], a category Petri of *unmarked* nets, i.e., nets without initial markings, whose transitions have finite pre- and post-sets. However, since the unfolding of a net is considered with respect to an initial marking, PTNets and all the categories of nets considered in [12] (and in related works) are categories of *marked* nets whose transitions, because of technical reasons, are forced to have possibly infinite pre- and post-sets and nonempty pre-sets. In order to solve this discrepancy, we simply restrict our attention to the subcategory of PTNets, say MPetri*, consisting of the nets whose transitions have finite pre- and post-sets, i.e., the nets with nonempty pre-sets in Petri equipped with an additional initial marking. Therefore, summing up, our result is that the following diagram commutes up to isomorphism

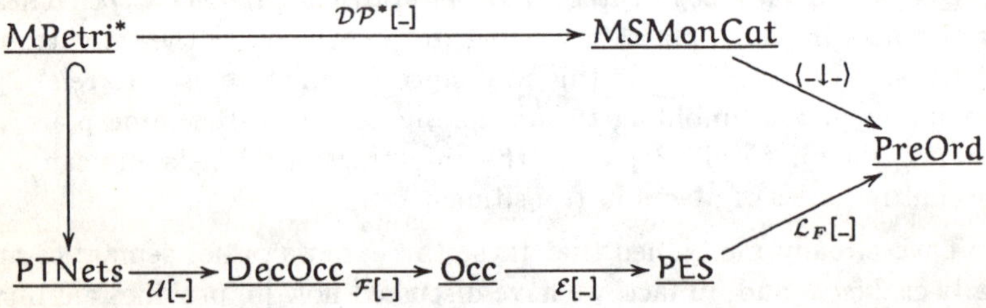


where $\hookrightarrow$ is the inclusion of MPetri* in PTNets, MSMonCat is the category of the "marked" symmetric strict monoidal categories, i.e., symmetric strict monoidal categories $\underline{C}$ with a distinguished object $c \in \underline{C}$, $\mathcal{DP}^*[\_]$ maps the marked net $(N, u_N)$ to $(u_N, \mathcal{DP}[N])$, PreOrd is the category of preorders, $\langle\_\downarrow\_\rangle$ is the comma category functor $(c, \underline{C}) \mapsto \langle c\downarrow\underline{C}\rangle$ (see Definition 3.11), and $\mathcal{L}_F$ returns the finite configurations of prime event structures ordered by inclusion. It should be stressed that our concern here is at the level of a *single* net, which means that the diagram above is defined only at the object level, i.e., the correspondence we establish is not functorial. Nevertheless, we think that this is an interesting result, since it provides a natural and unified account of the *algebraic*, the *process-oriented*, and the *denotational* views of net semantics. We remark that a similar approach has been followed in [16] in the case of elementary net systems—a particular class of safe nets without self-looping transitions—for unfoldings and non-sequential processes.

To conclude this discussion, we would like to mention that the correspondence of semantics presented here can be lifted smoothly to *infinite computations.* In [24], the present authors show that the symmetric monoidal category $\mathcal{P}[N]^\omega$ obtained as the completion of $\mathcal{P}[N]$ by colimits of $\omega$-diagrams can be understood as the category of possibly *infinite* concatenable processes of $N$. Working analogously, one can see that the arrows of the symmetric strict monoidal category $\mathcal{DP}[N]^\omega$ are possibly *infinite* decorated concatenable processes. Then, one can prove the commutativity (up to equivalence) of a diagram analogous to the one above involving all the configurations of $\mathcal{EFU}[N]$ and the comma category $\langle u_N \downarrow \mathcal{DP}[N]^\omega \rangle$. Unfortunately, due to lack of space, we shall not say more about this extension here. The details of this construction can be found in [22].

Concerning the organization of the paper, in Section 1 we recall the basic facts about the algebraic approach to Petri nets as given in [11] and [6]. Then, in Section 2 we give a brief overview of the formal development concerning the unfolding semantics introduced in [12]. In Section 3 we introduce the decorated processes and we illustrate their relationships with the unfolding semantics. Finally, we study the decorated concatenable processes of $N$ and their axiomatization as the arrows of the symmetric monoidal category $\mathcal{DP}[N]$.

Due to the extended abstract nature of this paper, we omit all proofs; the interested reader is referred to [14]. Some of the results presented here appear also in [22].

# 1 Petri Nets and their Processes

In this section we briefly recall the basic definitions about Petri nets (see [21] for a thorough introduction) and their processes.

NOTATION. Throughout the paper, $S^\oplus$ denotes the set of finite multisets on the set $S$. We recall that $S^\oplus$ is a commutative monoid, actually the free commutative monoid on $S$, under the operation of multiset union. For $\mu \in S^\oplus$ we write $[\![\mu]\!]$ to indicate the subset of $S$ consisting of those elements $s$ such that $\mu(s) > 0$. We shall represent a nonempty multiset $\mu$ as a formal sum $\bigoplus_{i\in I} n_i s_i$ where $\{s_i \mid i \in I\} = [\![\mu]\!]$ and $n_i = \mu(s_i)$. The empty multiset, i.e., the unit of the monoid, will be written as $0$. Finally, given a finite subset $S' \subseteq S$, we shall write $\bigoplus S'$ for the (multi)set $\bigoplus_{s\in S'} s$.

DEFINITION 1.1 *(Petri Nets and Marked PT Nets)*
*A Place/Transition (PT) Petri net is a structure $N = (\partial_N^0, \partial_N^1 \colon T_N \to S_N^\oplus)$, where $T_N$ is a set of transitions, $S$ is a set of places, and $\partial_N^0$ and $\partial_N^1$ are functions such that $\partial_N^0(t) \neq 0$.*

*A marked PT net is a pair $(N, u_N)$, where $N$ is a PT net and $u_N \in S_N^\oplus$ is the initial marking.*

This describes a Petri net precisely as a graph whose set of nodes is a free commutative monoid, i.e., the set of *finite multisets* on a given set of *places*. The source and target of an arc, here called a *transition*, are meant to represent, respectively, the *marking* consumed by the transition, i.e., the minimum multiset of tokens which allows the transition to fire, and the marking produced by the firing of the transition. The restriction to nets in which $\partial_N^0(t) \neq 0$ for each transition $t$ is due to the fact that such transitions are highly degenerated. In particular, the firing of any number of parallel instances of them is enabled at any marking, and this represents a serious problem for the unfolding semantics.

The formalization of nets as graphs with structure suggests considering graph morphisms which respect such structure as morphisms of nets.

Definition 1.2 *(PT Nets Morphisms)*
*A PT net morphism* $f\colon N_0 \to N_1$ *consists of a pair of functions* $\langle f_t, f_p\rangle$, *where* $f_t\colon T_{N_0} \to T_{N_1}$ *and* $f_p\colon S^{\oplus}_{N_0} \to S^{\oplus}_{N_1}$ *is a monoid homomorphism such that* $\langle f_t, f_p\rangle$ *respects source and target, i.e., it makes the two diagrams below commute.*

$$
\begin{array}{ccc}
T_{N_0} & \xrightarrow{\partial^0_{N_0}} & S^{\oplus}_{N_0} \\
{\scriptstyle f_t}\downarrow & & \downarrow{\scriptstyle f_p} \\
T_{N_1} & \xrightarrow[\partial^0_{N_1}]{} & S^{\oplus}_{N_1}
\end{array}
\qquad\qquad
\begin{array}{ccc}
T_{N_0} & \xrightarrow{\partial^1_{N_0}} & S^{\oplus}_{N_0} \\
{\scriptstyle f_t}\downarrow & & \downarrow{\scriptstyle f_p} \\
T_{N_1} & \xrightarrow[\partial^1_{N_1}]{} & S^{\oplus}_{N_1}
\end{array}
$$

*A morphism of marked PT nets from* $N_0$ *to* $N_1$ *is a PT net morphism* $f\colon N_0 \to N_1$ *which preserves the initial marking, i.e., such that* $f_p(u_{N_0}) = u_{N_1}$.

Notation. To simplify notation we shall almost always omit the subscripts $t$ and $p$ which distinguish the components of a morphism $f$. The type of the argument will identify which component we are referring to.

Processes provide a causal explanation of net behaviours, which is achieved by decorating the step sequences defined by the "token game" with explicit information about the *causal links* which ruled the firing of the transitions in the sequence. Usually one assumes that such links can be expressed faithfully as a partial order of transitions, the ordering being considered a cause/effect relationship. Thus, roughly speaking, a process of a net $N$ consists of a partial order built on a multisubset of transitions of $N$. The formalization of this gives the following notion of deterministic occurrence net.

Notation. In the following, in case of nets without multiplicities, we use the standard notation ${}^\bullet a$, for $a \in S_N$, to mean the *pre-set* of $a$, that is ${}^\bullet a = \{t \in T_N \mid a \in [\![\partial^1_N(t)]\!]\}$. Symmetrically, $a^\bullet$ indicates $\{t \in T_N \mid a \in [\![\partial^0_N(t)]\!]\}$, the *post-set* of $a$. These notations are extended in the obvious way to the case of sets of places. Recall that the terminology pre- and post-set is used also for transitions to indicate, respectively, ${}^\bullet t = [\![\partial^0_N(t)]\!]$ and $t^\bullet = [\![\partial^1_N(t)]\!]$. As usual, $|\_|$ indicates the cardinality of sets.

Definition 1.3 *(Occurrence and Process Nets)*
*An (nondeterministic) occurrence net is a PT net* $\Theta$ *such that:*

*i) for all* $t \in T_\Theta$, *for all* $a \in S_\Theta$ *one has* $\partial^0_\Theta(t)(a) \le 1$ *and* $\partial^1_\Theta(t)(a) \le 1$;

*ii) for all* $a \in S_\Theta$, $|{}^\bullet a| \le 1$;

*iii)* $\prec$ *is irreflexive, where* $\prec$ *is the transitive closure of the relation*

$$\prec^1 = \{(a,t) \mid a \in S_\Theta,\ t \in a^\bullet\} \cup \{(t,a) \mid a \in S_\Theta,\ t \in {}^\bullet a\};$$

*moreover,* $\forall t \in T_\Theta$, $\{t' \in T_\Theta \mid t' \prec t\}$ *is finite;*

*iv) the binary "conflict" relation* $\#$ *on* $T_\Theta \cup S_\Theta$ *is irreflexive, where*

$$\forall t_1, t_2 \in T_\Theta,\ t_1 \#_m t_2 \Leftrightarrow {}^\bullet t_1 \cap {}^\bullet t_2 \neq \varnothing \quad \text{and} \quad t_1 \neq t_2,$$

$\forall x, y \in T_\Theta \cup S_\Theta,\ x \# y \Leftrightarrow \exists t_1, t_2 \in T_\Theta :\ t_1 \#_m t_2$ *and* $t_1 \preccurlyeq x$ *and* $t_2 \preccurlyeq y$, *where* $\preccurlyeq$ *is the reflexive closure of* $\prec$.

*Given* $x, y \in T_\Theta \cup S_\Theta$, *we say that* $x$ *and* $y$ *are concurrent, in symbols* $x\ co\ y$, *if it is not the case that* ($x \prec y$ *or* $y \prec x$ *or* $x \# y$). *A set* $X \subseteq T_\Theta \cup S_\Theta$ *is concurrent, in symbols* $Co(X)$, *if* $\forall\ x, y \in X,\ x\ co\ y$ *and* $|\{t \in T_\Theta \mid \exists x \in X,\ t \preccurlyeq x\}| \in \omega$.

*We say that an occurrence net* $\Theta$ *is deterministic if for all* $a \in S_\Theta$, $|a^\bullet| \leq 1$. *Observe that, in this case, we have* $\# = \varnothing$. *We shall refer to deterministic occurrence nets also as process nets.*

Thus, in an occurrence nets each place belongs at most to one post-set and, if the net is a process net, at most to one pre-set. This makes the "flow" relation $\preccurlyeq$ be a pre-order. Thus, requiring $\prec$ to be irreflexive, which is equivalent to requiring that the net be acyclic, identifies a partial order on the transitions. The constraint about the cardinality of the set of predecessors of a transition is then the fairly intuitive requirement that each transition be *finitely caused*. (See [28] for a discussion in terms of event structures of this issue.)

We stipulate that occurrence nets are to be considered also as *marked* nets whose minimal (wrt. $\prec$) places constitute the initial marking. Observe that this matches exactly with the standard definition, according to which occurrence nets can be marked only by assigning a single token to each of its minimal places. In the following, therefore, we shall use occurrence nets both in contexts in which marked nets are expected and in contexts in which unmarked nets are.

DEFINITION 1.4 *(Non-Sequential Processes [8])*
*Given a net* $N$, *a process of* $N$ *is a PT net morphism* $\pi: \Theta \rightarrow N$ *which maps places to places (as opposed to morphisms which map places to markings), where* $\Theta$ *is a finite process net.*

*Similarly, a process of a marked net* $N$ *is a morphism* $\pi: \Theta \rightarrow N$ *of marked PT nets which maps places to places, for a finite process net* $\Theta$.

For the purpose of defining processes at the right level of abstraction, we need to make some identifications among process nets. Of course, we shall consider as identical process nets which are isomorphic and, consequently, we shall make no distinction between two processes $\pi: \Theta \rightarrow N$ and $\pi': \Theta' \rightarrow N$ for which there exists an isomorphism $\varphi: \Theta \rightarrow \Theta'$ such that $\pi' \circ \varphi = \pi$. Observe that the particular form of $\pi$ is relevant, since we certainly want process morphisms to be total and to map a single component of the process net to a single component of $N$. Otherwise said, process morphisms are nothing but *labellings* of $\Theta$ with an appropriate element of $N$. Moreover, as usual, in the case of marked nets, we want to consider only processes whose source is the initial marking.

Inspired by the current trends in the development of the theory of computation, one would certainly like to describe the processes of a net $N$ as an algebra whose operations model a minimal set of combinators on processes which capture the essence of concurrency. Clearly, in the present case the core of such an algebra must consist of the operations of *sequential* and *parallel* composition of processes. The problem which arises immediately is that non-sequential processes cannot be concatenated when multiplicities are present: in order to support such an operation one must *disambiguate* the identity of all the tokens in the multisets source and target of processes. In other words, one must recognize that process concatenation has to do with tokens rather than with places. This is the approach followed in [6], which led to the introduction of the *concatenable processes* of $N$. These are, as already sketched above, non-sequential processes enriched by total orderings of the minimal and maximal places carrying the same label. Then, exploiting the additional information, it is easy to define an operation of concatenation of such processes, and thus to organize them as the arrows of a category $\mathcal{CP}[N]$. In particular, since concatenable processes also admit an operation of parallel composition, $\mathcal{CP}[N]$ is a symmetric monoidal category. In addition, [6] shows that $\mathcal{CP}[N]$ can be axiomatized by means of an abstract symmetric monoidal category $\mathcal{P}[N]$ (see also [23]).

# 2 Unfolding Place/Transition Nets

In this section we sketch the basic notions concerning the unfolding of PT Petri nets as defined in [12, 13]. In order to keep the exposition of the background material as short as possible, we limit ourselves to the definitions of the object components of the functors $\mathcal{U}[\_]$, $\mathcal{F}[\_]$, $\mathcal{E}[\_]$ and $\mathcal{L}[\_]$. Moreover, to avoid details which are unnecessary in the present setting, we introduce only those notions that are strictly necessary for this paper. In particular, we shall not introduce explicitly the categories involved. The reader interested in the details is referred to [12, 28]. A complete survey of the topic is also given in [22].

As a first step, we define the *decorated occurrence nets*, a type of occurrence nets in which places are grouped into families. They allow a convenient treatment of multiplicity issues in the unfolding of PT nets. We shall use $[n]$ to denote the segment $\{1,\ldots,n\}$ of $\omega$.

DEFINITION 2.1 *(Decorated Occurrence Nets [12])*
*A decorated occurrence net is an occurrence net $\Theta$ such that:*

*i)* *$S_\Theta$ is of the form $\bigcup_{a\in A_\Theta}\{a\}\times[n_a]$, for some set $A_\Theta$, where the set $\{a\}\times[n_a]$ is called the family of $a$. We will use $a^F$ to denote the family of $a$ regarded as a multiset;*

*ii)* $\forall a \in A_\Theta,\ \forall x,y \in \{a\}\times[n_a],\ {}^\bullet x = {}^\bullet y.$

A family is thus a collection of finitely many places with the same pre-set, and a decorated occurrence net is an occurrence net where each place belongs to exactly one family. Families, and therefore decorated occurrence nets, are capable of describing *relationships* between places by grouping them together. We

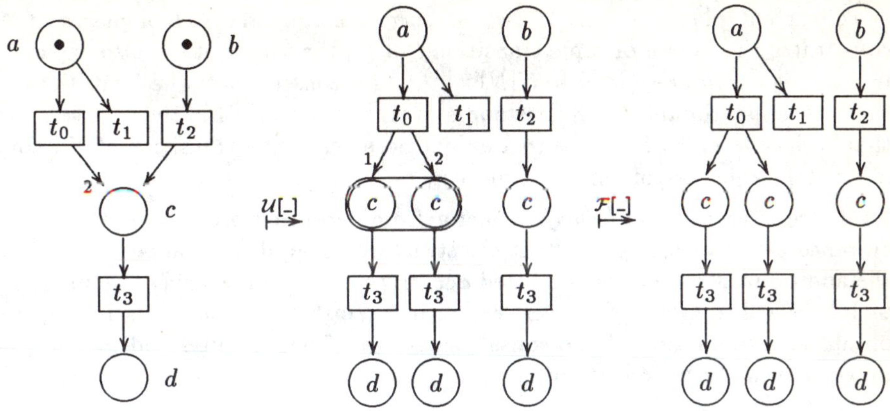


Figure 1: A net $N$, its unfolding $\mathcal{U}[N]$ and $\mathcal{FU}[N]$

will use families to relate places which are *instances* of the same place obtained in a process of unfolding.

Next, we define an unfolding procedure which maps marked PT nets to decorated occurrence nets.

DEFINITION 2.2 *(PT Nets Unfoldings: $\mathcal{U}[\_]$ [12])*
*Let $N = \big(\partial^0_N, \partial^1_N : T_N \to S_N^{\oplus}, u_N\big)$ be a marked net. We define the decorated occurrence net $\mathcal{U}[N]$ to be $\big(\partial^0, \partial^1 : T \to S^{\oplus}\big)$ , where $T$, $S$ and $\partial^0$ are generated inductively by the following inference rules.*

$$\frac{u_N(b) = n}{\Big\{\big(\varnothing, b\big)\Big\} \times [n] \subseteq S}$$

$$\frac{B = \Big\{\Big(\big(\epsilon_j, b_j\big), i_j\Big) \;\Big|\; j \in J\Big\} \subseteq S,\;\; Co(B),\;\; t \in T_N,\;\; \partial^0_N(t) = \bigoplus_{j \in J} b_j}{(B,t) \in T \quad \text{and} \quad \partial^0(B,t) = \bigoplus B}$$

$$\frac{x = (B,t) \in T,\;\; \partial^1_N(t)(b) = n}{\Big\{\big(\{x\}, b\big)\Big\} \times [n] \subseteq S}$$

*and for $x \in T$, $\partial^1(x) = \bigoplus_{b,i}\Big(\big(\{x\}, b\big), i\Big)$.*

Informally speaking, the definition above can be explained as follows, where we use $\mathcal{U}[N]^{(n)}$, $n \in \omega$, to denote the $n$-th approximation of $\mathcal{U}[N]$, i.e., the subnet of $\mathcal{U}[N]$ consisting of the elements at depth not greater than $n$. The net $\mathcal{U}[N]^{(0)}$ is obtained by exploding in families the initial marking of $N$, and $\mathcal{U}[N]^{(n+1)}$ is obtained, inductively, by generating a new transition for each possible subset of concurrent places of $\mathcal{U}[N]^{(n)}$ whose corresponding multiset of places of $N$ constitutes the source of some transition $t$ of $N$; the target of $t$ is also exploded in families which are added to $\mathcal{U}[N]^{(n+1)}$. As a consequence, the transitions of the $n$-th approximant net are instances of transitions of $N$, in the precise sense that each of them corresponds to a unique occurrence of a transition of $N$ in one of its step sequences of length at most $n$.

There is an obvious *forgetful* functor from decorated occurrence nets to occurrence nets which forgets about the structure of families. It allows us to drop the additional structure of decorated occurrence nets and to bring the unfolding of PT nets into Occ. Moreover, exploiting Winskel's coreflections in [28], we obtain an explanation of the causal behaviour of nets in PES and in Dom as illustrated in the introduction.

DEFINITION 2.3 *($\mathcal{F}[\_]$: from DecOcc to Occ [12])*
*Given a decorated occurrence net $\Theta$ define $\mathcal{F}[\Theta]$ to be the occurrence net underlying $\Theta$.*

Figure 1 shows a simple example of unfolding of PT nets. To make explicit the nature of the elements of $\mathcal{U}[N]$ and $\mathcal{FU}[N]$, in the picture we label them with the corresponding element $a, b, \ldots, t_3$ of $N$. In particular, the places of the unfolding labelled by $a$ and $b$ are respectively $(\varnothing, a)$ and $(\varnothing, b)$, the transitions labelled by $t_0$ and $t_2$ are $\bar{t}_0 = (\{(\varnothing, a)\}, t_0)$ and $\bar{t}_2 = (\{(\varnothing, b)\}, t_2)$, and thus the three instances of $c$ are $((\{\bar{t}_0\}, c), 1)$, $((\{\bar{t}_0\}, c), 2)$ and $((\{\bar{t}_2\}, c), 1)$. A family is represented by enclosing its elements into an oval. The numbers which label the outgoing arcs from $\bar{t}_0$ take into account the ordering of the elements in the family $(\{\bar{t}_0\}, c)^F$; since $\mathcal{U}[N]$ is an occurrence net, no confusion is possible with arc multiplicities. Families of cardinality one are not explicitly indicated. We call $\mathcal{U}[N]$ and $\mathcal{FU}[N]$ respectively the unfolding of $N$ in DecOcc and in Occ. However, in the following we shall avoid explicit reference to DecOcc and Occ.

Prime event structures [15, 28] are the simplest event based model of concurrency. They consist of a set of events, intended as indivisible *quanta* of computation, which are related to each other by two binary relation: *causality*, modelled by a partial order relation $\leq$, and conflict, modelled by an irreflexive, symmetric and hereditary relation $\#$.

DEFINITION 2.4 *(Prime Event Structures)*
*A prime event structure is a structure* $\mathrm{E} = (E, \#, \leq)$ *consisting of a set of events $E$ partially ordered by $\leq$, and a symmetric, irreflexive relation $\# \subseteq E \times E$, the conflict relation, such that*

$$\{e' \in E \mid e' \leq e\} \text{ is finite for each } e \in E$$

$$e \# e' \leq e'' \text{ implies } e \# e'' \text{ for each } e, e', e'' \in E.$$

The computational intuition behind event structures is really simple: an event $e$ can occur when all its *causes* have occurred and no event that is in

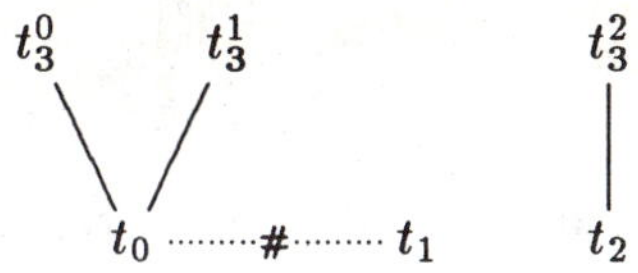


Figure 2: The event structure $\mathcal{EFU}[N]$ for the net in Figure 1

conflict with the given event has already occurred. This is formalized by the following notion of *configuration*.

DEFINITION 2.5 *(Configurations)*
*Given a prime event structure $(E, \#, \leq)$, define its configurations to be those subsets $x \subseteq E$ which are*

$$\textit{Conflict Free: } \forall e_1, e_2 \in x,\ \textit{not}(e_1 \# e_2)$$

$$\textit{Left Closed: } \forall e \in x\ \forall e' \leq e,\ e' \in x$$

*Let $\mathcal{L}(\mathrm{E})$ denote the set of configurations of the prime event structure* E *and $\mathcal{L}_F(\mathrm{E})$ the set of finite configurations of* E.

The following definition recalls how to translate occurrence nets into prime event structures. An example of this translation is shown in Figure 2, where, using the standard graphical representation of event structures, $\leq$ is indicated by (bottom-up) solid lines and # by a dotted line; we use superscripts to distinguish between the three instances of $t_3$ in $\mathcal{FU}[N]$.

DEFINITION 2.6 *($\mathcal{E}[\_]$: from <u>Occ</u> to <u>PES</u> [28])*
*Let $\Theta$ be an occurrence net. Then, $\mathcal{E}[\Theta]$ is the event structure $(T_\Theta, \preccurlyeq, \#)$, where $\preccurlyeq$ and # are the restrictions to the set of transitions of $\Theta$ of, respectively, the flow ordering and the conflict relation implicitly defined by $\Theta$.*

Finitary prime algebraic domains or dI-domains—introduced by G. Berry while studying sequentiality of functions [1]—are particular Scott's domains which are distributive and in which each finite element is preceded only by a finite number of elements of the domain. Here we are interested in their *"coherent"* version, i.e., in the version in which the underlying partial order is pairwise complete.

DEFINITION 2.7 *(Finitary (Coherent) Prime Algebraic Domains)*
*Let $(D, \sqsubseteq)$ be a partial order. Recall that a set $X \subseteq D$ is directed if all the pairs $x, y \in X$ have an upper bound in $X$, is compatible if there exists $d \in D$ such that $x \sqsubseteq d$ for all $x \in X$ and is pairwise compatible if $\{x, y\}$ is compatible for all $x, y \in X$. We say that $D$ is a (coherent) domain if it is pairwise complete, i.e., if for all pairwise compatible $X \subseteq D$ the least upper bound $\bigsqcup X$ of $X$ exists.*

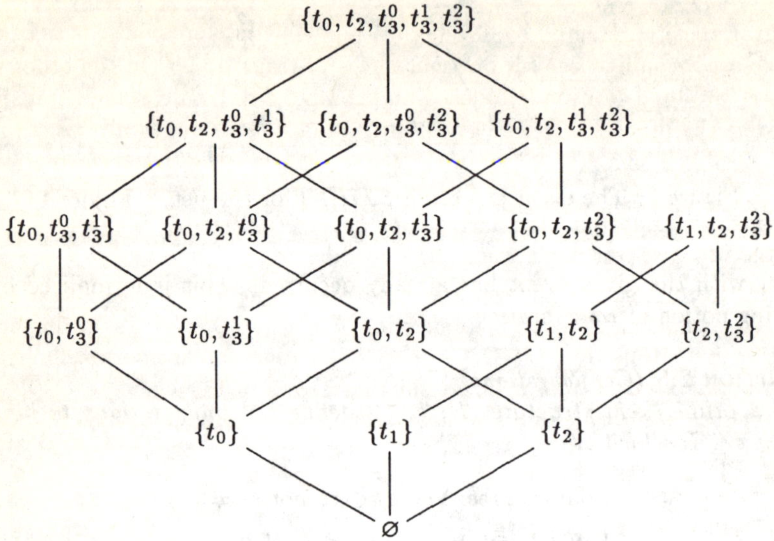


Figure 3: The Hasse diagram of the domain $\mathcal{LEFU}[N]$ for the net in Figure 1

*A complete prime of D is an element $p \in D$ such that, for any compatible $X \subseteq D$, if $p \sqsubseteq \bigsqcup X$, then there exists $x \in X$ such that $p \sqsubseteq x$. We say that a domain D is prime algebraic if for all $d \in D$ we have $d = \bigsqcup\{p \sqsubseteq d \mid p$ is a complete prime$\}$. Moreover, an element $e \in D$ is finite if for any directed $S \subseteq D$, if $e \sqsubseteq \bigsqcup S$, then there exists $s \in S$ such that $e \sqsubseteq s$. We say that D is finitary if for all finite elements $e \in D$, $|\{d \in D \mid d \sqsubseteq e\}| \in \omega$.*

Finitary prime algebraic domains can be equipped with a notion of morphism in such a way that the category Dom so obtained is *equivalent* to PES (see [28]). We conclude this section by recalling the object component of the equivalence functor $\mathcal{L}[\_]$: PES $\rightarrow$ Dom. An example is provided by Figure 3.

PROPOSITION 2.8 *($\mathcal{L}[\_]$: from PES to Dom [28])*
*Let E be a prime event structure. Then, $\mathcal{L}(E) = (\mathcal{L}(E), \subseteq)$, i.e., the set of configurations of E ordered by inclusion is a finitary (coherent) prime algebraic domain.*

# 3 Process vs. Unfolding Semantics for Nets

The semantics obtained via the unfolding yields an explanation of the behaviour of nets in terms of event structures, that is, in terms of domains. Domains can be unambiguously thought of as partial orderings of computations, where a computation is represented by a configuration, which, in our context, is a "downward" closed, conflict free set of occurrences of transitions. On the other hand, processes are by definition left closed and conflict free (multi)sets of transitions. Moreover, the processes from a given initial marking are naturally organized in a preorder-like fashion via a comma category construction which formalizes the usual notion of prefix ordering of processes. The question which therefore arises spontaneously concerns the relationships between these two notions; this is the question addressed in this section.

It is worth noticing that in the case of safe nets the question is readily answered exploiting Winskel's coreflection $\langle \hookrightarrow, \mathcal{U}[\_] \rangle : \underline{\mathrm{Occ}} \rightharpoonup \underline{\mathrm{Safe}}$. In fact, by definition an adjunction $\langle \mathsf{F}, \mathsf{G} \rangle : \underline{\mathsf{C}} \rightharpoonup \underline{\mathsf{D}}$ determines an isomorphism between arrows of the kind $\mathsf{F}(c) \to d$ in $\underline{\mathsf{D}}$ and the arrows of the kind $c \to \mathsf{G}(d)$ in $\underline{\mathsf{C}}$. Then, in the case of safe nets, we have a one-to-one correspondence

$$\pi : \Theta \to N \quad \Longleftrightarrow \quad \pi' : \Theta \to \mathcal{U}[N]$$

for each safe net $N$ and each occurrence net $\Theta$. Therefore, since such correspondence is easily seen to map processes to processes, in this special case, the correspondence between process and unfolding semantics of $N$ is very tidy: they are the same notion in the precise sense that there is an isomorphism between the processes of $N$ and the processes of $\mathcal{U}[N]$, i.e., the deterministic finite subnets of the unfolding of $N$, i.e., the finite configurations of $\mathcal{EU}[N]$.

In our context, however, we have that the unfolding of $N$ is strictly more concrete than the processes of $N$. For example, consider again the net $N$ and its unfolding $\mathcal{FU}[N]$ shown in Figure 1. Clearly, there is a unique process of $N$ in which $t_0$, $t_2$ and a single instance of $t_3$ caused by $t_0$ has occurred. Nevertheless, there are two deterministic subnets of $\mathcal{FU}[N]$ which correspond to such process, namely those obtained by choosing respectively the left and the right instance of $t_3$ below $t_0$. It is worth noticing that such subnets are isomorphic and that this is not a fortunate case, since it is easy to show that two finite deterministic subnets of $\mathcal{FU}[N]$ correspond to the same process of $N$ if and only if they are isomorphic via an isomorphism which sends instances of an element of $N$ to instances of the same element. More interestingly, the results of this paper will prove that this is the exact relationship between the two semantics of $N$: the unfolding contains several copies of the same process which, as illustrated in [12, 13], are needed to provide a fully *causal* explanation of the behaviour of $N$, i.e., to obtain an occurrence net whose transitions represent exactly the instances of the transitions of $N$ in all the possible causal contexts and which can therefore account for concurrent multiple instances of the same element of $N$, that is for *autoconcurrency*. More precisely, we shall see that the finite deterministic subnets of the unfolding of $N$ can be characterized by appropriately *decorating* the processes of $N$, which shows directly that the difference between process and unfolding semantics of $N$ is due only to the replication of data needed in the latter.

Of course, the appropriate decoration of processes is immediately suggested by the notion of family in decorated occurrence nets: a *decorated process* is simply a process whose underlying process net is a decorated occurrence net. This yields a *process-oriented* account of the unfolding construction. In addition, we shall also identify an abstract *symmetric strict monoidal category* $\mathcal{DP}[N]$ whose arrows axiomatize the decorated *concatenable* processes of $N$, and which therefore provides both the *algebraic* and the *process-oriented* account of the unfolding construction. In particular, for each marked PT net $(N, u_N)$ we have $\mathcal{L}_F\mathcal{EFU}[(N, u_N)] \cong \langle u_N \downarrow \mathcal{DP}[N] \rangle$, where the role of the comma category construction is to consider only the decorated concatenable processes from the initial marking $u_N$. Therefore, decorated (*deterministic*) occurrence nets, which at first seem to be just a convenient technical solution to establish the adjunction from PT nets to occurrence nets, provide *both* the process and the algebraic counterpart of the unfolding semantics.

The following proposition remarks the intuitive relationship between finite configurations and processes in the unfolding semantics.

PROPOSITION 3.1
*Let $N$ be a marked net. There is an isomorphism between the set of finite configurations of $\mathcal{EFU}[N]$ and the set of (marked) processes of $\mathcal{FU}[N]$.*

Our next task is to characterize the processes of $\mathcal{FU}[N]$ in terms of processes of $N$. We shall do it by means of the following notion of decorated process.

DEFINITION 3.2 *($f$-indexed orderings)*
*Given sets $A$ and $B$ together with a function $f\colon A \to B$, an $f$-indexed ordering of $A$ is a family $\{\ell_b \mid b \in B\}$ of bijections $\ell_b\colon f^{-1}(b) \to \{1, \ldots, |f^{-1}(b)|\}$, with $f^{-1}(b)$ being as usual the set $\{a \in A \mid f(a) = b\}$.*

Therefore, an $f$-indexed ordering of $A$ is a family of total orderings, one for each of the partitions of $A$ induced by $f$.

DEFINITION 3.3 *(Decorated Processes)*
*Let $N$ be a marked net. A decorated process of $N$ is a triple $DP = (\pi, \ell, \tau)$*

- *$\pi\colon \Theta \to N$ is a (marked) process of $N$;*
- *$\ell$ is a $\pi$-indexed ordering of $\min(\Theta)$, the minimal (wrt. $\preccurlyeq$) places of $\Theta$;*
- *$\tau$ is a family $\{\tau(t)\}$ indexed by the transitions $t$ of $\Theta$, where each $\tau(t)$ is a $\pi$-indexed ordering of the post-set of $t$ in $\Theta$.*

*The decorated processes $(\pi\colon \Theta \to N, \ell, \tau)$ and $(\pi'\colon \Theta' \to N, \ell', \tau')$ are isomorphic, and then identified, if their underlying processes are isomorphic via an isomorphism $\varphi$ which respects all the orderings, i.e., $\ell'_{\pi'(\varphi(a))}(\varphi(a)) = \ell_{\pi(a)}(a)$ for all $a \in \min(\Theta)$, and $\tau'(\varphi(t))_{\pi'(\varphi(a))}(\varphi(a)) = \tau(t)_{\pi(a)}(a)$ for all $t \in T_\Theta$ and for all $a \in t^\bullet$.*

*We say that $(\pi\colon \Theta \to N, \ell, \tau) \leq (\pi'\colon \Theta' \to N, \ell', \tau')$ if there exists $\varphi\colon \Theta \to \Theta'$ which preserves all the orderings and such that $\pi = \pi' \circ \varphi$. Since we identify isomorphic processes, the set of decorated processes is partially ordered by $\leq$. We shall write $DP[N]$ to indicate such ordering.*

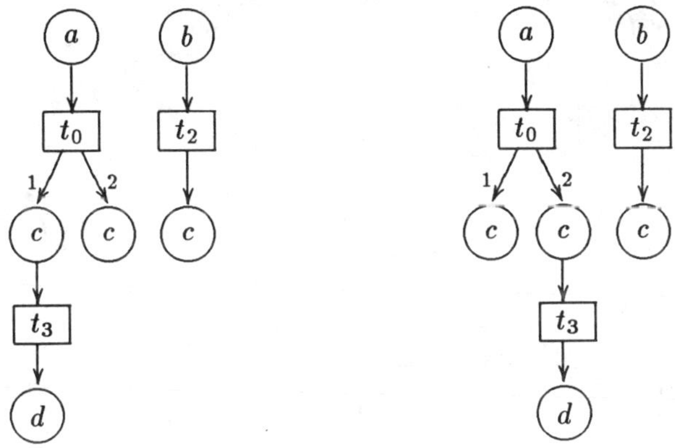


Figure 4: Two decorated processes of the net in Figure 1

Figure 4 shows the two decorated processes of the net $N$ in Figure 1 corresponding to the (unique) process of $N$ in which $t_0$, $t_2$ and an instance of $t_3$ caused by $t_0$ have occurred. In the pictures, we represent a process $\pi: \Theta \to N$ by drawing $\Theta$ and labelling its element $x$ by $\pi(x)$. Observe that Figure 4 also gives a hint about the announced correspondence.

PROPOSITION 3.4
*The set of decorated processes of N is isomorphic to the set of (marked) processes of* $\mathcal{FU}[N]$.

Since $\mathcal{FU}[N]$ is obtained from a decorated occurrence net via a forgetful functor, a process $\pi$ of $\mathcal{FU}[N]$ still contains information about the families of $\mathcal{U}[N]$. Informally speaking, the proof of the previous proposition consists of showing that it is possible to move such information from $\pi$ to $\ell$ and $\tau$ of a decorated process of $N$ and back in $\pi$ from $\ell$ and $\tau$, so obtaining an isomorphism. The correspondence above can be easily lifted to the partial orders of decorated processes and finite configurations of $\mathcal{EFU}[N]$.

PROPOSITION 3.5
$DP[N]$ *is isomorphic to* $\mathcal{L}_F\mathcal{EFU}[N]$.

Exploiting further the idea of decorated processes, the same conceptual step which led from non-sequential processes to concatenable processes suggests the following definition.

DEFINITION 3.6 *(Decorated Concatenable Processes)*
*A decorated concatenable process of the (unmarked) net* $N$, *is a quadruple* $(\pi, \ell, \tau, L)$ *where* $\pi$ *is a process of* $N$, $\ell$ *and* $L$ *are, respectively,* $\pi$*-indexed orderings of* $\min(\Theta)$ *and* $\max(\Theta)$, *i.e., the minimal and the maximal places of* $\Theta$,

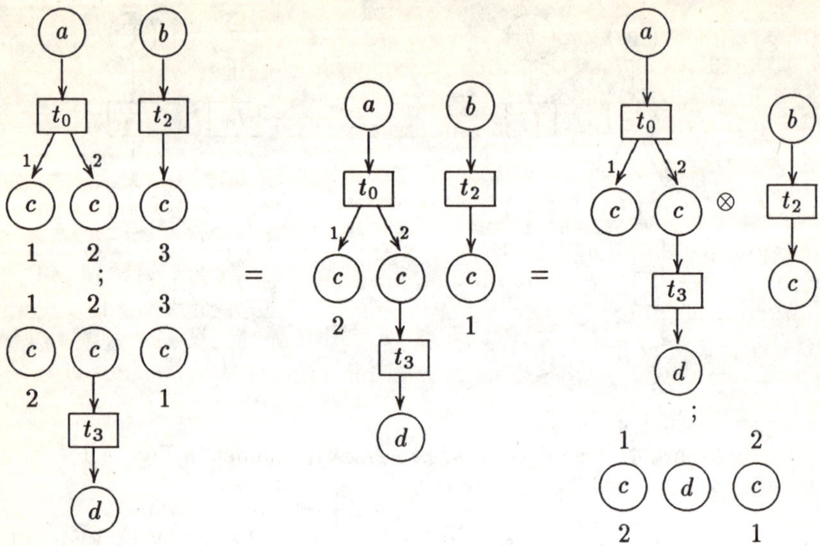


Figure 5: An example of the algebra of decorated concatenable processes

*and $\tau$ is a family $\{\tau(t)\}$ indexed by the transitions $t$ of $\Theta$, where each $\tau(t)$ is a $\pi$-indexed ordering of the post-set of $t$ in $\Theta$.*

*An isomorphism of decorated concatenable processes is an isomorphism of the underlying processes which, in addition, preserves all the orderings given by $\ell$, $\tau$ and $L$.*

So, a decorated concatenable process $(\pi, \ell, \tau, L)$ is a concatenable process $(\pi, \ell, L)$ where the post-sets of all transitions are $\pi$-indexed ordered. Observe that $(\pi, \ell, \tau)$ is different from a decorated process since $\pi$ is *unmarked*. It follows from the definitions that a place in $\max(\Theta)$ has in every case a double ordering: derived from $L$ and $\ell$ if it is also minimal, and derived from $L$ and $\tau$ otherwise.

It is clearly possible to define an operation of concatenation of decorated concatenable processes. We can associate a source and a target in $S_N^\oplus$ to any concatenable process $DCP$, namely by taking the image through $\pi$ of, respectively, $\min(\Theta)$ and $\max(\Theta)$, where $\Theta$ is the underlying process net of $DCP$. Then, the concatenation of $DCP_0 \colon u \to v$ and $DCP_1 \colon v \to w$ is defined in the obvious way exploiting the informations given by the labellings in order to merge the maximal places of the process nets underlying $DCP_0$ with the minimal places of the process net underlying $DCP_1$. (See also Figure 5, where the elements of the nets are labelled according to $\pi$, $\ell$, and $L$, $\tau(t)$ is represented by decorating the arcs outgoing from $t$ and all the trivial orderings are omitted.) Therefore, we can consider the category $\mathcal{DCP}[N]$ whose objects are the finite multisets on $S_N$ and whose arrows are the decorated concatenable processes.

PROPOSITION 3.7
*Under the above defined operation of sequential composition, $\mathcal{DCP}[N]$ is a category with identities those decorated concatenable processes consisting only of places, which therefore are both minimal and maximal, and such that $\ell = L$.*

Decorated concatenable processes admit also a tensor operation $\otimes$ which represents the parallel composition of processes. Given $DCP_0: u \to v$ and $DCP_1: u' \to v'$, $DCP_0 \otimes DCP_1: u \oplus u' \to v \oplus v'$ is the decorated concatenable process which may be graphically represented by putting side by side, from left to right, the graphical representations of $DCP_0$ and $DCP_1$ and reorganizing the labellings appropriately as shown in Figure 5. It is easy to see that the concatenable processes consisting only of places are the *symmetries* [10] which make $\mathcal{DCP}[N]$ into a *symmetric strict monoidal category.*

PROPOSITION 3.8
*$\mathcal{DCP}[N]$ is a symmetric strict monoidal category.*

Recalling that decorated concatenable processes are a refinement of concatenable processes and that the concatenable processes of $N$ correspond to the arrows of a category $\mathcal{P}[N]$, we are led to the following definition of the symmetric monoidal category $\mathcal{DP}[N]$ which captures the *algebraic essence* of decorated (concatenable) processes, and thus of the unfolding construction. It is worth remarking that the definition of $\mathcal{DP}[N]$ can be obtained simply by dropping one axiom from the definition of $\mathcal{P}[N]$. This shows that the difference between processes and decorated processes is simply axiomatizable in terms of monoidal categories.

DEFINITION 3.9 *(The category $\mathcal{DP}[N]$)*
*Let $N$ be a PT net. Then $\mathcal{DP}[N]$ is the monoidal quotient of the free symmetric strict monoidal category on $N$ modulo the axioms*

$$\begin{aligned} c_{a,b} &= id_{a\oplus b} && \text{if } a,b \in S_N \text{ and } a \neq b \\ s;t &= t && \text{if } t \in T_N \text{ and } s \text{ is a symmetry.} \end{aligned}$$

*where $c$ is the symmetry natural isomorphism. Explicitly, the category $\mathcal{DP}[N]$ is the category whose objects are the elements of $S_N^{\oplus}$ and whose arrows are generated by the inference rules*

$$\frac{u \in S_N^{\oplus}}{id_u: u \to u \text{ in } \mathcal{DP}[N]} \qquad \frac{u,v \text{ in } S_N^{\oplus}}{c_{u,v}: u \oplus v \to u \oplus v \text{ in } \mathcal{DP}[N]} \qquad \frac{t: u \to v \text{ in } T_N}{t: u \to v \text{ in } \mathcal{DP}[N]}$$

$$\frac{\alpha: u \to v \text{ and } \beta: u' \to v' \text{ in } \mathcal{DP}[N]}{\alpha \otimes \beta: u \oplus u' \to v \oplus v' \text{ in } \mathcal{DP}[N]} \qquad \frac{\alpha: u \to v \text{ and } \beta: v \to w \text{ in } \mathcal{DP}[N]}{\alpha;\beta: u \to w \text{ in } \mathcal{DP}[N]}$$

*modulo the axioms expressing that $\mathcal{DP}[N]$ is a strict monoidal category, namely,*

$$\begin{aligned} \alpha; id_v = \alpha = id_u;\alpha \quad &\text{and} \quad (\alpha;\beta);\delta = \alpha;(\beta;\delta), \\ (\alpha\otimes\beta)\otimes\delta = \alpha\otimes(\beta\otimes\delta) \quad &\text{and} \quad id_0 \otimes \alpha = \alpha = \alpha \otimes id_0, \qquad (1) \\ id_u \otimes id_v = id_{u\oplus v} \quad &\text{and} \quad (\alpha\otimes\alpha');(\beta\otimes\beta') = (\alpha;\beta)\otimes(\alpha';\beta'), \end{aligned}$$

*the latter whenever the lefthand term is defined, the following axioms expressing that $\mathcal{DP}[N]$ is symmetric with symmetry isomorphism c*

$$\begin{aligned} c_{u,v\oplus w} &= (c_{u,v} \otimes id_w); (id_v \otimes c_{u,w}), \\ c_{u,u'}; (\beta \otimes \alpha) &= (\alpha \otimes \beta); c_{v,v'} \quad \text{for } \alpha: u \to v, \ \beta: u' \to v', \\ c_{u,v}; c_{v,u} &= id_{u\oplus v}, \end{aligned} \tag{2}$$

*and the following axioms*

$$\begin{aligned} c_{a,b} &= id_{a\oplus b} \quad \text{if } a, b \in S_N \text{ and } a \neq b \\ (id_u \otimes c_{a,a} \otimes id_v); t &= t \qquad\quad \text{if } t \in T_N. \end{aligned} \tag{3}$$

It is worthwhile to remark that in the definition above axioms (1) and (2) define $\mathcal{F}(N)$, the free symmetric strict monoidal category on $N$. Observe that, exploiting the coherence axiom, i.e., the first of (2), a symmetry in $\mathcal{F}(N)$ can always be written as a composition of symmetries of the kind $(id_u \otimes c_{a,b} \otimes id_v)$ for $a, b \in S_N$. Then, since we have $c_{a,b} = id_{a\oplus b}$ if $a \neq b$, the axiom $s; t = t$ takes the particular form stated in (3).

For $\mathcal{DP}[N]$ and $\mathcal{DCP}[N]$ we have the following result which matches the analogous one for concatenable processes [6].

PROPOSITION 3.10
*$\mathcal{DCP}[N]$ and $\mathcal{DP}[N]$ are isomorphic.*

*Proof.* (Sketch.) Consider the following mapping F from the arrows of $\mathcal{DP}[N]$ to decorated concatenable processes.

- A transition $t$ of $N$ is mapped to the decorated concatenable processes with a unique transition and two layers of places: the minimal, in one-to-one correspondence with $\partial^0_N(t)$, and the maximal, in one-to-one correspondence with $\partial^1_N(t)$. The decoration, of course, consists in taking $\tau(t) = L$.
- A symmetry $c_{u,v}$, for $u = n_1a_1 \oplus \cdots \oplus n_ka_k$ and $v = m_1b_1 \oplus \cdots \oplus m_hb_h$ is mapped to the concatenable process having as many places as elements in the multiset $u \oplus v$ mapped by $\pi$ to the corresponding places of $N$ and such that $L_{a_i}(x) = v(a_i) + \ell_{a_i}(x)$ and $\ell_{b_i}(x) = L_{b_i}(x) - u(b_i)$.
- F is extended inductively to a generic term $\alpha$ of $\mathcal{DP}[N]$, i.e., $\alpha_0 \otimes \alpha_1$ is mapped to $\mathsf{F}(\alpha_0) \otimes \mathsf{F}(\alpha_1)$ and $\alpha_0 ; \alpha_1$ to $\mathsf{F}(\alpha_0); \mathsf{F}(\alpha_1)$.

Then, defining F to be the identity on the objects gives the required isomorphism $\mathsf{F}: \mathcal{DP}[N] \cong \mathcal{DCP}[N]$.

We conclude the section by getting back to the diagram discussed in the introduction. We recall the following simple notion from category theory.

DEFINITION 3.11 *(Comma Categories)*
*Let $\underline{\mathbf{C}}$ be a category and c an object of $\underline{\mathbf{C}}$. Then, the comma category $\langle c \downarrow \underline{\mathbf{C}} \rangle$, also called the category of elements under c, is the category whose objects are*

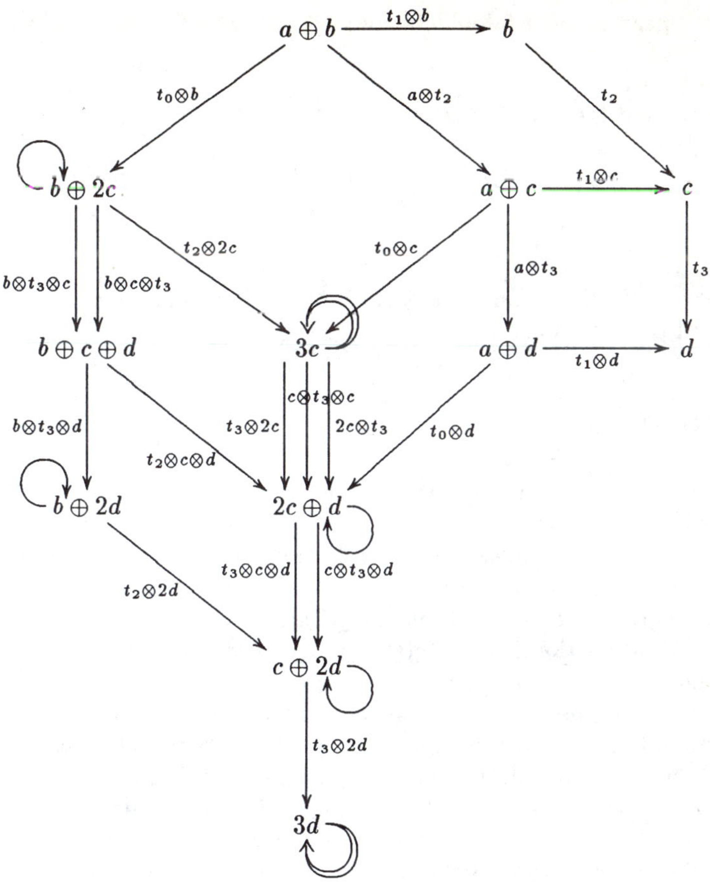


Figure 6: Some of the arrows with source $a \oplus b$ in $\mathcal{DP}[N]$ for the net of Figure 1

*the arrows* $f: c \rightarrow c'$ *of* $\underline{C}$ *and whose arrows* $h: (f: c \rightarrow c') \rightarrow (g: c \rightarrow c'')$ *are the commutative diagrams*

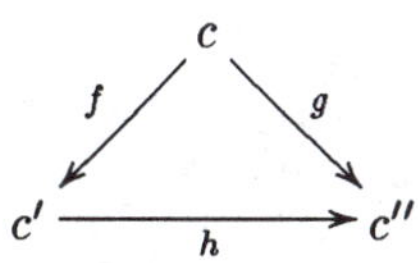


*Identities and arrow composition are inherited in the obvious way from* $\underline{C}$.

The next proposition essentially shows that the canonical partial order associated to $\langle u_N \downarrow N \rangle$ is $DP[(N, u_N)]$, and concludes our exposition. The proof follows easily from Proposition 3.10, and the intuition behind it can be grasped from Figure 6, where the self-looping arrows represent the non-identity symmetries. We warn the reader that not all the symmetries are shown in the picture;

this is the meaning of the double arrows which stand for several of them.

PROPOSITION 3.12
*For any marked PT net* $(N, u_N)$,

$$\langle u_N \downarrow \mathcal{DP}[N] \rangle \cong DP[(N, u_N)] \cong \mathcal{L}_F \mathcal{EFU}[(N, u_N)].$$

Observe that the second equivalence above is actually an isomorphism, as shown by Proposition 3.5.

# Conclusions

In this paper we have shown how the unfolding semantics given in [12, 13] can be reconciled with a process-oriented semantics. Moreover, we have seen that the algebraic structure of the processes of a net can be faithfully expressed by a symmetric monoidal category. The key of these formal achievements is the notion of *decorated occurrence net.* Although DecOcc arose from the need of factorizing the involved adjunction from PTNets to Occ, and, thus, decorated occurrence nets might at first seem to be just a convenient technical solution, we have shown that in fact useful insights on the semantics of nets are suggested by the present unfolding construction and the associated notion of decorated occurrence net. In fact, decorated *deterministic* occurrence nets, suitably axiomatized as arrows of the symmetric monoidal category $\mathcal{DP}[N]$, provide both the process-oriented and the algebraic counterpart of the unfolding semantics. Moreover, they can be characterized as the minimal refinement of Goltz-Reisig processes which guarantees the identity of all tokens, i.e., as the minimal refinement of occurrence nets which guarantees the existence of an unfolding for PT nets. In fact, in order to achieve this it is necessary to disambiguate both the tokens in the same place of the initial marking and the tokens which are multiple instances of the same place, and, therefore, to introduce the notion of *families*.

# References


[1] G. BERRY. Stable Models of Typed λ-calculi. In Proceedings *ICALP '78* , LNCS n. 62, pp. 72–89, 1978.

[2] E. BEST, AND R. DEVILLERS. Sequential and Concurrent Behaviour in Petri Net Theory. *Theoretical Computer Science*, n. 55, pp. 87–136, 1987.

[3] C. BROWN, AND D. GURR. A Categorical Linear Framework for Petri Nets. In *Proceedings of the 5th LICS Symposium*, pp. 208–218, 1990.

[4] C. BROWN, D. GURR, AND V. DE PAIVA. *A Linear Specification Language for Petri Nets.* Technical Report DAIMI PB-363, Computer Science Department, Aarhus University, 1981.

[5] P. DEGANO, R. DE NICOLA, AND U. MONTANARI. A Distributed Operational Semantics for CCS based on Condition/Event Systems. *Acta Informatica*, n. 26, pp. 59–91, 1988.

[6] P. DEGANO, J. MESEGUER, AND U. MONTANARI. Axiomatizing Net Computations and Processes. In *Proceedings of the 4th LICS Symposium*, pp. 175–185, IEEE, 1989.

[7] J. ENGELFRIET. Branching Processes of Petri Nets. *Acta Informatica*, n. 28, pp. 575–591, 1991.

[8] U. GOLTZ, AND W. REISIG. The Non-Sequential Behaviour of Petri Nets. *Information and Computation*, n. 57, pp. 125–147, 1983.

[9] P.W. HOOGERS, H.C.M. KLEIJN, AND P.S. THIAGARAJAN. Local Event Structures and Petri Nets In *Proceedings of CONCUR '93*, LNCS n. 715, pp. 462–476, Springer-Verlag, 1993.

[10] S. MACLANE. *Categories for the Working Mathematician.* Springer-Verlag, 1971.

[11] J. MESEGUER, AND U. MONTANARI. Petri Nets are Monoids. *Information and Computation*, n. 88, pp. 105–154, Academic Press, 1990.

[12] J. MESEGUER, U. MONTANARI, AND V. SASSONE. On the Semantics of Petri Nets. In proceedings of *CONCUR '92*, LNCS, n. 630, pp. 286–301, Springer-Verlag, 1992.

[13] J. MESEGUER, U. MONTANARI AND V. SASSONE. *On the Semantics of Place/Transition Petri Nets.* Technical Report TR 27/92 Dipartimento Informatica, Università di Pisa, 1992, and Technical Report SRI-CSL-92-09, SRI International, Computer Science Laboratory, 1992.

[14] J. MESEGUER, U. MONTANARI AND V. SASSONE. *Process versus Unfolding Semantics for PT Place/Transition Petri Nets* To appear as Technical Report of the Dipartimento di Informatica, Università di Pisa, and Technical Report of the Computer Science Laboratory, SRI International, 1994.

[15] M. NIELSEN, G. PLOTKIN, AND G. WINSKEL. Petri Nets, Event Structures and Domains, Part 1. *Theoretical Computer Science*, n. 13, pp. 85–108, 1981.

[16] M. NIELSEN, G. ROZENBERG, AND P.S. THIAGARAJAN. Behavioural Notions for Elementary Net Systems. *Distributed Computing*, n. 4, pp. 45–57, 1990.

[17] E.R. OLDEROG. A Petri Net Semantics for CCSP. In *Advances in Petri Nets*, LNCS, n. 255, pp. 196–223, Springer-Verlag, 1987.

[18] C.A. PETRI. *Kommunikation mit Automaten*. PhD thesis, Institut für Instrumentelle Mathematik, Bonn, Germany, 1962.

[19] C.A. PETRI. Concepts of Net Theory. In proceedings of *MFCS '73*, pp. 137–146, Mathematics Institute of the Slovak Academy of Science, 1973.

[20] C.A. PETRI. *Non-Sequential Processes*. Interner Bericht ISF–77–5, Gesellschaft für Mathematik und Datenverarbeitung, Bonn, FRG, 1977.

[21] W. REISIG. *Petri Nets*. Springer-Verlag, 1985.

[22] V. SASSONE. *On the Semantics of Petri Nets: Processes, Unfoldings and Infinite Computations*. PhD Thesis, TD 6/94 Dipartimento di Informatica, Università di Pisa, March 1994.

[23] V. SASSONE. *Some Remarks on Concatenable Processes*. Technical Report TR 6/94, Dipartimento di Informatica, Università di Pisa, April 1994.

[24] V. SASSONE, J. MESEGUER, AND U. MONTANARI. *ω-Ind Completion of Monoidal Categories and Infinite Petri Net Computations*. Presentation at the *Workshop on Topology and Completion in Sematics*, Institute Blaise Pascal, November 1993, Chartres, France. To appear as Technical Report of the Dipartimento di Informatica, Università di Pisa, and Technical Report of the Computer Science Laboratory, SRI International, 1994.

[25] D. SCOTT. Outline of a Mathematical Theory of Computation. In proceedings of *4th Annual Princeton Conference on Information Science and Systems*, pp. 169–176, 1970.

[26] R. VAN GLABBEEK, AND F. VAANDRAGER. Petri Net Models for Algebraic Theories of Concurrency. In proceedings of *PARLE*, LNCS, n. 259, pp. 224–242, Springer-Verlag, 1987.

[27] G. WINSKEL. Event Structure Semantics of CCS and Related Languages. In proceedings of *ICALP '82*, LNCS n. 140, pp. 561–567, Springer-Verlag, 1982. Expanded version available as technical report DAIMI PB-159, Computer Science Department, Aarhus University.

[28] G. WINSKEL. Event Structures. In proceedings of *Advanced Course on Petri Nets*, LNCS, n. 255, pp. 325–392, Springer-Verlag, 1987.

[29] G. WINSKEL. Petri Nets, Algebras, Morphisms and Compositionality. *Information and Computation*, n. 72, pp. 197–238, 1987.

# B-W Analysis: a Backward Reachability Analysis for Diagnostic Problem Solving Suitable to Parallel Implementation

**Cosimo Anglano, Luigi Portinale**

Dipartimento di Informatica - Universita' di Torino
C.so Svizzera 185 - 10149 Torino (Italy)
e-mail: {mino,portinal}@di.unito.it

**Abstract.** Backward reachability on Petri net models has been proposed since the beginning of the development of net theory without giving it a suitable motivation. For this reason, reachability analysis has been successively developed essentially by taking into account forward reachability. In this paper backward reachability analysis is motivated by showing its suitability to diagnostic problem solving. A particular technique for backward reachability analysis (B-W analysis) is defined for a net model called Behavioral Petri Net (BPN), intended to model the behavior of a system to be diagnosed. Such a technique is based on the use of two different types of tokens (normal and inhibitor tokens) whose aim is to represent the truth or falsity of the condition associated to a marked place. A distributed approach for parallel B-W analysis is also proposed. Such an approach is based on a set of processes derived from the structure of a BPN by means of a formal partitioning technique; this allows us to address the problem of mitigating the high inefficiency diagnostic algorithms usually have.

## 1 Introduction

Backward reachability on Petri net models has been proposed since the beginning of the development of net theory (e.g. in the Condition/Event systems [1]) without giving it a suitable motivation. For this reason, reachability analysis has been successively developed essentially by taking into account forward reachability. However, progresses in fields like Artificial Intelligence (AI) showed that a lot of different practical tasks require inference to be performed in backward fashion [13].

One of the most important task requiring backward reasoning is certainly represented by diagnostic problem solving. In this paper, we aim at discussing a particular technique for backward reachability analysis on a Petri net model, called B-W Analysis, that can be actually applied to diagnostic reasoning. This technique relies on a net model called BPN (Behavioral Petri Net) whose aim is to model the behavior of a system to be diagnosed. Formal aspects of B-W Analysis are introduced and its applicability to diagnostic problem solving is shown. The method exploits two different types of tokens (i.e. normal and inhibitor tokens) aimed at modeling the truth or falsity of the condition associated to the marked place. This allows us to point out inconsistencies when looking for the possible "explanation" of a given marking in a BPN.

One of the peculiarities of B-W analysis is its suitability to parallel implementations. In this paper we present a distributed approach for parallel B-W analysis, in which a set of processes is derived starting from the structure of a BPN. Given a BPN, its transitions are partitioned into independent subsets, from which both processes and interactions among them are derived. A simplification technique, whose purpose is that of producing a deadlock-free concurrent program, is also presented. The above technique exploits the information provided by the the marking from which the analysis starts.

The paper is organized as follows: in section 2, basic notions about Petri nets are reviewed; in section 3, the BPN model is introduced; in section 4, B-W Analysis is formally defined and in section 5 its application to diagnosis is discussed; in section 6, an approach to the parallel implementation of B-W Analysis is proposed; in section 7 an example of the parallel execution of B-W Analysis is discussed; final conclusions and comparison with related works are then reported in section 8.

## 2 Petri Nets: Fundamentals

This section outlines some basic definitions on which we will rely throughout the paper.

**Definition 1.** A (Petri) net is a triple N=(P,T,F) where

- $P \cap T = \emptyset$
- $P \cup T \neq \emptyset$
- $F \subseteq (P \times T) \cup (T \times P)$
- $dom(F) \cup cod(F) = P \cup T$

$P$ is the set of places, $T$ is the set of transitions and $F$ is the flow relation represented by means of directed arcs. For each $x \in P \cup T$ we will use the classical notations ${}^\bullet x = \{y|yFx\}$ and $x^\bullet = \{y|xFy\}$[1]. A *marking* is a function $\mu : P \rightarrow \mathbb{N}$ from places to nonnegative integers represented by means of *tokens* into places;

**Definition 2.** Given a Petri net $N = (P, T, F)$ and a marking $\mu$, a marked Petri net is defined as $\Sigma = (P, T, F, \mu)$ (we can also use $\Sigma = (N, \mu)$).

**Definition 3.** Let $\Sigma = (P, T, F, \mu)$ be a marked Petri net; a transition $t \in T$ has concession in $\mu$ if and only if $\forall p \in^\bullet t\ \mu(p) \geq 1$; if $t$ has concession in $\mu$, then $t$ may occur (fire) yielding a new marking $\mu'$ (we write $\mu[t > \mu')$ such that for every place $p \in P$ we have

$$\mu'(p) = \mu(p) - inp(p, t) + out(p, t)$$

where:

$inp(p, t) = 1$ if $(p, t) \in F$; 0 otherwise
$out(p, t) = 1$ if $(t, p) \in F$; 0 otherwise

[1] This notation can be trivially extended to sets as follows: ${}^\bullet X = \bigcup_{x \in X} {}^\bullet x$ and $X^\bullet = \bigcup_{x \in X} x^\bullet$.

The reachability set of a Petri net $N$ from a marking $\mu_0$, indicated as $R(N,\mu_0)$, is the smallest set of markings such that:
1) $\mu_0 \in R(\mu_0)$;
2) if $\mu_1 \in R(N,\mu_0)$ and $\mu_1[t > \mu_2$ for some $t \in T$, then $\mu_2 \in R(N,\mu_0)$.

**Definition 4.** Given a marked Petri net $\Sigma = (P,T,F,\mu_0)$ a place $p \in P$ is *k-bounded* if and only if $\forall \mu \in R(N,\mu_0)\ (\mu(p) \leq k)$; $\Sigma$ is *k-bounded* if and only if $\forall p_i \in P$ ($p_i$ is $k_i$-bounded and $k = max_i\ k_i$); $\Sigma$ is *safe* if and only if it is 1-bounded.

A *safe marking* is a marking such that $\forall p \in P(\mu(p) \leq 1)$. Two transitions are said to be *concurrent* if and only if each time they have both concession, the firing of one does not prevent the other from having concession.

**Definition 5.** A marked Petri net $\Sigma = (N,\mu_0)$ is *deterministic* if and only if

$$\forall \mu \in R(N,\mu_0)\ \forall t_1,t_2 \in T(\textit{if } \mu[t_1 > \textit{ and } \mu[t_2 > \textit{then } t_1 \textit{ and } t_2 \textit{ are concurrent}).$$

## 3 The Behavioral Petri Net Model

In [11], a Petri net model called *Behavioral Petri Net* (*BPN*) was introduced in order to address the problem of model-based diagnosis [5] in a Petri net framework; indeed, a Behavioral Petri Net is intended to model the behavior of a system to be diagnosed, while reachability analysis on the model can be used to perform diagnostic reasoning.

**Definition 6.** A *Behavioral Petri Net* is a 4-tuple $N = (P, T_N, T_{OR}, F)$ where:

1. $T_N \cap T_{OR} = \emptyset$
2. $(P, (T_N \cup T_{OR}), F)$ is a net
3. $F^+$ (transitive closure of $F$) is irreflexive
4. $\forall p \in P(|{}^\bullet p| \leq 1 \wedge |p^\bullet| \leq 1)$
5. $\forall p_1, p_2 \in P\ (({}^\bullet p_1 = {}^\bullet p_2) \wedge (p_1^\bullet = p_2^\bullet) \rightarrow p_1 = p_2)$
6. $\forall t \in T_N\ (|{}^\bullet t| = 1 \wedge |t^\bullet| > 0) \vee (|{}^\bullet t| > 0 \wedge |t^\bullet| = 1)$
7. $\forall t \in T_{OR}\ (|{}^\bullet t| = 2 \wedge |t^\bullet| = 1)$

The set of transitions of a BPN is partitioned into two subsets $T_N$ and $T_{OR}$. Transitions in $T_N$ are intended in the usual way, and can be classified as:

- *Linear Transitions*: $t \in T_N : |{}^\bullet t| = |t^\bullet| = 1$;
- *Fork Transitions*: $t \in T_N : |t^\bullet| > 1$;
- *Join Transitions*: $t \in T_N : |{}^\bullet t| > 1$;

Transitions in $T_{OR}$ are intended to represent the logical connective $OR$; a transition $t \in T_{OR}$ (graphically represented as an empty thick bar) is said to be an *OR-transition* and it represents a *macro transition* whose semantics can be given in terms of a Petri net with inhibitor arcs (see [11]). Informally, a transition $t \in T_{OR}$ has concession in a marking $\mu$ if and only if at least one of its input places is marked. An obvious consequence of the irreflexivity of $F^+$ (axiom 3 of definition 6) is that it defines a partial order, denoted as "$\prec$", over transitions of a BPN; given two transitions $t_1$ and $t_2$:

$$t_1 \prec t_2 \Leftrightarrow t_1 F^+ t_2$$

By considering such a partial order we can define the following *enabling rule*.

**Definition 7.** Given a Behavioral Petri Net $N$ and a marking $\mu$, a transition $t$ is *enabled* at $\mu$ (and so it may fire) if and only if it has concession in $\mu$ and $\nexists t' \prec t$ such that $t'$ has concession in $\mu$.

Definition 7 corresponds to imposing a priority ordering on transitions. Figure 1 shows an example of a BPN; as we will discuss in section 5, such a BPN is intended to represent a partial fault model of a car engine. For instance, transition $t1$ models the fact that an "increased oil consumption" (modeled by place $oil\_cons(incr)$) can be caused by either a "worn state of piston rings" (modeled by place $pist\_ring\_state(worn)$) or a "worn state of pistons" (modeled by place $pist\_state(worn)$).

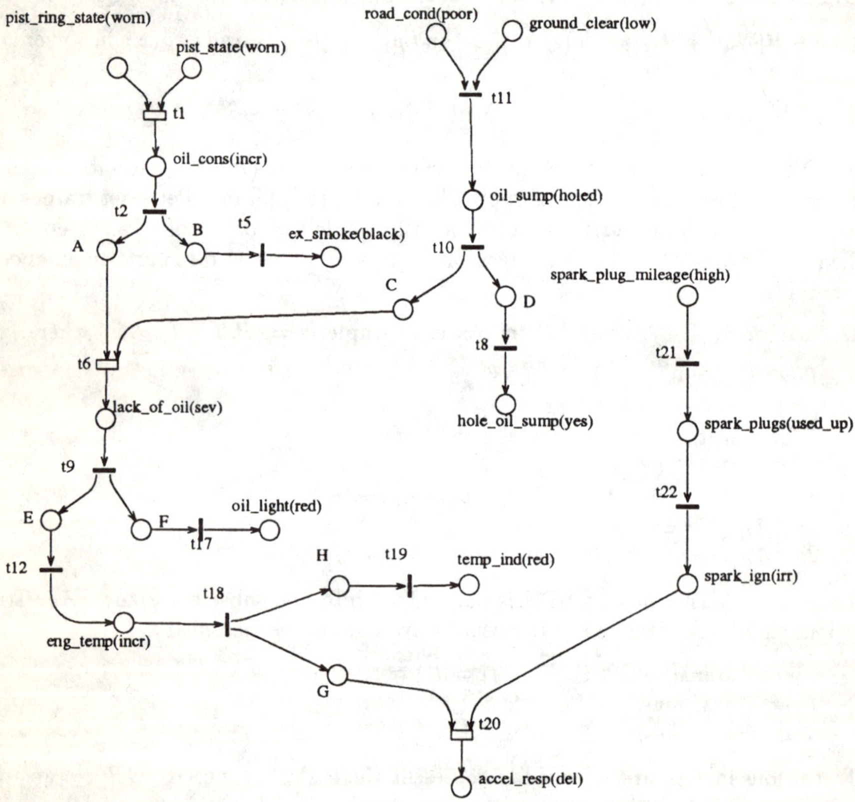


**Fig. 1.** Example of a BPN

Notice also that given a marking where $spark\_ign(irr)$ and $lack\_of\_oil(sev)$ are the only marked places, $t20$ and $t9$ have both concession, however $t9$ is the only enabled transition since $t9 \prec t20$.

**Definition 8.** An *initial marking* of a BPN is a safe marking $\mu_0$ such that if $\mu_0(p) = 1$ then $p$ is a source place (i.e. ${}^\bullet p = \emptyset$).

**Definition 9.** A *marked Behavioral Petri Net* is a pair $(N, \mu)$ where $N$ is a Behavioral Petri Net and $\mu$ is either an initial marking or a marking such that there exists an initial marking $\mu_0$ and $\mu \in R(N, \mu_0)$.

A marking $\mu$ such that no transition is enabled at $\mu$ is called a *final marking*. The following theorems have been proved in [11].

**Theorem 1.** *A marked BPN is* safe *and* deterministic*;*

**Theorem 2.** *In a marked BPN there exists exactly one final marking.*

Safeness property is a consequence of the priority imposed on transition firing and allows us to interpret in a simple way a marked place, i.e. a condition can be associated to the place and a token into the place means that the condition is satisfied; determinism is a trivial consequence of net structure and of the semantics of *OR-transitions*. The existence of a unique final marking in a BPN will be exploited in section 5 to characterize diagnoses on a BPN model.

## 4 The B-W Analysis

Given a Behavioral Petri Net, a particular analysis technique, called *B-W Analysis*, can be defined. Such a technique consists in a backward reachability analysis performed with two different types of tokens called respectively *normal tokens* (black) and *inhibitor tokens* (white). The meaning of normal tokens is as usual: we can associate a condition with a place and a normal token into a place means that such a condition is satisfied. On the contrary, inhibitor tokens represent conditions which are certainly not satisfied in the case under examination (i.e. the inhibition of a particular condition). If a place with associated condition $C$ is marked with this kind of token, then $\neg C$ holds. As a consequence, no constraint is imposed on condition $C$ when the corresponding place is empty. From a logical point of view, this corresponds to considering a three-valued logic whose truth value are $\{\mathbf{true}, \mathbf{false}, \mathbf{unknown}\}$. In order to formally discuss B-W Analysis, we have to generalize the concept of marking.

**Definition 10.** Given a BPN $N = (P, T_N, T_{OR}, F)$, a *b-w marking* is a function $\mu : P \rightarrow \{b, w, 0\}$.

If $\mu(p) = b$ then the place $p$ is marked with a normal token (black), if $\mu(p) = w$ then the place $p$ is marked with an inhibitor token (white) and if $\mu(p) = 0$ then the place $p$ is empty. Let us now introduce the backward enabling rule for this kind of analysis. Notice that in order to define such a rule, we need to consider the inverse relation of the partial order $\prec$ previously introduced; given two transitions $t_1$ and $t_2$ we define $t_1 \succ t_2 \Leftrightarrow t_2 \prec t_1$.

**Definition 11.** Let $N = (P, T_N, T_{OR}, F)$ be a BPN, $\mu$ a b-w marking and $t \in T_N \cup T_{OR}$ a transition:

- $t$ is *b-enabled* at $\mu$ if $\forall p \in t^\bullet (\mu(p) = b \wedge p^\bullet = \emptyset)$;
- $\neg t$ is *b-enabled* at $\mu$ if $\forall p \in t^\bullet (\mu(p) = w \wedge p^\bullet = \emptyset)$;

- $t$ is *b-enabled* at $\mu$ if $\forall p \in t^\bullet(\mu(p) = b \wedge p^\bullet \neq \emptyset \wedge \nexists t' \succ t(t'\ or\ \neg t'\ are\ b-enabled\ at\ \mu))$;
- $\neg t$ is *b-enabled* at $\mu$ if $\forall p \in t^\bullet(\mu(p) = w \wedge p^\bullet \neq \emptyset \wedge \nexists t' \succ t(t'\ or\ \neg t'\ are\ b-enabled\ at\ \mu))$.

The basic concept underlying this definition is that $t$ *b-enabled* means that we are reconstructing an evolution of the system where $t$ must fire, while $\neg t$ *b-enabled* means that we are reconstructing an evolution where $t$ must not fire.
Particularly relevant for B-W Analysis is the concept of *inconsistent marking.*

**Definition 12.** Given a BPN $N = (P, T_N, T_{OR}, F)$, a b-w marking $\mu$ is said to be *inconsistent* if and only if $\exists t \in T_N$ and $\exists p_i, p_j \in t^\bullet$ such that $\mu(p_i) = b$ and $\mu(p_j) = w$.

This definition reflects the fact that, given a fork transition, it is impossible to have its output places consistently marked with different types of tokens. Another feature of B-W Analysis is the possibility of forcing the backward firing of fork transitions, by performing an *abstraction step* as defined below.

**Definition 13.** Given a BPN $N = (P, T_N, T_{OR}, F)$, a transition $t \in T_N$ such that $t^\bullet = \{p_1 \ldots p_r\}$ is *forced* at the b-w marking $\mu$ if and only if:

1. $t$ is not b-enabled at $\mu$;
2. $\mu$ is not inconsistent;
3. $\exists p_i(1 \leq i \leq r)\mu(p_i) \neq 0$;
4. $\nexists t' \succ t(t'\ or\ \neg t'\ are\ b-enabled\ or\ forced\ at\ \mu)$.

The abstraction step can be formalized in the following way:

**if** $t$ **is forced then** $\forall p_i \in t^\bullet : \mu(p_i) = 0 \Rightarrow (\mu(p_i) = b) \vee (\mu(p_i) = w)$ **in such a way that** $\mu$ **is not inconsistent.**

This means that we have to mark empty places in the output of $t$ with the same type of tokens contained in the marked places.

**Definition 14.** Given a BPN and a b-w marking $\mu$, a *b-step* at $\mu$ is the maximal set $s = \{t_{i_1} \ldots t_{i_r}, \neg t_{j_1} \ldots \neg t_{j_s}\}$ having the property that each $t_{i_k}$ $(1 \leq k \leq r)$ and each $\neg t_{j_h}$ $(1 \leq h \leq s)$ is b-enabled at $\mu$.
Every $s' \subseteq s$ is said to be a *b-substep.*

Notice that in defining a b-step a condition of maximality is required. In order to show how a new b-w marking is produced from an old one in B-W Analysis, we introduce some *backward firing rules.* In the definition of the backward firing rules, given a place $p$ and a b-w marking $\mu$, we have to distinguish two possible cases:

1. ${}^\bullet p = \{t\} \wedge t \in T_N$ $({}^\bullet t = \{p_1 \ldots p_r\},\ r \geq 1)$
2. ${}^\bullet p = \{t\} \wedge t \in T_{OR}$ $({}^\bullet t = \{p_0, p_1\})$

In each case we can further distinguish two alternatives; let $\mu'$ be a marking resulting from the b-step $s$ at $\mu$:

1.a) $\mu(p) = b \Rightarrow \{t\}$ is a b-substep of $s$, $\mu'(p_1) = \ldots = \mu'(p_r) = b$ and $\mu'(p) = 0$.
1.b) $\mu(p) = w \Rightarrow \{\neg t\}$ is a b-substep of $s$ and we have a non-deterministic choice such that $< \mu'(p_1), \ldots \mu'(p_r) >$ is any combination of values from $\{b, w\}$ different from that corresponding to $\forall p_i$ $(1 \leq i \leq r)$ $\mu'(p_i) = b$.
2.a) $\mu(p) = w \Rightarrow \{\neg t\}$ is a b-substep of $s$, $\mu'(p_0) = \mu'(p_1) = w$ and $\mu'(p) = 0$.
2.b) $\mu(p) = b \Rightarrow$ non-deterministic choice among the following alternatives:
- $\{t\}$ is a b-substep of $s$, $\mu'(p_0) = b, \mu'(p_1) = w$ and $\mu'(p) = 0$;
- $\{t\}$ is a b-substep of $s$, $\mu'(p_0) = b, \mu'(p_1) = b$ and $\mu'(p) = 0$;
- $\{t\}$ is a b-substep of $s$, $\mu'(p_0) = w, \mu'(p_1) = b$ and $\mu'(p) = 0$.

In figure 2 the above backward firing rules are graphically represented. B-W Analysis

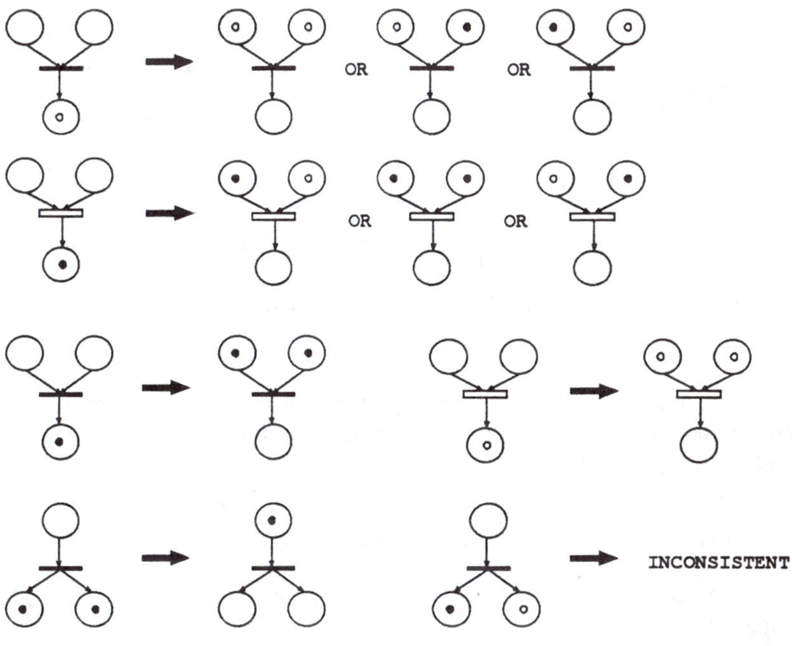


**Fig. 2.** Backward Firing Rules of a BPN

starts with a b-w marking $\mu$ such that $\mu(p) \neq 0 \rightarrow p^\bullet = \emptyset$ (a place $p$ such that $p^\bullet = \emptyset$ is said to be a sink place); a given alternative ends when either an initial b-w marking $\mu_0$ (i.e. $\mu_0(p) \neq 0 \rightarrow p$ is a source place) or an inconsistent b-w marking is reached. In the first case empty source places represent initial conditions that are not significant for the case under examination, while source places marked with normal and inhibitor tokens represent initial conditions that have been proved true and false respectively.

A backward reachability graph can be obtained in B-W Analysis as shown in figure 3. Notation $p[b]$ means that place $p$ is marked with a black token and $p[w]$ that $p$ is marked with a white token; arcs are labeled with b-steps and underlined

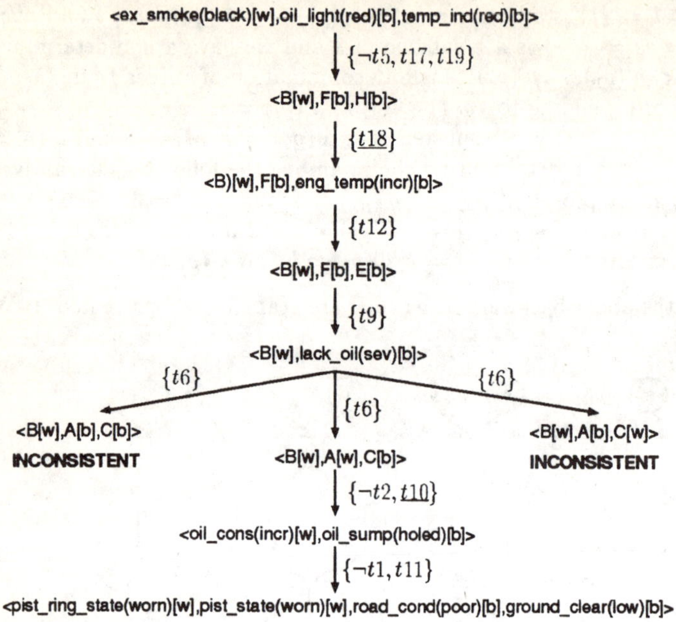


**Fig. 3.** Backward Reachability Graph for B-W Analysis

transitions represent forced transitions. The figure represents the result of B-W analysis on the BPN shown in figure 1 when the starting marking is $ex\_smoke(black)[w], oil\_light(red)[b], temp\_ind(red)[b]$.

## 5 Applying B-W Analysis to Diagnostic Problem Solving

One of the most important aspects of the BPN model concerns its suitability in modeling the behavior of a system to be diagnosed. In [11] the equivalence between the BPN model and a particular class of logical models quite popular in model-based diagnosis (i.e. hierarchical definite logic programs) is discussed and formally shown. In particular, following an approach pioneered in [17], by suitably introducing a notion of "derivation" in the net model, it is possible to show its equivalence with respect to classical derivation in the logical model. As a consequence, formal notions concerning diagnostic problems and their solutions can be re-formulated in the framework of Behavioral Petri Nets.

Following [3], a *diagnostic problem* can be suitably described by the 4-tuple $DP =< S, Exp, \Psi^+, \Psi^- >$ where $S$ is a set of logical formulae representing the model of the system to be diagnosed, $Exp$ is a set of ground atoms in terms of which the solutions have to be computed, $\Psi^+$ is a set of ground atoms representing observations to be covered in the case under examination and $\Psi^-$ is a set ground atoms representing the values of parameters conflicting with the observations. Observations are usually represented by manifestations of internal states of the modeled system, while solutions are computed in terms of initial causes of a malfunction. We can then think to $Exp$ as the set of possible initial causes of the behavior of the system under examination. If we abstract from time, we can assume that at most one instance of a manifestation can be observed in a given diagnostic problem; this means that every admissible value of a manifestation different from the observed one is conflicting with the observations. Given a diagnostic problem $DP =< S, Exp, \Psi^+, \Psi^- >$, a set $E \subseteq Exp$ is a solution to $DP$ or a *diagnosis* for DP if and only if:

$$\forall m \in \Psi^+ \; S \cup E \vdash m\,, \qquad \forall n \in \Psi^- \; S \cup E \not\vdash n$$

where $\vdash$ is the logical derivation symbol. Notice that $\Psi^+$ is in general a suitable subset of the set of all the observations $OBS$ (see also the example at the end of this section) and that different choices of $\Psi^+$ give rise to different definitions of diagnosis (see [3] for a more detailed discussion).

Let us introduce the labeling $\Phi : P \rightarrow B_S$ where $P$ is the set of places of a BPN and $B_S$ is the *Herbrand base* [6] of the logical model $S$ (i.e. the set of ground atoms of $S$). Since a BPN can play the role of model of the system to be diagnosed and since a notion of derivation equivalent to the logical one can be defined, then a diagnostic problem can be soundly captured in terms of a BPN.

**Definition 15.** Given a diagnostic problem $DP =< S, Exp, \Psi^+, \Psi^- >$ and a Behavioral Petri Net $N_S$ corresponding to $S$, we can define the diagnostic problem in terms of the BPN model in the following way: $DP =< N_S, P_E, P^+, P^- >$ where: $P_E = \{p \in P | \Phi(p) \in Exp\}$, $P^+ = \{p \in P | \Phi(p) \in \Psi^+\}$ and $P^- = \{p \in P | \Phi(p) \in \Psi^-\}$.

**Definition 16.** A marking $\mu$ of a Behavioral Petri Net *covers* a set of places $Q$ if and only if $\forall p \in Q \rightarrow \mu(p) = 1$.

**Definition 17.** A marking $\mu$ of a Behavioral Petri Net *zero-covers* a set of places $Q$ if and only if $\forall p \in Q \rightarrow \mu(p) = 0$.

The notion of diagnostic solution can then be captured with the following theorem (that has been proved in [11]).

**Theorem 3.** *Given a diagnostic problem $DP =< N_S, P_E, P^+, P^- >$, an initial marking $\mu_E$ is a* solution *to $DP$ if and only if the* final marking *$\mu$ of $N_S$* covers *$P^+$ and* zero-covers *$P^-$.*

It can be shown that, given this formalization, B-W Analysis can be actually used to compute diagnostic solutions. Let us consider the BPN of figure 1; we will consider the $\Phi$ labeling to be the *identity* function for every place except for places indicated as capital letters whose labeling is undefined. As already mentioned, this BPN represents a partial model concerning the faulty behavior of a car engine. Let us also consider the diagnostic problem characterized by the following observations: $OBS = \{ex\_smoke(normal), oil\_light(red), temp\_ind(red)\}$;
in case we require abnormal observations to be covered by a solution we obtain the following set of places:

$P_E = \{pist\_ring\_state(worn), pist\_state(worn), road\_cond(poor), ground\_clear(low), spark\_plug\_mileage(high)\}$;

$P^+ = \{oil\_light(red), temp\_ind(red)\};$
$P^- = \{ex\_smoke(black)\}.$

Diagnostic solutions can be computed by performing B-W Analysis starting from the following b-w marking $\mu$:

$$\mu(p) = \begin{cases} b \text{ if } p \in P^+ \\ w \text{ if } p \in P^- \\ 0 \text{ otherwise} \end{cases}$$

We can get all the possible solutions by exploring all the possible non-deterministic alternatives. In case we get an inconsistent marking, the explored alternative is not correct and we have to examine other possible search paths. By considering initial markings obtained at the end of B-W Analysis, the obtained solutions will be formed by the ground atoms corresponding to places marked with normal tokens and by the negation of the ground atoms corresponding to places marked with inhibitor tokens.

Graph of figure 3 represents the B-W Analysis leading to the computation of the solutions to the diagnostic problem $DP$ defined above; in particular the only possible diagnosis is: $E = \{road\_cond(poor), ground\_clear(low), \neg pist\_ring\_state(worn), \neg pist\_state(worn)\}$.

Notice that nothing has to be said about *spark_plug_mileage* since it is irrelevant for the case under examination.

## 6 Exploiting Parallelism in Backward Reachability Analysis

In this section we describe how B-W Analysis can be performed by means of a set of asynchronous processes, executing in a concurrent fashion and interacting through exchange of messages. The concurrent implementation of B-W Analysis corresponds to a concurrent implementation of the token game executed in a backward fashion.

Two different kinds of approaches have been proposed in the literature for the concurrent implementation of the token game, usually classified as *centralized* or *decentralized* ([2, 14, 15]). In the former approach the token game is carried out by a set of processes (usually identified with the transitions of the net) whose execution is controlled by a coordinator process. On the other hand decentralized approaches are based on a set of sequential processes (each of them managing a set of transitions) running asynchronously and interacting by a communication/synchronization mechanism.

As pointed out in [2] the centralized approach is not adequate when performance is the major concern, because of the intermediate coordinator process presence. Moreover if the message passing model is adopted, then the number of messages needed to implement the centralized approach is much larger than needed to implement the decentralized solution, resulting in a considerable overhead. The approach we choose in this paper is then the decentralized one.

The use of a set of asynchronous processes produces several advantages that can be better understood by comparing this execution model with a sequential one. In the latter, B-W analysis is performed step-by-step and transitions are considered as *static* entities that need to be tested for their enabling; the modifications in the marking produced by the fired step are computed looking at the type of transitions fired. On the contrary, in the proposed decentralized approach transitions are considered as *active* entities that decide to fire independently, according to the availability of the tokens they need. Moreover each transition knows the actions it has to perform when firing, so the produced modifications in the partial marking related to it are defined locally to transitions; global marking modifications are computed only when needed and only for meaningful situations. Finally when several alternative markings can be generated from a given one through different steps, all of them can be explored in the same execution.

Given a BPN model, a set of sequential processes implementing B-W analysis is automatically derived from the net structure. The basic idea underlying our approach is that independent transitions should be managed by different processes so that concurrency is effectively exploited.

Our approach works as follows: the first step is the derivation of a partition of the transitions of a given BPN model and then a process is associated with each subset of transitions. In this paper we will not deal with any specific implementation language, but rather we will present a general technique to obtain the parallel program implementing B-W analysis. The language we use to describe processes is a "meta-language", based on explicit message passing, easily translatable in one of the classical parallel languages such as OCCAM.

### 6.1 Net partitioning

As said above, the first step is that of deriving a partition of the set of transitions of a given BPN. To partition the set of transitions we use a binary relation $\asymp$ defined over $(T_N \cup T_{OR}) \times (T_N \cup T_{OR})$. In the following let $N = (P, T_N, T_{OR}, F)$ be a BPN and $T = T_N \cup T_{OR}$.

**Definition 18.** Let $t_i, t_j$ be two transitions of a BPN $N$. The relation $\asymp \subseteq T \times T$ is defined as:

$$t_i \asymp t_j \Leftrightarrow \mid {}^\bullet t_i \mid = 1 \text{ and } \mid t_j^\bullet \mid = 1 \wedge {}^\bullet t_i = t_j^\bullet$$

The following property can be easily proved.

**Property 1** *Let $t_i, t_j, t \in T$ be transitions of a BPN $N$. If both $t_i \asymp t$ and $t_j \asymp t$, then $t_i = t_j$.*

**Definition 19.** Given a BPN $N$, a transition $t \in T$ is said *starting* iff $\mid t^\bullet \mid > 1$ or $(\mid t^\bullet \mid = 1 \wedge t^\bullet \cap {}^\bullet t_j = \emptyset)$, $\forall t_j \in T$.

The above definition corresponds to say that a starting transition has either several output places or exactly one output sink place.

Given a starting transition $t$, it is possible to define a subset of transitions, called the *partitioning subset* generated by $t$, in the following way.

**Definition 20.** Let $t \in T$ be a starting transition of a BPN $N$. The partitioning subset $T_i \subseteq T$ generated by transition $t$ is defined in the following way:

1. $t \in T_i$
2. if $t_i \in T_i \wedge \exists t_j \in T$: $t_i \asymp t_j$ then $t_j \in T_i$.

Given a starting transition $t$, its corresponding partioning subset $T_i$ is built by computing the transitive closure $\asymp^+$ of the $\asymp$ relation. This process stops when the first transition $t_j$, such that $\forall t_k \in T\ t_j \not\asymp t_k$, is added to $T_i$. Such a transition $t_j$ is said to be the *ending* transition of $T_i$. By definition of the $\asymp$ relation, an ending transition $t$ can be characterized in the following way: either (a) $\mid {}^\bullet t \mid > 1$ or (b) ${}^\bullet t = \{p\}$ and p is a source place or (c) ${}^\bullet t = \{p\}$ and ${}^\bullet t \subseteq t_j^\bullet$ and $t_j$ is a starting transition.

We now prove that the partitioning subsets of a BPN (i.e. the set of partitioning subsets generated by all the starting transitions of the BPN) determine a partition of the set of transitions of the net.

**Lemma 4.** *Let $Q = \{T_i\}$ be the set of partitioning subsets of a BPN $N$. Then:*

1. $\forall T_i, T_j \in Q,\ T_i \cap T_j = \emptyset$ *and*
2. $\cup_{i=1}^{|Q|} T_i = T$

**Proof of 1**
By contradiction suppose that $\exists T_i, T_j \in Q,\ T_i \neq T_j$ such that $T_i \cap T_j \neq \emptyset$ and consider a transition $t \in T_i \cap T_j$. If $t$ is a starting transition, then, by definition 20, $T_i = T_j$ which is a contradiction.

On the other hand, if $t$ is not a starting transition then $\exists t_i \in T_i, t_j \in T_j, t_i \neq t_j$ such that both $t_i \asymp t$ and $t_j \asymp t$. By property 1, $t_i = t_j$ which is a contradiction.
**Proof of 2**
By contradiction suppose that $\exists t \in T$ such that $t \notin \cup_{i=1}^{|Q|} T_i$. If $t$ is a starting transition, then, by definition 20, $t$ belongs to the partioning subset it generates, so a contradiction arises. Suppose consequently that $t$ is not a starting transition. In this case if $t \notin \cup_{i=1}^{|Q|} T_i$ then

$$\forall T_i \in Q, \forall t_i \in T_i, t_i \not\asymp t \tag{1}$$

It should be noted that if $t$ is not a starting transition, then $\mid t^\bullet \mid = 1$. Denoting as $p$ the unique output place of $t$, condition (1) implies that $p$ is not in the input set of any transition belonging to any partitioning subset, i.e. $p$ is a sink place. This in turn implies that $t$ is a starting transition, since its output set comprises only one sink place, and this contradicts the hypothesis that $t$ is not a starting transition.
□

Given a partition of the set of transitions, it is possible to generate independent subnets called *transition-generated subnets.*

**Definition 21.** Let $N = (P, T_N, T_{OR}, F)$ be a BPN and let $T = T_N \cup T_{OR}$. Given a partition $\{T_1, \ldots, T_n\}$ of $T$, we define the *transition-generated subnets* (TGS) of $N$ w.r.t. the partition to be the nets $N_i = (P_i, T_i, F_i)$ $(i = 1, \ldots, n)$ where:

- $P_i = P \cap ({}^\bullet T_i \cup T_i^\bullet)$
- $F_i = F \cap ((P_i \times T_i) \cup (T_i \times P_i))$

Given a TGS $N_i$, we denote as $S(N_i)$ its starting transition and as $E(N_i)$ its ending transition.

Figure 4(a) shows the partitioning subsets $T_i$, obtained from the net of figure 1. In figure 4(b), for the sake of clarity, only some of the transition generated subnets (in particular those corresponding to $T_1, T_2$ and $T_8$) have been highlighted.

$T_1 = \{t1, t2\}$

$T_2 = \{t6, t9\}$

$T_3 = \{t5\}$

$T_4 = \{t10, t11\}$

$T_5 = \{t8\}$

$T_6 = \{t21, t22\}$

$T_7 = \{t17\}$

$T_8 = \{t12, t18\}$

$T_9 = \{t20\}$

$T_{10} = \{t19\}$

**(a)**

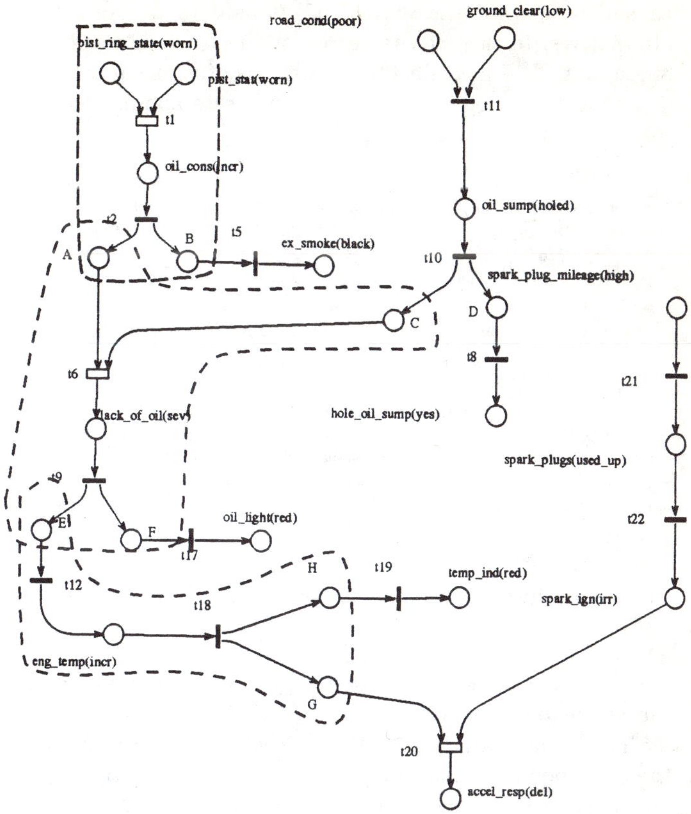


**(b)**

**Fig. 4.** An example of the application of the partitioning technique

## 6.2 Processes derivation

Given a BPN model, a process is associated with each TGS generated from it. Each process is constructed by providing, for each transition in a given TGS, an *activity* implementing its backward firing rule, by sequentially composing the above activities beginning from the starting transition and consecutively adding those related to transitions in the same TGS. An additional process, called the *Assembly Process* ($AP$), is used to collect partial results from processes holding source places and it is responsible for the assembling of final solutions.

According to the definitions given is section 3, each transition $t$ of a BPN $N = (P, T_N, T_{OR}, F)$ can be classified as Fork, Join, Or and Linear transition. To each transition class a simple activity, performing the backward firing rule for that class, can be associated as depicted in figure 5, where the following notation has been used: the input and output parameter lists have been specified using the keywords

**in** and **out**, while the symbol '□' is used to denote different mutually exclusive alternatives. In the activity corresponding to Join transitions we denote as $A_i$ one among the $2^n - 1$ possible instantiations of values to output variables (except that assigning b to all of them). The activity corresponding to linear transitions has been

```
procedure Join(in p ;out p1, p2, ..., pn );
begin
  if p = b   then
    p1 := b; p2 := b; ... ; pn := b
  else
    nondeterministically do
      □ A1
      □ A2
      ⋮
      □ A2^n-1
    enddo
endif
end
```

```
procedure Or(in p ; out p1, p2);
begin
if  p = w  then
  p1 := w, p2 := w
else
  nondeterministically do
    □ p1 := b, p2 := w
    □ p1 := b, p2 := b
    □ p1 := w, p2 := b
  enddo
endif
end
```

```
procedure Fork(in p1, p2, ..., pn ;out p );
begin
if ( p1, p2, ..., pn ) inconsitent then quit
endif
if ∃ i : pi = b  then p := b
else  p := w
endif
end
```

**Fig. 5.** Activities associated to transition classes

omitted since it corresponds to a copy of the contents of its unique input variable to the unique output variable.

The behavior of an activity is derived from the backward firing rule, defined for the corresponding class of transitions, in the following way: given a transition $t$, we introduce in the related activity an output variable for each places in $^\bullet t$ and one input variable for each place in $t^\bullet$. The above variables contain either black (b) or white (w) tokens; if no token is contained in a place, the content of the corresponding variable is undefined. In order for an activity to start, the contents of all its input variables must be defined and when it ends, the contents of its output variables are updated. An activity corresponding to a transition having more than one output places is called *synchronization point*, since all its input variables must be defined in order for the activity to start. Since the set of output variables of an activity $t_1$ may not be disjoint from the set of input variables of another activity $t_2$, the end of $t_1$ may imply the start of $t_2$; $t_2$ receives as input the data items produced by $t_1$. In this case we say that $t_1$ and $t_2$ are interacting activities. The content of variables corresponding to sink places is considered as a data item provided by the external environment, while in other places the data items are provided by other activities.

It should be noted that the set of TGSs does not form a partition over the set of places of a given BPN. This implies that some processes have variables corresponding to the same place of the BPN. In this case we consider the variables as input and output buffers of a communication channel among processes. More precisely if $N_i, N_j$ are two TGS such that $^\bullet E(N_i) \cap S(N_j)^\bullet \neq \emptyset$, then there is a communication channel between processes $\Pi_i$ and $\Pi_j$, respectively corresponding to $N_i$ and $N_j$, and the flow of messages occurs *asynchronously* (i.e. there is no rendez-vous) from $\Pi_i$ to $\Pi_j$.

The direct use of TGSs to derive the set of processes performing B-W analysis can lead to deadlock problems. In fact it may happen that some processes holding sink places are never active due the particular starting marking. Consequently some other processes, waiting messages from the inactive ones, are blocked. This effect is propagated through the set of processes, producing thus a deadlock. In the sequential execution model transitions which are never active are forced to fire (i.e. they are forced transitions). Forced transitions could also be used in the concurrent execution model, but a more efficient solution can be devised. This problem can indeed be solved by observing that if the starting transition of a TGS cannot receive tokens in some of its output places, then these places can be safely removed. If all the output places of the starting transition of a TGS are removed, then the overall TGS can be removed. Considering the concurrent program, this step corresponds to remove input channels and processes. To obtain a deadlock free concurrent program a simplification step over the set of TGSs is performed. The simplification step is defined as follows:

1. Given a TGS $N_i$, remove all the places $p \in S(N_i)^{\bullet}$ which are unmarked sink places;
2. if $S(N_i)^{\bullet}$ becomes empty then remove $N_i$ from the set of TGSs
3. if $N_i$ has been removed, then remove all the places in ${}^{\bullet}E(N_i)$.

The simplification step is iteratively performed considering each TGS in turn. After the simplification step, the set of processes corresponding to the remaining TGSs is built. In this way no process will wait for a message that cannot be sent, so the resulting program is deadlock free.

The set of TGSs corresponding to a given BPN can then be considered as a template, for the set of processes performing B-W analysis, which can be instantiated according to the particular starting marking provided for B-W analysis. When such a marking is specified, a simplification step (as described above) is performed and only the processes corresponding to the remaining TGSs are built and scheduled for execution. The construction of the set of TGS need to be performed only once for a given BPN model, while the simplification step has to be performed each time a new starting marking for B-W analysis is specified.

### 6.3 Multiple solutions and inconsistency checking

As said above, one of our design goals is that of analyzing all the possible evolutions of the system. In order to perform this task the behavior of the elementary activities needs to be modified as discussed below. When a nondeterministic choice is present, each possible alternative is generated and then propagated among activities, allowing the computation of the overall set of solutions within a single execution. It should be noted that exhaustive generation of alternatives implies that each place can simultaneously contain more tokens, each of them being related to a particular alternative. To correctly manage this situation it is necessary to keep track of the different evolutions generated. This can be accomplished by modifying the interpretation of variables, considering them as memory buffers which can contain several items of the type $< c, \nu >$ where $c \in \{b, w\}$ represents either a black ($c = b$) or white ($c = w$) token and $\nu$, called *evolution identifier* (**eid**), is a string denoting a path in the backward reachability graph. In a similar way the content of each message exchanged among processes can be defined. Nondeterministic choices are then eliminated generating all the possible alternatives, and to each of them a different eid is assigned. Using this method all the possible alternatives, generated during B-W analysis, can be considered in the same execution; this means that all possible solutions to a diagnostic problem can be collected at the end of the analysis. It should be noted that in this way, at the end of the analysis, the backward reachability graph is not available to the Assembly Process. Nevertheless it can be easily built by *AP* if data items are modified to include also information about the firing of transitions.

The choice of a set of processes running asynchronously implies that there is no explicit representation of the marking, which instead is split among different processes. This situation implies that a protocol must be devised for inconsistency checking. In fact a process discovering the inconsistency of a given marking should signal the inconsistent situation to all the other processes managing different parts of the same marking. However, because of the lack of a global state, it may be the case that the process discovering the inconsistency does not know the identities of the other processes holding part of the same marking. The solution we propose, called *i-propagation*, is based on the propagation of the inconsistency towards synchronization points. The process discovering the inconsistent marking produces a data item of the type $< i, \nu >$ where $i$ means "inconsistent" and $\nu$ is the current **eid**; such a data item is then propagated by means of messages to other processes. Inconsistency is discovered by means of a merge operation, defined on the above mentioned data items, performed at each synchronization point. The merge operation is a commutative operation defined as follows:

**Definition 22.** Let $\nu'$ be a prefix string of $\nu$ and $c, c' \in \{b, w\}$ such that $c \neq c'$.

- $\mathbf{merge}(< c, \nu >, < c, \nu' >) =< c, \nu >$
- $\mathbf{merge}(< c, \nu >, < c', \nu' >) =< i, \nu >$
- $\mathbf{merge}(< c, \nu >, < i, \nu' >) = \mathbf{merge}(< i, \nu >, < c, \nu' >) =< i, \nu >$

Since the first operation after a synchronization point corresponds to a Fork transition, the merge result is put into the location corresponding to the input place of such a transition. The merge operation is performed on data items with the *same* eid, contained into *different* memory buffers (corresponding to output places of the transition). At the end of the analysis the Assembly Process checks all the data items contained in the locations corresponding to source places. If one of these places contains a data item of the kind $< i, \nu >$, then from each source place the data items $< c, \nu >$, which are related to the same evolution $\nu$, are removed and not considered as final solutions. It should be noted that the introduction of the merge operation requires a slight modification of the activities associated to transitions (see [11] for more details).

## 7 An Example

In this section we describe, as an example, the derivation of the parallel program implementing B-W analysis for the BPN depicted in figure 4 and its execution on a particular starting marking. In what follows we denote as $N_i$ the transition generated subnet corresponding to the partitioning subset $T_i$, and with $\Pi_i$ the process corresponding to $N_i$. Suppose that the starting marking for B-W analysis is $ex\_smoke(black)[w], oil\_light(red)[b], temp\_ind(red)[b]$. The simplification step over the set of TGSs is performed as follows:

- $N_9$ is removed from the set of TGS since $S(N_9)^\bullet$ contains only one unmarked sink place;
- places $G$ and *spark_ign(irr)* are removed as a consequence of the removal of $N_9$;
- $N_6$ is removed since the unique place in $S(N_6)^\bullet$ (*spark_ign(irr)*) has been removed as a consequence of the removal of $N_9$;
- $N_5$ is removed since $S(N_5)^\bullet$ contains only one unmarked sink place.

After the simplification step described above, the set of processes corresponding to the remaining TGSs is built. In figure 6 the interaction among processes is represented through a graph in which nodes correspond to processes and directed arcs corresponds to communications whose flow is represented by arc directions.

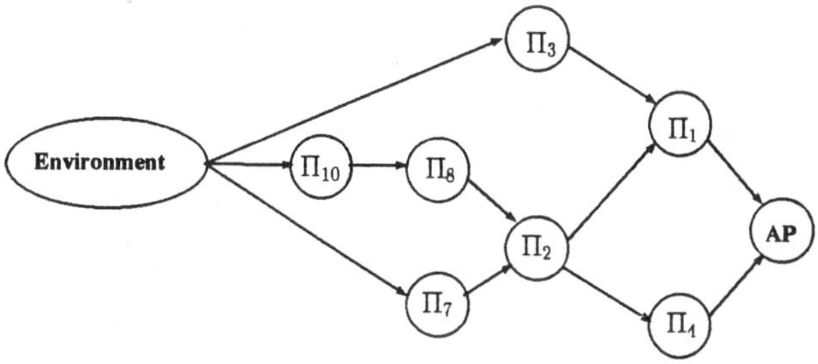


**Fig. 6.** Interaction structure of processes corresponding to TGSs in Fig. 4

Let us now present an example of the execution of the concurrent program on the starting marking above defined. For the sake of simplicity in the description we suppose that if more alternatives are generated by a process, then all the corresponding messages are sent in a single bunch to each destination process. Moreover we do not consider the fact that different processes may need different times to perform internal computations before sending messages. This implies that after receiving a message, the next action performed by a process is the sending of the messages corresponding to the results of its internal computation. This expedient allows us to describe the execution of the example program in a step-by-step fashion. This assumption is made only for description purposes; actually in real executions all the described actions occur asynchronously.

In the description we use the notation $\Pi_i \xrightarrow{Q} \Pi_j : \{m_1, \ldots, m_k\}$ to denote that process $\Pi_i$ sends to process $\Pi_j$ the bunch of messages $m_1, \ldots, m_k$ through channel $Q$. The notation $\Pi_j \overset{Q}{\rightsquigarrow} \Pi_i$ is used to denote that process $\Pi_i$ is waiting for a message from process $\Pi_j$ through channel $Q$. The specification of the channel on which messages are sent is needed if a process holds more than one source place (in our example $\Pi_1$ and $\Pi_4$). In this way the Assembly process can build final solutions. It should be noted that a process may have more than two input channels. The **merge** operation should be therefore applied in a pipeline fashion, i.e. if $\{m_1, \ldots, m_n\}$ is a set of messages such that $m_i$ is received from channel $p_i$, then the resulting sequence of applications of the **merge** operation is

$$merge(m_n, merge(m_{n-1}, merge(m_{n-2}, merge(\ldots merge(m_2, m_1)\ldots))))$$

The function **merge***($\{m_1, \ldots, m_n\}$) is used to denote the merge operation applied to a set of messages. This function is recursively defined as

- **merge***($\{m_1, \tau\}$)=$m_1$
- **merge***($\{m_1, m_2, \ldots, m_n\}$)=**merge**(**merge***($\{m_1, \ldots, m_{n-1}\}$), $m_n$)

where $m_1, \ldots, m_n$ are messages and $\tau$ denotes the empty message.
In the following we denote as $C_1, C_2, C_3, C_4$ respectively the communication channels *pist_ring_state(worn)*, *pist_state(worn)*, *road_cond(poor)* and *ground_clear(low)*. The generation of a new eid is performed through the concatenation of the previous eid with an integer representing a new evolution generated, separated through the character '#'. Each data item provided by the external environment is assumed to have the eid set to the empty string $\epsilon$.

**step 0:** Initially all the processes are waiting for input. The external environment performs the following actions: $Env \longrightarrow \Pi_3 :< w, \epsilon >$, $Env \longrightarrow \Pi_7 :< b, \epsilon >$, $Env \longrightarrow \Pi_{10} :< b, \epsilon >$.

**step 1:** $\Pi_3 \longrightarrow \Pi_1 :< w, \epsilon >$, $\Pi_{10} \longrightarrow \Pi_8 :< b, \epsilon >$, $\Pi_7 \longrightarrow \Pi_2 :< b, \epsilon >$. $\Pi_2 \rightsquigarrow \Pi_1$, $\Pi_8 \rightsquigarrow \Pi_2$, $\Pi_2 \rightsquigarrow \Pi_4$, $\Pi_1 \overset{C_1}{\rightsquigarrow} AP$, $\Pi_1 \overset{C_2}{\rightsquigarrow} AP$, $\Pi_4 \overset{C_3}{\rightsquigarrow} AP$, $\Pi_4 \overset{C_4}{\rightsquigarrow} AP$.

**step 2:** $\Pi_8 \longrightarrow \Pi_2 :< b, \epsilon >$. $\Pi_2$ performs merge*($\{< b, \epsilon >, < b, \epsilon >\}$), producing the message $< b, \epsilon >$;
$\Pi_2 \rightsquigarrow \Pi_1$, $\Pi_2 \rightsquigarrow \Pi_4$, $\Pi_1 \overset{C_1}{\rightsquigarrow} AP$, $\Pi_1 \overset{C_2}{\rightsquigarrow} AP$, $\Pi_4 \overset{C_3}{\rightsquigarrow} AP$, $\Pi_4 \overset{C_4}{\rightsquigarrow} AP$.

**step 3:** $\Pi_2$ produces the three messages corresponding to different assignments, since the output place of its ending transition ($t6$) contains $< b, \epsilon >$.
$\Pi_2 \longrightarrow \Pi_1 : \{< b, \#1 >, < b, \#2 >, < w, \#3 >\}$
$\Pi_2 \longrightarrow \Pi_4 : \{< w, \#1 >, < w, \#2 >, < b, \#3 >\}$
$\Pi_2 \rightsquigarrow \Pi_1$, $\Pi_2 \rightsquigarrow \Pi_4$ $\Pi_1 \overset{C_1}{\rightsquigarrow} AP$, $\Pi_1 \overset{C_2}{\rightsquigarrow} AP$, $\Pi_4 \overset{C_3}{\rightsquigarrow} AP$, $\Pi_4 \overset{C_4}{\rightsquigarrow} AP$.

**step 4:** $\Pi_1$ performs the following merge* operation:
merge*($\{< w, \epsilon >, < b, \#1 >\}$) $=< i, \#1 >$
merge*($\{< w, \epsilon >, < b, \#2 >\}$) $=< i, \#2 >$
merge*($\{< w, \epsilon >, < w, \#3 >\}$) $=< w, \#3 >$
$\Pi_1 \xrightarrow{C_1} AP : \{< i, \#1 >, < i, \#2 >, < w, \#3 >\}$
$\Pi_1 \xrightarrow{C_2} AP : \{< i, \#1 >, < i, \#2 >, < w, \#3 >\}$
$\Pi_4 \xrightarrow{C_3} AP : \{< w, \#1 >, < b, \#2 >, < b, \#3 >\}$
$\Pi_4 \xrightarrow{C_4} AP : \{< w, \#1 >, < b, \#2 >, < b, \#3 >\}$

**step 5:** $AP$ discards from all the source places the data items related to evolutions #1 and #2 since there are data items of the kind $< i, \#1 >$ and $< i, \#2 >$. The final result is thus:

$< pist\_state\_ring(worn)[w], pist\_state(worn)[w], road\_cond(poor)[b],$
$ground\_clear(low)[b] >$

## 8 Conclusions and Related Works

The approach proposed in this paper gives a suitable motivation to the usefulness of backward reachability analysis on Petri net models, by applying it to diagnostic problem solving; it can be seen as the complement of the approach proposed in [10], where T-invariant analysis is exploited to perform diagnosis.

The possibility of using Petri nets in AI applications is recently receiving a lot of attention [7, 16]; the present work shows how Petri net-based and AI-based techniques can be successfully merged in diagnostic problem solving.
Other attempts to address the problem of diagnostic reasoning using Petri nets can be found in [8], where algebraic analysis is used to detect single-fault diagnoses using a set of production rules modeled as a Petri net, and more recently in [12], where the correspondence between *linear logic* [4] and Petri nets is exploited in order to diagnose flexible manufacturing systems. We rely on a general and formal theory of diagnosis where multiple fault can be naturally addressed and we consider systems for which underlying logic is much simpler than linear one. However, when cyclic processes has to be modeled (as in the case of flexible manufacturing systems) the use of more complex frameworks is needed.

Notice that B-W Analysis can be applied also to tasks different than diagnostic one; indeed, in [9], the use of this type of analysis is proposed as a tool for the validation of a causal model (represented through a Petri net) following some correctness criteria and by detecting some errors that can happen when constructing the end-user model.

In the present paper, we also proposed a parallel implementation of B-W analys, based on the net structure and on the marking used to start analysis. The problem of obtaining, in a single execution, all the solutions to a diagnostic problem has been solved introducing the concept of evolution identifier (**eid**). This concept is also the basis for the proposed approach for inconsistency checking, namely *i-propagation*. The possibility of parallel implementation makes B-W analysis a good candidate for the mitigation of the relatively high inefficiency diagnostic algorithms usually have.

Future works are planned toward the actual evaluation of the proposed approach on a multiprocessor architecture and to the comparison of the parallel approach with a sequential one.

## Acknowledgements

The authors would like to thank Prof. Gianfranco Balbo and Dr. Matteo Sereno for fruitful discussions and comments during the preparation of the first draft of this paper.

## References

1. L. Bernardinello and F. De Cindio. A survey of basic net models and modular net classes. In G. Rozenberg, editor, *Advances in Petri Nets 1992, LNCS 609*, pages 304–351. Springer Verlag, 1992.

2. J.M. Colom, M. Silva, and J.L. Villarroel. On software implementation of Petri nets and Colored Petri nets using high-level concurrent languages. In *Proceedings of the* $7^{th}$ *International Conference on Application and Theory of Petri Nets*, 1986.
3. L. Console and P. Torasso. A spectrum of logical definitions of model-based diagnosis. *Computational Intelligence*, 7(3):133–141, 1991.
4. J.Y. Girard. Linear logic. *Theoretical Computer Science*, 50, 1987.
5. W. Hamscher, L. Console, and J. de Kleer. *Readings in Model-Based Diagnosis.* Morgan Kaufmann, 1992.
6. J.W. Lloyd. *Foundations of Logic Programming.* Springer-Verlag, 1987.
7. J. Martinez, P.R. Muro, M. Silva, S.F. Smith, and J.L. Villaroel. Merging artificial intelligence techniques and Petri nets for real-time scheduling and control of production systems. In *Proc. 12th IMACS World Congress on Scientific Computation*, pages 528–531, Paris, 1988.
8. T. Murata and J. Yim. Petri-net deduction methods for propositional-logic rule-based systems. Technical Report UIC-EECS-89-15, University of Illinois at Chicago, 1989.
9. L. Portinale. Verification of causal models using Petri nets. *International Journal of Intelligent Systems*, 7(8):715–742, 1992.
10. L. Portinale. Exploiting T-invariant analysis in diagnostic reasoning on a Petri net model. In *Proc. 14th Int. Conf. on Application and Theory of Petri Nets, LNCS 691*, pages 339–356. Springer Verlag, Chicago, 1993.
11. L. Portinale. *Petri net models for diagnostic knowledge representation and reasoning.* PhD Thesis, Dip. Informatica, Universita' di Torino, 1993.
12. B. Pradin-Chezalviel and R. Valette. Petri nets and linear logic for process oriented diagnosis. In *Proc. IEEE Int. Conference on Systems Man and Cybernetics*, Le Touquet, France, 1993.
13. E. Rich and K. Knight. *Artificial Intelligence (2nd Edition).* Mac Graw Hill, 1991.
14. M. Silva and R. Valette. Petri nets and flexible manifacturing. In *Advances in Petri Nets 1989.* Springer-Verlag, 1989. LNCS 424.
15. D. Taubner. On the implementation of Petri nets. In *Proceedings of the* $8^{th}$ *International Conference on Application and Theory of Petri Nets*, 1987.
16. R. Valette and M. Courvoisier. Petri nets and artificial intelligence. In *Proc. Int. Workshop on Emerging Technologies for Factory Automation*, North Queensland, Australia, 1992.
17. D. Zhang and T. Murata. Fixpoint semantics for Petri net model of definite clause logic programs. Technical Report UIC-EECS-87-2, University of Illinois at Chicago, 1987. also to appear in *Advances in the Theory of Computation and Computational Mathematics*, Ablex Publ.

# Probabilistic Validation of a Remote Procedure Call Protocol

N. Bennacer G. Florin C. Fraize S. Natkin
CEDRIC - Centre d'Etudes et de recherche
en Informatique CNAM
292 rue Saint-Martin 75141 Paris Cedex 03

**Abstract.** Classical validation approach tries to prove that failed events (events that do not verify an user property) will never occur. Probabilistic validation relies on a partial analysis on a system model and tries to prove that the failed event occurrences, have a sufficiently low probability. An incorrect behavior is a very rare event: it is the consequence of a complex unknown operation sequence. The system to be validated is modeled by a stochastic Petri net. The sequence of transition firings which may lead to critical (failed) Petri net markings are characterized at the Petri net level. An efficient travel through the reachability graph must be able to visit these sequences to reach as quickly as possible critical markings. From the incidence matrix of the net and the specification of the properties to be fulfilled, we derive a linear system called the "decision system". The set of the decision system solutions includes all characteristic vectors of sequences leading to critical markings. In this paper, we present the probabilistic validation of a Remote Procedure Call protocol . The goal is to evaluate the probability that the protocol satisfies the required "at most once" semantic. This model is solved using two probabilistic validation algorithms using the above principles. The first one is related to acyclic Markov graphs. A traversal technique combining a breadth and a depth first search traversal techniques is used considering only the critical trajectories. The second algorithm is a worst event driven and importance sampling simulation.

## 1 Probabilistic Validation

The validation of highly dependable and complex systems is a fundamental goal of their design methodology. It is generally a complex problem as an incorrect behavior (from an assertion correctness point of view) is a very rare event. If such an event happens it is generally the consequence of a complex unknown operation sequence which is outside the standard behavior of the system. Finding these sequences is a highly combinatorial task. Moreover, it is often verified that the relative number of these sequences and the corresponding probability of occurrences are very small. An exact proof (an assertion is always true) performed using an exhaustive analysis (state enumeration or formula rewriting techniques) is limited by the exponential growth of the model complexity [15] and [14]. Simulation of the behavior is not able to quantify the validation level [5]. Thus, it cannot be considered as a proof of a system property, but, like testing, it can increase the confidence in the system.

The main principle of probabilistic validation is the partial analysis of a system model. The model and the analysis method must allow to prove that assertions are verified with an acceptable probability level on transient or stationary horizon. Probabilistic validation uses the following proposition: "to prove an assertion with a very low probability, in a given period, is sometimes simpler that to prove that an event never occurs".

From a complexity point of view, this approach and behavioral simulation appear to be similar, but a probabilistic validation gives an operational estimate of the validation level.

The principles of probabilistic validation are presented in detail in [10]. It uses a state transition model which includes a description of stochastic operation duration and event frequency. The property to be validated is defined by an assertion on state space or more generally on the trajectory space. Our approach of probabilistic validation is based on stochastic Petri net models. From the structural analysis of the Petri net we derive a linear system of equations called the decision system. Solutions of the decision system allow to travel efficiently through a large state graph and to reach as quickly as possible the class of failed markings (markings which do not verify the required property). Using the above principles two methods (algorithms) have been developed. The first one is a partial breadth and depth first search reachability analysis, with probability evaluation in each visited marking. It can be used to solve acyclic Markov graphs. The probability evaluation is defined by the product of the probability to fire each transition [11]. The second one is a worst event driven and importance sampling simulation approach. This method is able to solve more general systems. It tries to guide the simulation to sample critical trajectories more frequently and to build an accurate estimate of a measure related to the assertion to be validated [3].

The goal of this paper is to show on a complex and real example that probabilistic validation is an efficient approach. The algorithms are applied to a model of a remote procedure call protocol (RPC) formerly used by CHORUS systems in micro-kernel distributed operating systems. The goal is to evaluate the probability that this protocol does not violate the required "at most once" semantic. In the following section, we present the probabilistic validation model. Then, due to the lack of space, we just recall in section 3 the basic principles of the algorithms. The reader can find a more formal and complete description of these algorithms in referenced papers. In section 4, we describe the remote procedure call protocol model. The results obtained are presented in section 5. The concluding section presents work in progress.

## 2 Probabilistic Validation Model

The formal model used in our approach of probabilistic validation is a stochastic Petri net. This model is first a Petri net, whose places represent resources and transitions represent operations. Random variables attached to transitions are associated with random operation durations. The stochastic behavior must be completely defined by a set of rules associated with the choice of the next transition to be fired in a given marking and the memory properties of the time already spent for a given operation.

The concepts and the notation used in this paper are detailed in [9]. We just recall in this section some important aspects.

## 2.1 Petri Nets

The underlying Petri net is a place-transition net. It is denoted by $R(\mathcal{P},\mathcal{T},\mathcal{V})$ where $\mathcal{P}$ is the set of places (with cardinal $|\mathcal{P}|$), $\mathcal{T}$ is the set of transitions (with cardinal $|\mathcal{T}|$) and $\mathcal{V}$ is the set of valued arcs between places and transitions. The incidence matrices define the valuation of each arc. The backward incidence matrix (relations between places and transitions) is denoted by $C^-$. The incidence matrix is denoted C. We denote the jth column of an incidence matrix (for instance $C^-$) by $C^-(.,j)$. A marking of the net, $M_k$, is an integer column vector. Its kth component $M_k(p_i)$ is the mark of place $p_i$ (number of tokens in the place) in marking $M_k$. Let $M_0$ be the initial marking of the Petri net. The behavior of the net is defined by:
**Condition:** a transition $t_j$ is firable from a marking $M_i$ if: $M_i \geq C^-(.,j)$.
**Action:** the firing of $t_j$ from $M_i$ leads to a new marking $M_k$ such that $M_k = M_i + C(.,j)$.

We denote by $\mathcal{S}_k = \{t_j : C^-(.,j) \leq M_k\}$ the set of firable transitions in $M_k$. Let $s = (t_{j_1},...,t_{j_n})$ be a sequence of transitions fired from a marking $M_0$ and $M_1$, $M_2$... $M_n$ be the sequence of successive reached markings in $s$. The characteristic vector of a sequence $s$ is an integer vector $\overline{s}$. The jth component of $\overline{s}$ is equal to the number of transition $t_j$ firings in the sequence $s$. According to this definition we can write the firing equation (or fundamental equation) of the net. For any marking $M_n$ reached from $M_0$: $M_n = M_0 + C\overline{s}$.

## 2.2 Stochastic Petri Nets

The definition of stochastic Petri nets (see [9], [1] and [16]) is introduced as follows:

• The marking $M(\tau) = M_n$ where $\tau \in [\tau_n, \tau_{n+1}[$. $\tau_n$ is the date of the nth transition firing leading to the marking $M_n$, with $\tau_0 = 0$.
• The operation duration associated with the transition $t_j$, $Z^j$, is a random variable and its probability distribution (respectively its density function) denoted by $F^j(a)$ (respectively $f^j(a)$).
• In the marking $M_n$, the time already spent by an operation associated with transition $t_j$ is denoted by $W_n^j$.
• In the marking $M_n$, the remaining time to finish an operation associated with transition $t_j$ is denoted by $TR_n^j = Z^j - W_n^j$.
• The system state $E_n$ is defined by a mix of a discrete state variable (marking) and a continuous state variable (the memory or the remaining time vector): $E_n = (M_n, TR_n)$.
• The state evolution is defined by the race model. The race model asserts that the next transition to be fired is the one associated with the shortest delay to finish. In each marking a race is done between operations associated with enabled transitions: $\tau_{n+1} - \tau_n = MIN_{t_j \in \mathcal{S}_n}\{TR_n^j\}$.
• The classes of stochastic Petri nets based on the race model differ in their ability to memorize the amount of time already spent by each concurrent operation:
- In age memory policy the memory of the time already spent is preserved on all successive reached markings until the corresponding transition is fired, even if the transition is not enabled in the current marking.
- Enabling memory policy keeps track of the time as long as a transition is contin-

uously enabled. But a transition may be disabled before its associated operation ends. In this case, the memory of the time already spent is lost.
- In the resampling policy the memory of all the different times spent are lost after the firing of any transition.
• We denote by $f_n^j(a)$ (respectively $F_n^j(a)$) the probability density function (respectively the distribution function) of the (conditional) "remaining lifetime" $TR_n^j$ of the operation transition $t_j$.

$$F_n^j(s) = \mathbf{P}[TR_n^j \leq s/E_n] = \frac{\mathbf{P}[W_n^j < Z^j \leq s + W_n^j]}{\mathbf{P}[Z^j > W_n^j]}$$

$$\overline{F}_n^j(s) = 1 - F_n^j(s) = \frac{\overline{F}^j(W_n^j + s)}{\overline{F}^j(W_n^j)} \quad \textit{and} \quad f_n^j(s) = \frac{f^j(s + W_n^j)}{\overline{F}^j(W_n^j)}$$

If $f^j(s)$ is an exponential distribution or more generally if $W_n^j = 0$: $F_n^j(s) = F^j(s)$ and $f_n^j(s) = f^j(s)$.
• For given state of the discrete event system $E_i$ after the i-th transition firing, we denote by $p(E_i, E_{i+1})$ the probability density (likelihood) that the next state is $E_{i+1}$ at the (i+1)th transition by firing the transition $t_{a_i}$. Then $\tau_{i+1} - \tau_i = TR_i^{a_i}$ is the sojourn time in state $E_i$.
It can be shown ([18], [19] and [13]) that the probability density (the likelihood) of a sample path $(E_0, \ldots, E_n)$ can be expressed as a function of the complementary probability distribution and probability density functions of remaining operation times $TR_i$.

$$(1) \qquad p(E_i, E_{i+1}) = f_i^{a_i}(TR_i^{a_i}) \prod_{t_{k \neq a_i} \in S_i} \overline{F}_i^k(TR_i^{a_i})$$

Let $(E_0, E_1, ..., E_n)$ be a sample path. Then the probability density of a sample path $(E_0, \ldots, E_n)$ is:

$$(2) \qquad p(E_0, E_1, \ldots, E_n) = \prod_{i=0}^{i=n-1} p(E_i, E_{i+1})$$

*Remarks:* To simplify the notations we have not introduced the two following extensions to the basic SPN definition.
1- The operation durations associated with transitions can have non continuous distributions (for example an instantaneous or a deterministic duration). Then several transitions can be fired at the same time. The race model does not provide any information to choose among them. Thus, a preselection policy based on a specified probability distribution must be defined over conflicting transitions.
2- For each enabled transition in a marking $M_i$ the time needed to perform an operation may depend on the amount of available resources. The local resource configuration is modeled by the marking $M_i$. This dependency is taken into account using either the notion of the clock rate [16] or, more generally, the notion of work [9] associated with a transition.

These extensions can be taken into account in the results presented in this paper. In counterpart, the SPN definition does not allow several simultaneous enabling of a given transition. Such feature, similar to parallel servers in queuing networks, changes radically the semantic of SPN definition.

# 3 Probabilistic Validation Algorithms

## 3.1 Qualitative Techniques

This section is devoted to the analysis of qualitative techniques able to allow a partial and efficient scanning of a large state graph in order to validate a state assertion. Basic principles of qualitative structural analysis are:
-A decision system characterizes the set of possible critical event sequences that might lead to failed markings to avoid the complete examination of the state space.
-A simulation or a traversal strategy methods is designed in order to visit these event sequences and to evaluate measures related to these sequences.

**Decision System.** We assume that the successful markings are defined by a boolean function (for example $g(M)$ : $M(p_1) > 0$ means that some resource like a processor modeled by place $p_1$ is available). Hence, failed markings are characterized by $\neg g(M)$. We assume that g(M) can be written as:

$$\neg g(M) = \bigvee_i \bigwedge_j (G_{ij}(M) \leq 0)$$

Where $G_{ij}(M)$ is a real function of the marking M. We denote by $G_i(M) \leq 0 = \bigwedge_j (G_{ij}(M) \leq 0)$ the ith minterm of the disjunctive form. $G_i(M) \leq 0$ is a set of simultaneous inequalities $G_{ij}(M) \leq 0$. Each set is studied independently.

Let us consider a marking $M_k$ reached from the initial marking $M_0$. We denote by $\mathcal{R}_k$ any subset of the set of firable transitions, $\mathcal{S}_k$, in $M_k$ ($\mathcal{R}_k$ is associated with the transitions not already fired in a given traversal of the reachability graph). The method must find, in the set $\mathcal{R}_k$, the subset of transitions leading to the set of failed markings. If a marking M such that $G_i(M) \leq 0$ is reachable from $M_k$, then the characteristic vector $X \geq 0$ of a firing sequence leading from $M_k$ to $M$ is a solution of the following decision system:

$$\begin{cases} -C.X \leq M_k \\ (G_i(M_k + C.X) \leq 0) \\ X \geq 0 \\ \sum_{\mathcal{R}_k} X_j \geq 1 \\ X \in I\!N^{|T|} \end{cases}$$

We assume in this paper that $G_{ij}(M)$ are linear functions of M. Many interesting properties of Petri net models can be expressed under this linear assumption [20]. From a practical point of view, all the correctness assertions we have encountered can be expressed as a linear function of the marking. In this case the decision system can be solved using a linear programming method. Moreover, as long as the $G_{ij}(M)$ functions are convex, there are still efficient numerical methods to solve this system.

When the decision system has no solution, none of the sequences starting by a transition in $\mathcal{R}_k$, can lead to a failed marking. Thus the generation of these sequences can be avoided. Unfortunately the existence of a solution X to the decision system does not imply the existence of a firing sequence leading from $M_k$ to $M$. Hence a partial generation algorithm must be used to derive from the linear programming solutions the sequences leading to critical markings.

### 3.2 Breadth and Depth First Search Traversal Algorithm

This algorithm is designed to deal with acyclic Markov Petri nets. An acyclic Markov Petri net is such that:

$$(\neg\exists)X \in I\!N^{|T|}, X \neq 0, \text{ such that } CX \geq 0$$

The following principles are used for the exploration of a large state space. The detailed algorithm and proof are given in [12].

• From a marking $M_i$ and a solution X to the decision system in $M_i$, the algorithm uses this solution and tries to reach by a depth first search traversal the failed markings. All the feasible sequences, the characteristic vector of which is X, are to be explored.
• Thus, a subgraph corresponding to all feasible combinations of characteristic vector is generated. For each last visited marking of the subgraph which is not failed, the algorithm tries to extend a sequence to a failed one by solving a new decision system in the current marking.
• A qualitative elimination method is associated with these principles. The exploration of a trajectory is stopped when the decision system has no positive solution in a given marking.
• The probability of the related sequences are computed during the subgraph generation.
• The algorithm must back track to a non evaluated marking $M_i$ to find new solutions to the decision system.
• The marking classification and the analytical probability computation of visited trajectories is performed. At a given step of the scanning of the state graph, the behavior of the system may be partitioned in three classes. The set of visited sequences such that the property is true has a probability P1, the set of visited sequences such that the property is false has a probability P2 and the set of non-visited sequences has the probability (1-P1-P2).
• Probabilistic rules to direct the traversal of a graph are used when the qualitative analysis does not define precisely the transition to fire in a given marking.
• The exploration is performed until either $P1 > (1-\epsilon)$ (the property is $\epsilon$ sure) or $P2 > \epsilon$ (the property is $\epsilon$ false).

When the probabilistic stopping criterion is not used ($\epsilon = 0$) the algorithm visits all sequences leading from $M_0$ to critical markings.

### 3.3 Probabilistic Validation Using Simulation

As stated above in this paper, we consider highly dependable systems modeled by stochastic Petri nets in order to validate an user assertion. So, from an assertion proof point of view, these systems can be characterized by the following properties:

- a failure is a rare event in the life of the system;
- failure occurrences result from complex unknown event sequences the probabilities of which are very small;
- the system can be correct so a failure is an event that will never happen.

Hence, it is clear that simulation analysis of highly dependable systems samples rarely (or never) sequences leading to failed markings and requires a large amount of computation time to obtain (if possible) significant estimates of measures related to these sequences. In the last ten years this kind of problem was encountered in dependability evaluation and was solved using importance sampling simulation (see [17], [18] and [7]).

The very low probability of an incorrect behavior can be the consequence of two phenomena:

1. Differences in the orders of magnitude of event rates.
   This problem is frequently encountered in availability and reliability studies. The fundamental problem with direct simulation is shown in [7] by considering a simple system model with three states (failure rate << repair rate).

2. Complex and long event sequences.
   These sequences are realized only after a very long time compared to the system or mission lifetime. Most of the probabilistic validation models includes such sequences.

The main difference between importance sampling applied to reliability analysis and to probabilistic validation is related to the knowledge of trajectories (event or event sequence) which lead to critical markings. In reliability analysis these events are generally defined as equipment failures. In probabilistic validation these events are unknown. It is even possible that such sequences of events may not exist. Hence, it is necessary to use some properties of the model to guide the simulation and select the worst possible event. The static analysis of the Petri net allows to find necessary conditions of worst event sequences. The frequency of these events is then increased according to the importance sampling technique.

**Importance Sampling Technique.** The basic idea of importance sampling (see for the basic concepts [19] and for the extension to stochastic systems [13]) is to simulate the system under a different probability distribution to appropriately and quickly move the system towards critical markings and to reduce significantly run lengths. In other words, the original operation time probability distributions are modified in order to increase the firing frequency of transitions that possibly lead to critical markings. Simulation of the modified stochastic Petri net distributions introduces a bias into the interesting measures. Importance sampling technique allows to maintain unbiased estimates of the desired measures by computing a correction factor (likelihood ratio) to compensate the resulting bias.

Consider a simulation system modeled by a stochastic Petri net and an user assertion g(M) over a period [0, T]. $T$ can be either a fixed delay or a stopping time corresponding for example to the end of a mission. We need to estimate the probability Y that the user assertion is not verified. An importance sampling simulation is performed as follows:
- the operation durations associated with transitions identified by the qualitative analysis are generated from new probability distributions;
- the transition remaining and memory times are updated according to their memory policy and the transition to fire is defined by the race model.

The estimated measure Y can be considered as a function of generated sample path ($E_0$, $E_1$,..., $E_n$) where n is a stopping time $n = MIN\{i/\tau_i \geq T$ and

$\forall k \leq i \ g(M_k)\}$ [6]:

$$Y(T) = Y(E_0, E_1, \ldots, E_n) \quad = 0 \text{ if g(M) is true} \\ \quad = 1 \text{ otherwise}$$

The expectation of Y can be expressed as follows:

$$E_p[Y] = \sum_{\forall(M_0,\ldots,M_n)} \int_0^{+\infty} \cdots \int_0^{+\infty} \\ Y(E_0,..,E_n)L(E_0,..,E_n)p^*(E_0,..,E_n)dTR_0^{a_0}\ldots dTR_n^{a_n} = E_{p^*}[YL]$$

$$E_p[Y] = \sum_{\forall(M_0,\ldots,M_n)} \prod_{i=0}^{n-1} \int_0^{+\infty} Y(E_i, E_{i+1})L(E_i, E_{i+1})p^*(E_i, E_{i+1})dTR_i^{a_i}$$

$E_p$ et $E_{p*}$ are respectively the expectation operators under the original and modified distributions. $p^*(E_0, E_1, ..., E_n)$ is the density probability of the generated sample path from the modified operation duration distributions and $L(E_0, ..., E_n)$ is the likelihood ratio. L is defined for a given sample path as the ratio of the probability density of the sample path under the original distributions over the probability density of the same sample path under the new distributions. This last equation shows how the importance sampling technique works. Instead of generating sample paths of the original Petri net, we can generate sample paths with the modified Petri net distributions. Estimating our measure with these sample paths we get biased results, which must be corrected by a factor $L(E_0, ..., E_n)$. A remaining key problem is, how an original model must be modified to make importance sampling efficient. An importance sampling simulation is efficient, if (for a given number of replications), we obtain confidence intervals as small as possible. Optimal sample paths are generated according to: $p* = Yp/E_p[Y]$.
This lead to an estimator (with zero variance) that requires the knowledge of $E_p[Y]$ which is unknown. However, the above equations indicate how the modified distributions should be selected to obtain estimates with reduced variance. We must consider as important in the sampling procedure the intervals where the contributions to the sum (in the expression of $E_p[Y]$) are the greatest. The choice of the modified distributions is an open problem. An application of this technique to a specific problem must use any knowledge related to the stochastic structure of the studied system.

**Worst Event Driven and Importance Sampling Simulation Algorithm.** From the algorithmic principles discussed previously we develop an efficient Monte Carlo simulation algorithm able to estimate with an acceptable accuracy, in a reasonable time, measures in discrete event dynamic systems strongly influenced by rare events. Two main procedures: structural analysis procedures and importance sampling procedure are used. The first one partitions the events (transition firings in a stochastic Petri net) in those leading to failed markings and those leading to successful markings. Simultaneously, by altering probability distributions, the critical sequences are visited as frequently as the uninteresting ones during simulation.

• **Structural analysis procedures:** In the simulation, the decision system is not used as a systematic way to explore critical sequences. It is used as an heuristic method to find the transition the frequency of which must be increased. Even if the transition which may lead to critical markings is not found using the decision system it can be fired according to the probabilistic sampling. The efficiency of the heuristical use of the decision system is measured by the accuracy of the estimates obtained at the end of the simulation.

Let $\overline{s}(M_i)$ be an integer solution to the decision system in $M_i$. Let $s_{ij}$ be a sequence leading from $M_i$ to $M_j$ with $\overline{s}_{ij} < \overline{s}(M_i)$. We define the updated characteristic vector solution in $M_j$ to the decision system in $M_i$ as the residual characteristic vector: $\overline{s}(M_i) - \overline{s}_{ij}$.

We have experimented the following heuristics:

1. The decision system is solved in $M_0$. In each marking $M_j$ the enabled transitions which are in the support of the updated characteristic vector solution $\overline{s}(M_0) - \overline{s}_{0j}$ are considered. The corresponding distributions are modified to increase the frequency of these transitions during simulation run. This forcing procedure is used until a transition outside $\overline{s}(M_0) - \overline{s}_{0j}$ is fired or until all enabled transitions in $\overline{s}(M_0) - \overline{s}_{0j}$ are exhausted. Then the simulation switches back to the original distributions (the forcing procedure is disabled and the original distributions are used).

2. If this first approach does not guide the simulation sufficiently we solve the decision system in a current marking $M_j$ each time a transition which is not in the support of the previous residual characteristic vector solution is fired or each time all enabled transitions are exhausted.

3. At last the decision system is solved in each reached marking.

If the decision system in $M_i$ has no solution then no failed markings can be reached from $M_i$. In this case, the elimination principle is helpful. Thus, the generation of the successors of $M_i$ and the prolongation of the considered sequence is avoided and the current simulation run is stopped.
• **Importance sampling procedure:** Importance sampling is integrated into the simulation algorithm to choose the fired transition (race model). Events are rescheduled by rescaling the remaining times associated with transitions of the updated characteristic vector solution. This is equivalent to generate the remaining times from new accelerated distributions. The likelihood ratio for a given sample path (simulation run) is obtained in an incremental computation using equations (1,2).

## 4 Remote Procedure Call Protocol Model

This section is not an exhaustive description of the RPC protocol. We give only the main aspects of this protocol to allow an understanding of the RPC protocol stochastic Petri net model.

### 4.1 RPC Protocol Semantics

In the case of a local procedure call, the caller procedure sends the request. The called procedure carries out systematically the work and sends back the reply. Unlike local procedure calls, remote procedure calls are subject to lost messages, server and client crashes. These features influence the semantics of the RPC and the goal of making it transparent. As a result of these problems, the exact semantics of RPC systems can be categorized in various ways [21]. The most interesting semantics are:

- **Exactly once:** Every call is carried out exactly once, no more and no less. This goal is unachievable because after a server crash it is not always possible to tell whether the operation was performed or not.
- **At most once:** This semantic is aimed by the modeled protocol. In this case, in the absence of faults, the control always returns to the client and the operation is performed exactly once. However, if a server crash is detected, the client stub gives up and returns an error code. Retransmission is attempted. In this case, the operation is performed either zero or one time, but no more. Further recovery is up to the client. Most of the protocols tend to implement this semantic. In our example, we show that this semantic is satisfied with a high probability. The uncertainty is related to the use of RPC sequence numbers and to transmission delays.
- **At least once:** The client stub just keeps trying over and over until it gets a proper reply. When the client gets an answer, he knows that the operation has been performed one or several times.

To satisfy the "at most once semantic" is similar to perform a release graceful of a transport connexion and in fact the two army problem [21]. Without any assumption on the maximum time needed to transmit a message or to execute the remote procedure, the problem has no deterministic solution. In fact, any efficient solution is a probabilistic algorithm which must be probabilistically validated. Hence, probabilistic validation seems to be an appropriate approach to cope with this kind of protocol. It takes naturally into account that a correct solution has a non null probability to fail.

### 4.2 Protocol Description

The remote procedure call protocol relies on the following basic principles [2] and [8].

**1-** A client initiates a request, the server carries out the request and sends back the reply; the client sends an acknowledgement message once it receives the reply. If the client has another request to submit to the same server, it inserts the acknowledgement in the request (piggyback acknowledgement), else it transmits an explicit acknowledgement message.
**2-** Once the client or the server initiates its message it starts a timer and waits for the reply (the client waits for the reply to its request and the server waits for the acknowledgement message). When the timers expire, the messages (request, reply) are retransmitted until a maximum number of attempts is reached. In this case, it is considered that the remote entity is failed.
**3-** A complete message exchange (request, reply, acknowledgement) is a "sequence" or a "transaction". For the client, a transaction starts when it initiates a request and finishes when it acknowledges the reply. For the server, a transaction starts when it carries out a request and finishes when it receives the acknowledgement of its reply or when it knows that it cannot receive it.

**To Satisfy the "at Most Once" Semantic.** The server must be able to identify a transaction. The transaction identifier contains the transmitter site identifier and an increasing sequence number allocated to each transaction. Moreover, to satisfy the "at most once" semantic the client counts the number of times it sends the same request in a given transaction. The current value of this number is included in the acknowledgement message. The server also counts the number of times it receives the same request in a given transaction. When the server receives an acknowledgement it compares the current value of its counter to the one included in the acknowledgement message. If these two counters are equal, there is no transiting message with the same transaction identifier. If the acknowledgement counter is greater than the server counter at least one message is still not received. In this case, the server must memorize the number of unreceived requests until it receives all of them or until a given delay. This delay must be sufficiently long to guarantee with a high probability that the unreceived messages are lost.

Figures 1 and 2 represent respectively a correct and an incorrect scenario.

**Timeout Delays.** To handle lost client requests, lost client acknowledgements, lost server replies and long requests the following timeout delays are used:

1. First-transmitting timer: The client starts a timer and waits for the reply to the first request. If it receives the reply before the timer expires it acknowledges the reply (explicit or piggyback acknowledgement) else it retransmits its request.

2. Retransmitting timer: The client starts a retransmitting timer after each retransmission. It waits either for the reply to its request or the message meaning that the request is currently carried out. After this timer expires the client retransmits until a maximum number of attempts is reached. In this case, the remote entity is considered to be failed.

3. Watchdog timer: The client starts a watchdog timer if, after several sent requests, it receives a message meaning that the submitted request is currently carried out. This timer is particularly long and if it expires before the reply arrives the server is considered to be failed.

4. Insertion timer: The client starts an insertion timer when it receives the reply. If there is another request to submit before the timer expires it sends a piggyback acknowledgement message else it sends an explicit acknowledgement message.

5. Acknowledgement timer: The server starts an acknowledgement timer if it sends back the reply and waits for the acknowledgment message. If this message arrives before the timer expires, the server compares its counter with the one included by the client in the acknowledgement message. If the timer expires the server retransmits the reply until a maximum number of attempts is reached.

6. Error timer: The server starts an error timer when the client counter included in the acknowledgement message is different from its own counter. This timer is stopped when all unreceived requests have been received. If the timer expires, the remaining unreceived messages are considered as lost. The

server counter is reset. Hence incoming requests will be accepted as new ones. As a consequence a correct value of this timer is critical to satisfy the "at most once" semantic. Unfortunately the protocol must be able to work simultaneously in different timing contexts (local and wide area networks) and a perfect tuning of the error timer is not always possible.

This protocol illustrates the probabilistic validation concept. The protocol violates "at most once" semantic since some scenarios exist such that the server executes the request more than once. For instance, a request can be received after the error timer expires. Then, in an exact validation study, the property "the request is executed at most once" is false. This kind of errors can be partly avoided using connexion management techniques such as three ways handshake and frozen references. But this classical approach induces a high overhead and also has a probabilistic behavior [21]. The probabilistic validation will try to estimate for each remote procedure call the probability to execute the procedure more than once is less than a given threshold $\epsilon$. Unlike an exact validation which proves that the property is true for all the trajectories, for an operational probabilistic validation it is sufficient that the property remains true with a probability greater than $(1-\epsilon)$.

### 4.3 Modelling

The stochastic Petri net is a one execution model [2]. It describes a behavior of the RPC from the beginning of the call until the last event related to this call. In fact, due to the piggyback acknowledgement of the call in the next one the model includes two successive executions of the RPC.
Figure 3 describes the different client states. Error cases are taken into account. The server is considered to be failed when the maximum request number is reached and when the watchdog timer expires without reply of the server. Figure 4 describes the different server states.

**Stochastic Petri net:** The stochastic Petri net corresponding to the RPC protocol has 36 transitions and 37 places. The places of the net represent client, server, network, timer states and counter values. The Petri net transitions allow to describe the synchronization between the place mark evolution. We present here a simple version of the client Petri net model (Figure 5) (there are no retransmission and no lost messages).

The following places are initially marked:
-the place which describes the client process ready to submit a request;
-the places which describe the timeout delays;
-the place which represents the maximum transmitted request number in a given transaction;
-the place which represents the maximum transmitted reply number in a given transaction;
-the place which represents the idle state of the server;
A place PS1-VALIDE allows the validation of the "at most once" semantic. A token is deposited in this place if the server finishes a request execution. Thus the PS1-VALIDE place tokens indicate the request execution number in a given transaction. The assertion to validate is " M(PS1-VALIDE) is less than two".

**Transition rates:** The transition rates associated with lost messages are such that the probability to lose a message is of 0.9999e-06. The request transmission duration is 5ms, the error timer duration is set to 15mn and the acknowledgement timer to 2mn. Timers related to the retransmitted request or reply are set to 100ms. For both validation algorithms, an exponential distribution of delays has been assumed. The simulation allows also to cope with deterministic and Erlang distributed delays.

# 5 Solving the RPC protocol model

This model was solved with different assumptions on the number of attempts to transmit a request or a reply. For a maximum of six attempts to transmit a request or a reply the model generates more than 800,000 markings. The size of the reachability graph generated by this model can easily be multiplied by increasing these numbers. In this section we present the results obtained by the two probabilistic validation algorithms on the RPC protocol model.

## 5.1 Breadth and Depth First Search Traversal Algorithm Results

This algorithm is able to solve efficiently models which are not tractable using a standard Markovian solver. Several traversal methods have been compared [4] and [12] (Figures 6 and 7). The results show that the breadth and depth first search traversal algorithm which uses most efficiently the solution to the decision system is the best for large state reachability graphs (> 100,000 markings). With the current workstation technology, we think that we are able to solve models having several million of reachable markings (Table 1).

## 5.2 Simulation Results

Using standard Monte Carlo simulation more than 20 million simulation runs (approximately 330 million of generated events) do not lead to the failed markings. The worst event driven and importance sampling algorithm allows to visit failed markings [3].

In section 3.3.2, we have briefly presented three possible uses of the decision system. These methods have an increasing complexity in terms of the number of decision systems solved. It appears that the second solution is an optimal compromise in terms of accuracy related to the number of generated events.

The simulation allows to validate the RPC model having more than 800,000 markings and to use deterministic durations for the timers (Table 2). Comparing the results of the two validation methods leads to the following conclusion.

## 5.3 Comparison of the two Validation Results

As long as the acyclic Markov hypothesis is fulfilled and the reachability subgraph generated by the breadth and depth first search traversal algorithm can be stored in main memory this approach can be used. In the contrary, for large non Markovian stochastic Petri nets, worst event driven and importance sampling simulation algorithm seems to be the most promising one.

| | Markovian solver | First algorithm |
|---|---|---|
| generated marking number for 5 reply or request attempts | 341830 | 73507 |
| P1 | $0.3025*10^{-9}$ | $0.2995*10^{-9}$ |
| 1-P1-P2 | 0.0 | $1*10^{-11}$ |

Table 1

| Second algorithm for 6 reply or request attempts | run number | estimate with 99% confidence interval | relative accuracy |
|---|---|---|---|
| standard simulation | 20 million | no estimate | - |
| first approach | 800000 | $1.1684 * 10^{-10} \pm 2.5754 * 10^{-10}$ | 220% |
| Second approach | 290000 | $2.7905 * 10^{-10} \pm 4.2136 * 10^{-11}$ | 15% |
| third approach | 250000 | $2.8971 * 10^{-10} \pm 3.1868 * 10^{-11}$ | 11% |

Table 2

## 6 Conclusion

From a practical point of view, the preceding results show that the RPC "at most than once" semantic is fulfilled with an acceptable probability. The probability to perform twice or more a given call is of $10^{-10}$ order. Compared with events such as the failure of the RPC server the "at most than once" semantic violation is still a very rare event.
These experiments show that the probabilistic validation is a promising technique in the field of distributed and safety systems able to deal with models which are untractable by the classical validation techniques.

We are now improving our tool in terms of performance and functionalities (general distributions of delays). From a theoretical point of view our work is focuses on two points:
- the ability to handle more complex assertions (trajectory assertions and non linear assertions);
- the failure probability estimation when no failed event occurred during the graph traversal.

# References

[1] M. Ajmone-Marsan, G. Balbo, G. Chiola, G. Conte and A. Cumani. On Petri nets with stochastic timing. PNPM85, IEE CN 85CH2 187-3, Turin, July, 1985.

[2] J. Barancourt. Modélisation et validation d'un protocole de RPC. Rapport de stage de DEA de systèmes informatiques, Université de Paris VI, 1991.

[3] N. Bennacer and G. Florin S. Natkin. Probabilistic validation using worst event driven and importance sampling simulation. CEDRIC Report 93-14 submitted to the 13th symposium on Reliable Distributed Systems, October, 1994.

[4] K. Barkaoui G. Florin C. Fraize B. Lemaire S. Natkin. Reliability analysis of non repairable systems using stochastic Petri nets. 18th Symposium on Fault Tolerant and Computing Systems, Tokyo, June 1988.

[5] A. R. Cavalli and E. Paul. Exhaustive analysis and simulation for distributed systems, both sides of the same coin. Distributed computing, n 2, 1988.

[6] E. Cinlar. Introduction to stochastic processes. Prentice Hall 1975.

[7] A. E. Conway and A. Goyal. Monte Carlo simulation of computer system availability/reliability models. Proceeding of the seventeenth symposium on Fault Tolerant Computing vol 6, pp 230-235, 1987.

[8] Chorus distributed operating systems C. Delorme. Spécifications de l'IPC Chorus V3.3, documentation interne, CS/TN-89-18, February, 1990.

[9] G. Florin C. Fraize S. Natkin. Stochastic Petri nets: properties, applications and tools. Micro-electronics and reliability, vol 31, n 4, pp 669-698, 1991.

[10] G. Florin C. Fraize S. Natkin. A new approach of formal proof: probabilistic validation. International working conference on Dependable Computing for Critical Applications, Tucson, Arizona, pp 18-20, February, 1991.

[11] G. Florin C. Fraize S. Natkin. Searching best paths to worst states. International working conference on Petri Nets Performance Models, Melbourne, pp 204-209, December, 1991.

[12] C. Fraize. Validation Probabiliste des systèmes sécuritaires ou distribués modélisés en termes de réseaux de Petri Stochastiques. Thèse de Doctorat, CNAM Paris, February, 1993.

[13] P. W. Glynn D. L. Iglehart. Importance Sampling for stochastic simulations. Management Science, vol 35, n 11, November, 1989.

[14] J. Gorski. Design for safety using temporal logic. Safecomp, Sarlat, France, 1986.

[15] B. T. Halpern and S. Oowicki. Modular verification of computer communication protocols. IEEE transactions on communications, vol com-31. n 1, 1983.

[16] J. P. Haas and G. S. Shedler. Modelling power of stochastic Petri nets for simulation. Probability in the engineering and information sciences, n 2, pp 135-159, 1988.

[17] E. E. Lewis F. böhm. Monte Carlo simulation of Markov unreliability models. Nuclear Engineering and Design, vol 77, pp 49-62, 1984.

[18] V. F. Nicola M. K. Nakayama P. Heidelberger A. Goyal. Fast simulation of dependability models with general failure, repair and maintenance processes. Proceedings of the twentieth symposium on fault- tolerant computing, pp 491-498, June, 1990.

[19] J. P. C. Kleijnen. Statistical Techniques in simulation. Part 1 Marcel Dekker, 1974.

[20] M. Silva. Structural analysis of Petri nets. Tutorials PNPM93, Toulouse, October, 1993.

[21] A. Tanenbaum. Communication networks. Second edition. Prentice-Hall editions.

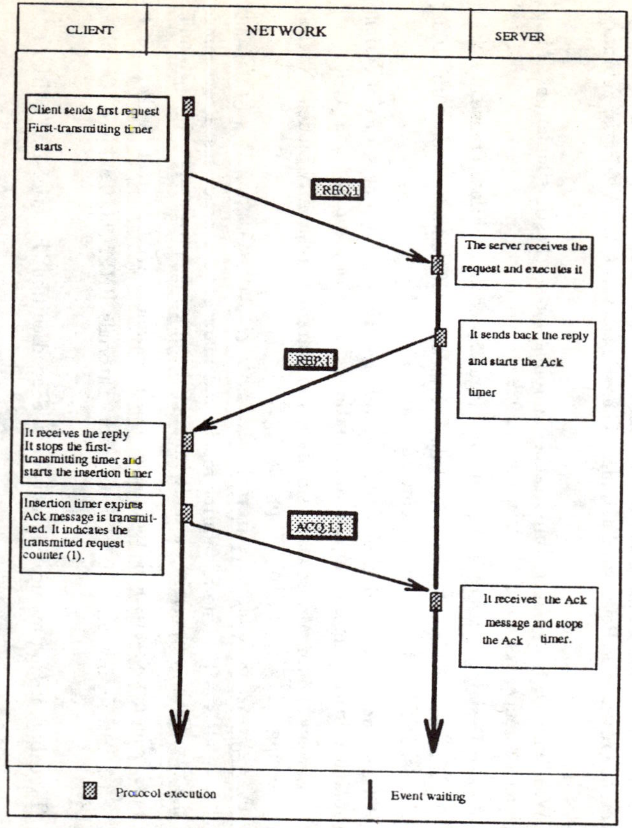


Figure 1: Correct scenario example

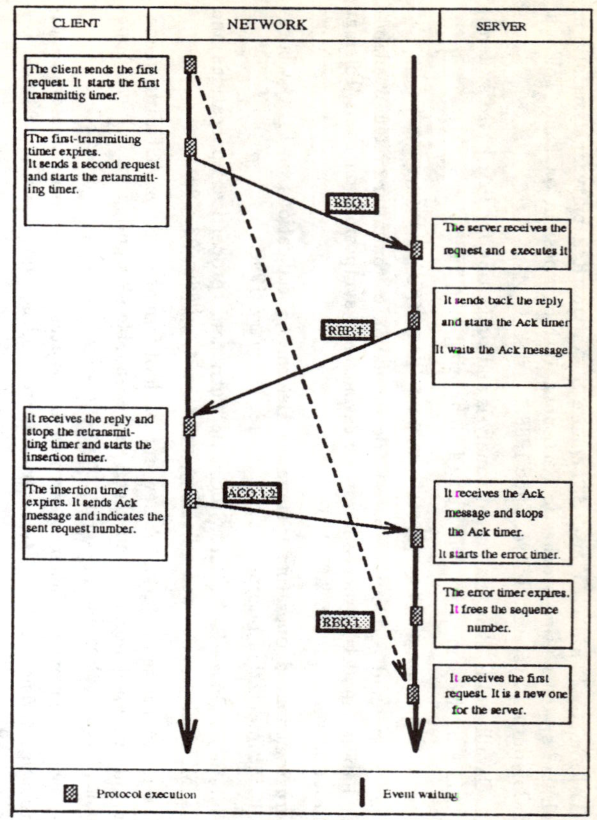


Figure 2: Incorrect scenario example

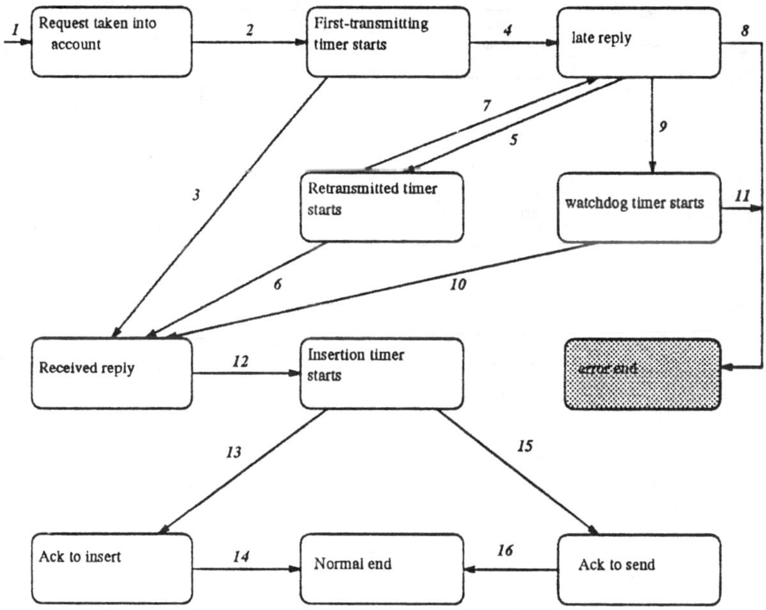


Figure 3: Client state machine

<u>Events</u>

1: A request is deposited by the client process.

2: This request is transmitted to the server. The first-transmitting timer is started.

3: The reply is received by the server. The first-transmitting timer is stopped.

4: The first-transmitting timer expires before the reply arrives. The maximum attempt number is initialized.

5: The maximum attempt number is not reached. It is decremented. The request is retransmitted. The retransmitting timer is started.

6: The reply is received by the client. The retransmitting timer is stopped.

7: The retransmitting timer expires.

8: The maximum attempt number is reached. The server is considered as failed.

9: A message of type "request is currently carried out" is received by the client. The retransmitting timer is stopped. The watchdog timer is started.

10: The server reply is received. The watchdog timer is stopped.

11: The watchdog timer expires. The server is considered as failed.

12: The insertion timer starts.

13: A request (following sequence) is submitted to the same server. The insertion timer is stopped.

14: The acknowledgement is inserted in the following request sequence.

15: The insertion timer expires.

16: The explicit acknowledgement is transmitted to the server.

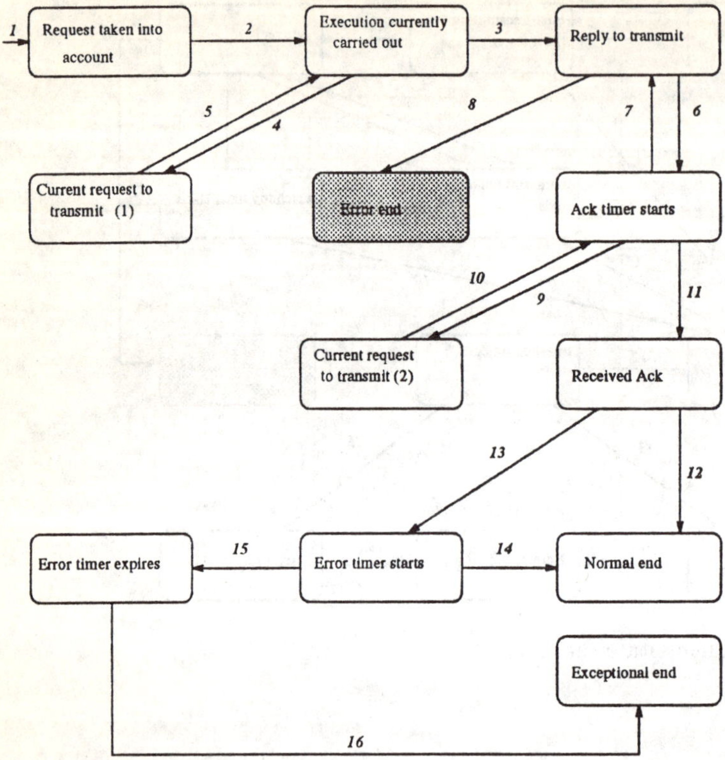


Figure 4: Server state machine

**Events**

1: A request is received by the server. The provided sequence identifier is unknown. A sequence is initialized. The maximum transmitted request number is initialized.

2: A server process is activated to carry out the request.

3: The request execution finishes.

4: The request is retransmitted by the client during the server request execution.

5: The "request is currently carried out" message is transmitted to the client.

6: The transmitted reply number is not reached. It is decremented. The reply is transmitted to the client. The acknowledgement timer starts.

7: The acknowledgement timer expires before the acknowledgement message arrives.

8: The transmitting reply number is reached. The client is considered as failed.

9: The server waits for the acknowledgement message, it receives the same request (retransmitted by the client).

10: The "request is currently carried out" message is sent to the client.

11: The acknowledgment arrives. The acknowledgement timer is stopped.

12: The transmitting request counter is equal to the received request counter.

13: The transmitting request counter is different from the received request counter. The error timer starts.

14: All the requests transiting in the network are received. The error timer is stopped.

15: The error timer expires.

16: The sequence is in "exceptional end".

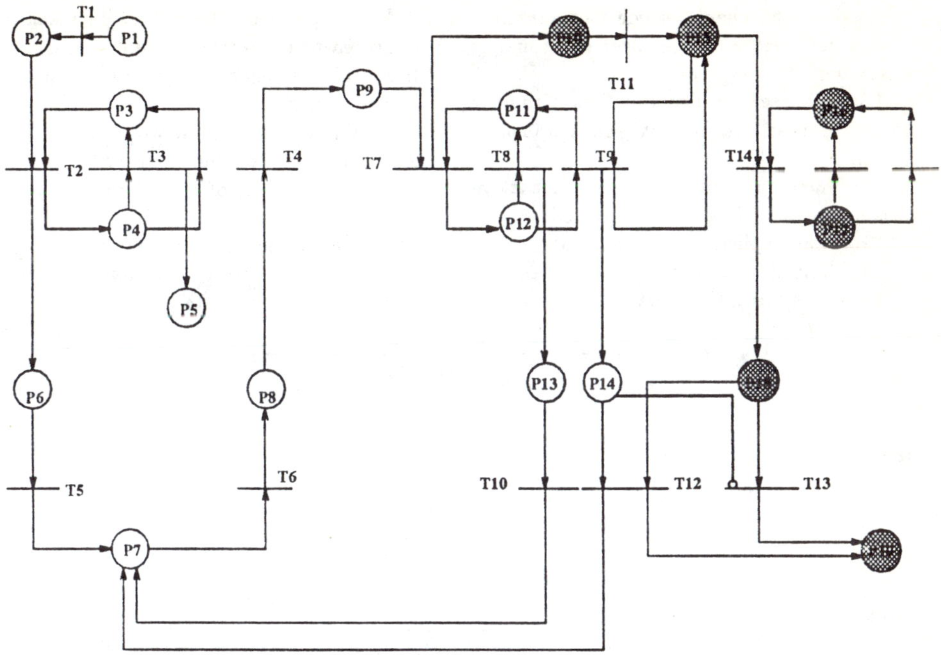


Figure 5: Petri net of client sequence

**Places:**

P1: Request process.

P2: The first request is submitted to the client.

P3: First-transmitting timer is timed out.

P4: First-transmitting timer is started.

P5: The first-transmitting timer is expired.

P6: The RPC request is transmitted.

P7: The first transaction server.

P8: The reply to the first request is transmitted.

P9: The reply to the first request is received.

P10: The request process can submit another request.

P11: The insertion timer is stopped.

P12: The insertion timer is started.

P13: An explicit Ack of the first transaction is transmitted.

P14: A piggyback Ack of the first transaction is transmitted.

P15: The second request is submitted to the client.

P16: The first-transmitting timer related to the second transaction is stopped.

P17: The first-transmitting timer related to the second transaction is started.

P18: The request of the second transaction is transmitted.

P19: The server of the second transaction.

P1, P3, P11 and P16 are initially marked.

**Fired transitions**

T1: A request is provided to the client.

T2: The first-transmitting timer is started and the request is transmitted.

T3: The first-transmitting timer is expired without receiving reply.

T4: The reply is received by the server. The first-transmitting timer is stopped.

T5: The request of the first transaction is received by the server.

T6: The server transmits the reply to the first request.

T7: The reply to the first request is pro-

vided to the client (which can submit a new request) and the insertion timer is started.
T8: The insertion timer expires without a new request. An explicit Ack is transmitted.
T9: The request of the second transaction is submitted. The Ack is inserted.
T10: The explicit Ack is received.
T11: The client submits a second request.
T12: The second request with inserted Ack is received by the server. The Ack is delivered to the server of the first transaction and the request to the server of the second transaction.
T13: The second request without inserted Ack is received by the server. It is delivered to the server of the second transaction.
T14: The second request is transmitted. The first-transmitting timer related to the second request is started.

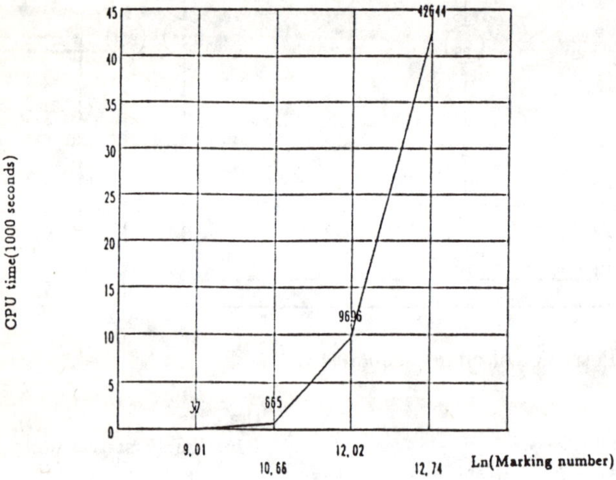


Figure 6: Markovian Solver

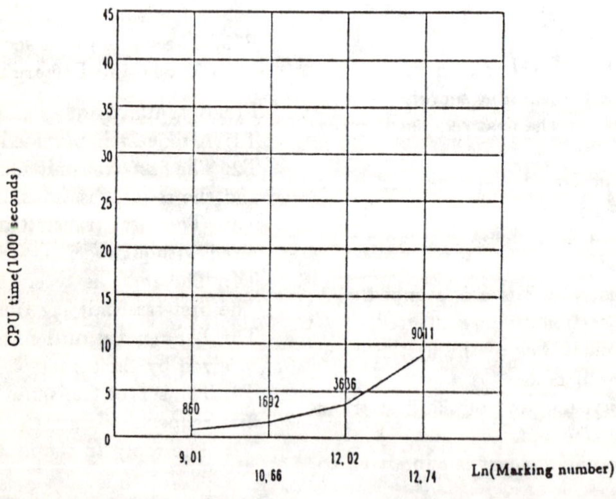


Figure 7: Breadth and depth fisrt search traversal algorithm

# OCCURRENCE GRAPHS FOR INTERVAL TIMED COLOURED NETS

Gérard BERTHELOT, CEDRIC-IIE-CNAM Paris
Hanifa BOUCHENEB, CEDRIC-IIE-CNAM and USTHB Alger

IIE-CNAM, 18 allée J.Rostand, 91025 EVRY CEDEX, FRANCE,
email : berthelot@iie.cnam.fr

**Abstract.** We present an approach to construct the occurrence graph for ITCPN (Interval Timed Coloured Petri Nets). These models, defined by Van Der Aalst in [VAN] can simulate other timed Petri nets and allow to describe large and complex real-time systems. We define classes as sets of states between two occurrences, and we use these classes to define the occurrence graph of an ITCPN. Then an equivalence relation based on time is defined for classes, and we show that occurrence graphs reduced using this equivalence relation are finite if and only if the set of reachable markings is finite. These graphs can be used to verify all the dynamic properties such as reachability, boundedness, home, liveness and fairness properties but also performance properties : minimal and maximal bounds along a occurrence sequence or a cycle. Finally we complete delay based equivalence with a colour based equivalence in order to achieve further reduction.

**Keywords:** interval timed coloured Petri nets, occurrence graph.

## 1 - Introduction

As formal verification techniques are wide spreading the interest for timed models increases constantly [SIF91], either in real times area with synchronous language like ESTEREL ([BERR]) or the protocol area with various timed LOTOS [QUE]. This seems quite normal since, in the common opinion, a model is correctly functioning not only if it has good behavioural properties such liveness and fairness but also if he has good time performances. Interest for time in Petri nets is quite old (Ramchandany paper [RAM] appears in 1973), but until now the various proposed models are not considered as totally adequate to model real systems. First timed Petri nets of [RAM], associated deterministic time to transitions or places [SIF,77] and were applicable only to few production systems or electronic devices where the time for performing an operation is constant. The behavioural properties of these models can be investigated only if they allow finite firing sequences or if there is a steady state when infinite firing sequences are allowed. To allow modelisation of a larger set of systems, Florin and Natkin have defined Stochastic Petri nets [FLO] where stochastic delays are associated with transitions. If these delays are distributed according to negative exponential probability laws, it is possible to translate the net into a continuous Markov chain. This Markov chain may then be analysed to gives probability laws of various quantities like mean firing period, mean residence time in a given state and so on. Besides these various performance measures, this model allows also to investigate usual behavioural properties. Despite its obvious interest, this model is still unsatisfactory for systems where strict time limits must be respected, as real time systems or protocols. By the way, negative exponential probability distribution allows to fire a transition after a delay ranging from 0 to infinity. So it is impossible to prove, for instance that a timer will send a signal before a given delay or not. Conversely a maximum firing delay for a transition may change the behavioural properties of the untimed underlying net, changing an unbounded but subject to deadlock net into a bounded and live one, since liveness is not monotonic with marking. So a model with time constraints defined by intervals is needed, and it must be analysed without removing time constraints. Such a model has been defined by Merlin [MER] which has proposed to associate an interval, specifying the minimal and maximal firing delays, with each transition. Berthomieu and Menasche in [BERT,83] have defined a procedure to obtain an occurrence graph for this model but this procedure is slow since it implies at each step (i.e. each firing), the resolution of a system of inequations. Moreover, Merlin's model is "uncoloured" and is difficult to use for modelisation of complex systems.

Van der Aalst ([VAN]) recently proposed Interval timed coloured Petri Nets, also called ITCPN, which are Coloured Petri Net (CPN) extended with time. A time stamp is associated with each token, specifying the minimal time from which this token could be consumed. When a token is created, its time stamp is defined by adding to the firing time of the creating transition, a value chosen randomly in an interval depending on three parameters : colours of consumed tokens, colour of the created token and arc used. ITCPN are extremely general since they can simulate other existing models, deterministic one's with intervals reduced to one value, as well as Merlin model with intervals depending only on the transition. They can also simulate Stochastic Petri nets since bounds of intervals may range from 0 to $\infty$, and one could specify a probability distribution for each interval. Consequently they allow to describe large and complex real-time systems.

The same author has proposed an analysis method, called the Modified Transition System Reduction Technique (MTSRT), which is "sound" i.e. any occurrence sequence in the initial model is possible also in the modified one, but not "complete", i.e. the converse is not true (some occurrence sequence in the modified model does not reflect any occurrence sequence of the initial model). Moreover, for nets allowing infinite occurrence sequences, MTSRT would lead to infinite graphs.

We present here an approach to built ITCPN occurrence graphs which have not these drawbacks. Firstly, we give basic definitions on ITCPN and ITCPN behaviour in section 2. In the following section we point out a simple condition for the occurrence of an event after n event occurrences. Then we define classes as sets of states between two occurrences and we define the occurrence graph of an ITCPN. In section 4 we define an equivalence relation based on delays, and we show that an occurrence graph, using this equivalence relation to fuse classes, is finite if and only if the set of reachable markings is finite (i.e. even if the net allows infinite occurrence sequences). It can be used to verify all the dynamic properties such as reachability, boundedness, home, liveness and fairness properties. Fifth section is devoted to the calculus of minimal and maximal delays for an occurrence sequence, cycling or not. In section 6, we show how to complete delay based equivalence with a colour based equivalence in order to achieve further reductions of an occurrence graph. We end with an example in section 7.

## 2 - Interval Timed Coloured Petri Nets

We will introduce here only necessary definitions and notations. For further details we refer to [JENa] for coloured Petri nets and to [VAN] for Interval timed coloured Petri nets. Let's just recall that $A_{MS}$ denotes the set of all multi-sets over a set A

An interval timed coloured Petri net is a coloured net where time stamps are attached to tokens when they are created. Time stamps must be chosen inside an interval associated with the arc creating each token.

**Definition 1**
TS is the **time set**, TS = $\{x \in \mathbb{R} \mid 0 \leq x\}$, i.e. the set all non-negative reals. INT = $\{[y,z] \in TS \times TS \mid y \leq z\}$ represents the set of all closed intervals. If $x \in TS$ and $[y,z] \in INT$, then $x \in [y,z]$ iff $y \leq x \leq z$.

An interval timed coloured Petri net is defined as follows:

**Definition 2**
An **Interval Timed Coloured Petri Net** is a five tuple ITCPN = $(\Sigma,P,T,C,F)$ where
(i) $\Sigma$ is a finite set of types, called colour sets.
(ii) P is a finite set of places.
(iii) T is a finite set of transitions.
(iv) $C \in P \rightarrow \Sigma$. C(p) is the type of $\Sigma$ which specifies the set of allowed values (or colours) for tokens of place p.
(v) $CT = \{(p,v) \mid p \in P \text{ and } v \in C(p)\}$ is the set of all possible coloured tokens
(vi) F is the transition function: F(t) specifies which tokens are consumed and produced by firing transition t and also the interval in which their time stamps must be chosen.
$F(t) \in CT_{MS} \rightarrow (CT \times INT)_{MS}$

### 2.1 ITCPN behaviour

A state of an ITCPN is a multi-set of tokens, each bearing a time stamp, describing a marking of the net. An event e = (t, bin, bout) represents the firing of t while consuming the tokens specified by bin and producing tokens specified by bout.

**Definition 3**
An event (t,bin,bout) is **enabled** in state s iff:
(i) bin ≤ s, i.e.: the tokens to be consumed are present in this state
(ii) M(bin) ∈ dom F(t), i.e. : there are enough tokens on each of its input place, as specified by dom(F(t))
(iii) F(t)(M(bin)) may be "specialised" into bout, i.e. : for each token in F(t)(M(bin)) there is the same token in bout and for every token of bout, its time stamp is inside the interval of the corresponding token of F(t)(M(bin)).

**Definition 4**
The **enabling time** of an event (t,bin,bout) is the maximum of all the time stamps of consumed tokens i.e.:

$$ET(t,bin,bout) = \max_{((p,v),x) \in bin} \{x\}$$

An enabled event is **time enabled** iff no other event have a smaller enabling time.

**Definition 5**
If an event (t,bin,bout) is time enabled in a state $s_1$, it **occurs** at its enabling time $\theta$ = ET((t,bin,bout), i.e. : transition t fires while removing tokens specified by bin and adding token specified by bout. This event occurrence is denoted by $s_1[e,\theta>s_2$. The new state $s_2$ is defined by

$$s_2 = (s_1 - bin) + (SC(bout,\theta))$$

where SC is a function which age every token by adding $\theta$ to their time stamps.

In other words, tokens must not be used "before the time" of their time stamps, but events must occur as soon as possible, i.e. if no event can occur before.

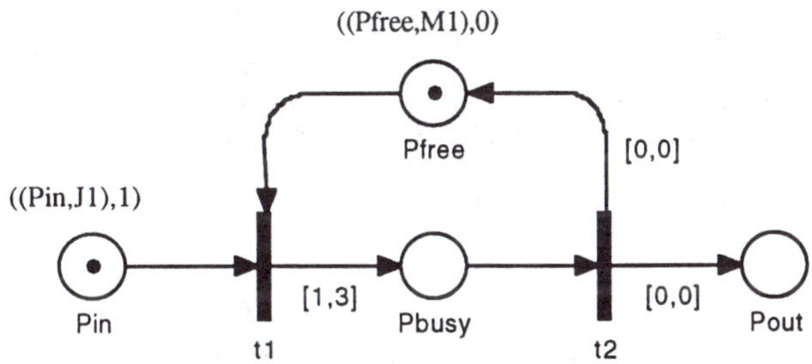

Figure 1

We will now explain the behaviour of an ITCPN, using an example given in [VAN] and reported here in figure 1. This figure is the graphic representation of an ITCPN which is composed of four places $\{p_{in}, p_{busy}, p_{free}, p_{out}\}$ and two transitions $\{t_1, t_2\}$. It represents a jobshop, where jobs arrive via place $p_{in}$ and leave the system via place $p_{out}$. The jobshop is composed of a number of machines. Each machine is represented by a token which is either in place $p_{free}$ or in place $p_{busy}$. There are three colour sets: Σ contains $M = (M_1, M_2,...)$, $J = (J_1, J_2,...)$ and $M \times J$. $M$ is attached to place $p_{free}$, $J$ is attached to places $\{p_{in}, p_{out}\}$ and colour set $M \times J$ is attached to place $p_{busy}$.

Suppose that 0 is the initial time. The initial state is $((p_{free}, M_I), 0) + ((p_{in}, J_I),1)$. Function $F(t_1)$ specifies the tokens consumed and produced by occurrence of $t_1$. It is defined as follows:

dom $(F(t_1)) = \{(p_{in},j)+(p_{free},m) \mid j \in J \text{ and } m \in M\}$

for $j = J_I$ and $m = M_I$, we have $F(t_1)\ ((p_{in}, j) + (p_{free}, m)) = ((p_{busy}, (M_I, J_I), [1,3]);$

Or also $t_1$ consumes one token from place $p_{in}$ and one token from place $p_{free}$ and it produces one token for place $p_{busy}$.

Because of token $((p_{in}, J_I),1)$, transition $t_1$ occurs at time 1. Occurrence of transition $t_1$ causes state change: the previous tokens are consumed and a new token is produced with a time stamp inside [1+1, 1+3], for instance 2, resulting in : $((p_{busy}, (M_I, J_I), 2)$.

From the obtained state, $t_2$ can occur at time 2, and the state will change according to the following:

for $j \in J$ and $m \in M$, we have $F(t_2)\ ((p_{busy}, (m,j)) = ((p_{free}, m), [0,0]) + ((p_{out}, j), [0,0])$

it consumes the token from place $p_{busy}$ and it produces two tokens (one for $p_{free}$ and one for $p_{out}$): $((p_{free}, M_I), 2) + ((p_{out}, J_I), 2)$.

**2.2 ITCPN state classes**

Even if events must occur as soon as possible, an ITCPN is not deterministic because of intervals specifying time stamps. So a given occurrence sequence may occur according to different delays and consequently, may results in different states. So the number of states, differing only with respect of time stamps of tokens is infinite. In order to be able to make some overall analysis, it is necessary to group all states having similar characteristics (i.e. identical untimed marking) in one state class.

To do so, Van Der Aalst have proposed a method, called the Modified Transition System Reduction Technique (MTSRT for short) in which time stamps of tokens are replaced by intervals reflecting the range of their possible time stamps (see figure 2).

Unfortunately, if this method is "sound" i.e. any occurrence sequence according to the initial model is possible also in the modified one, it is not "complete", i.e. the converse is not true (some occurrence sequence in the modified model does not reflect any occurrence sequence of the initial model). This is due to the fact that this method "forget" the occurrence time to memorise only intervals. Consequently, for events producing several tokens, the dependencies (relations binding intervals) are lost and classes including unreachable states are produced. In figure 2 for instance, the firing of transition t from a token $((p_1,e),[0,2])$ determine a class $((p_2,e),[1,4])+((p_3),[3,6])$ where states $((p_2,e),1)+((p_3),6)$ but also $((p_2,e),4)+((p_3),3)$ are represented but not reachable. Due to these represented but unreachable states, the state class occurrence graph obtained by MTSRT has not exactly the same properties as the initial model and a deadlock in the former does not imply a deadlock in the latter. For the same reasons the MTSRT can only be used to obtain larger bounds for time performances.

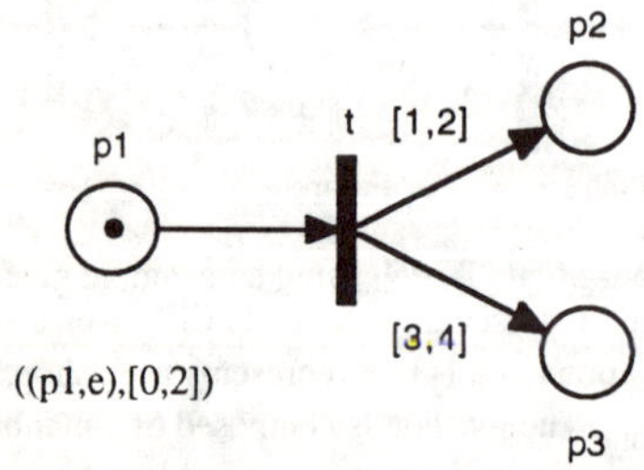


Figure 2

We present now an other technique which is both "sound" and "complete" and allows exact analysis both for behaviour properties and time performances.

## 3 - Class occurrence graph.

### 3.1 Tokens with interval and creation time stamp

Instead of associating only time stamps or intervals with tokens, we associates both time stamps (of creation time), and intervals.

**Definition 6**
A **interval timed token** is a 4-uple $((p_i,v_i),\theta_i,a_i,b_i)$ where
- $p_i \in P$ is the place where it stays,
- $v_i \in C(p_i)$, is its colour,
- $\theta_i \in TS$, is the time at which the token was produced and
- $(a_i,b_i) \in TS \times (TS \cup \{\infty\})$, $[a_i,b_i]$ is the interval used when the token was created.

This means that for a given value of $\theta_i$ a token becomes available at any arbitrary time between $\theta_i+a_i$ and $\theta_i+b_i$. $a_i$ and $b_i$ are called respectively the minimal and maximal delay of the token.

If we go back to figure 1, we can have the following behaviour. From the initial state $((p_{free}, M_1), \theta_0, 3, 5) + ((p_{in}, J_1),\theta_0, 0, 3)$, the model progress with occurrences of transitions. A transition t may occur at a given time θ if all input places contain at least all tokens which will be consumed by transition t and if all these tokens are available. Function $F(t_1)$ specifies the consumed and the produced tokens occurrence of $t_1$ as previously.

Transition $t_1$ may occur at any time $\theta_1$ between values $\theta_0+\max(3,0)$ and $\theta_0+\max(5,3)$. The first value represent the minimal time at which the two tokens become available. It is the minimal occurrence time of $t_1$. The second value is the maximal time at which the two tokens become available. $t_1$ must occur not later that time. So it is the maximal occurrence time of $t_1$.

Occurrence of transition $t_1$ causes state change: the previous tokens are consumed and the following token is produced: $((p_{busy}, (M_1, J_1)), \theta_1, 1, 3)$.

From the obtained state, $t_2$ can occur at any time $\theta_2$ between $\theta_1+1$ and $\theta_1+3$. Since $\theta_1$ is between $\theta_0+3$ and $\theta_0+5$, the minimal occurrence time of transition $t_2$ is $\theta_0+3+1$ and the maximal occurrence time of transition $t_2$ is $\theta_0+5+3$. If transition $t_2$ occurs, it consumes one token from place $p_{busy}$ and it produces two tokens:

$((p_{free}, M_1), \theta_2, 0, 0) + ((p_{out}, J_1), \theta_2, 0, 0)$.

### 3.2 Occurrence conditions when occurrence instants are not fixed.

Since we have settled into tokens, the necessary and sufficient information to choose a time stamp, it is clear that there is no difference, at the semantic level, between the time stamp tokens model and the time of creation and interval tokens model. From a practical point of view, the only difference is that time stamp can be chosen when a token is consumed and not when a token is created. But the upper bound and lower bound being the same, this has no influence for reachable states. We may conclude that any possible evolution in the initial model is possible also in the modified model and conversely. However, even with fixed $\theta_i$, a marking with intervals represents a set of markings with time stamps and it is necessary to precise the occurrence rules.

**Notations 1**
Let's denote by E the set of all events which are enabled by a marking M of a state s. The minimal and maximal occurrence time of an event (t, bin, bout) of E are:

$$FT_{min}((t, bin, bout)) = \max_{((p_i,v_i),\theta_i,a_i,b_i) \in bin} (\theta_i + a_i)$$

($FT_{min}$( (t, bin, bout) ) is the smallest time at which all tokens of bin become available).

$$FT_{max}( (t, bin, bout) ) = \max_{((p_i,v_i),\theta_i,a_i,b_i) \in bin} (\theta_i + b_i)$$

($FT_{max}$( (t, bin, bout) ) is the time at which all tokens of bin have reached their maximal delays).

**Proposition 1: condition for an enabled event to occur at time θ**

An event e can occur at time θ ⇔ (i) e is enabled and

(ii) $FT_{min}(e) \leq \theta \leq \min_{e' \in E} (FT_{max}(e'))$

**Remarks:**

- An enabled event e can occur at any time θ between its minimal and maximal occurrence time provided that it remains enabled and at time θ, no other enabled event e has exceeded its maximal occurrence time.
- An enabled event which reaches its maximal occurrence time must occur immediately without any delay.

This new presentation allows to keep dependencies between tokens, but its main advantage is that we can now represent a class of states simply by replacing fixed $\theta_i$ by variables $\tau_i$. We must also specify relations between these variables because they are not independent, since the maximal occurrence time of an event depends not only of the token it consumes but also of the other enabled events. To achieve this, we shall point out which conditions must be fulfilled for the occurrence of an event after the occurrence of any occurrence sequence $e_1, e_2,.., e_n$.

**Notations 2**

Let us assume that an occurrence sequence $e_1, e_2,.., e_n$ has occurred. The initial marking is denoted by $M_0$ and $M_i$ denotes the marking obtained by the ith occurrence (see figure 3):

– $\tau_i$ is a variable representing instants when the event $e_i$ may occurs.

- $x_i$ is a variable representing the difference between the (i-1)th occurrence and the ith occurrence, i.e. $x_i = \tau_i - \tau_{i-1}$. So we have $\tau_i = x_1 + x_2 + ... + x_i$.

For the sake of clarity, $x_0 = 0$ conventionally. Conventionally also, the minimum of an empty set will be ∞, and the maximum of an empty set will be 0.

We need to introduce some notations used in the next definition to determine occurrence instants.

Let $e_n = (t_n, bin_n, bout_n)$ be an enabled event at time $\tau_n$, we define the following sets for i ∈ [0,n-1] :

- **bin(i,n)** = $\{ ((p, v), \tau, a, b) \in bin_n \mid \tau = \tau_i \}$.

(bin(i,n) is the set of token created by the ith occurrence and consumed by the nth occurrence)

- $\mathbf{maxa}_{i,n}(e_n)$ = if bin(i,n) = Ø then 0 else $\max_{((p, v), \tau, a, b) \in bin(i,n)} a$

($maxa_{i,n}(e_n)$ is the value for which all tokens created by the ith occurrence and consumed by the nth occurrence have reached their minimal delay, i.e. are available)

- $\mathbf{maxb}_{i,n}(e_n)$ = if bin(i,n) = Ø then 0 else $\max_{((p, v), \tau, a, b) \in bin(i,n)} b$

($maxb_{i,n}(e_n)$ is the value for which all tokens created by the ith occurrence and consumed by the nth occurrence have reached their maximal delay)

- $\mathbf{E_i}$ will denotes the set of all events which are enabled for the marking $M_i$ (i.e. without taking care of time).

```
+-----x1--+----x2-----+-----------...--------+----------xi---+--------......--------+------xn----+------------>
τ0     τ1       τ2               τi-1            τi               τn-1           τn   occurrence times
+----M0--+-----M1--+----------....--------+----------Mi--+----------.....-------+------Mn---+  markings
       e1       e2               ei-1            ei               en-1           en   occurred events
```

Figure 3

In the following we investigate only time conditions. Marking conditions are the usual one's and are omitted.

**Proposition 2: conditions for the occurrence of an event $e_n$ after a sequence $e_1,e_2,\ldots,e_{n-1}$**

Let us suppose that $e_1, e_2,\ldots, e_{n-1}$ have occurred successively in that order.

An event $e_n$ may be the nth occurring event at time $\tau_n$ $\Leftrightarrow$ (i) $e_n$ enabled and (ii) it exists $x_n \geq 0$ such that :

$$\max_{i=0,n-1} \{\mathrm{maxa}_{i,n}(e_n)+ x_0+ \ldots +x_i \} \leq x_1+\ldots+x_n \leq \min_{e \in E_n} \{\max_{i=0,n-1} \{\mathrm{maxb}_{i,n}(e) + x_0+\ldots+x_i \}\}$$

**Proof :**

It suffices to sum up conditions specified by the previous proposition to reach $FT_{min}(e_n)$ without reaching $FT_{max}$ of any event enabled by some $e_i$.

**Proposition 3: conditions for the occurrence of a sequence $e_1, e_2,\ldots, e_n$**

Events $e_1, e_2,\ldots, e_n$ may occur respectively at the first, the second, .. and the nth occurrence

$\Leftrightarrow$

(i) $e_1, e_2,\ldots, e_n$ are enabled respectively by $M_0, M_1, M_2,\ldots,M_n$ and

(ii) there exist $x_1,\ldots, x_n$ such that:

$$I(n)\left|\begin{array}{l} \mathrm{maxa}_{0,1}(e_1) \leq x_1 \leq \min_{e \in E_1} \{\mathrm{maxb}_{0,1}(e)\} \text{ and} \\ \max_{i=0,1} (\mathrm{maxa}_{i,2}(e_2)+x_0 +\ldots+x_i) \leq x_1+x_2 \leq \min_{e \in E_2} \{ \max_{i=0,1} \{\mathrm{maxb}_{i,2}(e)+x_0 +\ldots+x_i\}\} \text{ and} \\ \quad\vdots \\ \max_{i=0,n-1} (\mathrm{maxa}_{i,n}(e_n)+x_0+\ldots+x_i ) \leq x_1+\ldots+x_n \leq \min_{e \in E_n} \{ \max_{i=0,n-1} \{\mathrm{maxb}_{i,n}(e)+x_0+\ldots+x_i \}\} \end{array}\right.$$

**Proof :**

It suffices to sum up conditions specified by the previous proposition for every event to reach its $FT_{min}$ without reaching $FT_{max}$ of any enabled event.

**Definition 7 : occurrence space at the nth occurrence**

The **occurrence space** for the nth occurrence, denoted by FS(n), is the set of solutions of the system I(n) given in proposition 3 above.

### 3.3 Simplification of the occurrence condition

The condition given above leads to solve a system of linear inequations for every occurrence and is rather difficult to use. In order to define a simpler condition , w
e will study the system I(n) and look for a condition implying the existence of solutions.
Intuitively, the idea behind the simplification is the following. Since occurrence times $\tau_i$ and $\tau_k$ are not independent we shall use conditions on delays between these occurrences. More precisely we shall prove with theorem 1 that in order to make it possible the occurrence of event $e_n$, for every i and j the delay $x_{i+1}+...+x_j$ must range between fixed values, i.e. we must have $Smin(i,j,n) \leq x_{i+1}+...+x_j \leq Smax(i,j,n)$. Smin and Smax depend not only on (i,j) but also on n, hence the following recursive definition.

**Notations 3**
Let $e_1, e_2, ..., e_n$ an enabled sequence of events. We define recursively the following for $i \in [0,n-1]$ :
**Smin(i,n,n)=**

$$\max\{\max_{k=1,i-1}\{maxa_{k,n}(e_n)-Smax(k,i,n-1)\}, maxa_{i,n}(e_n), \max_{k=i+1,n-1}\{maxa_{k,n}(e_n)+Smin(i,k,n-1)\}\}$$

**Smax(i,n,n)=**

$$\min_{e \in E_n}\{\max\{\max_{k=1,i-1}\{maxb_{k,n}(e) - Smin(k,i,n-1)\}, maxb_{i,n}(e), \max_{k=i+1,n-1}\{maxb_{k,n}(e)+Smax(i,k,n-1)\}\}$$

and for $i \in [0,j-1]$ and $j \in [1,n-1]$
**Smin(i,j,n)** = max(Smin(i,j,n-1), Smin(i,n,n) - Smax(j,n,n))
**Smax(i,j,n)** = min( Smax(i,j,n-1), Smax(i,n,n) - Smin(j,n,n))

The above notations are quite hard to read but can be understood with the help of some figures.

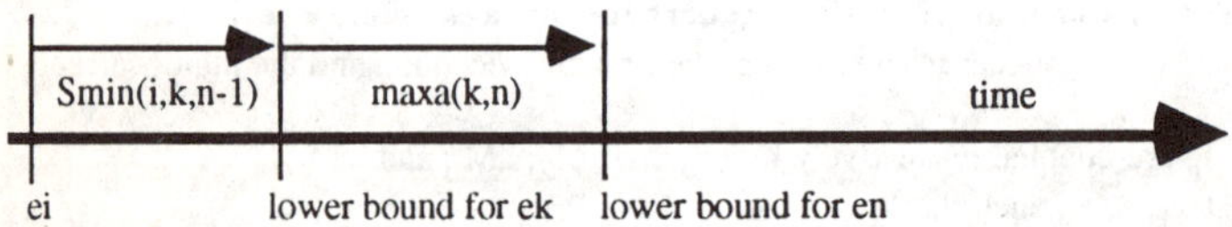


Figure 4

Explanations
Let's assume that $e_n$ consumes some tokens produced by $e_k$. By definition, they are not available during $maxa_{k,n}$. Let's assume that $e_i$ occurred before $e_k$ and let also assume that for the occurrence of $e_{n-1}$, the minimum delay between $e_i$ and $e_k$ is Smin(i,k,n-1). Then the minimum delay between $e_i$ and $e_n$ (Smin(i,n,n)) is greater or equal to the sum $maxa_{k,n}(e_n)$+Smin(i,k,n-1). Obviously, if $e_k$ has occurred before $e_i$, we must substract Smax(k,i,n-1) instead of adding Smin(i,k,n-1). We must repeat this for every event producing token consumed by $e_n$ and keep the maximum value.

The calculus for Smax(i,n,n) follows the same idea, but is a little bit more complicated since we must also check if other enabled event must not occur before $e_n$.

Concerning Smin(i,j,n), it can be seen easily on figure 5 that it may be bigger than Smin(i,j,n-1) if it is less than Smin(i,n,n)-Smax(j,n,n). The converse also may happen for Smax(i,j,n). We can conclude that Smin(i,j,n) may only increase with n, while Smax(i,j,n) may only decrease. In other words the range for the sum $x_{i+1}+...+x_j$ may only "decrease" as events occur. This means that in order to be able to fire $e_n$ latter we cannot use the full range of possibilities for occurrences of $e_i$ and $e_j$ but only a part of them.

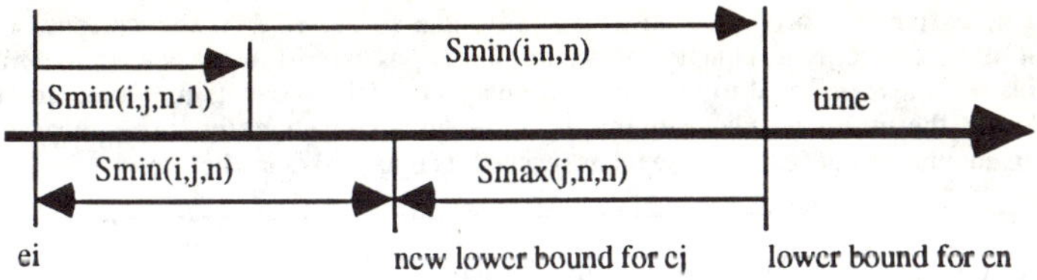


Figure 5

The first values of Smin and Smax are used in the following proposition which will be also the basis of proof of theorem 1 .

**Proposition 4:**
An event $e_1$ may be the first occurring event $\Leftrightarrow$
(i) $e_i$ is enabled for the initial marking and
(ii) $Smin(0,1,1) = maxa_{0,1}(e_1) \leq min\ \{maxb_{0,1}(e) \mid e \in E_1\} = Smax(0,1,1)$

**Proof:**
From notation 3 we have $Smin(0,1,1) = maxa_{0,1}(e_1)$ and $Smax(0,1,1) = min\ \{maxb_{0,1}(e) \mid e \in E_1\}$. Using proposition 2, it suffices to eliminate the variable $x_1$: $x_1$ exists $\Leftrightarrow$ $maxa_{0,1}(e_1) \leq min\ \{maxb_{0,1}(e) \mid e \in E_1\}$

QED

We are now ready to prove that Smin and Smax are respectively upper lower bound and lower upper bound of sums of $x_i$'s.

**Theorem 1:**
If FS(n-1) is not empty, then for every solution $(x_1,...,x_{n-1})$ of (FS(n-1), minimal and maximal values of the sum $x_{i+1}+..+x_j$ are respectively Smin(i,j,n-1) and Smax(i,j,n-1).

**Proof :** by induction on number of variables
**First step:** (n=1) see proof of proposition 4
**Second step:** Suppose that theorem 1 is true at step n-2 and show that it is true at step n-1.
From proposition 3, the occurrence space FS(n-1) can be expressed from FS(n-2) as follows:

$$FS(n-1) = \{ (x_1,...,x_{n-1}) \mid (x_1,...,x_{n-2}) \in FS(n-2) \text{ and}$$
$$\max_{k=0,n-2} \{maxa_{k,n-1}(e_{n-1})+x_0+...+x_k\} \leq x_1+...+x_{n-1} \leq \min_{e\in E_{n-1}} \{ \max_{k=0,n-2} \{ maxb_{k,n-1}(e)+x_0+...+x_k\}\}$$

To deduce bounds of $x_{i+1}+...+x_{n-1}$ , we substract the sum $x_0+...+x_i$ from the three parts of the above inequalities. Then we replace $(-x_{k+1} -...- x_i)$ by -Smax(k,i,n-2) in the left part and by -Smin(k,i,n-2) in the right part. Similarly the sum $(x_{i+1} +...+ x_k)$ is replaced by Smin(i,k,n-2) in the left part and by Smax(i,k,n-2) in the right part. Then we have obtained the expressions of Smin(i,n-1,n-1) and Smax(i,n-1,n-1) given in theorem 1.

Concerning Smin(i,j,n-1) and Smax(i,j,n-1), we have $x_{i+1}+...+x_j = (x_{i+1}+...+x_{n-1}) - (x_{j+1}+...+x_{n-1})$, then $Smin(i,j,n-1) \geq Smin(i,n-1,n-1) - Smax(j,n-1,n-1)$ and
$Smax(i,j,n-1) \leq Smax(i,n-1,n-1) - Smin(j,n-1,n-1)$.

QED

Remark : if a vector $\{x_1,...,x_{n-2},x_{n-1}\}$ belongs to FS(n-1), $\{x_1,...,x_{n-2}\}$ belongs to FS(n-2). So we have $Smin(i,j,n-1) \geq Smin(i,j,n-2)$ and $Smax(i,j,n-1) \leq Smax(i,j,n-2)$.

Theorem 2 hereafter uses properties of Smin and Smax to derive a new occurrence condition for a sequence of events. Rougly a sequence of occurences is possible iff at every step, the minimal delay to consume tokens is less or equal to the maximal delay for enabled events to occur. In other word it is possible to wait the minimal delay required by the tokens consumed by chosen events without to be forced to fire an other event first because it has reached its maximal delay before.

**Theorem 2:**
FS(n) is not empty $\Leftrightarrow \forall j \in [1,n], \forall i \in [0,j-1]\ maxa_{i,j}(e_j) \leq Smax(i,j,j)$

**Proof:** by induction
**First step:** (n=1) from proposition 4, $maxa_{0,1}(e_1) \leq Smax(0,1,1)$ .
**Second step:** By eliminating $x_n$ in system I(n), we obtain a system I'(n-1) which contains all inequalities of I(n-1) plus the following set of inequalities :

$$(1)\ \max_{i=0,n-1} \{maxa_{i,n}(e_n) + x_0 + \dots + x_i\} \leq \min_{e \in E_n}\{\ \max_{i=0,n-1}\{ maxb_{i,n}(e) + x_0 + \dots + x_i \}\}$$

Then system I(n) has a solution iff system I(n-1) has a solution and there exists $(x_1,\dots,x_{n-1})$ in FS(n-1) such that the set of inequalities (1) are satisfied. But the set of inequalities (1) may be organized as :
$\forall i \in [0,n-1]$

$$maxa_{i,n}(e_n) \leq \min_{e \in E_n}\{\ \max_{k=0,i-1}\{ maxb_{k,n}(e)-(x_{k+1}+\dots+x_i)\}, maxb_{i,n}(e),\ \max_{k=i+1,n-1}\{ maxb_{k,n}(e)+x_{i+1}+\dots+x_k\}\}$$

Using recurrence hypothesis, I(n) has a solution iff for j=1,n-1 and i=0,j-1 $maxa_{i,j}(e_j) \leq Smax(i,j,j)$ and $\forall i \in [0,n-1]$

$$maxa_{i,n}(e_n) \leq \min_{e \in E_n}\{ \max\{\max_{k=1,i-1}\{maxb_{k,n}(e)-Smin(k,i,n-1)\}, maxb_{i,n}(e),\ \max_{k=i+1,n-1}\{ maxb_{k,n}(e)+Smax(i,k,n-1)\}\}$$

QED

The following corollary defines the simplified occurence condition for the nth occurrence. It uses the result of theorem 2 but take advantage from the fact that when event $e_n$ occurs, events $e_0, e_1, \dots, e_{n-1}$ have occured before, so the corresponding inequalities have been yet verified and it suffices to verify only those corresponding to $e_n$.

**Corollary 1: simplified condition at the nth occurrence**
An event $e_n$ may be the nth occurred event $\Leftrightarrow$ (i) $e_n$ is enabled and
(ii) $\forall i \in [0,n-1]\ maxa_{i,n}(e_n) \leq Smax(i,n,n)$.

**Proof:** since at the nth occurrence, FS(n-1) is not empty, the simplified occurrence condition at the nth occurrence is obtained from theorem 2 by eliminating relations: $\forall j \in [1,n-1], \forall i \in [0,j-1]\ maxa_{i,j}(e_j) \leq Smax(i,j,j)$.

QED

Intuitively this condition can be interpreted as "in order to make it possible the occurrence of an event, it must be possible to wait until all the tokens consumed by that event are available without being forced to fire an other enabled event before". Quite surprisingly, this is basicaly the same idea as for the first firing, the difference being only in the calculus of the maximum delay.

We will use this result to define state classes and built the occurrence graph.

### 3.4 Building the class occurrence graph

We shall now introduce our notion of state class in ITCPN. A state could be seen as a pair (marking, time), defining the marking at a given time. Then a class could be the set of all states between two successive occurrences. Unfortunately, the bounds for occurrence depends strongly of times of preceding occurrences. Rather, we shall define a class as a marking and a period of time when this marking is possible. Informally, the class $CL_{n-1}$ obtained from $CL_0$ after successive occurrence of events $e_1$,

$e_2,...,e_{n-1}$ will be :

$CL_{n-1} = \{ (M_{n-1}, \tau_n) \mid (\tau_{n-1} \leq \tau_n \leq \tau_{n-1}+x_n$ and $x_n \in FS(n) \}$

The range of $\tau_n$ could be characterised using a set of inequations which must be solved for every occurrence as in [BER83], but instead we shall use inequalities of corollary 1 to achieve a faster method. From class $CL_{n-1}$, an event $e_n$ may occur iff event $e_n$ is enabled by the marking $M_n$ and it exists at least a state $(M_{n-1},\tau_n)$ such that at time $\tau_n$, $e_n$ reaches its minimal occurrence time and no other enabled event has exceeded its maximal occurrence time. In other words, iff the occurrence condition given in corollary 1 is verified.

We have shown that the occurrence condition for an event $e_n$ depends upon the marking, upon times at which tokens have been produced and upon minimal and maximal delays between previous occurrences. This information is sufficient to compute next classes and this leads us to the following definitions.

**Definition 8** :
**A state class** $CL_{n-1}$ is a pair $(M_{n-1}, TE_{n-1})$ where
- $M_{n-1}$ is a marking function,
- $TE_{n-1}$ is a function: $[0,n-1] \times [0,n-1] \dashrightarrow \mathbb{R}$ such that

$TE_{n-1}(i,j) = Smax(i,j,n-1)$ if $i < j$
$TE_{n-1}(i,j) = - Smin(j,i,n-1)$ if $i > j$
$TE_{n-1}(i,j) = 0$ if $i=j$

The initial state class is $CL_0 = (M_0, NULL)$ where $M_0$ is the initial marking and $TE_0$ is NULL.

**Remark**
Since this definition contains a function TE which defines lower and upper bounds for delays between every couple of previous occurrences, it can be thought that we must keep for every class an ever growing record of these bounds. This is partially true since we need only to keep bounds for occurrences which have created still existing tokens. For the other occurrences, this information is useless and can be forgotten. So if the net is bounded, then records for TE is also bounded.

We adapt now occurrence rule given in corollary 1 to this definition.

**Definition 9 : occurrence rules from a class**
From a class $CL_{n-1}=(M_{n-1}, TE_{n-1})$, an event $e_n = (t_n, bin_n, bout_n)$ **may occur** iff
- $e_n$ is enabled by $M_{n-1}$ and
- for each token $((p,v), \tau_i, a_i, b_i)$ of $bin_n$ : $a_i \leq TE_n(i,n)$.)

The occurrence of $e_n$ from state class $CL_{n-1}$ gives a state class $CL_n$. This is denoted by **$CL_{n-1}[[e_n>>CL_n$ .**

Remark (2) is equivalent to : for each enabled event $e_m = (t_m, bin_m, bout_m)$

$a_i \leq \max \{ b_j + TE_{n-1}(i,j), TE_{n-1}(i,n-1) \mid ((p,v), \tau_j, a_j, b_j) \in bin_m \}$

The new class CLn = (Mn, TEn) can be computed using this equivalent condition.

**$M_n$** : it is obtained in the same way as in coloured Petri nets.

**$TE_n$**: $TE_n(i,j)$ are defined as follows (and according to notation 3):

for $i \in [0,n-1]$ :

$$TE_n(n,i) = \min_{((p,v)\ \tau_j,a_j,b_j) \in bin(e_n)} \{TE_{n-1}(j,i) - a_j, TE_{n-1}(n-1,i)\}$$

$$TE_n(i,n) = \min\{ \max_{e \in E_n \; ((p,v)\, \tau_j, a_j, b_j) \in bin(e)} \{b_j + TE_{n-1}(i,j), TE_{n-1}(i, n-1)\}\}$$

for $i \in [0,n-1]$ and $j \in [1,n-1]$

$$TE_n(i,j) = \min \{TE_{n-1}(i,j), TE_n(i,n)+TE_n(n,j)\}$$

**Definition 10 :**
The **class occurrence graph** of an ITCPN is a couple COG = (S,X) where S is the set of nodes and X is the set of arcs. There is exactly one node $s_{CL}$ associated with each reachable state class CL and an arc labelled e leads from $s_{CL}$ to $s_{CL'}$ iff e may occur in CL and leads to CL'.

**Proposition 5 : soundness and completeness of state class occurrence graph.**
For any occurrence sequence in the initial model we have the same occurrence sequence in the state class occurrence graph and conversely.

**Proof**
This proposition is obviously true for marking (i.e. if we neglect time conditions). Moreover, starting from definition 5, we have used only equivalent occurrence conditions to obtain occurrence condition used in definition 9 above.

More precisely, let $e_1, e_2, \dots, e_n$ be an occurrence sequence and let $x_1, x_2, \dots, x_n$ be the corresponding delays. From proposition 3, we know that $\{x_1, x_2, \dots, x_n\}$ is a solution of I(n). From corollary 1, we deduce that conditions for $CL_0[[e_1>>CL_1$, $CL_1[[e_2>>CL_2$, ..., $CL_{n-1}[[e_n>>CL_n$, are successively satisfied. Hence this path exists in the state class occurrence graph.
Conversely, let $CL_0[[e_1>>CL_1[[e_2>>CL_2 \dots CL_{n-1}[[e_n>>CL_n$ be a path in the class occurrence graph, and let $x_1, x_2, \dots, x_n$ be n values such that for every $e_j$ : $\forall\, i \in [0,j-1]$ $maxa_{i,j}(e_j) \le x_{i+1} + \dots + x_j \le Smax(i,j,j)$
The conjunction of all these conditions for every event in the sequence gives the right hand side condition of theorem 2 and it can be shown that $x_1, x_2, \dots, x_n$ is a solution of I(n). So it exists an occurrence sequence with the corresponding delays.

QED

The class occurrence graph and the graph obtained by MTSRT method cannot be compared since the latter may contain nodes representing unreachable states. Hence one can think that the former is smaller, but this is not true since two or more of our classes can be fused by MTSRT, due to the larger definition of its classes. However we conjecture that when, for a given ITCPN, MTSRT is complete (i.e. does not lead to unreachable but represented state classes), the two graphs are isomorphic.

## 4 - Equivalent classes

For ITCPN allowing only finite occurrence sequences, the class occurrence graph, as MTSR, allows to obtain a finite graph to represent infinite number of states. However numerous real systems have cyclic behaviours and, consequently, allow infinite occurrence sequences. We shall see now how to regroup classes in order to have finite class occurrence graph when number of reachable markings is finite. First of all, we must define a criterion to agglomerate two classes. Intuitively we would like to fuse two state classes when they are equivalent from the behaviour point of view or, in other words, when they allow the same finite or infinite occurrence sequences. Such equivalence relations are well known and can be obtained formally as in [MIL] for instance.

**Definition 11 : Bisimulation relations**
A relation $\mathcal{R} \subseteq \mathcal{C} \times \mathcal{C}$ where $\mathcal{C}$ is the set of all possible state classes is a bisimulation if $(CL_1,CL_2) \in \mathcal{R}$ implies, for all event e.
(i) Whenever $CL_1[[e{>>}CL_1'$, then for some $CL_2'$, $CL_2[[e{>>}CL_2'$ and $(CL_1',CL_2') \in \mathcal{R}$
(ii) Whenever $CL_2[[e{>>}CL_2'$, then for some $CL_1'$, $CL_1[[e{>>}CL_1'$ and $(CL_1',CL_2') \in \mathcal{R}$

**Definition 12 : Equivalence of classes**
$CL_1$ and $CL_2$ are equivalent, written $CL_1 \sim CL_2$, if $(CL_1,CL_2) \in \mathcal{R}$ for some bisimulation.

It can easily be shown that $\sim = \bigcup \{\mathcal{R} \mid \mathcal{R}$ is a bisimulation$\}$ is reflexive, symmetric and transitive and hence is an equivalence relation. The following theorem defines a sufficient condition for equivalence of classes.

**Theorem 3**
Two classes $CL_{n-1}= (M_{n-1},TE_{n-1})$ and $CL_{n'-1}= (M_{n'-1},TE_{n'-1})$ are equivalent if
(i) it exists a bijective mapping g from $M_{n-1}$ to $M_{n'-1}$: $g( ((p_i,v_i),\tau_i,a_i,b_i) ) = ((p_i,v_i),\tau_i',a_i',b_i')$
(ii) and we have for each pair of tokens $((p_i,v_i), \tau_i, a_i, b_i)$ and $((p_j,v_j),\tau_j,a_j,b_j))$ of $M_{n-1} \times M_{n-1}$
(let $((p_i,v_i),\tau_i',a_i',b_i')$ and $((p_j,v_j),\tau_j',a_j',b_j')$ be their images by g)
(1) $\max(0,b_j+\min(TE_{n-1}(n-1,j),TE_{n-1}(i,j)-a_i))=\max(0,b_j'+\min(TE_{n'-1}(n'-1,j'),TE_{n'-1}(i',j')-a_i'))$
(2) $\min(0,\max(b_j+TE_{n-1}(i,j),TE_{n-1}(i,n-1))-a_i)=\min(0,\max(b_j'+TE_{n'-1}(i',j'),TE_{n'-1}(i',n'-1))-a_i')$
(iii) and for each token $((p_i,v_i), \tau_i, a_i, b_i)$ of $M_{n-1}$ (lets $(p_i,v_i), \tau_i', a_i', b_i'$) be its image in $M_{n'-1}$
(3) $\min( 0, TE_{n-1}(i,n-1) - a_i) = \min( 0, TE_{n'-1}(i',n'-1) - a_i')$
(4) $\max( 0, b_i + TE_{n-1}(n-1,i)) = \max( 0, b_i' + TE_{n'-1}(n'-1,i'))$

Intuitively, relations (1) and (2) mean that the occurrence of an event e consuming token $((p_i, v_i), \tau_i, a_i, b_i)$ in class $CL_{n-1}$ or consuming token $((p_i, v_i), \tau_i', a_i', b_i')$ in class $CL_{n'-1}$, , gives two classes such that each token will have the same minimal residual delay and the same maximal residual delay in both classes.
Relations ((3) and (4)) mean that each token has the same minimal residual delay and the same maximal residual delay in both classes.

**Proof** see appendix

We will now show in theorem 4 that if delays associated with tokens are rational constants and the number of reachable markings is finite, this equivalence relation implies a finite occurrence graph. But firstly we need the following proposition used to prove theorem 4.

**Proposition 6**
Let Q be a finite linear combination i.e.: $Q = n_1 \times q_1 + n_2 \times q_2 + ... + n_n \times q_n$ where $n_1,n_2,...,n_n$ are integer numbers and $q_1,q_2,..., q_n$ are rational constants. If Q is bounded by rational constants (i.e.. $a \leq Q \leq b$) then the number of different linear combinations is finite. In other words, Q may have a finite number of values.

**Proof:** the proof is given in [BER91].

**Theorem 4:**
An interval timed coloured Petri net has a finite occurrence graph iff its class occurrence graph contains a finite number of different markings and delays associated with tokens are rational constants or $\infty$.

**Proof:**

Since the number of different markings is finite, it suffices to prove that for a given marking, we have a only finite number of classes not satisfying at least one of relations (1), (2), (3), (4) given in theorem 3. More precisely,we shall show that for a marking $M_n$ the number of possible values of $H_1$, $H_2$, $H_3$ and $H_4$ defined as follows, are finite:

$$H_1 = \max(0, b_j + \min(TE_n(n,j), TE_n(i,j) - a_i)), \qquad H_2 = \min(0, \max(b_j + TE_n(i,j), TE_n(i,n)) - a_i)$$
$$H_3 = \min(0, TE_n(i,n) - a_i) \qquad H_4 = \max(0, b_j + TE_n(n,j))$$

$H_1$, $H_2$, $H_3$ and $H_4$ are finite combinations of rational constants (delays) with integer coefficients (finite because the number of different delays in the model is finite).

$H_1$, $H_2$, $H_3$ and $H_4$ are bounded or equal to $\infty$ :

$$0 \leq H_1 \leq b_j, \qquad -a_i \leq H_2 \leq 0, \qquad -a_i \leq H_3 \leq 0, \qquad 0 \leq H_4 \leq b_j.$$

Using proposition 6, they can take a finite numbers of values.

The converse is obvious.

QED

This necessary and sufficient condition to have a finite graph may be difficult to use since we have not a general procedure to decide whether or not any ITCPN has a finite number of different markings. However, we have a straightforward sufficient condition using the underlying CPN, and we know several methods to decide this property on CPN, namely the invariant method.

**Corollary 2:**

An interval timed coloured Petri net has a finite occurrence graph if the underlying coloured Petri net is bounded and delays associated with tokens are rational constants or $\infty$.

This last result is quite valuable since the hypothesis of the preceding corollary ( rationals bound for intervals and bounded underlying coloured net) are not real limitations, due to the following considerations. Firstly, only bounds of intervals must be rationals and events still may occur continuously, for any rational or real value. Secondly one can always find a rational number as close as necessary of a rational real number. Last but not least, computers deal only with rational numbers. About the requirement of a bounded underlying net, practical experiences of modelisation show that it is a general requirement for any practical systems of interest. From a programming point of view, similar programs for a slightly different model ([BOU]) have shown that an occurrence graph of several hundreds of thousands of classes may be constructed in half an hour delay.

## 5 - Computation of the minimal and maximal times along paths and cycles

In numerous systems, oftenly rewarded time informations are minimal and maximal bounds for executing a sequence of actions. Other important time informations are minimal and maximal cycle times. Using relation of equivalence given above, several classes which allow same occurrence sequences are represented by one node in the occurrence graph. Before computing minimal and maximal times, we must show that for two equivalent classes, the minimal and maximal delays for executing an occurrence sequence does not depend on the starting class.

**Theorem 5:**

Let $CL_{n-1}$ and $CL_{n'-1}$ be two equivalent classes, and let $CL_{n+m}$ and $CL_{n'+m}$ be the two classes obtained by occurrence sequence $e_n,\ldots,e_{n+m}$ starting respectively from $CL_{n-1}$ and from $CL_{n'-1}$. Then we have:

(i) $TE_{n+m}(n-1,n+m) = TE_{n'+m}(n'-1,n'+m)$ and

(ii) $TE_{n+m}(n+m,n-1) = TE_{n'+m}(n'+m,n'-1)$

**Proof:** see appendix.

**Corollary 3**
Let $CL_{n-1}=(M_{n-1},TE_{n-1})$, $CL_n=(M_n,TE_n)$,...,$CL_{n+m}=(M_{n+m},TE_{n+m})$ be a path and let $e_n$, ...., $e_{n+m}$ be the corresponding occurrence sequence. Then the minimal and maximal delays for this path are respectively:
- $TE_{n+m}(n+m,n-1)$ and $TE_{n+m}(n-1,n+m)$

**Proof:** From theorem 5, the minimal and maximal time for a path are not dependent of the class representing the first node of that path. Consequently, they can be computed for any class of that node by using definition 9.

**Remarks**
With $CL_{n-1}$ equivalent to $CL_{n+m}$, this result can be used to compute the minimal and maximal cycle times.

## 6 - Example

The example used to apply our approach is given in figure 6. This approach gives the occurrence graph shown in figure 7 where the set of reachable classes are computed as follows:

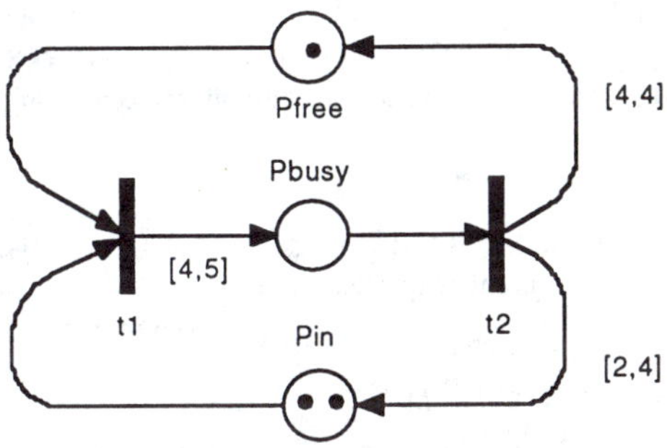


Figure 6

The initial marking is $M_0 = \{ ((p_{in}, j_1), \tau_0,1,3), ((p_{in}, j_2), \tau_0,4,4), ((p_{free}, M_1), \tau_0,4,4)\}$.
For the sake of simplicity $\tau_i$ are replaced by i in the sequel.

**Remark:** $F(t_1)$ and $F(t_2)$ are those given for ITCPN of figure 1.

**First occurrences:** (see figure 7)
- From $CL_0 = (M_0, NULL)$
we have two enabled events:

$e_{11}=(t_1; \{((p_{in}, j_1), 0,1,3), ((p_{free}, M_1), 0,4,4)\}; \{((p_{busy}, (j_1,M_1)), 1,4,5)\} )$

$e_{12}=(t_1; \{((p_{in}, j_2), 0,4,4)\}, ((p_{free}, M_1), 0,4,4)\}; \{((p_{busy}, (j_2,M_1)), 1,4,5)\} )$

The minimal and maximal occurrence times of events $e_{11}$ and $e_{12}$:

$FT_{min}(e_{11}) = max(1,4) = 4$; $FT_{max}(e_{11}) = max (3,4) = 4$

$FT_{min}(e_{12}) = max(4,4) = 4$; $FT_{max}(e_{12}) = max (4,4) = 4$

Events which can occur from $CL_0$ are:

$e_{11}$ may occur because $FT_{min}(e_{11}) \leq \min(FT_{max}(e_{11}), FT_{max}(e_{12})) = 4$

$e_{12}$ may occur because $FT_{min}(e_{12}) \leq \min(FT_{max}(e_{11}), FT_{max}(e_{12})) = 4$

Occurrence of $e_{11}$ gives class $CL_1=(M_1, TE_1)$ where

$M_1 = \{ ((p_{in}, j_2), 0,4,4), ((p_{busy}, (j_1, M_1)), 1,4,5)\}$ and

$TE_1(0,1) = \min(FT_{max}(e_{11}), FT_{max}(e_{12})) = 4$ and $TE_1(1,0) = - FT_{min}(e_{11}) = -4$

Occurrence of $e_{12}$ gives class $CL_1' = (M_1', TE_1')$ where

$M_1' = \{ ((p_{in}, j_1), 0,1,3), ((p_{busy}, (j_2, M_1)), 1,4,5)\}$

$TE_1'(0,1) = \min(FT_{max}(e_{11}), FT_{max}(e_{21})) = 4$ and $TE_1'(1,0) = - FT_{min}(e_{12}) = -4$

**Second occurrences:**

- From $CL_1 = (M_1, TE_1)$

we have one enabled event:

$e_{21}=(t_2; \{((p_{busy}, (j_1, M_1)), 1,4,5)\}; \{((p_{free}, M_1), 2,4,4), ((p_{in}, j_1), 2,2,4)\})$

minimal and maximal occurrence time of event $e_{21}$:

$FT_{min}(e_{21}) = \max(4 - TE_1(1,0)) = 8;$ $FT_{max}(e_{21}) = \max(5+TE_1(0,1)) = 9$

Occurrence of $e_{21}$ gives class $CL_2 =(M_2, TE_2)$ where

$M_2 = \{ ((p_{in}, j_2), 0,4,4), ((p_{in}, j_1), 2,4,4), ((p_{free}, M_1), 2,4,4)\}$ and

$TE_2(0,2) = FT_{max}(e_{21}) = 9$ and $TE_2(2,0) = - FT_{min}(e_{21}) = -8$

(remark : $TE_1(0,1)$ and $TE_1(1,0)$ are no longer usefull and can be dropped)

- From $CL_1' = (M_1', TE_1')$

we have one enabled event:

$e_{22}=(t_2; \{((p_{busy}, (j_2, M_1)), 1,4,5)\}; \{((p_{free}, M_1), 2,4,4), ((p_{in}, j_2), 2,2,4)\})$

minimal and maximal occurrence time of event $e_{22}$:

$FT_{min}(e_{22}) = \max(4 - TE_1'(1,0)) = 8;$ $FT_{max}(e_{22}) = \max(5 + TE_1'(0,1)) = 9$

Occurrence of $e_{22}$ gives class $CL_2' =(M_2', TE_2')$ where

$M_2' = \{ (p_{in}, j_1), 0,1,3), (p_{in}, j_2), 2,2,4), (p_{free}, M_1), 2,4,4)\}$ and

$TE_2'(0,2) = FT_{max}(e_{22}) = 9$ and $TE_2'(2,0) = - FT_{mim}(e_{22}) = -8$

**Third occurrences:**

- From class $CL_2 = (M_2, TE_2)$

we have two enabled events:

$e_{11}=(t_1; \{((p_{in}, j_1), 2,4,4), ((p_{free}, M_1), 2,4,4)\}; \{((p_{busy}, (j_1,M_1)), 3,4,5)\})$

$e_{12}=(t_1; \{((p_{in}, j_2), 0,4,4), ((p_{free}, M_1), 2,4,4)\}; \{((p_{busy}, (j_2,M_1)), 3,4,5)\})$

minimal and maximal occurrence times of events $e_{11}$ and $e_{12}$:

$FT_{min}(e_{11}) = \max(4,4) - TE_2(2,0) = 12;$ $FT_{max}(e_{11}) = \max(4,4) + TE_2(0,2) = 13$

$FT_{min}(e_{12}) = \max(4, 4-TE_2(2,0)) = 12;$ $FT_{max}(e_{12}) = \max(4, 4+TE_2(0,2)) = 13$

Events which can be fired from $CL_2$ are:

$e_{11}$ may occur because $FT_{min}(e_{11}) \leq \min(FT_{max}(e_{11}), FT_{max}(e_{12})) = 13$

$e_{12}$ may occur because $FT_{min}(e_{12}) \leq \min(FT_{max}(e_{11}), FT_{max}(e_{12})) = 13$

Occurrence of $e_{11}$ gives class $CL_3 = (M_3, TE_3)$

$M_3 = \{ ((p_{in}, j_2), 0,4,4), ((p_{busy}, (j_1, M_1)), 3,4,5)\}$ and

$TE_3\,(0,3) = \min(FT_{max}(e_{11}), FT_{max}(e_{12})) = 13$ and $TE_3(3,0) = -FT_{min}(e_{11}) = -12$

Occurrence of $e_{12}$ gives class $CL_3' = (M_3', TE_3')$ where

$M_3' = \{ ((p_{in}, j_1), 2,4,4), ((p_{busy}, (j_2, M_1)), 3,4,5)\}$ and

$TE_3'(2,3) = \min(\max(4,4), \max(4+TE_2(2,0), 4)) = 4$ and $TE_3'(3,2) = -\max(2- TE_2(0,2), 4) = -4$

- From class $CL_2' = (M_2', TE_2')$

we have two enabled events:

$e_{12}=(t_1;\ \{((p_{in}, j_2), 2,4,4), ((p_{free}, M_1), 2,4,4)\};\ \{((p_{busy}, (j_2,M_1)), 3,4,5)\}\ )$

$e_{11}=(t_1;\ \{((p_{in}, j_1), 0,1,3), ((p_{free}, M_1), 2,4,4)\};\ \{((p_{busy}, (j_1,M_1)), 3,4,5)\}\ )$

minimal and maximal occurrence times of events $e_{11}$ and $e_{12}$:

$FT_{min}(e_{12}) = \max(4, 4) - TE_2(2,0) = 12;$ $FT_{max}(e_{12}) = \max(4,4) + TE_2(0,2) = 13$

$FT_{min}(e_{11}) = \max(1,4-TE_2(2,0)) = 12;$ $FT_{max}(e_{11}) = \max(3, 4+TE_2(0,2)) = 13$

Events which can occur from $CL_2'$ are:

$e_{12}$ may occur because $FT_{min}(e_{12}) \le \min(FT_{max}(e_{11}), FT_{max}(e_{12})) = 13$

$e_{11}$ may occur because $FT_{min}(e_{11}) \le \min(FT_{max}(e_{11}), FT_{max}(e_{12})) = 13$

Occurrence of $e_{12}$ gives class $CL_3'' = (M_3'', TE_3'')$ where

$M_3'' = \{ ((p_{in}, j_1), 0,1,3), ((p_{busy}, (j_2, M_1)), 3,4,5)\}$ and

$TE_3''(0,3) = \min(FT_{max}(e_{11}), FT_{max}(e_{12})) = 13$ and $TE_3''(3,0) = -FT_{min}(e_{12})) = -12$

Occurrence of $e_{11}$ gives class $CL_3''' = (M_3''', TE_3''')$ where

$M_3''' = \{ ((p_{in}, j_2), 2,4,4), ((p_{busy}, (j_1, M_1)), 3,4,5)\}$ and

$TE_3'''(2,3) = \min(\max(4,4), \max(3+TE_2'(2,0), 4)) = 4$ and

$TE_3'''(3,2) = -\max(1 - TE_2'(0,2), 4) = -4$

**Equivalent classes:**

Using theorem 3, we can show that classes $CL_3$, $CL_3''$, $CL_1$ are equivalent and also that classes $CL_3'$, $CL_3'''$, $CL_1'$ are equivalent. For instance, class $CL_1'$ is equivalent to class $CL_3'$ because it exists mapping g from $M_1'$ to $M_3'$ such that:

$g(((p_{in}, j_1), 0,1,3)) = ((p_{in}, j_1), 2,4,4)$ and

$g(((p_{busy}, (j_2, M_1)), 1,4,5)) = ((p_{busy}, (j_2, M_1)), 3,4,5)$

for each pair of tokens we have relations (1) and (2) and (3) and (4) of theorem 3 which are satisfied:

- $((p_{in}, j_1), 0,1,3)$ and $((p_{busy}, (j_2, M_1)), 1,4,5)$ --> $((p_{in}, j_1), 2,4,4)$ and $((p_{busy}, (j_2, M_1)), 3,4,5)$

  $\max(0, 5 + \min(TE_1'(1,1), TE_1'(0,1)- 1)) = \max(0, 5 + \min(TE_3'(3,3), TE_3'(0,3)-4)) = 5$

  $\min(0, \max(5 + TE_1'(0,1), TE_1'(0,1)) - 1) = \min(0, \max(5 + TE_3'(0,3), TE_3'(0,3)) - 4) = 0$

- $((p_{busy}, (j_2, M_1)), 1,4,5)$ and $((p_{in}, j_1), 0,1,3)$ --> $((p_{busy}, (j_2, M_1)), 3,4,5)$ and $((p_{in}, j_1), 2,4,4)$

  $\max(0, 3 + \min(TE_1'(1,0), TE_1'(1,0)- 4)) = \max(0, 4 + \min(TE_3'(3,2), TE_3'(3,2)-4)) = 0$

  $\min(0, \max(3 + TE_1'(1,0), TE_1'(1,1)) - 4) = \min(0, \max(4 + TE_3'(3,2), TE_3'(3,3)) - 4) = -4$

- $((p_{in}, j_1), 0,1,3)$ --> $((p_{in}, j_1), 2,4,4)$

  $\max(0, 3 + TE_1'(1,0)) = \max(0, 4 + TE_3'(3,2)) = 0$

  $\min(0, TE_1'(0,1) - 1) = \min(0, TE_3'(2,3) - 2) = 0$

- $((p_{busy}, (j_2, M_1)), 1,4,5)$ --> $((p_{busy}, (j_2, M_1)), 3,4,5)$

  $\max(0, 5 + TE_1'(1,1)) = \max(0, 5 + TE_3'(3,3)) = 5$

  $\min(0, TE_1'(1,1) - 4) = \min(0, TE_3'(3,3) - 4) = -4$

Then, we have the following occurrence graph

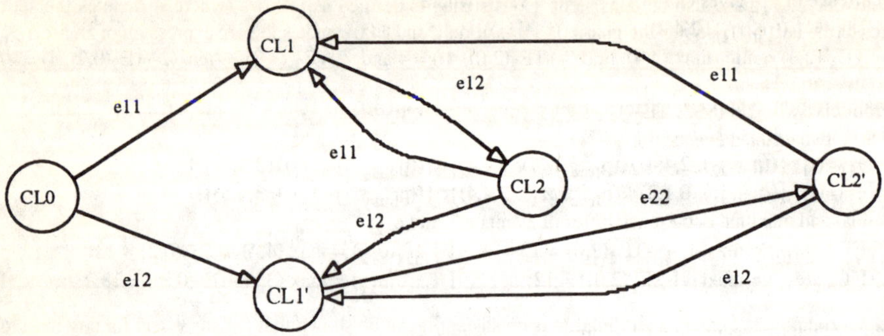


Figure 7

## 7 - Equivalencies and symmetries on colours.

Until now we have not taken advantage of equivalencies and symmetries on colour, but it can be done easily. The intuitive idea is to use the preceding delay based equivalence plus a colour based equivalence to fuse classes in larger classes. Colour based equivalence are investigated in details in [JENb] from which we take the three following definitions. Due to the fact that we need a one-to-one correspondence between tokens, we are restricted to permutation symmetries on colour sets $\Sigma$ (i.e. : renaming of colours). Permutation symmetries are defined only for atomic colour sets i.e. colour sets which are defined without reference to other existing colour sets. Permutation symmetries for structured colour sets (constructed from atomic colour sets ) are derived from permutation on atomic colour sets.
In the remainder of this section $\mathbb{M}$ denotes the set of all markings of a CP-net, and BE denotes the set of all binding elements of a CP-net , i.e. the set of all pairs (t,b) where t is a transition and b a binding (an association of a value to each variable) of t.

**Definition 13**
A **permutation symmetry specification** is a function SG which maps each atomic colour set S into a subgroup SG(S) of the set of all permutations of S.

A **permutation symmetry** for SG is a function $\phi$ which maps each atomic colour set S into a permutation of $\phi_S \in$ SG(S).

SG(S) is called the **symmetry group** of S while $\phi_S$ is called a colour symmetry. The set of all permutation symmetries for SG is denoted by $\Phi_{SG}$.

Permutation symmetries can be used to derive functions for markings, binding, and binding elements. $\phi_{C(p)}$ is the linear extension of the colour symmetry of C(p), while Type(v) is the type of the variable v.

**Definition 14**
Let a permutation symmetry $\phi \in \Phi_{SG}$ , a marking $M \in \mathbb{M}$ , a binding b, and a binding element (t,b) $\in$ BE be given. Then $\phi$(M), $\phi$(b) and $\phi$(t,b) are defined as follows :

(i) $\forall p \in P : \phi(M)(p) = \phi_{C(p)}\,(M(p))$.

(ii) $\forall v \in Var(t) : \phi(b)(v) = \phi_{Type(v)}\,(b(v))$.

(iii) $\phi(t,b) = (t,\phi(b))$.

A permutation symmetry $\Phi_{SG}$ defines equivalence relations on markings but to guarantee equivalent behaviour, $\Phi_{SG}$ must also be consistent. Consistency is defined as follows (where A denotes the set of Arcs of the net, G(t) denotes the guard of a transition t, and E(a) denotes the arc expression of an arc a):

**Definition 15**
A permutation symmetry specification $\Phi_{SG}$ is **consistent** iff the following properties are satisfied for all $\phi \subset \Phi_{SG}$, all $t \in T$ and all $a \in A$:

(i) $\phi(M_0) = M_0$.

(ii) $\forall b \in B(t) : G(t)\langle\phi(b)\rangle = G(t)\langle b\rangle$.

(iii) $\forall b \in B(t(a)) : E(a)\langle\phi(b)\rangle = \phi(E(a)\langle b\rangle)$.

If a permutation symmetry specification is consistent it can be shown that for any $\phi$ and for any marking $M_1$, for each step $M_1[b> M_1'$, we have also $\phi(M_1)[\phi(b)>\phi(M_1')$. We can now complete permutation symmetry specifications with the time equivalence relation defined in section 4.

**Definition 16**
Let an ITCPN and consistent permutation symmetry specification of permutation $\Phi_{SG}$ be given. Two classes $CL_{n-1}= (M_{n-1},TE_{n-1})$ and $CL_{n'-1}= (M_{n'-1},TE_{n'-1})$ of the occurrence graph are equivalent iff for all $\phi \in \Phi_{SG}$

(i) $M_{n'-1} = \phi(M_{n-1})$ and

(ii) for each pair of tokens $((p_i,v_i), \tau_i, a_i, b_i)$ and $((p_j,v_j), \tau_j, a_j, b_j))$ of $M_{n-1} \times M_{n-1}$
( let $(p_i,\phi(v_i)), \tau_{i'}, a_{i'}, b_{i'})$ and $((p_j,\phi(v_j)), \tau_{j'}, a_{j'}, b_{j'}))$ be their images by $\phi$) we have

(1) $\max(0, b_j + \min(TE_{n-1}(n-1,j), TE_{n-1}(i,j) - a_i)) = \max(0,b_{j'}+ \min(TE_{n'-1}(n'-1,j'), TE_{n'-1}(i',j') - a_{i'}))$

(2) $\min(0,\max(b_j + TE_{n-1}(i,j), TE_{n-1}(i,n-1)) - a_i) = \min(0,\max(b_{j'}+ TE_{n'-1}(i',j'), TE_{n'-1}(i',n'-1)) - a_{i'})$
and for each token $((p_i,v_i), \tau_i, a_i, b_i)$ of $M_{n-1}$ (let $((p_i,\phi(v_i)), \tau_{i'}, a_{i'}, b_{i'})$ its image by $\phi$)

(3) $\min( 0, TE_{n-1}(i,n-1) - a_i) = \min( 0, TE_{n'-1}(i',n'-1) - a_{i'})$.

(4) $\max( 0, b_i + TE_{n-1}(n-1,i)) = \max( 0, b_{i'} + TE_{n'-1}(n'-1,i'))$.

This equivalence relation may be used to reduce the state class occurrence graph. Since two equivalent classes have the same markings and allows the same occurrence sequences exactly as two untimed marking do, results proved in [JEN] remain true and the reduced graph may be used to investigate dynamic properties about boundedness, liveness, home and fairness. It can also be used for investigating minimal and maximal time along a path but not for a cycle since the initial and the final markings of the cycle may be different, even if they are equivalent.

For the net of figure 6 we can define a consistent permutation symmetry specification as $\Phi_{SG} = \{\phi_1, \phi_2\}$, with $\phi_1$: $J_1 \rightarrow J_2$ and $\phi_2$: $J_2 \rightarrow J_1)\}$. This symmetry allows to fuse first $CL_1$ and $CL_{1'}$, and then, $CL_2$ and $CL_2'$.

## 8 - Conclusion

We have presented here an approach to build the occurrence graph for interval timed coloured Petri nets (ITCPN). After studying the condition for an event to be the nth occurred event in the model, we have shown that this condition depends upon the marking, upon the number of occurrences which have created tokens, and upon delays between previous occurrences. This result leads both to an attractive definition of classes and to simple computations for reachable classes.

However, for models having infinite occurrence sequences, this definition implies infinite reachable classes. So, we have defined a delay based equivalence relation for classes. This equivalence leads to finite set of reachable classes iff the interval timed coloured Petri net is bounded. These occurrence graphs can be used to verify all the dynamic properties such as reachability, boundedness, home, liveness and fairness properties but also performance properties : minimal and maximal bounds along an occurrence sequence or a cycle. From a practical point of view, similar programs for a slightly different model ([BOU]) have shown that an occurrence graph of several hundreds of thousands of classes may be constructed in half an hour delay. Finally we have defined how to complete delay based equivalence with a colour based equivalence in order to further reduce occurrences graphs.

## REFERENCES


[AND,89] C.André "Synchronized Elmentary Net Systems". Advances in Petri Nets 1989, Grzegorz Rozenberg editor, LNCS 424, Springer -Verlag.

[BERR,88] G.Berry , G.Gonthier "The synchroneous Programming Language ESTEREL : Design, Semantic, Implementation". INRIA report 842 ,1988.

[BERT,83] B.Berthomieu, M.Menasche "An enumerative approach for analysing time Petri nets". IFIP Congress 1983, Paris, North-Holland.

[BERT,91] B.Berthomieu, M. Diaz "Modeling and verification of time dependent systems using time Petri nets". IEEE Transactions on Software Engineering vol 17, N°3, March 91.

[BOU,93] H.Boucheneb, G.Berthelot " Towards a Simplified building of time Petri Net Reachability graphs". Proc.of Petri Nets and Performance Models PNPM93, Toulouse France, October 1993, IEEE Computer Society Press.

[FLO,82] G.Florin, S.Natkin "Evaluation based upon Stochastic Petri nets of the maximum troughput of a full duplex protocol". Second European Workshop on Application and Theory of Petri Net, Springer-Verlag 1982.

[JENa] K.Jensen "Coloured Petri Nets : Basic concepts, Analysis Methods and Practical use. volume 1: Basic Concepts". To appear in EATCS Monographs on Theoritical Computer Science, Springer-Verlag.

[JENb] K.Jensen "Coloured Petri Nets : Basic concepts, Analysis Methods and Practical use. volume 2 : Analysis Methods"? To appear in EATCS Monographs on Theoritical Computer Science, Springer-Verlag.

[MEN,82] M.Menasche, "Analyse des réseaux de Petri temporisés et application aux systèmes distribués". Thèse de docteur ingénieur, université de Paul Sabatier, Toulouse, Novembre 82.

[MER,76] P.Merlin, D.J.Farber, "Recovability of communication protocols". IEEE Trans. on Communications , 24 (1976).

[MIL] R.Milner "Communication and Concurrency". Prentice Hall international series in computer science.

[QUE,89] J.Quemada, A.Azcorra, D.Frutos, "A timed Calculus for LOTOS". Proc. of Formal Description Techniques FORTE 89, S.Vuong editor, Vancouver Canada December 89.

[RAM,74] C.Ramchandani "Analysis of Asynchroneous Concurrent Systems by Timed Petri Nets"; Project MAC, TR 120, MIT, 1974.

[RAZ,84] R.R.Razouk, "The derivation of performance expressions for communication protocols for timed Petri nets". Computer Communication review, vol. 14, n° 12, pp 210- 217, 1984.

[SIF,77] J.Sifakis "Use of Petri Nets for Performance Evaluation". Measuring, Modeling and Evaluating Computer Systems, H.Beilner and E.Gelenbe editors, North-Holland, 1977.

[SIF,91] J.Sifakis, "An Overview and Synthesis on Timed Process Algebras". Proc of the 3rd International Workshop CAV'91, Alborg, Denmark, july 91, LNCS 575, Springer-Verlag.

[VAN,93] W.M.P. Van Der Aalst "Interval Timed Coloured Petri Nets and their Analysis", Application and Theory of Petri Nets 1993, 14th International Conference, Chicago, Illinois, USA, LNCS 691, Springer -Verlag.

# Implementation of Weighted Place / Transition Nets based on Linear Enabling Functions *

J.L. Briz and J.M. Colom

Depto. Ingeniería Eléctrica e Informática
Universidad de Zaragoza, c/ María de Luna, 3 (Pol. Actur)
50.015 Zaragoza, Spain
e-mail: briz@etsii.unizar.es

**Abstract.** Petri Nets should be implemented in an efficient and reliable way, specially when they are going to be used for critical problems, like that of giving support to Discrete Event Systems Simulation, whichever sequential or parallel strategies are adopted. One of the critical points while implementing a Petri Net, is that of determining whether a transition is enabled. In this contribution we classify transitions in several classes. The enabling of a transition is characterized by means of a Linear Enabling Function (LEF), that depends on the class. For some classes a transformation must be applied, preserving the behavior of the net. We show how LEFs can be applied to build a Simulation Engine that uses as data structure a DES described in terms of a Timed Petri Net, taking benefit of the properties of LEFs.
**Keywords.** Simulation of Weighted Place-Transition Systems, Timed Petri Nets, Linear Enabling Functions, Structure and Behavior of nets.

## 1 Introduction

The interest in parallel and distributed systems grows constantly as they are introduced in new domains. Their complex nature makes essential simulation and verification techniques during the design process, to prevent the occurrence of bad behaviors, to ensure that certain good properties hold, and to evaluate the performance. Petri Nets (PNs) have been pointed out as a good modelling tool, since many properties may be easily analyzed in a great number of cases. Moreover, when formal analysis becomes impracticable, the model may be simulated. As simulation tool, PNs allow the formulation of models with realistic features (as the competition for passive resources) absent in other paradigms (as nude queueing networks).

Given a PN model of a discrete physical system, we *simulate* the system by *playing the token game* on that PN, i.e. by firing transitions as a result of the available tokens. This is also referred as *implementing* the PN. If a deterministic or stochastic time interpretation is associated to transitions – Timed PNs (TPNs)

* Partially supported by grants CICYT TIC-91-0354 of the Plan Nacional de Investigación and P IT-6/91 of the CONAI (Diputación General de Aragón)

or Stochastic PNs (SPNs) –, the implementation of the TPN or SPN yields, actually, a Discrete Event Simulation system. In fact, SPNs have been proposed as the minimal discrete event notation [13], and both TPNs and SPNs are in the scope of recent works on Discrete Event Systems (DES) and Parallel DES (PDES) simulation [1, 11, 5, 27, 12].

PDES has arisen as a promising way of reducing simulation costs. However, sequential DES is still more efficient in a great number of cases, because of the overhead produced by conservative and optimistic strategies, which avoid causality errors or detect and recover from them. Whichever approach is used – sequential DES or PDES –, the final goals are always *efficiency* and *reliability*. Sequential event-driven simulation methods try to improve the list processing capabilities, and both conservative and optimistic approaches to PDES try to reduce communication costs [2, 8]. When PNs are used as the underlying frame, some of the heavy simulation components may be lightweighted, but the problem of how the token game can be efficiently carried out in a reliable way becomes critical. Particularly, the enabling test of the transitions plays a paramount role, since it is a rather time-consuming operation. This PN implementation problem, that has been present in the PN community since more than ten years ago (see [20, 16, 7, 24, 22, 25, 10] as some significant different approaches), recovers now a new meaning.

According to different criteria, the implementation of a PN may be *compiled* or *interpreted*, *sequential* or *concurrent*, *centralized* or *distributed*, *synchronous* or *asynchronous* [25]. For sequential, centralized methods, the solutions addressed to reduce the costs of the enabling test broadly fall into one of the following classes:

1. **Place-driven approaches.** Only the output transitions of some representative marked places are tested for firability. This gives a characterization of the partial enabling of transitions.
2. **Transition-driven approaches.** A characterization of the enabling of transitions is supplied, and only enabled transitions are considered. The firing of a given transition modifies the enabling conditions of the transitions connected to its input and output places. In general, it is not necessary an explicit representation of the marking.

Place-driven approaches have been extensively considered under different names [6, 21, 18, 24, 25]. Its main problem is the selection of a *good* set of representative places. A representative place is *good* when, if it is marked, its output transition is enabled. Today there is no practical way to compute this set nor to measure its quality. Nevertheless, very few works concern the transition-driven approach. These works either deal only with binary PNs [21, 18], or make a very inefficient characterization of the enabling, e.g. based on counters, which represents a clear disadvantage regarding to the place-driven approach [6].

For distributed implementations of PNs, the enabling test highly depends on the way in which places and transitions are represented. Thus, it may be carried out by means of complex reservation protocols, as in [22], or by following a client-server approach [3, 10].

We present here a new method that characterizes the enabling of each transition in a structurally bounded Weighted P/T-$\Sigma$ by means of a *Linear Enabling Function* (LEF). This greatly simplifies the enabling test, allowing a transition-driven approach for sequential simulations. The application of LEFs requires to classify transitions into five classes. We derive LEFs for classes 1, 2 and 3; for Class 1, the structural bound restriction is not strictly required. Classes 4 and 5 require previous transformations, based on the structure, to obtain transitions of the first three classes. This new characterization of the enabling of transitions, generalizes the previous works based on this approach, given that it can be applied to any net and it is simpler than [6], who uses numerous counters. As an example, we introduce the use of LEFs in a sequential, discrete event-driven simulation engine using TPNs.

The approach considered here to linearly characterize the enabling of a transition, has been also used in [23] in a different context to that of simulation: the structural analysis of the deadlock-freeness property. There, a linear characterization for a disabled transition at a marking is supplied in order to reduce the number of linear systems of inequalities needed to verify the property. The technical basis of these linear characterizations are very close to those used here in the definition of LEFs for transitions of Class 1, 2 and 3 and in the transformation of transitions of Class 4. Moreover, the characterization of the enabling of transitions might be used for the computation of the reachability graph, in order to speed up the process of finding enabled transitions.

This paper is organized as follows. Section 2 introduces the LEFs for Classes 1, 2 and 3. Section 3 and 4 present the transformations for Classes 4 and 5 with their resulting LEFs. Section 5 shows the algorithm of a Simulation Engine for TPNs based on LEFs. Finally, some conclusions are given, and an Appendix introduces some basics on Petri Nets and notation.

## 2 Linear Enabling Functions

In this section we define the so called *Linear Enabling Functions (LEF)* of a transition $t$, $f_t : R(\mathcal{N}, M_0) \to \mathbb{Z}$ with the following properties:

1. For a given marking, $M \in R(\mathcal{N}, M_0)$, $f_t(M) \leq 0$ characterizes the firability of the transition $t$ and $f_t(M) > 0$ characterizes the non-firability of $t$.
2. If $M[t'\rangle M'$, $f_t(M')$ can be computed from the value of $f_t(M)$ and some static parameters known before the execution.

The use of LEFs in the simulation of a PN, allows to simplify the part devoted to the enabling test, because we reduce it to a scanning for transitions with LEFs less than or equal to zero. Moreover, the LEF of a transition $t$ only must be updated if $t$ is fired or some transition belonging to ${}^{\bullet\bullet}t$ or $({}^{\bullet}t)^{\bullet}$ has been fired. The updating only requires the addition of a predefined constant to the previous value of the function. In some cases, a special technique is needed to preempt some enabled transitions.

Let $(\mathcal{N}, M_0)$ be a P/T-$\Sigma$ such that $\mathcal{N} = (P, T, Pre, Post)$. Then we define the following sets: $\Theta = \{p \in {}^\bullet t \mid SB(p) = Pre(p,t)\}$; $\Psi = \{p \in {}^\bullet t \mid SB(p) > Pre(p,t)\}$, i.e., ${}^\bullet t = \Theta \cup \Psi$ and $\Theta \cap \Psi = \emptyset$; $\Pi = \{p \in \Psi \mid p^\bullet = \{t\}\}$; $\Phi = \{p \in \Psi \mid p^\bullet \supset \{t\}\}$, i.e., $\Psi = \Phi \cup \Pi$. Accordingly, and in order to define LEFs, the set of transitions is partitioned into five Classes:

1. **Class 1.** Transitions with a single input place: $\mid \Theta \cup \Psi \mid = \mid {}^\bullet t \mid = 1$.
2. **Class 2.** Transitions with several input places, all of them with structural marking bound (SB) equal to the weight joining them to the transition: $\mid \Theta \mid > 1$, and $\mid \Psi \mid = 0$.
3. **Class 3.** Transitions with several input places, but only one with SB greater than the weight of its respective arc: $\mid \Theta \mid \geq 1$, and $\mid \Psi \mid = 1$.
4. **Class 4.** Transitions with at least two input places with only one output transition, and a SB greater than the weight of their arcs; the rest of the input places have a SB equal to the weight of their arcs: $\mid \Psi \mid = \mid \Pi \mid > 1$, hence $\Psi^\bullet = \{t\}$.
5. **Class 5.** Transitions that do not belong to precedent classes: $\mid \Psi \mid > 1$, and $\mid \Phi \mid \geq 1$.

The following algorithm carry out the partition of the set of transitions T:

**Algorithm 1.** Partition of the transitions of a weighted P/T-$\Sigma$.
**Input** Let$(\mathcal{N}, M_0)$ be a structurally bounded weighted P/T-$\Sigma$ where
$\mathcal{N} = (P, T, Pre, Post)$.
**Output** A partition of T into the classes defined above.

```
begin{
  for all ( p ∈ P) Compute SB(p)¹ at M0;
  Class1 := ∅; Class2 := ∅; Class3 := ∅; Class4 := ∅; Class5 := ∅;
  for all ( t ∈ T) {
     Θ = ∅; Ψ := ∅;
     for all( p ∈ •t)
        if (SB(p) = Pre(p,t)) Θ := Θ ∪ {p}; else Ψ := Ψ ∪ {p};
     case
        | Θ | + | Ψ | = 1: Class1 := Class1 ∪ {t};
        | Θ | > 1 and | Ψ | = 0: Class2 := Class2 ∪ {t};
        | Θ | ≥ 1 and | Ψ | = 1: Class3 := Class3 ∪ {t};
        | Ψ | > 1 and Ψ• = {t}: Class4 := Class4 ∪ {t};
        otherwise: Class5 := Class5 ∪ {t};
  }
}end
```

In the Classes defined above, SB$(p) < Pre(p,t)$ is not considered. This is because in such case $t$ is dead for any reachable marking, and we do not consider this kind of nets for simulation purposes.

[1] SB(p) is computed by solving a Linear Programming Problem[19]: $SB(p) = \max\{M(p) \mid M = M_0 + C \cdot \bar{\sigma}, M \geq 0, \bar{\sigma} \geq 0\}$. Therefore, this computation is of polynomial time complexity[9].

## 2.1 LEFs for Class 1 Transitions

Let $(\mathcal{N}, M_0)$ be a weighted P/T-$\Sigma$ and $t$ a transition such that ${}^\bullet t = \{p\}$.

**Definition 1.** The Linear Enabling Function (LEF) of a Class 1 transition $t$, $f_t : R(\mathcal{N}, M_0) \rightarrow \mathbb{Z}$ is defined as: $f_t(M) = Pre(p,t) - M(p)$; where $Pre(p,t) - \mathrm{SB}(p) \leq f_t(M) \leq Pre(p,t)$ □

For a given marking $M \in R(\mathcal{N}, M_0)$, we can use $f_t(M)$ to discriminate when a transition $t$ is enabled and, therefore, firable. The following obvious result characterizes this situation.

**Proposition 2.** *Transition $t$ of Class 1 is firable at marking $M$ iff the value of its LEF for $M$ is less than or equal to zero (i.e. $f_t(M) \leq 0$).*

Given a marking $M \in R(\mathcal{N}, M_0)$, we can know when the transition $t$ is firable by its LEF. Nevertheless, with the results presented until now, if we reach a new marking we must reevaluate all the LEFs. This reevaluation has three disadvantages: (1) We must explicitly have represented the marking of the net; (2) We must update the marking; (3) We must recompute the LEFs. Thus, it is clear that the use of LEFs increases the overhead of the simulation cycle, mainly referred to the reevaluation process.

We propose a method to update the LEFs, that does not require to keep the marking of the net, and the updating of a LEF is made only if the number of tokens of some input place of the transition has been modified.

**Proposition 3.** *Let $t$ be a transition of Class 1, and $t'$ a transition firable from a marking $M \in R(\mathcal{N}, M_0)$, i.e. $M[t'\rangle M' : f_t(M') = f_t(M) - C(p,t'), p \in {}^\bullet t$.*

In order to simulate a net system, we must compute previously $f_t(M_0)$ for all $t \in T$. After that, and during the simulation process, if we fire a transition we must update only the LEFs of those transitions whose input places have seen modified their number of tokens (in particular the fired transition itself). The updating is quite easy, because it is based on constants known before the simulation process (deduced mainly from the structure and weights of the arcs of the net).

Observe that the updating for the case in which $t' \neq t$ and $t' \in p^\bullet$, can be seen as the *preemption* of transition $t$ by the firing of transition $t'$ (this situation arises because $t$ and $t'$ are in a conflict relation). Observe also that in the case of weighted state machines, all transitions fall into the class defined in this section.

In this subsection we have presented with some generality the implications of the results stated for transitions in Class 1. In the following subsections, we will present in a more synthetic way the results, and all these comments can also be applied.

### 2.2 LEFs for Class 2 Transitions

Let suppose $(\mathcal{N}, M_0)$ be a weighted P/T-$\Sigma$ and $t$ a transition of Class 2.

**Definition 4.** The LEF of $t$ in Class 2, $f_t : R(\mathcal{N}, M_0) \to \mathbb{Z}$, is defined as:

$$f_t(M) = \sum_{p_i \in {}^\bullet t} (Pre(p_i, t) - M(p_i))$$

where $0 \leq f_t(M) \leq \sum_{p_i \in {}^\bullet t} Pre(p_i, t)$ □

The two following results trivially hold.

**Proposition 5.** *A transition $t$ of Class 2 is firable at marking $M$ iff the value of its associated LEF for $M$ is equal to zero (i.e.: $f_t(M) = 0$).*

**Proposition 6.** *Let $t$ be a transition of Class 2, and $t'$ a transition firable from a marking $M \in R(\mathcal{N}, M_0)$, i.e. $M[t'\rangle M' : f_t(M') = f_t(M) - \sum_{p_i \in {}^\bullet t} C(p_i, t')$*

As in Class 1, when ${}^\bullet t \cap {}^\bullet t' \neq \emptyset$, transition $t$ can be preempted by the firing of transition $t'$ (they are in conflict relation). For safe P/T-$\Sigma$, all transitions are of Class 2, and the LEFs take a simplified form:

$$f_t(M) = \sum_{p_i \in {}^\bullet t} (Pre(p_i, t) - M(p_i)) = |{}^\bullet t| - \sum_{p_i \in {}^\bullet t} M(p_i) \qquad (1)$$

This simplified LEF has been presented in [21] and [18] as *down-counter of a transition*, in the simulation method called *Enabled Transition Method.*

### 2.3 LEFs for Class 3 Transitions

It is easy to verify that the LEF of transitions of Class 2 do not characterize the enabling of transitions of Class 3. Let $(\mathcal{N}, M_0)$ be a weighted P/T-$\Sigma$ and $t$ a transition of this net such that $|\,\Theta\,| \geq 1$, and $\Psi = \{\pi\}$.

**Definition 7.** The LEF of a transition of Class 3, $f_t : R(\mathcal{N}, M_0) \to \mathbb{Z}$, is defined as: $f_t(M) = \mathrm{SB}(\pi) \cdot \sum_{p \in \Theta} (Pre(p, t) - M(p)) + Pre(\pi, t) - M(\pi)$; where $Pre(\pi, t) - \mathrm{SB}(\pi) \leq f_t(M) \leq \mathrm{SB}(\pi) \cdot \sum_{p \in \Theta} Pre(p, t) + Pre(\pi, t)$ □

Observe that the input places of this kind of transitions can be shared by other transitions of the net.

**Proposition 8.** *A transition $t$ of Class 3 is firable at a marking $M$ iff the value of its LEF for $M$ is less than or equal to zero (ie, $f_t(M) \leq 0$).*

*Proof.* $\Rightarrow$) If $t$ is firable at $M$, $\forall p \in {}^\bullet t, M(p) \geq Pre(p, t)$. Therefore $f_t(M) \leq 0$.

$\Leftarrow$) If $f_t(M) \leq 0$ and $t$ is not firable two cases arise for $t$ disabled:

**Case 1:** $\exists p' \in \Theta$ such that $M(p') \leq Pre(p', t) - 1$. Using also that $\forall p \in P : M(p) \leq \mathrm{SB}(p)$, and $\forall p \in \Theta : \mathrm{SB}(p) = Pre(p, t)$, we reach a contradiction: $f_t(M) = \mathrm{SB}(\pi) \cdot \sum_{p \in \Theta \setminus \{p'\}} (Pre(p, t) - M(p)) + \mathrm{SB}(\pi)\,(Pre(p', t) - M(p')) + Pre(\pi, t) - M(\pi) \geq \mathrm{SB}(\pi) \cdot \sum_{p \in \Theta \setminus \{p'\}} (Pre(p, t) - \mathrm{SB}(p)) + \mathrm{SB}(\pi) + Pre(\pi, t) - M(\pi) \geq Pre(\pi, t) > 0$.

**Case 2:** $M(\pi) < Pre(\pi, t)$. Using also that $\forall p \in P, M(p) \leq \mathrm{SB}(p)$, and $\forall p \in \Theta : \mathrm{SB}(p) = Pre(p, t)$ we reach a contradiction. □

**Proposition 9.** *Let $t$ be a transition of Class 3, and $t'$ a transition firable from $M \in R(\mathcal{N}, M_0)$, i.e. $M[t'\rangle M'$: $f_t(M') = f_t(M) - SB(\pi) \cdot \sum_{p \in \Theta} C(p,t') - C(\pi,t')$*

*Proof.* $f_t(M') = \mathrm{SB}(\pi) \cdot \sum_{p \in \Theta}(Pre(p,t) - M'(p)) + Pre(\pi,t) - M'(\pi) = \mathrm{SB}(\pi) \cdot \sum_{p \in \Theta}(Pre(p,t) - M(p)) + Pre(\pi,t) - M(\pi) - \mathrm{SB}(\pi) \cdot \sum_{p \in \Theta} C(p,t') - C(\pi,t') = f_t(M) - \mathrm{SB}(\pi) \cdot \sum_{p \in \Theta} C(p,t') - C(\pi,t')$. □

## 3 LEFs for Class 4 Transitions

Given a transition of Class 4, it is not possible to characterize its enabling by means of a LEF. In order to illustrate this we proceed by an intuitive geometric reasoning. Let us suppose that the transition $t_3$ of figure 1(a) belongs to Class 3. If we project the marking reachability space with respect to the marking of the input places ($p_a$ and $p_b$) of $t_3$, all the markings under which transition $t_3$ is enabled are projected into a segment (*potential firability region*). This segment is defined by the extremal points ($M_b = w_b, M_a = w_a$) and ($M_b = w_b, M_a = SB_a$).

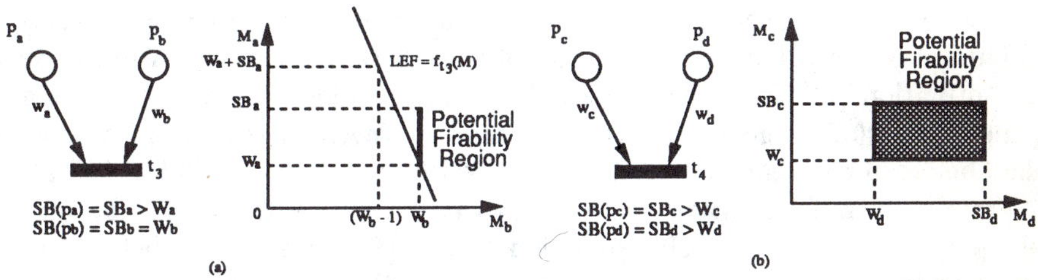


**Fig. 1.** A geometric interpretation of LEFs for transitions of Class 3 and Class 4

The LEF of transition $t_3$ splits the region of reachable markings $M_a$ and $M_b$ into two parts by means of the line depicted in the figure. The right region, including the line, contains only points with integer coordinates corresponding to the potential firability region (the LEF takes a value less than or equal to zero). The left region (LEF greater than zero) contains integer coordinates corresponding to markings of $p_a$ and $p_b$, under which transition $t_3$ is not firable.

Let us consider now the transition $t_4$ of Fig. 1 (b), of Class 4. Its potential firability region is a rectangle, and it is not possible to determine this region with a unique linear function.

In order to maintain a simulation scheme based on LEFs, we transform transitions belonging to Class 4 in such a way that: (1) The firing sequences of transitions of the original net are preserved in the transformed net; (2) The transitions resulting from the transformation, belong to classes whose LEFs are already known. In the sequel we present the transformation of transitions of Class 4 and the property of preservation of the firing sequences.

A transition $t$ of Class 4 verify: $^\bullet t = \Theta \cup \Pi$ where $\forall p_i \in \Theta, \mathrm{SB}(p_i) = Pre(p_i,t)$, $\Pi = \{\pi_i \in {}^\bullet t \mid i = 1 \ldots s, s > 1, \mathrm{SB}(\pi_i) > Pre(\pi_i,t)\}$, and $\Pi^\bullet = \{t\}$. The local

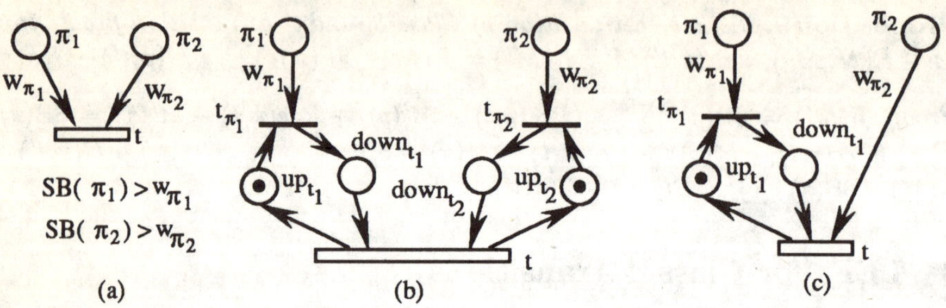


**Fig. 2.** (a) Transition of Class 4; (b,c) transformations preserving the firing language

transformation defined below affect to the places $\pi_i$ and the arcs joining them to transition $t$. In essence, this transformation is based on the addition of a new transition $t_{\pi_i}$ (called *silent transition*) for each place $\pi_i$ in order to take from place $\pi_i$ a number of tokens equal to the number of tokens needed to fire transition $t$. Only after the firing of transition $t$, new tokens can be taken from places $\pi_i$ for new firings of $t$. The transformation sequentializes the firings of transition $t$. Figure 2 presents a simple transition of this Class (a) and the corresponding transformation (b). The places $up_{t_i}$ and $down_{t_i}$ are added for taking from place $\pi_i$ the number of tokens needed to fire $t$, and to prevent new tokens from being taken before the firing of $t$.

From this transformation, we have two new transitions, $t_{\pi_1}$ and $t_{\pi_2}$, both belonging to Class 3. The old transition $t$ of Class 4 has been substituted by a new transition $t$ belonging to Class 2. Using this transformation, for each transition of Class 4 we must add only $\mid \Pi \mid$ new transitions. Nevertheless, we can add only $\mid \Pi \mid -1$ new transitions, according to the above transformation (see Fig. 2). In this case, the new transition $t$ will belong to Class 3, because of the non transformed place of $\Pi$. The following algorithm implements the local transformation of each transition of Class 4.

**Algorithm 2.** Local transformation of Class 4 transitions.
**Input** A P/T-$\Sigma$ $(\mathcal{N}, M_0)$, $\mathcal{N} = (P, T, Pre, Post)$ with transitions of Class 4.
**Output** A new P/T-$\Sigma$ $(\mathcal{N}', M_0')$ without transitions of Class 4.

**begin** {
  $T_\pi := \emptyset; T' := T; P' := P;$
  $\forall p \in P, \forall t \in T$: $Pre'(p,t) := Pre(p,t); Post'(p,t) := Post(p,t);$
  $\forall p \in P$: $M_0'(p) := M_0(p);$
  **for all** ($t \in$ Class 4 ){
    Compute $\Pi = \{\pi_i \in {}^\bullet t \mid i = 1 \dots s, s > 1, \mathrm{SB}(\pi_i) > Pre(\pi_i, t)\}$;
    **for** $(i := 0)$ **to** $(s-1)$ {
      $T' := T' \cup \{t_{\pi_i}\}; P' := P' \cup \{up_{t_i}, down_{t_i}\};$
      $Pre'(\pi_i, t_{\pi_i}) := Pre'(\pi_i, t);\ Pre'(up_{t_i}, t_{\pi_i}) := 1; Post'(up_{t_i}, t) := 1;$
      $Pre'(down_{t_i}, t) := 1; Post'(down_{t_i}, t_{\pi_i}) := 1;\ Pre'(\pi_i, t) := 0;$
      $M_0'(\mathrm{up}_{t_i}) := 1; M_0'(\mathrm{down}_{t_i}) := 0;$

```
        Tπ := Tπ ∪ {tπi}; Class 3 := Class 3 ∪ {tπi};
      }
      Class 3 := Class 3 ∪ {t}; Class 4 := Class 4\{t};
    }
}end
```

The following result states that the language of firing sequences of the P/T-$\Sigma$ $(\mathcal{N}, M_0)$, $L(\mathcal{N}, M_0)$, coincides with the language of firing sequences of the transformed net system obtained by the given algorithm, $(\mathcal{N}', M_0')$, after the removing all appearances of silent transitions $t_{\pi_i}$. If $L(\mathcal{N}', M_0')$ denotes the language of firing sequences of $(\mathcal{N}', M_0')$, the resulting language after removing the silent transitions will be denoted $L(\mathcal{N}', M_0')_{|T_\pi}$, where $T_\pi$ denotes the set of silent transitions.

**Proposition 10.** *Let* $(\mathcal{N}, M_0)$ *be a* P/T-$\Sigma$ *and* $(\mathcal{N}', M_0')$ *the transformed net system obtained from* $(\mathcal{N}, M_0)$ *by applying the algorithm* 2. $L(\mathcal{N}, M_0) = L(\mathcal{N}', M_0')_{|T_{\pi_i}}$.

*Proof.* $L(\mathcal{N}, M_0) \subseteq L(\mathcal{N}', M_0')_{|T_\pi}$: Let $\sigma \in L(\mathcal{N}, M_0)$ such that $\sigma = \sigma_0 t \sigma_1$, where the prefix $\sigma_0$ does not contain transformed transitions, and $t$ is the first transformed transition of $\sigma$. $\sigma_0 \in L(\mathcal{N}', M_0')$, because the firing of $\sigma_0$ in the transformed net affects to a part of the net equal to the original one, and $M_0 = M'_{0_{|P}}$. Moreover, if $M_0[\sigma_0 > M_1$ and $M_0'[\sigma_0 > M_1'$, then $M_1 = M'_{1_{|P}}$. From $M_1$ we can fire $t$, therefore from $M_1'$ we can fire concurrently all the silent transitions $t_{\pi_i}$ obtained in the transformation of $t$. After that, we can fire $t$: if $M_1[t > M_2$, then $M_1'[t_{\pi_1} > M_{11}'[t_{\pi_2} > \ldots [t_{\pi_{(s-1)}} > M'_{1(s-1)}[t > M_2'$ and $M_2 = M'_{2|P}$. From those new markings, $M_2$ and $M_2'$, and considering $\sigma_1$ instead of $\sigma$, we are in the same initial conditions of the proof. Applying iteratively the earlier reasoning we can build a firable sequence in $(\mathcal{N}', M_0')$ that, removing the silent transitions, matches the sequence $\sigma$.

$L(\mathcal{N}', M_0')_{|T_\pi} \subseteq L(\mathcal{N}, M_0)$: Let $\sigma' \in L(\mathcal{N}', M_0')$. Before the appearing of a transformed transition $t$ in $\sigma'$, all transitions $t_{\pi_i}$ obtained from the transformation of $t$ must appear exactly once (since $M_0'(down_{t_i}) = 0$). Moreover, all input places of transitions $t_{\pi_i}$ are not shared by other transitions. This means that if a $t_{\pi_i}$ is enabled it remains enabled until it fires. Therefore, we can reorder $\sigma'$ in such a way that all transitions $t_{\pi_i}$, needed to fire $t$, appear before a transformed transition $t$. The reordered sequence after removing all silent transitions, is also firable in $(\mathcal{N}, M_0)$. To prove this, let $M'$ be the marking that concurrently enables all the silent transitions of a transition $t$. Then, $M'$ projected w.r.t. the original places ($M'_{|P}$) is reachable in $(\mathcal{N}, M_0)$, and $M'_{|P}[t >$. □

In order to simulate a net system $(\mathcal{N}, M_0)$without transitions of Class 5, we previously transform all transitions of Class 4 (as described in this subsection), and some silent transitions are added to the net. The simulation of this new net allows to obtain the same firing sequences that the simulation of the original one (Proposition 10), ever since before the firing of any original transition we fire all enabled silent transitions. Observe that, in this new net, all transitions belong to Classes 1, 2 or 3. If $t$ is a transition of Class 4, as was defined above, after the transformation, the new generated transitions introduce the following LEFs:

$$f_t(M) = \mathrm{SB}(\pi_s) \cdot \left( \sum_{p_i \in \Theta} (Pre(p_i, t) - M(p_i)) + s - 1 - \sum_{i=1}^{s-1} M(down_{t_i}) \right) +$$

$$Pre(\pi_s, t) - M(\pi_s) \tag{2}$$

$$f_{t_{\pi_i}}(M) = \mathrm{SB}(\pi_i)\,(1 - M(up_{t_i})) + Pre(\pi_i, t_{\pi_i}) - M(\pi_i); \forall i = 1 \ldots (s-1). \tag{3}$$

The LEFs presented above, correspond to transitions of Class 3, therefore the updating rules are those stated in the previous section.

## 4 LEFs for Class 5 Transitions

It is not possible to characterize the enabling of a transition $t$ of Class 5 by a unique LEF, because there exist at least two places with SB greater than the weight of the input arc to $t$, as occurred in transitions of Class 4. Therefore, the net needs some transformations in order to integrate transitions of Class5 in a simulation scheme based on LEFs. These transformations add new transitions whose LEFs are already known, and preserve the behavior of the net. Two cases arise depending on the number of places in $\Phi$: $\mid \Phi \mid = 1$, and $\mid \Phi \mid > 1$.

If $|\Phi| = 1$, we transform each place of $\Pi$ as in Alg. 2. Places in $\Theta$ and $\Phi$ remain unchanged. Thus, we obtain a set of LEFs analogous to those obtained for transitions of Class 4. But in this case: 1) All places of the set $\Pi$ are expanded (in Class 4, only $|\Pi|-1$ places are expanded); 2) The input places of the transformed transition t are: the set $\Theta$, all places *down* arising from the expansion of places of $\Pi$ and the place of $\Phi$. This place of $\Phi$ plays the role of the place of the set $\Pi$ not expanded in transitions of Class 4. Obviously, all the added transitions belong to Class 3.

If $|\Phi| > 1$, then we cannot apply to places of $\Phi$ the same transformation we apply for Class 4, because the firing sequences are not preserved. Moreover, deadlock situations could appear, as it is shown in Fig. 3. The transition $t_2$ of Fig. 3 (a) is of Class 5. If we transform $t_2$ as a Class 4 transition, we obtain the net of Fig. 3 (b), where a deadlock can be reached that is not present in the original net.

The problem pointed out in Fig. 3, arises from the new silent transitions corresponding to the shared input places of the transition to be transformed: once the transition $t_a$ is fired, the tokens removed from $a$ cannot be used to fire $t_1$. To cope with this problem, we define a new transformation based on that of Class 4 transitions, according to the following principles:

1. A silent transition corresponding to a place of $\Phi$, does not remove tokens from this place (i.e. the place is an input and output place of the silent transition with equal weight than in the original net). Doing so, the tokens of a place of $\Phi$ can enable any of its output transitions. Moreover, the firing of the silent transition can be seen as the propagation of the enabling condition –concerning the corresponding place of $\Phi$– to the transformed transition .

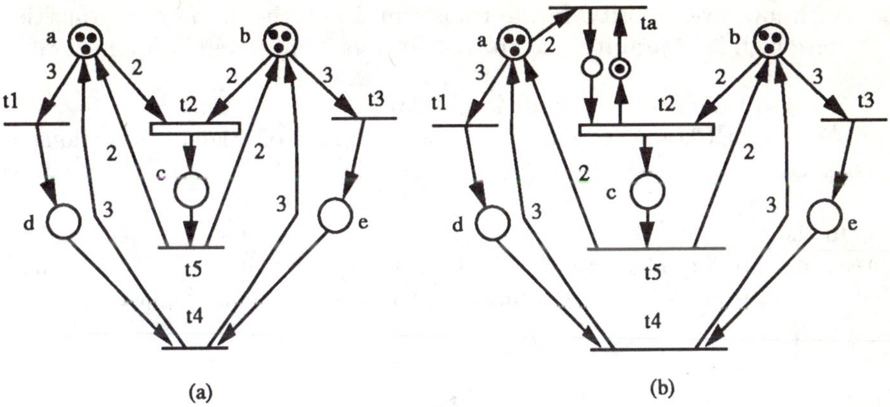


**Fig. 3.** (a) Live and bounded net system where $t_2$ is of Class 5. (b) Net system after applying the transformation of Class 4 to transition $t_2$. Observe that the sequence $t_a, t_3$ leads to a total deadlock.

2. The firing of a transformed transition $t$, must modify the LEFs of the transitions $({}^\bullet t)^\bullet$ and $(t^\bullet)^\bullet$ of the transformed net. We have to update also the LEFs for all the transitions $t' \in p^\bullet, p \in \Phi$, since the firing of a transformed transition $t$ must remove $Pre(p,t)$ tokens from each place $p \in \Phi$. To visualize this fact, we add a special arc from a place $p \in \Phi$ to the transformed transition, named *captor arc*, with an square head, meaning that place $p$ does not contribute to the enabling of $t$, but the firing of $t$ must take from $p$ a number of tokens equal to the weight of the arc.
3. An additional transition is needed for each expanded place $p \in \Phi$. Let us consider that, once a silent transition $t_p$, corresponding to a place $p \in \Phi$, is fired, an output transition of $p$, different from $t_p$, is also fired before the transformed transition $t$, making $M(p) < Pre(p, t_p)$. Then, $t_p$ must be *back-fired* (fired backward) because place $p$ does not enable $t$ any longer. This back-firing of the silent transition $t_p$ is implemented through a *preemptive transition*: its firing moves the token from the *down* place to the *up* place, and it has a special input arc from the place $p \in \Phi$ with a weight equal to the weight of the arc from $p$ to the original one of the transformed transition. This arc is a weighted inhibitor arc [4], and we call it *preemptor arc*. The firing semantics of this transition is as follows. The transition is enabled *iff* the *down* place contains at least one token, and the place $p \in \Phi$ contains a number of tokens lower than the weight of the preemptor arc. The firing moves one token from the *down* place to the *up* place, and the tokens of the place $p \in \Phi$ remain unchanged.
4. In order to preserve the firing sequences of transitions of the original net, we must assign static priorities to the firing of the transitions in the transformed net. There exist a high level and a low level of priority. If a transition of high level priority is enabled, then it must be fired before any enabled transition

with low level priority. In the transformed net, the preemptive transitions have high level priority, and the rest of transitions have low level priority.

Fig. 4 presents the transformation stated above, for the net of Fig. 3 (a), where $t_2$ is of Class 5. Observe that the sequences of original transitions are preserved in the transformed net, and that the deadlock of Fig. 3 (b) will not occur.

In the following we present: 1) The LEF corresponding to transitions with preemptor arcs; 2) The algorithm for transforming transitions of Class 5; and 3) The LEFs resulting from the transformation of a transition of Class 5.

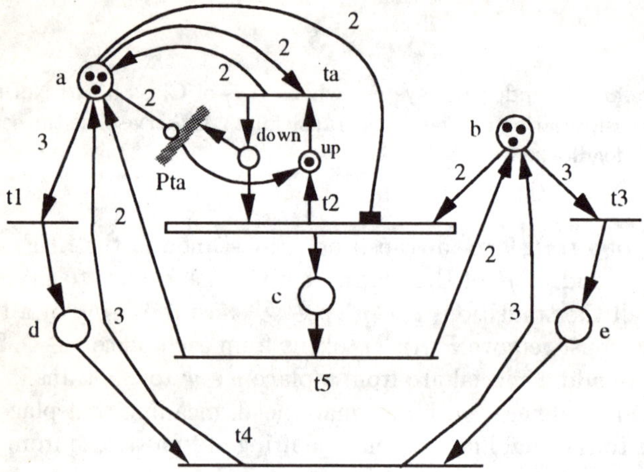


**Fig. 4.** Transformation of the transition $t_2$, of Fig. 3 (a), of Class 5

### 4.1 LEFs for Preemptive Transitions

In the description of the transformation of Class 5 transitions presented above, we have introduced a spcial class of transitions with a threshold inhibitor arc that we call *preemptor arc*. This arc checks that the contents of tokens of the connected place is lower than a threshold specified by the weight of the arc (denoted by $Pre^H(p,t)$ [4]). When $Pre^H(p,t) = 1$, we have a zero-testing inhibitor arc. Because of the lack of space, we only define here the LEF of a particular case of transitions with threshold inhibitor arcs, that we need for simulation purposes: the *preemptive transitions*. A transition $t$ is said to be a preemptive transition *iff* it has an input place $p$ such that $\mathrm{SB}(p) = 1$ and $Pre(p,t) = 1$, and an input place $\pi$ with a preemptor arc with a weight equal to $Pre^H(\pi,t)$, such that $\mathrm{SB}(\pi) \geq Pre^H(\pi,t)$.

**Definition 11.** The LEF of a preemptive transition t is defined as: $f_t(M) = M(\pi) + 1 - Pre^H(\pi,t)M(p)$, where $-(Pre^H(\pi,t) - 1) \leq f_t(M) \leq \mathrm{SB}(\pi) + 1$ □

Observe that for a preemptive transition with a zero-testing inhibitor arc, the LEF defined above becomes $f_t(M) = M(\pi) - M(p) + 1$.

**Proposition 12.** *A preemptive transition is firable at a marking $M$ iff the value of its LEF for $M$ is less than or equal to zero (i.e. $f_t(M) \leq 0$).*

*Proof.* $\Rightarrow$) If $t$ is firable at $M$, $M(p) = 1$ and $M(\pi) < Pre^H(\pi, t)$. Therefore, $f_t(M) \leq 0$.
$\Leftarrow$) If $f_t(M) \leq 0$ and $t$ is not firable, two cases arise for $t$ disabled, against the hypothesis: **Case 1:** $M(p) = 0$: $f_t(M) = M(\pi) + 1 > 0$; **Case 2:** $M(\pi) \geq Pre^H(\pi, t)$: $f_t(M) \geq 1 > 0$. □

**Proposition 13.** *Let $t$ be a preemptive transition, and $t'$ a transition firable from $M \in R(\mathcal{N}, M_0)$, i.e. $M[t' > M'$:*

1. $t' = t$: $f_t(M') = f_t(M) - Pre^H(\pi, t) \cdot C(p, t')$.
2. $t' \neq t$: $f_t(M') = f_t(M) - Pre^H(\pi, t) \cdot C(p, t') + C(\pi, t')$

*Proof.* $f_t(M') = M'(\pi) + 1 - Pre^H(\pi, t) \cdot M'(p)$. In any case, $M'(p) = M(p) + C(p, t')$. If $t' = t$, then $M'(\pi) = M(\pi)$ and the first equation is obtained, else $M'(\pi) = M(\pi) + C(\pi, t')$ and the second equation is obtained. □

## 4.2 Transformation of Transitions of Class 5

In order to transform Class 5 transitions, according to the previous considerations, we apply the following algorithm.

**Algorithm 3.** Local transformation of Class 5 transitions.

**Input** A P/T-$\Sigma$ $(\mathcal{N}, M_0)$, $\mathcal{N} = (P, T, Pre, Post)$, with transitions of Class 5.
**Output** A new P/T-$\Sigma$ $(\mathcal{N}', M_0')$ without transitions of Class 5. $T_\pi$ is the set of silent transitions and $T_{PT}$ is the set of preemptive transitions.

**begin** {
  $T_\pi := \emptyset$; $T_{PT} := \emptyset$; $P' := P$; $T' := T$; $\forall p \in P$: $M_0'(p) := M_0(p)$;
  $\forall p \in P, \forall t \in T$: $Pre'(p,t) := Pre(p,t)$; $Post'(p,t) := Post(p,t)$;
  **for all** ($t \in$ Class 5) {
    Compute $\Pi = \{\pi_i \in {}^\bullet t \mid {\pi^\bullet}_i = \{t\}, \mathrm{SB}(\pi_i) > Pre(\pi_i, t)\}$;
    Compute $\Phi = \{\phi_j \in {}^\bullet t \mid {\phi^\bullet}_j \supset \{t\}, \mathrm{SB}(\phi_j) > Pre(\phi_j, t)\}$;
    $\Psi = (\Pi \cup \Phi) \backslash \{\phi_r\}$;
    **for all** ( $p \in \Psi$ ) {
      $T' := T' \cup \{t_p\}$; $P' := P' \cup \{up_{tp}, down_{tp}\}$;
      $Pre'(p, t_p) := Pre'(p, t)$; $Pre'(up_{tp}, t_p) := 1$; $Post'(up_{tp}, t) := 1$;
      $Pre'(down_{tp}, t) := 1$; $Post'(down_{tp}, t_p) := 1$; $Pre'(p, t) := 0$;
      $T_\pi := T_\pi \cup \{t_p\}$; Class 3 := Class 3 $\cup \{t_p\}$;
      $M_0'(up_{tp}) := 1$; $M_0'(down_{tp}) := 0$;
      **if** ( $p \in \Phi$ ){
        $T' := T' \cup \{t_p^\pi\}$; $T_{PT} := T_{PT} \cup \{t_p^\pi\}$;
        $Post'(p, t_p) := Pre'(p, t_p)$; $Pre'(down_{tp}, t_p^\pi) := 1$;
        $Post'(up_{tp}, t_p^\pi) := 1$;

Add a captor arc from $p$ to $t$ with weight $Pre'(p, t_p)$;
Add a preemptor arc from $p$ to $t_p^\pi$ with weight $Pre'(p, t_p)$;
}
}
Class 5 := Class 5$\setminus\{t\}$; Class 3 := Class 3 $\cup \{t\}$;
}
}**end**

We call $L(\mathcal{N}, M_0)$ to the language of firing sequences of $(\mathcal{N}, M_0)$, and $L(\mathcal{N}', M_0')$ the language of firing sequences of the net transformed from $(\mathcal{N}, M_0)$by the Alg. 3. Let us consider the special firing semantics of transitions with preemptor and captor arcs, and the static priorities assigned to the firing of transitions. $L(\mathcal{N}', M_0')_{|T_\pi \cup T_{PT}}$ denotes the firing sequences of $(\mathcal{N}', M_0')$, after removing all appearances of silent and preemptive transitions.

**Proposition 14.** *Let* $(\mathcal{N}, M_0)$*be a* P/T-$\Sigma$ *and* $(\mathcal{N}', M_0')$ *the transformed net system obtained from* $(\mathcal{N}, M_0)$*by applying the Alg. 3. Then,* $L(\mathcal{N}, M_0) = L(\mathcal{N}', M_0')_{|T_\pi \cup T_{PT}}$.

*Proof.* $L(\mathcal{N}, M_0) \subseteq L(\mathcal{N}', M_0')_{|T_\pi \cup T_{PT}}$: Let $\sigma \in L(\mathcal{N}, M_0)$. Substitute any appearing of a transition $t$ of Class 5 for the firing of all silent transitions generated in the transformation of $t$, followed by the firing of the transformed transition $t$. If $t$ is firable in $(\mathcal{N}, M_0)$ at $M$, then $M$ is reachable in $R(\mathcal{N}', M_0')_{|P}$ and all silent transitions are concurrently enabled at $M$. After the firing of all the silent transitions concurrently enabled at $M$, the transformed transition $t$ is enabled (it can fire). Therefore, for all $\sigma \in L(\mathcal{N}, M_0)$ we can build a sequence in $(\mathcal{N}', M_0')$ that, removing the silent transitions, matches the sequence $\sigma$. Observe that preemptive transitions are not enabled, because the silent transitions are fired only when needed.

$L(\mathcal{N}', M_0')_{|T_\pi \cup T_{PT}} \subseteq L(\mathcal{N}, M_0)$: Let $\sigma' \in L(\mathcal{N}', M_0')$. If a preemptive transition $t_{PT}$ appears in $\sigma'$, then a silent transition $t_\pi$ must previously appears, such that ${}^\bullet t_{PT} \cap {}^\bullet t_\pi \in \Phi$. We can remove from $\sigma$ the transition $t_{PT}$ and $t_\pi$, because they do not modify the contents of tokens of the place ${}^\bullet t_{PT} \cap {}^\bullet t_\pi \in \Phi$, and the rest of transitions remain firable. If we apply the same procedure for all appearances of preemptive transitions, we obtain a new sequence $\sigma'$ where preemptive transitions do not appear. Now, we can reorder $\sigma'$ in such a way that, if a transformed transition $t$ appears in the sequence, then all the silent transitions needed for firing $t$ appear immediately before it. This is possible because $M'_{0|P} = M_0$ and only places *up* are marked in the initial marking. Therefore, removing all silent transitions of the reordered sequence $\sigma'$ (in a similar way to transitions of Class 4), the resulting sequence is firable in $(\mathcal{N}, M_0)$. □

In order to simulate a P/T-$\Sigma$, we previously transform all transitions of Classes 4 and 5, as described in section 3 and here. This transformations add some silent transitions and preemptive transitions. The simulation of this new net allows to obtain the same firing sequences that the simulation of the original one, ever since before the firing of any original transition, we fire all enabled silent transitions and when a preemptive transition is enabled, it is immediately fired. Observe that, in this new net, all transitions belong to Classes 1, 2, 3 or are preemptive transitions. The corresponding LEFs and their updating rules, have been presented in earlier sections.

## 5 Example: a Simulation Engine Based on LEFs

Timed Petri Nets (TPNs) may be used as a framework for Discrete Event Simulation. We present here a simulation algorithm working on a TPN structure. We associate a time interpretation to transitions, as in [17]. Thus, events are represented by transitions, and the structure of the TPN mirrors the causal relations among the events. From the PN point of view, the algorithm follows the general scheme of a synchronous centralized broad-first method, with a *central task*, *coordinator* or *interpreter* that plays the token game on a data structure representing the net, with *single server* semantics. LEFs allow an efficient implementation of a *Multiple server* semantics, but this is not included here by the lack of space. Transitions must have already been classified, as was exposed in section 2. Those transitions belonging to Classes 4 and 5 should be properly transformed. We use a *firing time* execution policy in the sense of [17, 4]. Silent and preemptive transitions are immediate transitions, i.e. they have no time associated. Therefore, every firing sequence in an untransformed P/T-$\Sigma$, holds in a transformed P/T-$\Sigma$ in which Class 4 and Class 5 transformations have been applied, and the behavior is preserved. From the DES simulation point of view, the algorithm is a sequential event-driven Simulation Engine [26, 14]. Firstly, we present the data structures used to describe the TPN and carry on the simulation, and secondly we give the algorithm.

### 5.1 Data Structures

The TPN is described by means of a list linking all the transitions in the net. To describe the transitions, we use *transition nodes*. The transition node for $t'$ has the following fields:

1. **Identifier** of $t'$.
2. **Priority** $Pr(t')$. *Low* for real transitions, *high* for preemptive transitions, and *null* for the silent ones.
3. $\tau(t')$. Deterministic firing time associated to transition $t'$. It stands for the time that the event modeled by $t'$ takes to occur.
4. **Counter.** The $f_{t'}(M)$,initialized with $f_{t'}(M_0)$, and updated whenever the transition –or a transition affecting it– fires, according to the equations stated in the earlier sections.
5. **Immediate Updating List** ($IUL(t')$) List of *updating nodes* covering the set $({}^\bullet t')^\bullet$ (Note that $({}^\bullet t')^\bullet$ includes $t'$).
6. **Projected Updating List** ($PUL(t')$) List of *updating nodes* covering the set $(t'^\bullet)^\bullet$.
7. **Next** Pointer to the next transition node (*null* in the last one).

The *updating nodes* have the following fields:

1. **Pointer to Transition**. Points to the transition node affected by the updating.

2. **Updating Factor** ($UF(t' \to t)$). As defined below.
3. **Next**. Pointer to the next updating node in the list (*null* in the last one).

Given two transitions $t$ and $t'$, the *Updating Factor* of $t'$ over $t$, $UF(t' \to t)$ represents the influence of firing $t'$ on the LEF of $t$, and can be defined as a function $UF(t' \to t) : \{Pre[t'] \cup Post[t']\} \to \mathbb{Z}$ such that:

$$UF(t' \to t) = \begin{cases} -C(p,t') & \text{if } t \in \text{ Class 1} \\ -\sum_{p_i \in {}^\bullet t} C(p_i,t') & \text{if } t \in \text{ Class 2} \\ -\text{SB}(\pi) \cdot \sum_{p \in \Theta} C(p,t') - C(\pi,t') & \text{if } t \in \text{ Class 3} \\ -Pre^H(\pi,t) \cdot C(p,t') + C(\pi,t') & \text{if } t \text{ is preemptive} \\ & \text{and } t \neq t' \\ -Pre^H(\pi,t) \cdot C(p,t) & \text{if } t \text{ is preemptive} \\ & \text{and } t = t' \end{cases}$$

Note that these are the only Classes of transitions we have after applying the transformations.

Using this definition, the updating equation of any transition $t$ affected by the firing of a transition $t'$ can be reformulated as $f_t(M') = f_t(M) + UF(t' \to t)$.

The algorithm uses four lists more. *SEL* has all the silent transitions enabled at the same step. *EL* contains the non-silent enabled transitions, and the function *insert()* maintains it ordered by priority. *FUL* contains *Future Updating Nodes* (FUNs), and the function *insert-FUL()* maintains them ordered by time. *FUL* plays the role of the Future Event List in an event-driven simulation algorithm for DES. A FUN holds:

1. a pointer (*pt*) to the transition to be updated;
2. the updating factor $UF(t' \to t)$ delivered by each fired transition ($t \in (t'^\bullet)^\bullet$);
3. the time (*time*) at which the updating must take effect.
4. **Next**. Pointer to the next updating node in the list (*null* in the last one).

Variable *head-FUL* is a pointer to *FUL*; *pop(FUL)* pops and returns the head of *FUL*. We access the fields of *FUN*s using the dot notation. The variable *clock* holds the current simulation time.

## 5.2 The Algorithm

**Algorithm 4.** Simulation Engine
**Input** A P/T-$\Sigma(\mathcal{N}', M_0')$ in which transitions have been classified using the Alg. 1, transformed according to the Algs. 2 and 3, and described with the exposed data structures.

**begin** {
  **for ever**{
    **for all** ($t' \in SEL$){
      **for all** ($t \in \{IUL(t') \cup PUL(t')\}$){
        $f_t(M) := f_t(M) + UF(t' \to t)$; **if** ($f_t(M) \leq 0$) insert($EL, t, Pr(t)$);

```
      }
    }
    for all (t' ∈ EL){
      if (f_{t'}(M) ≤ 0){
        for all (t ∈ IUL(t')){
          f_t(M) := f_t(M) + UF(t' → t);
          if (Pr(t) = High and f_t(M) ≤ 0)insert(EL, t, Pr(t));
          else if (Pr(t) = Low and t = t' and f_t(M) ≤ 0)
            insert-FUL (t, 0, τ(t) + clock);
        }
        for all (t ∈ PUL(t'))insert-FUL (t, UF(t' → t), τ(t') + clock);
      }
    }
    if (head-FUL.time > clock) clock := head-FUL.time;
    while (head-FUL.time = clock){
      t := head-FUL.pt; f_t(M) := f_t(M) + head-FUL.UF;
      if (f_t(M) ≤ 0){
        if (t is silent ) insert(SEL, t, null); else insert(EL, t, Pr(t));
      head-FUL := pop(FUL);
    }
  }
}end
```

The first **for all** fires all silent enabled transitions. The firing of a silent transition never prevents other silent ones transitions from firing. The LEFs of all the affected transitions are updated. Only real transitions can be enabled in this loop, and therefore included in *EL*.

The third **for all** visit *EL* and test transitions before firing, just in case a transition enabled when starting the cycle has been disabled by the firing of a previous one, in effective conflict relation. For the sake of clarity, here we solve conflicts by firing the first transition found in *EL*, but other preselection policies can be adopted. If a preemptive transition becomes enabled here, it is inserted in the head of *EL*, and fired immediately. If a fired transition remains enabled, then a *null* updating is inserted in *FUL*, in order to be considered at the next step. The updating of the LEFs of $(t^{\bullet})^{\bullet}$ are projected in the time, including the constant in FUL.

Finally, *clock* is advanced to the minimum time held in *FUL*, and all the updating factors projected for the current time take effect.

## 6 Conclusions

We have developed a new method to simulate PNs under a transition-driven approach. The method applies to general PNs, even with inhibitor arcs, and it is based on the definition of Linear Enabling Functions (LEFs). The LEF of a transition characterizes linearly the enabling of a transition at a given marking, reducing the cost of solving a complex equation system. This is an interesting property when simulating a system modeled by means of a Petri Net. We classify transitions into five Classes, and show how LEFs can be derived for Classes 1, 2 and 3. We give two transformations for Classes 4 and 5 that preserve the behavior of the system. Such transformations introduce new (*silent*) transitions, and a special class of transitions named *preemptive transitions*. All the transitions which appear in a transformed system, belongs to Classes 1, 2 and 3, or are *preemptive transitions*, and their enabling conditions can be characterized by means of a LEF. In particular, the LEF of *preemptive transitions* allows using threshold inhibitor arcs.

Finally, we have proposed a simulation algorithm of Timed Petri Nets with deterministic firing times. This algorithm is an event-driven Simulation Engine in the classic sense. The enabling is quickly characterized, and only the enabled transitions are tested for enabling. Further work includes the application of LEFs for distributed simulation of DES. Moreover, LEFs easily characterize the *k-firability* (k simultaneous firings of a transition in a same simulation cycle) of transitions of Classes 1, 2 and 3, and of preemptive transitions; k-firability of transitions of Classes 4 and 5 can be also characterized with minor changes in the simulation algorithm. Performance measures will be given in short, for both sequential and distributed DES.

## A Basic Concepts and Notations on Place/Transition Systems

We give here a brief recall of the basic Petri Net terminology and notation. The reader is referred to [15] for a nice tutorial on Petri nets.

A P/T-$\mathcal{N}$ $\mathcal{N}$ is defined as a 4-tuple $\mathcal{N} = (P, T, Pre, Post)$, where $P$ is the set of *places* ($|P| = n$), $T$ is the set of *transitions* ($P \cap T = \emptyset, P \cup T \neq \emptyset$) ($|T| = m$), $Pre$ ($Post$) is the pre- (post-) incidence function representing the input (output) arcs $Pre : P \times T \rightarrow \mathbb{N}$ ($Post : P \times T \rightarrow \mathbb{N}$). The pre- and post-incidence functions can be represented as $n \times m$ matrices $Pre$ and $Post$ with elements $Pre(p_i, t_j)$ and $Post(p_i, t_j)$, respectively. The *incidence matrix* $C$ of the net is defined by $C(p_i, t_j) = Post(p_i, t_j) - Pre(p_i, t_j)$. The *pre-* and *post-set* of a transition $t \in T$ are defined respectively as ${}^\bullet t = \{p | Pre(p, t) > 0\}$ and $t^\bullet = \{p | v(p, t) > 0\}$. The *pre-* and *post-set* of a place $p \in P$ are defined respectively as ${}^\bullet p = \{t | Post(p, t) > 0\}$ and $p^\bullet = \{t | Pre(p, t) > 0\}$.

A function $M : P \rightarrow \mathbb{N}$ is called a *marking*. A marking $M$ can be represented in vector form, with the $i^{th}$ component associated with the $i^{th}$ element of $P$. A P/T-$\Sigma$ $(\mathcal{N}, M_0)$ is a net $\mathcal{N}$ with an *initial marking* $M_0$. A transition $t \in T$ is *enabled* in marking $M$ iff $\forall p \in P$, $M(p) \geq Pre(p, t)$. A transition $t_j$ enabled in $M$ can *fire* yielding a new marking $M'$ defined by $M'(p) = M(p) - Pre(p, t_j) + Post(p, t_j) = M(p) + C(p, t_j)$. We denote as $M[t\rangle M'$ that transition $t$ is enabled in $M$ and that $M'$ is *reached* from $M$ by firing $t$. A sequence of transitions $\sigma = t_1 t_2 \ldots t_n$ is a *firing sequence* of $(\mathcal{N}, M_0)$ iff there exists a sequence of markings such that $M_0[t_1\rangle M_1[t_2\rangle M_2 \ldots [t_n\rangle M_n$. In this case, marking $M_n$ is said to be *reachable* from $M_0$ by firing $\sigma$, and this is denoted by $M_0[\sigma\rangle M_n$.

A place $p \in P$ is *bounded* iff $\max\{M(p), \forall M \in R(\mathcal{N}, M_0)\}$ is bounded, and this value is denoted B($p$). A P/T-$\Sigma$ $(\mathcal{N}, M_0)$is said to be bounded iff each one of its places is bounded. A place $p \in P$ is *structurally bounded* iff $\max\{M(p), \forall M = M_0 + C \cdot \bar{\sigma}, M \geq 0, \bar{\sigma} \geq 0\}$ is bounded, and this value is SB($p$). A transition $t \in T$ is *live* in $(\mathcal{N}, M_0)$ iff $\forall M \in R(\mathcal{N}, M_0)$ $\exists M' \in R(\mathcal{N}, M)$ such that $M'$ enables $t$. The marked net $(\mathcal{N}, M_0)$ is live iff all its transitions are live. A net $\mathcal{N}$ is *structurally live* iff $\exists M_0$ such that the marked net $(\mathcal{N}, M_0)$ is live. The marked net $(\mathcal{N}, M_0)$ is *deadlock-free* iff $\forall M \in R(\mathcal{N}, M_0)$ $\exists t \in T$ such that $M$ enables $t$.

## References


1. HH Ammar and Su Deng. Time Warp Simulation of Stochastic Petri Nets. In *Proc. of the 4th Int. Workshop on Petri Nets and Performance Models*, pages 186–195, Melbourne, Australia, December 1991. IEEE-CS Press.
2. J Banks and JS Carson. *Discrete-Event System Simulation.* Prentice Hall, Inc., 1984.
3. F Bréant. Rapid prototyping from petri net on a loosely coupled parallel architecture. In *Transputer applications 1991*, pages 28–30, Glasgow, Scotland (UK), August 1991.
4. G Chiola, S Donatelli, and G Franceschinis. Priorities, Inhibitor Arcs and Concurrency in P/T nets. In *Proc. of the 12th Int. Conference in Application and Theory of Petri Nets*, pages 182–205, Aarhus, June 1991.
5. G Chiola and A Ferscha. Distributed Simulation of Timed Petri Nets: Exploiting the Net Structure to Obtain Efficiency. In Marco Ajmone Marsan, editor, *Proc. of the 14th Int. Conf. on App. and Theory of Petri Nets*, pages 186–195, Melbourne, Australia, December 1993. Springer-Verlag.
6. D Chocron. Un Système de Programmation par RdP de Controleurs Industriels. Master's thesis, Montreal, Canada, 1980.
7. JM Colom, M Silva, and JL Villarroel. On software implementation of Petri Nets and Colored Petri Nets using high level concurrent languages. In *Proc. of 7th European Workshop on Application and Theory of Petri nets*, pages 207–241, Oxford, England, January 1986.
8. RM Fujimoto. Parallel Discrete Event Simulation. *Communications of the ACM*, 33(10):30–53, October 1990.
9. N Karmarkar. A New Polynomial Time Algorithm for Linear Programming. *Combinatorica*, (4):373–395, 1984.
10. F Kordon. *Prototypage de systèmes parallèles à partir de rèseaux de Petri colorés.* PhD thesis, Institut Blaise Pascal, Univ. Paris VI, 4, Place Jussieu 75252 PARIS CEDEX 05, 1992.
11. D Kumar. Systems with low distributed simulation overhead. *IEEE Transactions on Parallel and Distributed Systems*, 3(2):155–165, March 1992.
12. Y Li and WM Wonham. Control of Vector Discrete-Event Systems. I The Base Model. *IEEE Transactions on Automatic Control*, 38(8):1214–1227, August 1993.
13. J Miguel and M Graña. Towards the Distributed Implementation of Discrete Event Simulation Languages. In *Proc. Int. Conf. on Decentralized and Distributed Systems ICDDS'93*, pages 273–285, Mallorca, España, September 1993.
14. J Misra. Distributed discrete-event simulation. *ACM Computing Surveys*, 18(1), March 1986.

15. T Murata. Petri nets: properties, analysis, and applications. *Proceedings of the IEEE*, 77(4), April 1989.
16. RA Nelson, LM Haibt, and PB Sheridan. Casting Petri Nets into programs. *IEEE Transactions on Software Engineering*, 9(5):590–602, September 1983.
17. C Ramchandany. *Analysis of Asynchronous Concurrent Systems by Timed Petri Nets*. PhD thesis, Massachusetts Institute of Technology, Massachusetts 02139, USA, 1974.
18. M Silva. *Las Redes de Petri en la Informática y en la Automática*. AC, Madrid, 1985.
19. M Silva and JM Colom. On the Computation of Structural Synchronic Invariants in P/T Nets. In G Rozenberg, H Genrich, and G Roucairol, editors, *Advances in Petri Nets 1988*, volume 340 of *Lecture Notes in Computer Sciences*, pages 387–417. Springer-Verlag, Berlin, Germany, 1988.
20. M Silva and R David. Synthese programmée des automates logiques décrits par réseaux de petri: Une méthode de mise en oeuvre sur microcalculateurs. *Rairo-Automatique*, 13(4):369–393, november 1979.
21. M Silva and S Velilla. Programmable logic controllers and Petri Nets: A comparative study. In *Proc. of the Third IFAC/IFIP Symposium*, Software for Computer Control 1982, pages 83–88. Pergamon Press, 1982.
22. D Taubner. On the implementation of Petri Nets. In G. Rozenberg, H. Genrich, and G. Roucairol, editors, *Advances in Petri Nets 1988*, volume 340 of *Lecture Notes in Computer Sciences*, pages 418–439. Springer-Verlag, Berlin, Germany, 1988.
23. E. Teruel, J.M. Colom, and M. Silva. Linear analysis of deadlock-freeness of Petri net models. In *Proc. of the European Control Conference, ECC'93*, pages 513–518, Groningen, The Netherlands, June 28 - July 1 1993.
24. R Valette. Nets in production systems. In G Goos and Hartmann, editors, *Petri Nets: Applications and Relationships to Other Models of Concurrency*, volume 255 of *Lecture Notes in Computer Sciences*, pages 191–217. Springer-Verlag, Berlin, Germany, 1986.
25. JL Villarroel. *Integración Informática del Control en Sistemas Flexibles de Fabricación*. PhD thesis, Universidad de Zaragoza, María de Luna 3 E-50015 Zaragoza, España, 1990.
26. BP Zeigler. *Theory of Modelling and Simulation*. Wiley-Interscience, 1976.
27. BP Zeigler and WH Sanders. Frameworks for Evaluating Discrete Event Dynamic Systems. *Discrete Event Dynamic Systems: Theory and Applications*, (3):113–118, 1993.

# Hierarchical High Level Petri Nets for Complex System Analysis

Peter Buchholz

Informatik IV, Universität Dortmund, D-44221 Dortmund, Germany

**Abstract.** The class of Hierarchical High Level Petri Nets (HHPNs) is introduced. HHPNs provide a framework for a modular specification of complex parallel and concurrent systems which supports top-down and bottom-up design. Apart from specification convenience HHPNs can be analysed very efficiently according to functional aspects of the modelled system. Analysis is based on a divide and conquer approach combined with behaviour preserving reduction techniques on subsets of the reachability set. The complexity of an analysis on the reachability set of the net can often be reduced by several orders of a magnitude by considering only small parts of the complex reachability set in a single analysis step and performing behaviour preserving reductions before composing subsets of the reachability set.

## 1 Introduction

Petri Nets (PNs) are a well known formalism for the specification and analysis of parallel and concurrent systems [16, 17]. This is caused in particular by the combination of an expressive graphical specification with a well formalised mathematical model. Nevertheless, the limitations of ordinary PNs for the specification of complex and, unfortunately, realistic systems are also well known for the last 15 years. This motivates the development of High Level Petri Nets (HPNs) like coloured nets [12] or predicate transition nets [8] in the early eighties. Since this time HPNs have been extended in several directions by the integration of inhibitor arcs, transition priorities, place capacities etc. (see [1]). More recently it has been recognized that even HPNs are not adequate to specify complex systems since they describe a single view on a complex system. This yields the development of structuring mechanisms, in particular the introduction of hierarchies [7, 11, 13]. The approaches allow the substitution of places and transitions by more complex nets which describes a refinement of the view. However, the hierarchy is only used for net specification, only the most detailed net is really executable, thus any kind of analysis is performed on the flat detailed net.

Analysis of complex nets usually has to be performed on the reachability set of the net which is very large for realistic models and grows often exponentially with the size and complexity of the net. Analysis becomes a kind of state exploration, which is a common approach for automatic verification of parallel and concurrent systems regardless of the used specification technique. Exhaustive state exploration requires a finite number of states to yield exact results. Contemporary computer equipment allows the analysis of state spaces (reachability sets) with about $10^7$ markings. The dimension of reachability sets for realistic nets is often finite but usually several orders of a magnitude larger than the analysable number of states. Hierarchical specification approaches for PNs as mentioned above do not help here since analysis still has to be performed on the flat net.

The complexity of a state based analysis can be reduced significantly, if a structuring mechanism supporting also analysis is available (see e.g., [9, 15, 19, 20]). Common structuring mechanisms allow to describe a system by composing less complex process specifications in a well defined way. The central idea of all analysis approaches based on this paradigm is to define local equivalence relations. Processes can be substituted by less complex but equivalent behaving substitutes reducing the size of the state space significantly. Thus, divide and conquer is the major way to handle complexity.

In the area of PNs there are, of course, several approaches going in a similar direction (see e.g., [3, 6, 18]). However, in all cases the restrictions which have to be put on subnets or the composition medium are rather hard and limit the applicability of the approaches. In [20] synchronous composition of PNs is considered, based on an equivalence notation as given in CSP. The reachability set of the composed net is generated by composing reachability sets of subnets combined with behaviour preserving reductions between the steps. Such an approach is only possible if we can generate efficiently the reachability set of an isolated subnet without knowing the environment (i.e., for all possible environments). The problem is that the reachability set of an isolated subnet becomes very huge (in extreme cases infinite) although embedded in a concrete environment the number of reachable markings is rather small. This problem is most times not considered in compositional state space generation, limiting approaches to subnets with a small reachability sets in all (or a large class) of environments.

In this paper we also introduce hierarchies in PNs and use compositional state space generation, but the approach differs significantly from the former according to specification and analysis. In particular our hierarchy describes asynchronous interaction between subnets in contrast to the synchronous hierarchy which is normally used in composition based paradigms and the environment behaviour is partially considered during subnet reachability set generation. Nets are specified using different levels, a level consists of one or several subnets. Going down the hierarchy, the specification becomes more detailed and the part of the net which is considered in detail becomes smaller. For each subnet we define a detailed view and an aggregated view including the behaviour relevant for other parts. Consistency between detailed and aggregated view is assured by defining equivalence relations between both. Instead of restricting the allowed refinements by defining special classes of subnets, we allow arbitrary subnets and test the behaviour of a subnet against the aggregated view which is used for the subnet in a higher level. If both views are not equivalent, one has to be modified. This approach requires possibly iterations, however, each analysis step is performed on a rather small set of markings.

Subnets have a transition and a place border at input and output and allow the specification of asynchronous interactions with their environment. Synchronous communication between subnets can be explicitly modelled in the parent net of the subnets to be synchronized. In contrast to other classes of hierarchical HPNs (e.g., [13]) each subnet is executable, the dynamic behaviour is described by a local reachability graph including transitions in the detailed local part and aggregated non-local parts. Several properties of a net can be verified on small parts of the net or reachability graph which is very efficient even for nets with a large overall reachability set. Other properties require the computation of the complete reachability set, it is shown how this can be done by combining subspaces using appropriate operations and how to reduce subspaces during reachability set generation yielding a reduced reachability set of the complete net for the verification of the required properties.

The remainder of the paper is organized as follows. In the next section the class of HHPNs is introduced as an extension of a rather general class of HPNs. Afterwards in section 3 equivalence between different views on a subnet or the environment of a subnet are defined and algorithms to test equivalence are introduced. Section 4 describes several analysis algorithms for HHPNs.

# 2 Hierarchical Specification of HPNs

## 2.1 Basic High Level Petri Nets

**Definition 1.** A High Level Petri Net (HPN) is defined as a 7-tuple: $HPN = (P, T, C, W^+, W^-, \phi, m^0)$, where $P$ and $T$ are finite and disjoint sets (of places and transitions), $C$ is a function mapping $P \cup T$ on a non-empty set of colours ($C(p)$ and $C(t)$ are the colour domains of $p$ and $t$), $W^-$, $W^+$ are the input and output function where $W^-(p,t)$ ($W^+(p,t)$) is a function from $C(p) \times C(t) \rightarrow \mathbb{N}$, $\phi$ is a transition guard such that $\phi_t$ is function from $C(t) \times m \rightarrow \{TRUE, FALSE\}$, where $m$ is a marking of the net, and $m^0$ is the initial marking.

Let $m$ be a specific marking, $m(p,c)$ denotes the number of colour $c$ token on place $p$. $m(p)$ and $M(p)$ specify a marking or a set of markings for place $p$. We assume the usual interleaving semantic for firing transitions. Transition $t \in T$ is enabled for colour $c \in C(t)$ in marking $m$, if

$$\forall p \in P, \forall c' \in C(p) : W^-(p,t)(c',c) \leq m(p,c') \text{ and } \phi_t(c,m) = TRUE;$$

after firing transition $t$ for colour $c$ a new marking $m'$ is given by

$$\forall p \in P, \forall c' \in C(p) : m'(p,c') = m(p,c') - W^-(p,t)(c',c) + W^+(p,t)(c',c).$$

If a transition from marking $m$ to $m'$ is possible, then this is related to the firing of exactly one transition for one colour and each firing of a transition modifies the marking of the net. Marking $m'$ is directly reachable from $m$, if a transition $t$ is enabled in $m$ for some colour $c \in C(t)$ and the firing of $t$ in $c$ yields $m'$. The reachability set of markings $M$ includes all markings $m$ for which a sequence $\{m^0, \ldots, m^n\}$ exists such that $m^0$ is the initial marking and $m^k$ is directly reachable from $m^{k-1}$ for each $0 < k \leq n$. We use the notation $m \gg m'$, if $m'$ is reachable from $m$ by a finite sequence of transitions in the net and $m > m'$, if $m'$ is directly (by firing one transition) reachable from $m$. $m > t > m'$ indicates that $m'$ is reached from $m$ by firing transition $t$ . Let $R$ be the reachability graph of the net which is a directed graph with the set of nodes $M$. Two nodes $m$ and $m'$ are connected if $m > m'$. $R$ is an unlabelled graph, latter we define reachability graphs for subnets which are labelled graphs.

## 2.2 From HPNs to HHPNs

A HHPN, very roughly spoken, is a set of HPNs which are related in a hierarchical way. The overall structure forms a tree, each node of the tree is a HPN which describes a detailed local behaviour and an aggregated view on the nodes above which form the environment and the nodes below which are childs to be specified in more detail in the corresponding HPNs. HPNs forming a HHPN are numbered consecutive from top to bottom and left to right. We identify the HPNs in the sequel by their number ranging from 0 to $J$. The tree structure of the HPNs defines a function $h$ which determines the parent (or environment) of a node, we have $h(i) < i$ and $h(0) = \emptyset$. The inverse function $h^{-1}(i)$ defines the childs of HPN $j$, i.e., all $i$ with $h(i) = j$. The key point is to describe the behaviour of the environment and the children of a specific HPN, which represents a node in the tree, using a specification which is equivalent according the influences on the local behaviour but hides as much as possible of the details of non-local parts. We will denote such a description of a behaviour as an aggregated view and define it below.

**Definition 2.** A HHPN is a set of HPNs numbered from 0 to $J$ and a function $h$ as defined above. Each HPN $j$, denoted as a subnet of the HHPN, includes an aggregated description of HPN $h(j)$ and an aggregated description for each HPN in $h^{-1}(j)$.

Thus we are faced with two problems, namely to define aggregated views on non-local parts and to assure equivalence between aggregated and detailed specification of a part of the net. We start with the specification of aggregated views and introduce in the following section equivalence relations. Each HHPN implicitly defines a flat HPN which can be generated by substituting each subnet by the most detailed specification, the resulting net is a HPN without aggregated views.

Since we consider asynchronous interaction between subnets all influences are realized by tokens moving between the different subnets. An aggregated view on a subnet is realized by a set of input- and output ports to embed it in an environment. Input- and output ports can be divided into place- and transition border. Ports of the place border can only be connected with transitions or ports of the transition border, ports of the transition border can only be connected with places or ports of the place border. Using an appropriate colouring, we can assume that at most one port of each type exists (i.e., input place, input transition, output place, output transition). The input transition port fires tokens in the subnet which are offered by the environment, the input place port accepts tokens which are fired from the environment into the subnet, the output transition port fires tokens out of the subnet and the output place port offers tokens to the environment.

A subnet is described by the ports and a marking which specifies an aggregated view on the internal marking of the subnet realizing the detailed view. Each marking of the aggregated view covers a set of markings of the detailed view which observe specific conditions as defined below. The aggregated marking of a subnet is modified by internal events, input of tokens through an input port or output of tokens through an output port. We assume that the firing of a transition either adds tokens to, or removes tokens from a subnet. An aggregated view can be specified by a transition and a corresponding place, the marking of the place describes the state of the non-local part, abstracting from details, and the transition defines activities in the non-local part, abstracting from the details. In the graphical representation of a subnet we will describe aggregated views as boxes including a place and the ports (see the example below).

Let $P_j, T_j$ be the set of places and transitions and $\phi_j$ be the predicate function in the detailed description of subnet $j$. Assume that the HHPN includes subnets $0, \ldots, J$. Ordinary places in each subnet specification are numbered consecutive starting with $J+1$. Aggregated views, which are realized by a place and corresponding transition (see the definition below), receive the number of the corresponding subnet. E.g., $p_i^j$ and $t_i^j$ denote the aggregated view on subnet $i$ in subnet $j$ for $0 \leq i, j \leq J$ and denote for $J < i$ an ordinary place and transition in subnet $j$. Of course, $p_i^j$ and $t_i^j$ with $0 \leq i \leq J$ are only part of subnet $j$, if $h(j) = i$ or $i \in h^{-1}(j)$.

**Definition 3.** An aggregated view on subnet $i$ in subnet $j$ is defined by a place $p_i^j \in P_j$ including an aggregated view on the internal subnet marking with colour domain $C(p_i^j)$, for all $t_k^j \in T_j$ and $c \in C(t_k^j)$ $W^+(p_i^j, t_k^j)(c) = \emptyset$ or $W^-(p_i^j, t_k^j)(c) = \emptyset$, a transition $t_i^j \in T_j$ with colour domain $C(t_i^j)$.

Since aggregated views on non-local parts are realized by places and transitions, the reachability graph of each subnet can be generated by standard means (i.e., playing the token game). Let $M_j$ be the reachability set of subnet $j$, $m(i)$ ($m \in M_j$) is the aggregated marking of subnet $i$ included in the detailed view of $j$, $M_j(i)$ is the set of all reachable markings on $p_i^j$ and $R_j$ is the reachability graph of $j$.

We describe the concepts by a very simple example as shown in Fig. 1. The static structure of the HHPN is shown in i). For the function $h$ we get, $h(0) = \emptyset$, $h(1) = h(2) = 0$ and $h^{-1}(0) = \{1, 2\}$. The boxes in the nets ii)-iv) represent aggregated views on non-local parts, dynamic behaviour of aggregated views can be specified using a place and a transition as defined above. A detailed description of the dynamic of aggregated views will be given below. For the lack of space no functions at the arcs connecting ordinary transitions and places are given, however, the behaviour of all transitions, very roughly spoken, is to remove one token per arc from an input place and add a token of the same colour to the output place, additionally resource tokens (with colour black) are removed from and returned to the places $p_7^0$, $p_7^1$, $p_8^1$ and $p_{10}^2$ in the different subnets. In what follows we will come back from time to time to this example to clarify newly introduced concepts.

## 3 Consistent Views on Subnets

What we have done up to now is a specification of a net composed from several isolated parts, but consistency between different views on one subnet has not been considered yet. In particular it has to be assured that the aggregated view and the detailed view of a subnet are indistinguishable from the embedding environment. This characterisation depends, of course, on the concrete environment,

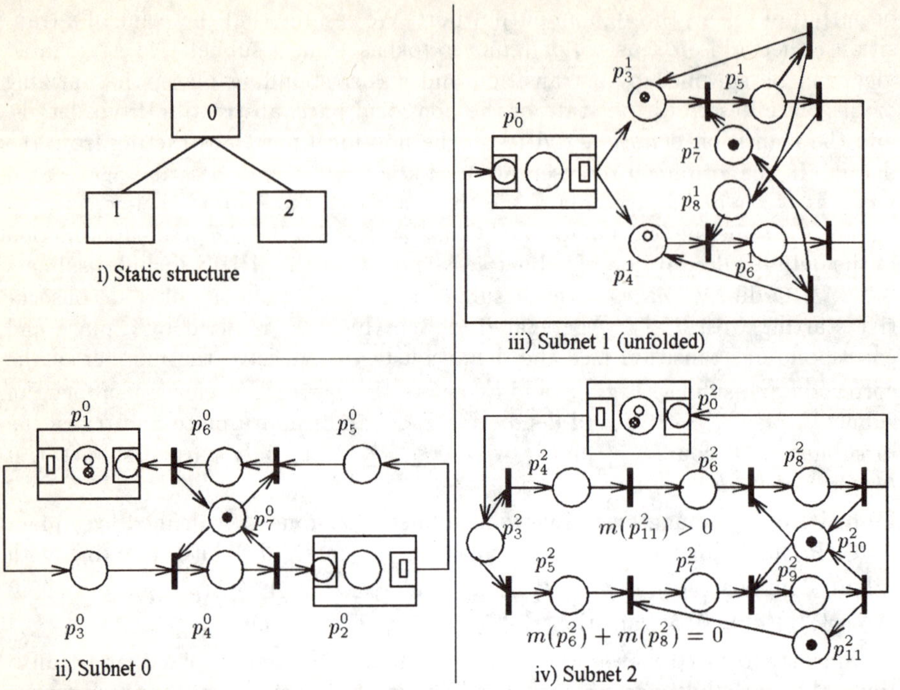


**Fig. 1.** Example HHPN

however, using stepwise refinement the environment is known during the detailed specification of a subnet. Bottom-up design combining pre-defined subnets is also possible, since in this case the subnet behaviour is known in detail and specifies the environment for other subnets. In this section we describe relations between isolated subnets and define consistency and equivalence relations between different representations and views. We consider first detailed specification of a subnet and afterwards the relation between a detailed and an aggregated view on a subnet.

First aggregation of markings has to be introduced. Aggregation is fundamental for specification and analysis of HHPNs. The idea is to represent several markings of a more detailed view by a single marking in a coarser view. A set of markings $\mathcal{A}$ is aggregated into a compound marking $\tilde{m}$ by substituting all incoming transitions $m' > m$ ($m \in \mathcal{A}, m' \notin \mathcal{A}$) by a transition $m' > \tilde{m}$ and all outgoing transitions $m > m'$ by a transition $\tilde{m} > m'$. If the reachability graph is represented by the incidence matrix, then aggregation is realized by adding the rows and columns belonging to the markings to be aggregated and setting the corresponding diagonal element to zero. Arbitrary sets of markings can be aggregated, but the resulting reachability graph usually will not preserve results for the original net, thus the key point is to find conditions allowing aggregation preserving the relevant behaviour.

The following notations are needed in the sequel. Assume that markings from a reachability set $M$ should be aggregated, $\tilde{M}$ is the resulting aggregated reachability set and $\tilde{R}$ the aggregated reachability graph. Let $f(m) \in \tilde{M}$ ($m \in M$) be a function defining the unique aggregated marking which belongs to $m$. Function $f$ can also be used for sets of markings yielding the corresponding set of aggregated markings (e.g., $f(M) = \tilde{M}$). $f^{-1}(\tilde{m})$ is the set of markings which is aggregated to $\tilde{m}$. Define $f_{in}^{-1}(\tilde{m}) \subseteq f^{-1}(\tilde{m})$ as the set of entry markings of the set $f^{-1}(\tilde{m})$, i.e.,

$$f_{in}^{-1}(\tilde{m}) = \{m | m \in f^{-1}(\tilde{m}) : \exists \bar{m} \notin f^{-1}(\tilde{m}) \; \bar{m} > m \text{ or } m = m^0\}.$$

We sometimes use the notation $\tilde{m}$ to describe the aggregated marking which belongs to marking $m$, i.e., $\tilde{m} = f(m)$. The aggregation of markings hides some of the transitions from the original reachability graph we use the notation $m > \tau^* > m'$ for $f(m) = f(m')$, if $m$ and $m'$ are connected via a path which includes only transitions between markings that are aggregated into the same compound marking. For transitions which are visible in the aggregated reachability set we use the notation $> t >$ (e.g., $m > \tau^* t \tau^* > m'$ denotes that $m'$ is reachable from $m$ by a number of non-visible transitions in $f^{-1}(\tilde{m})$ followed by a visible transition between $\tilde{m}$ and $\tilde{m}'$ followed by a number of non-visible transitions in $f^{-1}(\tilde{m}')$). The aggregation assure that a transition in the detailed reachability set is represented by a transition in the aggregated reachability set but not generally vice versa (i.e., if $m > m' \Rightarrow f(m) > f(m')$ for $f(m) \neq f(m')$).

## 3.1 Detailed subnet specification

The reachability set of a subnet obviously depends on the embedding environment. We consider here two different ways to describe an environment. The first is based on top-down design (stepwise refinement), the second is more related to bottom-up design by generating subnet reachability sets for a class of environments. We start with a stepwise refinement approach.

In a stepwise refinement approach the parent subnet is completely specified at the point when the children are specified in detail. Assume that subnet $j$ has been completely specified, the reachability set has been generated and subnet $i$, which is a child of $j$, has to be specified in detail. In the detailed view of $i$ the environment, which is defined by $j$, is represented by an aggregated view, i.e., a place $p_j^i$ and transitions $t_j^i$. The generation of the environment representation is performed in two steps. The first step is the transformation of the reachability set $M_j$ of $j$ in a set of markings for the place $p_j^i$ and the definition of transitions to realize dynamic behaviour of the environment. The second step is the aggregation of environment markings and transitions to get a smaller representation for the environment of a subnet. Let the aggregated view on $i$ in $j$ be represented by place $p_i^j$ and the corresponding transition $t_i^j$. The transformation of a marking $m \in M_j$ in a marking of the place $p_j^i$ is performed by introducing new colour domains. Each token of colour $c \in C(p_k^j)$ on place $p_k^j$ is represented by a token of colour $\langle c, p_k^j \rangle$ on $p_j^i$[1]. The colour domain of $p_j^i$ is given by $C(p_j^i) = \cup_{p \in P_j / p_i^j} \cup_{c \in C(p)} \langle c, p \rangle$. The transformation defines an equivalence relation between markings. Let $g(m) \in M(p_j^i)$ be the representation of marking $m \in M_j$.

[1] $\langle c_1, c_2 \rangle$ describes a new colour which is composed from the colours $c_1$ and $c_2$.

In a similar way the transitions $t_j^i$ can be specified. If transition $t_k^j$ ($k \neq i$) fires for colour $c \in C(t_k^j)$ transforming marking $m$ in $m'$, then $t_j^i$ fires for colour $\langle c, t_k^j \rangle$ transforming $g(m)$ in $g(m')$. The transformation of transitions is performed automatically on reachability set level.

The previous transformation yields a net which consists of two aggregated views on subnets, represented by the places $p_j^i$, $p_i^j$ and the transitions $t_j^i$, $t_i^j$, respectively[2]. However, $p_j^i$, $t_j^i$ and $p_i^j$, $t_i^j$ still include a complete specification of subnet $j$ since all markings are available. For the detailed specification of $i$ we need only to distinguish those markings of the environment which differ in their effects on the internal behaviour of $i$. Thus, the next step is the reduction of the set of markings and related transitions on $p_j^i, t_j^i$. This reduction of markings is nothing else then the generation of an aggregated view, an algorithm to compute aggregated views is given in the next subsection.

After generating the environment the aggregated view on subnet $i$ has to substituted by a detailed subnet specification. Obviously the detailed specified subnet has to behave according to the environment as defined by the aggregated view. The user has, of course, during specification the aggregated view in mind which describes the behaviour to meet by the detailed specification. Afterwards detailed and aggregated view are tested according to consistency as described below. For the moment we will assume that both are consistent.

From the detailed specification including the aggregated view of the environment and the initial marking $m_i^0$ (see below), the detailed reachability set of the subnet can be generated by standard means. What is still missing is the definition of a relation between a marking in the detailed view and a marking in the aggregated view. We assume that markings in the aggregated view describe disjoint subsets of markings in the detailed view. Thus the aggregated view describes an aggregation of the detailed view. The corresponding aggregation function $f(.)$ is usually defined on net level of the detailed view and should be independent from the marking of the environment $m(p_j^i)$. The following relation has to hold for all $m$ and for all $c$ for which $t_j^i$ can fire in $f(m)$

$$m > t_j^i(c) > m' \Rightarrow f(m) > t_j^i(c) > f(m')$$

The initial marking of the detailed subnet $m_i^0$ is a refinement of the initial marking of the parent net $m_j^0$. Thus, $m_i^0(p_j^i)$ is uniquely defined by transforming $m_j^0$ in the marking of place $p_j^i$ as described above and representing $m_i^0(p_j^i)$ as the aggregated description of the resulting marking. The initial marking $m_i^0$ of the detailed subnet $i$ has to be chosen such that $f(m_i^0) = m_j^0(p_i^j)$.

Obviously the aggregation function $f(.)$ is not unique and might range between a single marking and the representation of all markings of the detailed view in the aggregated view. The latter view destroys all advantages of the hierarchy since all details are visible. The former view, although often used,

---

[2] It should be noticed that this net is only an intermediate result which will not really be specified. All operations will be performed on the reachability graph which equals the reachability graph of HPN $j$ and is transformed to yield automatically a specification for $p_j^i$, $t_j^i$. Thus it does not matter that $p_j^i$, $t_j^i$ and $p_i^j$, $t_i^j$ belong to different HPNs

hides usually too much of the subnet behaviour which is relevant for the rest of the system. Normally $f$ can be defined according to some physical state of the modelled system (e.g., the number of outstanding requests, the number of active processes, the state of a counter, ...). According to $f$ $M_i$ is decomposed in disjoint subsets $M_i(n) = f^{-1}(n)$. Markings from a subset $M_i(n)$ are indistinguishable in the aggregated view. Similar to the reachability set the reachability graph $R_i$ is decomposed according to the aggregated view. Since the reachability graph describes transitions between markings we have to distinguish transitions which are visible in the aggregated view and transitions which are not visible in the aggregated view. Thus, let $R_i^n$ be the portion of the reachability graph which includes all transition between two markings $m, m' \in M_i(n)$. Transitions from $R_i^n$ are not visible in the parent net of subnet $i$. Transitions in subnet $i$ which are related to the firing of $t_j^i$ $(h(i) = j)$ for some colour $c$ are collected in $R_{i+}^{n,c}$. Transitions related to the firing of $t_i^j$ for some colour $c$ are collected in $R_{i-}^{n,c}$. Each subgraph $R_{i-}^{n,c}$ $(R_{i+}^{n,c})$ includes all transitions from the detailed view between markings from $M_j(n)$ and $M_j(n')$ with $n > t_i^j(c) > n'$ $(n > t_j^i(c) > n')$.

We come back to our example in Fig. 1. The initial marking of the HHPN consists of one grey and one white token originally in subnet 1. An aggregation of markings can be defined for both subnets 1 and 2 by counting the number of white and grey tokens actually in a subnet. However, we will see below that this view is not detailed enough to specify exactly the subnet behaviour. For the moment we assume that tokens enter a subnet and eventually leave the subnet without changing the colour. If a grey and a white token are in a subnet, then both can leave as the next one. This behaviour is very simple but not unusual for subnets of a complex net. With the specification of aggregated views for subnet behaviour we have completely specified the dynamic behaviour of subnet 0 and can compute the corresponding reachability graph. From these reachability graph the environment for subnet 1 and 2 can be computed and prepresented by $p_0^1, t_0^1$ and $p_0^2, t_0^2$, respectively.

### 3.2 Detailed and aggregated views on subnets

In this subsection aggregated and detailed views are formally related. There are two possible situations, the aggregated view is specified before the detailed view, then we have to check consistency between detailed and aggregated view, or the detailed view is available and we need an aggregated description to lower the complexity.

We first establish conditions between detailed and aggregated view to assure consistency. Interactions between subnet and environment have to be preserved when a detailed specification is substituted by an aggregated view. Therefore all transitions internal in a subnet (i.e., inside a subset $M_j(n)$) become invisible. A sequence of such invisible transitions is denoted by $\tau^*$.

Assume that we have a detailed description of some subnet $i$ and that the marking space of the subnet should be aggregated according to some subnet $j$ represented by $p_j^i$ and $t_j^i$ in $i$. This description covers both situation, that we want to check consistency between detailed and aggregated view for $i$, then $h(i) = j$, or that we want to generate an environment for $j$, then $h(j) = i$.

We introduce the following concept to assure equivalence between aggregated and detailed view. A transition is invisible in subnet $j$, if it equals the firing of $t_k^i$ for some colour $c$ such that $W^-(p_j^i, t_k^i)(c) = W^+(p_j^i, t_k^i)(c) = \emptyset$. The aggregated view has to preserve all possible interaction with subnet $j$, i.e.,

$$m^0 > \tau^* t_1 \tau^* t_2 \tau^* \ldots t_n > \Leftrightarrow \tilde{m}^0 > \tau^* t_1 \tau^* t_2 \tau^* \ldots t_n > \ .$$

The above condition is not sufficient for the analysis of liveness aspects (e.g., the reachability or non-reachability of markings). In this case we have to assure that after observation of a sequence of interactions between subnet and environment the possible future interactions between subnet and environment are the same using the aggregated or the detailed view, i.e.,

$$\tilde{m} > \tau^* t_1 \tau^* t_2 \tau^* \ldots t_n > \Leftrightarrow \forall m \in f_{in}^{-1}(\tilde{m})\, m > \tau^* t_1 \tau^* t_2 \tau^* \ldots t_n > \ .$$

The above condition considers only markings from $f_{in}^{-1}(\tilde{m})$ for the remaining markings in$f^{-1}(\tilde{m})$ we have to assure that no deadlocks that are not visible in the aggregated view occur. Such a situation is given when the detailed description is in a marking $m$ such that no $m'$ with $m \gg m'$ and $f(m') \neq f(m)$ exists, but the aggregated view allows a transition $\tilde{m} \gg \tilde{m}'$ for some $\tilde{m}' \neq \tilde{m}$.

The conditions are not easy to prove, however, since the reachability set is finite we can formulate local conditions which are easier to analyse and assure identical behaviour of aggregated and detailed view.

**Theorem 4.** *The aggregated and the detailed view of the behaviour of subnet i according to subnet j is completely consistent, if*

- $\forall m, m' \in f^{-1}(\tilde{m})\; m(p_j^i) = m'(p_j^i)$,
- $\tilde{m} > t > \tilde{m}' \Rightarrow \forall m \in f_{in}^{-1}(\tilde{m}) \exists m' \in f_{in}^{-1}(\tilde{m}')\; m > \tau^* t > m'$, *and*
- $\tilde{m} > t > \tilde{m}' \Rightarrow \forall m \in f^{-1}(\tilde{m}) \exists \bar{m} \notin f^{-1}(\tilde{m})\; m > \tau^* t' > \bar{m}$ *with* $t, t' \neq t_j^i$.

*Proof.* $m(p_j^i) = m'(p_j^i)$ has to hold since subnet $j$ has to be left unchanged . $\tilde{m}^0 > \tau^* t_1 \tau^* t_2 \tau^* \ldots t_n > \Rightarrow m^0 > \tau^* t_1 \tau^* t_2 \tau^* \ldots t_n >$ holds since for each path $\tilde{m}^1 > \tilde{m}^2 > \ldots > \tilde{m}^n$ in the aggregated marking space exists a path $m^{1_1} > \ldots > m^{n_1} > m^{2_1} > \ldots > m^{n_n}$ such that $f(m^{x_y}) = \tilde{m}^y$ and $m^{1_y} \in f_{in}^{-1}(\tilde{m}^y)$. By assumption for each marking $m \in f_{in}^{-1}(\tilde{m}^y)\; m > \tau^* t > m'$ with $m' \in f_{in}^{-1}(\tilde{m}^{y+1})$ holds, if $\tilde{m}^y > \tilde{m}^{y+1}$ holds in the aggregated model.
For the same reasons $\tilde{m} > \tau^* t_1 \tau^* t_2 \tau^* \ldots t_n > \Rightarrow m > \tau^* t_1 \tau^* t_2 \tau^* \ldots t_n >$ holds for all $m \in f_{in}^{-1}(\tilde{m})$.
The third condition assure that the detailed environment cannot deadlock, if the aggregated view allows a transition out of the actual marking. □

Aggregated and detailed view are defined by the user a priori during model specification and are afterwards tested according to the above conditions or the aggregated view is computed from the detailed view by an algorithm which should be presented now.

**Stepwise refinement algorithm to compute an aggregated view**
Choose an initial aggregation (e.g., aggregate $m, m' \in M$ with $m(p_j^i) = m'(p_j^i)$).
*repeat*
 *for* all $\tilde{m} \in \tilde{M}$ *do*

*if* the conditions of theorem 4 are not observed for all $m \in f^{-1}(\tilde{m})$ *then*
refine the aggregation by splitting $f^{-1}(\tilde{m})$ into new groups such that the conditions are observed
*until* theorem 4 holds for all $m$

The refinement step can be implemented using algorithms similar to those used to find the relational coarsest partition in a graph (see [2]). The result is an aggregated view with the smallest number of markings. Aggregation of markings can be performed completely algorithmic. If an aggregated view has to be modified after it has been used in the specification of other subnets, this might generate different environments for these subnets requiring an iterative specification.

If we now consider the aggregated views defined for the subnets 1 and 2 in our example, we have to notice that the straightforward specification, allowing a token eventually to leave whenever it resides in a subnet, is not completely consistent. Let us start with subnet 1. White and grey tokens might cycle via the places $p_4^1$, $p_6^1$ and $p_3^1$, $p_5^1$, respectively. However, such cycles have to be alternating, i.e., if a white token ends its cycle, it has to wait for a grey and vice versa. This behaviour is caused by the resource token on $p_7^1$ or $p_8^1$. Thus we might have the situation that a token cannot leave before the other token enters the subnet. Using the above algorithm to refine the aggregated view we get a completely consistent specification which includes one marking when both tokens are in the subnet and two markings if only one token is in the subnet, in the first aggregated marking the token possibly can leave the subnet or can enter the second aggregated marking, in the second aggregated marking it is blocked until the other token arrives in the subnet. The situation is similar for subnet 2. In this subnet a token might have reached a point, namely the places $p_7^2$, $p_8^2$, $p_9^2$, where it cannot be overtaken and is therefore definitely the next token to leave. The refinement algorithm generates an aggregated view which distinguishes whether a token and which token has reached such a state. Using the algorithm to generate an environment for the subnets 1 and 2 we get in both cases a behaviour which is completely equivalent to the aggregated view for subnet 2. Although the example is very simple it should be obvious that the approach works for more complex nets, as long as we can relate the inputs and outputs of subnets in some way.

It is sometimes convenient or necessary to define some weaker relations between aggregated and detailed view which also allow the analysis of the net but reduce the number of markings in the aggregated view yielding a more efficient analysis. This is especially important for subnets with a rather strong connection to their environment, such that the aggregated view has to include a lot of the details of the detailed view to be completely consistent according to the conditions in theorem 4 and it is also necessary in bottom up design where environments for subnets have to be defined without knowing an exact specification of their behaviour.

The idea is to define two aggregated views for a detailed specification, the first includes all transition which can occur between subnet and environment starting in one of the markings from a set of aggregated markings and a second view including those transitions which occur in each marking from a set of aggregated markings. Both aggregated views are formally defined as follows.

**Definition 5.** According to the aggregation of the detailed reachability set of subnet $i$ using aggregation function $f$, an aggregated view is

- almost consistent, if it results from the normal aggregation (i.e., $m > m' \Rightarrow \tilde{m} > \tilde{m}'$), and
- it is at least consistent, if $\tilde{m} > \tilde{m}'$ requires that for all $m \in f_{in}^{-1}(\tilde{m})$ a marking $m' \in f_{in}^{-1}(\tilde{m}')$ with $m > \tau^* t > m'$ exists and no marking $m \in f^{-1}(\tilde{m})$ exists such that $m > \tau^* t > m'$ is not possible for some $m' \notin f^{-1}(\tilde{m})$ and $t \neq t_i^j$ with $h(i) = j$.

If the aggregated view is completely consistent, then the almost and at least consistent view are identical. Often it is possible to include some additional information in the at least or almost consistent view, like the information that a marking is no deadlock marking although the at least consistent view includes no outgoing transitions, but one of the outgoing transitions of the almost consistent view will eventually be realized. Obviously consistent views are not unique. However, for a given function $f$ an almost consistent and an at least consistent aggregated view can be defined automatically, whereas a completely consistent view needs not to exist. An almost and an at least consistent view belonging to a specific aggregation will be denoted as weakly consistent aggregated view.

In our example we can use the original idea, counting the white and grey tokens for aggregation, to define at least and almost consistent aggregated views. For subnet 1 with one token, in the almost consistent view the token can leave and in the at least consistent view the token cannot leave. If two tokens are in subnet 1, then both colours can eventually leave, the at least and almost consistent view are identical. For subnet 2 with two tokens both tokens can leave in the at least consistent view, in the almost consistent view none of the tokens can leave, since we cannot assure whether there has been already the decision that a specific token will leave next. But we additionally get the information that no deadlock occurs, i.e., one of the tokens will eventually leave. For one token in subnet 1, at least and almost consistent view are identical and allow the token to leave the subnet. Weakly consistent environments for subnets 1 and 2 can be defined equivalently to the weakly consistent view for subnet 2.

If environment and children of a subnet are defined using weakly consistent aggregated view, then the almost consistent aggregated view results from using almost consistent aggregated views and the at least consistent view from using at least consistent views for the children and environment.

Let $M_j^+$ ($M_j^-$) be the reachability set of $j$ where all aggregated views are completely or at least (almost) consistent. $R_j^+$ and $R_j^-$ are the corresponding reachability graphs. If we define $M_j$, $R_j$ as the projection of the reachability set/graph of the complete (flat) net on the reachability set of subnet $j$, then $M_j^- \subseteq M_j \subseteq M_j^+$ and $R_j^- \subseteq R_j \subseteq R_j^+$. Thus the reachability set and reachability graph of the detailed subnet $j$ embedded in its environment can at least be bounded using weakly consistent aggregated views for the behaviour of children and environment. We denoted a HHPN as completely consistent if all included aggregated views are completely consistent, if at least one aggregated view is weakly consistent, then the HHPN is weakly consistent. If $M_j^- = M_j^+$, then also

$M_j$ is exactly known, additionally we know that whenever $M_j^+$ is finite also $M_j$ is finite.

If we use weakly consistent views defined above in our example, then $M_0^- = M_0^+$, $M_1^- = M_1^+$ and $M_2^- = M_2^+$.

## 4 Analysis of HHPNs

Analysis of HHPNs considers dynamic properties based on the reachable markings of the net. The usual distinction is between liveness (some subset of markings will eventually be reached) and safety (a specific subset of markings will never be reached). More formally we consider the following dynamic properties of a $HPN$ with reachability set $M$.

- A HPN is safe according to a set of safety properties $S = \{s_1, \ldots, s_n\}$, $s_i : M \to \{TRUE, FALSE\}$, if $s_i(m) = TRUE$ for all $m \in M$ and $s_i \in S$.
- A HPN is live for a set of liveness properties $L = \{l_1, \ldots, l_n\}$, $l_i : M \to \{TRUE, FALSE\}$, if for each $m \in M$ and each $l_i \in L$ a marking $m' \in M$ exists such that $m \gg m'$ and $l_i(m') = TRUE$.

The properties have been defined for flat HPNs, but each subnet of a HHPN is a flat HPN, if we consider aggregated views on subnets like ordinary places and transitions. Furthermore the complete HHPN or a part of it can be interpreted as a flat HPN. Two situations are distinguished in HHPNs. First, the definition of properties local to a specific view and, second, conditions which are global for all or some views.

### 4.1 Local Properties

We define a properties to be local, if it can be verified by analysing subnets without computing the combined reachability set of all or some subnets. If aggregated views are completely consistent with the detailed specification according to the externally observable behaviour of a specific part, a property which can be completely specified in one view can also be verified in this view.

Thus, liveness on net level in a completely consistent HHPN can be proved using only the subnet reachability sets, since the enabling of ordinary transitions depends only on markings visible in a subnet. The decision whether the initial marking is a home marking according to the most detailed view belonging to the flat specification of the HHPN depends on the complete reachability set of the net and is therefore a global liveness property. Local properties in a completely consistent HHPN are verified efficiently on the local reachability graph possibly during generation. Furthermore global properties might be negated locally. We define the restriction of a safety property $s(m_i)$ on the reachability set $M_i$ of subnet $i$ as $FALSE$, if no marking $m' \in M$ with $m'_i = m_i$ and $s(m') = TRUE$ exists, otherwise it is $TRUE$. The restriction of a liveness property $l(m_i)$ on $M_i$ is $FALSE$, if no $m' \in M$ with $m'_i = m_i$ and $l(m') = TRUE$ exists. A safety property $s$ restricted on $M_i$ does not hold, if a marking $m_i \in M_i$ with $s(m_i) = FALSE$ exists, otherwise it potentially holds. A liveness property $l$ restricted on $M_i$ does not hold, if a marking $m_i \in M_i$ exists such that no marking $m'_i \in M_i$ with $l(m'_i) = TRUE$ and $m_i \gg m'_i$ exists, otherwise it potentially holds. A liveness or safety property for a HHPN does not hold, if it does not hold in one of the subnets. If it potentially holds in all subnet, it might or might not hold for the complete HHPN.

The previous results are all based on complete consistency between aggregated and detailed view. If only weakly consistent aggregated views are known for non-local parts of the subnet, then the following theorem applies.

**Theorem 6.** *Let $M_j^+$ ($M_j^-$) and $R_j^+$ ($R_j^-$) be the reachability set and reachability graph for subnet j where all aggregated views are weakly consistent views for the corresponding detailed specifications, then the following results hold.*

- *If $M_j^- = M_j^+$ and a local liveness property l is TRUE on $R_j^-$, then l is also TRUE for the complete net.*
- *If a safety property s is FALSE for $M_j^-$, then it is also FALSE for the complete net.*
- *If $M_j^- = M_j^+$ and a liveness property l is FALSE on $R_j^+$, then l is FALSE for the complete net.*

*Proof.* The central idea of the proof is that $R_j^-$ includes a minimum number of transitions and $R_j^+$ a maximum number. Liveness cannot be destroyed by adding transitions without generating new markings, a safety property that does not hold cannot become valid by new transitions and by removing transitions a non-reachable marking cannot become reachable. □

In our example we can prove liveness on net level for all subnets using the isolated subnets and weakly consistent views, as defined above, for all non-local parts.

### 4.2 Global Properties

The analysis of global properties or local properties in weakly consistent HHPNs requires the complete or partial generation of the reachability set/graph from the isolated parts. On net level this is done by substituting each subnet by the most detailed view and generating the reachability set of the resulting net. However, it is much more convenient and efficient to consider the hierarchical structure. The idea is to generate the detailed reachability set bottom-up and reduce it in each step without loosing significant results. We assume that the properties which should be verified require the complete reachability set, a partial generation, which is often sufficient, follows immediately from the introduced concept. The negation of a global property during generation of the reachability set implies that the property does not hold and needs not to be considered in the subsequent steps. Since the complete HHPN is analysed bottom-up, the only aggregated view included in a subnet which is analysed describes the environment behaviour. If the aggregated view of the environment is weakly consistent, then the results of theorem 6 apply. In a completely consistent environment results can be verified or negated without further constraints, if they are defined locally in the actual view.

Let $S = \{s_1, \ldots, s_n\}$ be a set of safety properties and $L = \{l_1, \ldots, l_m\}$ be a set of liveness properties which have to be verified. Let $\mathcal{M}_0$ be the reachability set of the flat HPN underlying the HHPN and $\bar{\mathcal{R}}_0$ the reachability graph. $\bar{\mathcal{M}}_j, \bar{\mathcal{R}}_j$ are projections of the detailed reachability set/graph on the reachability set/graph of subnet $j$. $\mathcal{M}_j, \mathcal{R}_j$ are detailed reachability set/graph generated hierarchically as shown below. Each marking $m \in M_j$ describes a set of marking from $\mathcal{M}_j$ which are indistinguishable in the coarser view. The set $\mathcal{M}_j$ can be decomposed into subsets $\mathcal{M}_j(n)$ including all markings $m'$ which belong to a marking $m \in M_j(n)$. The sets $\mathcal{M}_j$ are computed recursively as

$$\mathcal{M}_j = \begin{cases} M_j & \text{if } h^{-1}(j) = \emptyset \\ \cup_{m \in M_j} \times_{i \in h^{-1}(j)} \mathcal{M}_i(m) & \text{else} \end{cases} \tag{1}$$

A detailed reachability graph $\mathcal{R}_j$ can be defined on $\mathcal{M}_j \times \mathcal{M}_j$. It is easy to verify that $\bar{\mathcal{M}}_j \subseteq \mathcal{M}_j$ and $\bar{\mathcal{R}}_j \subseteq \mathcal{R}_j$ holds for a completely consistent HHPN. In the sequel we assume that non-reachable markings are eliminated during or after generation of the composed reachability graph.

Like the original graph $R_j$ $\mathcal{R}_j$ is decomposed into disjoint subgraphs $\mathcal{R}_j^n$, $\mathcal{R}_{j+}^{n,c}$ and $\mathcal{R}_{j-}^{n,c}$. For computation of the graphs we have to assume a convenient representation amenable to appropriate operations. If $h^{-1}(j) = \emptyset$, then $R_j$ is the incidence matrix of the graph and $\mathcal{R}_j = R_j$. In the sequel we identify $\mathcal{R}_j$ with the incidence matrix of the graph. Reachability graphs $R_j$ of subnets in the higher levels of the hierarchy have labelled arcs. Each arc between two markings $m$ and $m'$ is labelled with an element $(i, c, \mathcal{T}_{i,c})$. $i$ denotes a transition which fires for colour $c$ transforming marking $m$ in $m'$. We use the notation $i \in h^{-1}(j)$ for a transition specifying the aggregated view on subnet $i$ which is a child of $j$. $\mathcal{T}_{i,c} = \{(k, c')\}$ is a set of subnets $k \in h^{-1}(j)$ $(k \neq i)$ which are affected by the firing of $i$ for colour $c$ and the effect is realized by firing $t_j^k$ for colour $c'$.

**Definition 7.** Let $A_1, A_2$ be two $n_i \times m_i$ $(i = 1, 2)$ matrices with elements $\{0, 1\}$, then the following operations are defined.

- The sum $C = A + B$ is defined for $n_1 = n_2$, $m_1 = m_2$ and equals a $n_1 \times m_1$ matrix with $C(i,j) = 1$ if $A(i,j) + B(i,j) > 0$ and 0 else.
- The product $C = AB$ is defined for $m_1 = n_2$ and equals a $n_1 \times m_2$ matrix with $C(i,j) = A(i,j)B(j,i)$.
- The tensor product $C = A \otimes B$ equals a $n_1 n_2 \times m_1 m_2$ matrix with $C((i_1 - 1)n_1 + i_2, (j_1 - 1)n_1 + j_2) = A(i_1, j_1)B(i_2, j_2)$.

All operation are associative. For the generation of $\mathcal{R}_j$ additionally the following two functions are needed for each $i \in h^{-1}(j)$ and each marking $m \in M_j$.

$$l_i^j(m) = \prod_{k \in h^{-1}(j), k < i} |\mathcal{M}_k(m)| \qquad u_i^j(m) = \prod_{k \in h^{-1}(j), k > i} |\mathcal{M}_k(m)|$$

$|\ldots|$ denotes the cardinality of a set. $u_0^j(m)$ equals $|\mathcal{M}_j(m)|$. Let $\mathcal{I}_n$ be the identity matrix of order $n$. Since $\mathcal{R}_j$ is computed bottom-up, the reachability graphs $\mathcal{R}_i$ $(i \in h^{-1}(j))$ are known during the computation of $\mathcal{R}_j$. The next step is to map transitions $M_j$ onto $\mathcal{M}_j$, this is done by definition of appropriate matrices (graphs) $\mathcal{S}_i^m$, $\mathcal{S}_{i-}^{m,c}$ and $\mathcal{S}_{k+}^{m,c'}$ (for $i \in h^{-1}(j)$ and $(k, c') \in \mathcal{T}_{i,c}$) describing the transformation of transitions from the local reachability set of a subnet into the detailed reachability set of the parent net (see [4, 5] for further details).

$$
\begin{aligned}
\mathcal{S}_i^m &= \mathcal{I}_{l_i^j(m)} \otimes \mathcal{R}_i^m \otimes \mathcal{I}_{u_i^j(m)} \\
\mathcal{S}_{i-}^{m,c} &= \mathcal{I}_{l_i^j(m)} \otimes \mathcal{R}_{i-}^{m,c} \otimes \mathcal{I}_{u_i^j(m)} \\
\mathcal{S}_{k+}^{m,c} &= \mathcal{I}_{l_k^j(m')} \otimes \mathcal{R}_{k+}^{m,c'} \otimes \mathcal{I}_{u_k^j(m^{*i})}
\end{aligned} \tag{2}
$$

where $i, k \in h^{-1}(j)$, $(k, c') \in \mathcal{T}_{i,c}$, $m'(l) = m(l) + W^+(p_l^j, t_i^j)(c) - W^-(p_l^j, t_i^j)(c)$, $m^{*i}(i) = m(i) + W^+(p_i^j, t_i^j)(c) - W^-(p_i^j, t_i^j)(c)$ and $m^{*i}(l) = m(l)$ for $l \neq i$.

For the generation of $\mathcal{R}_j$ the above matrices have to be combined appropriately. $\mathcal{R}_j$ can be decomposed into subgraphs $\mathcal{R}_j(m, m')$ $(m, m' \in M_j)$, each subgraph includes all transitions which describe a transition from $m$ to $m'$ in $R_j$.

$$
\mathcal{R}_j(m,m') = \begin{cases} 0 & \text{if } R_j(m,m') = 0 \\ \prod\limits_{(k,d)\in\mathcal{T}_{i,c}} \mathcal{S}_{k+}^{m,d} & \text{if } R_j(m,m') = (i,c,\mathcal{T}_{i,c}) \text{ and } i \neq h^{-1}(j) \\ \mathcal{S}_{i-}^{m,c} \prod\limits_{(k,d)\in\mathcal{T}_{i,c}} \mathcal{S}_{k+}^{m,d} & \text{if } R_j(m,m') = (i,c,\mathcal{T}_{i,c}),\ i \in h^{-1}(j), m \neq m' \\ \sum\limits_{i\in h^{-1}(j)} \mathcal{S}_i^m & \text{if } m = m' \end{cases} \tag{3}
$$

With the above equation the detailed reachability graph can be computed step by step, non-reachable markings are eliminated. Each $\mathcal{R}_j$ $(j \neq 0)$ can be decomposed into disjoint subgraphs $\mathcal{R}_j^n$, $\mathcal{R}_{j-}^{n,c}$ and $\mathcal{R}_{j+}^{n,c}$ according to the decomposition of $R_j$ (i.e., if $R_j(m, m')$ belongs to $R_j^n$ $(R_{j+}^{n,c}, R_{j-}^{n,c})$, then $\mathcal{R}_j(m, m')$ belongs to $\mathcal{R}_j^n$ $(\mathcal{R}_{j+}^{n,c}, \mathcal{R}_{j-}^{n,c})$). $\mathcal{R}_0$ equals the reachability graph of the flat HPN underlying the HHPN after elimination of unreachable markings.

The reachability sets of the subnets in our example include 32, 12 and 51 markings, respectively. Combining them according to the above formulas yields the complete reachability set of the HHPN with 278 markings.

### 4.3 Behaviour preserving aggregation of subspaces

The size of the reachability graph grows rapidly with an increasing number of subnets and levels. But the computation of the huge reachability graph $\mathcal{R}_0$ is only necessary if it is needed for the verification of the required properties. Usually it is possible to reduce the size by aggregation of markings during the recursive generation without loosing the possibility to verify the required properties.

In a reachability graph $\mathcal{R}_i$ the reachability set of the HHPN represented by the subtree with root node $i$ is considered in detail, the rest of the hierarchical net is represented by an aggregated view. Markings from $\mathcal{M}_i$ can only be aggregated, if they belong to the same subset $\mathcal{M}_i(n)$ otherwise the marking of the aggregated view for $i$ in the parent net would not be unique. Let $\tilde{\mathcal{M}}_i$ be the aggregated reachability set of subnet $i$ which can be decomposed into subsets $\mathcal{M}_i(n)$. Aggregation of markings is performed as described in section 3, the notation is also defined there.

Since markings are aggregated during computation of the detailed reachability graphs for a subnet we use unlabelled reachability graphs. The incidence matrix of the aggregated reachability graph $\tilde{\mathcal{R}}_i$ can be computed from $\mathcal{R}_i$ by adding all rows and columns belonging to markings which are represented by the same aggregated marking and setting the diagonal element to zero. $\tilde{\mathcal{R}}_j$ can be partioned in subgraphs like $\mathcal{R}_j$ and these subgraphs are used in the upper levels of the hierarchy to generate aggregated versions of reachability graphs. $\tilde{\mathcal{R}}_0$ is the aggregated reachability graph of the complete net. Due to the construction of the reachability graph in an upper level using reachability graphs from the lower levels (see (2) and (3)), a reduction of the reachability set of subnet $i$ by a factor $r$ normally yields a reduction of the reachability set for the parent net by a similar factor. This immediately shows the benefits of aggregation, reduction can be performed on small portions of the original reachability graph yielding reductions for the overall reachability graph. It is often worth to put some effort in the reduction of a subnet reachability set instead of analysing the huge reachability set of the overall net.

Assume that a safety property $s$ should be verified using the aggregated reachability graph. Let $n \in \mathcal{M}_0$ and $\tilde{n} \in \tilde{\mathcal{M}}_0$, $\tilde{\mathcal{M}}_0$ is the reachability set of the flat net where subnet $i$ has been aggregated. Let $n(i)$ be the part of the marking which describes the marking in the detailed subnet $i$ (i.e., all children of $i$ are represented by their most detailed specification). Then $n \in f^{-1}(\tilde{n})$, if and only if $n(i) \in f^{-1}(\tilde{n}(i))$ and $n(j) = \tilde{n}(j)$ for all $j \neq i$. In the following theorem we use the notation $m$ to describe markings of subnet $i$ which should be aggregated and prove the result for marking $n \in \tilde{\mathcal{M}}_0$ which is given by combining the aggregated subnet $i$ with the original subnets $j$. Since the aggregation preserves safety and/or liveness the results obviously hold if more than one subnet is aggregated according to the theorems below. A safety property $s$ is specified according to the aggregation as $s(\tilde{n}) = FALSE$, iff there exists a marking $n \in f^{-1}(\tilde{n})$ with $s(n) = FALSE$.

**Theorem 8.** *An aggregation on the reachability set of subnet i preserves a safety property s, if*

1. $m_i^0 \in f^{-1}(\tilde{m}) \Rightarrow \forall m \in f^{-1}(\tilde{m})\ m_i^0 > \tau^* > m$, *and*
2. $\tilde{m} > t > \tilde{m}'\ (\tilde{m} \neq \tilde{m}') \Rightarrow \forall m' \in f^{-1}(\tilde{m}')\ \exists m \in f^{-1}(\tilde{m})\ , m > t\tau^* > m'$.

*Proof.* $s$ does not hold for $\mathcal{R}_0 \Rightarrow s$ does not hold for $\tilde{\mathcal{R}}_0$:
A path $m^0 = n^0 > t > n^1 > \ldots > t > n^X$ exists such that $s(n^X) = FALSE$. Let $n^x \in f^{-1}(\tilde{n}^x)$. A path $\tilde{n}^0 > \tilde{t} > \tilde{n}^1 > \ldots > \tilde{t} > \tilde{n}^X$ exists in $\tilde{R}_0$ such that $\tilde{t} = \tau$ if $\tilde{n}^{x-1} = \tilde{n}^x$ and $t$ else. Furthermore $s(n^X) = FALSE \Rightarrow s(\tilde{n}^X) = FALSE$.
$s$ does not hold for $\tilde{\mathcal{R}}_0 \Rightarrow s$ does not hold for $\mathcal{R}_0$:
A path $\tilde{n}^0 > \tilde{t} > \tilde{n}^1 > \ldots > \tilde{t} > \tilde{n}^X$ exists such that $s(\tilde{n}^X) = FALSE$, $m^0 \in f^{-1}(\tilde{n}^0)$ and for some $n^X \in f^{-1}(\tilde{n}^X)\ s(n^X) = FALSE$ holds. The second condition implies that for each $n^x \in f^{-1}(\tilde{n}^x)\ (x = 1 \ldots X)$ a $n^{x-1} \in f^{-1}(\tilde{n}^{x-1})$ exists such that $n^{x-1} > \tau > n^{x_1} > \ldots > \tau > n^{x_Y} > t > n^x$ with $n^{x_1}, \ldots, n^{x_Y} \in f^{-1}(\tilde{n}^x)$. Furthermore for each $n \in f^{-1}(\tilde{n}^0)\ m^0 > \tau^* > n$ holds. Therefore a path $m^0 > \tau^* t \ldots t\tau^* > n^X$ with $s(n^X) = FALSE$ exists in $\mathcal{R}_0$. □

Aggregation according to liveness is more crucial since liveness requires that every possible path fulfills a specific condition. The liveness property $l$ is defined according to the aggregation as $l(\tilde{n}) = TRUE$, iff there exists a marking $n \in f^{-1}(\tilde{n})$ with $l(n) = TRUE$, otherwise $l(\tilde{n}) = FALSE$.

**Theorem 9.** *An aggregation of the reachability set of a subnet i with $h(i) = j$ preserves a liveness property l, if*

3. $\tilde{m} > t > \tilde{m}' \Rightarrow \forall m \in f_{in}^{-1}(\tilde{m}), \exists m' \in f_{in}^{-1}(\tilde{m}')$ *with* $m > \tau^* t > m'$,
4. $\tilde{m} > t > \tilde{m}'$ $(\tilde{m} \neq \tilde{m}',\ t \neq t_j^i)$ $\Rightarrow \forall m \in f^{-1}(\tilde{m})\ \exists t' \neq t_j^i\ \exists m' \notin f^{-1}(\tilde{m})$ *with* $m > \tau^* t' > m'$, *and*
5. $l(\tilde{m}) = TRUE \Rightarrow \forall m \in f^{-1}(\tilde{m})\ \exists m' \in f^{-1}(\tilde{m})$ *with* $m > \tau^* > m'$ *and* $l(m') = TRUE$.

*Proof.* $l$ holds for $\mathcal{R}_0 \Rightarrow l$ holds for $\tilde{\mathcal{R}}_0$:
Choose an arbitrary marking $n \in \mathcal{M}_0$ which is reachable from $m^0$. Since the original net is live a path from $n$ to a marking $n'$ with $l(n') = TRUE$ exists or $l(n) = TRUE$. Let $n \in f^{-1}(\tilde{n})$. By construction of the aggregated transitions, $\tilde{n}$ is reachable in $\tilde{\mathcal{R}}_0$ from $\tilde{m}_0$. If $l(n) = TRUE$ or $n' \in f^{-1}(\tilde{n})$, then $l(\tilde{n}) = TRUE$. Now let $n' \in f^{-1}(\tilde{n}')$, then a path between $n$ and $n'$ implies a path between $\tilde{n}$ and $\tilde{n}'$ in the aggregated reachability graph and $l(\tilde{n}') = TRUE$ since $l(n') = TRUE$.
$l$ holds for $\tilde{\mathcal{R}}_0 \Rightarrow l$ holds for $\mathcal{R}_0$:
Choose an arbitrary marking $\tilde{n} \in \tilde{\mathcal{M}}_0$ which is reachable from $\tilde{m}^0$. Since $m^0 \in f_{in}^{-1}(\tilde{m})$ it is connected to some $n^1 \in f_{in}^{-1}(\tilde{n}^1)$ (i.e., $m^0 > \tau^* t > n^1$). In the same way $\tilde{n}^{x-1} > t > \tilde{n}^x$ implies for $n^{x-1} \in f_{in}^{-1}(\tilde{n}^{x-1})$ that some $n_x \in f_{in}^{-1}(\tilde{n}^x)$ with $n^{x-1} > \tau^* t > n^x$ exists. Thus a marking $n \in f^{-1}(\tilde{n})$ is reachable.

Now assume that we have reached some $n \in f^{-1}(\tilde{n})$ and the aggregated net is live for $l$. If $l(\tilde{n}) = TRUE$, then there exists a $n' \in f^{-1}(\tilde{n})$ with $l(n') = TRUE$ and $n > \tau^* > n'$. If $l(\tilde{n}) = FALSE$, then $n > \tau^* t > n'$ with $n' \in f_{in}^{-1}(\tilde{n}')$ exists. Since the aggregated net is live, there exists a path from $\tilde{n}'$ to some $\tilde{n}''$ with $l(\tilde{n}'') = TRUE$. Using the arguments from above there has to exist a path from $n'$ to some $n'' \in f_{in}^{-1}(\tilde{n}'')$ and $n'' > \tau^* > \bar{n}$ with $l(\bar{n}) = TRUE$ which assures liveness of the original net. □

If combined safety and liveness properties should be analysed, then condition 5 in theorem 9 is automatically fulfilled using condition 2 of theorem 8. The above theorems describe conditions for a behaviour preserving aggregation of subnet markings spaces, they do not introduce a constructive method for aggregation. Such a method can be directly derived using a stepwise refinement approach. Stepwise refinement for behaviour preserving aggregation is not new, in [14] a similar idea is used to reduce the complexity of a protocol specification which results in an aggregated state space for the verification of safety and liveness properties.

**Stepwise refinement algorithm for aggregate construction**
Choose an initial aggregation (e.g., aggregate all markings from $\mathcal{M}_j(n)$).
*repeat*
  *for* all $\tilde{m} \in \tilde{\mathcal{M}}_j$ *do*
    *if* the conditions 1.-4. are not observed for all $m \in f^{-1}(\tilde{m})$ *then*
      split $f^{-1}(\tilde{m})$ into new groups such that 1.-4. are observed
*until* conditions 1.-4. are observed for all $m$

If only safety or only liveness properties should be proved, only the corresponding conditions are used in the algorithm. The algorithm obviously stops, in the worst case without any reduction of the reachability set. Aggregation of markings has been defined here for a completely consistent view of the environment. Since aggregation is performed according to local transitions which are completely defined, the approach is still valid using the almost consistent environment specification including a maximum number of transitions originated in the environment.

For the analysis of a HHPN the aggregation step is first performed for all subnets of the lowest level. Afterwards the reduced subnet reachability sets are combined in the level above, if this combination is necessary for the verification of the required properties. Then the combined marking spaces are aggregated and the resulting aggregated reachability sets can be used in the next higher level for the generation of the detailed reachability set. So it goes on until all properties have been checked, this can be done at least using $\tilde{\mathcal{M}}_0$ and $\tilde{\mathcal{R}}_0$. Aggregation and several steps of detailed reachability graph generation can be performed in parallel which is an important point for the analysis of complex systems. Identities and symmetries in the specification are exploited by analysing identical parts only once.

Applying the approach for our example we start with the reduction of the reachability sets for the subnets 1 and 2. In both cases the aggregated reachability graphs are equivalent to the completely consistent aggregated views defined for the subnets above, if we consider aggregation according to safety and liveness properties. The aggregated reachability graph $\tilde{\mathcal{R}}_0$ includes 87 markings instead of 278 markings in $\mathcal{R}_0$ without aggregation.

# 5 Conclusions

We have presented a class of hierarchical High Level Petri nets for the modular specification of complex parallel and concurrent systems. In contrast to other approaches for hierarchical PNs, each view on a net is executable and allowed refinements are not restricted a priori. Consistency between different views is analysed afterwards by testing detailed against aggregated views. Analysis of HHPNs is supported by the hierarchical structure since reachability set and reachability graph can be composed from much smaller parts, several analysis and reduction steps are performed locally on small parts reducing the size of the reachability sets in upper levels often significantly. The approach has been presented here in the framework of High Level Petri Nets, thus we can also use analysis approaches for HPNs on net level (e.g. reduction rules [10]) to decrease the complexity of subnets.

Obviously the underlying concept of hierarchical system specification and analysis is not restricted to HPNs, other paradigms for state based specifications can be handled similar and multi-paradigm specifications using different paradigms for different parts of a specification are also possible. The original motivation behind this research was the analysis of large Markov chains for quantitative system analysis [4, 5], in this area we have used the approach to combine stochastic Petri nets, queueing networks and stochastic automata in one model. We assume that in a similar way different, more functional oriented, specification techniques can be combined.

## References


1. J. Billington; Extensions to Coloured Petri Nets; *in Proc. of the Third Int. Workshop on Petri Nets and Performance Models, Kyoto 1989.*
2. T. Bolognesi, S.A. Smolka; Fundamental Results for the Verification of Observational Equivalence: A Survey; *in: H. Rudin, C. West (eds.), Protocol Specification, Testing and Verification VII, North Holland (1987) 165-179.*
3. W. Brauer, R. Gold, W. Vogler; A Survey of Behaviour and Equivalence Preserving Refinements of Petri Nets; *in G. Rozenberg (ed.), Advances in Petri Nets 90, LNCS 424, Springer 1990, pp. 1-46.*
4. P. Buchholz; A Hierarchical View of GCSPNs and its Impact on Qualitative and Quantitative Analysis; *Journal of Parallel and Distributed Computing, vol. 15, no. 2, July 1992, pp. 207-224.*
5. P. Buchholz; Hierarchies in Colored GSPNs; *in M. Ajmone Marsan (ed.), Application and Theory of Petri Nets 1993, LNCS 691, Springer 1993, pp. 106-125.*
6. G. Chehaibar; Use of Reentrant Nets in Modular Analysis of Colored Nets; *in K. Jensen, G. Rozenberg (ed.), High-Level Petri Nets. Theory and Application, Springer 1991.*
7. R. Fehling; A Concept of Hierarchical Petri Nets with Building Blocks; *in G. Rozenberg (ed.), Advances in Petri Nets 93, LNCS 674, Springer 1993.*
8. H.J. Genrich, K. Lautenbach; System modelling with high-level Petri nets; *Theoretical Computer Science, vol. 13, 1981, pp. 109-136.*
9. J.F. Groote, F. Moller; Verification of Parallel Systems via Decomposition; *in W.R. Cleaveland (ed.), CONCUR 92, LNCS 630, Springer 1992, pp. 62-76.*
10. M.S. Haddad; A Reduction Theory for colored Nets; *in G. Rozenberg (ed.), Advances in Petri Nets 90, LNCS 424, Springer 1990, pp. 209-235.*
11. P. Huber, K. Jensen, R.M. Shapiro; Hierarchies in colored Petri Nets; *in G. Rozenberg (ed.), Advances in Petri Nets 90, LNCS 483, Springer 1990, pp. 313-341.*
12. K. Jensen; Coloured Petri Nets and the invariant-method; *Theoretical Computer Science, Vol. 14, 1981, pp. 317-336*
13. K. Jensen; Coloured Petri Nets: A High Level Language for System Design and Analysis; *in G. Rozenberg (ed.), Advances in Petri Nets 90, LNCS 483, Springer 1990, pp. 342-416.*
14. S.S. Lam, A.U. Shankar; Protocol Verification via Projections; *IEEE Trans. on Softw. Eng., Vol. 10, 1984, pp. 325-342.*
15. R. Milner; Communication and Concurrency; *Prentice Hall, 1989.*
16. J.L. Peterson; Petri Net Theory and the Modeling of Systems; *Prentice Hall, Englewood Cliffs, 1981.*
17. W. Reisig; Petri Nets, An Introduction; *EATCS Monographs on Theoretical Computer Science, Springer 1985.*
18. Y. Souissi, G. Memmi; Composition of Nets via a Communication Medium; *in G. Rozenberg (ed.), Advances in Petri Nets 91, LNCS 483, Springer 1992, pp. 457-470.*
19. A. Valmari, M. Clegg; Reduced Labelled Transition Systems Save Verification Effort; *in W.R. Cleaveland (ed.), CONCUR 91, LNCS 527, Springer 1991, pp. 526-540.*
20. A. Valmari; Compositional State Space Generation; *in G. Rozenberg (ed.), Advances in Petri Nets 93, LNCS 674, Springer 1993, pp. 427-457.*

# Modeling Symmetric Computer Architectures by SWNs

G. Chiola, G. Franceschinis, and R. Gaeta

Dipartimento di Informatica
Università degli Studi di Torino
corso Svizzera 185
10149 Torino, Italy

**Abstract.** Timed and Stochastic Well-formed colored nets (SWNs) have been introduced as a good modeling tool for complex systems with inherently high degree of symmetry. Analysis and simulation algorithms allow the automatic exploitation of model symmetries to improve their efficiency. Fairly strong constraints are posed over the color definition syntax in order to support such automatic symmetry exploitation as compared to other high level Petri net formalisms. In this paper we derive several models of parallel computer architectures in order to show not only that the formalism is adequate for this class of applications, but also how the different types of symmetries can be mapped into the allowed specification formalism. From this set of case studies we conclude that SWNs are an "intermediate level" formalism, closer to the application domain than P/T nets, yet requiring some ingenuity and experience from the modeler in order to exploit their (high) potential.

## 1 Introduction

An ideal modeling tool for the study of complex systems such as parallel computer architectures has at least three main requirements that partially contradict one the other:

1. it should be based on a very simple and intuitive formalism, so that it can be used by people that are not expert modelers; a graphic representation is usually preferred by users if the formalism has the capability of conveying the main ideas in a few small pictures;
2. it should be powerful enough so that all basic problems and mechanisms in the selected application domain may be represented in a concise and efficient way;
3. it should be amenable to efficient analysis or (at least) simulation.

High level Petri nets have been pushed by several academic researchers as well as profit organizations as a good trade-off among these main requirements. In

[0] This work was supported by the ESPRIT–BRA project No.7269 "QMIPS" and the Italian MURST 40% project.

particular, the idea of folding the behavior representation expressed in net form by identifying several token types in the places is most effectively exploited for the representation of symmetric systems.

Although the proposed formalisms (Pr/T, CPN, etc.) have the power to handle the folding of inhomogeneous process behavior, up to the extreme case in which the complete system state is encoded in the data structure associated with a single token in a single place (manipulated by a single transition) [1], this power is never fully exploited in practice. The practical reason for stopping the folding process at a given point is that this folding in general shifts the complexity from the graphic structure to the textual description. Usually the only practical case in which the modeler feels him/herself confident to be able to master the complexity of color inscriptions in high level Petri net model is when a substantial amount of symmetry is inherent to the color definition part [2, 3]. Hence, well-formed colored nets (WNs) [4, 5] where only two basic forms of symmetries give the basis of color structures are not usually perceived as a severely restricted formalism for practical applications. On the contrary, the simplicity of their basic constructs might be seen as an advantage from the point of view of understanding the model specification by non experts. On the other hand, the syntactic restrictions of the formalism provide the condition for efficient analysis as well as simulation algorithms [6, 7] to deal with models of large systems.

Timed and stochastic WNs (SWNs) appear then as a very good modeling formalism for highly symmetric systems: their graphic representation can be kept small and easy to understand, their color structures do not "hide" complexity of the model behavior, the actual size of the system in terms of number of similarly behaving module components may be parametrized in the color class cardinality and may (in several cases) affect only marginally the complexity of the model analysis and/or simulation.

Most parallel computer architectures, on the other hand, tend to be extremely regular interconnections of (small or large quantities of) uniformly behaving components. The design and implementation is heavily based on such symmetry properties, so that it appears quite natural to base also the modeling techniques used for their behavioral and/or quantitative evaluation on the same symmetry properties for better efficiency.

The purpose of this work is to assess the adequacy of the SWN formalism for behavioral and performance modeling of typical problems and mechanisms found in parallel architectures. We carry out this assessment by developing and studying two models of classical problems in this domain. The rest of the paper is organized as a set of independent sections, each one devoted to the development of efficient SWN models for the study of different aspects and mechanism related to parallel processing. Each considered example is characterized by some intrinsic symmetry that is exploited using the SWN formalism in order to simplify both the model structure and its analysis/simulation. The last section resumes the main results obtained on these case studies and contains an organic discussion of the practical modeling power of the formalism for the particular application domain. For the sake of brevity we decided to sacrifice the self-containment of the paper by not including a definition of the WN/SWN formalism. The reader is referred to [4, 5] for such a definition.

## 2 Multilevel Fat Trees

The first model that we consider, represents an interconnection network similar to that used in the Connection Machine CM-5. This interconnection network, introduced by Leiserson in [8], is called *fat-tree.* The net structure is a $k$-ary tree, whose leaves are the processors while the internal nodes are switching elements. The tree structure is called *fat* because the branches of the tree closer to the root are "thicker" (i.e., they may support a larger number of communications in parallel).

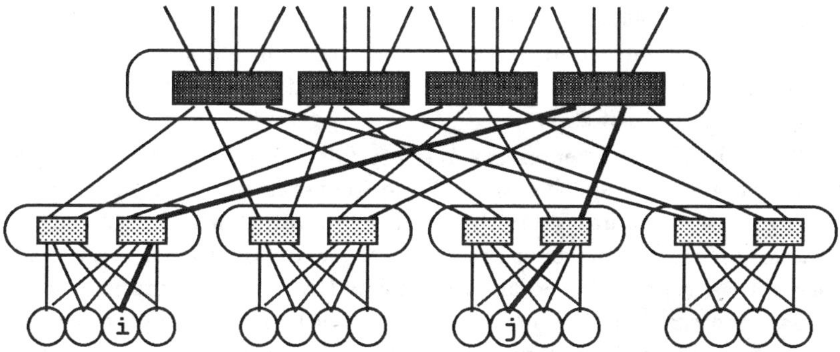


**Fig. 1.** A 4-ary fat tree (from [9], Fig. 8.29).

The CM-5 data network is implemented with a 4-ary fat tree. Figure 1 depicts such interconnection network: each internal node is made up of several router chips; when a process mapped on processor $i$ wants to communicate with a process mapped on processor $j$, the message climbs up towards the root of the tree until the closest common ancestor is found, then it descends towards the destination processor leaf.

The hierarchical structure of the network naturally suggests a hierarchical naming schema for the tree nodes: each node is uniquely identified by a pathname. Since we are dealing with a 4—ary tree, we can label all the nodes with common father with four different labels, e.g., $a$, $b$, $c$, and $d$, and then identify a generic node "n" with the sequence of labels associated with the nodes on the path from the root to "n". Consider the three-levels tree in Figure 2, the pathname for processor $j$ is $\langle a, a, d\rangle$, while the pathname for processor $i$ is $\langle a, b, c\rangle$. Their closest common ancestor is identified by the pathname $\langle a\rangle$. Notice that the pathname of the closest common ancestor of two nodes corresponds to the longest common prefix in their pathnames. This encoding of the net nodes eas-

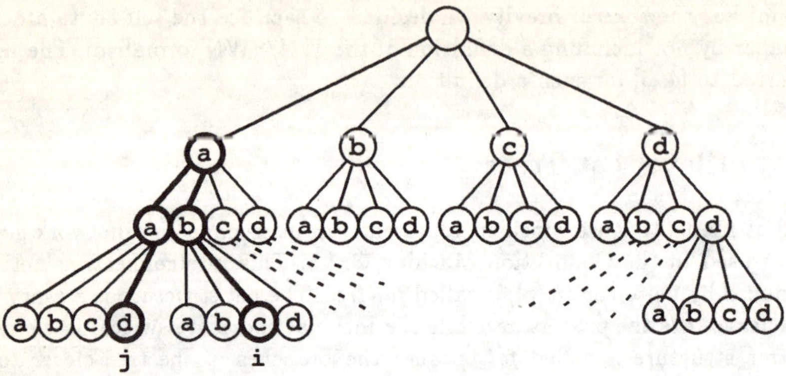


**Fig. 2.** Naming nodes through pathnames in a 3 level 4-ary tree.

ily translates in a specification for the color structure of the SWN model of the interconnection network: $k$-ary trees of depth $D$ require a single (non ordered) color class $C$ of cardinality $k$, and a token representing a node at depth level[1] $j$ $(1 \leq j \leq D)$ is a tuple of arity $j$.

The SWN in Figure 3 represents a 4—ary fat tree of depth three. The model includes a very simple specification of the messages workload: each processor "p" runs one process that performs local computation for an average time of $1/\lambda$ (transition *start* represents the end of a local computation), then it sends a message to a randomly chosen destination (the firing of immediate transition *choose* represents the choice of a destination); the probability of chosing a given destination is uniformly distributed among all remaining processors in the system.

The presence of a 6-tuple $\langle s_3, s_2, s_1, r_3, r_2, r_1 \rangle$ in place *level*3 means that a message is being sent from processor $\langle s_3, s_2, s_1 \rangle$ to processor $\langle r_3, r_2, r_1 \rangle$. The routing of a message is modeled by a three-level subnet. The level $i$ subnet contains two places $level_i$ and $comm_i$, and two transitions $sc_i$ and $ec_i$. Transitions $climb_j$ represent the search for the sender-receiver common ancestor (predicates are associated with these transitions that compare the prefixes of the two nodes up to the considered level). Places *rootlink*, *link*32, *link*21 and *leaves* represent the resources needed to set up the communication between the sender and receiver processors: the initial marking of these places is set to the maximum number of concurrent communications that can pass through a switch at a given level. For example the initial marking of place *rootlink* is 8, hence at most 8 communications that need to be routed through the root may be concurrently served. Place *link*21 initially contains two copies of each identifier $\langle s_3, s_2 \rangle$ representing a switch at level two meaning that at most two communications can be concurrently routed through the switch $\langle s_3, s_2 \rangle$. Finally, place *link*32 initially contains four copies of each identifier $\langle s_3 \rangle$ representing a switch at level one

[1] The depth level of the root is 0, while for the leaves it is $D$.

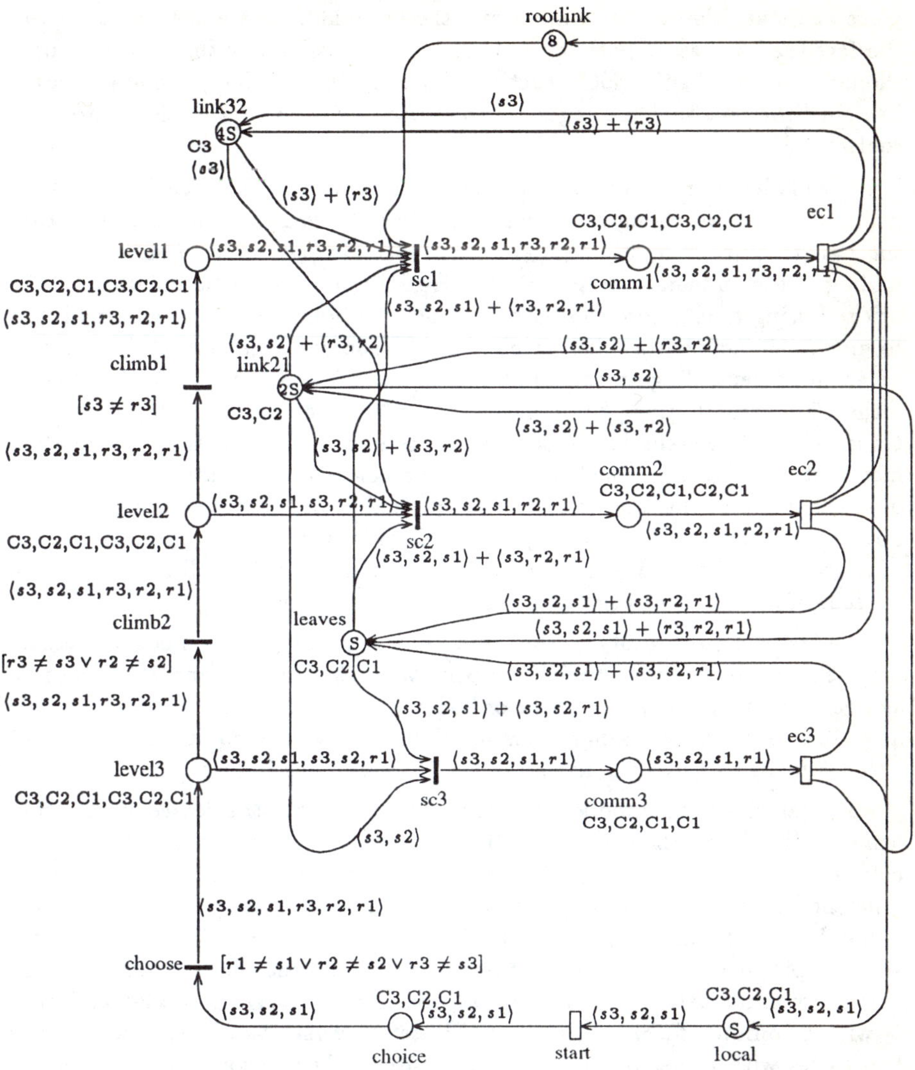


$$C1 = \{a, b, c, d\}; \quad C2 = \{a, b, c, d\}; \quad C3 = \{a, b, c, d\}$$

**Fig. 3.** First SWN model of the messages routing in a 3 level 4-ary fat-tree.

meaning that at most four communications can be concurrently routed through the switch $\langle s_3 \rangle$.

Observe that any $k$-ary fat tree with three levels can be modeled by the same net structure, changing only the definition of the basic color class $C$, so that $|C| = k$. On the contrary, $k$-ary fat trees with $d \neq 3$ levels require that the place and transition color domains and the arc functions are changed to reflect the fact that processor pathnames have length $d$; moreover the routing subnet should comprise $d$ submodels, each representing the climbing of one tree level, to model the fact that messages may need to climb up to $d$ levels towards the root.

The model is thus fully parametrized with respect to the arity $k$ of the tree, while its structure changes when the number of levels $d$ changes. Notice that the net structure is highly regular with respect to parameter $d$ as well, and a folding of the subnets representing the tree climbing could be obtained at the price of using much more complex arc functions (guarded functions are needed to model the fact that a different number of resources —i.e., switches— have to be acquired depending on the number of climbed levels). In the SWN in Figure 3 there are actually three color classes: $C1$, $C2$, $C3$ all of cardinality four: class $Ci$ is used to represent labels of nodes at level $d - i + 1$. This is to stress the fact that the algorithm has to consider the colors used to identify objects at different levels of the tree as independent. If the three color components were not considered as independent by the SRG construction algorithm, the degree of aggregation achieved would be substantially decreased. A discussion on the concept of independent color components can be found in [10].

We first ran our prototype SRG computation program on a model representing a 2-level binary tree, obtaining a SRG with 184 tangible and 731 vanishing symbolic markings. The size of the ordinary RG for this model was 505 tangible and 2020 vanishing markings (thus obtaining an average compaction factor of about 3 by using the SRG instead of the RG technique in this case).

By analyzing the symbolic markings we observed that the degree of reduction achieved was lower than expected. This was due to the fact that the specified color structure introduces an unnecessary dependence between nodes that have different fathers but same associated color characterizing the postfix of the pathname. Let us explain the problem on the following example. Consider the three states depicted in Figure 4: all of them represent a situation in which two concurrent communications are in progress, the former involving two leaves $l_1$ and $l_2$ with a common father $f_1$, while the latter traverses two levels and involves two leaves with father different from $f_1$. Intuitively we would like to consider these markings as equivalent. Instead the SRG keeps the first two in the same symbolic marking while it associates the third with a different symbolic marking. The equivalence relation that we have in mind relates any two states that can be obtained one from the other by permutation of the nodes at the same level with the only constraint of preserving the father-children connection. The equivalence relation used by the SRG generation algorithm, instead, adds a further constraint due to the fact that the permutations used to define the symbolic markings operate at the basic color class level. Since leaves with different fathers are associated with labels from the same class $C1$ in the model, an example of allowed permutation is the exchange of objects $a$ and $c$ in class $C1$, that has the effect of simultaneously permuting nodes $\langle x, y, a\rangle$ and $\langle x, y, c\rangle$ for all $\langle x, y\rangle \in C3, C2$, thus preventing the possibility of exchanging at the same time

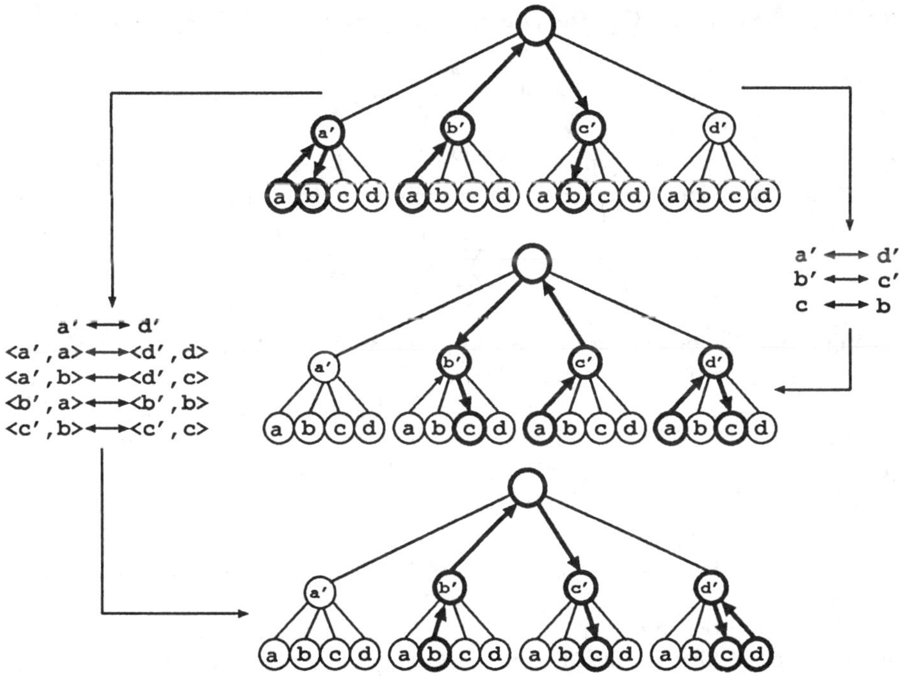


**Fig. 4.** Equivalent markings.

$\langle x', y', a\rangle$ and $\langle x', y', b\rangle$ for some $\langle x', y'\rangle$ in $C3, C2$.

An easy solution to this problem is obtained as follows. Define class $Ci$ (representing the nodes at level $d-i+1$) as a color set of cardinality $k^{d-i+1}$, and then represent the father-children relation of the tree structure in the initial (symbolic) marking. For example in the SWN model of a 3 level 4-ary tree depicted in Figure 5, color classes are defined as follows: $C1 : |C1| = 64$, $C2 : |C2| = 16$, $C3 : |C3| = 4$. Moreover places *conn*21 represents the father-children relation at the leaves level (i.e., a pair $\langle x, y\rangle, x \in C2, y \in C1$ is in this place iff $x$ is father of $y$) while place *conn*32 represents the father-children relation one level above. The initial marking of these places is defined in a symbolic form: class $C1$ is partitioned into 16 *dynamic subclasses* ($Z_1^i, i = 1, \ldots, 16$) of cardinality four[2] (objects in the same dynamic subclass represent siblings); 16 dynamic subclasses ($Z_2^i, i = 1, \ldots, 16$) of cardinality one are defined for class $C2$; 4 dynamic subclasses ($Z_3^i, i = 1, \ldots, 4$) of cardinality one are defined for class $C3$. The initial

[2] The cardinality completely characterizes a dynamic subclass; a partition of a class in dynamic subclasses denotes *any* partition satisfying the given subclass cardinality constraints.

marking of place *conn*21 is then defined as follows:

$$\mathcal{M}_0(conn21) = \sum_{i=1}^{16} \langle Z_2^i, Z_1^i \rangle$$

meaning that the object represented by subclass $Z_2^i$ is father of the four objects in subclass $Z_1^i$. Place *conn*32's initial marking is defined as:

$$\mathcal{M}_0(conn32) = \sum_{i=1}^{4} \sum_{j=1}^{4} \langle Z_3^i, Z_2^{4(i-1)+j} \rangle$$

meaning that the object represented by subclass $Z_3^1$ is father of the four objects represented by subclasses $\{Z_2^1, \ldots Z_2^4\}$, etc.

The initial symbolic marking represents several initial ordinary markings corresponding to all the possible ways of connecting labeled nodes in a father-children relation consistently with the desired tree structure. As a consequence, the SRG represents $|\mathcal{M}_0|$ ordinary reachability graphs (the different RGs are not connected with each other).

This new representation allows us to achieve the goal of allowing independent permutations within sets of siblings associated with different fathers. Applying the SRG computation algorithm to an SWN model of a 2-level binary tree represented in the new proposed form, 113 tangible and 301 vanishing symbolic markings are generated. The model can be further refined to reduce the number of vanishing markings by "fusing" immediate transitions *choose*, *climb*2 and *climb*1 into timed transition *start* using the transformations described in [11]. The reduced model of the two-level binary tree has 113 tangible and 73 vanishing symbolic states, thus yielding a compaction factor of about 4.5 on tangible markings.

The solution of a 2-level ternary tree was already beyond the possibilities of our SRG construction program prototype (the current implementation is rather inefficient). Even if the SRG generation did not complete for time and space limits, we could analyze the SRG portion generated (that included 9232 tangible and 5447 vanishing symbolic states) to evaluate the level of aggregation achieved. The initial symbolic marking represents 1680 ordinary markings, while the symbolic markings average cardinality is in the order of one million of states. Hence the average compaction factor (i.e., the ratio between number of states in the ordinary RG and number of symbolic states in the SRG), obtained by dividing the average symbolic marking cardinality by $|\mathcal{M}_0|$ gives about 700 (please compare with the compaction factor of about 4.5 for the 2-level binary tree). Similar levels of improvement can be obtained in simulation by using the symbolic simulation technique instead of the usual event-driven simulation for colored Petri nets.

## 2.1 Discussion

In this section we have developed a modeling exercise with the aim of highlighting an aspect of SWN color structure definition that may influence the degree

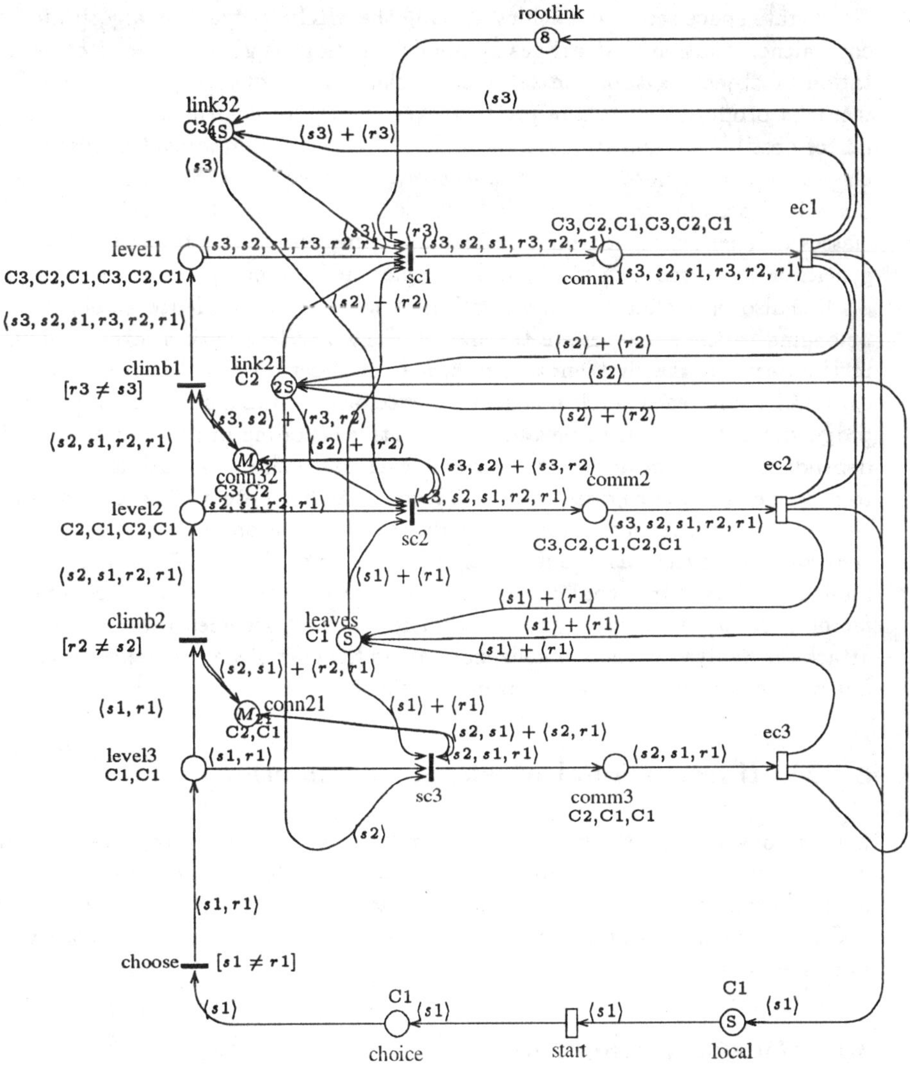


$$C1 = \{aaa, aab, \cdots, ddc, ddd\}; \quad C2 = \{aa, ab, \cdots, dc, dd\}; \quad C3 = \{a, b, c, d\}$$

**Fig. 5.** SWN model of 3 level 4-ary "fat tree" interconnection capturing all permutation symmetries of state space reduction achieved using the SRG construction algorithm. The equivalence relation that defines symbolic markings is given in terms of permutation of objects within the same color class, as a consequence the reduction factor is proportional to the product over the set of color classes of the number of possible permutations in each class. Hence it is important to distinguish classes used to denote different objects even if they have the same cardinality. Since it is possible to automatically check on the model structure for the color classes independence, it would be helpful to implement this feature in the SRG generator to overcome a source of incomplete reduction.

It is also important to avoid creating unnecessary dependence among objects belonging to the same set. In the model of a $k$-ary tree with $d$ levels described in this section, the first encoding chosen for describing nodes used classes of cardinality $k$ for each level. As a consequence only a maximal potential reduction proportional to $(k!)^d$ is achievable. Indeed, this encoding introduced an artificial dependence between nodes at the same level. By defining a cardinality $k^i$ class to describe the set of nodes at level $i$, a much higher potential reduction degree can be achieved. In our example, the degree of reduction achieved on a ternary tree with two levels was about 36 using the first model and 700 using the second model. For larger tree configurations (4-ary trees of depth $\geq 2$) the compaction factor becomes enormous, so that its exploitation becomes the only hope to attack the analysis of such models, not only in the case of state space enumeration but also in the case of discrete event simulation.

# 3 Multidimensional Mesh interconnection

Let us consider MIMD distributed-memory architectures with $n$-dimensional mesh interconnection structure, possibly having a different number of processors in each dimension. Figure 6 depicts a 3-dimensional structure with a number of nodes in the vertical direction different from the number of nodes in the other two directions.

## 3.1 Modeling assumptions

Let us assume that the workload is evenly distributed across all the processing elements and the computations assigned to each processor entail interaction with the local neighbors only [12]. The computation assigned to each processor consists of a set of $N \geq 1$ concurrent tasks performing a cyclic activity and one server task. The cyclic activity of the identical tasks includes a local operation phase (whose duration is a random, exponentially distributed variable with parameter $\lambda$), followed by a request for service addressed with equal probability to one of the neighbors. Service time is also a random, exponentially distributed variable with parameter $\mu$. Furthermore let us consider two interaction policies:

- *preemptive interaction*: requests coming from the neighbors are served with higher priority than the local activity, preempting any computation task; no

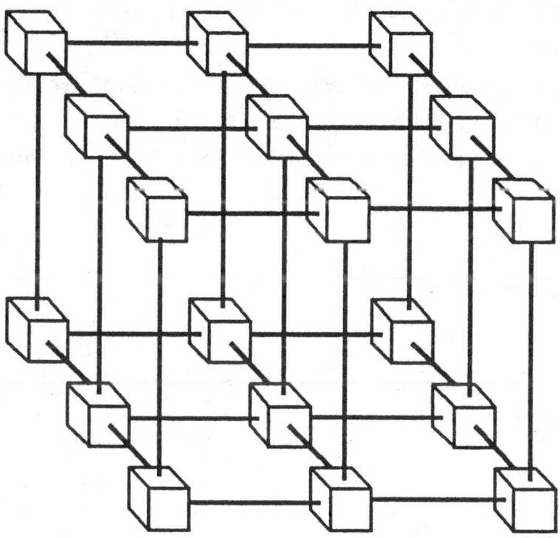

```
Computation task:
 while TRUE do
     local_computation;
     neighbour:= select(random);
     send_request(neighbour);
     receive(neighbour);
 od

Server task:
 while TRUE do
     receive(neighbour);
     compute;
     send_answer(neighbour);
 od;
```

(a) (b)

**Fig. 6.** (a) $3 \times 3 \times 2$ mesh (b) workload description

particular local task scheduling policy is specified: only the priority of server processes over local tasks is assumed;

- *non-preemptive interaction*: requests are not preemptive, but are served at the end of the local operation with higher priority with respect to the local computation tasks; this policy implies that local tasks are run by the processor up to their completion, so that for instance round-robin scheduling is excluded; on the other hand, no further scheduling policy assumption is made besides the fact that local tasks are run one at a time, without interruption up to their completion.

We assume that communication delays are negligible in our model. The addition of two delay transitions to take them into account would be however straightforward.

As an example of performance measure we can consider the node utilization for different levels of multiprogramming $N$.

### 3.2 Modeling the interconnection symmetry

The mesh structure depicted in Figure 6 looks extremely regular. Despite its regularity it is not fully symmetric. Internal nodes in the structure show indeed the same interconnection pattern with their nearest neighbor nodes. However part of the regularity is lost when we consider the "boundary" nodes, which show a reduced number of connections compared to the internal nodes.

A "slight" modification of this interconnection schema is the folding obtained by addition of one link connecting the last node to the first one in each dimension.

In this way we obtain an interconnection schema called ($n$-dimensional) torus, which is instead completely symmetric.

For the sake of simplicity let us first consider the case of 1-dimensional torus, i.e., a set of $m$ processing units connected in a ring. Let us consider a very simple WN model that represent the choice of one neighbor from a node that wants to communicate. This (trivial) WN model is depicted in Figure 7. Each

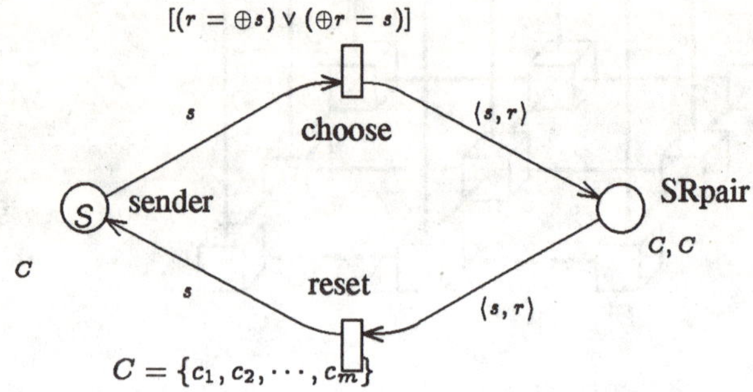


**Fig. 7.** Basic color definitions for nearest neighbor identification in 1-dimensional torus (ring).

processing node is uniquely associated with one element of the ordered basic class $C$. The predicate associated with the enabling of transition "choose" restricts the identity of the receiver to be either the previous or the next element with respect to the sender.

The symmetry of the structure is captured by the WN model that allows a reduction of $1/|C|$ in the number of symbolic markings with respect to the number of ordinary markings. The use of SWN models for $n$-dimensional torus topologies (composed as explained later on by Cartesian product of the color class $C$) yields a compaction factor of $1/\prod_1^n |C_i|$, and can thus be considered natural and efficient at the same time; indeed it results in models that are parametric on the number of nodes, whose color structure is very easy to understand, and that are efficient to analyze using either the SRG or the symbolic simulation technique.

The correct treatment of boundary conditions for mesh interconnections requires the introduction of predicates that distinguish inner node identities from boundary node identities. A careless and direct introduction of such predicates completely destroys all symmetry properties of the WN model, thus giving rise to models that are still easy to draw and describe, somewhat cumbersome to understand for what concerns the treatment of boundary conditions, and impossible to analyze efficiently. Of course, one might "approximate" the behavior of a mesh interconnection by ignoring the anomalies due to the boundary nodes and using the torus model instead of the mesh model. In this example, however, we are interested in representing such boundary anomalies in an accurate way in order to compare the performance of mesh versus torus interconnection. We first introduce a direct representation of the structure in terms of color structure that has all the problems just mentioned, and then refine the model in order to be able to capture and exploit those symmetries that are anyhow present in the system.

The case of 1-dimensional mesh can be obtained by refining the predicate associated with transition "choose" in order to distinguish the two boundary nodes from the others. This WN model is depicted in Figure 8. The ordered

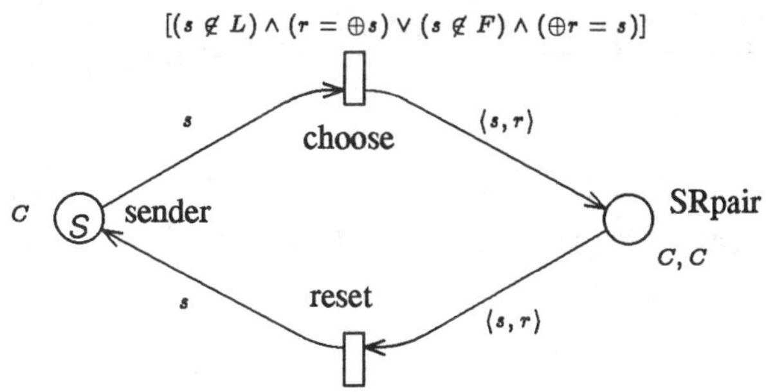


$C = F \cup M \cup L; \quad F = \{c_1\}; \quad M = \{c_2, \cdots, c_{m-1}\} \quad L = \{c_m\}$

**Fig. 8.** Basic color definitions for nearest neighbor identification in 1-dimensional mesh.

class $C$ is now partitioned in three subsets, so that it is possible to identify the first (the only element of subclass $F$) and the last (the only element of subclass $L$) elements. The predicate associated with the enabling of transition "choose" restricts the identity of the receiver to be either the previous or the next element with respect to the sender identity for internal nodes. Boundary nodes have only one neighbor (the next for node $c_1$ and the previous for node $c_m$).

The extension to higher dimension levels is trivial both from a conceptual and a practical point of view. Figure 9 depicts the color structures and functions for the 3-dimensional case. The predicate restricting the enabling of transition "choose" increases in complexity in order to handle the boundary condition in each dimension. The mapping of the interconnection structure into color structures remains however straightforward to understand.

Unfortunately, the partitioning of the ordered color classes (that is necessary to distinguish boundary conditions by means of predicates) destroys all useful model symmetries: symbolic marking representations collapse to ordinary marking representation, thus making the complexity of the model analysis to increase dramatically with both the number of dimensions and the size in each dimension. On the other hand, some kind of symmetries are still enjoyed by our interconnection schema, even if they are not directly captured by the WN formalism in Figures 8 and 9. The kind of symmetry that we should model is, in each dimension, the symmetry with respect to the central point of a segment.

$$[((s1 \not\in L1) \wedge (r1 = \oplus s1) \vee (s1 \not\in F1) \wedge (\oplus r1 = s1)) \wedge (s2 = r2) \wedge (s3 = r3) \vee$$
$$((s2 \not\in L2) \wedge (r2 = \oplus s2) \vee (s2 \not\in F2) \wedge (\oplus r2 = s2)) \wedge (s1 = r1) \wedge (s3 = r3) \vee$$
$$((s3 \not\in L3) \wedge (r3 = \oplus s3) \vee (s3 \not\in F3) \wedge (\oplus r3 = s3)) \wedge (s1 = r1) \wedge (s2 = r2)]$$

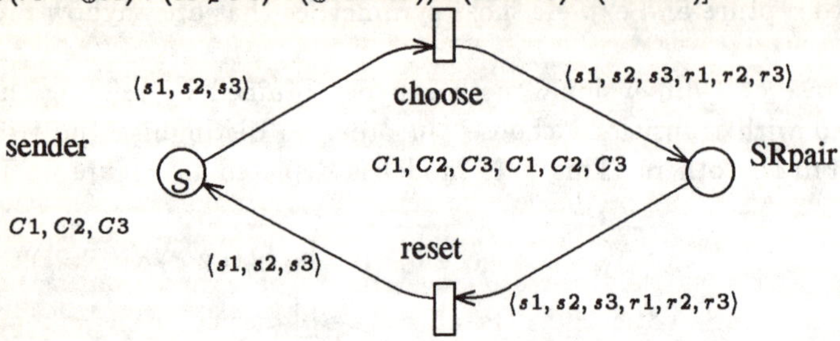


$$C1 = F1 \cup M1 \cup L1; \quad F1 = \{a_1\}; \quad M1 = \{a_2, \cdots, a_{m_a - 1}\} \quad L1 = \{a_{m_a}\}$$
$$C2 = F2 \cup M2 \cup L2; \quad F2 = \{b_1\}; \quad M2 = \{b_2, \cdots, b_{m_b - 1}\} \quad L2 = \{b_{m_b}\}$$
$$C3 = F3 \cup M3 \cup L3; \quad F3 = \{c_1\}; \quad M3 = \{c_2, \cdots, c_{m_c - 1}\} \quad L3 = \{c_{m_c}\}$$

**Fig. 9.** Basic color definitions for nearest neighbor identification in 3-dimensional mesh.

The "mirror" symmetry that we need in order to efficiently handle this case is not directly supported by the WN formalism, but it can be "simulated" as a special case of "rotation" (modeled by ordered classes) in which only two basic objects are considered. Figure 10 depicts a more sophisticated way of modeling the linear connection using the WN formalism. The "coordinate" of each

$$[(sd \not\in L) \wedge (rd = \oplus sd) \wedge (rr = sr) \vee (sd \not\in F) \wedge (\oplus rd = sd) \wedge (rr = sr) \vee (sd \in F) \wedge (rd = sd) \wedge (rr = \oplus sr)]$$

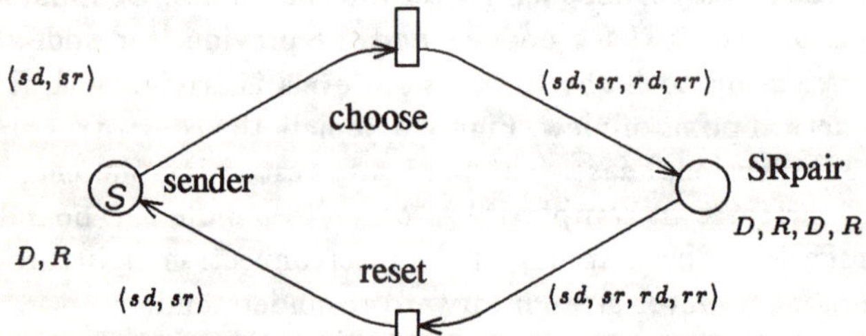


$$D = F \cup M \cup L; \quad F = \{d_1\}; \quad M = \{d_2, \cdots, d_{m-1}\} \quad L = \{d_m\}; \qquad R = \{l, r\}$$

**Fig. 10.** Basic color definitions for nearest neighbor identification in 1-dimensional mesh capturing specular symmetry.

processing node is encoded using a pair of ordered basic classes, the first one representing the distance from the center of the segment, and the second one the direction (left or right) with respect to the center. The class representing the distance is still partitioned in subclasses to identify the first and last elements. The class representing the direction is instead not partitioned, thus inducing actual symmetric folding of the symbolic marking representation.

The extension from 1-dimensional to $n$-dimensional mesh interconnections may be as straightforward in principle as in the first model in which symmetries were not captured: one may just replicate the pair of basic classes as many times as the number of dimensions (thus also expanding the complexity of the predicate associated with transition "choose").

The degree of compaction of the symbolic marking representation is only a factor of 2 in the case of 1-dimensional mesh. It increases however exponentially with the dimension, thus providing a factor $2^n$ of compaction in the case of $n$-dimensional mesh. Notice that the case of "$n$-dimensional hypercube" in which the number of nodes per dimension is fixed to 2 is particularly advantageous from the symmetric folding point of view: the color class $D$ can be omitted and the coordinate of a node reduces to a set of $n$ colors of type $R$ for which complete symmetry is available.

Notice also that, even in the case of completely homogeneous local computation workload that we are examining, the compaction factor $2^n$ is the best result that we may obtain. Indeed, processing nodes at different distances from the center (or, equivalently, from the nearest boundary) are subject to different communication loads (due to the different number of neighbors on the boundary), thus making the exact aggregation of nodes at different distances impossible from a performance evaluation point of view.

In the case of equal number of components in two or more dimensions, however, another form of permutation symmetry with respect to the ordering of dimensions may be observed. In other words, if we focus on a 2-dimensional case with $2m$ homogeneous nodes in each dimension, we can observe a new form of symmetry corresponding to a flip of the structure with respect to either main diagonal. This additional symmetry can be properly taken into account at the price of some additional complication in the encoding of the coordinates of each node and in the predicate that establishes the nearest neighbor connection. In particular, the identity of a node may be encoded by means of a 5-tuple of type $R, R, R, D, D$ of the two following basic color classes:

$$D = F \cup M \cup L; \quad F = \{d_1\}; \quad M = \{d_2, \cdots, d_{m-1}\} \quad L = \{d_m\};$$

$$R = Z \cup NZ; \quad Z = \{0\}; \quad NZ = \{-1, 1\};$$

the first one ordered (with a successor function defined among elements), and the second one not ordered. The first two elements in the encoding identify one among four "quadrants" in the mesh. The third element is zero for the nodes laying on the main diagonals, and either 1 or $-1$ for the other nodes (see Fig. 11). Hence the three components of type $R$ are used to encode one out of eight "sectors". The last two elements of type $D$ identify a specific node in a sector, by encoding its "latitude" and "longitude"; for instance in Fig. 11 the encoding of node **a1** is $\langle 1, 0, 0, 3, 1\rangle$, that of node **f2** is $\langle 0, 1, 1, 2, 2\rangle$, and finally that of node **b6** is $\langle 0, -1, -1, 3, 2\rangle$.

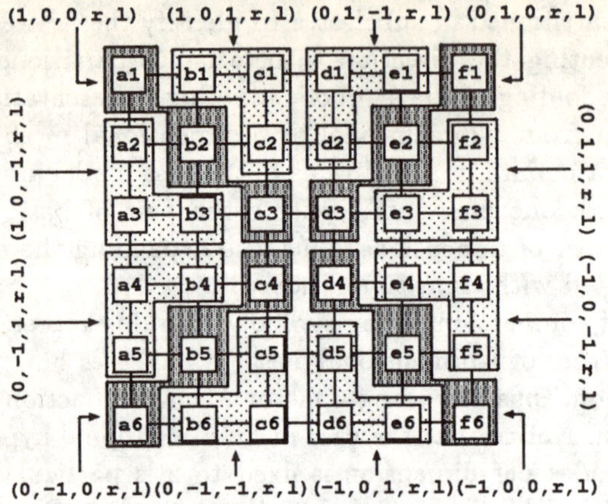


**Fig. 11.** Encoding of nodes of a 2-dimensional structure.

A valid identification of a node is subject to the following restrictions: call $\langle dx, dy, dd, r, l\rangle$ a color instantiation corresponding to a node,

$$dx \neq 0 \text{ iff } dy = 0 \quad \wedge$$

$$\text{if } dd = 0 \text{ then } l = d_1 \quad \wedge$$

$$l \leq r \text{ (i.e. if } r = d_i \text{ then } l = d_j \ \wedge\ j \leq i \text{ )}$$

The resulting model is cumbersome to draw and to explain, so that it is not reported here. The compaction factor in this case of 2-dimensional mesh varies with number of nodes per dimension from a minimum of 4 in the case of $m = 1$ (i.e., a 2-dimensional hypercube) to a limiting value of 8 in the case $m \to \infty$ (which corresponds to the number of symmetric sectors defined in the color encoding).

### 3.3 Modeling non-preemptive interaction

The first interaction policy considered is the non-preemptive one, in which external requests are not preemptive, but are served at the end of the local operation with higher priority with respect to computation tasks. As already pointed out, this model assumes a non interruptible, single server type scheduling of computation tasks.

In Fig. 12 the SWN model corresponding to the non-preemptive policy for a $2m$ nodes, 1-dimensional (1D) mesh is depicted. The processor identities are modeled as shown in the previous section using a pair of order classes (one without symmetry, and the other one with 2 symmetric elements).

Place ActTask models the presence of $N$ indistinguishable tasks on the different processors, each one characterized by a pair representing the identity of the processor on which it is allocated. Immediate transition start_loc_comp models

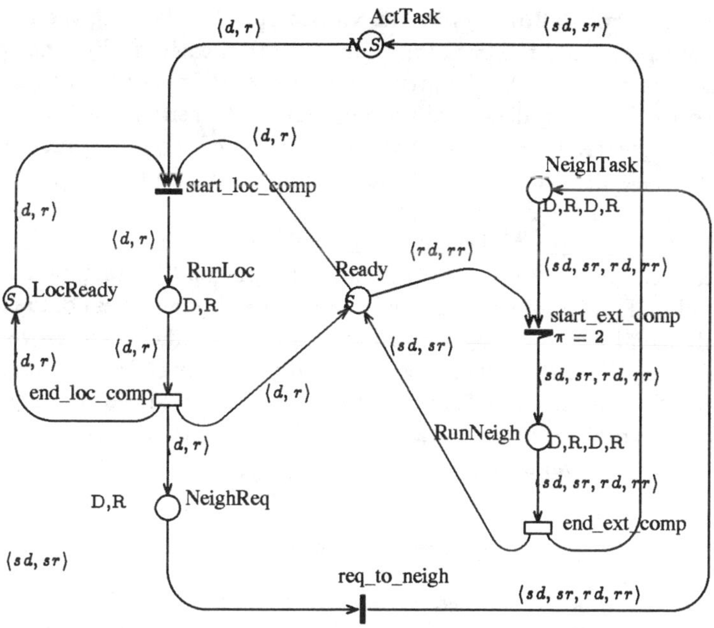


$$[(sd \notin L) \wedge (rd = \oplus sd) \wedge (rr = sr) \vee (sd \notin F) \wedge (\oplus rd = sd) \wedge (rr = sr) \vee (sd \in F) \wedge (rd = sd) \wedge (rr = \oplus sr)]$$

**Fig. 12.** $2m$, 1D mesh with non-preemptive policy

the scheduling of such tasks one at a time on each processor by synchronizing the tokens in place ActTask with the ones in place LocReady, initially containing only one token per coordinate pair. Place Ready initially containing one token per processor identity represents the availability of processors for the execution of computation and server tasks. Transition end_loc_comp (whose rate is $\lambda$) models the activity of processors that are not currently involved in a communication. Its firing represents issuing a communication request to another processor. The selection of the destination processor is made at random through the firing of immediate transition req_to_neigh, that "invents" the identity of the destination by means of the additional components $rd$ and $rr$ in the function labeling its output arc. The values that can be bound to functions $rd$ and $rr$ are restricted by the transition predicate in order to correctly implement the nearest neighbors connection on the 1D mesh topology as illustrated in Figure 10. Immediate transition start_ext_comp models the scheduling of the activity related to the message handling at the destination processor. The completion of the activity related to the communication is modeled by the timed transition end_ext_comp (whose rate is $\mu$), The non-preemptive priority policy is simply obtained by giving immediate transition start_ext_comp higher priority than immediate transition start_loc_comp. Indeed once a processors has chosen to start a local computation phase, the token with processor identity is removed from place Ready and it is not available for the enabling of transition start_ext_comp until the local operation completes.

**Structural model analysis and validation** By checking the incidence functions of the net and the initial marking, one may easily realize that place LocReady is behaviorally as well as structurally implicit. Indeed its marking may never reduce the enabling degree of its only output transition start_loc_comp since it is always greater than or equal to the marking of the implicant place Ready. It may then be deleted together with its input and output arcs without affecting the behavior of the model in any way.

After the erasure of the implicit place LocReady the net exhibits two colored, minimal support P-semiflows that cover all places. The fact that all places are covered by P-semiflows is sufficient to guarantee boundedness of the model, while an intelligent use of the two invariant relations is sufficient to prove liveness and reversibility (i.e., the ergodicity of the corresponding Markovian process). The details of the formal proofs are omitted here for the sake of brevity. Notice that all these formal proofs are parametric with respect to the cardinality of the color classes (i.e., are valid $\forall m \geq 1$).

### 3.4 Modeling preemptive interactions

The second model proposed is relative to the preemptive interaction policy. Please remember that in the preemptive interaction policy requests coming from the neighbors are served with higher priority than the local activity, immediately preempting any computation task. However nothing is assumed concerning the scheduling of computation tasks.

The model is not reported here for the sake of brevity. The definition of the basic color class is exactly the same as for the model of the non-preemptive case. The main difference with respect to the SWN depicted in Figure 12 is that the arc from place Ready to transition start_loc_comp has been moved to transition end_loc_comp. In this way the token representing the processor is used for the enabling of timed transition end_loc_comp, but is not removed during the execution of computation tasks. Immediate transition start_ext_proc may be enabled by an incoming communication request (if no other communication request is currently being served) even during the execution of a computation task. The fact that start_ext_proc is immediate forces its firing right away, thus possibly disabling end_loc_comp. The memoryless property of the firing time distribution of transition end_loc_comp makes it unnecessary the specification of an age memory policy to correctly handle the resume of activity after preemption. The completion of the activity related to the communication is still modeled by the timed transition end_ext_comp (whose rate is $\mu$). Its firing may resume the enabling of transition end_loc_comp by returning the corresponding processor identity token in place Ready (if no other communication requests are pending).

In this case, place LocReady is not behaviorally redundant with respect to the enabling degree of transition start_loc_comp. However, the latter is an immediate transition, for which the enabling degree is not semantically relevant. The enabling degree of timed transition end_loc_comp that is semantically crucial in the behavior of the timed model, instead is not affected by the presence of place LocReady, since it is already limited to the maximum value 1 by the marking of place Ready. Therefore, removing place LocReady does not change the timed measurable behavior of the model.

From the model point of view, after the erasure of place LocReady the net may be further reduced without affecting its qualitative and performance properties by removing immediate transition start_loc_comp and merging places ActTask and RunLoc into a single place.

Boundedness, liveness, and reversibility may be formally proven in a parametric way as in the non-preemptive case. The details of the proofs are omitted here.

### 3.5 Discussion

The formalism provides a direct mapping of the interconnection structure into the model color structure. In case of torus and hypercube topologies the direct mapping into the color structures and functions captures all the symmetries of the model, while in the case of mesh interconnection it does not. Less trivial WN representations may be found that capture system symmetries of mesh interconnection at different levels at the expense of an increasing model complexity. Some conceptually simple extensions of the formalism such as the direct support of the "mirror" type symmetry or the "flip" across color components (permutation of elements in a tuple) would allow the treatment of such more difficult cases in a natural way.

## 4 Conclusions

We have developed a set of examples of use of the SWN formalism for the description of problems arising in the study of parallel computer architectures. The common denominator of these cases that makes the use of SWNs practically advantageous is their intrinsic high degree of symmetry. In some cases the system symmetry directly matches the basic ones provided by the formalism, while in other cases the basic color and function definitions must be composed in a non trivial way in order to capture the desired system symmetries.

In this sense, the WN formalism can be seen as an intermediate level of abstraction in which some more complex symmetric structures can be reproduced only at the cost of some model complexity. From a practical point of view, it would be better to have the possibility of adding (possibly problem or domain specific) color structures and associated manipulation functions in order to match more closely the modeling needs.

On the other hand, the basic principle of restricting the color complexity so that it is mainly related to the symmetry of the system to be modeled seems to be the key for the success of this modeling approach. Future works on the subject should include the development of a clean method to define a library of ready-to-use symmetry structures together with their appropriate manipulation functions that allow efficient implementation of the symbolic firing technique.

## References


1. H.J. Genrich. Equivalence transformations of Pr/T nets. In *Proc. 9$^{th}$ Europ. Workshop on Application and Theory of Petri Nets*, Venezia, Italy, June 1988.
2. Chuang Lin and Dan C. Marinescu. Stochastic high level Petri nets and applications. *IEEE Transactions on Computers*, 37(7):815–825, July 1988.
3. J. A. Carrasco. Automated construction of compound Markov chains from generalized stochastic high-level Petri nets. In *Proc. 3rd Intern. Workshop on Petri Nets and Performance Models*, pages 93–102, Kyoto, Japan, December 1989. IEEE-CS Press.
4. G. Chiola, C. Dutheillet, G. Franceschinis, and S. Haddad. On well-formed coloured nets and their symbolic reachability graph. In *Proc. 11$^{th}$ International Conference on Application and Theory of Petri Nets*, Paris, France, June 1990. Reprinted in *High-Level Petri Nets. Theory and Application*, K. Jensen and G. Rozenberg (editors), Springer Verlag, 1991.
5. G. Chiola, C. Dutheillet, G. Franceschinis, and S. Haddad. Stochastic well-formed coloured nets for symmetric modelling applications. *IEEE Transactions on Computers*, 42(11), November 1993.
6. G. Chiola, G. Franceschinis, and R. Gaeta. A symbolic simulation mechanism for well-formed coloured Petri nets. In *Proc. 25th SCS Annual Simulation Symposium*, Orlando, Florida, April 1992.
7. G. Chiola, R. Gaeta, and M. Ribaudo. Designing an efficient tool for Stochastic Well-Formed Coloured Petri Nets. In R. Pooley and J. Hillston, editors, *Proc. 6$^{th}$ Int. Conference on Modelling Techniques and Tools for Computer Performance Evaluation*, pages 391–395, Edinburg, UK, September 1992. Antony Rowe Ltd.
8. C.E. Leiserson. Fat-trees: Universal networks for hardware efficient supercomputing. *IEEE Transactions on Computers*, C-34(10), October 1985.
9. Kay Hwang. *Advanced Computer Architecture with Parallel Programming.* Mc Graw Hill, 1993.
10. G. Chiola and G. Franceschinis. A structural colour simplification in Well-Formed coloured nets. In *Proc. 4th Intern. Workshop on Petri Nets and Performance Models*, pages 144–153, Melbourne, Australia, December 1991. IEEE-CS Press.
11. G. Chiola, S. Donatelli, and G. Franceschinis. GSPN versus SPN: what is the actual role of immediate transitions? In *Proc. 4th Intern. Workshop on Petri Nets and Performance Models*, pages 20–31, Melbourne, Australia, December 1991. IEEE-CS Press.
12. S. Caselli and G. Conte. GSPN models of concurrent architectures with mesh topology. In *Proc. 4th Intern. Workshop on Petri Nets and Performance Models*, pages 280–289, Melbourne, Australia, December 1991. IEEE-CS Press.

# Coloured Petri Nets Extended with Channels for Synchronous Communication

Søren Christensen
Niels Damgaard Hansen
Computer Science Department, Aarhus University
Ny Munkegade, Bldg. 540
DK-8000 Aarhus C, Denmark
Phone: +45 89 42 31 88
Telefax: +45 89 42 32 55
E-mail: schristensen@daimi.aau.dk, ndh@daimi.aau.dk

**Abstract**

This paper shows how Coloured Petri Nets (CP-nets) can be extended to support synchronous communication. We introduce coloured communication channels through which transitions are allowed to communicate complex values. Small examples show how channel communication is convenient for creating compact and comprehensive models.

The concepts introduced in this paper originate from the practical use of Petri nets for modelling, and they are formally defined in such a way that they preserve the basic properties of CP-nets. We show how a CP-net with channels can be transformed into a behaviourally equivalent CP-net. This allows us to deduce properties of CP-nets with channels from well-known properties of CP-nets. As an example, we extend the concept of place invariants to cope with CP-nets with channels and show how place invariants can be found. This is done without transforming the CP-nets with channels into their equivalent CP-nets.

The reader is assumed to be familiar with the notion of CP-nets.

**Keywords:**
Coloured Petri nets, synchronous communication, channels, modular specifications, re-usable models, invariant analysis.

The work presented in this paper has been supported by a grant from the Danish Research Programme for Informatics—grant number 5.26.18.19.

## Introduction

During our involvement in modelling projects using hierarchical CP-nets [HJS90], [Jen91], [Jen92], it has become clear that it would be valuable to include constructs making it easy to model synchronisation and synchronous communication. Without such constructs it is necessary to model explicitly the synchronous communication through additional places and transitions often resulting in a complex net structure. This means that the modeller has to devote much attention to the model, instead of focusing on the problem being modelled. We propose to extend CP-nets to support communication through channels. The concept of channels is influenced by CCS, CSP and communication constructs found in high level programming languages, e.g., [Hoa85] and [Mil89]. For the sake of simplicity we show how to extend CP-nets and not hierarchical CP-nets. It should, however, be obvious that the extension can be generalised to hierarchical nets.

Other Petri net models have used the notation of communication channels, e.g., [HT91], but the concept has, to the best of our knowledge, never been formally defined before and never fully integrated into the Petri net framework.

The paper is organised as follows: first, the new constructs are informally introduced. Then we show how CP-nets with channels can be formally defined and transformed into behaviourally equivalent CP-nets. In section 3 we show how analysis methods—especially place invariants—can be extended to cope with CP-nets with channels without having to transform the models into the equivalent CP-nets. Finally, we present a number of small examples illustrating the convenience of CP-nets with channels for modelling.

In this paper we have left out the technical details of some of the proofs. The detailed proofs can be found in [CD92].

## 1. Informal Introduction to CP-nets with Channels

In this section we introduce CP-nets with channels. We informally describe the new concepts by means of small examples. These examples illustrate how channels may be used for widely different purposes.

Extending CP-nets with channels allow transitions to communicate through named and coloured communication channels. Transitions, which use channel communication, are called communication transitions and for each channel they are divided into !?-transitions and ?!-transitions. A communication between two transitions is only possible if one of the transitions is a !?-transition and the other is a ?!-transition—and they use the same channel. No direction of communication is intended, this is the reason for using !? and ?! instead of just ! and ? used in CSP. !? and ?! is merely used to divide the transitions into two groups and it would make no difference if all !? were exchanged by ?!.

An example of a CP-net with channels is shown in Fig. 1.1. The inscriptions next to the communication transitions Send and Receive are called communication expressions specifying the channel and the actual value communicated through the channel.

We explicitly declare the colour set, i.e., the type, of the communication channel. This makes static type checking of the communication expressions possible, since we, based on the net-inscriptions, are able to determine if the communication expressions evaluate to values of the same type as the associated communication channel.

Intuitively, a communication through a channel ch is enabled if and only if there exist two communication transitions $t_1$ and $t_2$ with communication expressions $expr_1$ and $expr_2$ and two bindings $b_1$ and $b_2$ such that:

- transition $t_1$ is enabled for the binding $b_1$ and $t_2$ is enabled for the binding $b_2$, i.e., there are sufficient tokens of the correct colours on the input places,
- $t_1$ has a communication expression of the form $expr_1$ !?ch,
- $t_2$ has a communication expression of the form $expr_2$ ?!ch,
- $expr_1$<$b_1$> = $expr_2$<$b_2$>, i.e., $expr_1$ and $expr_2$ evaluate to the same value when they are evaluated in the bindings for which the two transitions occur.

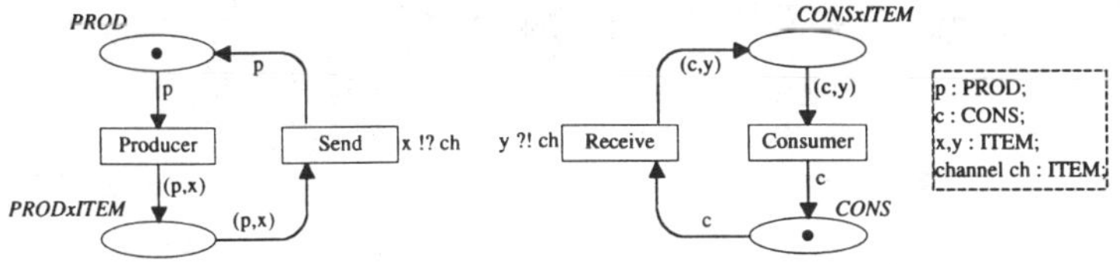


Figure 1.1: CP-net with two communication transitions using a channel

Please note that although most of the examples in this paper show transitions communicating on one channel only, our formalism are able handle transitions communicating on multiple channels.

The semantics of a CP-net with channels may be illustrated by constructing the behaviourally equivalent CP-net. Fig. 1.2 shows the behaviourally equivalent CP-net of the CP-net with channels shown in Fig 1.1. Intuitively the equivalent CP-net is constructed by merging the !?-transition and the ?!-transition so that the arcs of the merged transition are the union of the arcs of the original transitions. The guard of the merged transition is formed by the conjunction of the original guards and an expression stating that the communication expressions must evaluate to the same value. Section 2.3 describes this transformation in more detail.

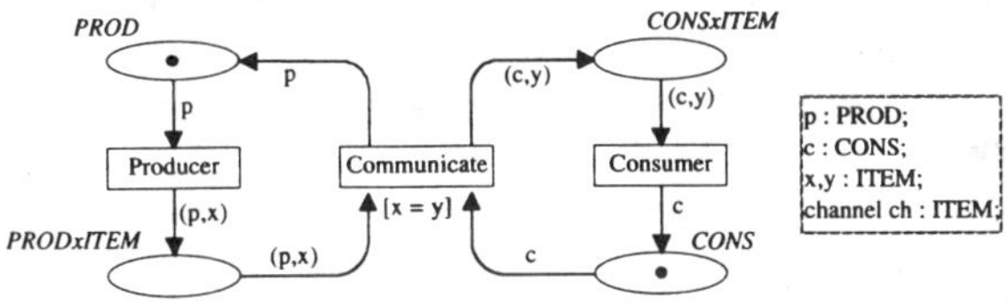


Figure 1.2: Equivalent CP-net of the CP-net with channels in Fig. 1.1

It should be noted that information during a communication may be passed in both directions, i.e., from a !?-transition to a ?!-transition, or vice versa. In the example above the variable x appears in the input arc expression of Send, while the variable y appears in the output arc expression of Receive. This means that the colour of the input token of Send determines the colour of the output token of Receive, and thus information is sent (through the channel) from Send to Receive. In the general case, two communication transitions may have communication expressions specifying bi-directional exchange of information. Throughout this paper we have chosen to use the !?-transition to denote the communication transition determining the value communicated through the associated channel. This convention can of course not always be used, e.g., when the communication is bi-directional.

If we extend the CP-net of Fig. 1.1 to consist of two producers and two consumers we get the CP-net with channels presented in Fig. 1.3. Often, we would fold the producers and consumers into a single structure and distinguish the processes by means of the associated colours—but let us assume that all four processes have internal structures that make it difficult to make this folding. All processes communicate using the channel called ch. In this way each producer is able to send to any of the consumers ready to receive data.

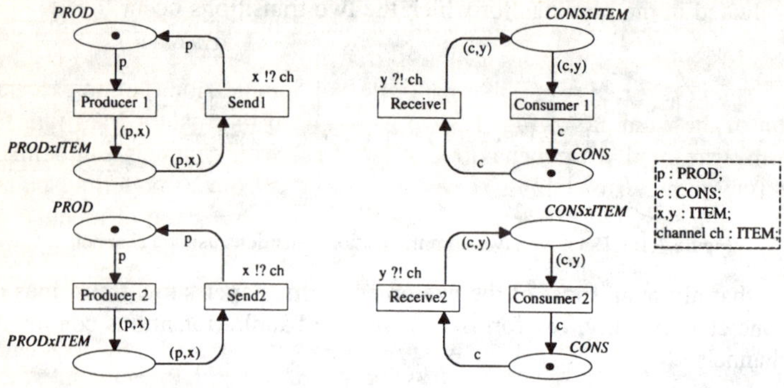


Figure 1.3: Producer-consumer modelled using CP-nets with channels

Fig. 1.4 illustrates that the behaviourally equivalent CP-net becomes rather complex. If we have n !?-transition and m ?!-transition for a given channel we will need n*m transitions to model the same patterns of communication. This observation illustrates a reason why the introduction of channels may result in a dramatic simplification of the net structure. When applied to nodes in a single net, e.g., in Fig. 1.3, channel communication is mainly a drawing convenience allowing the user to avoid too many crossing arcs. Fig. 1.4 illustrates that this is a valuable quality. Channel communication becomes a strong description primitive of its own right especially when it is combined with structured net descriptions such as hierarchical CP-nets, where a model is described as a set of related subnets, e.g., drawn on separate pages.

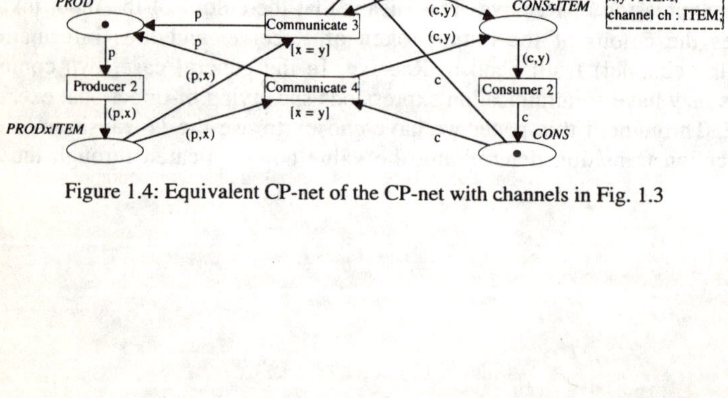


Figure 1.4: Equivalent CP-net of the CP-net with channels in Fig. 1.3

The example of Fig. 1.3 also illustrates one of the differences between channel communication and transition fusion described in [HJS90] and [CP92]. When we compare the concept of channel communication to the concept of transition fusion we find two major differences: first, the bindings of transitions involved in a channel communication are independent except for the restrictions expressed in the communication expressions. This is in contrast to transition fusion where the transitions share a common binding. Secondly, the occurrence of a transition fusion set always involves all members of the fusion set. For a CP-net with channels an occurrence involves pairs of !?-transitions and ?!-transitions. This means that some communication transitions using a given channel may occur without involving the other communication transitions using the same channel. So, if we would model channel communication by means of transition fusion, we would need to create a transition fusion set for each possible pattern of communication, i.e., to express the communication of the example of Fig. 1.3 we would need to create 4 transition fusion sets. Although this means that the behaviour of a CP-net with channels can be expressed by means of transition fusion, this is often not feasible from a modelling point of view. Transition fusion in [CP92] is defined to facilitate the discussion of modular analysis of CP-nets while CP-nets with channels are defined to ease modelling. The main advantage of CP-nets with channels is the possibility to create and analyse independent and re-usable sub-models.

The model shown in Fig. 1.3 is structured in such a way that the producers and the consumers are described in separate nets, while the channel defines the interface between the otherwise independent sub-models. It is possible, e.g., to add another consumer or change the internal description of a consumer independently of the other descriptions—as long as the channel specifying the interface between the descriptions is not changed.

Until now, we have only considered data communication. It is, however, also easy to model communication without exchange of data, i.e., mere synchronisation. This can be done by using an elementary channel colour E = {e} containing only one possible value. It is analogous to the way in which we handle tokens carrying no information in their colour.

In the last example of this section we use CP-nets with channels to model an abstract data type. We use the term abstract data type to stress that the internal representation of the information is accessed through a fixed interface. Abstract data types may in a CP-net be expressed as a colour set having a number of operations. It can, however, also be convenient to model data types through small subnets allowing us to reason about the net structure, e.g., to use invariants.

In Fig. 1.5, we describe a queue and its operations. This queue could for instance be inserted between the producers and the consumers describing that items from the producers are buffered before arriving at the consumers.

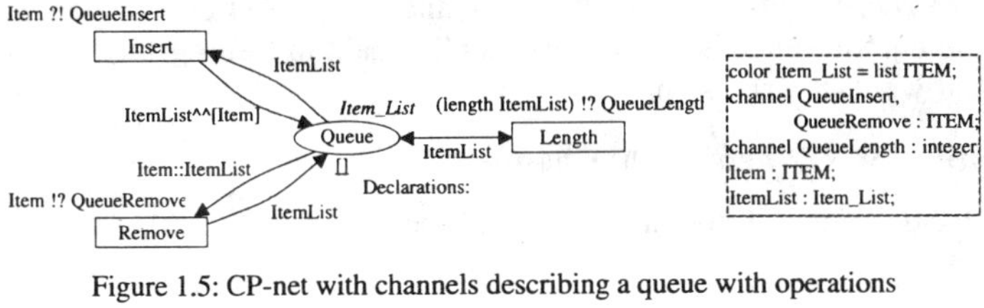


Figure 1.5: CP-net with channels describing a queue with operations

We have introduced a channel for each of the basic operations: Insert, Remove and Length. The queue itself is a token on the Queue place consisting of a list with the buffered elements. Initially the queue is empty as indicated by []. Elements are inserted at the end of the list using ^^ for list concatenation. The list constructor :: is used to split the list into its head element and tail.

The communication transition modelling an operation is enabled by communicating through the associated channel, e.g., the Insert transition is enabled by communicating an item through the Insert channel.

An alternative way of implementing the queue would be to use only one channel and communicate both operation and data through the channel to let the queue implementation distinguish the operations. Similar to this we could implement multiple queues without duplicating the net structure simply by extending the channel communication to contain information on the queue we want to access.

If we want to add a queue between the producers and the consumers we just have to change their use of channels, see Fig 1.6.

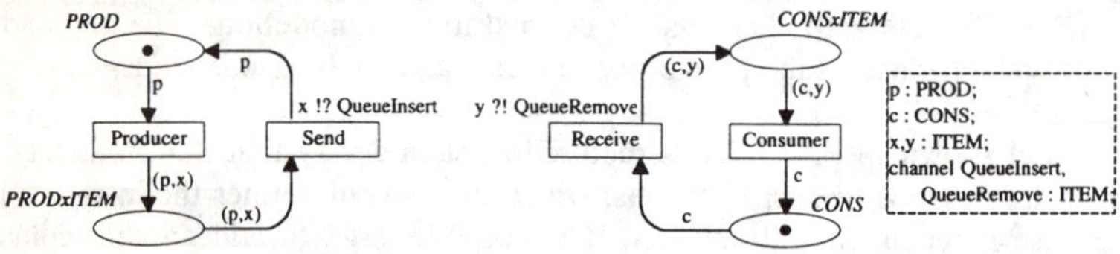


Figure 1.6: Producer and consumer using a queue

The producer uses the QueueInsert channel to insert an element into the queue, whereas the consumer uses the QueueRemove channel to obtain an element from the queue. In this example we do not use the Length operation.

# 2. Formal Definition of CP-nets with Channels

In this section we formalise how channel communication can be added to CP-nets. The notation used in this chapter is consistent with the one used in [Jen91] and [Jen92]. First we recall the definition and notation for multi-sets.

A **multi-set** m, over a non-empty set S, is a function $m \in [S \to \mathbb{N}]$. The non-negative integer $m(s) \in \mathbb{N}$ is the number of appearances of the element s in the multi-set m. We usually represent the multi-set m by a formal sum:

$$\sum_{s \in S} m(s)`s$$

By $S_{MS}$ we denote the set of all multi-sets over S. The non-negative integers $\{m(s) \mid s \in S\}$ are called the **coefficients** of the multi-set m, and the number of appearances of s, m(s), is called the coefficient of s. An element $s \in S$ is said to **belong** to the multi-set m iff $m(s) \neq 0$ and we then write $s \in m$.

## 2.1 Definition of CP-nets with Channels

We recall the definition of non-hierarchical CP-nets:

**Definition 2.1** ([Jen92], Def. 2.5)
A **CP-net** is a tuple CPN = $(\Sigma, P, T, A, N, C, G, E, I)$ satisfying the requirements below:

(i) $\Sigma$ is a finite set of types, called **colour sets**.
(ii) P is a finite set of **places**.
(iii) T is a finite set of **transitions**.
(iv) A is a finite set of **arcs** such that:
 • $P \cap T = P \cap A = T \cap A = \emptyset$.
(v) N is a **node** function. It is defined from A into $P \times T \cup T \times P$.
(vi) C is a **colour** function. It is defined from P into $\Sigma$.
(vii) G is a **guard** function. It is defined from T into boolean expressions such that:
 • $\forall t \in T$: $[\text{Type}(G(t)) = \mathbb{B} \wedge \text{Type}(\text{Var}(G(t))) \subseteq \Sigma]$.
(viii) E is an **arc expression** function. It is defined from A into expressions such that:
 • $\forall a \in A$: $[\text{Type}(E(a)) = C(p(a))_{MS} \wedge \text{Type}(\text{Var}(E(a))) \subseteq \Sigma]$
 where p(a) is the place of N(a).
(ix) I is an **initialization** function. It is defined from P into closed expressions such that:
 • $\forall p \in P$: $[\text{Type}(I(p)) = C(p)_{MS}]$.

A detailed explanation of the requirements of this definition can be found in [Jen91] and [Jen92]. For an arc, a, we use p(a), t(a), s(a) and d(a) to denote the place, transition, source and destination respectively. For a node x we use A(x) to denote the set of arcs connected to x. We furthermore use $\mathbb{B}$ to denote the boolean type containing the elements {false, true} and having the standard operations from propositional logic.

We use the set $\Delta = \{!?, ?!\}$ to denote the two relations a communication transition can have to a channel. Moreover, we use Var(expr) to denote the variables appearing in an expression expr and we use Type(v) and Type(expr) to denote the type of a variable v and an expression expr, respectively. For more information about the use of types, expressions and variables in CP-nets see [Jen91] or [Jen92]. We are now ready to give a formal definition of CP-nets with channels. A short explanation is given below the definition. It is recommended to read this explanation in parallel with the definition.

**Definition 2.2**
A CP-net with channels is a tuple CCPN = (CPN, CS) satisfying the requirements below:

(i) CPN is a non-hierarchical CP-net $(\Sigma, P, T, A, N, C, G, E, I)$.
(ii) CS is a **channel specification** (CH, CT, CE)
(iii) CH is a finite set of **channels** such that:
 • $(P \cup T \cup A) \cap CH = \emptyset$.
(iv) CT is a **channel type** function. It is defined from CH into $\Sigma$.
(v) CE is a **communication expression** function. It is defined from T into finite sets of communication expressions. All elements of CE(t) are triples (expr, #, ch) where expr is an expression, $\# \in \Delta$ and $ch \in CH$ such that:
 • $\forall t \in T$: $\forall (expr,\#,ch) \in CE(t)$:
  $[\text{Type}(expr) = CT(ch) \wedge \text{Type}(\text{Var}(expr)) \subseteq \Sigma]$.

(i) CPN is a CP-net, see Def. 2.1.
(ii) The channel specification is a triple containing a set of channels, a channel type function and a channel expression function.
(iii) CH is a finite set of channels, which can be distinguished from the net elements of the CP-net. In Fig. 1.1 we have: CH = {ch}.
(iv) The channel type specifies the kind of information that can be passed through a given channel. In Fig. 1.1 we have: CT(ch) = ITEMS.
(v) The expression of a communication expression must have a type identical to the type of the channel. Moreover, all variables must be of a known type. In Fig. 1.1 we have: CE(Producer) = { }, CE(Consumer) = { }, CE(Send) = {(x, !?, ch)} and CE(Receive) = {(y, ?!, ch)}. We could generalise the communication expression function to return multi-sets of communication expressions, but since it is not important for the results shown in this paper we have chosen to use plain sets.

For $t \in T$, $\# \in \Delta$ and $ch \in CH$ we use Expr(t,#,ch) to denote the set of communication expressions connecting t and ch in direction #:

$$Expr(t,\#,ch) = \{expr' \mid (expr',\#,ch) \in CE(t)\}.$$

We generalise the expression function E to handle channel expressions. E(t,#,ch) denotes the multi-set sum of all expressions in Expr(t,#,ch):

$$E(t,\#,ch) = \sum_{expr \in Expr(t,\#,ch)} 1`expr.$$

The multi-set sum in E is well-defined since all expressions of a given channel evaluate to value of the channel type.

## 2.2 Behaviour of CP-nets with Channels

Having defined the static structure of CP-nets with channels we are now ready to consider their behaviour. Adding communication to a given transition constrains its enabling, but does not change the effect of an occurrence.

First we re-define the set of variables for a transition so that it also includes the variables of the communication expressions:

$$\forall t \in T: Var(t) = \{v \mid v \in Var(G(t)) \vee \exists a \in A(t): v \in Var(E(a)) \vee \exists (expr,\#,ch) \in CE(t): v \in Var(expr)\}.$$

Our definitions of bindings, B(t), token elements, TE, binding elements, BE, markings, $\mathbb{M}$, and steps, $\mathbb{Y}$, are identical to the definitions given in [Jen92], which except for technical details are identical to those of [Jen91]. However, the enabling rule of CP-nets must be extended to take into account the communication expressions.

**Definition 2.3**
A step $Y \in \mathbb{Y}$ is **enabled** in a marking $M \in \mathbb{M}$ iff the following properties are satisfied:

(i) $\forall p \in P: \sum_{(t,b) \in Y} E(p,t)<b> \leq M(p).$

(ii) $\forall ch \in CH: \sum_{(t,b) \in Y} E(t,!?,ch)<b> = \sum_{(t,b) \in Y} E(t,?!,ch)<b>.$

If Y fulfils (ii) it is said to be **communication enabled**.

(i) This is the enabling rule of CP-nets, See [Jen91] Def. 2.6 or [Jen92] Def 2.8.

(ii) For each channel we demand that the multi-set of values obtained by evaluating the !?-expressions must match the multi-set of values obtained by evaluating the ?!-expressions. This means that a step is enabled iff all binding elements are enabled and the communication can take place.

Our definitions of occurrence, occurrence sequence and reachability are identical to those of [Jen91] and [Jen92].

## 2.3 Behaviourally Equivalent CP-net

In this section we show, that although adding channels to CP-nets increases the possibility for creating compact and comprehensive models it does not increase the computational power. We show that a large class of CP-nets with channels, which we call well-formed, can be transformed into behaviourally equivalent CP-nets. By a behaviourally equivalent CP-net, we mean a CP-net without channels behaving like the original CP-net with channels—that is, there is a one to one correspondence between markings and enabled steps of the two models (see Theorem 2.5). The existence of an equivalent CP-net is extremely useful, because it tells us how to generalise the basic concepts and the analysis methods of CP-nets to CP-nets with channels. We simply define these concepts in such a way that a CP-net with channels has a given property iff the equivalent CP-net has the corresponding property. It is important to understand that we never make the transformation for a particular CP-net with channels. When we describe a system we directly use CP-nets with channels without constructing the equivalent CP-net. Similarly, we directly analyse a CP-net with channels without constructing the equivalent CP-net.

In Def. 2.3 we defined the enabling rule of CP-nets with channels. This definition gives an operational semantics that is easy to understand and use. In this section we show how to transform the restrictions imposed by the additional enabling restriction into a structural property of the resulting CP-net. The formal proof involves a number of definitions and mathematical proofs. In this paper we have chosen to give the intuition behind the transformation and only the most central definitions and proofs. The formal definitions and proofs are found in [CD92].

The basic idea behind the transformation of a CP-net with channels to an equivalent CP-net is illustrated in Fig. 1.2 and Fig. 1.4: all transitions involved in a channel communication are merged together so that the arcs of the merged transition are the union of the arcs of the original transitions and the guard of the merged transition is formed by the conjunction of the original guards and an expression stating that the communication expressions must evaluate to the same value.

Since the bindings of the communication transitions involved in a channel communication are independent except for the restrictions expressed in the communication expressions we need to make sure that set of variables of the involved communication transitions are disjoint before we merge the transitions. Therefore, for each transition, t, we replace each variable $v \in Var(t)$ with a new variable $v_t$, of the same type as v. We assume that the names of the new variables are different.

Since the transformation is based only on the net structure and does not take the inscriptions into account, the transformation obviously fails to be finite if the communication contains cycles. A simplified example of a net with a cyclic communication is shown in Fig. 2.1. The t1 transition initiates the communication by sending an integer value on channel ch1. If the value is negative t2 just receives the value. Otherwise t3 sends the decremented value on ch2. t4 just retransmits on ch1 the value received on ch2. The transformation of this net fails to be finite because it is not possible from the net structure alone to determine how many instances of t3 and t4 should be merged. It depends on the actual value of the variable v, e.g., if v=3 the transformation would include one instance of t1 and t2, but 3 instances of t3 and t4.

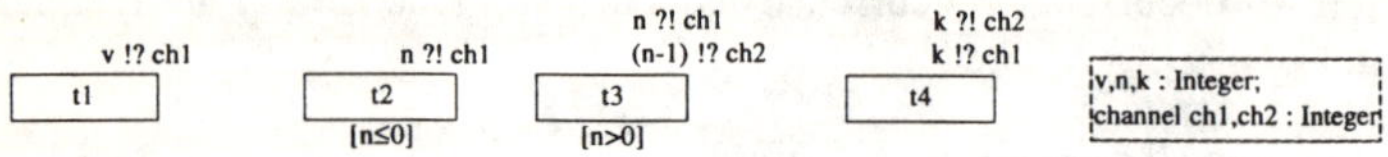


Figure 2.1: Cyclic channel communication

In the rest of this paper we only consider what we call well-formed CP-nets with channels. We say that a CP-net with channels is **well-formed** iff the channel communication has no directed cycles. Well-formed CP-nets with channels have behaviourally equivalent CP-nets. It should be noted that not well-formed CP-nets with channels may have a reasonable meaning—although they are forbidden in order to be able to transform CP-nets with channels to CP-nets.

In the definition of the equivalent CP-net we need to define which transitions are merged together. We define the term **transition groups**. Intuitively a set of transitions form a transition group iff they fulfil:

- if the transition is an ordinary transition it constitutes a transition group of its own,
- for each channel involved in the communication between transitions in the set, we must have exactly one !? and one ?!-transition,
- the transition group is minimal. That is, a transition group contains only one ordinary transition or the communicating transitions involved in one specific communication. Since a communication transition may communicate on several channels a transition group may include transitions communicating on a number of channels.

In Fig. 1.3 we have eight communication groups: one for each of the ordinary transitions, i.e., the producer and consumer transitions and four groups consisting of the possible communication paths, i.e. {Send1, Receive1}, {Send1, Receive2}, {Send2, Receive1} and {Send2, Receive2}.

For each transition group, we are now able to define the conditions that communication enables the transitions. The **communication guard** function CG is defined from the set of transition groups, TG, into boolean expressions:

$$\forall T'' \in TG: \; CG(T'') = \bigwedge_{ch \in CH} \left[ E(T'',!?,ch) = E(T'',?!,ch) \right].$$

Having this notion of transition groups and communication guard, we are now able to define the transformation from CP-nets with channels into a behaviourally equivalent CP-nets:

**Definition 2.4** (Def. 2.8 in [CD92])
Let a CP-net with channels CCPN = (CPN, CS) be given, where CPN = (Σ, P, T, A, N, C, G, E, I) and CS = (CH, CT, CE). Then we define the **equivalent CP-net** to be CPN* = (Σ*, P*, T*, A*, N*, C*, G*, E*, I*) where:
(i) $\Sigma^* = \Sigma$.
(ii) $P^* = P$.

(iii) $T^* = TG$.
(iv) $A^* = \{(a,t^*) \in A \times T^* \mid t(a) \in t^*\}$.
(v) $\forall a^*=(a, t^*) \in A^*$:
$$[\ s(a) \in P \Rightarrow N^*(a^*) = (p(a),t^*) \wedge$$
$$s(a) \in T \Rightarrow N^*(a^*) = (t^*,p(a))].$$
(vi) $C^* = C$.
(vii) $\forall t^* \in T^*: G^*(t^*) = \bigwedge_{t \in t^*} G(t) \wedge CG(t^*)$.
(viii) $\forall a^*=(a,t^*) \in A^*: E^*(a^*) = E(a)$.
(ix) $I^* = I$.

(i) The set of colour sets is unchanged.
(ii) The set of places is unchanged.
(iii) We have a transition for each of the transition groups. We recall that transitions, which do not communicate, are represented as singleton transition groups, i.e., by communication groups containing only a single transition.
(iv) The arcs are duplicated such that for an arc, a, we get a copy of a for each communication group having the transition of a as a member.
(v) The node function is changed to match (iv).
(vi) The colour function is unchanged.
(vii) The guard consists of two independent parts: the conjunction of the guards in the original transitions and the communication guard defined above.
(viii) The arc expression function is changed to match (iv).
(ix) The initialization expression function is unchanged.

The equivalent CP-net has a set of steps that is different from the CP-net with channels. This is because each binding element in the equivalent CP-net by definition fulfils the guard. Thus we only have a binding element when the involved communication expressions have matching values. A similar property is not satisfied for the CP-net with channels, where bindings of the communication transitions may be defined in such a way that the values of the communication expressions do not match. A step with such bindings will of course not be enabled—but the binding elements do exist. All concepts with a star refer to CPN*, while those without refer to CCPN.

**Theorem 2.5** (Theorem 2.9 in [CD92])
Let CCPN be a CP-net with channels and let CPN* be the equivalent CP-net. Then we have the following properties:
(i) $\mathbb{M} = \mathbb{M}^* \wedge M_0 = M_0^*$.
(ii) There exists a bijective function, $\varphi$, which maps communication enabled steps of CCPN onto steps of CPN*.
(iii) $\forall M_1, M_2 \in \mathbb{M}\ \forall Y \in \mathbb{Y}$:
$$M_1\ [Y\rangle_{CCPN} M_2 \Leftrightarrow M_1\ [\varphi(Y)\rangle_{CPN^*} M_2$$

***Proof***: The proof (shown in [CD92]) can be derived from the previous definitions.

## 3. Analysis of CP-nets with Channels

It is important that our extension do not invalidate the analysis methods already known for CP-nets. In practice, it is not sufficient to be able to map a CP-net with channels into a behaviourally equivalent CP-net. We must also be able to perform analysis directly on the CP-nets with channels. The present section discusses how this can be done. We consider simulation, occurrence graphs and place invariants.

## 3.1 Simulation and Occurrence Graph Analysis

The main problem of computer support for simulation of CP-nets with channels is to find a good way to represent the channel communication to the user. Especially the representation of locally enabled transitions that are restricted by their communications, can be difficult to present. However, if a simulator can handle CP-nets we see no major conceptual problems in extending it to handle CP-nets with channels.

In occurrence graphs we usually only consider steps containing a single binding element, i.e., a single transition, with a single binding. For CP-nets with channels this is not sufficient, because we need at least two binding elements to have a communication. Thus we must instead consider steps corresponding to transition groups in the equivalent CP-net. However, again we see no conceptual problems.

## 3.2 Place Invariants

In this section we show how the concepts of place invariant and place flow can be extended to CP-nets with channels. Place invariants can be used in the proofs of properties of a CP-net, e.g., absence of dead markings. In this paper, we focus on the concepts of place invariants and place flows more than on the use of invariants and flows in the proof of properties of CP-nets. For an introduction to invariants for CP-nets see [Jen81], [Jen86], [Jen91] and [Jen92].

### 3.2.1 Formal Definition of Place Invariants for CP-nets

In the following, we formally define the concepts of place invariants and place flows for CP-nets. Before doing this it is necessary to define the notion of weighted set.

A **weighted set** over a non-empty set S, is defined in exactly the same way as a multi-set—except that we now also allow negative coefficients. This means that we can always subtract two weighted sets over the same set S, from each other, and it also means that scalar-multiplication with negative integers is allowed. The set of all weighted sets over S is denoted by $S_{WS}$. Weighted sets have properties that are analogous to those of multi-sets. In particular, we say that a function $W \in [A_{WS} \rightarrow B_{WS}]$ is **linear** iff:

$$W(w_1+w_2) = W(w_1)+W(w_2)$$

for all weighted-sets $w_1, w_2 \in A_{WS}$. The set of linear functions in $[A_{WS} \rightarrow B_{WS}]$ is denoted by $[A_{WS} \rightarrow B_{WS}]_L$.

**Definition 3.1**

Let CPN = (Σ, P, T, A, N, C, G, E, I) be a CP-net. We then define:

(i) W is a **place weight** function, with range $R \in \Sigma$:
$\forall p \in P$: $W(p) \in [C(p)_{WS} \rightarrow R_{WS}]_L$.

(ii) For each marking M the place weight function W determines $W(M) \in R_{WS}$:
$\forall M \in \mathbb{M}$: $W(M) = \sum_{p \in P} W(p)(M(p))$.

(iii) The place weight function W determines the **place invariant** $W(M) = W(M_0)$ iff the weighted marking is constant for all reachable markings:

$$\forall M \in [M_0\rangle: \ W(M) = W(M_0).$$

(iv) The place weight function W is a **place flow** iff:

$$\forall (t,b) \in BE: \sum_{p \in P} W(p)(E(p,t)\langle b\rangle) = \sum_{p \in P} W(p)(E(t,p)\langle b\rangle).$$

We use $\mathbb{W}_P$ to denote the set of all place weight functions, while we use $\mathbb{W}_{PF} \subseteq \mathbb{W}_P$ and $\mathbb{W}_{PI} \subseteq \mathbb{W}_P$ to denote the set of those place weight functions that are place flows and which determine place invariants.

All weights are linear functions. This means that any linear combination of two place flows is a place flow, e.g., if $W_1$ and $W_2$ are place flows, with identical range, and $z_1, z_2 \in \mathbb{Z}$ then $z_1 * W_1 + z_2 * W_2$ is a place flow. A zero weight is a function mapping any weighted-set to the empty set. The weight function assigning zero weights to all places is always a place flow. We say that a place p is included in W if W(p) is a non-zero function. Similar remarks apply to place invariants. The main reason for introducing place flows is the difficulty to check place invariants on the total set of reachable states whereas a place flows is a static property, which can be checked on the structure of the CP-net.

The following theorem describes the relationship between place invariants and place flows. A binding element is said to be dead when it can never occur.

**Theorem 3.2**
Let a CP-net be given. We then have:

(i) $\mathbb{W}_{PF} \subseteq \mathbb{W}_{PI}$.

(ii) No dead binding elements $\Rightarrow \mathbb{W}_{PF} = \mathbb{W}_{PI}$.

***Proof:*** The theorem is part of the classical theory for invariant analysis. For CP-nets a proof of (i) can be found in [Jen81] and [Jen86]. The proof of (ii) is straightforward. ♦

### 3.2.2 Extending Place Invariants to CP-nets with Channels

For CP-nets with channels we define place weight functions, weighted markings, place invariants, $\mathbb{W}_P$ and $\mathbb{W}_{PI}$ in exactly the same way as we did for CP-nets. Moreover, we define place/channel weight functions and place/channel flows as follows:

**Definition 3.3**
Let CCPN = (CPN, CS) be a CP-net with channels, where CPN = (Σ, P, T, A, N, C, G, E, I) and CS = (CH, CT, CE). We then define:

(i) W is a **place/channel weight** function, with range $R \in \Sigma$:

$$\forall p \in P: W(p) \in [C(p)_{WS} \rightarrow R_{WS}]_L \ \wedge \ \forall ch \in CH: W(ch) \in [CT(ch)_{WS} \rightarrow R_{WS}]_L.$$

(ii) For each marking M the place/channel weight function W determines $W(M) \in R_{WS}$:

$$\forall M \in \mathbb{M}: \ W(M) = \sum_{p \in P} W(p)(M(p)).$$

(iii) The place/channel weight function W determines the **place invariant** $W(M) = W(M_0)$ iff the weighted marking is constant for all reachable markings:

$$\forall M \in [M_0\rangle: \ W(M) = W(M_0).$$

(iv) The place/channel weight function W is a **place/channel flow** iff:
$\forall (t,b) \in BE$:

$$\sum_{p \in P} W(p)(E(p,t)<b>) + \sum_{ch \in CH} W(ch)(E(t,!?,ch)<b>) =$$
$$\sum_{p \in P} W(p)(E(t,p)<b>) + \sum_{ch \in CH} W(ch)(E(t,?!,ch)<b>).$$

We use $\mathbb{W}_{PC}$ to denote the set of all place/channel weight functions, while we use $\mathbb{W}_{PCF} \subseteq \mathbb{W}_{PC}$ and $\mathbb{W}_{PCI} \subseteq \mathbb{W}_{PC}$ to denote the set of those place/channel weight functions that are place/channel flows and which determine place invariants.

The domain of a weight function is extended to cover both places and channels, i.e., each place and each channel are mapped into a weight. A weight of a place is a linear function from weighted sets over the colour set of the place to weighted sets over R. A weight of a channel is a linear function from weighted sets over the channel type to weighted sets over R. A place/channel weight function determines a place/channel flow if the weighted sum of the tokens consumed together with the values of the !?-expressions are equal to the weighted sum of the tokens produced together with the values of the ?!-expressions. This is the only non-symmetric use of !? and ?!. We have chosen to group !?-expressions with the input arcs, and ?!-expressions with the output arcs. We could just as well do the opposite. The definition of place invariants is equivalent to the definition given for CP-nets.

For a weight function $W \in \mathbb{W}_{PC}$ we use $W|P$ to denote its restriction to P and we use $\mathbb{W}_{PCI}|P$ to denote $\{(W|P) \mid W \in \mathbb{W}_{PCI}\}$. This means $\mathbb{W}_{PCI}|P = \mathbb{W}_{PI}$.

In the following we show how the concepts of place invariant and place/channel flow of CP-nets with channels match the corresponding concepts for CP-nets. A star is used to indicate that a symbol refers to the equivalent CP-net. As an example we use $W^* \in \mathbb{W}_P^*$ to denote a place weight function of CPN*.

**Theorem 3.4**
Let CCPN be a CP-net with channels, having CPN* as the equivalent CP-net. We then have:
(i) $\mathbb{W}_{PCF}|P \subseteq \mathbb{W}_{PF}^*$.
(ii) $\mathbb{W}_{PCF} \subseteq \mathbb{W}_{PCI}$.

***Proof:***

**Property (i):** This property is shown by reformulating the place/channel flow of Def. 3.3 (iv) in terms of the behaviourally equivalent CP-net using the transformation specified in Def. 2.4. The proof can be found in [CD92].

**Property (ii):** We have that $\mathbb{W}_{PCF}|P \subseteq \mathbb{W}_{PF}^* \subseteq \mathbb{W}_{PI}^*$. The first inclusion follows from Th. 3.4 (i) and the second from Th. 3.2 (i). We conclude the proof of the property by noting that any extension of $\mathbb{W}_{PI}^*$ to cover channels is a place/channel invariant. ♦

As we will illustrate by the examples in Sec. 4 it is possible to determine the interesting place invariants of a CP-net with channels from the place/channel flows, it is however the case that certain use of channel communication can make it impossible to find place/channel invariants corresponding to all place invariants.

In the example of Fig. 3.1 and 3.2 we can specify three weight functions which determines place invariants:

$W_1$: $P_1(A + B)$
$W_2$: $P_2(A + B)$
$W_3$: $Id(A + B)$.

$W_1$ and $W_2$ specifies that the multi-set sum of the projection of the two places is constant for all reachable markings. $W_3$ specifies that the multi-set sum of the markings of the two places are constant. It can be verified from fig. 3.2 that $W_1$, $W_2$ and $W_3$ are place flows of the behaviourally equivalent CP-net.

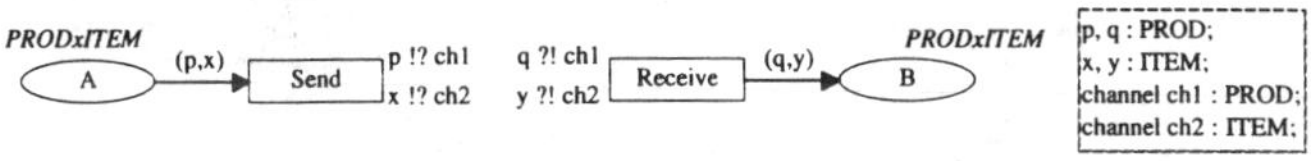


Figure 3.1: CP-net with channel communication

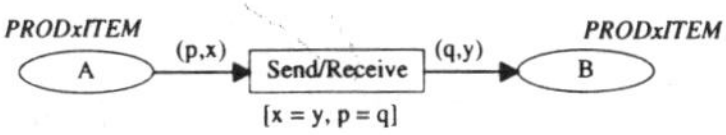


Figure 3.2: Behaviourally equivalent CP-net

It is possible to find place/channel flows which corresponds to $W_1$ and $W_2$:

$W_1$: $P_1(A + B) + Id(ch1)$
$W_2$: $P_2(A + B) + Id(ch2)$.

But it is not possible to find any weights of ch1 and ch2 corresponding to $W_3$.

# 4. Examples of CP-nets with Channels

In this section we illustrate the modelling convenience of CP-nets with channels by means of small examples. We focus on how channels can be used to glue together different sub-models into larger models.

The straightforward use of CP-nets with channels is to describe the synchronous communication between separate processes. Channels are then used to define the interface through which the processes communicate by means of communication transitions. This approach has a number of advantages. The most obvious advantage is that it is possible to structure the model into communicating sub-models in such a way, that you can change parts of the model without changing other parts of the model as long as the interface specified by means of channels is not changed.

## 4.1 A Resource Sharing Example

In the following we show how the resource allocation system in Fig. 4.1, [Jen92], can be re-structured using channel communication. The resource allocation example has a set of processes that share a common pool of resources. There are two different kinds of processes, called p-processes and q-processes, and three different kinds of resources, called r-resources, s-resources and t-resources. Each process is cyclic and during the individual parts of its cycle, the process needs to have exclusive access to a varying amount of resources. For each process, we have an integer value counting the number of process cycles. We use the following definition of colours: P = {p,q}, I = Integer, U = P×I and R = {r,s,t}. We use a variable x of type P and a variable i of type I. The p-processes can be in four different states, while q-processes can be in five different states. In the initial state, there are 2 p-processes, 3 q-processes, 1 r-resource, 3 s-resources and 2 t-resources. The CP-net is presented in Fig. 4.1 is so small, that we would not decompose it in practice, but it can still be used to illustrate the convenience of CP-nets with channels.

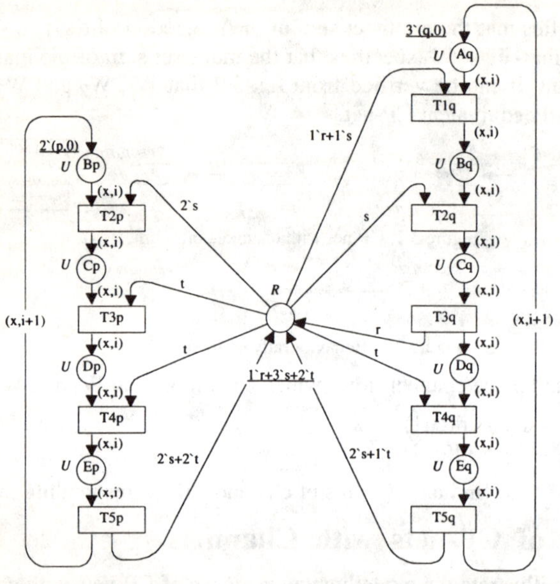


Figure 4.1: Resource allocation modelled with CP-nets

The idea is to re-structure the CP-net model of Fig. 4.1 into three separate sub-models communicating through channels—sub-models describing the p-processes, the q-processes and the resource allocation respectively. We use channels instead of ordinary arcs to reserve and release resources. Hereby the model is broken up into three separate submodels with interface described by means of channels. The sub-models are shown in Fig. 4.2.

The resource allocation sub-model consists of the resource place and two communication transitions GetResource and PutResource communicating through channels called Reserve and Release. The channel type of Reserve and Release is multi-sets over the colour set R and res is a multi-set variable over the colour set R.

In the subnets modelling the processes we replace the transitions normally connected to the resource place with communication transitions. Transitions reserving a resource are replaced with communication transitions using the Reserve channel and transitions releasing a resource are replaced with communication transitions using the Release channel.

Note that we allow a multi-set to be communicated through a channel, but we do not allow a multi set from one sender to be split up into pieces received by different receivers.

Given the model of Fig. 4.2 consisting of three separate sub-models, it is now possible to, e.g., redefine the resource allocation strategy without changing the process descriptions.

On the other hand the figure also illustrates that breaking a small example into sub-models by replacing some of the arcs with channels does not always enhance the readability of the model. The real advantage of CP-nets with channels will be gained when modelling large systems where it is impossible to model the whole system on a single page.

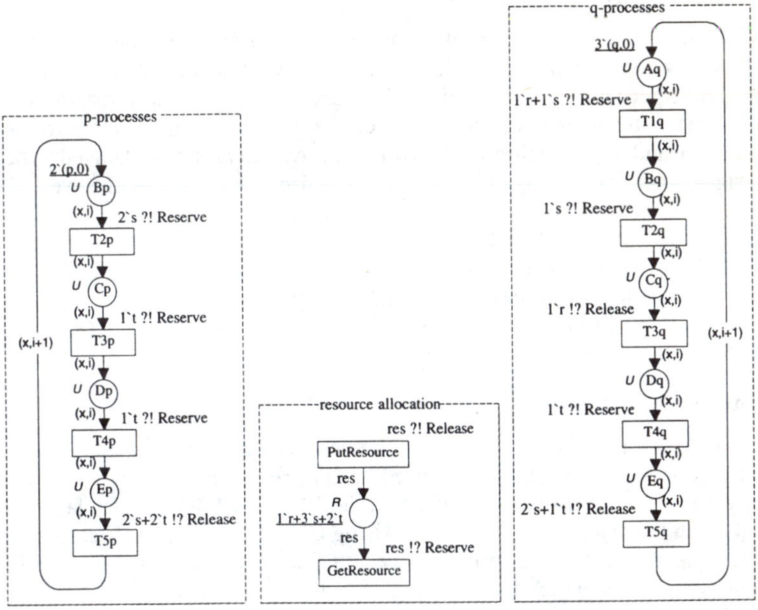


Figure 4.2: Resource allocation modelled with channels

Before we look at the place/channel flows of the CP-net with channels shown in Fig. 4.2 we look at the place flows of the CP-net in Fig. 4.1: We use the following notation for weight functions: places having a zero weight are simply left out, the identity function is implicit, the Pr (Projection) function maps multi-sets of pairs of P×I into multi-sets of P, the Ig (Ignore colour) function maps a multi-set of size s into s\`e, the indicator functions $I_r$, $I_s$, $I_t$ map multi-sets of resources into the size of the respective resources, i.e., $I_r$(2\`r + 1\`s + 3\`t) = 2\`e, $I_s$(2\`r + 1\`s + 3\`t) = 1\`e and $I_t$(2\`r + 1\`s + 3\`t) = 3\`e. Finally, we use the names of the places to refer to the marking of the place, e.g., we write Bp instead of M(Bp). For the example in Fig. 4.1 we then have the following five place weight functions. By checking that the weighted sum of tokens consumed by each binding for each transition is equal to the weighted sum of tokens produced it can easily be shown that the weight functions are place flows and therefore determine place invariants.

$W_1$: Pr(Bp + Cp + Dp + Ep)
$W_2$: Pr(Aq + Bq + Cq + Dq + Eq)
$W_3$: $I_r$(R) + Ig(Bq + Cq)
$W_4$: $I_s$(R) + Ig(Bq) + 2*Ig(Cp + Dp + Ep + Cq + Dq + Eq)
$W_5$: $I_t$(R) + Ig(Dp + Eq) + 2*Ig(Ep).

We can construct other place flows, but all of the above place flows can easily be interpreted in terms of the CP-net: as an example, $W_1$ shows that all the p-processes are in one of the states represented by Bp, Cp, Dp or Ep. $W_3$ shows that the r resources are either free (i.e., in state R) or occupied by a q-process in state Bq or Cq. Using the information from the five place flows above, it is straightforward to prove that the system is deadlock free and similar behavioural properties.

Lets now consider the place/channel flows of the CP-nets with channels in Fig. 4.2. A place/channel weight function is a place/channel flow if the weighted sum of the tokens consumed together with the values of the !?-expressions are equal to the weighted set of tokens produced together with the values of the ?!-expressions. Therefore the weight functions, $W_1$ and $W_2$, which do not involve the resources are unchanged, whereas the other weight functions, $W_3$, $W_4$ and $W_5$, are extended to include the Reserve and Release channels.

$W_1$: Pr(Bp + Cp + Dp + Ep)
$W_2$: Pr(Aq + Bq + Cq + Dq + Eq)
$W_3$: $I_r$(R + Reserve + Release) + Ig(Bq + Cq)
$W_4$: $I_s$(R + Reserve + Release) + Ig(Bq) + 2∗Ig(Cp + Dp + Ep + Cq + Dq + Eq)
$W_5$: $I_t$(R + Reserve + Release) + Ig(Dp + Eq) + 2∗Ig(Ep).

## 4.2 Protocol Modelling

Although ordinary Petri nets has shown to be very useful in the area of protocol specification, CP-nets with channels may be used in this area with advantage. Often protocols are divided into a number of layers like, e.g., the OSI model, where each layer communicates only with the layer above and below. Using CP-nets with channels each layer is then modelled separately and the interfaces between the layers are modelled by means of channels. Hereby it is possible to model each layer independently and possibly at different abstraction levels. At the lowest level of details a layer can be modelled by means of two communication transition as shown in Fig. 4.3, where each communication transition handles the communication in one direction. At this abstraction level, the layer is transparent—messages are sent unchanged to the layers above and below. If it is later on decided to detail the description of this protocol layer, the communication transitions are just replaced with a subnet describing the internal behaviour of the layer at an appropriate level of abstraction.

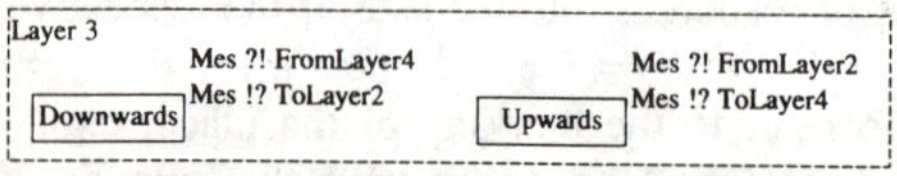


Figure 4.3: Transparent protocol layer

Using channels to describe the interface between two layers furthermore eases the description of, e.g., an unreliable communication. If we want to describe that data may be lost on the channel 'ToLayer2' in Fig 4.3 we do not need to change the original model, but just add a single transition occasionally removing data transmitted via the channel, as illustrated in Fig. 4.4.

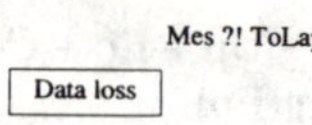


Figure 4.4: Modelling data loss on a channel

In this section we have illustrated how CP-nets with channels provide useful concepts for generating structured net models. We have illustrated how channels are useful to describe synchronous communication between separate sub-models and to glue separate sub-models together into larger models with extended functionality.

## 5. Conclusion

In this paper we have shown how the concept of synchronous communication channels can be introduced in the framework of CP-nets. CP-nets with channels facilitates the creation of compact and comprehensive models of systems with synchronous communication.

Channel communication is a valuable concept for structuring net models. Defining the interface between different subnets by means of communication channels makes it easy to create expandable and tailorable sub-models, where all dependencies between sub-models are expressed in the communication expressions. The advantage of such a modular approach has been known in the area of traditional programming languages for a long time, and has resulted in a number of different concepts, e.g., abstract data types and modules. CP-nets with channels have been developed to support similar efforts inside the area of Petri nets.

CP-nets with channels are highly influenced by needs originating from practical use of hierarchical CP-nets for modelling, and we have shown how CP-nets with channels fit nicely into the Petri net framework. We have shown that each CP-net with channels can be mapped into an equivalent CP-net, i.e., a CP-net with identical behaviour. We have defined place flows and place invariants for CP-nets with channels, in such a way that they have similar properties as in CP-nets.

## Acknowledgement

We would like to thank Kurt Jensen for valuable discussions and Peter Huber, Mogens Nielsen and Laura Petrucci for many helpful comments on earlier versions of this paper.

## References


[CD92] S. Christensen, N. Damgaard Hansen: **Coloured Petri nets extended with channels for synchronous communication.** Daimi PB–390, ISSN 0105–8517, April 1992. Available as: Daimi PB–390, ISSN 0105–8517, April 1992.

[CP92] S. Christensen, L. Petrucci: **Towards a modular analysis of coloured Petri nets**. In: K. Jensen (ed.): Application and Theory of Petri Nets 1992. Lecture Notes in Computer Science, vol. 616, Springer-Verlag, 1992, 113-133.

[HT91] T. Hildebrand and N. Trèves: **S-CORT: A method for the development of electronic payment systems.** In: G. Rozenberg (ed.): Advances in Petri Nets 1989, Lecture Notes in Computer Science vol. 424, Springer-Verlag 1990, 262-280.

[Hoa85] C. A. R. Hoare: **Communicating sequential processes** ISBN 0-13-153289-8, Prentice Hall, 1985.

[HJS90] P. Huber, K. Jensen and R. M. Shapiro: **Hierarchies in coloured Petri nets.** In: G. Rozenberg (ed.): Advances in Petri Nets 1990. Lecture Notes in Computer Science, vol. 383, Springer-Verlag, 1990, 342-416. Also in [JR91], 215-243.

[Jen81] K. Jensen: **Coloured Petri nets and the invariant method.** Theoretical Computer Science 14 (1981), Springer-Verlag 1981, 317-336.

[Jen86] K. Jensen: **Coloured Petri nets.** In: W. Brauer, W. Reisig and G. Rozenberg (eds.): Petri Nets: Central Models and Their Properties, Advances in Petri Nets 1986 Part I, Lecture Notes in Computer Science, vol. 254, Springer-Verlag 1987, 248-299.

[Jen91] K. Jensen: **Coloured Petri nets: A high level language for system design and analysis**. In: G. Rozenberg (ed.): Advances in Petri Nets 1990. Lecture Notes in Computer Science, vol. 383. Springer-Verlag, 1990, 342-416. Also in [JR91], 44-119.

[Jen92] K. Jensen: **Coloured Petri nets. Basic concepts, analysis methods and practical use. Volume 1: Basic concepts.** EATCS monographs on Theoretical Computer Science, Springer-Verlag 1992.

[JR91] K. Jensen and G. Rozenberg (eds.): **High-level Petri nets: theory and application.** Springer-Verlag 1991. ISBN 3-540-54125-X/0-387-54125-X.

[Mil89] R. Milner: **Communication and concurrency**. ISBN 0-13-114984-9, Prentice Hall, 1989.

# Petri Nets with Marking-Dependent Arc Cardinality: Properties and Analysis

Gianfranco Ciardo

Department of Computer Science
College of William and Mary
Williamsburg, VA 23187-8795, USA
ciardo@cs.wm.edu

**Abstract.** We discuss P/T-nets where the arc cardinalities are allowed to be marking-dependent expressions of various types, resulting in a hierarchy of subclasses. Some of the language and decidability properties of these classes have been studied before, but we focus on the practical implications in systems modeling, adding some new insight to the known results about the relative expressive power of the subclasses.
We show how the p-semiflows of a P/T-net with marking-dependent arc cardinality can be obtained from the p-semiflows of a related ordinary P/T-net and how bounds on the relative throughputs of the transitions can be obtained, a weaker condition than t-semiflows.
Finally, we briefly discuss several modeling applications where these subclasses are used.

## 1 Introduction

Petri nets were introduced by Petri [22] to model concurrent behavior and, since then, they have been studied for their theoretical properties and adopted as an effective description formalism to model discrete-state systems.

Informally, a Petri net is a finite bipartite directed graph where the nodes are either places or transitions. Tokens reside in places, and move according to the firing rule. Three classes of Petri nets have been considered [23]: condition-event nets (C/E-nets), where at most one token can reside in each place, resulting in a finite state-space; place-transition nets (P/T-nets), which we consider in this paper, where this restriction is lifted, resulting in a possibly infinite state-space; and predicate-event nets (P/E-nets) and high-level nets, where tokens have individual identities.

Two main directions of Petri net research are well represented in two cycles of conferences: the International (formerly European) Conference on Applications and Theory of Petri Nets, held annually, with a focus on the theoretical computer science aspects of nets, and the International Workshop on Petri Nets and Performance Models, with a focus on the use of nets augmented with stochastic or deterministic timing for the quantitative analysis of systems (stochastic Petri nets, or SPNs). While these two areas span a wide range of interests in computer science, from formal languages, decidability, and complexity theory, to stochastic processes, numerical analysis, and optimization, there has been a good amount of cross-fertilization, as attested by the number of researchers active in both conferences.

The standard definition of Petri net assumes that the cardinality of the input and output arcs is constant but, at times, the reality being modeled behaves differently: a given event might require to remove from a place, or add to a place, a number of tokens which varies according to the marking of the net. Assuming that the behavior can be modeled at all, extraneous places and transitions must be used, whose only purpose is to model explicitly the movement of a variable number of tokens.

From a practical point of view this is both unpleasant, since it adds additional clutter to the model without adding any useful detail, and inefficient, since it requires to process sequentially what is essentially a bulk movement of tokens. For this reason, SPNs with marking-dependent arc cardinality were introduced in [8], adopted in several types of SPN definitions [9, 10, 11], and implemented in a computer-based analysis tool, SPNP [13]. More recently other SPN modeling tools have adopted them as well [18, 19].

Much earlier, though, researchers interested in the expressive power of (untimed) P/T-nets, introduced similar extensions: reset nets, where the firing of a transition can empty a place [3], and self-modifying nets (SM-nets) and their subclasses [24, 25], where the cardinality of input and output arcs can be any nonhomogeneous linear combination of the number of tokens in each place.

In this paper, we consider P/T-nets with various restrictions on the type of arc cardinality (Sect. 2), obtaining a hierarchy of nets. For these, we consider both the computation of minimal p-semiflows (Sect. 3), and the analysis of their expressive power (Sect. 4). Finally, we list some examples from the modeling literature where P/T-nets with marking-dependent arc cardinality have been used, or could have been used, advantageously (Sect. 5).

## 2 P/T-Nets with Marking-Dependent Arc Cardinality

A P/T-net with marking-dependent arc cardinality is a tuple

$$N = (P, T, D^-, D^+, \mu^{[0]})$$

where:

- $P = \{p_1, p_2, \ldots p_{|P|}\}$ is a finite set of places, which can contain tokens. A marking $\mu = (\mu_1, \mu_2, \ldots \mu_{|P|}) \in \mathbb{N}^{|P|}$ describes an assignment of tokens to each place[1]. A marking-dependent expression is a function of the marking, $f(\mu) = f(\mu_1, \mu_2, \ldots, \mu_{|P|})$.
- $T = \{t_1, t_2, \ldots t_{|T|}\}$ is a finite set of transitions, $P \cap T = \emptyset$.
- $\forall i, 1 \le i \le |P|, \forall j, 1 \le j \le |T|$, $D^-_{i,j} : \mathbb{N}^{|P|} \to \mathbb{N}$ and $D^+_{i,j} : \mathbb{N}^{|P|} \to \mathbb{N}$ are the marking-dependent cardinalities of the input arc from $p_i$ to $t_j$ and of the output arc from $t_j$ to $p_i$, respectively.
- $\mu^{[0]} \in \mathbb{N}^{|P|}$ is the initial marking.

[1] We use the symbols $\mathbb{N}$, $\mathbb{Z}$, and $\mathbb{Q}$ to indicate the non-negative integers, the integers, and the rational numbers, respectively

Places and transitions are drawn as circles and rectangles, respectively. The number of tokens in a place is written inside the place itself (default is zero). Input and output arcs have an arrowhead on their destination. The cardinality is written on the arc (default is the constant function one). A missing arc indicates that the cardinality is the constant function zero.

A transition $t_j \in T$ is enabled in marking $\mu$ iff $D^-_{\bullet,j}(\mu) \leq \mu$ ($A_{\bullet,j}$ indicates the $j$-th column of $A$). The set of transitions enabled in marking $\mu$ is denoted by $\mathcal{E}(\mu) = \{t_j \in T : D^-_{\bullet,j}(\mu) \leq \mu\}$. A transition $t_j \in \mathcal{E}(\mu)$ can fire, we denote the new marking $\mathcal{M}(t,\mu) = \mu - D^-_{\bullet,j}(\mu) + D^+_{\bullet,j}(\mu) = \mu + D_{\bullet,j}(\mu)$, where $D = D^+ - D^-$ is the incidence matrix. The reachability set is $\mathcal{R}(N) = \{\mu \in \mathbb{N}^{|P|} : \exists \sigma \in T^* \wedge \mu = \mathcal{M}(\sigma, \mu^{[0]})\}$, where the first argument of function $\mathcal{M}$ is extended to sequences of transitions.

## 2.1 Classes of P/T-Nets with Marking-Dependent Arc Cardinality

The class of P/T-nets just defined is Turing-equivalent, since an input arc from $p_i$ to $t_j$ with cardinality $2\mu_i$ is exactly equivalent to an inhibitor arc from $p_i$ to $t_j$ (see Fig. 1), and P/T-nets with inhibitor arcs can model Turing machines [15]: if $\mu_i = 0$, the arc has no effect on the enabling of $t_j$ and no token is removed from $p_i$ if $t_j$ fires; if $\mu_i > 0$, the cardinality of the arc is greater than the number of tokens in $p_i$, hence $t_j$ is disabled.

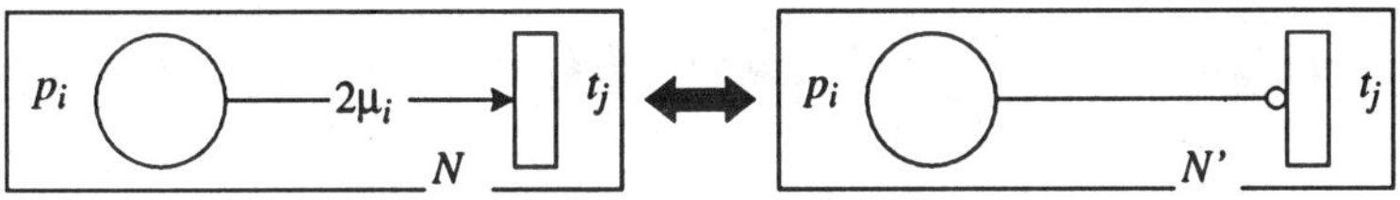


**Fig. 1.** Equivalence of inhibitor arcs and marking-dependent cardinality arcs.

Certain marking-dependent behaviors, though, such as removing all tokens from a place or moving all tokens from a place to another are often useful to model many systems of practical interest, hence we define the following subclasses of P/T-nets with restricted marking-dependent cardinality arcs (assume $\alpha^-, \alpha^+ \in \mathbb{N}^{|P|\times|T|}$ and $\beta^-, \beta^+ \in \mathbb{N}^{|P|\times|T|\times|P|}$):

$\mathcal{N}_o$: ordinary P/T-nets, with constant arc cardinalities,

$$D^-_{i,j} = \alpha^-_{i,j} \quad \wedge \quad D^+_{i,j} = \alpha^+_{i,j} \; .$$

$\mathcal{N}_r$: reset P/T-nets [3], with the addition of reset, or flushing arcs,

$$D^-_{i,j} = \alpha^-_{i,j} + \beta^-_{i,j,i}\mu_i, \; \beta^-_{i,j,i} \leq 1, \beta^-_{i,j,i} = 1 \Rightarrow \alpha^-_{i,j} = 0 \quad \wedge \quad D^+_{i,j} = \alpha^+_{i,j} \; .$$

$\mathcal{N}_p$: post self modifying P/T-nets [24], also called set nets in [25], where the cardinality of output arcs can be any nonhomogeneous linear combination of the marking (hence they allow duplication of the number of tokens in a place),

$$D^-_{i,j} = \alpha^-_{i,j} \ \wedge \ D^+_{i,j} = \alpha^+_{i,j} + \sum_{1 \le l \le |P|} \beta^+_{i,j,l}\mu_l \ .$$

$\mathcal{N}_t$: transfer P/T-nets, a new class, where the firing of a transition can move all the tokens from one place to another (but not duplicate them),

$$D^-_{i,j} = \alpha^-_{i,j} + \beta^-_{i,j,i}\mu_i,\ \beta^-_{i,j,i} \le 1, \beta^-_{i,j,i} = 1 \Rightarrow \alpha^-_{i,j} = 0$$
$$\wedge\, D^+_{i,j} = \alpha^+_{i,j} + \sum_{1 \le l \le |P|} \beta^+_{i,j,l}\mu_l \ ,$$

subject to $\beta^-_{i,j,i} = \sum_{1 \le l \le |P|} \beta^+_{l,j,i}$.

$\mathcal{N}_l$: linear transfer P/T-nets, also called reset-set nets in [25], which allow both reset and duplication, hence transfer, behavior,

$$D^-_{i,j} = \alpha^-_{i,j} + \beta^-_{i,j,i}\mu_i,\ \beta^-_{i,j,i} \le 1, \beta^-_{i,j,i} = 1 \Rightarrow \alpha^-_{i,j} = 0$$
$$\wedge\, D^+_{i,j} = \alpha^+_{i,j} + \sum_{1 \le l \le |P|} \beta^+_{i,j,l}\mu_l \ .$$

$\mathcal{N}_s$: self modifying P/T-nets [24], where the cardinality of both input and output arcs can be any nonhomogeneous linear combination of the marking,

$$D^-_{i,j} = \alpha^-_{i,j} + \sum_{1 \le l \le |P|} \beta^-_{i,j,l}\mu_l \ \wedge \ D^+_{i,j} = \alpha^+_{i,j} + \sum_{1 \le l \le |P|} \beta^+_{i,j,l}\mu_l \ .$$

The possible patterns of marking-dependent arc cardinalities are summarized in Fig. 2.

### 2.2 Observations

Often, a transition $t_1$ having $p_1$ as its only input place, with cardinality $\mu_1$, should not be considered enabled when $\mu_1 = 0$. In other words, the flushing of $p_1$ should occur only when $\mu_1$ reaches a minimum threshold $k$, often one (see $N$ in Fig. 3). This behavior is represented in $N' \in \mathcal{N}_r$ (hence $\mathcal{N}_t$, $\mathcal{N}_l$, and $\mathcal{N}_s$) by adding places $p_{run}$ and $p_{stop}$ and transition $t_{(1,p_1)}$, with $p_{run}$ being input and output with cardinality one to any other transition in the net (this is a standard technique which ensures that $t_{(1,p_1)}$ is the only enabled transition when the token is removed from $p_{run}$). Hence, given $\delta^- \in \mathbb{N}^{|P| \times |T|}$, we can allow $D^-_{i,j}$ to be of the form $\max\{\delta^-_{i,j}, \alpha^-_{i,j} + \beta^-_{i,j,i}\mu_i\}$ for nets is $\mathcal{N}_r$, $\mathcal{N}_t$, and $\mathcal{N}_l$, or $\max\{\delta^-_{i,j}, \alpha^-_{i,j} + \sum_{1 \le l \le |P|} \beta^-_{i,j,l}\mu_l\}$ for nets in $\mathcal{N}_s$, with the understanding that this is just a shorthand, not a real extension.

In the original definition [3], reset arcs were drawn from the transition to the place to be reset, with a circle drawn on them. Hence, the threshold behavior

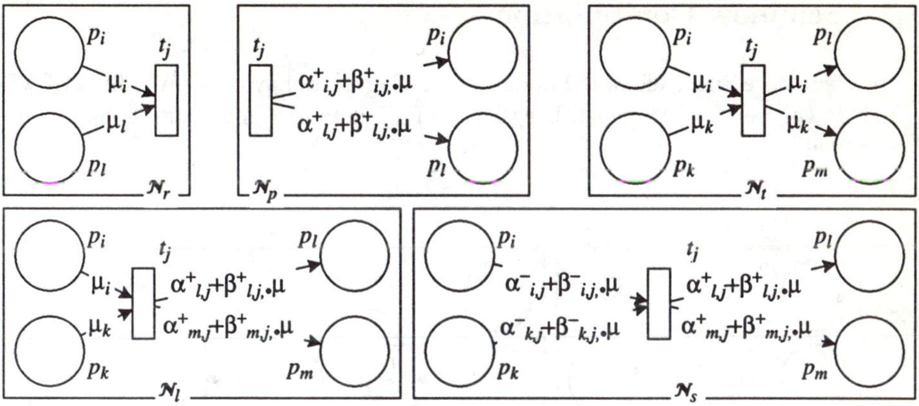


**Fig. 2.** Patterns of marking-dependent cardinality arcs for $\mathcal{N}_r$, $\mathcal{N}_p$, $\mathcal{N}_t$, $\mathcal{N}_l$, and $\mathcal{N}_s$.

could be represented naturally with a normal input arc (see net $N''$ in Fig. 3). An additional requirement could be that $m$ tokens be deposited in $p_1$ after this flushing, so that the entire semantic of $N$ is that $t_1$ fires only if $\mu_1 \geq k$, and, after the firing, $\mu_1 = m$. This can be accomplished with an arc from $t_1$, for $N$ and $N''$, or from $t_{(1,p_1)}$, for $N'$, back to $p_1$, with cardinality $m$.

It is also easy to show that a reset behavior can be modeled by $\mathcal{N}_t$, $\mathcal{N}_l$, and $\mathcal{N}_s$, in addition to $\mathcal{N}_r$. In the case of $\mathcal{N}_t$, it is necessary to add a dummy place where the tokens removed from the place to be reset can be deposited (or we could explicitly allow a reset behavior in addition to the transfer behavior in $\mathcal{N}_t$, by requiring $\beta^-_{i,j,i} \geq \sum_{1 \leq l \leq |P|} \beta^+_{l,j,i}$, instead of a strict equality).

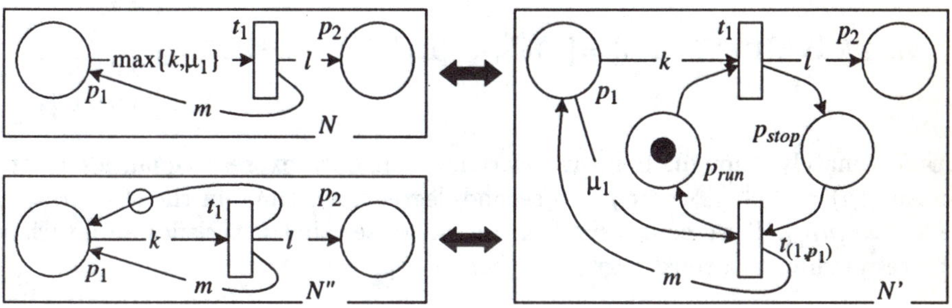


**Fig. 3.** Flushing $p_1$ only after a threshold $k$.

## 3 Semiflow Computation

Invariants are an excellent aid in understanding the behavior of an ordinary P/T-net. In this section, we generalize them to P/T-nets with marking-dependent arc cardinality.

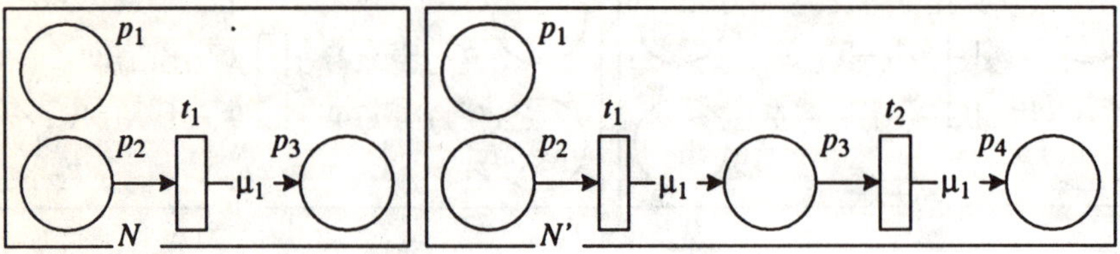


**Fig. 4.** P/T-nets with non-linear invariants.

### 3.1 P-Semiflows

In ordinary P/T-nets, a structural linear p-invariant could state that $\forall \mu \in \mathcal{R}(N), x \cdot \mu = x \cdot \mu^{[0]}$, where $x \in \mathbb{Q}^{|P|}$ is a p-flow, and satisfies $xD = 0$, and the operator "$\cdot$" indicates inner (scalar) product between two vectors. Specifically, we are interested in the minimal non-zero p-flows $x \in \mathbb{N}^{|P|}$, or minimal p-semiflows [14]. If $D$ contains arbitrary marking-dependent expressions, non-linear structural p-invariants might exist as well. For example, $N$ in Fig. 4 satisfies the non-linear invariant $\forall \mu \in \mathcal{R}(N), \mu_1\mu_2 + \mu_3 = \mu_1^{[0]}\mu_2^{[0]} + \mu_3^{[0]}$.

Following [25], this behavior can be described by a "bilinear invariant" $X \in \mathbb{Z}^{(1+|P|)\times(1+|P|)}$:

$$\forall \mu \in \mathcal{R}(N), [1,\mu]X \cdot [1,\mu] = [1,\mu^{[0]}]X \cdot [1,\mu^{[0]}] \quad \text{where} \quad X = \begin{bmatrix} 0&0&0&1 \\ 0&0&1&0 \\ 0&0&0&0 \\ 0&0&0&0 \end{bmatrix} .$$

Unfortunately, though, bilinear invariants can only express equalities of the form $p(\mu) = p(\mu^{[0]})$, where $p$ is a second-degree polynomial in the $|P|$ variables $\{\mu_1, \ldots, \mu_{|P|}\}$. If we consider $N'$ in Fig. 4, we see that it satisfies an invariant corresponding to a third-degree polynomial,

$$\forall \mu \in \mathcal{R}(N), \mu_1^2\mu_2 + \mu_1\mu_3 + \mu_4 = (\mu_1^{[0]})^2\mu_2^{[0]} + \mu_1^{[0]}\mu_3^{[0]} + \mu_4^{[0]} .$$

In this paper, we restrict ourselves to (linear) p-semiflows for the class $\mathcal{N}_s$ (hence $\mathcal{N}_r$, $\mathcal{N}_p$, $\mathcal{N}_t$, and $\mathcal{N}_l$), that is, solutions $x \in \mathbb{N}^{|P|}, x \neq 0$ to the equation $xD = 0$.

**Theorem 1.** The set of p-semiflows of $N = (P, T, D^-, D^+, \mu^{[0]}) \in \mathcal{N}_s$ equals the set of p-semiflows of $N' = (P, T', D'^-, D'^+, \mu^{[0]}) \in \mathcal{N}_o$, where $T' = T \cup \{t_{(j,l)} : 1 \le j \le |T|, 1 \le l \le |P|\}$ and $D'^-_{i,j} = \alpha^-_{i,j}$, $D'^+_{i,j} = \alpha^+_{i,j}$, $D'^-_{i,(j,l)} = \beta^-_{i,j,l}$, $D'^+_{i,(j,l)} = \beta^+_{i,j,l}$.

**Proof**: the entries of the incidence matrix $D$ of $N$ are

$$D_{i,j}(\mu) = \alpha^+_{i,j} - \alpha^-_{i,j} + \sum_{1 \le l \le |P|} (\beta^+_{i,j,l} - \beta^-_{i,j,l})\mu_l = \alpha_{i,j} + \sum_{1 \le l \le |P|} \beta_{i,j,l}\mu_l \ ,$$

where $\alpha = \alpha^+ - \alpha^-$, and $\beta = \beta^+ - \beta^-$. A p-semiflow for $N$, $x \in \mathbb{N}^{|P|}$, $x \ne 0$, satisfies $xD = 0$, resulting in the $|T|$ equations

$$\forall j, 1 \le j \le |T|, \sum_{1 \le i \le |P|} x_i \left( \alpha_{i,j} + \sum_{1 \le l \le |P|} \beta_{i,j,l}\mu_l \right) = 0 \ , \qquad (1)$$

which can be rearranged into

$$\forall j, 1 \le j \le |T|, \sum_{1 \le i \le |P|} x_i \alpha_{i,j} + \sum_{1 \le l \le |P|} \left( \sum_{1 \le i \le |P|} x_i \beta_{i,j,l} \right) \mu_l = 0 \ .$$

The above equations must be satisfied by $x$ independent of the marking $\mu$, hence, in particular, they must be satisfied by the empty marking, resulting in

$$\forall j, 1 \le j \le |T|, \sum_{1 \le i \le |P|} x_i \alpha_{i,j} = 0 \ , \qquad (2)$$

and by the $|P|$ markings having one token in place $l, 1 \le l \le |P|$, and no tokens elsewhere, resulting in

$$\forall j, 1 \le j \le |T|, \forall l, 1 \le l \le |P|, \sum_{1 \le i \le |P|} x_i \alpha_{i,j} + \sum_{1 \le i \le |P|} x_i \beta_{i,j,l} = 0 \ ,$$

which, given (2), imply

$$\forall j, 1 \le j \le |T|, \forall l, 1 \le l \le |P|, \sum_{1 \le i \le |P|} x_i \beta_{i,j,l} = 0 \ . \qquad (3)$$

But these are exactly the constraints satisfied by the p-semiflows of $N'$: (2) are imposed by the set of transitions $T \subseteq T'$ and (3) are imposed by the set of transitions $T' \setminus T$, hence $xD' = 0$. On the other hand, any p-semiflow $x$ for $N'$, solution of $xD' = 0$, satisfies (1) as well, and is therefore a p-semiflow for $N$. □

This result has an intuitively appealing interpretation. Consider, for example, Fig. 5, containing both $N \in \mathcal{N}_s$ and its corresponding $N' \in \mathcal{N}_o$ (when $\forall i, 1 \le i \le |P|, \beta^-_{i,j,l} = \beta^+_{i,j,l} = 0$, transition $t_{(j,l)}$ has empty input and output bags and can be removed from $N'$). Clearly, the effect of firing $t_1$ in $N$ when the marking

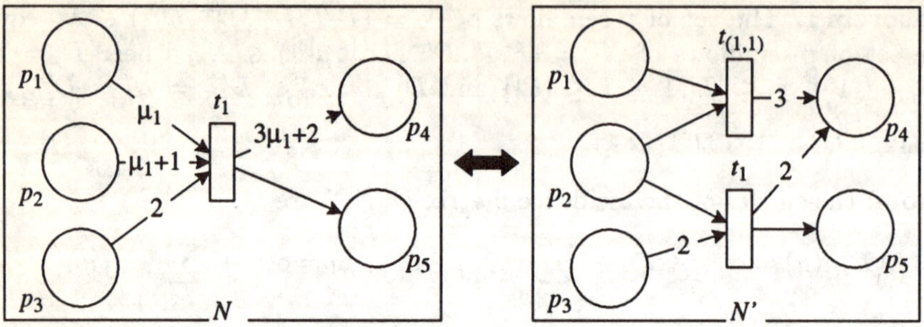


**Fig. 5.** Computing p-semiflows for a net $N \in \mathcal{N}_s$.

is $\mu$ can be simulated by firing $t_1$ once and $t_{(1,1)}$ $\mu_1$ times in $N'$. The constraints imposed on $x$ by this example are:

$$\text{from (2)}: \; 0x_1 + (-1)x_2 + (-2)x_3 + 2x_4 + 1x_5 = 0$$
$$\text{from (3)}: \; (-1)x_1 + (-1)x_2 + 0x_3 + 3x_4 + 0x_5 = 0$$

resulting in the four minimal p-semiflows shown in Fig. 6.

| $x_1$ | $x_2$ | $x_3$ | $x_4$ | $x_5$ |
|---|---|---|---|---|
| 1 | 2 | 0 | 1 | 0 |
| 3 | 0 | 1 | 1 | 0 |
| 0 | 3 | 0 | 1 | 1 |
| 0 | 0 | 1 | 0 | 2 |

**Fig. 6.** The minimal p-semiflows of $N$ in Fig. 5.

In ordinary P/T-nets, if, for a given $t_j$, $\alpha_{\bullet,j} \leq 0$ (or $\alpha_{\bullet,j} \geq 0$), no place $p_i$ for which $\alpha_{i,j} < 0$ (or $\alpha_{i,j} > 0$) can be covered by a linear structural p-invariant. In linear transfer and self modifying P/T nets, in addition, $p_i$ cannot be covered if, for some $l$, $\beta_{\bullet,j,l} \leq 0$ (or $\beta_{\bullet,j,l} \geq 0$) and $\beta_{i,j,l} < 0$ (or $\beta_{i,j,l} > 0$). This implies that a necessary condition for the net to be completely covered by linear structural p-invariants is that, for each $t_j \in T$, the set of places appearing in the marking dependent expressions of the input and output arc cardinalities must coincide. In particular, then, a place $p_i$ such that $\beta^-_{i,j} = 1$ in a reset P/T-net (or such that $\beta^+_{i,j,l} > 0$ for some $l$, in a post self modifying P/T-net) cannot be covered by a linear structural p-invariant.

The approach just used for the computation of minimal p-semiflows can still be applied even if arbitrary non-negative integer functions are used in the specification of the marking-dependent arc cardinalities, instead of just linear expressions. Consider, for example, the net $N$ in Fig. 7. Informally, the same approach still applies if we consider $\mu_3\mu_4$ and $\mu_1^2$ as "atomic terms", hence

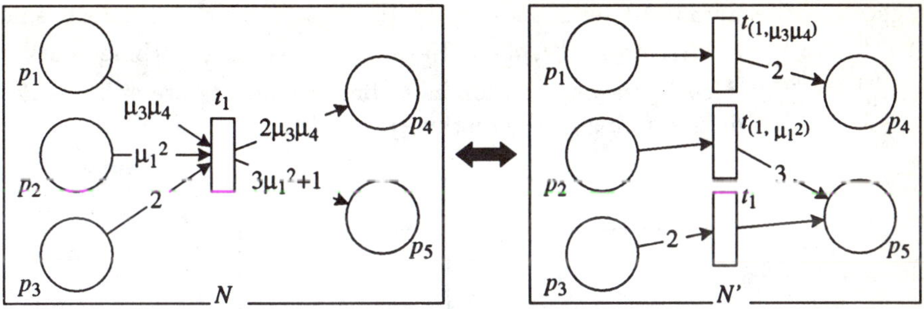


**Fig. 7.** P-semiflows in a net with nonlinear arc cardinalities.

the net $N'$ on the right in Fig. 7 has the same p-semiflows as $N$. A different transition can be defined for each atomic term, such as $t_{(1,\mu_3\mu_4)}$ and $t_{(1,\mu_1^2)}$ in $N'$. However, elaborate symbolic manipulations might be required to establish nonlinear integer equalities among atomic terms, (e.g., $\lfloor e^{\mu_1} \rfloor + 1 = \lceil e^{\mu_1} \rceil$, or $\mu_1^2 + \mu_2^2 + 2\mu_1\mu_2 = (\mu_1 + \mu_2)^2$). P/T-nets with such nonlinear behavior, though, are unlikely to be found in practical modeling applications, so we do not consider them further.

### 3.2 T-Semiflows

The existence of t-semiflows, solutions $y \in \mathbb{N}^{|T|}$, $y \neq 0$, for the equation $Dy = 0$ (where $y$ is a column vector), is also an important property of P/T-nets. If $y$ is interpreted as the firing vector of a sequence $\sigma \in T^*$, that is, $y_j$ is the number of times $t_j$ appears in $\sigma$, then $\forall \mu \in \mathbb{N}^{|P|}, \mu = \mathcal{M}(\sigma, \mu)$, provided $\sigma$ can be fired starting in $\mu$.

Unfortunately, with marking-dependent arc cardinalities, the definition of t-semiflows, let alone their existence, becomes a problem. If we naively carry on the product $Dy$, we obtain:

$$\forall i, 1 \leq i \leq |P|, \sum_{1 \leq j \leq |T|} \left( \alpha_{i,j} + \sum_{1 \leq l \leq |P|} \beta_{i,j,l}\mu_l \right) y_j \ ,$$

but these $|P|$ equations are semantically incorrect, since the usage of $\mu_l$ is ambiguous. When multiplied by $\beta_{i,j_1,l}$, $\mu_l$ refers to the number of tokens in $p_l$ when $t_{j_1}$ fires, but, when multiplied by $\beta_{i,j_2,l}$, it refers to the number of tokens in $p_l$ when $t_{j_2}$ fires, which can be a different quantity (this quantity can vary even between two firings of the same transition). The following example typifies the problem. Consider $N$ in Fig. 8 where the equation $Dy = 0$ results in

$$\begin{bmatrix} -\mu_1 & \mu_1 \\ \mu_1 & -\mu_1 \end{bmatrix} \begin{bmatrix} y_1 \\ y_2 \end{bmatrix} = 0 \ \Rightarrow \ -\mu_1 y_1 + \mu_1 y_2 = 0 \ \Rightarrow \ y_1 = y_2 \ ,$$

suggesting the minimal t-semiflow $y = [1,1]$. This is clearly incorrect, since $[0, \mu_1^{[0]} + \mu_2^{[0]}] = \mathcal{M}((t_1,t_2),[\mu_1^{[0]}, \mu_2^{[0]}])$. Indeed, $N$ has an absorbing marking $[0, \mu_1^{[0]} + \mu_2^{[0]}]$, which is reached as soon as $t_1$ fires ($t_1$ and $t_2$ are still enabled, but their firing does not change the marking).

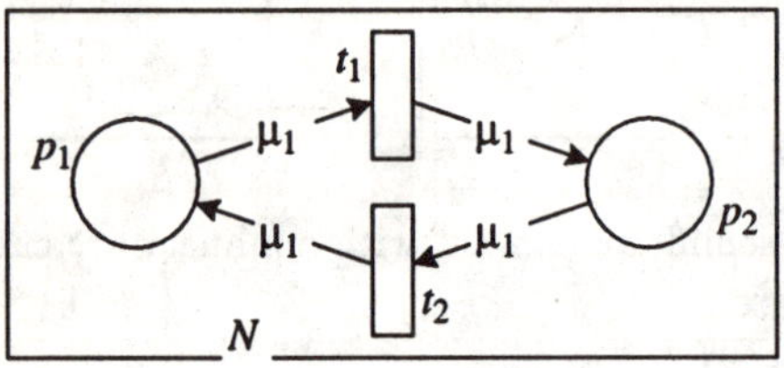


**Fig. 8.** A net with no t-invariants.

This shows how the existence of t-semiflows is irreparably compromised by marking-dependent arc cardinalities, unless the marking-dependent arc cardinalities themselves happen to be irrelevant, as in the case of Fig. 9, where the only difference between $N$ and $N'$ is that the recurrent markings are $[\mu_1^{[0]} + \mu_2^{[0]}, 0]$ and $[0, \mu_1^{[0]} + \mu_2^{[0]}]$ in the former, and $[1,0]$ and $[0,1]$ in the latter. Both nets have the minimal t-semiflow $y = [1,1]$.

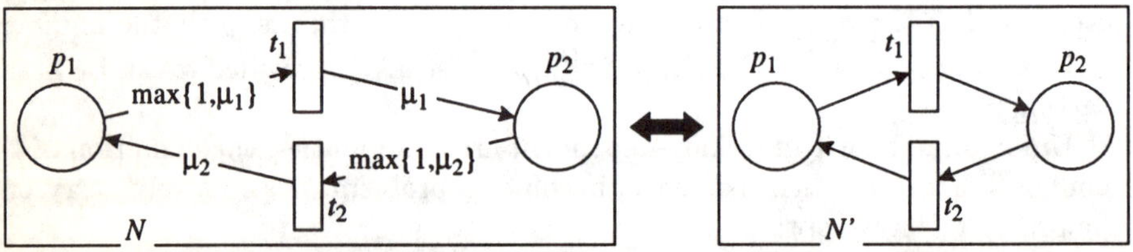


**Fig. 9.** A net with marking-dependent arc cardinalities and an equivalent one without.

Some useful information about firing sequences can still be obtained, though, even in the presence of marking-dependent arc cardinalities. An important reason to study t-semiflows when modeling a system is to establish relationships among the throughputs (firing rates) of various transitions. For example, assume that we are interested in performing a steady-state simulation of $N$ in Fig. 10, which has minimal t-semiflows $y^{(1)} = [1,1,1,0,1]$ and $y^{(2)} = [0,1,0,1,0]$. We can say that $y^{(1)}$ "fires" at the unknown rate $\phi^{(1)}$, and that $y^{(2)}$ "fires" at the unknown rate $\phi^{(2)}$. If $\phi_j$ is the throughput of transition $t_j$, we can then conclude that $\phi_1 = \phi_3 = \phi_5 = \phi^{(1)}$, $\phi_2 = \phi^{(1)} + \phi^{(2)}$, and $\phi_4 = \phi^{(2)}$, hence we could collect statistics only for $t_2$ and $t_4$, and still obtain the throughput of all five transitions ($\phi_1 = \phi_3 = \phi_5 = \phi_2 - \phi_4$).

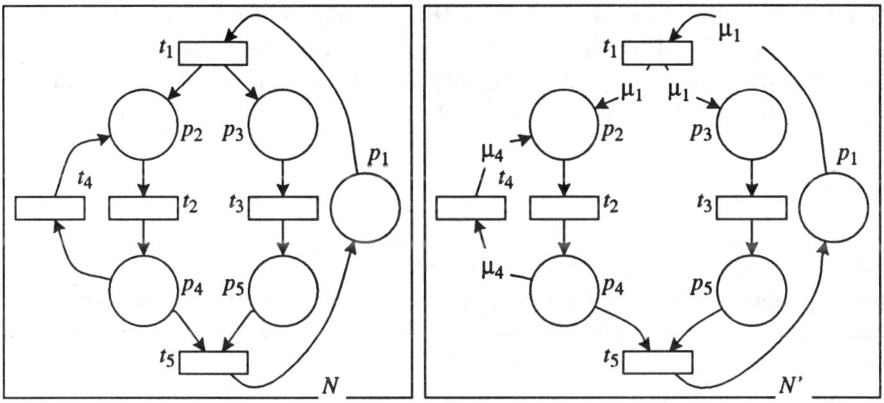


**Fig. 10.** An example.

Consider now the case of $N' \in \mathcal{N}_l$ in Fig. 10, where

$$D = \begin{bmatrix} -\mu_1 & & & & 1 \\ \mu_1 & -1 & & \mu_4 & \\ \mu_1 & & -1 & & \\ & 1 & & -\mu_4 & -1 \\ & & 1 & & -1 \end{bmatrix} \quad \text{and} \quad Dy = 0 \ \Rightarrow \ \begin{cases} -\mu_1 y_1 + y_5 = 0 \\ \mu_1 y_1 - y_2 + \mu_4 y_4 = 0 \\ \mu_1 y_1 - y_3 = 0 \\ y_2 - \mu_4 y_4 - y_5 = 0 \\ y_3 - y_5 = 0 \end{cases} .$$

Defining $y'_1 = \mu_1 y_1$ and $y'_4 = \mu_4 y_4$, a set of equations in $\{y'_1, y_2, y_3, y'_4, y_5\}$ results, with the same coefficients as in the constant case. Hence, we can compute the following "pseudo t-semiflows": $y^{(1)} = [1/\mu_1, 1, 1, 0, 1]$ and $y^{(2)} = [0, 1, 0, 1/\mu_4, 0]$. These can then be used to compute bounds on the ratio between throughputs, provided we know bounds on the minimum and maximum values that $\mu_1$ and $\mu_4$ can assume when $t_1$ and $t_4$ fire, respectively (these, of course, depend on $\mu^{[0]}$).

Given the initial marking and the fact that $N'$ is covered by p-semiflows, we can compute the maximum number of tokens in $p_1$ and $p_4$ as $M_1 = \mu_1^{[0]} + \min\{\mu_2^{[0]} + \mu_4^{[0]}, \mu_3^{[0]} + \mu_5^{[0]}\}$, and $M_4 = \mu_1^{[0]} + \mu_2^{[0]} + \mu_4^{[0]}$, respectively. Furthermore, the system might impose a minimum threshold $m_1 = \delta^-_{1,1}$ and $m_4 = \delta^-_{4,4}$ on the number of tokens required in $p_1$ and $p_4$ before $t_1$ and $t_4$ can fire, respectively. Then, the pseudo t-semiflows still fire at an unknown rate $\phi^{(1)}$ and $\phi^{(2)}$. We can then conclude that $\phi_1 \in [\phi^{(1)}/M_1, \phi^{(1)}/m_1]$, $\phi_2 = \phi^{(1)} + \phi^{(2)}$, $\phi_3 = \phi_5 = \phi^{(1)}$, and $\phi_4 \in [\phi^{(2)}/M_4, \phi^{(2)}/m_4]$.

If $p_1$ is unbounded ($M_1 = \infty$), we only obtain the upper bound $\phi_1 < \phi^{(1)}/m_1$. If $t_1$ can fire when $p_1$ is empty ($m_1 = 0$), we only obtain the lower bound $\phi_1 > \phi^{(1)}/M_1$ and, in the particular case of $N$ in Fig. 10, the throughput of $t_1$ is then simply equal to the inverse of its average firing time, since $t_1$ is always enabled.

Realistically, though, it is more likely that, in such a system, we are interested in computing the throughput of $t_1$ counting only the firings that change the marking, which is equivalent to saying $m_1 = 1$. A meaningful upper bound on $\phi_1$ as a function of $\phi^{(1)}$ can then be obtained.

## 4 Expressive Power

When modeling a system with a P/T-net, it is important to know whether the net is bounded or live, and whether certain markings are reachable. State-space-based analytical approaches, for example, require to generate and store the entire reachability graph, which must then be finite. Some partial answers might be obtainable from invariant analysis (a net is bounded if it is completely covered by p-semiflows), but more complex issues of decidability exist. Traditionally, these have been studied by addressing P/T-nets as language generators.

Given a P/T-net $N = (P, T, D^-, D^+, \mu^{[0]})$, a set of final markings $\mathcal{F} \subseteq \mathcal{R}(N)$, and a labelling function $\sigma : T \to \Sigma \cup \lambda$, where $\lambda \notin \Sigma$ denotes the empty string, we can define the language $\mathcal{L}(N, \mathcal{F}, \sigma) = \{\sigma(s) : s \in T^*, \mathcal{M}(s, \mu^{[0]}) \in \mathcal{F}\}$. Petri net languages can then be classified according to three parameters:

1. The type of P/T-net: $\mathcal{N}_o$, $\mathcal{N}_r$, $\mathcal{N}_p$, $\mathcal{N}_t$, $\mathcal{N}_l$, and $\mathcal{N}_s$.
2. The set of final states $\mathcal{F}$ [21]:
   $\mathcal{F} = \mathcal{R}(N)$: *P*-type, any reachable marking is a final state.
   $\mathcal{F} = \{\mu_1^f, \mu_2^f, \ldots, \mu_k^f\}$: *L*-type, a finite set of reachable markings.
   $\mathcal{F} = \{\mu \in \mathcal{R}(N) : \exists \mu^f \in \{\mu_1^f, \mu_2^f, \ldots, \mu_k^f\}, \mu \geq \mu^f,\}$: *G*-type, any reachable marking covering one in a finite set of markings.
   $\mathcal{F} = \{\mu \in \mathcal{R}(N) : \mathcal{E}(\mu) = \emptyset\}$: *T*-type, any terminal (absorbing, dead) marking is a final state.
3. The labelling function $\sigma$ [17]:
   $\forall t \in T, \sigma(t) \in \Sigma \wedge \forall t_1, t_2 \in T, \sigma(t_1) = \sigma(t_2) \Rightarrow t_1 = t_2$: *Free*, each transition has a different label in $\Sigma$.
   $\forall \mu \in \mathcal{R}(N), \forall t_1, t_2 \in \mathcal{E}(\mu), t_1 \neq t_2 \Rightarrow \sigma(t_1) \in \Sigma \wedge \sigma(t_2) \in \Sigma \wedge \sigma(t_1) \neq \sigma(t_2)$: *Deterministic*, each enabled transition has a different label, and, if its label is $\lambda$, it is the only one enabled.
   $\forall t \in T, \sigma(t) \in \Sigma$: *Non-$\lambda$*, each transition has a non-empty label.
   $\forall t \in T, \sigma(t) \in \Sigma \cup \{\lambda\}$: $\lambda$, no restrictions.

Using a notation similar to that of Peterson [21], $X_z^y$ indicates the class of languages which can be generated by a PN of type $z$ ($o$, $r$, $p$, $t$, $l$, and $s$) using a final state definition of type $X$ ($P$, $L$, $G$, or $T$), and a labelling function of type $y$ ($f$, $d$, $n$, $\lambda$).

From a modeling point of view, $\lambda$-transitions are in some way analogous to immediate transitions, which fire in zero time [1], and non-$\lambda$-transitions are analogous to timed transitions. One major difference, though, is that immediate transitions have an implicit priority over timed transitions, since they will fire before any timed transition with probability one. Since P/T-nets with priorities can model Turing machines, the introduction of immediate transitions could destroy our ability to analyze the net. Two possible solutions are:

- Introduce a new type of labelling, "exclusive-$\lambda$" that does not require an implicit priority: $\forall \mu \in \mathcal{R}(N), \forall t_1, t_2 \in \mathcal{E}(\mu), \sigma(t_1) = \lambda \Rightarrow \sigma(t_2) = \lambda$.
  This is more general than a non-$\lambda$ labelling, but less general than a $\lambda$-labelling. Unfortunately, to the best of our knowledge, this type of labelling has not been considered by researchers.
- Restrict ourselves to $k$-prompt nets [15], where no more than $k$ $\lambda$-transitions can fire consecutively. Such a net can be transformed into a non-$\lambda$ net, $\lambda$-transitions are just a shorthand. This is in particular true of a bounded net, where the only infinite sequences of $\lambda$-transitions must return infinitely often to the same marking. This fact was used in [7] to show that immediate transitions can be eliminated from GSPN families if the net is bounded. Unfortunately, there are practical systems which do not fall in this category. Consider for example the GSPN in Fig. 11, modeling a queue with batch arrivals of arbitrary size. The two immediate transitions $t_2$ and $t_3$, drawn with a thin line, cannot be eliminated, since the net is unbounded and not $k$-prompt. The "reduced reachability graph" shown on the right is the results of the elimination of the vanishing markings, that is, markings that enable only immediate transitions. Note that, in this net, immediate transitions are not required to have implicit priority over timed transitions, since place $p_{run}$ enforces mutual exclusion between timed and immediate transitions. This is an example of exclusive-$\lambda$ labelling.

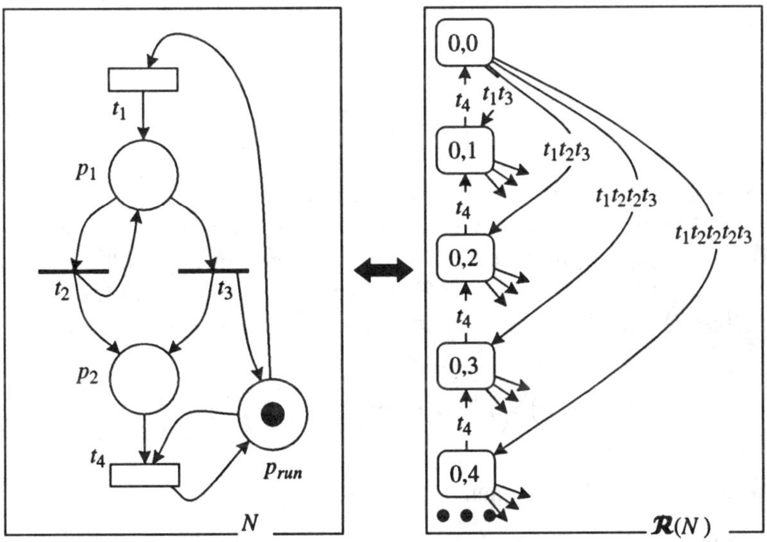


**Fig. 11.** A case where immediate transitions cannot be eliminated.

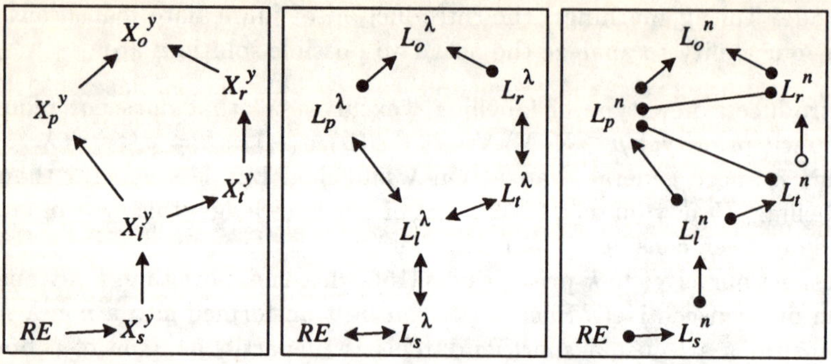


**Fig. 12.** Relations among some Petri net languages.

Given the definition of $\mathcal{N}_o$, $\mathcal{N}_r$, $\mathcal{N}_p$, $\mathcal{N}_t$, $\mathcal{N}_l$, and $\mathcal{N}_s$, we can immediately derive the relations shown on the left portion of Fig. 12, for any $X$ and $y$ (on an edge $(a, b)$, an arrowhead on $b$ signifies that any language in $b$ is also a language in $a$, a full, or empty, circle on $b$ signifies that there are, or we conjecture the existence of, languages in $b$ which are not in $a$, respectively).

For $L_\bullet^\lambda$ languages, it is known that $L_o^\lambda$ cannot generate some context-free language [17], while $L_r^\lambda$ or $L_p^\lambda$ can model a counter automaton, hence they allow to generate any recursive enumerable language ($RE$) [3, 24], resulting in the center portion of Fig. 12. On the other hand, it is also known that, for non-$\lambda$, hence $k$-prompt, languages, there is a strict hierarchy as shown in the right portion of Fig. 12 [25]. The only relations shown in Fig. 12 and not discussed before in the referenced literature are those regarding transfer nets. Since, as pointed out in Sect. 2.2, transfer behavior includes reset behavior, the only relations that need to be proven are:

$L_p^n$ •–$L_t^n$: consider the language

$$\mathcal{L}_1 = \{wcd^{f(w)} : w \in \{a,b\}^*\} ,$$

where $f$ is the value of $w$ in base 3 when $a = 1$ and $b = 2$: $f(\lambda) = 0$, $f(wa) = 3f(w){+}1$, $f(wb) = 3f(w){+}2$. Using an argument similar to that of Peterson's [21], we can show that $\mathcal{L}_1$ cannot be in $L_t^n$ (hence in $L_r^n$ or $L_o^n$). Assume that $\mathcal{L}_1 \in L_t^n$, that is, $\mathcal{L}_1 = \mathcal{L}(N, \mathcal{F}, \sigma)$, where $N = (P, T, D^-, D^+, \mu^{[0]})$ and $\forall t \in T, \sigma(t) \in \{a, b, c, d\}$. Then after $k = |w|$ firings in $N$, at most

$$n(k) = \mu^{[0]} + k \max_{1 \le j \le |T|} \left\{ \sum_{1 \le i \le |P|} \alpha_{i,j} \right\}$$

tokens can be in the net, hence at most

$$\binom{n(k) + |P| - 1}{n(k)} = \frac{(n(k) + |P| - 1) \cdots (n(k) + 1)}{(|P| - 1)!} < (n(k) + |P|)^{|P|}$$

different markings can be reached in $k$ firings. But $n(k)$ is linear in $k$, hence $(n(k)+|P|)^{|P|} < 2^k$ for a sufficiently large $k$, and, after $k$ firings, $N$ cannot distinguish between the $2^k$ possible $w \in \{a,b\}^k$, that is, $\exists w_1, w_2 \in \{a,b\}^k$, $w_1 \neq w_2$, such that $\mathcal{M}(w_1, \mu^{[0]}) = \mathcal{M}(w_2, \mu^{[0]})$. Then, $\mathcal{M}(w_1 cd^{f(w_1)}, \mu^{[0]}) = \mathcal{M}(w_2 cd^{f(w_1)}, \mu^{[0]})$, while we should have $w_1 cd^{f(w_1)} \in \mathcal{F}$ and $w_2 cd^{f(w_1)} \notin \mathcal{F}$, a contradiction.

The net $N' \in L_p^n$ in Fig. 13, instead, generates $\mathcal{L}_1$ with a non-$\lambda$ labelling and $\mathcal{F} = \{[0,0,1]\}$.

$L_p^n \multimap\!\!\bullet\, L_t^n$: follows from $L_p^n \multimap\!\!\bullet\, L_r^n$ [25] and $L_t^n \to L_r^n$. We conjecture that

$$\mathcal{L}_2 = \{a^{m_1} cb^{n_1} d \cdots a^{m_k} cb^{n_k} d : k \in \mathbb{N}, \forall i, 1 \leq i \leq k, m_i \geq n_i\}$$

is not in $L_p^n$, but it is generated by $N \in \mathcal{N}_r$ in Fig. 14, with $\mathcal{F} = \{[1,0,0]\}$.

$L_l^n \bullet\!\!\rightarrow L_t^n$: follows from $L_p^n \bullet\!\!- L_t^n$ and $L_l^n \to L_p^n$.

$L_t^n \bullet\!\!- L_r^n$: we conjecture that

$$\mathcal{L}_3 = \{a^{m_1} cb^{n_1} d \cdots a^{m_k} cb^{n_k} def^l : k \in \mathbb{N}, \forall i, 1 \leq i \leq k, m_i \geq n_i, l = \sum_{1 \leq i \leq k} (m_i - n_i)\}$$

is not in $L_r^n$, but it is generated by $N' \in \mathcal{N}_t$ in Fig. 14, with $\mathcal{F} = \{[0,0,0,1,0]\}$.

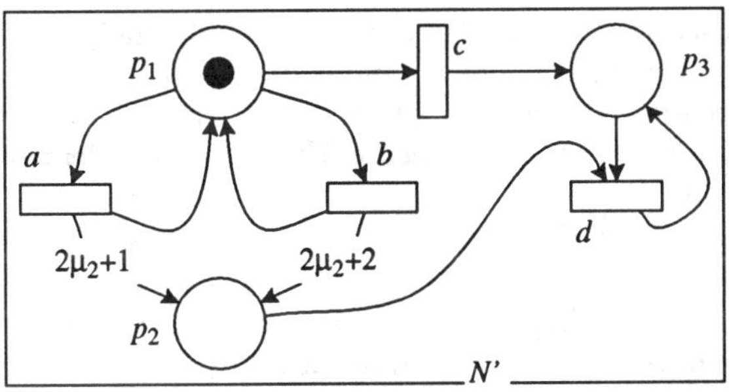


**Fig. 13.** A language in $L_p^n$ but not in $L_t^n$.

## 5 Practical Applications of Variable Cardinality Arcs

From the previous sections, it should be apparent that P/T-nets with marking-dependent arc cardinalities have the potential to express certain system behaviors more naturally than ordinary P/T-nets. This is of particular interest when solving large SPN models.

Many SPN models have exponentially distributed or zero firing times, hence an underlying continuous-time Markov chain, as in GSPNs [1, 10], or exponentially distributed or constant, including zero, firing times, hence an underlying

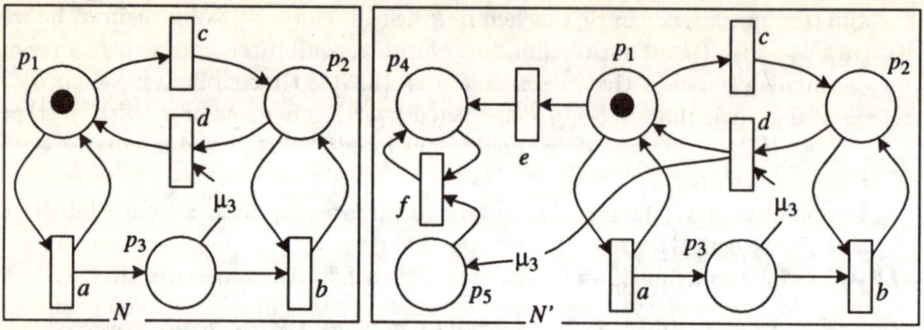


**Fig. 14.** Conjectures: a language in $L_r^n$ but not in $L_p^n$ and one in $L_t^n$ but not in $L_r^n$.

semi-regenerative process, as in DSPNs [2, 10]. With finite reachability sets, the marking-dependent arc cardinalities are not needed in theory, since the behavior of the system is described by a large finite automaton, but, in practice, they improve the description, resulting in a more concise model, and allow a more efficient generation of the state space.

In all cases, the behavior to be modeled is the transfer of all tokens from one or more places to one or more other places, often as a result of a some kind of reset action. A common pattern, then, is that of $N$ in Fig. 15. If ordinary P/T-nets are used, a subnet similar to that of $N'$ in Fig. 15 must be used, where, in the usual notation of SPNs, immediate transitions, having a zero firing time, are drawn with a thin bar. Alternative approaches could use transitions priorities or guards or, when the net is bounded, complementary places, but the amount of clutter would be analogous. Clearly, $N$ and $N'$ have the same stochastic behavior, but the readability of $N'$ is seriously hindered.

Furthermore, computer-based packages used for the numerical solution of these models [6, 13, 18, 19] generate the entire reachability graph. Structural simplifications aimed at removing immediate transitions before the generation of the reachability graph are sometimes possible [7]. These technique apply to GSPN families, that is, they are independent of the initial marking $\mu^{[0]}$, and substitute a timed transition $t_i$, plus one or more immediate transitions, with a set of timed transitions. They are not applicable to our case, though, since the number of timed transitions to be introduced is dependent on the maximum number of tokens to be removed, hence is not obtainable from a structural analysis of the net. If we are willing to perform simplifications that take into account the initial marking, then an equivalent SPN can always be obtained from a GSPN, provided the reachability set is finite, but the efficiency of this approach is questionable.

With $N'$ as input, then, a computer tool has little choice but generating long sequences of vanishing markings, which are then eliminated either during or after the generation of the reachability graph.

The resulting reduced reachability graph is the same as that generated by $N$, but the computational effort is larger. The total number of additional reachability graph nodes and arcs corresponding to the vanishing markings due to $t_{(1,0)}$, $t_{(1,1)}$, and $t_{(1,2)}$ could be as large as the number of "tangible" (enabling only timed transitions) markings. In fact, for each tangible marking $\mu$ with a token in $p_3$ there is a corresponding vanishing marking $\mu'$ with a token in $p_{(1,0)}$ with $\mu'_1 = \mu_1$, $\mu'_2 = \mu_2$, and $\mu'_3 = \mu_3$.

The number of additional reachability graph arcs is even larger if the inhibitor arc from $p_2$ to $t_{(1,1)}$ is omitted, since, in this case, all the possible interleavings of $t_{(1,1)}$ and $t_{(1,2)}$ are generated. The resulting stochastic process is still the same as when the firings of $t_{(1,2)}$ are considered before those of $t_{(1,1)}$, since both transitions have zero firing times, but more arcs will be generated. Fortunately, this situation, "compatible" immediate transitions, is recognizable at the structural level in this case [4].

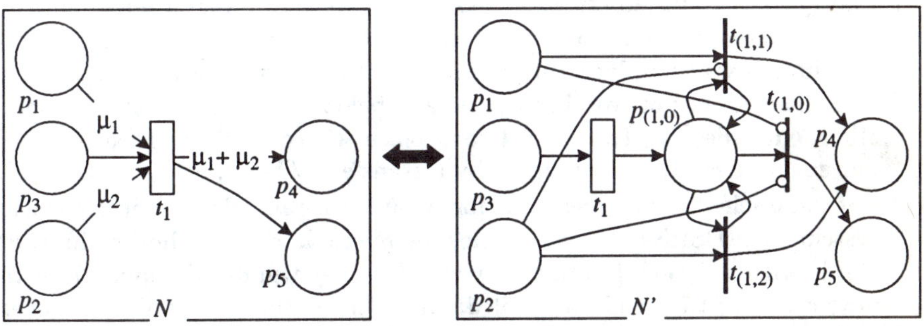


**Fig. 15.** A common pattern when modeling closed systems.

In the following, we give a few references to recent practical modeling applications where marking-dependent arc cardinalities are used, or could have been used, advantageously. In all cases, the stochastic behavior is that of a closed system and the net is covered by p-semiflows.

[5] models the availability of a replicated file system using CTMC-based SPNs. Various hosts can fail and be repaired, hence a read or write request for a file can be satisfied by any working host, as long as there is a quorum to certify that the host has an up-to-date copy. Both the quorum-gathering process and the mechanism by which hosts not included into the quorum for a write request become out-of-date require to move a marking-dependent number of tokens. As an ordinary P/T-net was used, the resulting model is quite complex.

[12] models the productivity of a flexible manufacturing system with CTMC-based stochastic reward nets. When finished parts are ready to leave the factory, they are gathered in a place, from where they are removed periodically, in bulk. This behavior closely reflects reality, where the means used

for transportation do not usually carry items individually. A finished part of type $P12$ is obtained by assembling two different parts, $P1$ and $P2$. To maintain a closed environment, $k$ raw parts $P1$ and $P2$ enter the system whenever $k$ finished parts $P12$ leaves. The resulting model is a linear transfer P/T-net.

[11] studies the performance of a simple protocol with DSPNs. In this protocol, if no acknowledgment is received within a given constant time, a timeout occurs and the current message is retransmitted. This can happen either because the message was not received properly at the other end or because of delays. In the resulting P/T-net, these two cases correspond to having a token in one of two places. When the timeout occurs, though, the token must be removed, independently of its position. This is easily accomplished using a linear transfer P/T-net with a single transition which resets both places. In this case, though, the same behavior could be represented by an ordinary P/T-net with two transitions, one to remove the token if it is in the first place, the other if it is in the second place.

[16] analyzes the availability of a VAXcluster system. When a node fails, the cluster attempts to reconfigure itself. Failed nodes with uncovered faults remain in a waiting place until either an automatic reboot isolates them, or until the number of working nodes fails below a threshold. At this point, all failed nodes are transferred to a place where they are repaired. This movement is easily accomplished by a transfer P/T-net.

[20] studies self-stability measures for a fault-tolerant clock synchronization system with DSPNs. At each cycle, the operable clocks "choose" independently with a Bernoulli probability whether they will be still operable at the next cycle, or they will have failed. After this decision is made, the tokens corresponding to the clocks which did not fail must be moved back to the place containing the operable clocks, in zero time. Since an ordinary P/T-net is used, these tokens are moved back one at a time with an immediate transition, with a negative effect on the readability of the model and on the execution time. A transfer P/T-net would have allowed to move all these tokens with the firing of a single transition.

## 6 Concluding Remarks

We have discussed several behaviors which, in the unbounded case, cannot be modeled by ordinary P/T-nets, yet are often needed when modeling real systems.

It is well known that *zero-testing*, the ability of firing a transition $t$ only when a given place $p$ is empty, for example by using inhibitor arcs, makes the formalism Turing-equivalent.

There are at least two other extensions, less known than zero-testing and complementary to each other, which are of practical interest: *zero-enforcing* ($\mathcal{N}_r$), the ability to ensure that, *after* the firing of a transition $t$, a given place $p$ becomes empty, and *duplication* ($\mathcal{N}_p$), that is, the ability to add to a place $p_1$, after the firing of a transition $t$, as many tokens as there are in a place $p_2$.

In addition, the ability to transfer all the tokens from a place to another ($\mathcal{N}_t$), a generalization of zero-enforcing, is conjectured to be more general than strict zero-enforcing, although it cannot model duplication. On one hand this could seem obvious, since the restriction imposed on $\mathcal{N}_t$ is not dissimilar to requiring, in ordinary P/T-nets, that the number of tokens generated by a firing is not greater than the number of tokens consumed. On the other hand, in $\mathcal{N}_l$, this rule only applies to the marking-dependent portion of the arc cardinalities, not to the constant portion, so a firing can still increase the number of tokens in the net.

We conclude by observing that, from a modeling point of view, the most important feature of each subclass considered is the type of reachability graphs that it can generate. Translated into the terminology of Petri net languages, this corresponds to $P$-type languages, for which, unfortunately, few results are know.

One important property shared by $\mathcal{N}_o$, $\mathcal{N}_r$, $\mathcal{N}_p$, $\mathcal{N}_t$, and $\mathcal{N}_l$, but not $\mathcal{N}_s$ or inhibitor nets, is *monotonicity* [25] (for both the enabling and the firing of a transition): $t \in \mathcal{E}(\mu) \wedge \mu \leq \mu' \Rightarrow t \in \mathcal{E}(\mu') \wedge \mathcal{M}(t,\mu) \leq \mathcal{M}(t,\mu')$. Note, though, that, strict inequality implies strict inequality in the above for $\mathcal{N}_o$ and $\mathcal{N}_p$, but not for $\mathcal{N}_r$, $\mathcal{N}_t$, and $\mathcal{N}_l$. This important property strongly characterizes the structure of the reachability graph.

## References


1. M. Ajmone Marsan, G. Balbo, and G. Conte. A class of Generalized Stochastic Petri Nets for the performance evaluation of multiprocessor systems. *ACM Trans. Comp. Syst.*, 2(2):93–122, May 1984.
2. M. Ajmone Marsan and G. Chiola. On Petri Nets with deterministic and exponentially distributed firing times. In G. Rozenberg, editor, *Adv. in Petri Nets 1987, Lecture Notes in Computer Science 266*, pages 132–145. Springer-Verlag, 1987.
3. T. Araki and T. Kasami. Some decision problems related to the reachability problem for Petri nets. *Theoretical Computer Science*, 3:85–104, 1977.
4. G. Balbo, G. Chiola, G. Franceschinis, and G. Molinari Roet. On the efficient construction of the tangible reachability graph of generalized stochastic Petri nets. In *Proc. of the Int. Workshop on Petri Nets and Performance Models*, Madison, Wisconsin, Aug. 1987.
5. J. Bechta Dugan and G. Ciardo. Stochastic Petri net analysis of a replicated file system. *IEEE Trans. Softw. Eng.*, 15(4):394–401, Apr. 1989.
6. G. Chiola. A Graphical Petri Net Tool for Performance Analysis. In *Proc. 3rd Int. Conf. on Modeling Techniques and Tools for Performance Analysis, Paris*, pages 323–333, 1987.
7. G. Chiola, S. Donatelli, and G. Franceschinis. GSPNs versus SPNs: what is the actual role of immediate transitions? In *Proc. of the Fourth Int. Workshop on Petri Nets and Performance Models (PNPM91)*, Melbourne, Australia, Dec. 1991.
8. G. Ciardo. *Analysis of large stochastic Petri net models.* PhD thesis, Duke University, Durham, North Carolina, 1989.
9. G. Ciardo, A. Blakemore, P. F. J. Chimento, J. K. Muppala, and K. S. Trivedi. Automated generation and analysis of Markov reward models using Stochastic Reward Nets. In C. Meyer and R. J. Plemmons, editors, *Linear Algebra, Markov*

*Chains, and Queueing Models*, volume 48 of *IMA Volumes in Mathematics and its Applications*, pages 145–191. Springer-Verlag, 1993.

10. G. Ciardo, R. German, and C. Lindemann. A characterization of the stochastic process underlying a stochastic Petri net. In *Proc. of the Fifth Int. Workshop on Petri Nets and Performance Models (PNPM93)*, Toulouse, France, Oct. 1993.
11. G. Ciardo and C. Lindemann. Analysis of deterministic and stochastic Petri nets. In *Proc. of the Fifth Int. Workshop on Petri Nets and Performance Models (PNPM93)*, Toulouse, France, Oct. 1993.
12. G. Ciardo and K. S. Trivedi. A decomposition approach for stochastic reward net models. *Perf. Eval.*, 18:37–59, 1993.
13. G. Ciardo, K. S. Trivedi, and J. Muppala. SPNP: stochastic Petri net package. In *Proc. of the Third Int. Workshop on Petri Nets and Performance Models (PNPM89)*, pages 142–151, Kyoto, Japan, Dec. 1989. IEEE Computer Society Press.
14. J. M. Colom and M. Silva. Convex geometry and semiflows in P/T nets. A comparative study of algorithms for the computation of minimal p-semiflows. In *10th International Conference on Application and Theory of Petri Nets*, Bonn, Germany, 1989.
15. M. Hack. Petri net languages. Technical Report 159, Laboratory for Computer Science, Massachusetts Institute of Technology, Cambridge, Massachussetts, Mar. 1976.
16. O. C. Ibe and K. S. Trivedi. Stochastic Petri net modeling of VAXCluster system availability. In *Proc. of the Third Int. Workshop on Petri Nets and Performance Models (PNPM89)*, Kyoto, Japan, Dec. 1989.
17. M. Jantzen. Language theory of Petri nets. In W. Brauer, W. Reisig, and G. Rozenberg, editors, *Adv. in Petri Nets 1986, Part 1, Lecture Notes in Computer Science 254*, pages 397–412. Springer-Verlag, 1986.
18. G. Klas and R. Lepold. TOMSPIN, a tool for modeling with stochastic Petri nets. In *CompEuro 92*, pages 618–623, The Hague, The Netherlands, May 1992.
19. C. Lindemann. DSPNexpress: A software package for the efficient solution of Deterministic and Stochastic Petri Nets. In *Proc. 6th Int. Conf. on Modelling Techniques and Tools for Computer Performance Evaluation*, pages 15–29, Edinburgh, Great Britain, 1992.
20. M. Lu, D. Zhang, and T. Murata. Analysis of self-stabilizing clock synchronization by means of stochastic Petri nets. *IEEE Trans. Comp.*, 39:597–604, 1990.
21. J. L. Peterson. *Petri Net Theory and the Modeling of Systems*. Prentice-Hall, 1981.
22. C. Petri. *Kommunikation mit Automaten*. PhD thesis, University of Bonn, Bonn, West Germany, 1962.
23. W. Reisig. *Petri Nets*, volume 4 of *EATC Monographs on Theoretical Computer Science*. Springer-Verlag, 1985.
24. R. Valk. On the computational power of extended Petri nets. In *Seventh Symposium on Mathematical Foundations of Computer Science, Lecture Notes in Computer Science 64*, pages 527–535. Springer-Verlag, 1978.
25. R. Valk. Generalizations of Petri nets. In *Mathematical foundations of computer science, Lecture Notes in Computer Science 118*, pages 140–155. Springer-Verlag, 1981.

# New Structural Invariants for Petri Nets Analysis

**J.M. Couvreur, E. Paviot-Adet**

Laboratoire MASI, Institut Blaise Pascal, Université Paris VI
4, Place Jussieu, 75252 Paris CEDEX 05, France

**Abstract**

This paper introduces two new kinds of invariant relations based on the Petri net structure and the already proved invariant properties. These invariants establish a relation between the marking of two place subsets A and B: *Exclusive Invariant* states that A and B cannot be simultaneously marked, *Implication Invariant* states that when A is marked, B is also marked. In order to show the applicability of the presented invariants to the validation of models, we apply these invariants to the validation of two classical distributed algorithms: alternate bit protocol and Peterson's algorithm for n processes

## 1. Introduction

Distributed programs are often described informally. Such a description offers the possibility to study a problem without the formal background of a theory. The major drawback of this approach is that it can easily lead to a wrong program [Chandy, Misra 88]. Therefore, the need of correctness proofs is especially great with distributed algorithms. The prevalence of programming errors and the enormous number of possible execution sequences that makes exhaustive testing impossible, have led to an interest in proving the properties of these algorithms [Lamport 77]. Invariant properties are generally formally written and are verified from axioms and inference rules of a formal system by using the induction technique [Hoare 83], [Nguyen et al. 86]. The drawback of this method is the difficulty to find the invariance properties.

Petri nets could avoid these drawbacks [Murata 89]. First, they offer a formal specification well-suited to help the programmer to avoid some obvious errors. Secondly, this model allows to obtain invariance properties by linear systems resolution. The purpose is to systematically find the invariance properties of a formally specified distributed algorithm.

---

This work has been supported by the Indo-French Centre for the Promotion of Advanced Research (IFCPAR), Project 302-1.

At the present time, invariance properties that can be automatically computed, help to build the proofs but are far from being sufficient: the deductions remain complex. The main reason, is that their deductive power is limited by the fact they only take into account a small portion of the Petri net semantic and never use the already known properties. For example, flows and semi-flows take into account only the incidence matrix of a Petri net and never the pre incidence function and the inhibitor arcs which allows to decide if a transition is enabled.

This paper introduces two new kinds of structural invariants which establish a relation between the marking of two place subsets A and B and are called *Exclusive Invariant* and *Implication Invariant:*

- the *Exclusive Invariant* states that A and B cannot be simultaneously marked,
- the *Implication Invariant* states that when A is marked, B is also marked

A verification algorithm based on linear programming techniques is also presented. It uses the already proved properties on the net behavior.

As we will see, these two classes of invariants, even being very particular, are useful for the proof of some very interesting properties of the model reachable states.

The paper is structured in three parts. In the first part, definitions and notations concerning Petri nets are briefly considered. In particular, section 2.2. defines drain transitions which is a basic definition for the new invariants. In the second part, exclusive and implication invariants are formally introduced after the exposition of a preliminary example. In the last part, the invariants are used to prove classical distributed algorithms: alternate bit protocol [Morgan, Razouk 87, Suzuki 90] and Peterson's algorithm for n processes [Peterson 81].

## 2. Definitions and Notation

This section introduces the notations and definitions related to Petri nets used in this paper. Section 2.2. introduces a new terminology used in section 3. to define the new invariants.

### 2.1. Basic terminology and notations

A Petri net is a bipartite graph. The nodes are places and transitions, modeling, respectively, process states and instructions. In this paper we consider inhibitor arcs, an extension of Petri nets which allows to test place emptiness.

**Definitions**

- A *Petri net* N, is a 5-tuple, N = <P, T, Pre, Post, H>, where:
  - * P is the place set,
  - * T is the transition set ($P \cap T = \emptyset$),
  - * Pre (resp. Post) is the Pre (resp. Post) incidence function:

    Pre: $P \times T \rightarrow \mathbb{N}$ (arcs from places to transitions),

    Post: $P \times T \rightarrow \mathbb{N}$ (arcs from transitions to places),

* H is the inhibitor relation between places and transitions, $H(p,t)$ is true iff an inhibitor arc exists between place p and transition t.

• A function $M: P \rightarrow \mathbb{N}$ is called a *marking*. A marked Petri net $<N,M_0>$ is a Petri net N with an initial marking $M_0$.

• A transition t is enabled for a marking M iff:

$$\forall\, p \in P,\ Pre(p,t) \leq M(p) \text{ and } H(p,t) \Rightarrow M(p) = 0.$$

• If transition t is enabled for a marking M, the transition firing yields to a new marking M':

$$\forall\, p \in P,\ M'(p) = M(p) + W(p,t)$$

where W= Post - Pre is the *incidence matrix*.

• $R(N,M_0)$ denotes the set of reachable markings for a Petri net N and an initial marking $M_0$.

• A *flow* is a function $F: P \rightarrow \mathbb{Q}$ solution of the incidence matrix equation $F^t.W = 0$. A flow gives a linear invariant property for any reachable marking M: $F^t.M = F^t.M_0$.

**Notations**

• The input set (resp. output set) of a place set A is:

$${}^{\bullet}A = \{t \in T \mid \exists\, p \in A,\ Post(p,t) \neq 0\} \text{ (resp. } A^{\bullet} = \{t \in T \mid \exists\, p \in A,\ Pre(p,t) \neq 0\})$$

• The exclusive input (resp. exclusive output ) set of place set A is ${}^{\bullet}A \setminus A^{\bullet}$ (resp. $A^{\bullet} \setminus {}^{\bullet}A$)

• For a reachable marking M, the marking of a place set A, denoted as M(A), is defined as $M(A) = \sum_{p \in A} M(p)$. A is marked when $M(A) \geq 1$. If $M(A)=0$, A is said to be unmarked.

• If A and B are two place sets, the notation $M(A) <> M(B)$ means that places of A and B are in mutual exclusion for marking M:

$$[M(A) <> M(B)] \equiv [M(A)=0 \text{ or } M(B)=0]$$

• If A and B are two sets of places, the notation $M(A) \Rightarrow M(B)$ means that if A is marked then B is marked for marking M:

$$[M(A) \Rightarrow M(B)] \equiv [M(A) \geq 1 \Rightarrow M(B) \geq 1]$$

• For invariant properties, the marking M is sometimes omitted in order to simplify the notation. So, "$\forall\, M \in R(N),\ M(A) <> M(B)$" (resp. "$\forall\, M \in R(N),\ M(A) \Rightarrow M(B)$") will be denoted "$A <> B$" (resp. "$A \Rightarrow B$").

### 2.2. Drain transitions

This part contains the definition of the drain transitions. This concept is the key point of the new invariant definitions introduced in this paper.

A transition t is said to be drain relatively to a place subset A if one of the following alternative is satisfied for all reachable marking:

- t is not fired when A is marked,
- if t fires when A is marked, after firing, A is unmarked for the reached marking.

**Definition**
Transition t is a *drain transition* for a place subset A iff:

$$\forall M \in R(N,M_0),\ M[t\rhd M' \Rightarrow (M(A) = 0 \text{ or } M'(A) = 0)$$

Drain(A) is the set of the drain transitions for place subset A.

**Drain transition verification**

• The only transitions t which are not drain transitions for a place subset A are the ones such that an accessible marking M exists which fulfills :
(C1) $M[t\rhd M'$, $M(A) \geq 1$ and $M'(A) \geq 1$.

• The draining property can only be wholly verified using the reachable marked graph. To avoid this problem, we verify the property over the set of markings satisfying the invariant properties already obtained. Most of the time, the invariants obtained are linear equations or inequations, thus it is possible to transform the draining property verification of transition t for set A into the feasibility of a linear program where the variables are the marking of the places (vector M) and the constraints come from the linear equations (LE), inequations (LI) and from (C1):

| | |
|---|---|
| $LE.M = U$ | {linear invariant equations. LE is either a vector or a matrix} |
| $LI.M \geq V$ | {linear invariant inequations. LI is either a vector or a matrix} |
| $M \geq Pre(.,t)$ | {t is enabled} |
| $M(I(t))=0$ | {I(t) is the set of places associated to the inhibitor arcs of t} |
| $M(A)\geq 1$ | {A is marked for M} |
| $(M+W(.,t))(A) \geq 1$ | {A is marked for the marking reached after t firing} |

If the system is not feasible, t is a drain transition for A. If the system is feasible, one of the following alternative is true:

- the informations already obtaines are not sufficient to prove that transition t is a drain transition for A,
- t is not a drain transition for A.

• More generally, the properties introduced in this paper are not necessarily expressed as linear invariant equations or inequations. However, they may be expressed as a disjunction of linear invariants:

$$\bigvee_i S_i(M), \text{ with } S_i(M) \equiv (LE_i.M = U_i) \wedge (LI_i.M \geq V_i)$$

Thus it is possible to transform the draining property verification for a transition into the feasibility of a set of linear programs:

$LE_i.M = U$
$LI_i.M \geq V$
$M \geq Pre(.,t)$
$M(I(t))=0$
$M(A)\geq 1$
$(M+W(.,t))(A) \geq 1$

If all the systems are unfeasible, t is a drain transition for A. If at least, one system is feasible, one of the following alternative is true:

- the informations already obtaines are not sufficient to prove that transition t is a drain transition for A,
- t is not a drain transition for A.

## 3. New invariants based on drain transitions

Two new invariants are defined in this section. They establish a link between the marking of two place subsets. They both are structural invariants, meaning that verifying such properties only involve Petri net structure and the previously proved properties. A preliminary example gives an informal introduction for each of them.

### 3.1 Exclusive invariants

Exclusive invariants ensure that two place subsets cannot be simultaneously marked.

#### 3.1.1. Preliminary example

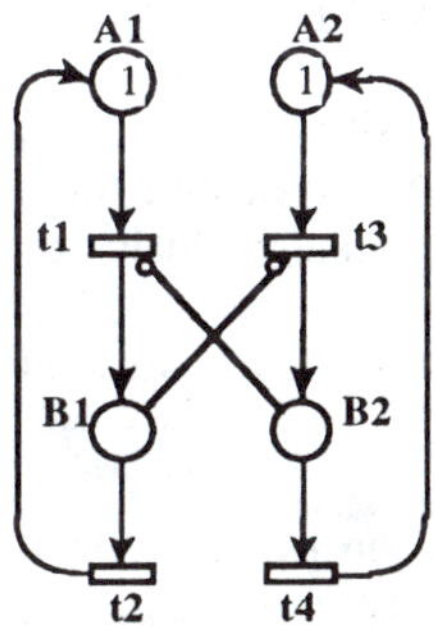


Figure 1: Simple example of mutual exclusion between two places

When analyzing a parallel algorithm, exclusion between process states, resources or variables is a property often sought for. Exclusive invariant is a structural invariant designed to handle this property in Petri nets.

Let A and B be two place subsets. The fact that A and B are structurally exclusive means that when A becomes marked after firing an exclusive input transition t, B is empty. This condition is verified if transition t don't marks B and belongs to Drain(B). The same argument can be applied to B.
Figure 1 is an example where t1 is a drain transition for {B2} and t3 is a drain transition for {B1}. Thus B1 <> B2 is an invariant.

### 3.1.2. Formal presentation of exclusive invariants

**Theorem**
Let A, B be two place subsets. If the following conditions hold:

(a) $M_0(A) <> M_0(B)$
(b) ${}^\bullet B \setminus B^\bullet \subseteq \text{Drain}(A)$
(c) ${}^\bullet A \setminus A^\bullet \subseteq \text{Drain}(B)$
(d) ${}^\bullet A \setminus A^\bullet \cap {}^\bullet B \setminus B^\bullet = \emptyset$

then M(A) <> M(B) is an invariant. It is called an *exclusive invariant.*

*Proof*
We prove by induction that M(A) <> M(B) is true for all reachable markings:
* M(A) <> M(B) is true for the initial marking (a).
* We have to prove [M[t>M' and (M(A) <> M(B))] ⇒ [M'(A) <> M'(B)]. We consider the three cases where M(A) <> M(B) hold:

1- M(A) = 0 and M(B) = 0

If set A becomes marked: M'(A)≥1, transition t is an exclusive input of set A: $t \in {}^\bullet A \setminus A^\bullet$
Using (d) condition, we obtain $t \notin ({}^\bullet B \setminus B^\bullet)$. Thus set B stays unmarked for M': M'(B)=0.
We have proved that set A and set B stay in mutual exclusion for M'.

2- M(A) ≥ 1 and M(B) = 0

If set B becomes marked: M'(B)≥1, transition t is an exclusive input of set B: $t \in {}^\bullet B \setminus B^\bullet$.
Using (b) condition, transition t is a drain transition for A: t ∈ Drain(A). From the drain definition, we deduce that A becomes unmarked: M'(A)=0. So, we have proved the mutual exclusion property for M'.

3- M(A) = 0 and M(B) ≥ 1

This case is symmetric with the previous one. ❑

*Remarks*

- Proving that M(A) <> M(B) is an *exclusive invariant.* can be done automatically over the invariant properties already computed:

  Conditions (a) and (d) are structural properties of the marked Petri net
  Conditions (b) and (c) are draining properties. They can be verified by solving linear programs (section 2.2.).

- Each exclusive invariant may induce new drain properties:

  ${}^\bullet A \cup A^\bullet \subseteq \text{Drain}(B)$
  ${}^\bullet B \cup B^\bullet \subseteq \text{Drain}(A)$

• An exclusive invariant is a disjunction of linear properties: M(A)=0 or M(B)=0.

• If $\bigvee_i S_i(M)$ express the invariants already obtained, adding an exclusive invariant leads to the new system: $\bigvee_i \left(S_i(M) \wedge M(A)=0\right) \vee \bigvee_i \left(S_i(M) \wedge M(B)=0\right)$

### 3.2. Implication invariants

Implication invariants ensure that, when a given place subset is marked, another place subset is also marked. The preliminary example is based on Dijkstra's proof [Dijkstra 81] of Peterson's mutual exclusion algorithm for two processes [Peterson 81] and highlights the close relationship between the two approaches.

#### 3.2.1. Preliminary example

Dijkstra's proof of the Peterson's mutual exclusion algorithm for two processes is built on insertion invariants, local to each process. These local invariants can be seen as implication invariants of the Petri net modeling the algorithm.

The algorithm (Figure 2) uses variable sharing as communication mechanism:

- variable Turn contains the number of the first process to request the critical section, the initial value is 1 or 2,
- Flag is an array of two Boolean variables, Flag[i] = true indicates that process i is requesting the critical section, the initial value is false.

| Process $P_1$ | | Process $P_2$ | |
|---|---|---|---|
| | Flag[1] ← false; | | Flag[2] ← false; |
| | While true | | While true |
| A1: | { Flag[1] ← true; | A2: | { Flag[2] ← true; |
| B1: | Turn ← 1; | B2: | Turn ← 2; |
| C1: | wait until Flag[2] = false or Turn = 2; | C2: | wait until Flag[1] = false or Turn = 1; |
| CS1: | <Critical Section> | CS2: | <Critical Section> |
| | Flag[1] ← false; | | Flag[2] ← false; |
| | } | | } |

Figure 2: Peterson's algorithm for two processes

To prove the algorithm, [Dijkstra 81] uses assertional invariants (Figure 3). A variable Li which contains the state of process i, is introduced.

```
Process Pi i ∈ [1..2]
    Flag[i] ← false; Li ← Ai;
    While true                                          {Li = Ai and Flag[i] = False}
    {  Flag[i] ← true; Li ← Bi,                         {Pi.1: Li = Bi and Flag[i] = true}
       Turn ← i; Li ← Ci,                               (Pi.2: Li = Ci and Hi}
       wait until (Flag[j] = false or Turn = j);Li ← CSi;  {Pi.3: Li = CSi and Hi}
       <Critical Section>
       Flag[i] ← false; Li ← Ai,;
    }

with
    Hi = (Turn = i or Lj = Cj)
    if i = 1 then j = 2 else j = 1
```

Figure 3: Assertional invariants of Peterson's algorithm

Dijkstra observes that process i cannot falsify Hj to establish the validity of the local assertions. Afterwards he verifies that Pi.3 and Pj.3 cannot be satisfied at the same time :

[Pi.3 and Pj.3] ≡ [(Li = CSi and Hi) and (Lj = CSj and Hj) ⇒ (Turn = i and Turn = j)] ≡ false

Pi.2 and Pi.3 are the key invariants of this proof. They can be slightly transformed to become an inductive invariant:

(I) (Li = Ci or Li = CSi) ⇒ (Turn = i or Lj = Cj)

Figure 4 is the model of the algorithm. Places Ai, Bi, Ci and CSi model the different states of process i, places Flag[i], Turn1 and Turn2 model the shared variables. In order to simplify the model Turn is initially valued to 2.

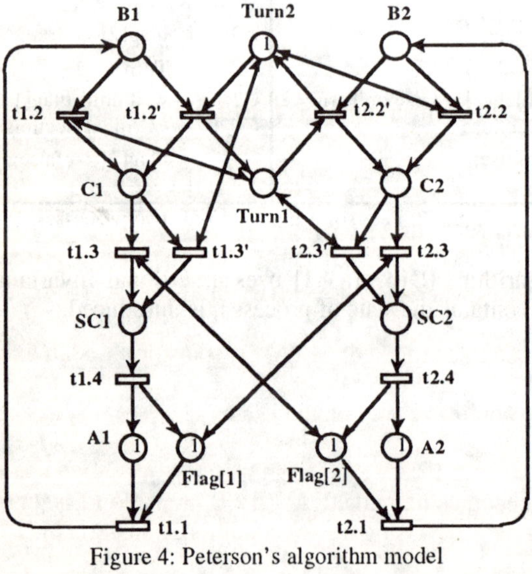


Figure 4: Peterson's algorithm model

Five flows can be found:

| (F1) $A1 + B1 + C1 + CS1 = 1$ |
|---|
| (F2) $Flag[1] + B1 + C1 + CS1 = 1$ |
| (F3) $Turn1 + Turn2 = 1$ |
| (F4) $A2 + B2 + C2 + CS2 = 1$ |
| (F5) $Flag[2] + B2 + C2 + CS2 = 1$ |

Invariant (I) can be translated in:

| (I1) $(Ci \geq 1 \text{ or } CSi \geq 1) \Rightarrow (Turni \geq 1 \text{ or } Cj \geq 1)$ |
|---|

It is equivalent to prove that when at least one of the places in A = {Ci, CSi} is marked, then at least one of the places in B = {Turni, Cj} is marked: A marked ⇒ B marked. Such an invariant is an implication invariant.

Intuitively, this property is valid if the following conditions are fulfilled:

(1) as soon as A becomes marked, B is also marked,

(2) while A is marked, B cannot be emptied.

The first point is structurally verified if the exclusive inputs of A are inputs of B. For invariant (I1) $^\bullet A \setminus A^\bullet = \{ti.2, ti.2'\} \subseteq {}^\bullet B$.

To structurally verify the second point, it is equivalent to say that the exclusive output of B are drain transitions for A. For invariant (I1) we must verify that tj.3 is a drain transition for A: when Ci or CSi is marked, Flag[i] is unmarked (from invariants (F2) and (F5)). Thus tj.3 cannot be fired while A is marked.

### 3.2.2. Formal presentation of implication invariants

**Theorem**

Let A, B be two place subsets. If the following conditions hold:

(a) $M_0(A) \Rightarrow M_0(B)$

(b) $^\bullet A \setminus A^\bullet \subseteq {}^\bullet B$

(c) $B^\bullet \setminus {}^\bullet B \subseteq Drain(A)$

then $M(A) \Rightarrow M(B)$ is an invariant. It is called an *implication invariant*..

*Proof*

We prove by induction $M(A) \Rightarrow M(B)$ is true for all reachable markings:

* $M(A) \geq 1 \Rightarrow M(B) \geq 1$ is true for the initial marking (a).

* We have to prove $[M[t{>}M' \text{ and } (M(A) \Rightarrow M(B))] \Rightarrow [M'(A) \Rightarrow M'(B)]$. We consider the two cases where $M(A) \Rightarrow M(B)$ holds:

1- $M(A) = 0$

If set A becomes marked: $M'(A) \geq 1$, transition t is an exclusive input of set A: $t \in {}^\bullet A \setminus A^\bullet$

Using (b) condition, we obtain $t \in {}^\bullet B$. Thus set B is marked for M': $M'(B) \geq 1$.

2- $M(A) \geq 1$ and $M(B) \geq 1$

If set B becomes unmarked: $M'(B)=0$, transition t is an exclusive output of set B: $t \in B^\bullet \setminus {}^\bullet B$. Using the (c) condition, transition t is a drain transition for

A: $t \in$ Drain(A). Thus set A becomes unmarked for M': M'(A)=0. ❑

*Remarks*

• Proving that $M(A) \Rightarrow M(B)$ is an *implication invariant* can be done automatically over the invariant properties already computed:

Conditions (a) and (b) are just structural properties of the Petri net

Condition (c) is a draining property. It can be verified by solving linear programs (section 2.2.).

• An implication invariant is a disjunction of linear properties: M(A)=0 or M(B)≥1.

• If $\bigvee_i S_i(M)$ express the invariants already obtained, adding an exclusive invariant leads to the new system: $\bigvee_i \left(S_i(M) \wedge M(A)=0\right) \vee \bigvee_i \left(S_i(M) \wedge M(B)\geq 1\right)$

# 4. Applications

In this section, more complete examples are analyzed: the alternate bit protocol proof uses exclusive invariants and the generalized Peterson's algorithm proof uses implication invariants. This last example highlights that these invariants can be easily used with high level formalisms.

## 4.1. Alternate bit protocol

In the alternate bit protocol [Morgan, Razouk 87, Suzuki 90] (Figure 5), a process 1 sends messages to a process 2 which returns acknowledgments. The communication medium is:

- asynchronous,
- FIFO, messages and acknowledgments are treated in the order they are sent,
- unsafe, thus messages and acknowledgments are repeated.

To ensure no confusion between messages, each message has a type: 0 or 1. When process 2 returns acknowledgments, it specifies the type of the message acknowledged. Process 1 changes the type for each new message. When process 2 (resp. 1) receives a message (resp. an ack) of the wrong type, the message (resp. ack) is lost by the process.

| Process P1 | Process P2 |
|---|---|
| While True | While True |
| A: { New(m); Send(m,0); | E: { Repeat Send(1) until Receive(m,0); |
| B: Repeat Send(m,0) until Receive(0); | F: Send(0); |
| C: New(m); Send(m,1); | G: Repeat Send(0) until Receive(m,1); |
| D: Repeat Send(m,1) until Receive(1); | H: Send(1) |
| } | } |

Figure 5: Alternate bit protocol

As proved further, the algorithm has the following property: when process 1 is ready to send a message of a given type, there is no message of this type left in the communication medium (invariants (a), (b) (c) and (d)). Thus, the only possible states of the buffer are:

- empty,
- messages of type 0,
- messages of type 1,
- messages of type 0 followed by messages of type 1,
- messages of type 1 followed by messages of type 0.

The communication medium is modeled like an infinite FIFO buffer. The buffer model (Figure 6) is cut in two parts to model the precedence of the different types of messages. The communication medium is modeled with four places (two places for each part of the buffer, one per type) mess0, messP0, mess1, messP1, where mess0 and mess1 model the low part and messP0 and messP1 the high part of the buffer. A message of type i in the low part can move to the high part iff all messages of type j≠i are removed of the high part.

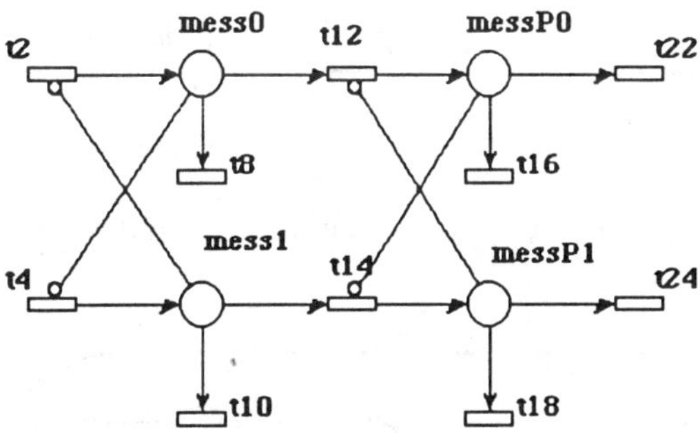

Figure 6: Communication buffer model

The same property will be proved for process 2 and the acknowledgments. Therefore the communication medium between process 2 and process 1 is modeled in the same way.

The correctness of the protocol is based on the following properties of the model (Figure 7):

|  |  |
|---|---|
| (a) | A=1 ⇒ E=1 and mess0 = messP0 = ack0 = ackP0 = 0 |
| (b) | C=1 ⇒ G=1 and mess1 = messP1 = ack1 = ackP1 = 0 |
| (c) | F=1 ⇒ B=1 and ack0 = ackP0 = mess1 = messP1 = 0 |
| (d) | H=1 ⇒ D=1 and ack1 = ackP1 = mess0 = messP0 = 0 |

When Process 1 is ready to send a message of type 0 (resp. type 1), Process 2 is ready to receive it and there is no message and no acknowledgment of type 0 (resp. type 1) in transit. Properties (a) and (b) prove that there is no confusion between messages of the same type. In the same way properties (c) and (d) prove that there is no confusion between acknowledgments of the same type.

In order to prove properties over the model, four places (Iack1, Imess0, Iack0 and Imess1) are added (figure 8). We will have to prove that these places are implicit ones.

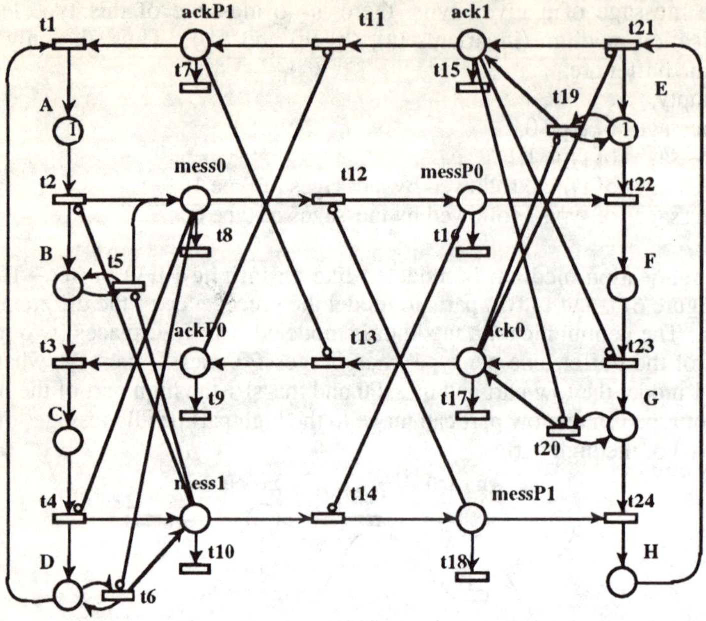


Figure 7: Alternate bit protocol model

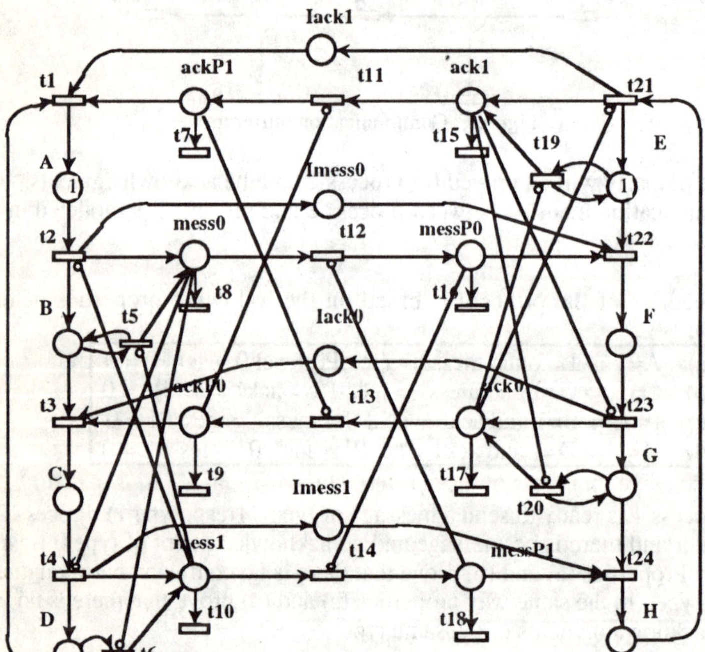


Figure 8: Alternate bit protocol model with implicit places

Five P-flows form a generative family which can easily be computed:

| | |
|---|---|
| (F1) | A + Imess0 + F + Iack0 + C + Imess1 + H + Iack1 = 1 |
| (F2) | E = Iack1 + A + Imess0 |
| (F3) | B = Imess0 + F + Iack0 |
| (F4) | G = Iack0 + C + Imess1 |
| (F5) | D = Imess1 + H + Iack1 |

Invariants (F1), (F2) (F3) and (F4) prove a part of correctness properties (a), (b), (c) and (b)

| | |
|---|---|
| (a) | A=1 ⇒ E=1 |
| (b) | C=1 ⇒ G=1 |
| (c) | F=1 ⇒ B=1 |
| (d) | H=1 ⇒ D=1 |

The completion of the model correctness is based on exclusive invariants. The two first classes of exclusive invariants (E1) and (E2) are just a step in proving the more interesting exclusion invariants (E3).

**Exclusive invariants (E1)**

| | |
|---|---|
| (E1.1) | (mess1 + A + D) <> (mess0 + B + C) |
| (E1.2) | (ack1 + E + F) <> (ack0 + G + H) |

*Proof*

Proof of (E1.2)

(a) The property holds for the initial marking.

(b) Transition t23 is the only exclusive input of (ack0+G+H). It is a drain transition for (ack1+E+ F):

If M[t23>M' then M(F) = 1 and M(ack1) = 0.
Thus M'(ack1) = 0 and M'(G) = 1. From the P-flows we easily obtain M'(ack1) = M'(E) = M'(F) = 0.

(c) Transition t21 is the only exclusive input of (ack1+E+F). It is a drain transition for (ack0+G+H):

The same proof as t23 can applies for transition t21.

(d) The exclusive input sets of (ack1+E+F) and (ack0+G+H) are separate.

By symmetry, the same proof can be applied to (E1.1). ❑

**Exclusive invariants (E2)**

| | |
|---|---|
| (E2.1) | (A + ackP1) <> (C + ackP0) |
| (E2.2) | (H + messP1) <> (F + messP0) |

*Proof*
Proof of (E2.1)
(a) The property holds for the initial marking.

(b) Transition t13 is the only exclusive input of (C + ackP0). Transition t13 is a drain transition for (A + ackP1):

If M[t13>M' then M(ackP1) = 0 and M(ack0) ≥ 1. From (E1.1) M(E) = 0 and from (F2) M(A) = 0. Thus M(ackP1) = M(A) = 0 is a condition to fire transition t13. This proves that transition t13 is a drain transition for (A + ackP1).

(c) Transition t11 is the only exclusive input of (A + ackP1). Transition t13 is a drain transition for (C + ackP0):

The same proof as t13 applies for transition t11.

(d) The exclusive input sets of (A + ackP1) and (C + ackP0) are separate.

By symmetry, the same proof can be applied to (E2.2). ❑

The following exclusive invariants complete the proof of correctness properties. When process 1 is in state A (resp state C), there is no message and no acknowledgment of type 0 (resp. type 1) in transit. The same arguments appear for Process 2.

**Exclusive invariants (E3)**

| | |
|---|---|
| (E3.1) | (ack1 + ackP1 + E) <> (C + Imess1 + H) |
| (E3.2) | (ack0 + ackP0 + G) <> (A + Imess0 + F) |
| (E3.3) | (mess1 + messP1 + D) <> (F + Iack0 + C) |
| (E3.4) | (mess0 + messP0 + B) <> (H + Iack1 + A) |

*Proof*
We just give the proof of (E3.1). The others can be obtain by symmetry.
(a) The property holds for the initial marking.

(b) Transition t21 is the only exclusive input of (ack1 + ackP1 + E). Transition t21 is a drain transition for (C + Imess1 + H):

If M[t21>M' then M'(E) ≥ 1. From (F1) and (F2) M'(C) = M'(Imess1) = M'(H) = 0. Thus transition t21 is a drain transition for (C + Imess1 + H).

(c) Transition t3 is the only exclusive input of (C + Imess1 + H). Transition t3 is a drain transition for (ack1 + ackP1 + E). We consider the three cases where (ack1 + ackP1 + E) is marked and prove that t3 is not enabled:

1- M(ackP1)≥1: From invariant (E2.1), M(ackP0)=0. So t3 is not enable.
2- M(ack1)≥1: From invariant (E1.2), M(G)=0. Using the flow property (F4), we deduce M(Iack0)=0. So t3 is not enabled.
3- M(E)≥1: From the flow properties (F1) and (F2), we deduce that M(Iack0)=0. So t3 is not enabled.

(d) The exclusive input sets of (ack1 + ackP1 + E) and (C + Imess1 + H) are separate. ❑

**Implicit place proof**
Now, it is easy to prove that Iack1 is an implicit place (and by symmetry, places Imess0, Iack0, Imess1). Since t1 is the only transition with Iack1 as precondition place, we must prove that:

$$\forall M \in R(R,M_0), (M(D) = 1 \text{ and } M(ackP1) \geq 1) \Rightarrow M(Iack1) = 1$$

*Proof*
From (F5): $M(D) = 1 \Rightarrow M(Imess1) + M(H) + M(Iack1) = 1$.
From (E3.1): $M(ackP1) \geq 1 \Rightarrow M(Imess1) = M(H) = 0$.
Thus from (F5), $M(D) = 1$ and $M(ackP1) \geq 1 \Rightarrow M(Iack1) = 1$. ❑

### 4.2. Generalized Peterson's algorithm

This algorithm [Peterson 81] (Figure 9) is the generalization to n processes of the mutual exclusion algorithm analyzed in section 3.1.2. It uses the same principles over (n-1) levels to eliminate one process each time. Shared variables are:

- Turn which is an array containing the number of the last process to request the critical section for each level,
- Flag which is an array containing the level of each process.

```
Process Px, x ∈ [0..n-1]
        Flag[x] ← 0;
        While true
Idle:{     <Critical section not requested>
          For j from 1 to n-1
A:        {    Flag[x] ← j;
B:             Turn[j] ← x;
C:             Wait until (∀ y ≠ x: Flag[y]<j or Turn[j] ≠x);
D:        }
CS:  <Critical Section>
E:        Flag[x] ← 0;
        }
```

Figure 9: Peterson's algorithm for n processes

Figure 10 is the colored net model of the algorithm. The following analysis of the net is done over the unfolded net, some notations are useful:

- variables x, y, z refer to processes,
- variables j, k refer to levels,
- the ordinary places obtained from a colored place p are denoted $p[i_1, ..., i_u]$ where $i_j$ is a process or a level according to the color domain of place p.
- the ordinary transitions obtained from a colored transition t are denoted $t[i_1, ..., i_u]$ where $i_j$ is a process or a level according to the color domain of transition t.

The color function Sum(x≠y, k≥ j) associated to the inhibitor arc between place Flag and transition t5 means that when firing t5[x,j], places Flag[y,k] with y≠x and k ≥ j must be unmarked.

Classes
  Process = {x1, ..., xn}
  Level = {0, ..., n-1}
Place domains
  Idle, CS, E: Process
  A, B, C, D, Flag: Process x Level
  Turn: Level1 x Process
Initialization
  Idle = <x1>+ ... +<xn>
  Flag = <x1,0>+ ... +<xn,0>
  Turn = <1,x1>+ ... + <n-1,x1>

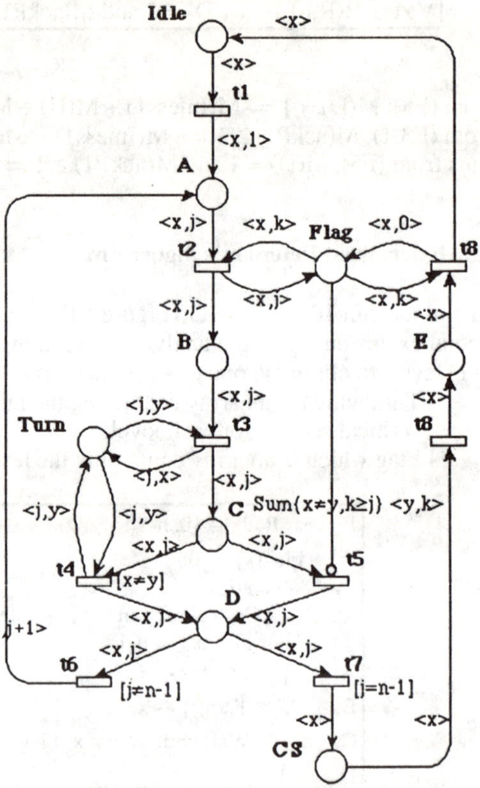


Figure 10: Peterson's algorithm model

From this net we obtain three flows. They correspond to the state of processes, variables Turn and Flag:

(F1) $\forall x \in$ Process,

$$Idle[x] + \sum_{j=1}^{n-1} \big(A[x,j] + B[x,j] + C[x,j] + D[x,j]\big) + CS[x] + E[x] = 1$$

(F2) $\forall x \in$ Process, $\sum_{j=0}^{n-1} Flag[x,j] = 1$

(F3) $\forall j \in \{1, ..., n-1\}$, $\sum_{x \in Process} Turn[j,x] = 1$

The following invariants are deduced from unmarked deadlocks (i.e unmarked place sets without exclusive input). They express that processes cannot be in states A, B, C, D with loop variable j equals 0 and the variable Turn[0] is not defined.

(D1) $\forall$ x $\in$ Process, A[x,0] + B[x,0] + C[x,0] + D[x,0] = 0

(D2) $\sum_{x \in Process}$ Turn[0,x] = 0

The following invariants give a relationship between the value of Flag and the processes state. Even if these invariants are linear ones, they cannot be considered as flows. One can notice invariant (I1) is not a flow since t2[x,j,0] $j \neq 1$ is an output of {Flag[x,0]}, this would means that Flag[x,0] could be emptied without changing the status of Idle[x] + A[x,1]. Fortunately, this transition is a dead one. The implication invariants notion allows to prove the invariants.

**Implication invariants (I)**

(I1) Flag[x,0] $\Leftrightarrow$ Idle[x] + A[x,1]

(I2) Flag[x,j] $\Leftrightarrow$ A[x,j+1] + B[x,j] + C[x,j] + D[x,j], j $\in$ {1, ..., n-2}

(I3) Flag[x,n-1] $\Leftrightarrow$ B[x,n-1] + C[x,n-1] + D[x,n-1] + CS[x] + E[x]

*Proof*

Proof of (I2):

1- Flag[x,j] $\Rightarrow$ A[x,j+1] + B[x,j] + C[x,j] + D[x,j] (*)

We want to prove that (*) is an implication invariant.

(a) It is obvious that the property is true for the initial marking.

(b) The exclusive inputs of {Flag[x,j]} are inputs of {A[x,j+1], B[x,j], C[x,j], D[x,j]}
$^{\bullet}${Flag[x,j]} = {t2[x,j,k] | k $\in$ {0, ..., n-1}}
Each transition t2[x,j,k] is an entry of place B[x,j] which proves (b)

(c) The exclusive outputs of U = {A[x,j+1], B[x,j], C[x,j], D[x,j]} are drain transitions of{Flag[x,j]})
The exclusive output set of U is A[x,j+1]$^{\bullet}$ = {t2[x,j+1,k] | k $\in$ {0, ..., n-1}}.
Each transition t2[x,j+1,k] is a drain transition for {Flag[x,j]}: firing this transition produces a token in the place Flag[x,j+1] and empties the place Flag[x,j] (deduced from flow (F2)).

2- Flag[x,j] $\Leftarrow$ A[x,j+1] + B[x,j] + C[x,j] + D[x,j] (**)

We want to prove that (**) is an implication invariant.

(a) It is obvious that the property is true for the initial marking.

(b) The exclusive inputs of U = {A[x,j+1], B[x,j], C[x,j], D[x,j]} is inputs of $^{\bullet}${Flag[j,x]}

The exclusive input set of U is ${}^{\bullet}B[x,j] = \{t2[x,j,k] \mid k \in \{0, ..., n-1\}\}$
Each transition t2[x,j,k] is an entry of {Flag[x,j]}.

(c) The exclusive outputs of {Flag[x,j]} are drain transitions of {A[x,j+1], B[x,j], C[x,j], D[x,j]}
The exclusive output set of {Flag[x,j]} is $\{t2[x,k,j] \mid k \neq j\} \cup \{t8[x,j]\}$.
Each transition t2[x,k,j] for k≠j is a drain transition for U: firing this transition produces a token in place B[x,k] with k≠j and empties U (deduced from flow (F1)).
Each transition t8[x,j] is a drain transition for U: firing this transition produces a token in place Idle[x] and empties U (deduced from flow (F1)).

Proof of (I1) and (I3) are similar and are let to the readers. ❑

**Implication invariant P(x,y,j)**

The following invariant is the generalization of the one used in the analysis section 3.1.2. It proves that a process is alone in its level j and in higher levels, or that the process which posses the Turn for level j is in place C:

$$\forall x,y \in \text{Process},\ x \neq y:\ P(x,y,j) \equiv$$
$$\underbrace{C[x,j]+D[x,j]+\sum_{k>j}(A[x,k]+B[x,k]+C[x,k]+D[x,k])+CS[x]+E[x]}_{A} \Rightarrow \underbrace{C[y,j]+\sum_{z\neq y} Turn[j,z]}_{B}$$

*Proof*

(a) It is obvious that the property is true for the initial marking.

(b) The exclusive inputs of A are inputs of B.
${}^{\bullet}A \backslash A^{\bullet} = \{t3[x,j,u] \mid u \in \text{Process}\}$. Each transition t3[x,j,u] is an entry of place Turn[j,x] which proves (b).

(c) The exclusive outputs of B belong to Drain(A).
The exclusive output set of B is {t5[y,j]}. The transition t5[y,j] is a drain transition for A: when A is marked, one of the places Flag[x,k] with $k \geq j$ is marked (deduced from the invariants I2, I3). Thus transition t5[y,j] is not enabled because of the inhibitor arcs. This proves that t5[y,j] is a drain transition for A. ❑

**Mutual exclusion proof**

Using the previous invariant, we can now prove that processes are in mutual exclusion for the critical section:

$$|CS| \leq 1$$

*Proof*

If CS[x] = CS[y] = 1, with x≠y:

$\forall j \in \{1, ..., n-1\}$, let $v_j$ such that Turn[j,vj] = 1

P(y,x,j) and (F3) $\Rightarrow \sum_{z \neq x} Turn[j,z] = 1 \Rightarrow v_j \neq x$

P(x,$v_j$,j) and (F3) $\Rightarrow C[v_j,j] = 1 \Rightarrow |C| \geq n-1$

Thus, $|CS| + |C| \geq n+1$. This is inconsistent with flow (F1). ❑

## 5. Conclusion

In this paper we have introduced the first structural invariants that take in account the invariants previously established. These new invariants constitute a great progress considering the complexity of classical proofs of distributed algorithm. As highlighted in the Peterson's algorithm proof, implication invariants express the inductive ones used in the insertion proof. The exclusive invariants are the ones that the Petri net modelers intuitively know using inhibitor arcs. Structural characterization of this type of invariants allows us to give a simple extension of Dijkstra's proof to n processes and an elegant proof of the alternate bit protocol.

The first attempts to compute exhaustively these invariants failed on combinatory explosion of space and time. However, we have designed some algorithms which seem to work in reasonable time but only compute a part of them for ordinary net. The invariants used in our examples belong to the ones automatically obtained. For future work, the automatic computation of these new invariants is a priority. Another way of research is the understanding of inductive invariants used in classical proofs that may lead to the structural definition of some other invariants.

## References


[Chandy, Misra 88] Chandy K.M. and Misra J. "Parallel Program Design: A Foundation". Addison-Wesley Publishing Compagny, 1988, 516 pages.

[Dijkstra 81] E.W. Dijkstra, "An assertional proof of a program by G.L. Peterson". EWD 779, Burroughs Corp. 1981.

[Hoare 83] Hoare C.A.R., "Toward a Theory of Parallel Programming". Operating Systems Techniques, Hoare and Perrots (eds), Academic Press New York, 1972, p. 61-71.

[Lamport 77] Lamport L., "Proving the Correctness of Multiprocess Programs". IEEE Transactions on Software Engineering, vol. SE-3:2, March 1977, p. 125-143.

[Nguyen et al. 86] Nguyen V., Demers A., Gries D. and Owicki S., "A model and Temporal Proof System for Network of Processes". Distributed Computing, Springer Verlag, 1986, p. 7-25.

[Morgan, Razouk 87] Morgan E.T. and Razouk R.R., "Interactive State-Space Analysis of Concurrent Systems". IEEE Transactions on Software Engineering, vol. 13, no 10, October 1987, p. 1080-1091.

[Murata 89] Murata T., "PetriNets: Properties, Analysis and Applications". Proceedings of the IEEE, vol. 77, no. 4, April 1989, p. 541-580.

[Peterson 81] G.L. Peterson, "Myths about the Mutual Exclusion Problem". Information Processing Letters, 12:3. June 1981, p. 115-116.

[Suzuki 90] Suzuki I., "Formal Analysis of the Alternating Bit Protocol by Temporal Petri Nets". IEEE Transactions on Software Engineering, vol. 16, no 11, November 1990, p. 1273-1281.

# Time Stream Petri Nets
# A Model for Timed Multimedia Information[1]

Michel Diaz°, Patrick Sénac ° *

° LAAS du CNRS
7, avenue du Colonel Roche
31077 Toulouse Cedex - France

* ENSICA
49 Avenue Léon Blum
31056 Toulouse Cedex - France

**Abstract.** This paper introduces a model for specifying synchronization constraints in distributed asynchronous multimedia systems and applications. The consistency and semantics of multimedia systems depend on the temporal behaviour of information streams, like audio and video streams, whose synchronization constraints in asynchronous environments need to be enforced. The promoted model, named Time Stream Petri Nets (TStreamPN) allows the timed behaviour of streams to be fully, accurately and formally described using an extension of time Petri nets. This model uses time intervals to label arcs that leave the places of the net. A complete set of firing rules is also proposed to accurately enforce actual synchronization policies between different and related multimedia streams. Therefore, this model allows a formal characterization and verification of time parameters in distributed asynchronous multimedia systems.

## 1. Introduction

Multimedia systems have to synchronize the access, communication and presentation of static (*i.e.* discrete data such as text and still images) and continuous streams of time constrained data (such as audio and video). Therefore, the synchronization requirements of multimedia applications may entail arbitrarily complex synchronization scenarios whose satisfaction is fundamental for the application semantics. Moreover, temporal non-determinism entailed by asynchronous distributed systems increases the complexity of reasoning about synchronization in distributed multimedia applications. Thus, distributed asynchronous multimedia systems would greatly benefit from the use of formal models that allow synchronization scenarios to be easily described, verified and validated. Starting from these considerations, Time Petri nets have been selected and extended by features that allow multimedia synchronization to be easily modeled. Thus, Time Stream Petri Nets (TStreamPNs) are proposed to precisely model the complex behaviour of synchronized sets of multimedia streams.

This paper is organized as follows: after a brief introduction, in Section 2, to

[1] This work has been developed as part of the CESAME project, a research collaboration between CNET, CCETT and CNRS, and under the grant 92 1B 178 from CNET - France Telecom.

multimedia synchronization issues, Section 3 presents how Timed Petri Nets have been used to model multimedia synchronization scenarios and indicates their limitations. Then, Section 4 overviews the Time Petri Net model which suppresses the modeling power limitations of the Timed Petri Net model; it is shown that this model suffers from modeling power limitations. Time Arcs Petri Nets (TAPNs) are introduced in Section 5, and it is discussed why TAPNs need to be extended to attain a full expressive power for describing and handling timed streams. The proposed TStreamPN model is given in Section 6. It is shown how it captures and represents the basic synchronization constraints of multimedia streams. The problems related to streams synchronization are presented and new firing rules are proposed for TStreamPNs. Then, in Section 7, the formal semantics of TStreamPN is given, and finally, some of their main properties are expressed in Section 8.

## 2. Multimedia Synchronization issues

Multimedia systems and applications have to manage the communication, processing, and presentation of not only traditional static data as text and still images but also dynamic data as audio, video and synthetic animations. Multimedia data (*i.e.* static and dynamic) can be defined as series of *information units*[2]. For instance a video is a serie of video frames. Therefore, the processing of multimedia data can be defined as a serie of time constrained tasks on information units. Such a synchronization structure is called a *stream* and tasks on information units are called *synchronization units*. For instance a video stream for presentation concern is a serie of video frames which must be displayed using at of 25 images per second. Note that a synchronization unit may be dedicated to the processing of several information units. Thus, multimedia systems need to manage the so called *intra-stream synchronization* to insure the respect of synchronization constraints inside streams. Moreover, by definition, multimedia applications induce synchronization scenarios that involve several streams. Therefore, multimedia systems have also to manage *inter-stream synchronization* constraints [Senac3][Rust].

Intra and inter-stream synchronization schemes involved in multimedia systems are usually specified in "digital production studios" by placing multimedia streams along a common time line. Figure 1 illustrates, using a "*digital production studio*" stream composition paradigm, intra and inter-stream synchronization constraints between an audio and a video streams. This synchronisation scheme specifies intra-synchronization constraints such as the duration and the relative dates of presentation of each video and audio information unit. Moreover, inter-stream synchronization constraints such as "the audio stream must start with the video stream" (i.e. discrete *event synchronization*) and "every two video information units are to be synchronized with their 4 related audio information units" (*i.e. continuous synchronization*) can be deduced from such a specification. Note that such a method does not have a well defined inter-stream synchronization semantics. Indeed, inter-stream synchronization constraints are defined indirectly from the time line, and thus are not explicitly and clearly specified. That is, using such an informal method, the author of complex multimedia scenarios can not master the consistency of the specified synchronization scheme.

---

[2]Note that a static data (e.g. text or still image) is a special case of stream with only one information unit

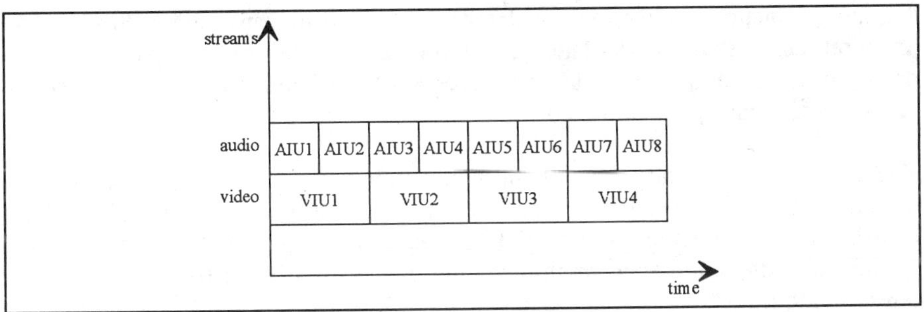


**Figure 1.** The digital production studio composition paradigm

Asynchronous distributed systems entail temporal non-determinism for the communication, processing and presentation of multimedia streams. This temporal non-determinism can be taken into account through the notion of admissible *jitter*, *i.e.* the admissible temporal variations of a synchronization unit, compared to its nominal duration. For instance the presentation of a video frame can support a minimum and a maximum jitters of 10 ms. Figure 2-a illustrates the minimum and the maximum jitter (denoted as j and J respectively) of a synchronization unit relative to its nominal duration (denoted as n).

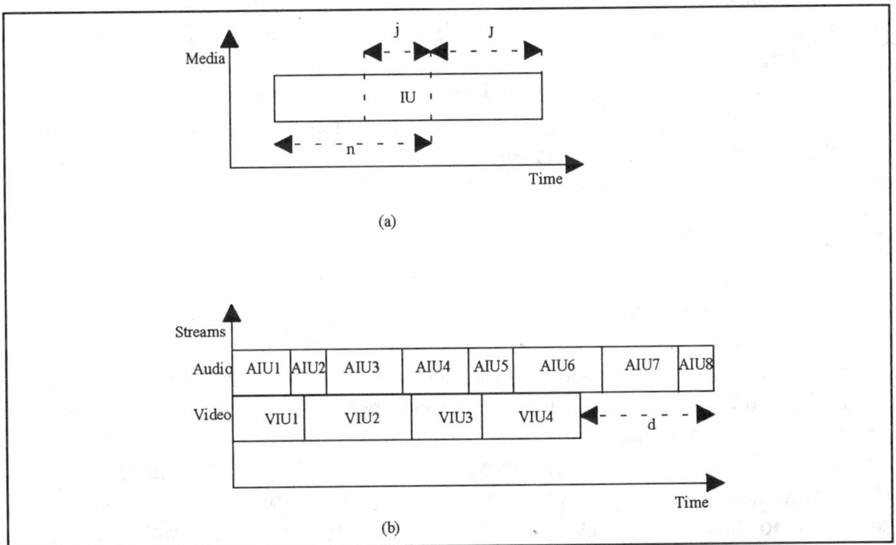


**Figure 2.** Temporal jitter (a) and drift (b) in multimedia streams

The introduction of jitter in multimedia processing entails temporal *drifts* between streams. For instance jitter on video and audio synchronization units can transform the synchronization scheme in Figure 1 into the one in Figure 2-b by introducing a temporal drift, denoted as d, between the two streams. Note that inter-stream synchronization schemes may be deeply modified because of jitter, and that the "digital production studio" multimedia stream composition paradigm do not tackle temporal non determinism. Thus, for asynchronous multimedia systems to have deterministic dynamic behaviour, it is necessary to use inter-stream synchronization semantics that take jitter into account.

Therefore, the commonly used "digital production studio" composition paradigm is not suitable for the specification of distributed asynchronous multimedia systems. However, the modeling power of this composition paradigm allows an easy specification of multimedia synchronization, and so is a good basis for new more powerful models.

From the above considerations, Petri Nets extended with time have been chosen for the formal modeling of multimedia systems. Indeed, Timed Petri Nets allow multimedia synchronization to be specified with a modeling power very close from the one of the "digital production studio" composition paradigm [Little1]. Moreover, the formal semantics of Petri nets allows the dynamic and temporal behaviour of multimedia systems to be simulated, verified and validated. Thus, the following sections investigate the usefulness of existing Petri nets (with time) models for specifying synchronization constraints in asynchronous multimedia systems. Two main evaluation criteria are used in the following:

- the *modeling power* of a model. That is, the capability to *easily* express a given problematic.
- the *expressive power* of a model. That is, the capability to *fully* express a given problematic.

It will be shown, through a modeling example of continuous synchronization between an audio and a video stream, that existing temporal extensions to Petri nets do not offer both satisfying modeling and expressive powers for the specification of asynchronous distributed multimedia systems. Therefore, a new model allowing a full, easy and accurate specification of multimedia synchronization, called Time Stream Petri Net (TStreamPN), will be introduced.

## 3. The Timed Petri Net Model

The timed Petri nets model (*i.e.* Petri net with time duration on places) [Ramchandani] has been proposed in [Little2], under the name Object Composition Petri Net (OCPN), for modeling multimedia synchronization. Timed Petri nets are proposed as a basic model to represent the dynamic behaviour of multimedia systems and to implement their synchronized behaviours. The OCPN model adopts a structuring and minimalist approach using a set of seven primitives that are sufficient to characterize all possible temporal relative positions between two intervals [Hamblin] [Allen]. These primitives define the set of composition rules that are used to build complex models out of simple ones. Indeed, in [Allen] the relative temporal locations of two intervals relating two processings Pi and Pj are described by seven constructs (six others symmetrical relate Pj to Pi). If ";" means sequence, "||" parallelism, and if "$\tau$", "$\tau_1$" and "$\tau_2$" represent any time positive delays, the seven constructs are :

- meets : $P_i;P_j$ where $P_j$ directly follows $P_i$ ;
- before : $P_i;\tau;P_j$ where $P_j$ follows $P_i$ after $\tau$ time units ;

- overlap : $(P_i;\tau_1)||(\tau_2;P_j)$ where $P_i$ starts first and $P_j$ ends last ;
- during : $(\tau_1;P_i;\tau_2)||P_j$ where $P_i$ starts later and ends before $P_j$ ;
- start : $(P_i;\tau)||P_j$ where $P_i$ and $P_j$ start at the same time, and $P_i$ ends before $P_j$;
- end : $(\tau;P_i)||P_j$ , where $P_i$ starts after $P_j$ and both end at the same time ;
- equality : $(P_i||P_j)$ , where $P_i$ and $P_j$ start and end at the same time.

Basis on these constructs, [Little2] shows that timed place Petri nets can represent any temporal relationship between two intervals. OCPN are obtained by composing in a serial-parallel way timed place Petri nets using these seven temporal constructs and associating multimedia types to each place. The exclusive use of these seven fundamental synchronization structures induces a marked graph structure and a safe (if the initial marking is safe) and L1-live Petri Net. A Petri Net build using these synchronization construct is called a *structured* one.

### 3.1. Timed Petri Nets Assessment for Multimedia Modeling

The modeling power of this model allows an easy and structured specification of intra and inter-synchronization constraints. Indeed, a stream is modeled, using sequential synchronization constructs (*i.e.* "meet" and "before"), as a sequence of timed places. Therefore, using Timed Petri Nets for stream modeling, synchronization units are modeled as timed places. The time value associated with each place specifies its nominal duration. For instance, Figure 3-a illustrates a video stream composed of four synchronization units with the same nominal duration denoted as n. Using Timed Petri Nets, inter-stream synchronization schemes are built easily, *i.e.* just by merging some transitions of the involved streams. Such a way to build synchronization scenarios offers a modeling power very close to the "digital production studio" composition paradigm.

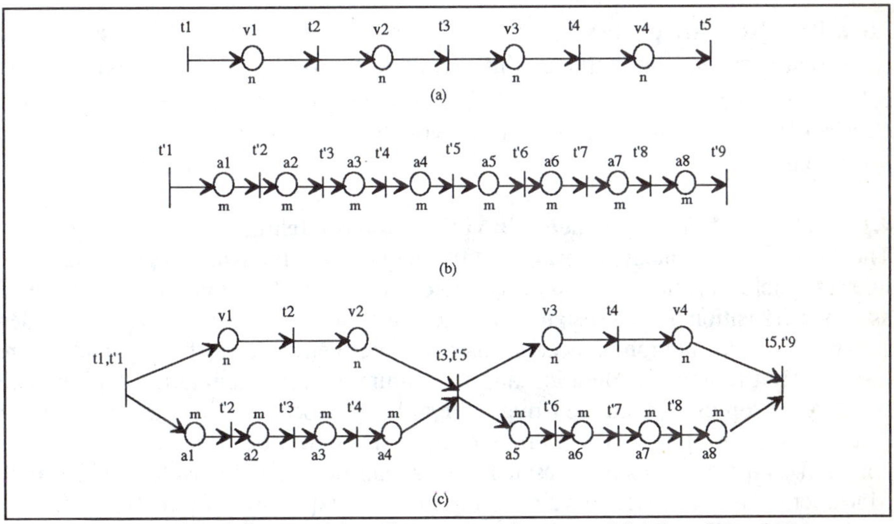


**Figure 3.** Modeling multimedia synchronization with Timed Petri Nets

Figure 3 gives an example of an inter-stream synchronization scheme between a video stream (figure 3-a) and an audio stream (figure 3-b). Note that the resulting

inter-stream synchronization scheme is obtained synchronizing iteratively two video synchronization units with their four related audio synchronization units. Note that this synchronization scheme uses the "equality" synchronization structure to specify the continuous synchronization between the audio and the video streams. Therefore, the nominal durations of the audio and video synchronization units (m and n respectively) must verify : 2×n=4×m.
Though the modeling power of the Timed Petri Net model allows an easy specification of multimedia synchronization scenarios by merging transitions, it can be easily noted that Timed Petri Nets use only nominal durations. Therefore, Timed Petri Nets do not allow the admissible temporal jitter of multimedia processings to be specified. So, Timed Petri Nets do not have an expressive power that allows a realistic specification of asynchronous distributed systems.

# 4. Time Petri Nets

The first model that defines imperfect timings is the Time Petri Net model [Merlin]. In Time Petri Nets, the time interval that labels all transitions is a potential interval of firing.
*Definition:* A Time Petri net is a tuple ( P, T, B, F, Mo, SI ) where:

- (P, T, B, F , Mo) defines a Petri net;
- SI is a mapping function called Static Interval, SI : T —> $Q^+ \times (Q^+ \cup +\infty)$

where, for verification [Berthomieu], $Q^+$ is the set of positive rational numbers. $SI(t_i) = (x_i, y_i)$, where $x_i$ and $y_i$ are rational numbers such that:

$$0 \leq x_i \leq y_i \leq +\infty \text{ and } x_i < +\infty$$

the interval $[x_i, y_i]$ is the firing interval of transition $t_i$;

Time Petri Nets firing rules are defined as follows: assuming that transition $t_i$ is enabled at absolute time $\tau$, then, while being continuously enabled, $t_i$ may not fire before time $(\tau + x_i)$, and $t_i$ must fire before or at the latest at time $(\tau + y_i)$. Note that, classical Petri nets are Time Petri Nets with $(x = 0, y = +\infty)$ associated to every transitions.

## 4.1. Time Petri Nets Assessment for Multimedia Modeling

The Time Petri Net model allows, with the help of time intervals on transitions, the jitter of synchronization units to be specified (*i.e.* synchronization units are modeled as timed transitions). For instance, Figure 4-a illustrates the modeling of a video stream (for presentation concerns) using Time Petri Nets. If n, j and J are respectively the nominal duration and the minimum and maximum jitter of each synchronization units, then the firing interval [x,y] associated with each transition is given by: [x,y]=[n-j,n+J]. Therefore, the expressive power of Time Petri Nets allows temporal non-determinism in asynchronous multimedia systems to be specified. Moreover, Time Petri Nets have a Turing machine expressive power. However, this model does not allow an easy specification of inter-stream synchronization by using a "digital production studio" paradigm (*i.e.* by just merging transitions of the involved streams). Indeed, let us note first that when merging two transitions their respective firing intervals have no more sense (i.e. the initial temporal semantics of the related synchronization units is lost). Moreover, using intervals on transitions induces a strong synchronism on the places that belongs to the prefix of the transition and involves a synchronization always driven by the latest stream. For example, Figure 4 illustrates the composition, by transition merging , of two streams described using Time Petri Net (figure 4-a). A worst case analysis (i.e. a scenario where we get the maximum delay for each synchronization unit of one stream and the minimum delay for each synchronization unit of the other stream) shows that it may be impossible to find an interval [u,v] for transition (t3,t5') allowing one to satisfy in every case the temporal semantics of both synchronization units modeled as t3 and t5' (and their related places v2 and a4). So the modeling power of Time Petri Nets is not satisfying for the specification of multimedia synchronization scenarios. Putting intervals on arcs instead of transitions would allow an easier specification of inter streams synchronization schemes.

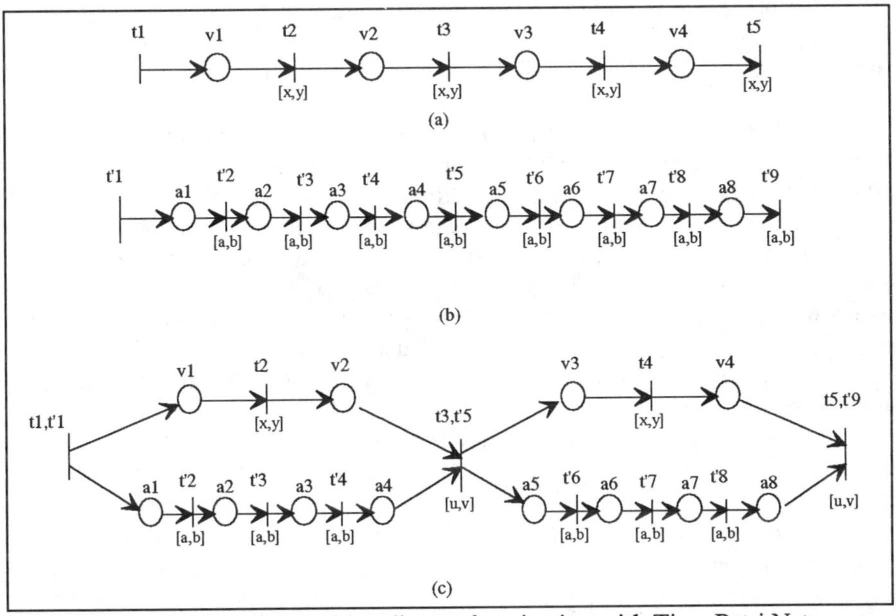


**Figure 4.** Modeling multimedia synchronization with Time Petri Nets

## 5. Petri Nets with Time Intervals on Arcs

A model with intervals on arcs, here called Time Arc Petri Nets (TAPNs), has been proposed in [Walter].

*Definition* : An TAPN is a tuple ( P, T, B, F, Mo, SIM ) where:

- (P, T, B, F, Mo) defines a Petri net;

Let A be the set of arcs entering transitions: $A = \{ (p_i,t_i) \mid B(p_i,t_i)>0 \}$

- IM is a Interval Mapping function,

  $IM : A \rightarrow Q^+ \times (Q^+ \cup +\infty)$

$IM(a_i) = (x_i,y_i)$, where $x_i$ and $y_i$ are rationals such that: $0 \le x_i \le y_i \le +$ and $x_i < +\infty$. Function IM defines the temporal semantics of arcs. The interval $[x_i,y_i]$ is called the *(relative) temporal validity interval* of arc $a_i$.

The difference between TAPN and Time Petri Nets appears when a place receives a token; then, a timer local to each outgoing arc of the place is started (*i.e.* arcs are said *enabled*). Assuming that place $p_i$ is marked at absolute time $\tau_i$, and assuming that interval $[x_i,y_i]$ labels arc $a_i$, then:

- t may not fire before time $(\tau_i + x_i)$ and
- t must fire before or at the latest at time $(\tau_i + y_i)$.

$[\tau_i + x_i, \tau_i + y_i]$ is called the *absolute temporal validity interval* of arc $(p_i, t)$.
Therefore, when multiple places are synchronized by one transition the firing condition must hold for all arcs. As a consequence, transition t is firable if all its related arcs lead to its firing. That is, transition t is firable if the intersection of all dynamic intervals related to all arcs is not empty. [Walter] considers that the behaviour of the net is undefined when there is a *time mismatch*, *i.e.* when the intersection is empty. Such a case must not occur in [Walter] but no efficient validation procedure is given to detect these inconsistent behaviours. In a different way, developing a process algebra basis on this approach, [Bolognesi] assumes that in this case not firable tokens are discarded and disappear from the related places.

### 5.1. TAPNs Assessment for Multimedia Synchronization Modeling

Neither of the two proposed solutions for tackling desynchronization in TAPNs is adequate for modeling and managing multimedia synchronization. Indeed, in case of desynchronization between several streams the firing rules must ensure the continuation of the multimedia scenario and the continuity of streams. For example, Figure 5 illustrates, using TAPN, the modeling of continuous synchronization between an audio and a video streams. Suppose that, the nominal durations of audio and video synchronization units, here modeled as timed arcs, are respectively equal to 2 and 4 time units and that their temporal validity intervals are [1,3] and [3,5] respectively. A worst case analysis (i.e. t2', t3' and t4' fire at the latest firing time of their related arc and t2 fires at its earliest firing time) shows that the absolute temporal validity interval of arcs (v2,(t3,t5')) and (a4,(t3,t5')) are then equal to [6,8] and [10,12] respectively. Thus, the intersection of the absolute temporal validity intervals is empty (*i.e.* $[6,8] \cap [10,12] = \emptyset$), though the nominal duration of the two composed streams are equal. Note that in case of such a time mismatch, a synchronization runtime using the synchronization semantics promoted in [Walter] or [Bolognesi] would stop the multimedia presentation of the two streams. Such desynchronization risks could be avoided using finer synchronization granularity between the two streams (by example by merging transitions t2 and t3' ), but this is not always feasible because some resource constraints may not allow a finer granularity. Thus, it would be excessive to forbid synchronization schemes on the pretext that they risk desynchronization in case of overload of the multimedia system. A compromise based on a "best effort" policy seems more realistic. Moreover, in asynchronous distributed systems, risks of desynchronization between distributed multimedia servers can not be avoided. Therefore, the expressive power of TAPN does not allow to fully tackle desynchronization problems between several multimedia streams. TAPN firing rules promote a too strict all or nothing way to synchronize multimedia streams. In case of desynchronization, TAPNs firing rules (*i.e.* their synchronization semantics) are not able to insure a possible degradation of the temporal constraints of the related streams. It seems unrealistic to stop a multimedia application just because of desynchronization between streams (the show must go on!).

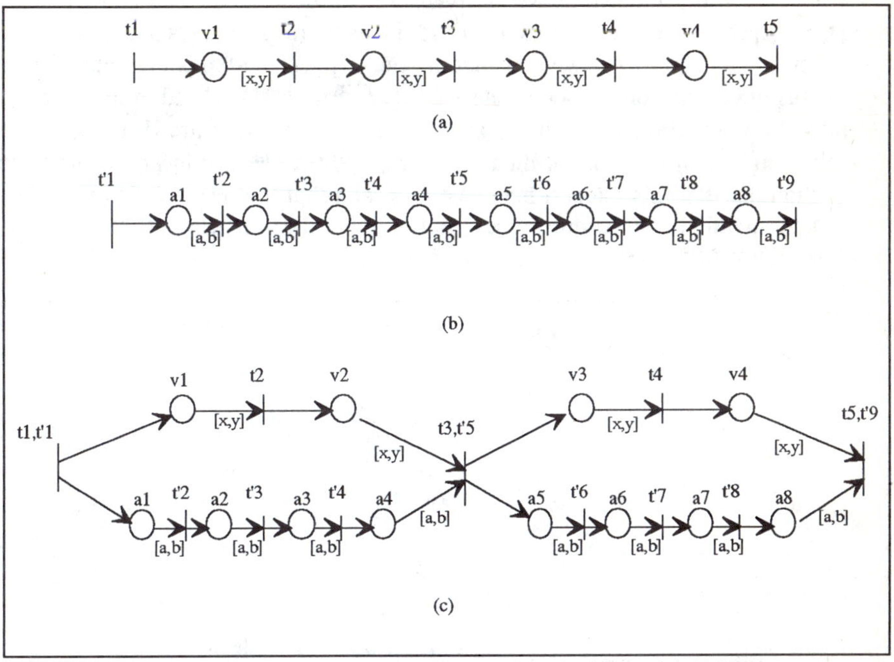


**Figure 5.** Modeling multimedia synchronization with TAPNs

To sum up, TAPN modeling power allow an easy specification of multimedia synchronization schemes by taking into account temporal non determinism of asynchronous systems, but their expressive power appears too limited for tackling risks of desynchronization between streams.

## 6. The Time Stream Petri Net Model

Time Stream Petri Nets (TStreamPNs) aim to encompass both the modeling power of TAPNs and the expressive power of TPNs [Diaz2][Sénac2]. This model like TAPN uses temporal intervals on arcs outgoing from places. The TStreamPN model introduces new firing rules in order to tackle desynchronization problems.

In order to formally define these new firing rules, for any transition t and any related arc $a_i=(p_i,t)$ with a temporal validity interval $[x_i,y_i]$, five time instants are defined :

- $\tau_i^{tok}$ : when the token enters place $p_i$ and enables arcs $(p_i,t)$,
- $\tau_i^{min}$: when the age of the token reaches $x_i$ ,
- $\tau_i^{max}$: when the age of the token reaches $y_i$ ,
- $\tau^{tra}$: when the transition becomes enabled : the first instant at which all its arcs are enabled, *i.e.* when the latest token reaches its $\tau_i^{tok}$ ,
- $\tau^{fir}$: when the transition is fired.

Note that $[\tau_i^{min},\tau_i^{max}]$ is, as defined in Section 5, the absolute temporal validity interval of arc $a_i$.
The firing rules must have certain properties.

1. Each token stays in an arc $a_i$ an amount of time, defining the *temporal semantics* of the arc, comprised between $\tau_i^{min}$ and $\tau_i^{max}$. It follows that $\tau_i^{fir} \in [\tau_i^{min},\tau_i^{max}]$ must be fulfilled to ensure the temporal semantics of arc $a_i$.
2. When a transition t has several ingoing arcs, $(p_i,t)$, the temporal semantics of the firing depends on the set of intervals $\{[\tau_i^{min},\tau_i^{max}]\}$ related to the different arcs. The condition given in 1) above must hold for any arc, leading to the following condition to be fulfilled : $\tau^{fir} \in \cap_i[\tau_i^{min},\tau_i^{max}]$. It has been shown in Section 5.1 that this interval may be empty, and that the respect of intra-stream synchronization constraints does not imply the respect of inter-stream synchronization constraints.

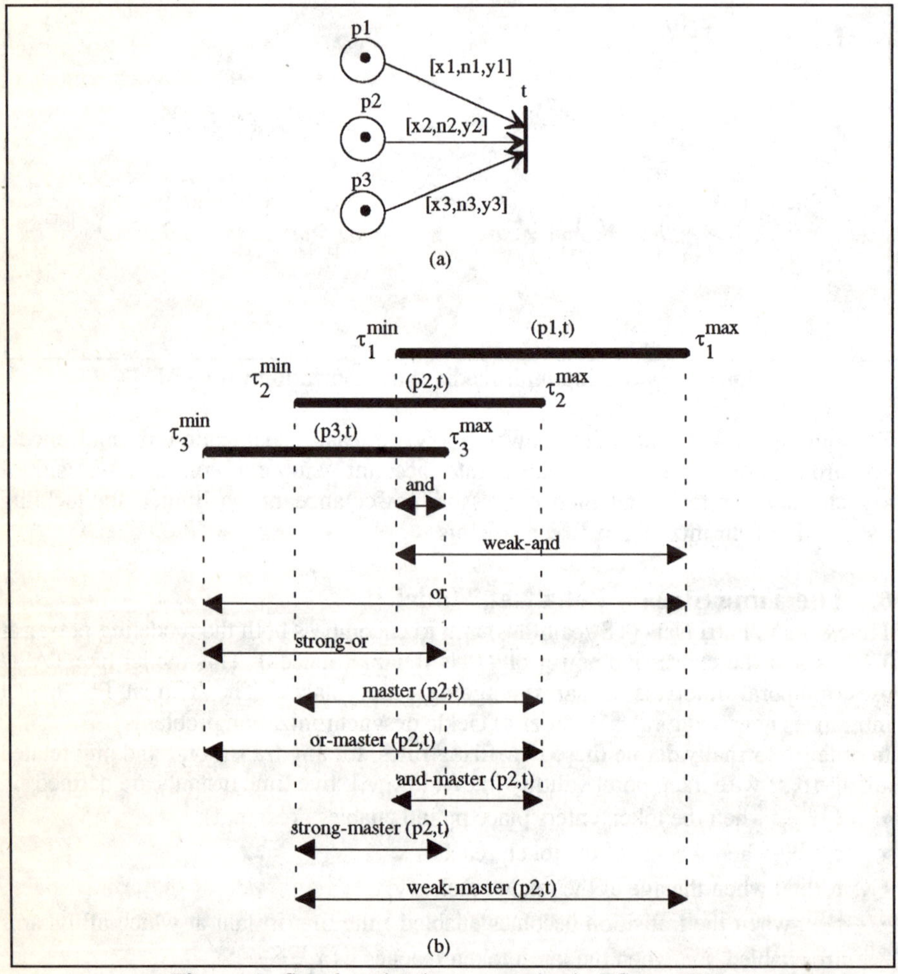


**Figure 6.** Synchronization semantics in TStreamPNs

TAPNs firing rules are not able to tackle inter-stream desynchronization, so, we need new deterministic firing rules ensuring stream continuity and scenario continuation even in presence of time mismatches. The main idea behind these new firing rules is to give priority, in case of desynchronization, to a given arc (belonging to the prefix of the considered transition) . This strategy induces 3 fundamental classes of firing rules:

- a *dynamic* synchronization firing rule driven by the *earliest arc*, *i.e.* the arc with the greatest $\tau_i^{max}$ ("strong-or firing rule"). This arc is defined dynamically in function of cumulative delays in the related streams.
- a *dynamic* synchronization firing rule driven by the *latest arc*, *i.e.* the arc with the lowest $\tau_i^{max}$ ("weak-and firing rule").
- a *static* synchronization firing rule driven by a predetermined *master* arc ("master firing rule").

More precisely, the absolute temporal validity intervals (i.e. $[\tau_i^{min}, \tau_i^{max}]$ intervals) of arcs relating to a given transition may be combined to give a temporally consistent and complete set of nine firing intervals that entail nine fundamental firing rules for TStreamPNs. The firing intervals of a given inter-stream synchronization transition t (Figure 6-a) are illustrated in Figure 6-b according to the predefined time instants and the firing rule associated with transition t (in this example, when the master synchronization is used, the master arc is arc (p2,t)). Therefore, the definition of a TStreamPN can now be given.

*Definition* : A TSPN is a tuple ( P, T, B, F, Mo, IM, SYN ) where:

- (P,T, B, F, Mo) defines a Petri Net.

Let A be the set of transition ingoing arcs : $A = \{ a_i=(p_i,t_i) \mid B(p_i,t_i)>0 \}$;

- IM is an Interval mapping function,

$$IM : A \rightarrow Q^+ \times Q^+ \times (Q^+ U + \infty), \quad a_i \rightarrow ( x_i^S, n_i^S , y_i^S )$$

where this 3-uple represents, the minimum, nominal ,and maximum static enabling durations for arc $a_i$, and verify: $x_i^s \leq n_i^s \leq y_i^s$. This 3-uple is called the *static temporal validity interval* of arc $a_i$ and defines its *temporal semantics*. The nominal duration has been added in order to be able to calculate, during a simulation, advance or delay with regard to the nominal duration. The firing rules between states defined in Section 7.3, transform these static value into dynamic values denoted as $x_i$, $n_i$ and $y_i$ respectively. The 3-uple ($x_i$, $n_i$ ,$y_i$) is called the *(dynamic) temporal validity interval* of arc $a_i$.

- SYN is a typing function defining different transition firings rules:

  SYN: T→{and,weak-and,or,strong-or,master,or-master,and-master,weak-master,strong-master},

The formal semantics of TStreamPNs and of five among nine of the evoked firing rules is given in the next section (namely the "and", "weak-and" "or", "strong-or", "master" firing rules).

# 7. Formal semantics of TStreamPNs

## 7.1. States in a TStreamPN

A state S of a TStreamPN can be defined as a pair (M,I) consisting of:

- the marking M, and
- the list I of the dynamic temporal validity intervals of enabled arcs. The number of entries in this list is the number of arcs enabled by marking M.

Let us consider the TStreamPN given in Figure 7. This scenario describes a multimedia coarse grain synchronization presentation scenario. The presentation starts by presenting a title in place $p_0$, during a nominal duration of 4 time units, and with a minimum and maximum durations equal to 2 and 7 respectively. Then, a voice comment starts in $p_5$. It lasts for the entire presentation, in parallel with two images, presented sequentially in $p_3$ and $p_4$. The first image is presented in parallel with 2 textual comments, sequentially presented in $p_1$ and $p_2$. If we have a form of semantics redundancy between the image and the textual comment, then we can associate with transition $t_2$ a "strong-or" firing rule, so the first ending object associated with this transition (*i.e.* text or image) triggers the synchronization and is not delayed. A "weak-and" type is associated with $t_3$ transition in order to ensure that each multimedia object (*i.e.* image associated with place $p_4$ and comment associated with place $p_5$) to be completely presented.

The initial state of this TstreamPN is given by $S_0=(M_0,I_0)$ with:

$M_0 = p_0(1)$, *i.e.* transition $t_0$ is the only enabled transition and,

$I_0=((2,4,7))$, *i.e.* $t_0$ can fire at any time between 2 and 7.

When $t_0$ fires, it leads to state $S_1=(M_1,I_1)$, with $M_1=(p_1(1),p_3(1),p_5(1))$, where arcs $(p_1,t_1)$, $(p_3,t2)$ and $(p_5,t_3)$ are enabled, and $I_1 = ((2,5,7),(5,10,15),(10,15,20))$.

## 7.2. Firing Rule between States

Let us assume that transition t be firable at time θ from state S. Then state S'=(M',I') reached from S by firing t at θ can be computed as:

1. M' is computed as:

   $(\forall\ p)\ M'(p) = M(p) - B(t,p) + F(t,p)$ , as usually in Petri nets,

2. I' is computed in three steps:
   a. Remove from the expression of I the 3-uples that are related to arcs disabled when t is fired; these disabled arcs are those enabled by M and not enabled by M(.) - B(t,.). They include the ones related to transition t.
   b. Shift of a value θ towards the origin of times all temporal validity intervals in I (*i.e.* the intervals that remain enabled and so remain in I) and truncate them, when necessary, to non negative values. So, for all arcs, $a_k = (p_k,t_n)$, which remain enabled, then replace their dynamic temporal validity interval $(x_k,n_k,y_k)$, in I, with $(\max(0,x_k-\theta),n_k-\theta,\max(0,y_k-\theta))$, where $x_k$ and $y_k$ denote respectively the minimum and maximum enabling durations for arc $a_k$ from state S, and $n_k$ denotes, relatively to the arc nominal duration, the advance, when positive, or delay, when negative.
   c. Introduce in the domain the static intervals of the new enabled arcs. The new enabled arcs are those not enabled by M(.) - B(t,.) and enabled by M'(.) = M(.) - B(t,.) + F(t,.). So, the static temporal validity intervals of all arcs newly enabled by M' (including those enabled by M', but already enabled by M and in conflict with t in M) are contained in I'.

**7.3. The "And " Firing Rule**

With the "*and*" firing rule all arcs satisfy, if there is no time mismatch, their temporal semantics. Therefore, if there is no time mismatch, the firing interval defined by the "and" firing rule is the same that the one defined by the firing rule of TAPNs (*i.e.* the intersection of the absolute temporal validity intervals). However, in case of time mismatch, this firing interval amounts to a time instant defined by the maximum date $\tau_i^{min}$ among the set of the related arcs (*i.e.* all arcs have to reach their minimum duration). Remark that structured TStreamPN allow one to apply static verification procedures to detect risks of time mismatch [Senac3]. In case of risk of time mismatch the specifier may modify the synchronization scheme or change the "and" firing rule by a less stringent one (*e.g.* the "weak-and" one). However, the "and" firing rule can be used even in case of time mismatch.

*Definition* : A transition $t_k$, such that SYN($t_k$)="and", is firable at time $\tau^{fir}=\tau^{tra}+\theta$, from state S=(M,I), iff the two following conditions hold:

1. $t_k$ is enabled by M at time $\tau^{fir}$
2. the relative firing time θ, such that $\tau^{tra}+\theta=\tau^{fir}$, verifies the following system of inequations that defines the firing interval:

$$\max_i\left(\tau_i^{tok}+x_i^s\right)\leq\tau^{tra}+\theta\leq\max\left(\min_i\left(\tau_i^{tok}+y_i^s\right),\max_i\left(\tau_i^{tok}+x_i^s\right)\right)$$

$$\text{with } i\in N_{t_k}=\left\{i\in N/a_i\in A_{t_k}\right\}\text{ and } A_{t_k}=\left\{a\in A/a=(p_n,t_k)\right\}$$

From the firing rules between states, it follows that an enabled "and" transition $t_k$ is firable from state S=(M,I) at relative time θ (*i.e.* θ is relative to the occurrence of state S) iff θ verifies the following inequalities:

$$MIN_k=\max_i\{x_i\}\leq\theta\leq\max(\min_i\{y_i\},\max_i\{x_i\})=MAX_k$$

where i ranges over $N_{t_k}$ and where $x_i$ and $y_i$ are the bounds of the temporal validity intervals in I related to arcs in $A_{t_k}$. Remark that this firing rule uses relative instants only.

**7.4. The "Weak-And" Firing Rule**

With the "*weak-and*" rule, if due to any condition, one arc becomes in advance with respect to the others, then this arc is forced to wait. This implies that the faster stream will be delayed to wait for the slowest one, up to a maximum value of time, this value being the greatest value $\tau_i^{max}$ among the related arcs .

*Definition:* A transition $t_k$, such that SYN(t)="weak-and", is firable at time $\tau^{tra}+\theta$ from state S=(M,I) iff both the following conditions hold :

1. $t_k$ is enabled by marking M
2. the relative firing time θ, such that $\tau^{tra}+\theta=\tau^{fir}$, is in the interval of time where

the following inequalities hold :

$$\max_i\left(\tau_i^{tok}+x_i^s\right)\le\tau^{tra}+\theta\le\max_i\left(\tau_i^{tok}+y_i^s\right)\text{ with } i\in N_{t_k}$$

From the firing rules between states, it follows that a "weak-and" enabled transition $t_k$ is firable from state S=(M,I), at relative time θ, iff θ verifies the following inequalities:

$$MIN_k = \max_i \{x_i\} \le \theta \le \max_i \{y_i\} = MAX_k$$

**7.5. The "Strong-Or" Firing Rule**

With the "strong-or" firing rule, if due to adverse conditions, an arc is in advance with respect to the others, the arc in advance imposes the firing. That is, the faster stream is not delayed, but rather speeds up the other streams.

*Definition:* A transition t, such that SYN(t)="strong-or", being enabled at $\tau^{tra}$, is firable at time $\tau^{tra}+\theta = \tau^{fir}$, from state S=(M,I), iff both the following conditions hold :

1. $t_k$ is enabled by marking M at time $\tau^{fir}$
2. the relative firing time θ, relative to the absolute time $\tau^{tra}$, is in the interval where the following inequalities hold :

$$\min_i\left(\tau_i^{tok}+x_i^s\right)\le\tau^{tra}+\theta\le\min_i\left(\tau_i^{tok}+y_i^s\right)\text{ with } i\in N_{t_k}$$

From the firing rules between states, it follows that a "strong-or" enabled transition $t_k$ is firable from state S=(M,I), at relative time θ, iff θ verifies the following inequalities:

$$MIN_k = \min_i \{x_i\} \le \theta \le \min_i \{y_i\} = MAX_k\text{, where i ranges over } N_{t_k}.$$

**7.6. The "Or" Firing Rule**

The "or" firing rule is the weakest one. It allows the transition to fire at any time inside the time interval defined by the largest temporal interval that may be defined from the bounds of the absolute temporal validity intervals of the related arcs. This seems to be the weakest possible synchronization constraint, in the sense that the firing is defined as acceptable as long as the temporal semantics of at least one arc is guaranteed. That is, this firing rule introduces a model of failure for arcs.

*Definition:* A transition $t_k$, such that SYN(t)="or", being enabled at $\tau^{tra}$, is firable at time $\tau^{tra}+\theta = \tau^{fir}$, from state S=(M,I), iff both the following conditions hold :

1. $t_k$ is enabled by marking M at time $\tau^{fir}$
2. the relative firing time θ, relative to the absolute enabling date $\tau^{tra}$, is in the interval where the following inequalities hold :

$$\min_i\left(\tau_i^{tok}+x_i^s\right)\le\tau^{tra}+\theta\le\max_i\left(\tau_i^{tok}+y_i^s\right)\text{ with } i\in N_{t_k}$$

From the firing rules between states, it follows that an enabled "or" type transition $t_k$ is firable at relative time θ, from occurrence of state S=(M,I), iff θ verifies the following inequalities:

$$MIN_k = \min_i \{x_i\} \le \theta \le \max_i \{y_i\} = MAX_k$$

where $x_i$ and $y_i$ are the bounds of the related temporal validity intervals in I.

### 7.7. The "Master" Firing Rule

The "master" firing rule defines one arc (and the subjacent stream) as the leading (the most important) arc. This arc enforces the firing of the related transition.

*Definition:* A transition $t_k$, such that SYN($t_k$)="master", with arc $a_m$ being the master arc, is firable at time $\tau^{tra}+\theta = \tau^{fir}$, from state S=(M,I), iff both the following conditions hold :

1. $t_k$ is enabled by marking M at time $\tau^{fir}$
2. the relative firing time θ, relative to the absolute enabling time $\tau^{tra}$, is in the interval where the following inequalities hold :

$$\left(\tau_m^{tok} + x_m^s\right) \leq \tau^{tra} + \theta \leq \left(\tau_m^{tok} + y_m^s\right)$$

From the firing rules between states, it follows that an enabled "master" type transition $t_k$ is firable at relative time θ, from the occurrence of state S=(M,I), iff θ verifies the following inequalities:

$$MIN_k = x_m \leq \theta \leq y_m = MAX_k$$

where $x_m$ and $y_m$ are respectively the minimum and maximum bounds of the temporal validity interval of master arc $a_m$ in I.

### 7.8. Enabledness and Firability Condition of a Set of Transitions

A transition $t_k$ is firable at the relative instant θ, from the occurrence of state S=(M,I), iff both the following conditions hold :

1. t is enabled by marking M: $(\forall\, p_i)\ (M(p_i) \geq B(t, p_i))$
2. θ satisfies the following inequalities: $MIN_k \leq \theta \leq \min_j\{MAX_j\}$, where j ranges over the set of transitions enabled by M.

Note that 2) is required because, at time $\min_j\{MAX_j\}$, the corresponding transition must fire, modifying the marking and so the state of the TStreamPN.

### 7.9. Example

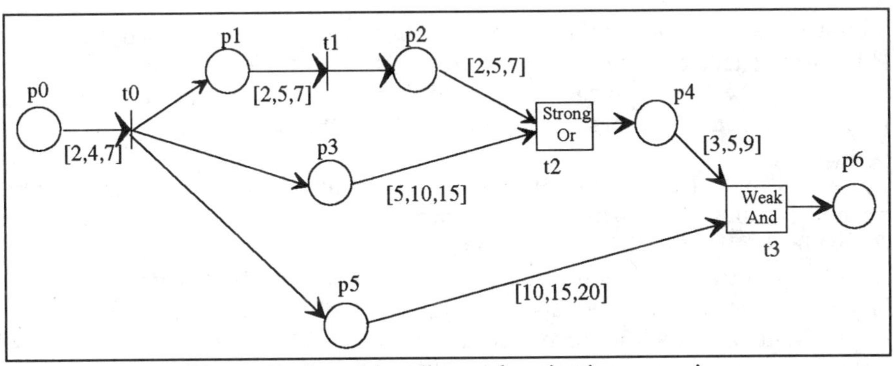


**Figure 7.** A multimedia synchronization scenario

Let's consider the TStreamPN in Figure 7. The initial state of this TStreamPN is

$S_0=(M_0,I_0)$ with:

- $M_0=(p_0(1))$
- $I_0=((2,4,7))$

From state $S_0$ only $t_0$ can fire. If $t_0$ fires between 2 and 7, say at 4 (*i.e.* the nominal duration), then we get the new state $S_0=(M_0,I_0)$ with:

- $M_1 = (p_1(1),p_3(1),p_5(1))$ where $(p_1,t_1)$, $(p_3,t_2)$ and $(p_5,t_3)$ are enabled and,
- I1 = ((2,5,7) (5,10,15) (10,15,20))

From the state $S_1=(M_1,I_1)$ only $t_1$ can fire. If $t_1$ fires between 2 and 7, say at $\theta_1 = 6$ (*i.e.* with a delay of 1 time unit compared to the nominal duration of arc $(p_1,t_1)$), then we get the new state $S_2=(M_2,I_2)$ with:

- $M_2 =(p_2\ (1),\ p_3\ (1),\ p_5\ (1))$ where $t_2$ is the only enabled transition
- $I_2 = ((2,5,7)$ for $(p_2,t_2)$,
  $(\max(0,5 - \theta_1),10-\theta_1,\max(0,(15 - \theta_1))) = (0,4,9)$ for $(p_3,t_2)$ and
  $(\max (0,10 - \theta_1),15-\theta_1, \max(0,(20 - \theta_1))) = (4,9,14)$ for $(p_5,t_3))$

As a "strong-or" transition, $t_2$ can fire between [min(2,0), min(7,9)] = [0,7]. If $t_2$ fires at $\theta_2=1$, then, note that the temporal semantic of arc (p2,t2) is not respected, *i.e.* the synchronization unit modeled as (p2,t2) is stopped by the termination of the synchronization unit modeled as arc (p3,t2). Thus, we get the new state $S_3=(M_3,I_3)$ with:

- $M_3=(p_4(1),\ p_5(1))$
- $I_3 = ((3,5,9)$ for $(p_4,t_3)$, (3,8,13) for $(p_5,t_3))$.

As a "weak-and" transition t3 can fire between max(3,3) and max(9,13), *i.e.* in the relative interval [3, 13]. Note that t3 may fires at relative date $\theta_3=10$ though at this time the temporal semantics of arc (p4,t3) is no longer satisfied.

## 8. Main Properties

### 8.1. Consistency With the Statically Defined Temporal Semantics of Arcs

*Definition* : A transition t of a Time Stream Petri Net is *well-designed* iff :

$$\forall (a_i,a_j) \in A_t \times A_t,\ \tau_i^{\min} \leq \tau_j^{\max}$$

$$\text{where } A_t = \{(p,t) \in A\}$$

Well-designed transitions are of importance because (cf. the next property) they help to insure arcs temporal semantics in TStreamPNs. The "well-designed" property can be statically checked in structured TStreamPNs by applying a reduction algorithm on the considered inter-stream synchronization scheme [Senac3].

*Property:* “and” (see Figure 6) well-designed transitions enforce the temporal semantics of arcs in inter-stream synchronization schemes.
*Proof* : by definitions of the “and” firing rule and of the temporal semantic. Indeed, well-designed transitions allow time mismatches to be avoided, *i.e.* the intersection of the absolute temporal validity intervals of a well-designed transition is never empty. Moreover, when no time mismatch, the and firing rule ensures the related transition to be fired within the intersection of the temporal validity intervals. Therefore, the temporal semantics of all related arcs is satisfied. Thus, preserving both streams continuity and temporal semantics of synchronization units, the “and”

firing rule and well-designed transitions are well adapted for tight synchronization between continuous streams (*i.e.* continuous synchronization).

**8.2. Boundedness**

*Property*: If the arcs of a structured Time Stream Petri Net are temporally bounded, that is if $\forall a \in A / IM(a) = (x,n,y)$ , $y < +\infty$, then the TStreamPN is temporally bounded (*i.e.* the modeled scenario has a finite bounded duration).

*Proof*: by giving an effective procedure for the calculus of the minimum and maximum bounds [Diaz2].

**8.3. Relationships with Time Petri Nets (Merlin's model)**

*Property:* A Time Petri Net (Merlin's model) is a particular case of Time Stream Petri Net where:

1. all transitions are of "weak-and" type, and
2. the interval [x,y] of each transition t in the Time Petri Net is the interval associated with each arc ingoing transition t in the equivalent TStreamPN.

*Proof*: We can give here a short "proof" of this property considering that, in a TstreamPN, a "weak-and" enabled transition, $t_m$, is firable at relative time $\theta$, from state S=(M,I), iff $\max_i\{x_i\} \leq \theta \leq \max_i\{y_i\}$, where i ranges over $N_t$ and where $(x_i,n_i,y_i)$ are the related dynamic time intervals in I. If $\forall\ a_k \in A_t$, $SYN(a_k)=(x,n,y)$ and if the transition is newly enabled in state S, then the above inequality becomes, $x \leq \theta \leq y$, which is the traditional set of inequations defining the firing interval of a just enabled transition, with an associated static interval [x,y], in a TPN. Figure 8 illustrates a Time Petri Net (Figure 8-a) and the equivalent TStreamPN (Figure 8-b).

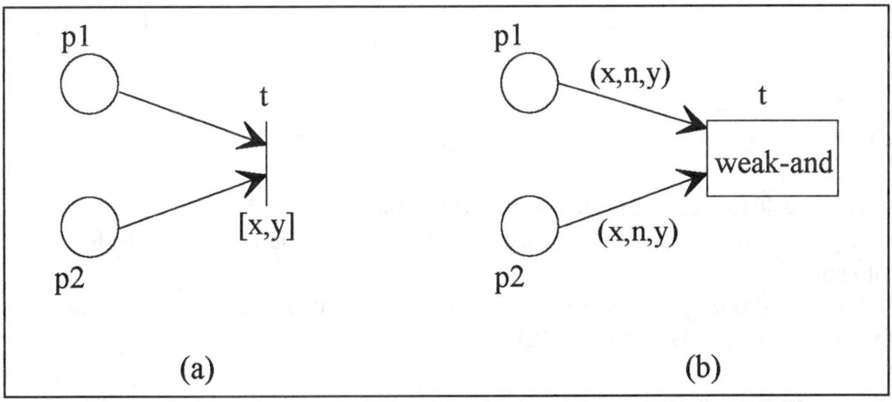


**Figure 8.** Translation of a Timed Petri Net (a) into a TStreamPN (b)

*Property:* Each Time Stream Petri Net has an equivalent Time Petri Net.

*Proof:* By construction. Figure 9 gives the translation of a TStreamPN "and" transition (Figure 9-a) into an equivalent Time Petri Net (Figure 9-b). This Figure illustrates that the TStreamPN model has a stronger modeling power for multimedia synchronization, than the Time Petri Net model. Indeed, as illustrated in Figure 9 the translation of a TStreamPN into a Time Petri Net entails an explosion of the number of places, transitions and arcs in the resulting Time Petri Net.

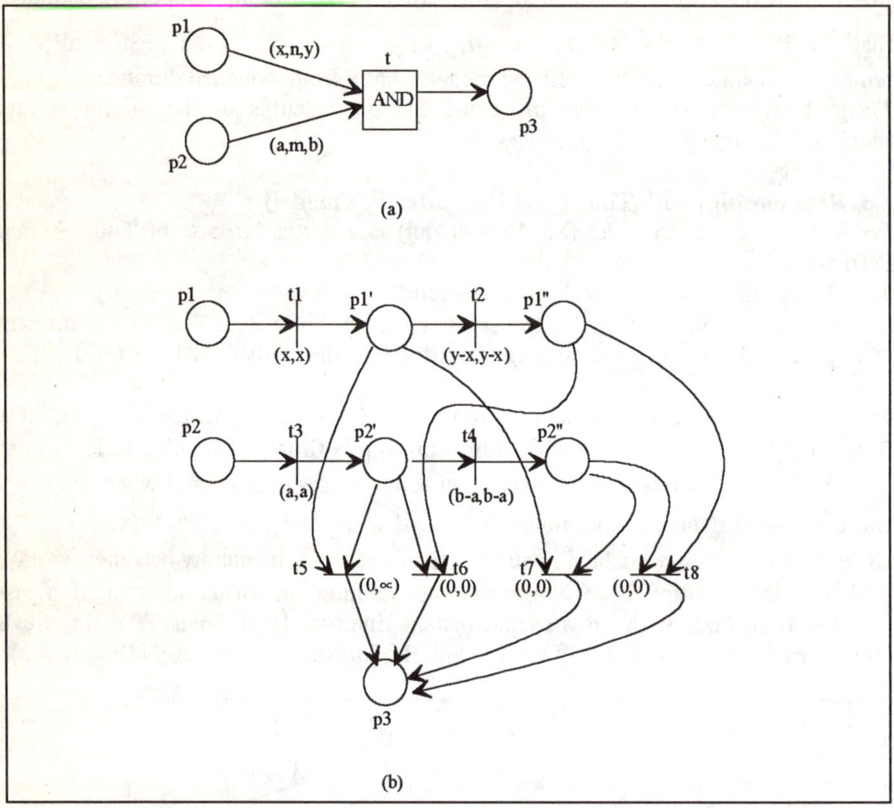


**Figure 9.** Expressing TStreamPNs (a) with Time Petri Nets (b)

### 8.4. Undecidability of Reachability and Boundedness Problems

*Property:* The reachability and boundedness problems are undecidable for Timed Stream Petri Nets.

*Proof*: The property comes from the fact that every Time Petri Net can be translated into an equivalent Time Stream Petri Net.

## 9. Conclusion

This paper has introduced TStreamPNs to describe synchronization constraints in asynchronous distributed systems, and mainly in multimedia distributed asynchronous systems. It has been shown that the TStreamPN model brings interesting contributions to Petri Nets as a model that encompasses both the modeling power of the TAPN model, and the expressive power of the Time Petri Net model. Therefore, TStreamPNs model allow an easy, complete and accurate

specification of synchronization constraints within asynchronous distributed systems. Furthermore, TStreamPNs can be used as a modeling basis for any multimedia objects and at different levels of granularity. Moreover, TStreamPNs are presently used for the specification of scheduling profiles for a synchronization controller in a multimedia project [Diaz3],[Diaz4][Senac1].
Analysis and verification techniques still need to be developed, mainly those concerning the correctness of time behaviours.

## 11. References


[Allen] J. F. Allen, *Maintaining Knowledge about Temporal Intervals*, Comm. of hte ACM, 26, (11), November 1983.

[Berthomieu] B. Berthomieu, M. Diaz, Modeling and Verification of Time Dependent Systems using Time Petri Nets, IEEE Transactions on Software Engineering, V 17, N3, March 1991.

[Bolognesi] T. Bolognesi, *From Timed Petri Nets to Timed LOTOS*, IFIP Int Conf on Protocol Specification, Testing and Verification, Ottawa, June 1990, North-Holland.

[Diaz1] M. Diaz, *Modeling and Analysis of Communication and Cooperation Protocols using Petri Net basisd Models*, Computer Networks, December 1982.

[Diaz2] M. Diaz, P. Senac, *Synchronization and Multimedia Objects*, Technical Report 92437, LAAS, November 1992.

[Diaz3] M. Diaz, P. de Saqui-Sannes, P. Senac, *Un Modèle Formel pour la Spécification de la Synchronization Multimédia en Environment Distribué*, Proceedings of CFIP'93, Montréal, September 1993.

[Diaz4] M. Diaz, P. Senac, *Time Stream Petri Nets, a Model for Multimedia Streams Synchronization,* Proceedings of MMM'93, Singapore, November 1993.

[Hamblin] C. L. Hamblin, *Instants and Intervals*, Proc. 1st Int. Conf. for the study of time, Springer Verlag, J. T. Fraser at al Editors, 1972.

[Little1] T. Little, A. Ghafor, *Multimedia Synchronization*, IEEE Data Engineering September 1991.

[Little2] T. Little, A. Ghafor, *Synchronization and Storage Models for Multimedia Objects*, IEEE Journal on Selected Areas in Communications, 8, (3), April 1990.

[Merlin] P. Merlin, *A Study of the Recoverability of Computer Systems*, Thesis, Computer Science Dept., University of California, Irvine, 1974.

[Murata] T. Murata, *Petri Nets: Properties, Analysis and Applications*, Proceedings of the IEEE, Vol 77, No 4, April 1989.

[Ramchandani] C. Ramchandani, *Analysis of Asynchronous Concurrent Systems by Timed Petri Nets*, Project MAC, TR 120, MIT, February 1974.

[Rust] L. F. Rust da Costa Carmo, P. de Saqui-Sannes, J. P. Courtiat, *Basic Synchronization Concepts in Multimedia Systems*, Proceedings of the 3rd International Workshop on Network and Operating Support for Digital Audio and Video, La Jolla, California, November 1992.

[Senac1] P. Senac, M. Diaz, P. de Saqui-Sannes, *A Formal Environment for the Specification an Design of Multimedia Synchronization Scenarios*, short Proceedings of the 4th International Workshop on Network and Operating Support for Digital Audio and Video, Lancaster, November 1993.

[Senac2] P. Senac, M. Diaz, P. de Saqui-Sannes, *Un Modèle Formel pour la Spécification de la Synchronization Multimédia en Environment Distribué*, Proceedings of RTS'94, Paris, January 1994.

[Sénac3] P. Senac, M. Diaz, *The Properties of Structured Time Stream Petri Nets*, Technical Report LAAS, April 1994.

[Senac4] P. Senac, M. Diaz, P. de Saqui-Sannes, *Toward a Formal Specification of Multimedia Synchronization,* published in "Annals of Telecommunications", May/June 1994.

[Walter] B. Walter, *Timed Petri Nets for Modeling and Analysing protocols with time*, Proc of the IFIP Conf on Protocol Specification, Testing and Verification, III, 1983, North Holland, H. Rudin & C. West eds.

# A Term Representation of P/T Systems

Cheryl Dietz , Gerlinde Schreiber*
C.v.O. Universität Oldenburg
Fachbereich Informatik
D-26111 Oldenburg
linde.schreiber@informatik.uni-oldenburg.de

Net semantics for CCS-like calculi have been developed in order to combine the explicit parallelism of Petri nets and their graphical intuition with the techniques for modular design and analysis of process algebras. In this paper we pursue the converse: starting from a place/transition system we construct an equivalent term representation. The resulting terms include the operators action, nondeterminism, parallelism, recursion and restriction. We prove the correctness of the term semantics according to bisimulation equivalence and investigate stronger correctness notions.

Topic: Relationship between net theory and CCS-like calculi

## 1 Introduction

Petri nets (for an overview see [Reisig85]) and process algebras like CCS ([Milner89]), CSP ([Hoare85]) or ACP ([BK85]) are formalisms for the specification and analysis of distributed systems. Both emphasize different aspects: while Petri nets provide a graphical representation of a system making its inherent parallelism explicit, CCS-like calculi support modular design techniques and compositional semantics.

In the last years several attempts have been made to combine the advantages of both formalisms. One topic of research has been the development of net semantics for CCS-like calculi by defining CCS-like composition operators for nets (see for example [Olderog87], [vGV87], [DDM88], [Goltz88] and [Taubner89]). This is a constructive method to support system design by nets, as these operators are equipped with the induced compositional semantics.

In our work we pursue the converse: is it possible to translate an arbitrary place/transition system into a term of a process algebra? This question seems to be much less investigated. In [BRS86] each node of a net is translated to a parallel component of the term. These parallel components communicate by playing the token game: they pass "send token"/"receive token"-messages to one another according to the flow relation of the net. The initial marking is converted to the number of "send token"-messages a process can communicate initially.

In contrast to this approach we are interested in a term representation of the net where the parallel components do not correspond to the static net structure (i.e. to its nodes) but where they reflect the parallelism in its dynamic behaviour. We want the parallel components of a term to represent the runs of tokens through the net keeping track of conflicts, and we want communication between these processes to model

* This work has been supported under grant Cl53/4 by the Deutsche Forschungsgemeinschaft.

synchronisation between those runs. This intuition of our term semantics for nets is very close to the notion of processes of a net [Reisig85], moreover to its branching processes as introduced by Engelfriet [Engelfriet91]. While the branching processes of a net build a generally infinite lattice with infinite elements the term representation of arbitrary finite place/transition nets is finite.

Thus we yield a finite representation of a kind of Petri net semantics that has up to now been expressed by infinite structures and we combine the nice intuition of Petri nets (that are not limited to certain well-structured subclasses) and the modular design and analysis techniques of CCS-like calculi.

The term representation of a Petri net and its correctness proof is gained as follows: In chapter 2 we list the necessary basic definitions on Petri nets and process terms. For technical reasons we restrict ourselves to nets with binary synchronisation and binary choice. Transitions are labelled with actions from an action alphabet Act. In the process algebra considered possible synchronisation between pairs of actions is denoted by a symmetric synchronisation function $\gamma$:ActxAct→Act similar to synchronisation in ACP. We introduce in chapter 3 a special class of Petri nets, the so-called synchronisation-free P/T systems (for short: SF nets). We need these nets in an intermediate step during our constructions. In SF nets all transitions have at most one ingoing arc: they are structurally free of synchronisation. We show that every P/T system can be transformed into a synchronisation-free P/T system by splitting those transitions with a larger preset as sketched in figure 1:

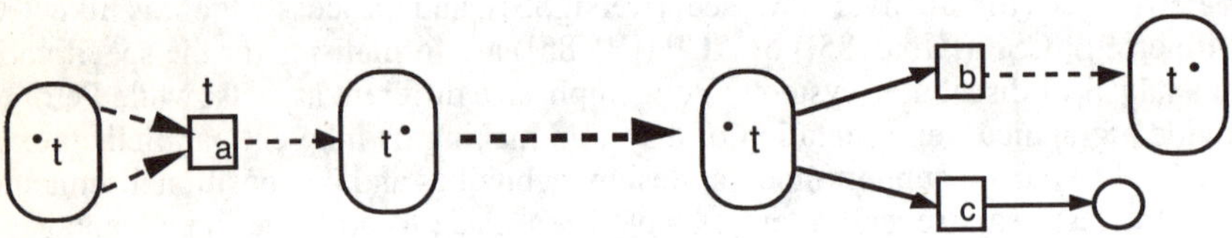


Fig. 1: Construction of synchronisation-free P/T systems

The synchronisation function $\gamma$:ActxAct→Act keeps book of this splitting of transitions (in the above figure: $\gamma$(b,c)=a). Every action may synchronise with at most one partner. We equip SF nets with a new occurrence rule: a transition is enabled only when both the transition and its partner according to $\gamma$ are enabled in the usual way. Then both transitions must occur jointly and synchronise as the one action given by the value of $\gamma$. Further constructions now only rely on SF nets. In the fourth chapter we present an algorithm that translates SF nets into terms of the process algebra. The term representation is defined by structural induction on the initial part of the net (precisely the marked places with their pre- and postsets). By the inductive reasoning a net is transformed into tree-like components with recursion possible inside the branches. The resulting terms include the operators action, nondeterminism, parallelism and recursion. Possible synchronisation is captured by the function $\gamma$ that is adopted from the SF net. To enforce synchronisation we finally add a global restriction on all those actions $\gamma$ is defined on.

The correctness of our term representation is investigated in chapter 5. The usual interleaving semantics of nets can be retrieved from the term semantics. The proof consists of three steps: We first state that an arbitrary P/T-system is bisimilar to its synchronisation-free version working according to the new occurrence rule for SF nets. In the second step we turn from SF nets where synchronisation is enforced (formally: they are equipped with the new occurrence rule) to SF nets where synchronisation is possible (formally: they are equipped with the union of the new and the standard occurrence rule). These nets are bisimilar to the unrestricted term. Then we show that enforcing synchronisation in terms (by restriction) and enforcing synchronisation in SF nets (by use of the new occurrence rule only) correspond. By transitivity of bisimulation equivalence we can finally conclude that the restricted term is bisimilar to the P/T system we started with.

Figure 2 illustrates the proof structure (where N denotes a P/T system, SF(N) its synchronisation-free version, T(SF(N)) the term representing SF(N) with the union of standard and modified occurrence rule and T(N) the restricted term representing SF(N) with the modified occurrence rule only ):

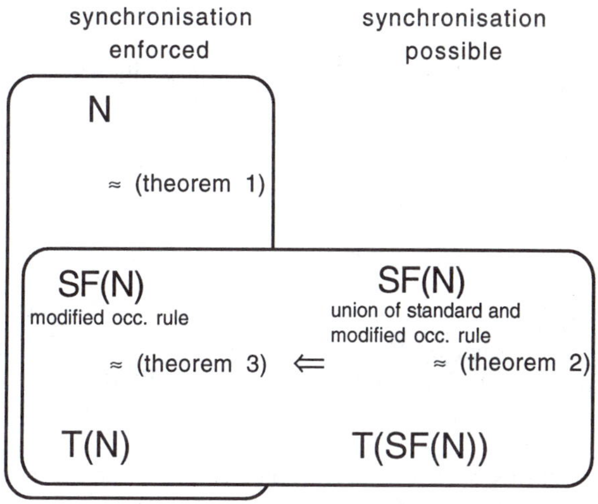


Fig. 2: Proof strucure

We discuss stronger correctness notions and illustrate the relationship to branching processes by an example. The paper ends with concluding remarks in chapter 6.

## 2 Basic definitions

In this chapter we provide the necessary basic definitions on syntax and semantics of petri nets and CCS-like terms.

### 2.1 Petri nets

Our definitions on Petri nets are based on multisets. A <u>multiset</u> M over a set X is a mapping $M:X \rightarrow \mathbb{N}_0$ (written as $M \in \mathbb{N}_0^X$). $x \in X$ with $M(x)>0$ is called <u>element</u> of M. $|M|$ is the sum of the multiplicities of the elements of M: $|M|=\Sigma_{(x\in X)}M(x)$. $M \in \mathbb{N}_+^X$ denotes a multiset including at least one element. $0_X$ determined by $0_X(x)=0\ \forall\ x \in X$

denotes the empty multiset over X. Addition and subtraction of multisets over the same set are defined componentwise, restriction of $M \in \mathbb{N}_0^X$ to a subset X' of X is denoted by $M|_{X'}$. $\mathbb{N}_0^X$ is partially ordered by "≤": $M \leq M' :\Leftrightarrow \forall x \in X: M(x) \leq M'(x)$. For $M \in \mathbb{N}_0^X$ and $M' \in \mathbb{N}_0^{X'}$ the multiset union $M \cup M' \in \mathbb{N}_0^{X \cup X'}$ is defined if $X \cap X' = \emptyset$. $M \in \mathbb{N}_0^X$ will often be written as $\{x \rightarrow n \mid x \in X, M(x)=n \neq 0\}$, where only its elements are listed.

We use the following definition of a labelled finite place/transition system.

**Definition**

$N=(S,T,M_0)$ is called a (labelled finite) place/transition system (P/T system) :⇔

- S is a nonempty finite set of places,
- $T \subseteq \mathbb{N}_+^S x Act x \mathbb{N}_+^S$ is a finite set of transitions and
- $M_0 \in \mathbb{N}_+^S$ is the initial marking.

Transitions are given by triples (preset, label, postset). The preset of a transition is denoted by $^\bullet t$, the label by l(t) and the postset by $t^\bullet$. Preset and postset of a transition are multisets, thus representing arc weights. By this definition we exclude empty pre- and postsets.

When dealing with nets we make use of the following notations: $s^\bullet$ and $^\bullet s$ denote the post- and preset of $s \in S$ defined as usual. Fin(N) is the set of places of N with empty postset (final places). In(N) is the set of initial places of N, precisely those places that are marked by the initial marking. $IP(N)=\{(^\bullet s,s,s^\bullet) \mid s \in In(N)\}$ denotes the initial part of N. The disjoint union is defined on nets with disjoint sets of places by componentwise union. Note that labels don't have to be unique. We consider nets as isomorphic if their graphs are isomorphic and the labels of the transitions are preserved. Their initial marking is not considered.

The dynamic behaviour of nets is given by the usual occurrence rule.

**Definition**

Let $N=(S,T,M_0)$ be a place/transition system.

- $M \in \mathbb{N}_0^S$ is called a marking of N.
- A transition $(^\bullet t,l(t),t^\bullet)$ is enabled at a marking M if $^\bullet t \leq M$ (written M[t> or M[l(t)>).
- $(^\bullet t,l(t),t^\bullet)$ enabled at M may occur yielding a new marking M' with $M'=M-^\bullet t+t^\bullet$ (written M[t>M' or M[l(t)>M'). This notion is extended to arbitrary sequences $w \in T^*$ as usual.
- $[M\rangle := \{M' \mid \exists w \in T^* \text{ with } M[w\rangle M'\}$ denotes the set of all reachable markings of N beginning at marking M.

For the comparison of the behaviour of nets and terms we introduce (labelled) transition systems.

**Definition**

$A=(S,\rightarrow,q_0)$ is called <u>transition system over</u> Act :$\Leftrightarrow$

- S is a set (states),
- $\rightarrow\subseteq S\times Act\times S$ (transition relation) and
- $q_0\in S$ (initial state).

$(p,a,q)\in\rightarrow$ will be written as $p\overset{a}{\rightarrow}q$.

With these notions we can now define the usual interleaving semantics for nets. Let N be a P/T system. The <u>interleaving semantics</u> of N is given by the transition system

$\mathcal{A}(N)=(\mathbb{N}_0^S,\rightarrow,M_0)$ where $M\overset{a}{\rightarrow}M'$ :$\Leftrightarrow$ M[a>M'.

We will prove the correctness of our term semantics with respect to bisimulation equivalence. Two transition systems $A_1=(S_1,\rightarrow,q_1)$ and $A_2=(S_2,\rightarrow,q_2)$ are called <u>bisimulation equivalent</u> (written $A_1\sim A_2$) if there is a bisimulation relation $R\subseteq S_1\times S_2$ with $(q_1,q_2)\in R$ and for all $(p,q)\in R$, $a\in Act$:

- $p\overset{a}{\rightarrow}p' \Rightarrow \exists q': q\overset{a}{\rightarrow}q'$ with $(p',q')\in R$ and
- $q\overset{a}{\rightarrow}q' \Rightarrow \exists p': p\overset{a}{\rightarrow}p'$ with $(p',q')\in R$.

P/T systems N and N' are called bisimulation equivalent (written N~N') if
$\mathcal{A}(N)\sim\mathcal{A}(N')$.

## 2.2 Process Terms

Let Act be a set of actions (with typical elements a, b, ...), Var a set of variables (with typical elements x, y, ...) and $\gamma$:ActxAct $\rightarrow$ Act a symmetrical function such that $\gamma(\gamma(.,.),.)$ is never defined. $\gamma$ represents synchronisation information.

**Definition**

The set PT is defined by the following production system
P ::= NIL | x | a.P | P+P | P|P | P\A | $\mu$x.P where $x\in Var$, $a\in Act$ and $A\subseteq Act$.

The restriction to binary choice is kept for better readability. The following definitions can naturally be extended to n-ary choice.

In the sequel we use some syntactical abbreviations:

$\overset{n}{\underset{i=1}{|}} T$ denotes the n-ary parallel composition of the term T, $\overset{0}{\underset{i=1}{|}} T := NIL$ for $T \in PT$ and $n \in \mathbb{N}$.

Informally the (interleaving) semantics of PT is as follows: NIL is not able to perform any action. a.P first engages in the action a and then behaves like P. P+Q either behaves like P or like Q, depending on the first action performed. P|Q denotes the parallel execution of P and Q that may be synchronised according to the synchronisation function γ: if γ(a,b)=c is defined for the actions a (in P) and b (in Q), then their joint execution will be observed as c. In P|Q synchronisation is possible, but not enforced. This can be achieved using the restriction operator P\A that forbids the execution of elements of A appearing as actions in P. The term μx.P denotes recursion. When μx.P is executed it behaves like the term P where every free occurrence of x in P is substituted by μx.P.

Formally the interleaving semantics of a process term in PT is given by the following transition system:

**Definition**

The <u>(interleaving) semantics</u> of $T \in PT$ is $\mathcal{A}(T)=(PT,\rightarrow,T)$, where $\rightarrow \subseteq PT \times Act \times PT$ satisfies the following axiom and rules for $P, Q \in PT$, $a,b,c \in Act$ and $A \subseteq Act$:

(action) $\quad a.P \xrightarrow{a} P$

(choice) $$\frac{P \xrightarrow{a} P'}{P+Q \xrightarrow{a} P',\ Q+P \xrightarrow{a} P'}$$

(synchrony) $$\frac{P \xrightarrow{a} P',\ Q \xrightarrow{b} Q',\ \gamma(a,b)=c}{P|Q \xrightarrow{c} P'|Q'}$$

(asynchrony) $$\frac{P \xrightarrow{a} P'}{P|Q \xrightarrow{a} P'|Q,\ Q|P \xrightarrow{a} Q|P'}$$

(restriction) $$\frac{P \xrightarrow{a} P',\ a \notin A}{P\backslash A \xrightarrow{a} P'\backslash A}$$

(recursion) $$\frac{P[\mu x.P/x] \xrightarrow{a} P'}{\mu x.P \xrightarrow{a} P'}$$

where P[μx.P/x] denotes the term resulting from the substitution of all free occurrences of x in P by μx.P.

The parallel composition operator is commutative, associative (as $\gamma(\gamma(.,.),.)$ is never defined) and has NIL as neutral element.

## 3 Synchronisation-free P/T systems

One step of our term representation can only be developed for those P/T systems where for all transitions t it holds that $|\cdot t| \leq 1$. In this chapter we present a way to transform an arbitrary P/T system into an equivalent one with this property.

We introduce a new net class that is needed in an intermediate step in our translation to process terms - the synchronisation-free P/T systems (for short SF systems). These nets are structurally free of synchronisation: $|\cdot t| \leq 1$ holds for all transitions t. Instead synchronisation information is given by the function $\gamma$, $\gamma$:ActxAct $\rightarrow$Act (already known from the set PT of terms). We will change the occurrence rule for SF systems appropriately: two enabled transitions labelled with actions a and b where $\gamma(a,b)$ is defined are enforced to occur jointly thus simulating the synchronisation of usual P/T-systems. We will associate to every P/T-system a set of its so-called synchronisation-free versions.

**Definition**

$N=(S,T,M_0,L,\gamma)$ is called <u>synchronisation-free P/T system</u> (SF system):$\Leftrightarrow$

- $(S,T,M_0)$ is a P/T system,
- $|\cdot t|=1 \; \forall \; t \in T$,
- $\gamma$:ActxAct$\rightarrow$Act is partially defined such that $\gamma(\gamma(.,.),.)$ is never defined, symmetrical, and

  (*) $\forall$ $a,b_1,b_2 \in$ Act with $\gamma(a,b_1)$ defined and $\gamma(a,b_2)$ defined $\Rightarrow b_1=b_2$,
- L:Fin(N)$\rightarrow$Var is partially defined.

Condition (*) on $\gamma$ ensures that possible partners for synchronisations are determined uniquely. The labelling of places according to L arises in the translation of SF systems into process terms similar to its use in [Goltz88].

A synchronisation-free version of an arbitrary P/T system is constructed by splitting all synchronisation transitions into two independent transitions. $\gamma$ keeps book of this splitting.(As explained in the introduction we restrict ourselves to P/T systems with binary synchronisation only. We might as well allow arbitrary finite synchronisation yielding an n-ary $\gamma$-function with $n = \max_{(t \in T)} |\cdot t|$. )

**Definition**

Let $N=(S,T,M_0)$ be a P/T system (with binary synchronisation).

$SF(N)=(S_{SF},T_{SF},M_0,L,\gamma)$ is called a <u>synchronisation-free version</u> of N :⇔ $S_{SF}=S\cup S^{split}$, $T_{SF}=T\backslash T^{syn}\cup T^{split}$ and L is totally undefined,

where:

$T^{syn}=\{\ t\in T \mid |{\cdot}t|=2\}$ is the set of synchronisation transitions of N,

$S^{split}:=\{s_t \ | t\in T^{syn}\}$ is a set of new places, one for each transition of $T^{syn}$,

$T^{split}$ is a set of new transitions such that each transition of $T^{syn}$ is represented by two transitions of $T^{split}$.

$T^{split} := \{(M_1(t_i), l_1(t_i), t_i{\cdot}) \mid t_i \in T^{syn}\}$
$\quad \cup \{(M_2(t_i), l_2(t_i), s_{t_i}\rightarrow 1) \mid t_i \in T^{syn}, s_{t_i}\in S^{split}\}$, where

$M_1(t_i) \cup M_2(t_i) = {\cdot}t_i$ and $\gamma(l_1(t_i),l_2(t_i)) = \gamma(l_2(t_i),l_1(t_i)) = l(t_i)$ for all $i\leq|T^{syn}|$

and $l_i(t_k)=l_j(t_m)$ for $i,j\leq 2$ and $k,m\leq|T^{syn}|$ implies $i=j$ and $k=m$,

and $l_i(t_k)\neq l(t_m)$ for $i\leq 2$ and $k,m\leq|T^{syn}|$.

$\mathfrak{SF}(N)$ denotes the set of all synchronisation-free versions of N.

In $SF(N) \in \mathfrak{SF}(N)$ a transition with two places in its preset is replaced as follows:

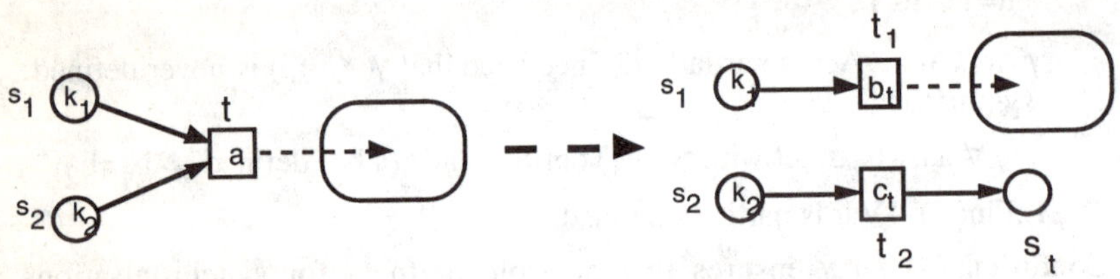


A transition with one ingoing arc with arc weight 2 is replaced by

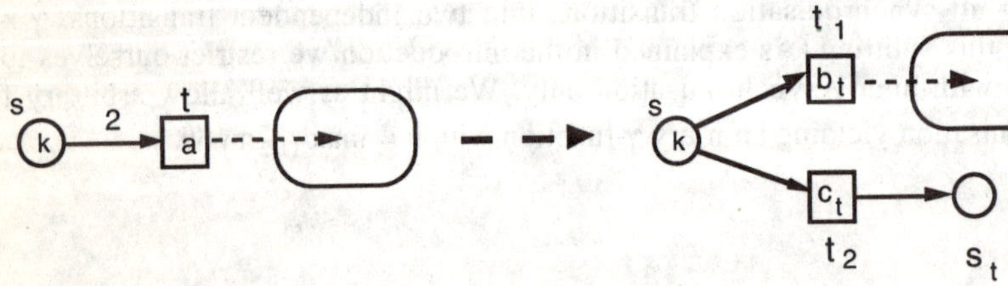


Each $SF(N) \in \mathfrak{SF}(N)$ is a synchronisation-free P/T system. Note that synchronisation transitions with equal names (we do not demand transition labellings to be injective) are replaced by pairs of transitions with different names.

We now define the modified occurrence rule for SF systems:

**Definition**

Let $N=(S,T,M_0,L,\gamma)$ be an SF system.

- A transition t with l(t)=a may occur at a marking M yielding a new marking M' defined by (a) or (b) :⇔
  (a) $\gamma(a,.)$ is undefined, $^\bullet t \leq M$ and $M'=M-{}^\bullet t+t^\bullet$. We write $M[t\geq M'$ or $M[a\geq M'$.
  (b) $\exists b\in Act$ with $\gamma(a,b)=c$, $\exists t'\in T$ with $l(t')=b$, $M\geq {}^\bullet t+{}^\bullet t'$ and $M'=M-{}^\bullet t+t^\bullet - {}^\bullet t'+t'^\bullet$. We write $M[t,t'\geq M'$ or $M[c\geq M'$.
  This notion is extended to arbitrary sequences $w\in(T\cup TxT)^*$ (respectively $Act^*$) as usual.
- $[M\geq:=\{M' \mid \exists w\in Act^* \text{ with } M[w\geq M'\}$ denotes the set of all reachable markings of N starting with M.

We define the interleaving semantics of SF systems by the modified occurrence rule. The interleaving semantics of an SF system N is given by the transition system

$$\mathbb{C}(N)=(\mathbb{N}_0^S,\rightarrow,M_0) \text{ where } M\xrightarrow{a}M' :\Leftrightarrow M[a\geq M'.$$

In chapter 5 we will show that a synchronisation-free version of a P/T system N equipped with the modified occurrence rule is equivalent to N.

# 4 A Term Representation for P/T Systems

In this chapter we define the term representation for P/T systems. First we introduce some notations for markings of SF systems needed in the following constructions. Then we show how to obtain a process term from an SF system by structural induction on the initial part of the system.

The s-marking $M_s$ of N only preserves the initial marking of the place s (formally: $M_s(s')=M_0(s)$ for s'=s, 0 otherwise). The t-marking $M_t$ simulates occurrence of t yielding the marking $t^\bullet$ where all places not included in the postset of t are set to zero (formally: $M_t(s)=t^\bullet(s)$ ). We need two special kinds of subnets of a synchronisation-free net N: the net $N_s$ being the subnet reachable from a place s and the net $N_t$ reachable after occurrence of a transition t.

**Definition**

Let $N=(S,T,M_0,L,\gamma)$ be an SF system , $s\in S$ and $t\in T$.

(1) $N_t=(S',T',M',L',\gamma)$ is the subnet reachable from t :⇔
$S'=\{s\in S \mid \exists\, M\in[M_t> \text{ with } M(s)>0\}$,
$T'=\{t'\in T \mid \exists\, M\in[M_t> \text{ with } M[t'>\}$,
$M'=M_t|_{S'}$ , $L'=L|_{S'}$.

(2) $N_s=(S',T',M',L',\gamma)$ is the <u>subnet reachable from s</u> :⇔
$S'=\{s'\in S \mid \exists\, M\in[M_s\rangle \text{ with } M(s')>0\}$,
$T'=\{t\in T \mid \exists\, M\in[M_s\rangle \text{ with } M[t\rangle\}$,
$M'=M_s|_{S'}$ , $L'=L|_{S'}$ .

We are now ready to define the translation of SF nets to process terms. We give graphical impressions of the handled nets and then define and explain the corresponding constructions on terms.

**Definition**

Let $N=(S,T,M,L,\gamma)$ be an SF system. The <u>process term semantics</u> T of N is a mapping from SF systems to PT as defined by (a) - (e), where $\gamma$ is given by the synchronisation function of N.

(a) <u>Inaction:</u> N is a net with initial part 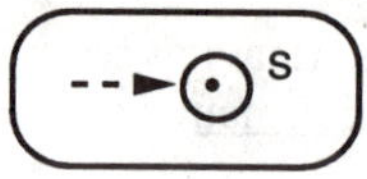

, formally

$IP(N)=\{({}^{\cdot}s,s,\emptyset)\}$, $M(s)=1$. Then we define $T(N):=\begin{cases} x & \text{if } L(s)=x, \\ NIL & \text{otherwise.} \end{cases}$

(b) <u>Action:</u> N begins with one transition:

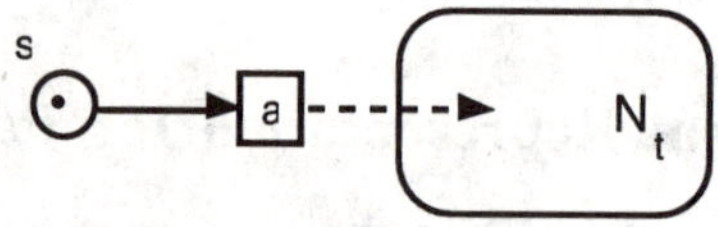


$IP(N)=\{(\emptyset,s,\{t\})\}$, $M(s)=1$, with $l(t)=a$. Then we define

$$T(N):=a.T(N_t)$$

(c) <u>Nondeterminism:</u> N starts with a conflict:

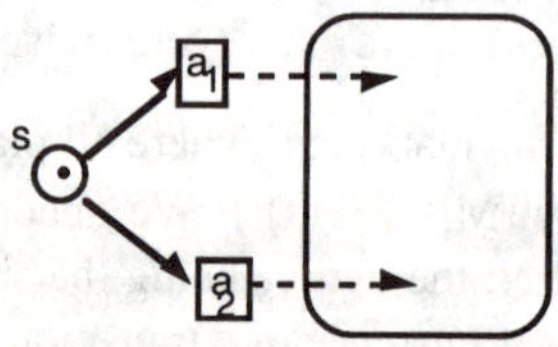


$IP(N)=\{(\emptyset,s,\{t_1,t_2\})\}$, $M(s)=1$, where $l(t_i)=a_i$ for $i\in\{1,2\}$. Let $N_i$ be isomorphic to $N_{t_i}$ with the same initial marking on isomorphic places and let $N_1$ and $N_2$ be disjoint.Then we define

$$T(N):=a_1.T(N_1)+a_2.T(N_2)$$

(d) Parallelism

(i) Structural: The initial part of the net consists of n>1 places:

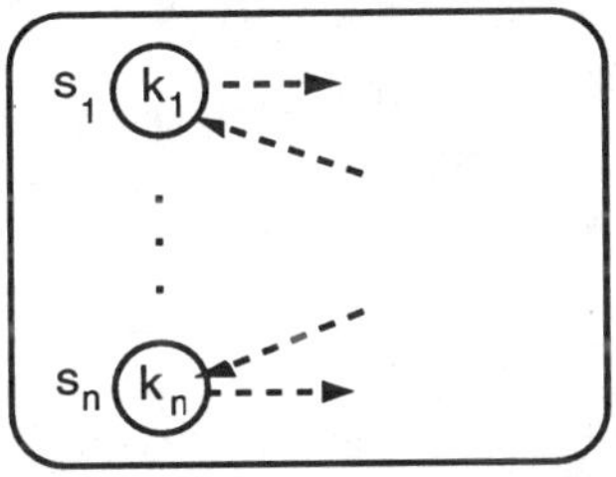


$IP(N)=\{(^{\bullet}s_i,s_i,s_i^{\bullet}) \mid i\in\{1,\dots,n\}\}$ with $M(s_i)=k_i$ for $i\in\{1,\dots,n\}$. Let $N_1,\dots,N_n$ be disjoint, isomorphic to $N_{s_1},\dots,N_{s_n}$ with the same initial marking on the isomorphic places. Then we define

$$T(N):=\mathop{|}_{i=1}^{n} T(N_i)$$

(ii) Dynamical: The initial place of N is marked with more than one token:

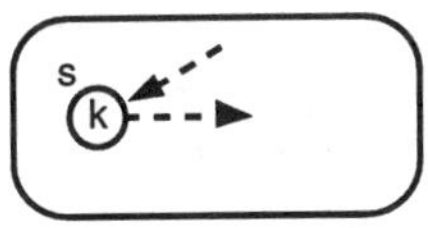


$IP(N)=\{(^{\bullet}s,s,s^{\bullet})\}$, $M(s)=k>1$. Then we define

$$T(N):=\mathop{|}_{i=1}^{k} T(N')$$

where N' is isomorphic to N with an initial marking of only one token on the place isomorphic to s. Thus we treat N like the k-ary parallel composition of N'.

(e) Recursion: The initial place of N has in- and outgoing arcs and a marking with one token:

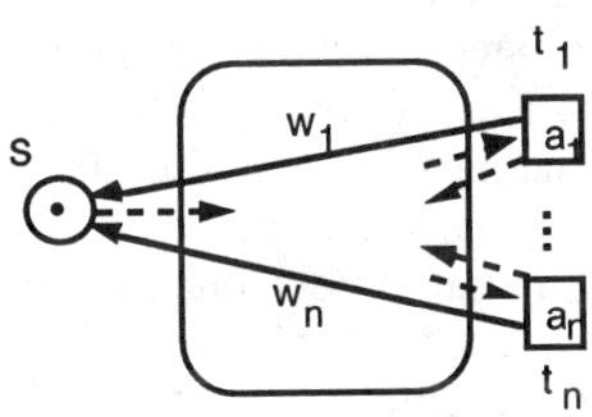


$IP(N)=\{(\{t_1,\dots,t_n\},s,\{u_1,\dots,u_r\})\}, n,r\in\mathbb{N}$ (the transitions $u_i$ are not shown in the figure); $M(s)=1$, $l(t_i)=a_i$ and $t_i^{\bullet}(s)=w_i$ for $i\in\{1,\dots,n\}$. Then we define

$$T(N):=\mu x.T(N_x) \text{ for } x\notin L(S),$$

$N_X=(S_X,T_X,M_X,L_X,\gamma)$ where

$S_X:=S\cup S'$ with $S':=\{s_{t_i} \mid s_{t_i}\notin S$ for $i\in\{1,\ldots,n\}\}$;

$T_X:=(T\setminus\{t_1,\ldots,t_n\})\cup T'$ with

$T'=\{({}^{\cdot}t_i,l(t_i),(t_i^{\cdot})') \mid t_i\in {}^{\cdot}s$ for $i\in\{1,\ldots,n\}\}$,

$$(t_i^{\cdot})'(s')=\begin{cases} t_i^{\cdot}(s) & \text{for } s'=s_{t_i}, \\ 0 & \text{for } s'=s \\ t_i^{\cdot}(s') & \text{otherwise} \end{cases}$$

$M_X:=M\cup 0_{S'}$,

$L_X:=L\cup\{s\to x \mid s\in S'\}$ and

The net constructed has the structure

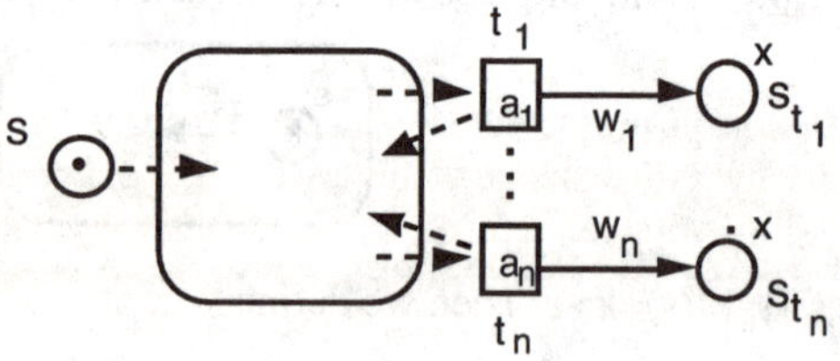


For each transition contained in the preset of the initial place we add a new place labelled with a so far unused variable that indicates the possibility of entering a recursion. The weights of the arcs to these new places are adopted from the corresponding arcs leading to the initial place in N, which are removed in $N_X$.

This construction is well-defined: every possible structure of the initial part of an SF system is treated in a unique way and the procedure terminates for finite SF systems after finitely many steps as shown in [DS93].

The style of the above definition of T(N) for SF systems N emphasizes the relation to work on net semantics for terms (see also the correctness proofs, especially theorem(2) in the appendix). An alternative definition of our term representation for SF systems can be given that points out the relation to automata and regular expressions. This definition can be sketched as follows:

Introduce a recursion variable $X_s$ for every place s of N. For a marking M let T(M) be the parallel composition of M(s) copies of $X_s$ for every place s. Then the term corresponding to N is $T(M_0)$ where $M_0$ is the initial marking of N and where the recursion variables are given as follows: if $s^{\cdot}=\{t_1,\ldots,t_n\}$ with $l(t_i)=a_i$ for $i=1,\ldots,n$ then $X_s=a_1.T(M_{t_1})+\ldots+a_n.T(M_{t_n})$ (where $M_{t_i}$ is the $t_i$-marking introduced before).

This definition generalizes the construction of a regular expression for a finite automaton.

The term semantics of a P/T system is now defined as the set of all terms representing the synchronisation-free versions of N that are restricted by all actions that may synchronise according to the synchronisation function $\gamma$.

**Definition**

The <u>process term semantics</u> of a P/T system N is defined by

$$T(N):=\{T(SF(N))\backslash\{a\in Act \mid \gamma(a,.) \text{ defined}\} \mid SF(N)\in \mathfrak{SF}(N)\}.$$

# 5 Correctness of the Term Representation

In this chapter we investigate the relation between the term semantics introduced and other semantic models for Petri nets. First we show that the interleaving semantics of a net can be retrieved from its term semantics. As indicated in the introduction the proof is achieved in three steps. We first state the bisimulation equivalence between a P/T system and its synchronisation-free version where synchronisation is enforced by the new occurrence rule (theorem(1) ). Then we consider SF systems where synchronisation according to $\gamma$ is possible. Formally this is captured by equipping the net with the union of the standard and the modified occurrence rule: an action may synchronise according to $\gamma$, but it may as well occur on its own. We state the equivalence to the associated process term (theorem (2) ). In the next step we show that enforcing synchronisation in nets with the modified occurrence rule and enforcing synchronisation in terms by restriction correspond to each other (theorem (3) ). For better readability we only state the theorems and add a proof outline in the appendix.

**Theorem**

For each SF system N let $\mathfrak{B}(N)$ denote the transition system associated to N by the union of the standard and the modified occurrence rule, and let $\mathfrak{C}(N)$ denote the transition system associated to N by the modified occurrence rule.

(1) Let $N=(S_N,T_N,M_N)$ be a P/T system and $SF(N)=(S_{SF(N)},T_{SF(N)},M_{SF(N)},L,\gamma)\in\mathfrak{SF}(N)$. Then $\mathcal{A}(N)\sim\mathfrak{C}(SF(N))$.

(2) Let N be an SF system. Then $\mathfrak{B}(N)\sim\mathcal{A}(T(N))$.

(3) Let N be an SF system, $\gamma$ the synchronisation function of N, $T\in PT$ and let $A=\{a\in Act \mid \gamma(a,.) \text{ defined}\}$. Then $\mathfrak{B}(N)\sim\mathcal{A}(T) \Rightarrow \mathfrak{C}(N)\sim\mathcal{A}(T\backslash A)$ holds.

(4) Let N be a P/T system and $T\in T(N)$. Then $\mathcal{A}(N)\sim\mathcal{A}(T)$.

The kind of semantics closest related to our term semantics are the branching processes of Engelfriet [Engelfriet91]: they represent the prefixes of the branching run (unfolding) of a system. The branching processes of a net build a lattice with the unfolding of the net [NPW81] as largest element.

The following figure shows a net:

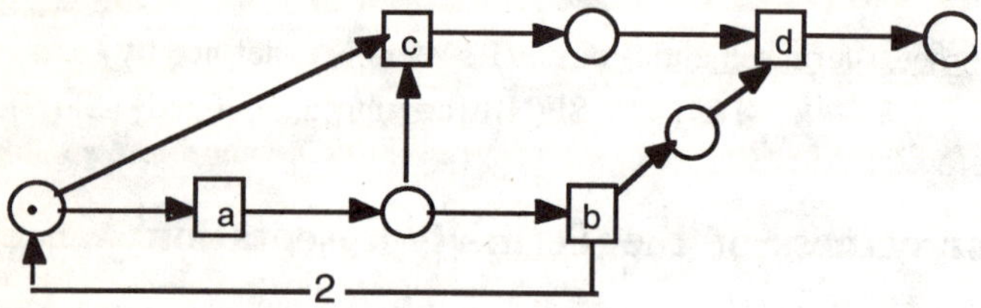

and a prefix of its unfolding:

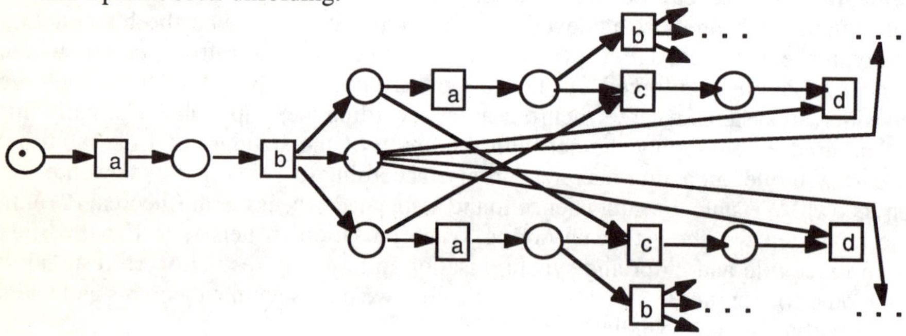

The term semantics of nets originates from the same intuition as the branching processes - we follow the branching run of the initial tokens through the net. One SF-version of the net is :

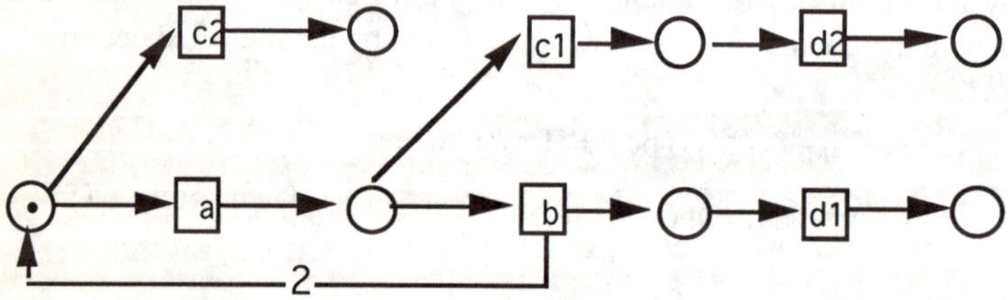

with the term representation $\mu x.(a.(c_1.d_2.NIL + b.(d_1.NIL \mid x \mid x)) + c_2.NIL)$ and $\gamma(c_1,c_2)=c$, $\gamma(d_1,d_2)=d$. While the unfolding is infinite the term representation gives a finite description of the branching behaviour of the net.

Formally the correctness of our term representation according to branching processes is gained as follows: A net semantics N for processes T(N) can be defined similarly to Goltz' net semantics for RCCS [Goltz88]. We strongly conjecture that the net N(T(N)) and the net N coincide in their branching processes.

## 6 Concluding remarks

One aim of the definition of net semantics for terms was to identify well-structured subclasses of nets (that are generated by the operators induced from CCS-like calculi). We view our work as an advancement of this research direction: we show that <u>any</u> net can be represented by an ACP-like term where the processes model the branching runs of the initial tokens of the net. Thus any net can be represented in a well-structured way (we prove: up to bisimulation equivalence, we conjecture: up to branching process equivalence).

Our work seems to be close to a recent result of Christensen [Christensen93].

We thank Volker Claus, Ernst-Rüdiger Olderog and an anonymous referee for valuable comments.

## References

[BK85] Bergstra, J.A.; Klop, J.W.: Algebra of communicating processes with abstraction; TCS 37, 1985

[BRS86] Boudol,G.; Roucairol,G.; de Simone,R.: Petri nets and algebraic calculi of processes; LNCS 222, Springer1986

[Christensen93] Christensen, Soren: Decidability and Decomposition in Process Algebras; Report ECS-LFCS-93-278, Univ. of Edinburgh, Dept. of Computer Science, 1993

[DDM88] Degano, Pierpaolo; De Nicola, Rocco; Montanari, Ugo: A Distributed Operational Semantics for CCS based on C/E-Systems; Acta Informatica 26, 1988

[DS93] Dietz, Cheryl; Schreiber, Gerlinde: A Term Representation for P/T Systems (in German); Report 1/93 ; Univ. Oldenburg, Dept. of Computer Science, 1993

[Engelfriet91] Engelfriet, Joost: Branching processes of Petri nets; Acta Informatica 28, 1991

[vGV87] van Glabbeek,R.; Vandraager, F.: Petri Net Models for Algebraic Theories of Concurrency; LNCS 259, Springer 1987

[Goltz88] Goltz, Ursula: On representing CCS programs by finite Petri Nets; LNCS 324, Springer 1988

[Goltz88b] Goltz, Ursula: On the representation of CCS programs by Petri Nets (in German); GMD-Bericht 172, Oldenbourg 1988

[Hoare85] Hoare, C.A.R.: Communicating Sequential Processes; Prentice Hall 1985

[Milner89] Milner, Robin: Communication and Concurrency; Prentice Hall 1989

[NPW81] Nielsen,M.; Plotkin,G.; Winskel,G.: Petri nets, event structures and domains; TCS 13, 1981

[Olderog87] Olderog, Ernst-Rüdiger: Operational Petri Net Semantics for CCSP; LNCS 266, Springer 1987

[Reisig85] Reisig, Wolfgang: Petri Nets; EATCS Monographs, vol.4, Springer 1985

[Taubner89] Taubner, Dirk: The Finite Representation of CCS and TCSP Programs by Automata and Petri Nets; LNCS 369, Springer 1989

# Appendix

We repeat the correctness theorem of chapter 5 and give a proof outline. The proof is executed in detail in [DS93] .

**Theorem (1)**

Let $N=(S_N,T_N,M_N)$ be a P/T system,
$SF(N)=(S_{SF(N)},T_{SF(N)},M_{SF(N)},L,\gamma)\in \mathcal{SF}(N)$ with the modified occurrence rule. Then $\mathcal{A}(N)\sim\mathcal{C}(SF(N))$.

Proof(outline): We define a bisimulation that relates the places of the nets. We presuppose a labelling h(s) of the "new" places $s \in S^{split}$ with NULL. All places in SF(N) not labelled with NULL are as well represented in N, so a bijection $\sigma$ between $S_N$ and $S_{SF(N)}\backslash\{s\in S_{SF(N)} | h(s)=NULL\}=:S_{SF}$ exists.

The relation $B\subseteq[M_N\rangle\times[M_{SF(N)}\rangle$ defined by

$B:=\{(M,M') | M(s)=M'(\sigma(s)) \ \forall\ s\in S_N\}$

is a bisimulation for $\mathcal{A}(N)$ and $\mathcal{C}(SF(N))$.

**Theorem (2)**

Let N be an SF system with the union of the standard and the modified occurrence rule. Then $\mathcal{B}(N)\sim\mathcal{A}(T(N))$.

Proof (outline): The proof is executed in detail in [DS93] . It is done by induction on the syntactical structure of the initial part of N; this includes exactly the cases (a)-(e) of the definition of T(N). For each case a bisimulation (similar to the bisimulations in Goltz' correctness proof for her net semantics for terms in [Goltz88b]) is stated. The cases (c):nondeterminism and (d):parallelism have to be handled carefully as in these steps the net has to be transformed slightly. The transformations are shown below. They preserve bisimulation equivalence. By the transitivity of ~ the desired result follows.

Ad (c): <u>nondeterminism</u>

$IP(N)=\{(\emptyset,s,\{t_1,t_2\})\}$, $M(s)=1$, $l(t_i)=a_i$ and $t_i^{\bullet}=\{s_k \rightarrow w_{i_k} \mid 1 \leq k \leq n\}$ for $i \in \{1,2\}$.

The net under consideration is

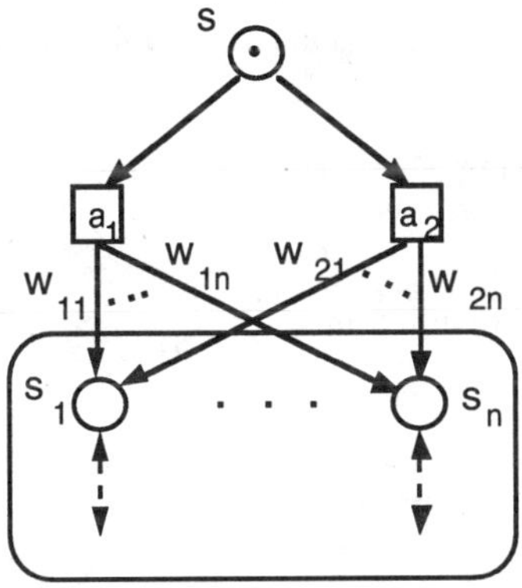


It is transformed as sketched into:

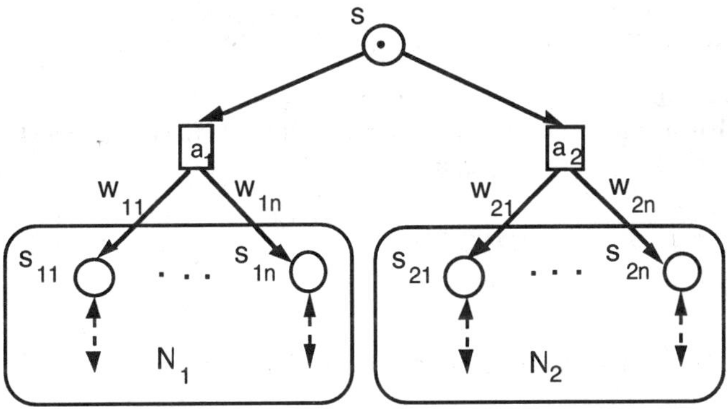


Substructures shared by different processes are separated by using disjoint copies of the affected subnets reachable from the $t_i$ (all with an empty initial marking).

Ad (d): <u>Parallelism</u>

(i) <u>structural</u>:

The net has $n > 1$ initial places each marked with respectively $k_i$ tokens: $In(N)=\{s_1,...,s_n\}$ with $M(s_i)=k_i$ for $i \in \{1,...,n\}$. Let $N_{s_1},...,N_{s_n}$ be disjoint, isomorphic to the subnets reachable from the places $s_1,...,s_n$ with $In(N_{s_i})=\{s_i\}$ and an initial marking $M_{s_i}(s_i)=k_i$ for $i \in \{1,...,n\}$. Then $N \sim \mathop{\mid}\limits_{i=1}^{n} N_{s_i}$ where $\mathop{\mid}\limits_{i=1}^{n} N_{s_i}$ denotes the disjoint union of $N_{s_1},...,N_{s_n}$. The picture below illustrates this construction:

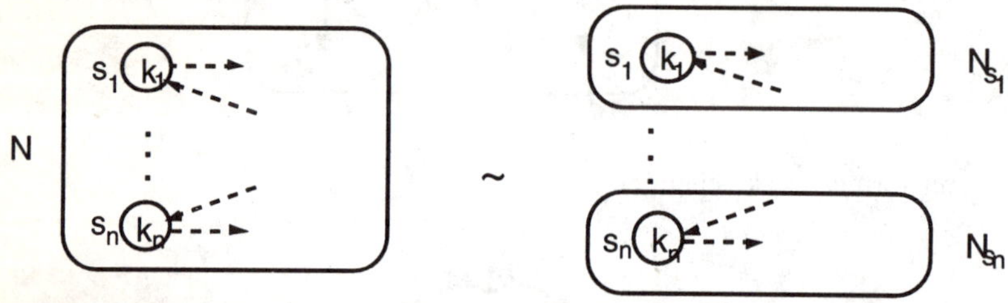


(ii) <u>dynamical</u>:

The net has one initial place with n tokens: $In(N)=\{s\}$ and $M(s)=n>1$. Let $N_1,...,N_n$ be disjoint, isomorphic to N with $In(N_i)=\{s_i\}$ and $M_i(s_i)=1$ for $i \in \{1,...,n\}$.

Then $N \sim \mathop{\mid}\limits_{i=1}^{n} N_i$.

This transformation can be illustrated by the following picture:

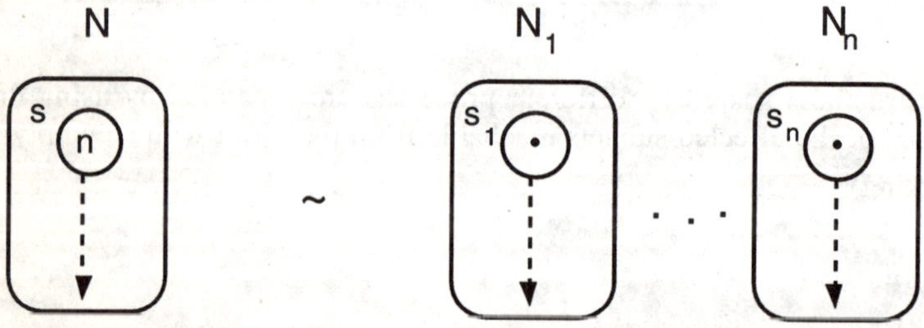


The correctness of the given transformations essentially depends on the restricted net structure of SF nets ($|{}^\bullet t|=1$ for all transitions t). By the transitivity of ~ the theorem is proved.

**Theorem (3)**

Let N be an SF system, $\gamma$ the synchronisation function of N, $T \in PT$ and let $A=\{a \in Act \mid \gamma(a,.) \text{ defined}\}$.
Then $\mathfrak{B}(N) \sim \mathcal{A}(T) \Rightarrow \mathfrak{C}(N) \sim \mathcal{A}(T \backslash A)$ holds.

Proof (outline): Let the symbol [» denote the occurrence rule that captures the union of both occurrence rules. As $\mathfrak{B}(N) \sim \mathcal{A}(T)$, a bisimulation $B \in [M_0\text{»} \times PT$ for N and T exists. Then

$$B' := \{(M,P) \mid \exists w \in (Act \backslash A)^* \text{ with } M_0[w \geq M,\ T\backslash A \xrightarrow{w} P,\ P = \overline{P} \backslash A \text{ and } (M, \overline{P}) \in B\}$$

is a bisimulation for $\mathfrak{C}(N)$ and $\mathcal{A}(T\backslash A)$.

**Theorem (4)**

Let N be a P/T system and $T \in T(N)$. Then $\mathcal{A}(N) \sim \mathcal{A}(T)$.

Proof: Let $A=\{a \in Act \mid \gamma(a,.) \text{ defined}\}$.

We summarize:
$\mathcal{A}(T) = \mathcal{A}(T(SF(N)) \backslash A)$ for an $SF(N) \in \mathfrak{SF}(N)$ by definition of T(N).
By theorem(1) it holds:
$\mathcal{A}(N) \sim \mathfrak{C}(SF(N))$ where SF(N) is equipped with the modified occurrence rule.

Allowing additionally the standard occurrence rule for SF(N) we get by theorem(2):
$\mathfrak{B}(SF(N)) \sim \mathcal{A}(T(SF(N)))$.

Now by theorem(3) we conclude:
$\mathfrak{C}(SF(N)) \sim \mathcal{A}(T(SF(N)) \backslash A)$ where SF(N) is equipped with the modified occurrence rule only.
The transitivity of ~ implies $\mathcal{A}(N) \sim \mathcal{A}(T)$.
The structure of the whole proof is illustrated in figure 2 in the introduction.

# Superposed Generalized Stochastic Petri nets: definition and efficient solution

Susanna Donatelli *

Dipartimento di Informatica, Università di Torino
corso Svizzera 185, 10149 Torino, Italy
phone: (+39-11)-7712002
fax: (+39-11)-751603
email: susi@di.unito.it

**Abstract.** In a previous paper we have defined Superposed Stochastic Automata (SSA) [13], a class of Stochastic Petri Nets (SPN) whose solution can be efficiently computed since it never requires the construction of the complete Markov chain of the underlying Markovian process. The efficient solution of SSA is based on a method proposed by Plateau in [23] for the analysis of stochastic processes generated by the composition of stochastic automata. Efficient analysis is there achieved (both in terms of space and time) with a technique based on Kronecker (tensor) algebra for matrices.

A SSA is basically a set of Stochastic State machines that interact through transition superposition: their application to real models is therefore limited. The technique defined for SSA is here extended to Superposed Generalized Stochastic Petri Nets (SGSPN), a set of GSPN nets that interact through transition superposition.

In this paper we define SGSPN, explain how the solution method proposed by Plateau in [23] and already used for SSA can be adapted to work for this larger class of SPN, and discuss the possibility of using SGSPN for the performance evaluation of concurrent processes. The solution is implemented by a set of programs that interact with the GreatSPN [8] package: a SGSPN net is specified through the GreatSPN graphical interface, so that also all classical analysis methods already available for GSPN in the package can still be applied.

## 1 Introduction

It is well known that the major draw-back of Stochastic Petri Nets (SPN) and of their generalization GSPN is the large state space that can be generated even by rather simple models. The solution of a SPN indeed requires the construction of the associated Continuous Time Markov Chain (CTMC), of its infinitesimal generator and the solution of a number of linear equations equal to the number of states of the SPN (GSPN).

---

* This work was supported by the CNR "Progetto Finalizzato Sistemi Informatici e Calcolo Parallelo", contract no. 92.01563.PF69, the Italian MURST "40%" project, and by the ESPRIT-BRA project No. 7269 "QMIPS"

There have been many attempts to simplify the SPN solution, in particular trying to identify subclasses of Petri nets that have a solution of the "product form" type [19, 17, 16], or trying to exploit symmetries to produce lumped Markov chains, as for example in Stochastic Well Formed Nets [7].

In [23, 22] Plateau presents a solution method for networks of Markovian stochastic automata that has two main advantages. The first is that there is *no need to store the infinitesimal generator of the complete Markov chain*, as it is expressed as Kronecker sum and product (see [10]) of the infinitesimal generators of the single automata and of a certain number of "correcting matrices": the solution process exploits the particular form of the Kronecker expression to avoid the computation of the infinitesimal generator of the global process. The second one is that *the solution process is easily amenable to parallel implementation.* For the time being only the sequential version of the method has been implemented in a tool called PEPS [24]. The approach has two main disadvantages: each component is a state machine (a flexible but low level model) and the user has to specify manually all the matrices involved. This last drawback has been solved in a later paper [25] where the authors introduce a language to describe interacting stochastic automata, from which the matrices are automatically defined.

In a previous paper [13] we present a subclass of SPN called Superposed Stochastic Automata (SSA). An SSA is basically a set of stochastic state machines (SSM) that interact through transition synchronization. From an SSA we can automatically derive a Kronecker expression for its infinitesimal generator of the type used by PEPS, while still retaining the advantages of being able to study and define the model using all facilities of packages like GreatSPN [8] and SPNP [15]. This include the possibility of automatic definition of performance indices, steady state as well as transient solution, approximate solution based on decomposition [9] and structural analysis techniques as whose based on the computation of P- and T-invariants.

SSA still suffers from the fact that each component is a state machine: in this paper we show instead how the same analysis technique can be adapted to work for a wider class of timed nets called Superposed Generalized Stochastic Petri Nets, where each component is a GSPN [2]. Given a SGSPN described in graphical form using the GreatSPN package [8], we automatically determine the Kronecker expression for its infinitesimal generator in a way that is completely transparent to the user.

There have been other attempts to exploit Kronecker algebra to solve complex stochastic Petri nets. In [5] Buchholz applies it to the study of hierarchical Markovian models described through queueing networks and Petri nets, and in [4] Beounes uses Kronecker algebra to compute the reachability sets of models composed by superposition on common places of submodels (that are marked graphs). In both cases the type of interactions considered is not of the synchronization type, as submodels interact through the exchange of tokens (asynchronous interaction).

The paper is organized as follows. Section 2 recalls the definition of SSA, the solution method defined in [13] and their applicability to real examples.

Section 3 is the core of the paper: it defines the SGSPN class and shows the extension of the method already applied for SSA to this more general class. Possible applications of SGSPN to different domains, and some ideas on how to enlarge the SGSPN class are also discussed here. Section 4 concludes the paper.

## 2 The previous work: SSA nets

In this section we recall the definition of the SSA class and we show the Kronecker expression of the infinitesimal generator of an SSA that allows steady state probabilities to be efficiently computed. The content of this section is a digest of the results presented in [13].

"Superposed Stochastic Automata," (SSA) is a subclass of stochastic Petri nets (SPN), the class of timed nets defined by Molloy in [20] and Natkin in [21]. The name for the subclass is due to the fact that SSA are the timed, stochastic counterpart of a class of untimed Petri nets called "Superposed Automata" nets (SA) introduced by DeMichelis et al. in [11]. As it is well known, the addition of stochastic time, with infinite support distribution, to Petri nets does not change the logical behavior of the model, and therefore all results obtained for SA nets still apply to SSA.

The distinctive feature of SSA nets is that they can be considered as a set of stochastic state machines (SSM) [2] combined through a synchronization operator of the rendez-vous type. In Petri net terminology an SSA net is a set of SPN nets of the state machine type composed by superposition of transitions. We now define Superposed Stochastic Automata.

**Definition 1** *We call Superposed Stochastic Automata the tuple* $\mathcal{S} = (P, T, F, \Pi, W, M0)$ *where*

- $P$ *is the non empty set of places*
- $T$ *is the non empty set of transitions*
- $F \subseteq P \times T \cup T \times P$ *with* $dom(F) \cup codom(F) = P \cup T$ *is the flow relation. It has to satisfy the following restriction:* $\forall t \in T : |\, {}^{\bullet}t \,| = |\, t^{\bullet} \,| \geq 1$
- $\Pi = \{\Pi_0, \cdots, \Pi_{N-1}\}$ *is a partition of* $P$ *such that* $\forall t \in T$ *and* $\forall i \in \{0, \cdots, N-1\} : |\Pi_i \cap {}^{\bullet}t| = |\Pi_i \cap t^{\bullet}| \leq 1$
- $W : T \rightarrow I\!R^+$*;* $W(t)$ *is the rate of the exponential distribution associated to transition* $t$
- $M0 \subseteq P$ *and* $\forall i \in [0, \ldots, N-1]\ |\Pi_i \cap M0| = 1$

The condition on the partition $\Pi$ identifies a set of stochastic state machines that compose the SSA. Observe that SSA allows only one type of synchronization: each transition with more than one input or output arc must have the same number of input and output arcs.

---

[2] An SSM is a Stochastic Petri Net (SPN) with the condition that all transitions are 1-in/1-out ( $\forall t \in T : |\, {}^{\bullet}t \,| = |\, t^{\bullet} \,| = 1$), where ${}^{\bullet}t$ ($t^{\bullet}$) the set of input (output) places of $t$.

SSA transitions that have $|{}^{\bullet}t| = |t^{\bullet}| > 1$ are called *synchronization (or synchronized) transitions*; we indicate with $TS$ the set of synchronized transitions ($TS = \{t \in T : |{}^{\bullet}t| > 1\}$).

The requirement on the initial marking is such that *there is exactly one place marked in each element of the partition* in any reachable marking. Therefore a SSA can be considered as the superposition on transitions of equal rate of a set SSM's.

Fig. 1 provides a pictorial representation of a SSA where the subset of places is partitioned into three subsets. The three elements of the partition, $\Pi_0, \Pi_1$ and $\Pi_2$ determine the three SSM components $SSM_0$, $SSM_1$, and $SSM_2$. $T3$ is the synchronization transition of $SSM_0$ and $SSM_1$, and transition $T9$ is the synchronization between $SSM_1$ and $SSM_2$. In general a transition can of course synchronize more than two SSMs.

$\Pi_0 = \{P1, P2, P3, P4, P5\}$
$\Pi_1 = \{P6, P7, P8\}$
$\Pi_2 = \{P9, P10, P11\}$

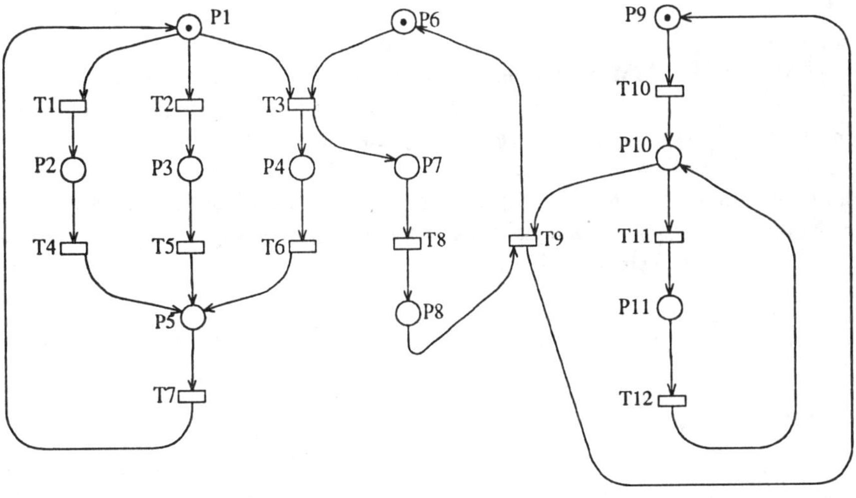


**Fig. 1.** An SSA system

### 2.1 Solution method

In this section we show the Kronecker expression of the infinitesimal generator of an SSA system, that is given in terms of the infinitesimal generators of some independent processes plus some adequate matrices that play the role of "correction matrices".

In [23] Plateau suggests and applies the idea of separating in each process the independent behavior from the dependent one and to express the infinitesimal generator of the composed process as the sum of two parts: one is the composition of the independent behaviors of the component automata, and the other one takes into account all dependencies. The same idea has been applied to SSA, and the results are shown in the following.

Let $TS \subseteq T$ be the set of synchronization transitions of an SSA composed of $N$ stochastic state machines $\mathrm{SSM}_i$. We say that an automata, component of an SSA, participates to the $t$ synchronization iff $| {}^{\bullet}t \cap \mathrm{SSM}_i| = 1$.

In [13] we derived the following formula that, given an SSA systems, computes its infinitesimal generator $Q$ as Kronecker expressions of matrices of the $n_i \times n_i$ type, where $n_i$ is the number of states of each $\mathrm{SSM}_i$.

$$Q = \Pi_{\oplus_{i=0}^{N-1}} Q_i' - \sum_{t \in TS} W(t) \Pi_{\otimes_{i=0}^{N-1}} K_i'(t) + \sum_{t \in TS} W(t) \Pi_{\otimes_{i=0}^{N-1}} K_i''(t) \tag{1}$$

$Q_i$ is the infinitesimal generator of $SSM_i$, while $Q_i'$ is the infinitesimal generator of $SSM_i$ derived from $Q_i$ by deleting all contributions due to synchronized transitions. In formula: $Q_i' = Q_i - \sum_{t \in TS} W(t) E({}^{\bullet}t, t^{\bullet}) + \sum_{t \in TS} W(t) E({}^{\bullet}t, {}^{\bullet}t)$, where $E({}^{\bullet}t, t^{\bullet})$ ($E({}^{\bullet}t, {}^{\bullet}t)$) is a matrix whose only non null element is the one corresponding to the state transition from ${}^{\bullet}t$ to $t^{\bullet}$ (from ${}^{\bullet}t$ to ${}^{\bullet}t$); remember that a state of a SSM is uniquely identified by a place, and that by definition of SSM each transition has exactly one input and one output place. The correcting factors $K_i'(t)$ and $K_i''(t)$ are identity matrices if the i-th automata does not participate in the synchronization; otherwise $K_i''(t)$ is a matrix whose only non null element is the one that represents the state transition caused by the firing of $t$ (that is equal to 1), also $K_i'(t)$ is a matrix with a single non null entry equal to 1: the diagonal element of state ${}^{\bullet}t$ (that is equal to 1). In more formal terms:

$$K_i'(t) = \begin{cases} \text{identity} & \text{if } \{ {}^{\bullet}t \cap SSM_i\} = \emptyset \\ \text{all zeros but } ( {}^{\bullet}t \cap SSM_i, {}^{\bullet}t \cap SSM_i) & \text{if } | {}^{\bullet}t \cap SSM_i| = 1 \end{cases}$$

$$K_i''(t) = \begin{cases} \text{identity} & \text{if } \{ {}^{\bullet}t \cap SSM_i\} = \emptyset \\ \text{all zeros but } ( {}^{\bullet}t \cap SSM_i, t^{\bullet} \cap SSM_i) & \text{if } | {}^{\bullet}t \cap SSM_i| = 1 \end{cases}$$

If the SSA system under study is composed of $N$ automata and $TS$ synchronization transitions, then the number of matrices involved in the expression is $N(1+2|TS|)$, although many of them may be identity matrices. It is important to observe that the expression of $Q$ is only a more compact way of writing the infinitesimal generator of an SSA system, but actually the number of elements of $Q$ is still $\Pi_{i=0}^{N-1} n_i \times \Pi_{i=0}^{N-1} n_i$.

It is nevertheless possible to find a solution of the equation $\nu Q = 0$ (the characteristic equation of a Markov chain) that does not require the computation and the storage of $Q$. We apply an iterative method, where at each step $k$ we compute the $k$-th approximation $\nu^k$ of the probability vector $\nu$ by applying the classical formula of the power method:

$$\nu^k = \nu^{k-1} + \frac{1}{\omega(1+\varepsilon)} \nu^{k-1} Q$$

where $\omega$ is the maximum of the diagonal elements of $Q$ and $\varepsilon$ is an arbitrarily small constant.

Observe that the matrix expressed by Eq. 1 may not be irreducible, even if the process associated to the SSA is actually ergodic. Indeed we consider as possible states of an SSA system all combinations of the states of the single SSMs, but clearly not all states are reachable from the initial marking of the system. But we should also consider that, starting from an initial value of $\nu_0$ in which we assign probability 1 to the state that corresponds to the initial marking, and probability 0 to all other states, then by computing at each iteration the formula $\nu^k = \nu^{k-1} + \frac{1}{\omega(1+\varepsilon)}\nu^{k-1}Q$, we never set to zero the probability of states that are not reachable from the initial marking.

In [13] we have shown how, following the method proposed by Plateau, we can rewrite Eq. 2.1 as:

$$\begin{aligned}\nu^k &= \nu^{k-1} + \frac{1}{\omega(1+\varepsilon)}( \\ &\sum_{j=0}^{N-1}(\nu^{k-1}\Pi_{i=0}^{N-1}S_{n_0\cdots n_i,n_{i+1}\cdots n_{N-1}}(Id_{q_j}\otimes A_{ij})S^T_{n_0\cdots n_i,n_{i+1}\cdots n_{N-1}}) - \\ &\sum_{t\epsilon TS} W(t)(\nu^{k-1}\Pi_{i=0}^{N-1}S_{n_0\cdots n_i,n_{i+1}\cdots n_{N-1}}(Id_{q_i}\otimes K_i'(t))S^T_{n_0\cdots n_i,n_{i+1}\cdots n_{N-1}}) + \\ &\sum_{t\epsilon TS} W(t)(\nu^{k-1}\Pi_{i=0}^{N-1}S_{n_0\cdots n_i,n_{i+1}\cdots n_{N-1}}(Id_{q_i}\otimes K_i''(t))S^T_{n_0\cdots n_i,n_{i+1}\cdots n_{N-1}}))\end{aligned} \quad (2)$$

where $A_{ij}$ is the $Q_{ij}$ matrix if $i = j$, and the identity otherwise, $S_{p,q}$ is a (p,q) perfect shuffle permutation, with $p = n_0 \cdots n_i$ and $q = n_{i+1} \cdots n_{N-1}$, $S^T$ indicates the inverse of the permutation realized by $S$, and the matrix multiplication by $S^T$ is at no cost if indices are appropriately stored.

*Complexity* By taking advantage of the special properties of the Kronecker product we can therefore solve the characteristic equation of an SSA system by simply using a vector of $\Pi_{i=0}^{N-1}n_i$ elements (number of (potential) states of the SSA), and matrices of size $(n_i \times n_i)$ $(0 \leq i < N)$ without being forced to explicitly compute and store the infinitesimal generator of the SSA, which has a (potential) size of $(\Pi_{i=0}^{N-1}n_i)^2$. The complexity of the computation of Equation 2 is of the order

$$N^2\alpha + 2|TS|N\alpha = O(N^2\alpha) \qquad \text{with } \alpha = (\Pi_{i=0}^{N-1}n_i)(\sum_{i=0}^{N-1} n_i)$$

where the term $\alpha$ expresses the complexity of the inner computation of Equation 2. In the product $\nu S_{n_0\cdots n_i,n_{i+1}\cdots n_{N-1}}(Id_{q_i}\otimes A_{ij})S^T_{n_0\cdots n_i,n_{i+1}\cdots n_{N-1}}$ the cost due to permutations is of the order $\Pi_{i=0}^{N-1}n_i$ while to execute the product $\nu S_{n_0\cdots n_i,n_{i+1}\cdots n_{N-1}}(Id_{q_i}\otimes A_{ij})$ it is necessary to execute $q_i$ $(q_i = \Pi_{j=0,j\neq i}^{N-1}n_j)$ products of order $n_j$.

*Applicability* The big limitation of SSA is due to the requirement that each component of an SSA has to be a SSM, that means that one, and exactly one, token can be going around in the net, and that there is no synchronization. An additional constraint is that all transitions have to be timed, so that even a transition that represents a logical event is modelled as a time consuming activity.

## 3 From SSA to SGSPN

In this section we define the class of Superposed Generalized Stochastic Petri nets (SGSPN), and we show how the SSA solution can be adapted to work for this larger class.

Informally a SGSPN is a set of Generalized Stochastic Petri nets (GSPN) that synchronize on a common subset of transitions of equal rate. GSPN is a class of Stochastic Petri nets (defined in [2]), characterized by having timed (exponentially distributed) transitions, as well as immediate (zero time) transitions. Immediate transitions fire with priority over timed ones, and it is possible to define different levels of priority also among immediate transitions (for simplicity we shall consider instead only two priority levels: priority zero for timed transitions and priority one for immediate ones).

**Definition 2** *A GSPN is an eight-tuple*

$$( P,\ T,\ \pi,\ I,\ O,\ H,\ W,\ M0 ) \tag{3}$$

*where*

$P$ *is the set of places;*
$T$ *is the set of transitions such that* $T \cap P = \emptyset$*;*
$\pi : T \rightarrow \{0, 1\}$ *is the priority function;*
$I, O, H : T \rightarrow Bag(P)$ *, are the input, output and inhibition functions, respectively, where* $Bag(P)$ *is the multiset on* $P$*;*
$W : T \rightarrow \mathbb{R}$ *is a function that assigns a weight to each transition;*
$M0 : P \rightarrow \mathbb{N}$ *is the initial marking: a function that assigns a nonnegative integer value to each place*

We shall indicate with $T_E$ the subset of timed transitions of $T$ and with $T_I$ the set of immediate transitions of $T$; they are defined as $T_E = \{t \in T \mid \pi(t) = 0\}$ and $T_I = \{t \in T \mid \pi(t) = 1\}$.

We now remind the classical definitions of reachability set (RS), reachability graph (RG), tangible reachability set (TRS) and tangible reachability graph (TRG).

**Definition 3** *The Reachability Set* $RS(M0)$ *of a GSPN system is defined as the smallest set of markings such that*

- $M0 \in RS(M0)$
- $M1 \in RS(M0) \wedge \exists t \in T : M1[t\rangle M2 \Rightarrow M2 \in RS(M0)$

where $M1[t\rangle M2$ indicates that the system changes state from $M1$ to $M2$ due to the firing of $t$; we shall indicate with $M[t\rangle$ the condition "transition $t$ is enabled in $M$" and with $M[\rangle$ the set of all transitions enabled in $M$.

**Definition 4** *The Reachability Graph $RG(M0)$ of a GSPN of initial marking $M0$ is a labelled directed multigraph whose set of nodes is $RS(M0)$ and whose set of arcs $A$ is defined as follows:*

$$A \subseteq RS(M0) \times RS(M0) \times T$$

$$\langle M, M', t\rangle \in A \longleftrightarrow M \in RS(M0), M' \in RS(M0) \text{ and } M[t\rangle M'$$

We shall say that a marking $M$ is *vanishing* iff $\exists t \in T_I \mid M[t\rangle$; a marking $M$ is instead called *tangible* iff $\nexists t \in T_E \mid M[t\rangle$.[3]

The reachability set and graph include both vanishing and tangible markings. From the point of view of the associated stochastic process we are interested only in tangible markings.

**Definition 5** *The Tangible Reachability Set $TRS(M0)$ of a GSPN system is the set of all tangible markings of $RS(M0)$*

The Tangible Reachability Graph ($TRG(M0)$) is a graph whose set of nodes is the set of tangible states, and there is an arc between two nodes *for each possible path in the corresponding RG between the same two nodes that passes only through vanishing states.* Each arc is labelled with the firing sequence of the path that is a timed transition followed by $n$ immediate transitions ($n \geq 0$).

**Definition 6** *Given a GSPN system and its reachability graph $RG(M0)$ we can define the Tangible Reachability Graph $TRG(M0)$ as a labelled directed multigraph whose set of nodes is $TRS(M0)$ and whose set of arcs $A$ is defined as follows:*

$$A \subseteq TRS(M0) \times TRS(M0) \times T_E \times T_I^*$$

$$\langle M, M', t_0, \sigma\rangle \in A \text{ iff } M[t_o\rangle M_1, \sigma = t_1, \ldots, t_n, (n \geq 0) \text{ and}$$

$$\exists\ M_2, \ldots, M_n :\ M_1[t_1\rangle M_2[t_2\rangle \ldots M_n[t_n\rangle M'$$

The CTMC associated to a GSPN of initial marking $M0$ is obtained from the $TRG(M0)$ by considering each node of the graph as a state of the CTMC; the weight of each arc can be computed as $W(t_0) \cdot W(\sigma)$ [4], while the transition rate in the CTMC between two states is determined as the sum of the weights of the arcs between the two corresponding nodes in the $TRG$.

We can now define SGSPN as a GSPN in which we identify a partition $\Pi$ of the set of places $P$, and a set of timed transitions $TS \subseteq T_E$.

[3] Due to priority, if there is one immediate transition enabled in $M$ then all transitions enabled in $M$ are actually immediate, vice-versa if there is one timed transition enabled in $M$, then all transitions enabled in $M$ are timed.

[4] See [1] for the computation of $W(\sigma)$, the weight of a sequence of immediate transitions

**Definition 7** *A SGSPN is a ten-tuple*

$$( P, T, \pi, I, O, H, W, M0, \Pi, TS ) \tag{4}$$

*where*

$( P, T, \pi, I, O, H, W, M0 )$ *is a GSPN,*
$\Pi = \{P_0, \cdots, P_{N-1}\}$ *is a partition of* $P$,
$TS \subseteq T_E$ *is the set of synchronized transitions, that are timed by definition.*

*Moreover* $\Pi$ *induces on* $T \backslash TS$ *a partition of transitions.*

Let us explain the last point of the definition. If $N$ is the number of elements of the partition we can derive from a SGSPN $N$ components $GSPN_i = ( P_i, T_i, \pi_i, I_i, O_i, H_i, W_i, M0_i)$ as follows. $P_i$ is element $i$ of partition $\Pi$, $T_i$ is the set of transitions that are connected by an input, output, or inhibitor arc to a place in $P_i$. $I_i, O_i, H_i, \pi_i, W_i, M0_i$ are the restrictions of $I, O, H, \pi, W, M0$, respectively, on the elements of $P_i$ and $T_i$. We require that the sets $T_i \backslash TS$ are a partition of $T \backslash TS$, that is to say, the only transitions that may belong to more than one GSPN are the one in $TS$.

We denote with $TRG_i(M0_i)$ the TRG of $GSPN_i$ and we impose the additional constraint that each $TRG$ of each $GSPN$ is strongly connected (a necessary condition for the associated CTMC to be ergodic), moreover we also require that the firing of a synchronized transition $t$ in each of the GSPN leads from a tangible state to a tangible state. That is to say we pose the additional restrictions:

- $TRG_i(M0_i)$ is strongly connected
- "Synchronized transitions lead to tangible states": $\forall t \in TS$, if $\langle M, M', t, \sigma \rangle$ is an arc of $TRG_i(M0_i)$ then $\sigma$ is the empty string ($M[t\rangle M'$ and $M' \in TRS_i(M)_i$)).

At the end of this section we shall discuss this last limitation, as well as the condition that all transitions in $TS$ have to be timed. Observe that there is no requirement on the number of input and output arcs for synchronized transitions (as there is instead for SSA): indeed a transition that synchronize two GSPNs may also express a synchronization (more than one input arc) inside the single GSPNs as well as a fork event (more than one output arc).

Fig. 2 with $TS = \{T_1, T_2\}$, and with the partition represented by the dotted lines, shows an example of SGSPN that is the composition of three GSPNs. Each GSPN can be considered as a model of a process: the upper left is a while statement nested into an infinite loop, the body of the inner loop is an if statement that leads to synchronization with either the upper right GSPN or with the bottom one. The other two processes are replicas of the same GSPN: each can be considered as a model of a process that executes in an infinite loop a for statement controlled by a variable initially set to the value 10. The body of the for activates four processes in parallel with a co-begin/co-end scheme, and one of the four synchronize with the first process.

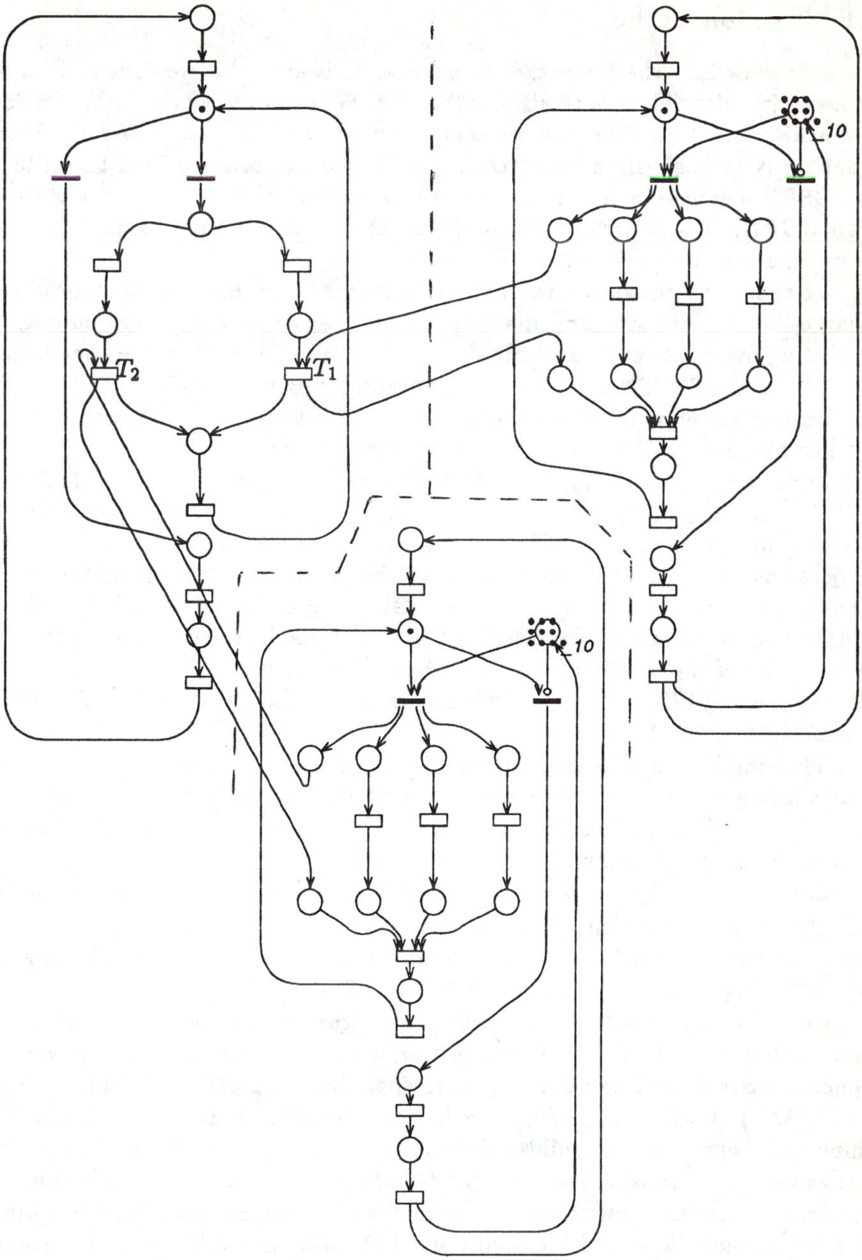


Fig. 2. A example of SGSPN

## 3.1 Solution method

We now show how the Kronecker expression derived in the previous section for SSA can be adapted to work also for SGSPN. We can observe that while in SSA a synchronized transition can be enabled in at most one state of each $SSM_i$ (that leads to basically a single correcting factor per synchronized transition), in SGSPN a synchronized transition can be enabled in more than one state of each $GSPN_i$ (and typically this is the case) so that we have to provide more complex correcting factors.

We shall first provide a rather trivial construction method, that produces a large number of correcting matrices, to then show how the large number of correcting matrices can be factorized into a few ones. In particular we shall show that also in the SGSPN case the number of correcting matrices is dependent only on the *number of synchronized transitions* and not on the *number of states* of each component GSPN that enables a synchronization transition.

As for SSA we investigate what is the relation between the CTMC of a SGSPN and the independent composition of the CTMCs of the component GSPNs. Let us term "global" the state $M$ of a SGSPN, and "local" the $N$ projections $M_i$ of $M$ over the elements of the partition. The independent composition of the CTMC of the component GSPNs produces a CTMC that allows a state transition due to the firing of $t'$ in any (global) state for which there is at least one of the local states that enables $t'$, independently of the local states of the other GSPN components, which is a correct behaviour if $t' \notin TS$, while it is not if $t' \in TS$.

Following the same approach pursued for SSA we can build from each infinitesimal generator $Q_i$ of each component GSPN, the matrix $Q'_i$, obtained from $Q_i$ by deleting all contributions due to the firing of all $t \in TS$ (of course also in the diagonal elements).

Matrix $Q' = \Pi_{\oplus_{i=0}^{N-1}} Q'_i$ differs from the infinitesimal generator $Q$ of the SGSPN only in the contribution due to $t$, that are completely absent. How many contributions due to $t$ are there in $Q$? There is one in each global state $M = M_0 \cdot M_1 \cdots M_{N-1}$ such that $\forall i \in [0, \ldots, N-1] : M_i[t\rangle$.

To correctly compute the correcting matrices we can perform the transformation illustrated in Fig. 3. Starting from a SGSPN we construct the $N$ components $GSPN_i$ and compute the corresponding tangible reachability graphs $TRG_i(M0_i)$. Each $TRG_i(M0_i)$ can be considered as a Stochastic State Machine and hence we can build $SSM_i = (P_{SM_i}, T_{SM_i}, F_{SM_i}, W_{SM_i}, p_{SM_i})$ that has as many places as nodes in the $TRG_i(M0_i)$, and as many transitions as the number of arcs; the flow relation is defined as follows. Let us establish a bijection $\gamma_i$ between $P_{SM_i}$ and the nodes of $TRG_i(M0_i)$ and a bijection $\delta_i$ between $T_{SM_i}$ and the arcs $A_i$ of $TRG_i$. Then the flow relation of $SSM_i$ is defined as: $(p, t) \in F_{SM_i}$ and $(t, p') \in F_{SM_i}$ iff $\delta_i(t) = \langle \gamma(p), \gamma(p'), t', \sigma \rangle \in A_i$ for some $t'$ and $\sigma$. $W_{SM_i}(t)$ is the weight of the arc $\delta_i(t)$, and $\gamma_i(p_{SM_i}) = M0_i$.

If $\mathrm{MAP}_i : T_{SM_i} \rightarrow T_E$ is the function that associates to each transition $t$ of

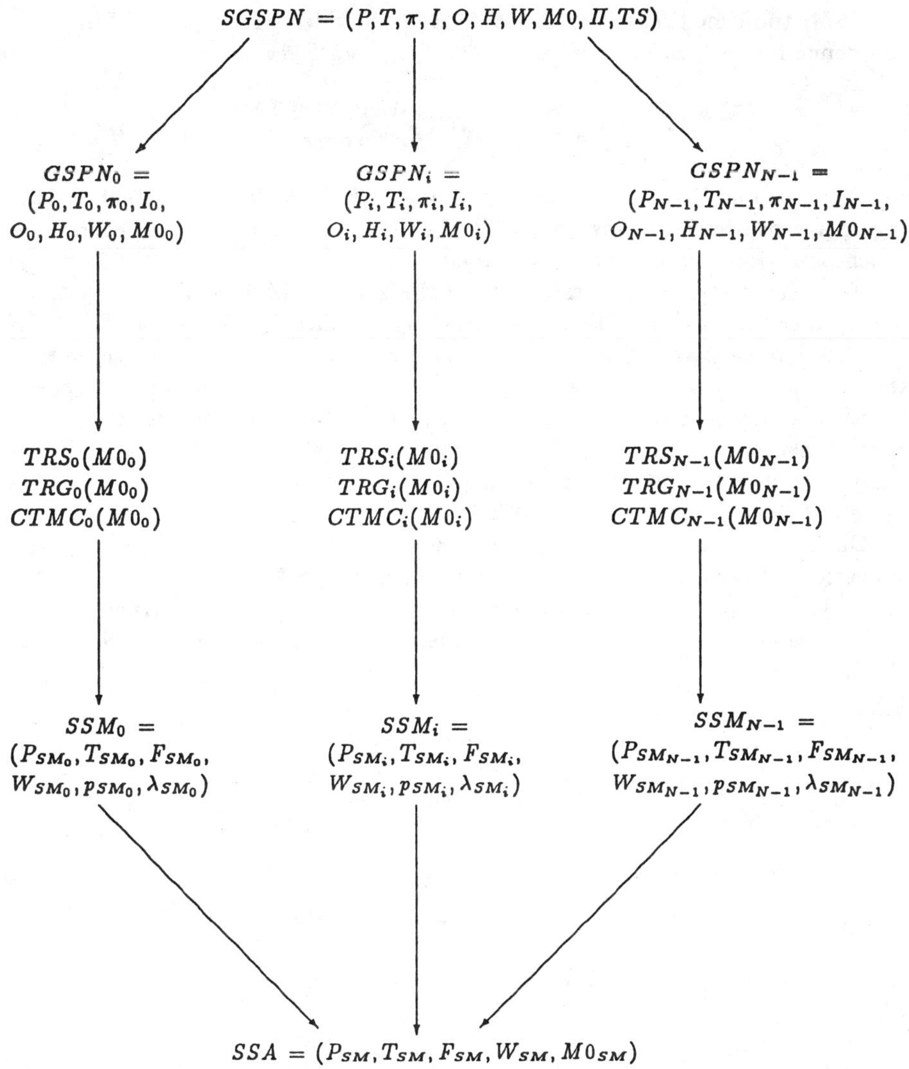


**Fig. 3.** A first solution method

$SSM_i$ the timed transition of the SGSPN of which $t$ represents a firing[5] we can define for each $SSM_i$ a labelling $\lambda_i : T_{SM_i} \rightarrow TS \cup \tau$ as

$$\lambda_i(t) = \begin{cases} \tau & \mathrm{MAP}_i(t) \notin TS \\ \mathrm{MAP}_i(t) & \textit{otherwise} \end{cases}$$

The labelling identifies the transitions of each $SSM_i$ that represent the firing of a synchronized transition, while all transitions that do not represent a synchronization among GSPNs are labelled $\tau$.

If we create the SSA obtained from the superposition of the $N$ SSM$_i$ on non-$\tau$ transitions of equal label, then each synchronized transition of label $t$ of the SSA represents the firing of $t$ in a given state of the SGSPN. Observe that the superposition makes sense only if the rates associated to the superposed transitions are equal, but this is indeed the case, as they represent the firing of the same transition $t'$ in the GSPN, and thanks to the restriction that "synchronized transitions lead to tangible states", all arcs in all $TRG_i(M0_i)$ that represent a firing of $t'$ have the same weight $W(t')$.

Observe that the SSA contains, for each transition $t' \in TS$, a synchronized transition for each element of the Cartesian product $\mathrm{MAP}_0^{-1}(t') \times \mathrm{MAP}_1^{-1}(t') \times \ldots \times \mathrm{MAP}_{N-1}^{-1}(t')$. We shall call $TS_{SSA}$ the set of synchronized transitions of the SSA, and $TS_{SM_i} = T_{SM_i} \cap TS_{SSA}$ the set of all synchronized transitions of $SSM_i$.

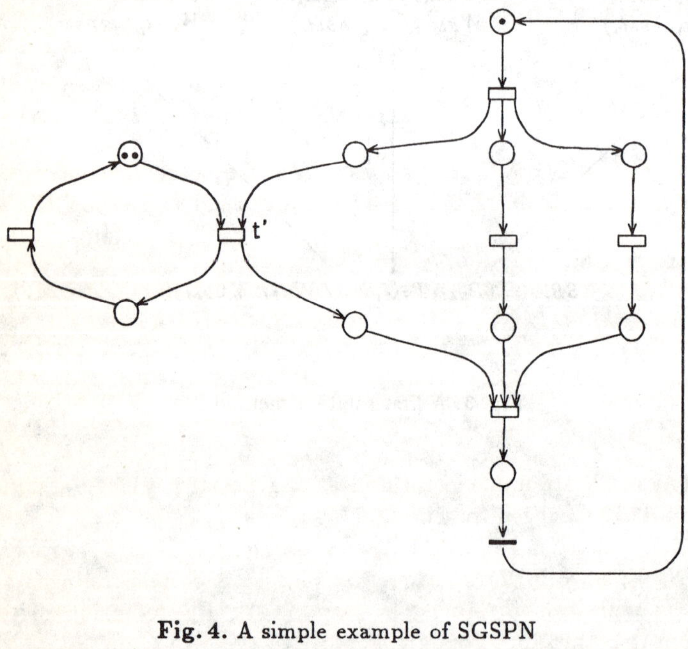


**Fig. 4.** A simple example of SGSPN

[5] If $\delta_i(t) = \langle \gamma(p), \gamma(p'), t', \sigma \rangle$, then $\mathrm{MAP}_i(t) = t'$.

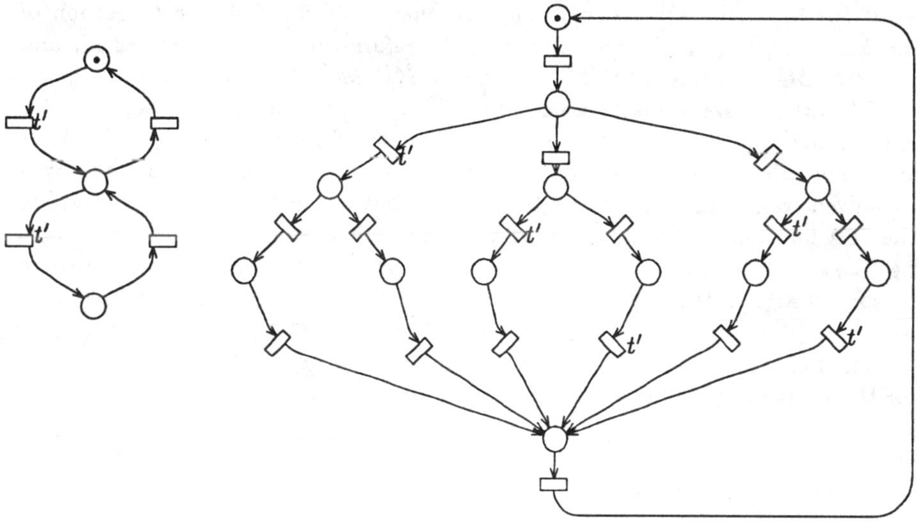


**Fig. 5.** The two corresponding SSMs

Fig. 5 shows the two SSMs that are produced by the SGSPN of Fig. 4. The SGSPN is composed by two GSPN that synchronize on transition $t'$. For the two SSM we have shown the associated label: there are two transitions of label $t'$ in the first SSM and five in the second one. This lead to the construction of an SSA that has ten synchronized transitions. The SSA is not shown in the picture as the net become quite cluttered with arcs.

We now prove that the infinitesimal generator of the SSA is also the infinitesimal generator of the SGSPN. To do this we first show that they have isomorphic TRGs, and that arcs in the TRGs have the same weight. We show this property only for synchronized transitions, as for regular one it is obvious.

**Property 1** *If $M$ and $M'$ are two elements of the TRS(M0), $M = (M_0, \ldots, M_{N-1})$, $M' = (M'_0, \ldots, M'_{N-1})$, $t' \in TS$, $t \in TS_{SSA}$, then*

$$M[t'\rangle M' \longleftrightarrow (p_0, p_1, \ldots, p_{N-1})[t\rangle(p'_0, p'_1, \ldots, p'_{N-1})$$

*with $\gamma(p_i) = M_i$, $\gamma(p'_i) = M'_i$ and $\lambda(t) = t'$*

For simplicity of notation, and with no loss of generality, we can assume that $t'$ is a synchronized transition for *all* $GSPN_i$.

$\rightarrow$ The relation is trivially proved by observing that $M[t\rangle M'$ implies that in each $GSPN_i$ we have $M_i[t\rangle M'_i$, therefore there is a corresponding arc in each $TRG_i$, and a transition of label $t$ in each $SSM_i$. By definition of transition superposition it is therefore possible to move in the SSA from $(p_0, p_1, \ldots, p_{N-1})$ to $(p'_0, p'_1, \ldots, p'_{N-1})$, by firing a transition of label $t'$.

$\leftarrow$ If $(p_0, p_1, ..., p_{N-1})[t\rangle(p'_0, p'_1, ..., p'_{N-1})$, then each $SSM_i$ has a transition of label $t$, input bag $p_i$ and output bag $p'_i$, therefore there exist $N$ states $M_i$ and $N$ states $M'_i$ such that $M_i[t\rangle M'_i$, and hence $M[t\rangle M'$.

From the above property we can only assume that the two infinitesimal generators have the same number of non null entries. To prove that they also have the same value we only have to remark that, due to the restriction that "synchronized transitions lead to tangible states" and to the particular way in which the SSA has been build by superposing transitions of equal label, all transitions $t \in T_{SSA}$ such that $\lambda(t) = t'$ have the same weight $W(t')$, and that for the same reason all arcs in the $TRG(M)$ that correspond to a firing of $t'$ have the same weight $W(t')$.

The following formula expresses the infinitesimal generator of the SSA, that, for the property above is also the infinitesimal generator of the SGSPN:

$$Q = \Pi_{\oplus_{i=0}^{N-1}} Q'_i - \sum_{t \in TS_{SSA}} W(t) \Pi_{\otimes_{i=0}^{N-1}} K'_i(t) + \sum_{t \in TS_{SSA}} W(t) \Pi_{\otimes_{i=0}^{N-1}} K''_i(t) \quad (5)$$

$Q_i$ is the infinitesimal generator of $SSM_i$, and $Q'_i$ is the infinitesimal generator of $SSM_i$ where the contributions of state transitions due to synchronized transitions in $TS_{SM_i}$ have been deleted. In formula: $Q'_i = Q_i - \sum_{t \in TS_{SM_i}} W(t)E({}^\bullet t, t^\bullet) + \sum_{t \in TS_{SM_i}} W(t)E({}^\bullet t, {}^\bullet t)$.

We can rewrite the equation above in terms of the $TS$ of the SGSPN by grouping terms that refer to transitions of equal label,as:

$$\begin{aligned} Q = \Pi_{\oplus_{i=0}^{N-1}} Q'_i &- \sum_{t' \in TS} \sum_{t \in \mathrm{MAP}^{-1}(t')} W(t) \Pi_{\otimes_{i=0}^{N-1}} K'_i(t) \\ &+ \sum_{t' \in TS} \sum_{t \in \mathrm{MAP}^{-1}(t')} W(t) \Pi_{\otimes_{i=0}^{N-1}} K''_i(t) \end{aligned} \quad (6)$$

Due to the restriction posed on the synchronized transitions of SGSPN, we have $W(t) = W(t')$, and hence it can be taken out of the inner sum.

This expression has two correcting matrices for each synchronized transition of the SSA, that is to say for each state of the SGSPN that enables a synchronized transition. But this expression can be factorized, by exchanging the inner matrix sum with the Kronecker product (a property that does not hold in general) to get:

$$\begin{aligned} Q = \Pi_{\oplus_{i=0}^{N-1}} Q'_i &- \sum_{t' \in TS} \mu'_t \Pi_{\otimes_{i=0}^{N-1}} \sum_{t \in \mathrm{MAP}_i^{-1}(t')} K'_i(t) \\ &+ \sum_{t' \in TS} W(t') \Pi_{\otimes_{i=0}^{N-1}} \sum_{t \in \mathrm{MAP}_i^{-1}(t')} K''_i(t) \end{aligned} \quad (7)$$

The factorization can be justified from the linearity of the $\Pi_{\oplus_{i=0}^{N-1}}$ function and observing that the term $\sum_{t \in \mathrm{MAP}^{-1}(t')}$ is the sum over all transitions that are obtained by the superposition of transition of equal label, and that the cardinality of this set is

$$|\,\mathrm{MAP}^{-1}(t')\,| = |\,\mathrm{MAP}_0^{-1}(t')\,| \cdot |\,\mathrm{MAP}_1^{-1}(t')\,| \cdot \ldots \cdot |\,\mathrm{MAP}_{N-1}^{-1}(t')\,|$$

therefore there is a correcting factor for each element of the Cartesian product of the states that enable, in each $SSM_i$, the transitions of $\mathrm{MAP}_i^{-1}(t')$, and this can be indifferently computed as the matrix sum of the Kronecker product of the correcting matrices with a single non null entry, or as the Kronecker product of the matrices obtained as matrix sum of the elementary correcting matrices.

## 3.2 Implementation

The extension of the SSA implementation to the SGSPN case has required a stronger interaction with the GreatSPN package. In the previous implementation an SSA is specified through the graphical interface as a single net, and then the user has to create a file that lists, for each component SSM, which places and transitions of the SSA belong to that SSM, as well as the set of synchronized transitions: quite an error prone work for the user. The current implementation allows instead each component to be identified at the graphical interface level by using the GreatSPN feature of the layers: each component GSPN is graphically assigned to a different layer, with synchronized transitions being assigned to all layers of the GSPNs of which they represent a synchronization.

Given a graphical representation of the SGSPN the solution program automatically produces the component GSPNs and identifies the set of synchronized transitions. It then calls, on each newly created $\mathrm{GSPN}_i$, the GreatSPN facility that computes the associated $\mathrm{TRG}_i$ and $\mathrm{CTMC}_i$. Note that the simpler case of SSA did not require the use of the GreatSPN facility to build the TRG since the TRG of an SSM can be very simply generated observing that the number of states is equal to the number of places. From the CTMC of each component GSPN, the program computes all the matrices involved in the Kronecker expression of the infinitesimal generator of the SGSPN, and the older solution can still be applied. We should also mention that the old implementation has been enhanced to treat all matrices in sparse form.

## 3.3 Applications and extensions

*Applications* A SGSPN is basically a set of components that interact through transition synchronization, and can therefore be applied to study all those systems that interact in a rendez-vous manner.

As a first example consider a set of $N$ systems that use a common global resource. This case can be very easily modelled by a SGSPN with $N+1$ component GSPNs: one component per system, plus a component that represents the resource manager.

Another class of problems that can fit into the SGSPN scheme is that of Petri net models of CSP-like programs. There have been a few attempts to use

Petri nets to study the performance of parallel programs of the CSP type, see for example [3], [12] and [18]. A tool is already available for the automatic translation of Occam programs into GSPN, in a format that is compatible with the GreatSPN package. A program is translated as a set of GSPNs, that in the tool are already assigned to different layers, and GSPNs interact by exchanging messages with a rendez-vous protocol. The rendez-vous is translated as a synchronization of two processes over an immediate transition, thus violating the SGSPN assumptions. The consideration that communication channels are 1-in/1-out allows an application of the algorithm for the elimination of immediate transition [6] that does not lead to an explosion in the number of synchronized timed transitions and that produces a model that can be solved with the method proposed in this paper.

Many of the models produced by the translation tool are of a size that is solvable by GreatSPN. This is because the tool does not translate program variables, which in certain cases has severe implications on the usefulness of the analysis performed. Indeed the reason that urged us to look into SGSPN is to be able to solve models that are produced by an enhanced version of the translation tool that explicitly include in the model a selected subset of the program variables.

*Synchronized transitions have to be timed?* Right now the SGSPN class has been defined with the major limitation of not allowing a synchronized transition to be immediate. In [6] the authors discuss the role of immediate transitions and conclude that in most cases (if there are no loops of immediate transitions) immediate transitions are not essential and an algorithm is given to eliminate them from the net at the expenses of introducing a higher number of timed transitions. We can not therefore consider this constraint as very limitative from a *theoretical* point of view, although it may become a serious one from a *practical* point of view: indeed to eliminate even a single immediate synchronized transition we may have to introduce a large number of timed transitions that are also synchronized transition, thus augmenting the number of correcting matrices we have to introduce. We think that is possible to extend the solution method, and hence the SGSPN class, to deal also with immediate transition synchronization, and this possibility is currently under study.

*Synchronized transitions have to lead to tangible states?* We have introduced the requirement that if $t$ is a synchronized transition, then $\forall i$, and for all arcs of the $TRG_i$ of the type $M[t\rangle M'$, $M'$ has to be tangible. This is equivalent to the requirement that all arcs in the $TRG_i$ that have $t$ has timed transition have an empty sequence of immediate transitions $\sigma$. This implies that all transitions of the corresponding $SSM_i$ of label $t$ have the same associated weight, and this is what allows us to easily build the SSA with a simple superposition operation, and to factorize the set of correcting matrices (inversion of the sum of matrices with the Kronecker product). It is surely possible to remove this condition, but then the number of correcting matrices may grow wildly again.

## 4 Conclusions

In this paper we have presented Superposed Generalized Stochastic Petri Nets (SGSPN), that can be considered as GSPN obtained by transition synchronization of a set of smaller GSPN components.

We have shown how the solution of the Markov chain associated to a SGSPN can be computed without the need to store the complete infinitesimal generator. This has been achieved by rewriting the generator as a Kronecker expression for which we know that an efficient solution method exists.

The work on SGSPN constitutes an extension of the work already done for SSA, a class of SPN obtained by superposition on transition of SPN of the state machine type. The trivial extension of the solution applied to SSA to this larger class of nets leads to a complex Kronecker expression where the number of terms may lead to very slow solution: we have shown instead that the number of correcting matrices is still dependent only on the *number of synchronized transitions* and not on the *number of states* of each component GSPN that enables a synchronization transition. It is clear that if there are a large number of correcting factors then part of, or all, the advantages introduced by the solution method proposed can be lost.

The method has been presented here in the context of steady state solution, but it can indeed be applied also to transient solution algorithm based on the product vector by infinitesimal generator type. The use of this technique for other transient analysis algorithm, as the one based on uniformization, requires instead some investigation.

The formula for the infinitesimal generator of SSA, as well as the one for SGSPN is easily amenable for parallel implementation, and indeed a parallel solution for SSA already exists, although there is no sparse matrix implementation available. The solution of SGSPN, that is interfaced with GreatSPN and that uses sparse matrices, is instead only sequential, although it is planned to bring it to our recently acquired new parallel machine.

Although the method presented does not require to store the infinitesimal generator, it does require to store the steady state probability vector $\nu$, so that if the number of state grows over a certain limit (limit that depends on the machine being used, but that on our Sun workstations is larger than a million) then we have to resort to lumpability techniques as whose associated to Stochastic Well Formed Nets, or to approximate solutions.

A lot of work is still ahead of us, in particular we are planning to adapt this solution method to solve the nets that are automatically produced from CSP-like programs by the EPOCA [14] tool, and to implement the solution of SGSPN in a parallel programming environment.

## 5 Acknowledgments

I would like to thank the anonymous referees for their careful reading of the paper and for their useful suggestions on how to improve it.

## References


1. M. Ajmone Marsan, G. Balbo, G. Chiola, G. Conte, S. Donatelli, and G. Franceschinis. An introduction to Generalized Stochastic Petri Nets. *Microelectronics and Reliability*, 31(4):699–725, 1991. Special issue on Petri nets and related graph models.
2. M. Ajmone Marsan, G. Balbo, and G. Conte. A class of generalized stochastic Petri nets for the performance analysis of multiprocessor systems. *ACM Transactions on Computer Systems*, 2(1), May 1984.
3. G. Balbo, S. Donatelli, and G. Franceschinis. Understanding parallel programs behaviour through Petri net models. *Journal of Parallel and Distributed Computing*, 15(3), 1992.
4. C. Beounes. Stochastic petri net modeling for dependability evaluation of complex computer system. In *Proc. of the International Workshop on Timed Petri nets*, Torino, Italy, July 1985. IEEE-CS Press.
5. P. Buchholz. Numerical solution methods based on structured descriptions of Markovian models. In *Proc. $5^{th}$ Int. Conf. Modeling Techniques and Tools for Computer Performance Evaluation*, Torino, Italy, February 1991.
6. G. Chiola, S. Donatelli, and G. Franceschinis. GSPN versus SPN: what is the actual role of immediate transitions? In *Proc. 4th Intern. Workshop on Petri Nets and Performance Models*, Melbourne, Australia, December 1991.
7. G. Chiola, C. Dutheillet, G. Franceschinis, and S. Haddad. Stochastic well-formed coloured nets and multiprocessor modelling applications. In K. Jensen and G. Rozenberg, editors, *High-Level Petri Nets. Theory and Application.* Springer Verlag, 1991.
8. Giovanni Chiola. *GreatSPN* 1.5 software architecture. In *Proc. $5^{th}$ Int. Conf. Modeling Techniques and Tools for Computer Performance Evaluation*, Torino, Italy, February 1991.
9. G. Ciardo and K.S. Trivedi. A decomposition approach for Stochastic Petri net models. In *Proc. 4th Intern. Workshop on Petri Nets and Performance Models*, Melbourne, Australia, December 1991.
10. M. Davio. Kronacker products and shuffle algebra. *IEEE Transactions on Computers*, 30(2):1099–1109, 1981.
11. F. De Cindio, G. De Michelis, L. Pomello, and C. Simone. Superposed automata nets. In C. Girault and W. Reisig, editors, *Application and Theory of Petri Nets.* IFB 52, New York and London, 1982.
12. F. DeCindio and O. Botti. Comparing Occam2 program placements by a GSPN model. In *Proc. 4th Intern. Workshop on Petri Nets and Performance Models*, Melbourne, Australia, December 1991.
13. S. Donatelli. Superposed stochastic automata: a class of stochastic Petri nets with parallel solution and distributed state space. *Performance Evaluation*, 18:21–36, 1993.
14. S. Donatelli, G. Franceschinis, N. Mazzocca, and S. Russo. Software architecture of the epoca integrated environment. In *Proc. of the Seventh International Conference on Modelling Techniques and Tools for Computer Performance Evaluation*, May 1994.
15. J.B. Dugan, A. Bobbio, and G. Ciardo K.S. Trivedi. The design of a unified package for the solution of stochastic Petri net models. In *Proc. Int. Workshop on Timed Petri Nets*, Torino, Italy, July 1985. IEEE-CS Press.

16. G. Florin and S. Natkin. Matrix product form solution for closed synchronized queueing networks. In *Proc. 3rd Intern. Workshop on Petri Nets and Performance Models*, pages 29–39, Kyoto, Japan, December 1989. IEEE-CS Press.
17. W. Henderson and P.G. Taylor. Aggregation methods in exact performance analysis of stochastic Petri nets. In *Proc. 3rd Intern. Workshop on Petri Nets and Performance Models*, pages 12–18, Kyoto, Japan, December 1989. IEEE-CS Press.
18. I.E. Jelly and J.P. Grey. Prototyping parallel systems: a performanve evaluation approach. In *Proc. ISMM Conference on Parallel and Distributed Computing Systems*, Pittsburgh, USA, October 1992.
19. A.A. Lazar and T.G. Robertazzi. Markovian Petri net protocols with product form solution. In *Proc. of IEEE INFOCOM'87*, San Francisco, CA, USA, March 1987. Also in *Performance Evaluation*, Vol.12, 1991, pp.67–77.
20. M. K. Molloy. Performance analysis using stochastic Petri nets. *IEEE Transaction on Computers*, 31(9):913–917, September 1982.
21. S. Natkin. Timed and stochastic Petri nets: from the validation to the performance of synchronization schemes. In *Proc. Int. Workshop on Timed Petri Nets*, Torino, Italy, July 1985. IEEE-CS Press.
22. B. Plateau. *Repartition, parallelisme et des elements de leur valutation.* PhD thesis, (in french), Paris, France, 1981.
23. B. Plateau. On the stochastic structure of parallelism and synchronization models for distributed algorithms. In *Proc. 1985 SIGMETRICS Conference*, pages 147–154, Austin, TX, USA, August 1985. ACM.
24. B. Plateau. Peps: a package for solving complex Markov models of parallel systems. In R. Puigjaner and D. Poiter, editors, *Modeling techniques and tools for computer performance evaluation*, pages 291–306. Plenum Press, New York and London, 1990.
25. B. Plateau and K. Atif. Stochastic automata network for modeling parallel systems. *IEEE Transactions on Software Engineering*, 17(10):1093–1108, 1991.

# Verification of Recipe-Based Control Procedures by Means of Predicate/Transition Nets

Hartmann J. Genrich[1], Hans-Michael Hanisch[2a] and Konrad Wöllhaf[3]

[1]Gesellschaft für Mathematik und Datenverarbeitung (GMD), PF 1316, Schloß Birlinghoven, D-53757 Sankt Augustin, Germany
[2]Otto-von-Guericke-Universität Magdeburg, Dept. of Electrical Engineering, PF 4120, D-39016 Magdeburg, Germany
[3]Universität Dortmund, Dept. of Chemical Engineering, D-44221 Dortmund, Germany

**Abstract.** The paper presents a method for the verification of control procedures which are based on so-called recipes. The concept of recipes is widely used in the control of multipurpose chemical batch plants which are very similar to Flexible Manufacturing Systems.
The recipes as well as the plant description are transformed into Predicate/ Transition nets (abbr.: Pr/T nets). The Pr/T net can be tested either by simulation or by analysis based on the computation of the occurence graph. Several formal net properties (conflicts, deadlocks etc.) indicate desired or critical behavior of the controlled batch plant.
An illustrative example shows the complexity of the problem as well as the transformation into Pr/T nets and the results of the analysis which was performed by Design/CPN.

## Preface

Batch production systems are used to manufacture chemical products with a high value (and a high degree of possible danger during the manufacturing process) such as pharmaceutical and dye products, catalysts etc.
The control problems which occur in batch processes are partially problems of the control of Discrete Event Dynamic Systems (abbr.: DEDS). However, the application of theoretical approaches for the description and analysis of DEDS to batch processes is very sparse. The concept of recipes developed by chemists and chemical engineers of the big chemical companies and described in standards [6] and recommendations [9] is far away from any theory of DEDS.
We present here an approach to cross the gap between the concept of recipes and Pr/T nets to make one of the most powerful concepts of DEDS applicable to recipe-

---

[a] Corresponding author, email: hami@infaut.e-technik.uni-magdeburg.de

controlled batch processes. We will show that recipes can be modeled by Pr/T nets in a very natural way. Based on these models, we can transfer all theoretical methods and the simulation and analysis tools for Pr/T nets to a new field of application with a large extension. We can get a deeper understanding of the processes which are described by recipes and we can prevent malfunctions before they occur in the real production system.

# 1 Modeling of Recipe-Controlled Multipurpose Plants

The initial state of the computer-aided investigation of technical systems is a model that contains all relevant aspects of a real system in a suitable data structure. The insight of the plant operator into the real system and the knowledge about the formal model are needed for the construction of such models. The knowledge of the process is represented by the description of recipes. An object oriented data structure and a graphical editor to create this standard description for batch processes were developed for a simulation tool that supports the simulation of the mixed, continuous and discrete dynamic of multipurpose batch plants [1]. The developed data structure will be used for a formal construction of appropriate Pr/T net models.

## 1.1 Structure of Multipurpose Batch Plants

The representation of the production process is based on the concept of recipe-controlled operation with basic and control recipes which has been developed by the big chemical companies [6,9,10] and is used in the automation of batch plants by all major suppliers of process control systems. The structure of the data model is based upon this recommendations and consists of the basic recipe, the production schedule, the multipurpose plant description and the executable control recipe.
Figure 1 shows the main components to create a control recipe. A customer orders an amount of a product, and the production schedule synchronizes this order with other production tasks. A basic recipe is selected, that describes the way the product is manufactured. This basic recipe is independent of a special plant. The control recipe is a copy of the basic recipe and the parameters of the basic recipe e. g. the lot and the devices are replaced by real numbers or components of the multipurpose batch plant. The control recipe can be executed on a real plant.

## 1.2 The Basic Recipe

The basic recipe describes - independent of a special plant - the production process of a standard lot of a product. It consists of (figure 2): the recipe header with general documentation data, the list of the products and the formula of involved substances (in the data model the formula is replaced by a reference to a component

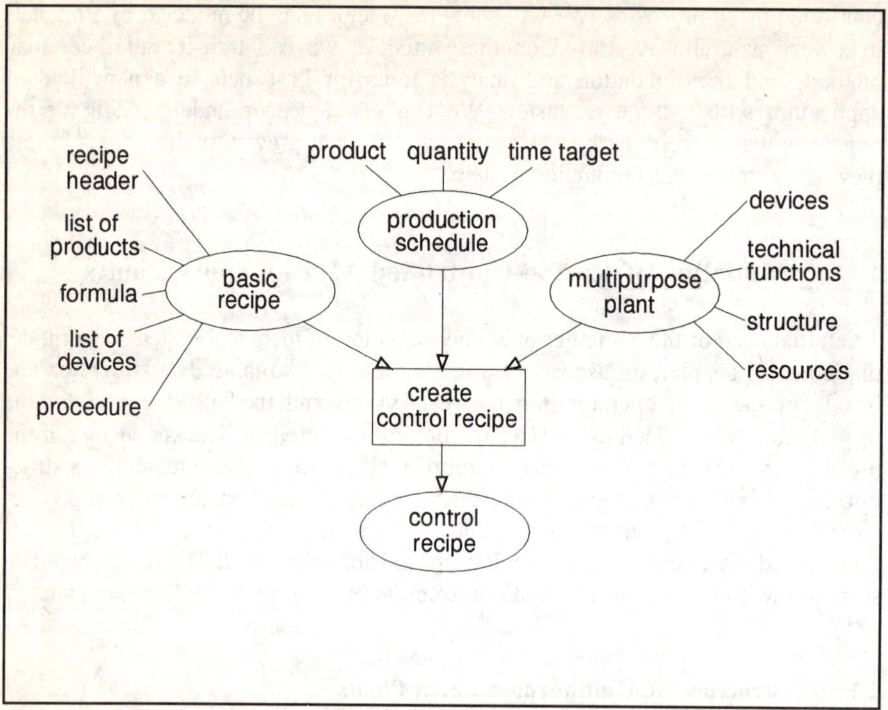


**Fig. 1.** Input Parameters to Create a Control Recipe

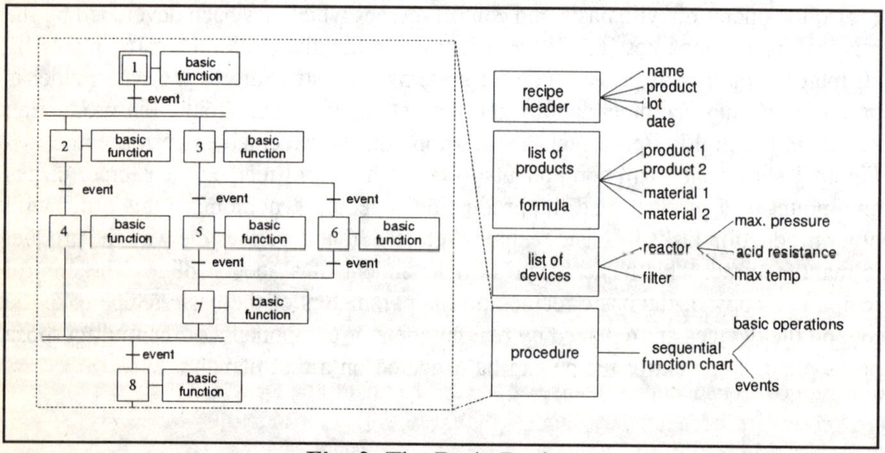


**Fig. 2.** The Basic Recipe

database) and a list of needed devices with their properties. The procedure contains the operation instructions, that is an event-controlled combination of actions, described by sequential function charts. The events may be time events caused by clocks or state events caused the crossing of a threshold by continuous process variables

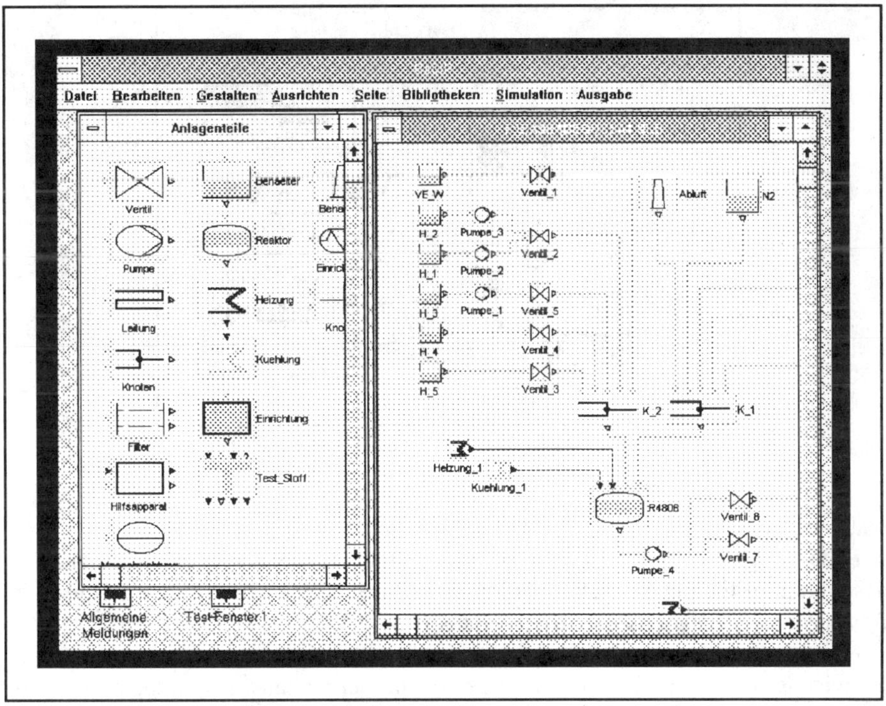


**Fig. 3.** Graphical Editor to Construct the Plant Model

like pressure, temperature or concentration. The actions are the basic functions like feeding a material, heating, cooling or basic operations that are the functional composition of different basic functions. The structure of sequential function charts is similar to Petri nets because they describe the causal order of states and state transitions.

## 1.3 Production Schedule

The production schedule includes specifications concerning the product, quantity, required quality, time target and the priority of the task. As a rule, the processing of a production schedule means the actual production of a lot. This few parameters are significant because they specify the target of the production process.

## 1.4 Multipurpose Plant

The model of the plant consists of the devices, their connections and the technical functions. The multipurpose plant is "the hardware" to execute the production process defined in the basic recipe and offers the technical functions that are the

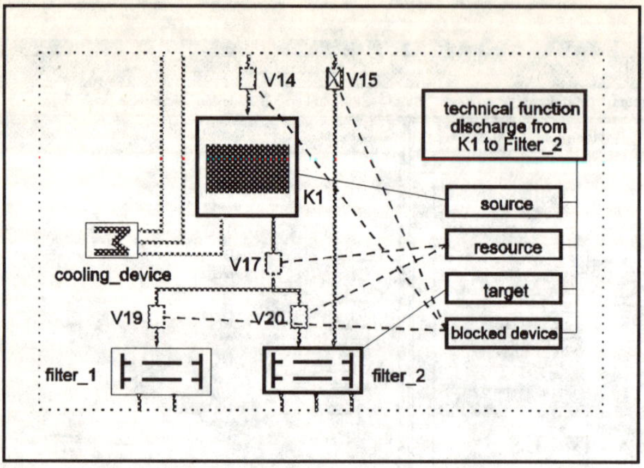


**Fig. 4.** Construction of a Technical Function

realization of the basic functions of the basic recipe.

For the design of a model of the plant, a library of devices is available and the description of the plant is supported by a graphical editor with the symbols used in flowsheets and diagrams of process plants (figure 3).

The properties and parameters of the devices are defined in parameter dialogues. If a device is selected in the control recipe, the properties must match the parameters given in the basic recipe. The operational functions of the plant are described by technical functions.

A part of a plant model (dotted rectangle) containing a crystallizer, two filters, a cooling device and different valves is shown in figure 4. A technical function of this unit is: "Discharge from K1 to filter_2". The crystallizer K1 is the source of the material, the filter is the target, and the valves V17 and V20 are resources, that must be used exclusively by this technical function. It must be guaranteed that no other technical function will open the valve V19 while this function is active, so this valve must be blocked. Another technical function for this unit may be "cooling of K1 with cooling_device". The cooling device may be used by different devices and the capacity of the cooling device may be exhausted. So the cooling device need not to be used exclusively but is a limited resource.

## 1.5 Example

Figure 5 shows the flowsheet of a multipurpose batch plant for the production of two similar products (substances D and E) which are denoted by "product I" and "product II " in the models. The manufacturing processes are performed as follows:

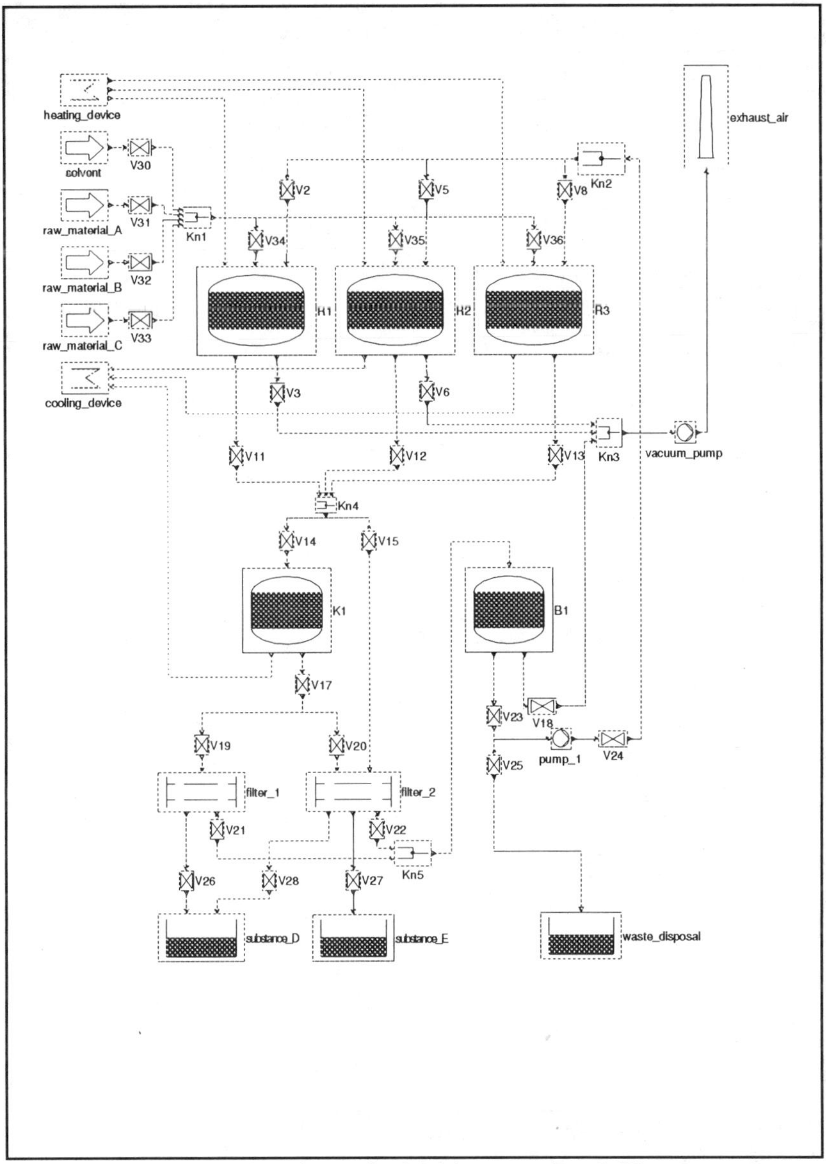


**Fig. 5.** Flowsheet of a Multipurpose Batch Plant

Product I

For the manufacturing of product I solvent has to be pumped into a reactor. Then the raw materials A and B are added in the stoichiometric proportion. Next the reactor is evacuated end heated to start the reaction. If the concentration of the final product I has reached the prescribed concentration, the contents of the reactor is

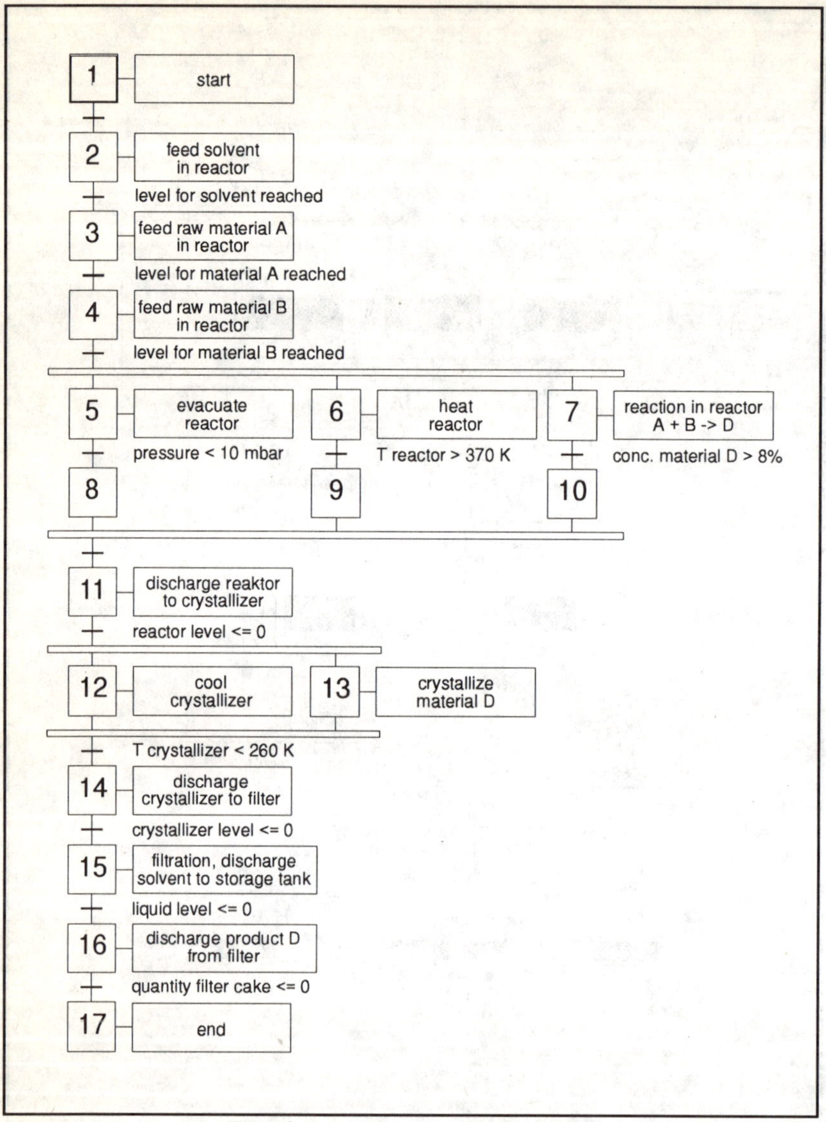


**Fig. 6.** Basic Recipe for Product I

discharged into a crystallizer. While the crystallizer is cooled down, product I crystallizes. If the contents of the crystallizer has reached a defined temperature, the crystallization process is assumed to be finished and the whole mixture is discharged into a filter, where product I is separated from the solvent. The solvent is gathered in a storage tank to be recycled in a subsequent batch.

Figure 6 shows the **basic recipe** for product I.

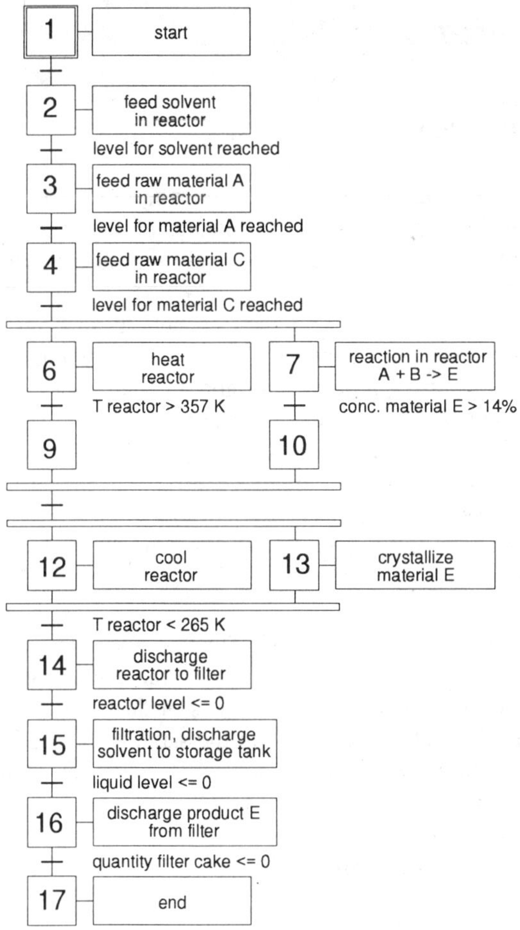


**Fig. 7.** Basic Recipe for Product II

Product II

The manufacturing process of product II also requires the same solvent as product I. The solvent is pumped into a reactor. Then the raw materials A and C are feeded into the reactor and the reaction is started by heating the reactor. If a defined concentration of product II is measured, the contents is cooled down and the crystallization process is started directly in the reactor. Then the contents is discharged

into a filter, product II and the solvent are separated, and the solvent can be recycled.
Figure 7 shows the **basic recipe** for product II.

## 2 The Verification Problem

Although the concept of recipes and appropriate control procedures for chemical batch plants has become a widely used approach for the design of process control systems in the process industry, the concept has some unsolved problems which are:

1. Recipes are designed by intuition and they cannot be analyzed in the original form to guarantee that the recipe corresponds to the requirements of the particular manufacturing process. Our goal is to analyze the recipes as they are prescribed by the chemical engineer and not to deal with formal synthesis techniques for controllers in batch processes. We assume that the basic functions and the appropriate technical functions are implemented correctly. However, if we once have a model, the verification of the single recipes can be supported by the excellent simulation facilities of Pr/T nets and the tool we use.

2. Each recipe describes the sequence of technological operations to manufacture a single product. There is no concept in the recipe description neither to describe nor to analyze the interactions of different or identical recipes which are performed concurrently on the same plant and are coupled by limited resources (metering tanks, auxiliary devices, mains for steam etc.).
   Petri nets are an ideal concept for modeling and analysis of such problems. Hence, the paper focuses on this problem.

3. Recipes usually describe the desired sequence of steps. The control operations which are necessary to reject disturbances are specified separately. Such control operations can change the interactions between several recipes fundamentally and can lead to very critical, dangerous situations.
   The example given in this paper only describes the normal mode of operation. However, we are able to model disturbances and the modification of resource allocation caused by disturbances.

The inherent complexity of the resource allocation problem is more or less hidden in the recipe description since the problem is split into different parts (see figure 1). If a real plant is running under control of recipes, all those parts interact in a way that can not be predicted by the knowledge of the several parts but only by a model that represents the complexity as a whole.
Hence, more and more of the users of process control systems that support recipe

control feel an increasing need for formal techniques which enable verification of control procedures to prove the correctness before the procedures are executed in a process control system that controls a real (dangerous) chemical process. For this purpose, the recipes as well as the description of the plant must be transformed into mathematical models which can be formally analyzed.

## 3 Transformation of Recipes into Predicate/Transition Nets

The basic idea of the modeling of recipe-controlled batch plants by means of Pr/T nets is to integrate the different parts of the description (see figure 1) in one model which describes the whole plant together with the control procedures. The model must therefore at least express the following items:

1. The causal structure of basic functions and events prescribed in the procedures of the basic recipe.
2. The devices which are part of the plant the production process is performed on (the "existing hardware").
3. The technical functions of the existing devices which must correspond to the basic functions of the basic recipe.
4. The specifications of the devices which are given as a list of devices of the basic recipe (the "hardware requirements" of a production process).
5. The resources and allocation strategies which are necessary to perform a production process.

Figures 8 and 9 show the Pr/T nets for the recipes of products I and II. Their relationship to figures 6 respectively 7 is obvious. Their syntactical form is that of CPN models (cf. [5]) as supported by Design/CPN, the software tool we use for constructing and analyzing higher level Petri net models [7].
The model is set up in such a way that the description of the devices can be easily varied according to different plants. In our example, however, we check only the proper technical functions of the devices.

### 3.1 The Basic Recipe

The causal structure of a basic recipe can be immediately transformed into the structure of a Pr/T net where

- The basic functions are represented by predicates.

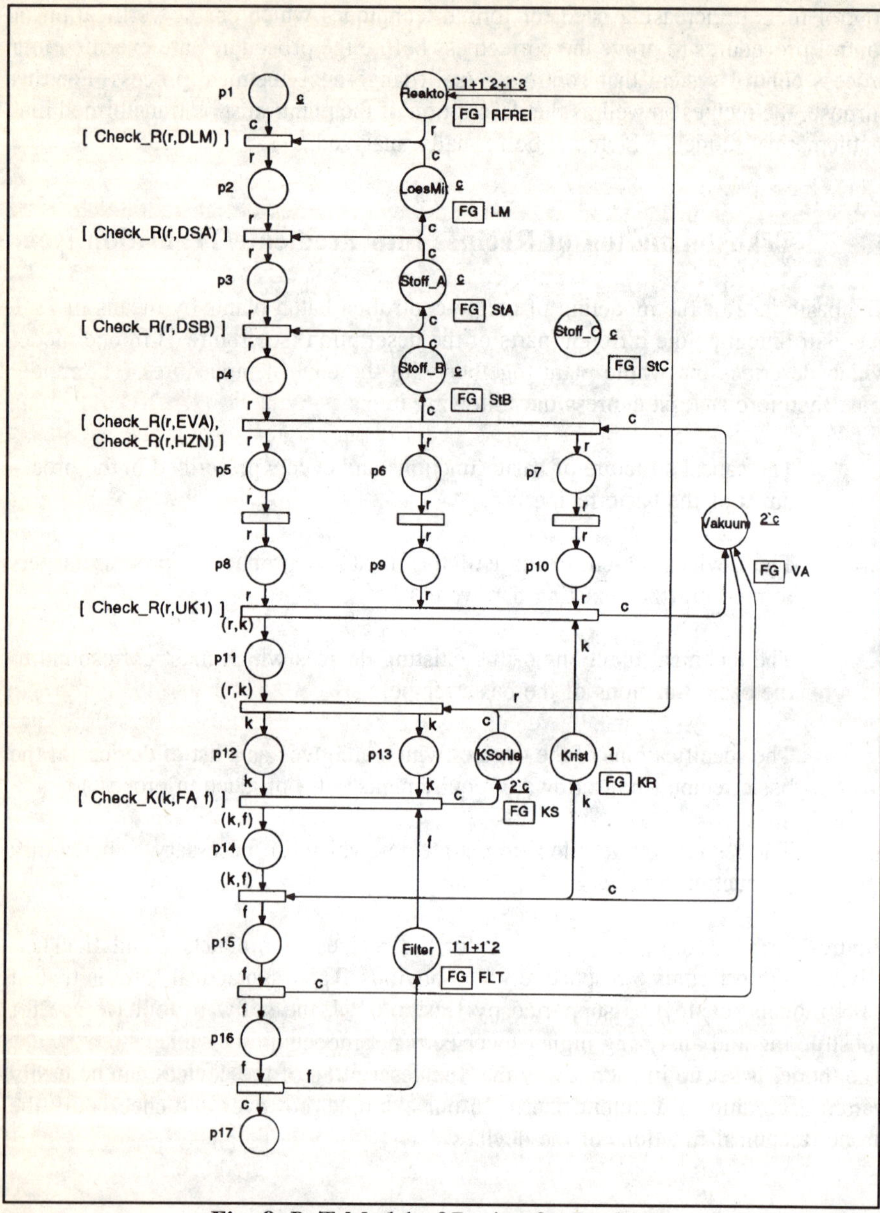


**Fig. 8.** Pr/T Model of Recipe for Product I

- The events which start the basic functions and the events which must be supervised to indicate the end of the basic functions are represented by transitions.

- The causal order is represented by the flow relation (the arcs).

Since the basic recipe is independent of the plant (see Section 1), the inscriptions at the net elements contain variables which correspond to the parameters of the appropriate basic functions. One can see easily that the representation of recipes by means of Pr/T nets is a very natural way to describe recipes by mathematical models.

*Example:* A comparison of figures 6 and 7 with figures 8 and 9 shows that the structures of recipes and Pr/T nets are identical.

### 3.2 The Devices of the Plant

The devices of the plant are modeled by means of individual tokens. The distribution of the tokens in the net represents the technological operations which are currently performed by the control recipes in the devices.
The tokens are structured (records) and contain all information described in figure 2 (list of devices). Additionally, a list of technical functions that contains all technical functions the process device is able to perform is part of each token (see below).

*Example:* The tokens representing the reactors, for instance, are records containing information about the volume (important for the size of a batch and the number of batches which must be manufactured), the maximal temperature and pressure which must not be exceeded, about the material (this can be important if very corrosive substances are manufactured) and about the technical functions the reactor is able to perform (see table 1).

### 3.3 The Technical Functions

The lists of the technical functions are based on the description of the technical functions in the simulation model (see figure 4) and on the structure of the plant.
The technical functions of the devices are part of the individual tokens the devices are modeled by. Each token of a device has a list of technical functions the device is able to perform. The guard expressions at the transitions which model the start of basic functions ensure that the technical functions of the devices (the tokens) correspond to the basic functions of the basic recipe (the predicates).

*Example:* In our example, reactor R1 has the following technical functions (see figure 5 and table 1):

DLM Dosing of solvent via pipe connection from sovent to reactor R1
DSA Dosing of substance A via pipe connection from raw_material_A to reactor R1
DSB Dosing of substance B via pipe connection from raw_material_B to reactor

```
update(R_Daten,1,
      {Volumen   = 6.3,  (*m^3*)
      MaxTemp    = 473.0,  (*Kelvin*)
      MaxDruck   = 10.0,  (*bar*)
      Material   = Mat0,
      TechFn     =[DLM,DSA,DSB,DSC,HZN,EVA,UK1,UF2]});
update(R_Daten,2,
      {Volumen   = 6.3,
      MaxTemp    = 473.0,
      MaxDruck   = 10.0,
      Material   = Mat0,
      TechFn     =
      [DLM,DSA,DSB,DSC,HZN,EVA,UK1,UF2,KUE]});
update(R_Daten,3,
      {Volumen   = 6.3,
      MaxTemp    = 473.0,
      MaxDruck   = 10.0,
      Material   = Mat0,
      TechFn     =
      [DLM,DSA,DSB,DSC,HZN,UK1,UF2,KUE]});
update(K_Daten,1,
      {Volumen   = 9.8,
      TechFn     = [UF1,UF2]});
update(F_Daten,1,
      {Volumen   = 9.8,
      TechFn     = [EFT]});
update(F_Daten,2,
      {Volumen   = 9.8,
      TechFn     = [EFT]});
```

**Table 1** Device Data

| | |
|---|---|
| | R1 |
| DSC | Dosing of substance C via pipe connection from raw_material_C to reactor R1 |
| HZN | Heating by means of heating_device (steam) |
| EVA | Evacuation via pipe connection from R1 to vacuum_pump |
| UK1 | Transfer into crystallizer K1 via pipe connection |
| UF2 | Transfer into filter_2 via pipe connection. |

Reactor R1 has no technical function F1 (transfer into filter_1), because there is no direct pipe connection from R1 to filter_1 (only via the crystallizer).

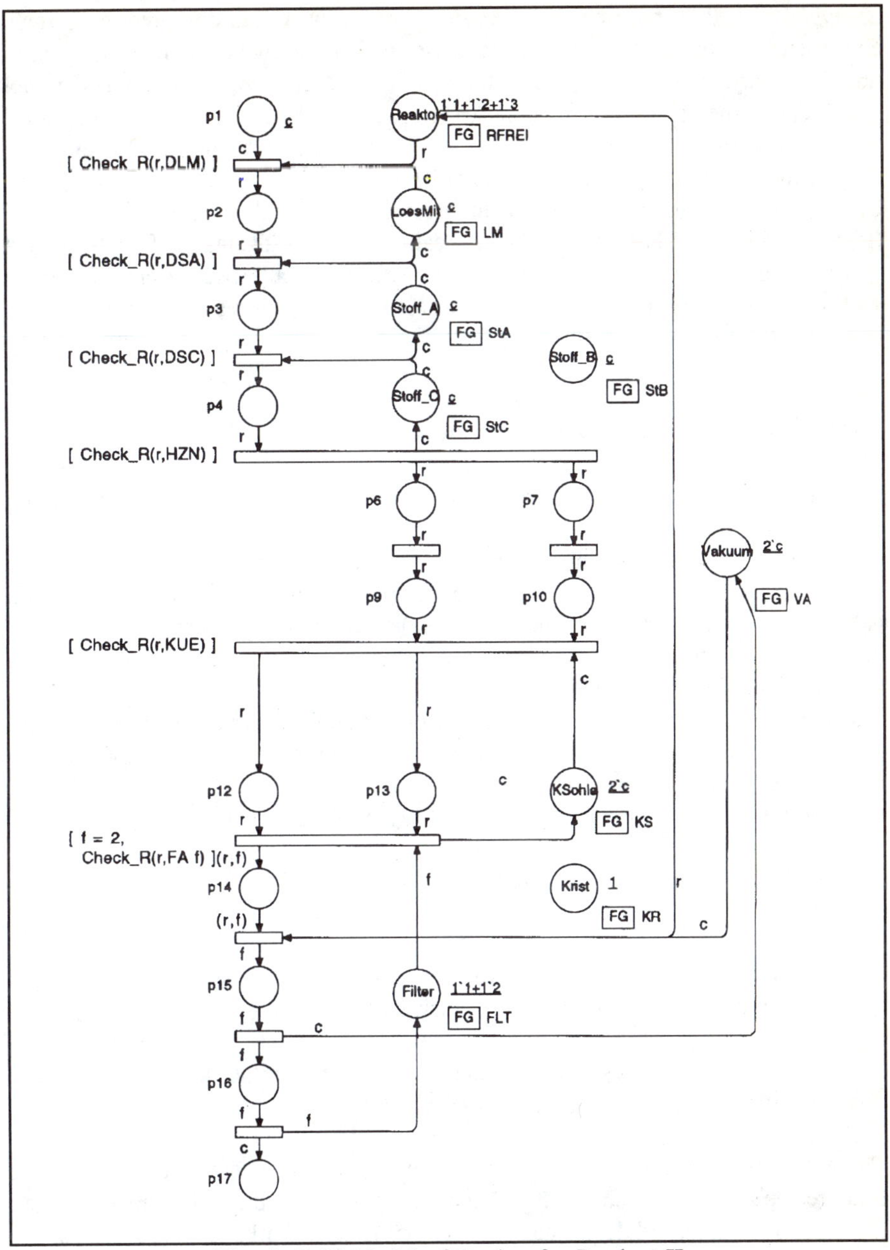


**Fig. 9.** Pr/T Model of Recipe for Product II

### 3.4 The Specification of the Devices

The requirements for the devices needed to perform the process which is described by the basic recipe are represented by the guard expressions the transitions are labeled with.

*Example:* The Pr/T net models in figures 8 and 9 have guard expressions which check whether the requirements of a process are fulfilled by the devices involved or not. The guard expression will have the value "FALSE" if the requirements are not fulfilled, and the value "TRUE" otherwise. Figure 10 shows two guard expressions. The declaration of the guard expressions is part of the CPN model. However, we will not describe this part in detail (see [7]).
In recipe I, for example, the reactor involved is tested at the beginning of phase 5 (reaction) whether it has evacuation and heating. (The disadvantage of this approach will show up in Section 4) In recipe II, only filter_2 is allowed for entering phase 14 (discharge from reactor to filter) because only filter_2 has a direct pipe connection from the reactors (see figure 5).

### 3.5 The Resource Allocation Strategy

Unfortunately, resource allocation requirements for multipurpose batch plants are not explicitly described neither in the basic recipe nor in the plant description and therefore such information must be added in a textual form. The same goes for the strategy of resource allocation to the single recipe-controlled processes. The decision is left to the plant-operator or (the worst case) to the intuition of the programmer of the process control system who is normally not familiar with the manufacturing processes and their particular requirements.
The modeling of resource allocation strategies is therefore the most complicated part of modeling of recipes. This part can not - at least up to now - be created formally from the recipes.
Generally, we have to distinguish different types of resources:

1. Process devices of the plant such as reactors, distillation columns, filters etc.

2. Metering and storage tanks for raw materials, products etc. and switching valves which must be used exclusively.

3. Auxiliary devices such as mains for steam and vacuum with a given capacity that must not be exceeded.

The allocation of all these different resources cannot be determined based on the basic recipe but only when a control recipe is created by adding information about the plant and the schedule to the basic recipe. Hence, we have first to model the devices and the structure of the plant by means of Pr/T nets as described above.

**The process devices.** The process devices are characterized by a set of variables which describe the physical properties (acid-resistance, maximal pressure, maximal temperature etc.) and by a set of technical functions the process device is able to perform (see above).

**Metering tanks and switching valves.** This kind of resources are used exclusively by the different production processes. Hence, each resource is modeled by a predicate which denotes that the resource is available (if the predicate contains a non-individual token) or not (if the predicate is clean).

*Example:* Resources of this type are modeled by predicates "LoesMit" (solvent), "Stoff_A" (raw_material_A), "Stoff_B" (raw_material_B) and "Stoff_C" (raw_material_C) in figures 8 and 9. One can see that the initial marking of these predicates is given by one non-individual token for each predicate. The predicates are fused globally, denoted by the rectangle with the inscription FG, to model the exclusive use of the corresponding resources if several control recipes are executed concurrently.

**Auxiliary devices.** Auxiliary devices have a limited capacity which must not be exceeded in order to prevent dangerous situations. Hence, such devices are modeled by means of predicates with a finite number of tokens that describe the available capacity of such a resource. If the appropriate predicate is clean, then the resource is not available.

*Example:* These resources are modeled by predicates "Vakuum" (vacuum_pump) and "KSole" (cooling_device) in figures 8 and 9. Each of this predicates carries two black tokens which denote that the resources can be used by two devices at the same time but never by three or more devices because this would exceed the capacity of the resource. These predicates are also fused globally.

Insufficient resource allocation strategies can cause hazardous states or even deadlocks of the production system. Hence, we apply the formal analysis techniques of Pr/T nets to detect such errors *before* the production process is performed. If a suitable control strategy is found and verified, this strategy can be used to control the real process.

## 4 Simulation and Analysis of the Pr/T Net Model of the Batch Plant

The two models can be executed and tested either separately or together as one system. In the latter case, they interact via the common resource places (process devices, metering tanks for the raw materials and mains for vacuum and cooling

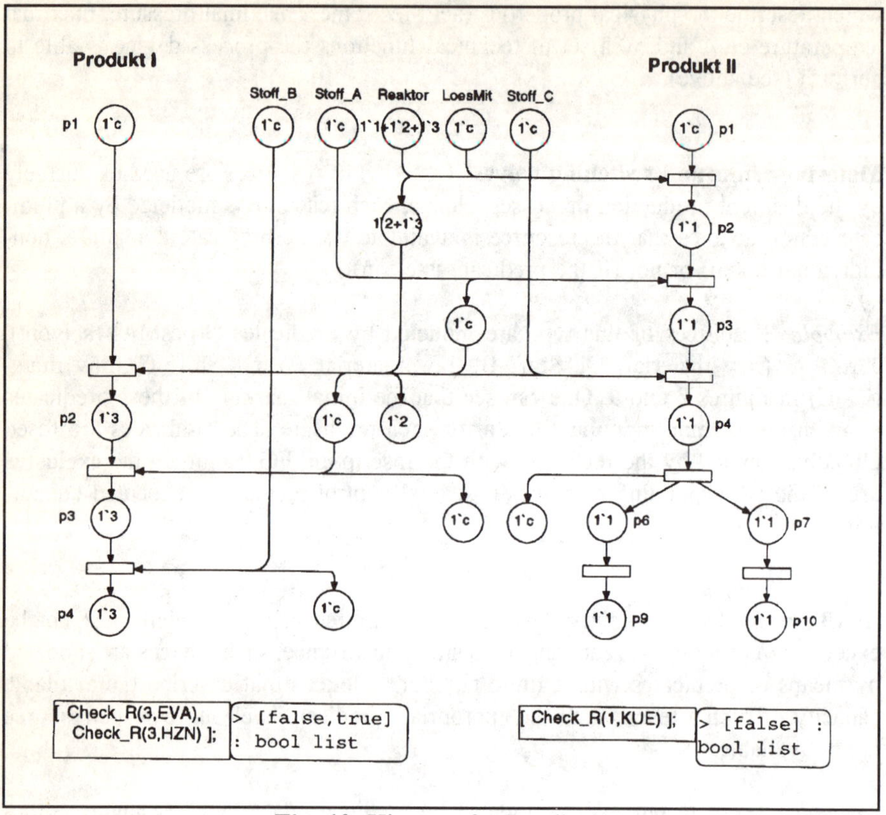


**Fig. 10.** History of a Deadlock

agent) that are 'fused' together.

The initial marking of p1 in figures 8 and 9 models the production target (manufacturing of one batch of both products). The conflicts in the models represent the decisions to select devices for the manufacturing processes described by the basic recipes. Suitable decisions will ensure that no deadlocks occur.

The models can be debugged by interactive simulation. We can study the execution of control recipes (which are created by binding individual tokens to variables of the arc inscriptions and guard expressions) and verify whether the sequences of technological operations prescribed by the recipes meet the requirements of the manufacturing processes or not. We can also study the flow of resources as well as the flow of batches through the production system.

For systematic detection of possible malfunctions like deadlocks, respectively for verifying their absence, we use the occurrence graph analyzer that is part of the software tool [8]. The understanding of a deadlock is facilitated by portraying its history. Adopting techniques developed in a different context (design of asynchronous circuits, cf. [2]) we construct and display these histories automatically. The complete occurrence graph of our example has 801 nodes (reachable markings) and 1857 arcs (transition occurrences).

*Example:* Figure 10 shows the history of a deadlock caused by choosing reactor R3 for manufacturing of product I and reactor R1 for manufacturing of product II. The graph is derived automatically from the occurence graph and drawn by using the graphics routines of the tool. Each transition occurence is shown with the markings of the input places before, and the markings of the output places after, that occurence. Empty markings are omitted. The layout of the history graph was changed afterwards by using the tool's interactive editing facilities. Also, the guard regions of the two transitions that would have to occur next were evaluated. They have both a false part which explains the premature halt. The deadlock results from the fact that reactor R1 can not be cooled down because there is no cooling provided and reactor R3 can not be evacuated (see guard regions in figure 10).

Obviously the reactors must be checked in the beginning for all technical functions needed during processing the recipe. But even if this is done, we can find inadequate resource allocation strategies which cause at least delays or blockings of one recipe by the other one. If for example filter_2 is chosen for the manufacturing of product I (this is possible, because filter_2 has all the technical functions for manufacturing of product I), then no batch of product II can be produced, because filter_1 has no technical function for the transfer of product from the reactors to the filter (only via the crystallizer, which is occupied by product I).

All these malfunctions are more or less obvious, but this is only for the sake of simplicity of our example here. Existing batch plants are much more complex, up to 100 different products and about 10 control recipes running concurrently are no exceptions. In such cases, a much more subtle interplay of several recipes may lead to occasional malfunctions that would be very hard to detect and analyze without occurrence graph analysis and history display.

Compared with the present practice in designing control systems for batch processes, the capability of an analysis (not only a simulation) is a substantial progress, and according to the experience one of us has with applying similar techniques to asynchronous circuit design [2], occurrence graphs with up to 10000 nodes are still quite manageable with the tools at hand.

This fact gives rise to the hope that our concept is applicable to even large plants in the process industry.

## 6 Conclusions

The paper describes a first approach for analyzing recipe-controlled processes by means of Pr/T nets. The approach can be easily extended to larger and more complicated problems, because all of the information for control procedures provided by the modeling and simulation tool in Section 1 is contained in the Pr/T models (excepted the blocking of valves). For instance, the guard expressions can be extended by checking proper materials of the devices, proper temperature and pressure limits etc.

The first subject of our further work will be the integration of all switching valves in the model. Then we will be able to verify completely all the details of the control procedures, not only the resource allocation strategies.
According to the experience one of us has with performance evaluation and optimization of the dynamic behavior of batch processes by means of timed Place/Transition nets [3,4], we also will include time into our models and try to solve problems of scheduling and performance evaluation.
It is obvious, that most of the information provided by the modeling and simulation tool in Section 1 can be formally (automatically) transformed into the Pr/T model. Hence, an interface between the modeling and simulation tool and Design/CPN is needed to do this really automatically. This will also be a subject of further studies.
The basic ideas described here may lead to a computer-aided verification tool for recipes to ensure more effective and safer control procedures in the chemical industry.

## Acknowledgement

The work was partially supported by the Deutsche Forschungsgemeinschaft under grant Ha 1886/1-1 and by the Ministry of Science and Technology of the Province of Nordrhein-Westfalen.

## References

1. Engell S.; Wöllhaf, K.: Dynamic simulation of batch plants, Proc. European Symposium on Computer Aided Process Engineering-3 (ESCAPE-3), Graz, 439-444, (1993).
2. Genrich, H.J.; Shapiro, R.M.: Hazard Analysis. Technical Report, Meta Software Corporation (Nov. 1993).
3. Hanisch, H.-M.: Petri-Netze in der Verfahrenstechnik. Muenchen/Wien: Oldenbourg-Verlag (1992).
4. Hanisch, H.-M.: Analysis of Place/Transition Nets with Timed Arcs and Its Application to Batch Process Control. LNCS 691, Berlin: Springer-Verlag (1993).
5. Jensen, K.: Coloured Petri Nets. Basic Concepts, Analysis Methods and Practical Use. Volume 1, Basic Concepts. Berlin: Springer-Verlag (1992).
6. Mergen, R. J.: Batch Control Systems Standards. An Update on ISA/SP88. Wickliffe, Ohio 1990.
7. Meta Software Corporation: Design/CPN Reference Manual. Version 2.0 (1993).
8. Meta Software Corporation: The Design/CPN Occurrence Graph Analyzer. Version 0.3 (1992).

9. NAMUR-Empfehlung: Anforderungen an Systeme zur Rezeptfahrweise (Requirements for Batch Control Systems). NAMUR AK 2.3 Funktionen der Betriebs- und Produktionsleitebene (1992).

10. Uhlig, R.J.: Erstellen von Ablaufsteuerungen für Chargenprozesse mit wechselnden Rezepturen (Development of sequential controllers for batch processes with variable recipes). Automatisierungstechnische Praxis 29, 17-23 (1987).

# Towards Comprehensive Support for the Dynamic Analysis of Petri Net based Models

Rudolf K. Keller* Marianne Ozkan** Xijin Shen**

Centre de recherche informatique de Montréal (CRIM)
1801 McGill College, Suite 800, Montréal, PQ, Canada H3A 2N4
e-mail: {keller, ozkan, shen}@crim.ca; ftp-site: petri.crim.ca

**Abstract.** For dynamic analysis to be a powerful and convenient instrument in system modelling, comprehensive support is required. Dynamic analysis should be embedded in an appropriate modelling environment and be supported by adequate visualization mechanisms. Such an environment may support model consistency, selective information display, hypertext-based documentation, model substitution, and complexity management; key mechanisms include flexible and comprehensive graphical simulation, a visual front-end to performance analysis, and an integrated user interface for the various dynamic analysis resolution methods. These concepts are being validated in Macrotec, a toolset for business modelling which is based on Macronets, a variation of the Petri net formalism. Preliminary results from using Macrotec indicate that the described concepts substantially facilitate dynamic analysis and may be carried over to system modelling in other domains.
**Keywords.** Dynamic analysis, Petri net, Macronet, environment, visualization, graphical simulation, performance analysis, business modelling, tool evaluation.

## 1 Introduction

Modelling and model analysis have become an important aspect of system engineering. Particularly, dynamic analysis, which is the study of the behavioral aspects of systems, is playing an increasingly important role in many areas of system design. One such area is business modelling and reengineering, where the overall goal is the identification, design and evaluation of opportunities for business improvement.

The dynamic analysis of system models comprises tasks such as the analysis of behavioral characteristics (deadlock situations, critical paths, etc.), the answering of what-if questions, and the calculation of performance indices. Depending on the nature of the models and the intent of the analysis, dynamic analysis is carried out with various analytical and simulation methods.

Various methodologies and formalisms have been suggested to support system modelling and dynamic analysis. Among the most promising are approaches based on

This research is in part supported by research grants from NSERC, Canada, and by the IT Macroscope Project (managed by DMR Group Inc., Montréal).

* Author carried out part of this research as an adjunct professor at McGill University.

**Authors' contributions are part of their Master thesis research at McGill University.

Petri nets and generalizations thereof. These approaches excel in their descriptive power and allow for the combination of classical resolution methods and powerful modelling formalisms into a single framework [18]. We suggest that such frameworks, at the tool level, be comprehensively supported by an elaborate modelling environment and expressive visualization mechanisms, incorporating state-of-the-art graphics and modern animation techniques.

Graphical model editing is a well-developed discipline, and there are some systems that support high-quality graphical simulation (cf. [12]). However, these systems provide only limited functionality for many other aspects of dynamic analysis [5]. They typically suffer from poor user interfaces, lack visual support for analytical resolution methods, and do not provide sufficient support at the environment level. However, we contend that, if dynamic analysis is to play a prominent role in system modelling and prototyping, comprehensive support is required.

In a joint research project, we have developed, in cooperation with DMR Group Inc., a new approach for the architectural modelling and high-level requirements specification of business processes and information systems [2]. To support and validate our approach, we have engineered the Macrotec[1] toolset, a CASE tool which will eventually support all facets of our approach [16]. As an integral part of this effort, we have developed the environmental and visualization concepts described in this paper, using Macrotec as a testbed to validate our work.

In the remaining sections of this paper, we first relate business modelling to Petri nets, review the algorithmic and functional background of dynamic analysis, and discuss the environmental and visualization aspects of existing work. Then, we provide an overview of our modelling approach and of the Macrotec toolset. Next, we discuss the requirements for comprehensive support of dynamic analysis, both at the environment level and in terms of visualization, and detail our solution approach. Furthermore, we report on the design and implementation of the toolset and our experience at using it. Finally, we discuss its evaluation in the course of ECORP, an ongoing project at CRIM for evaluating Petri net tools, and present some plans for future work.

## 2 Background

In this section, we shall first discuss the relevance of the Petri net formalism for business modelling. Then, we shall outline major functional and algorithmic aspects of dynamic analysis, and review how dynamic analysis is supported in existing approaches.

### 2.1 Business Modelling using Petri Nets

Business modelling is a technique for specifying, analyzing, and reengineering an enterprise's infrastructure. Its overall goal is to identify, design, and evaluate value-adding opportunities for business improvement. To this end, business modelling approaches should support the design and analysis of an enterprise's architectural

[1] "Macrotec" is a contraction of the words *Macro*scope, our project's name, and archi*tec*ture.

structures and information technology components as well as the analysis of their dynamic aspects.

To meet these requirements, a formalism is needed which allows for the description of such structures and components in terms of the activities involved, time and cost, required resources, created products, etc. Furthermore, such a formalism should be based on data flow rather than event flow, since in the business domain, data are generally more stable than the processes that act upon them.

The Petri net formalism lends itself to business modelling, given its descriptive power and flexibility [23]. Colored Petri nets, for instance, allow for the direct mapping of hierarchical system entities, i.e., data, into models, through instantiation of resource and product classes. Furthermore, stochastic and deterministic Petri nets enable the user to characterize the time duration of activities via exponentially distributed functions and fixed values, respectively. Also, since Petri net based models may be hierarchical, they are fit to capture typical, hierarchical business structures and reflect the frequently used top-down development methods.

The flexibility of Petri nets and their underlying resolution techniques make them tailorable to specific business behavior. For instance, model components may be characterized by functions and constraints, such as cost functions over activities and quantity restrictions over entities. This leads to detailed behavioral descriptions which can be analyzed according to various parameters.

### 2.2 Dynamic Analysis and its Visualization

Dynamic analysis (DA) should encompass logical and structural assessment, as well as analytical performance evaluation. Specifically, it should provide support for the analysis of causal relationships between processes, for the investigation of undesirable system properties such as deadlocks, and for the automatic generation of quantifiable timed performance measures.

Typical Petri net analysis methods include graphical simulation, performance analysis, and sensitivity analysis [26]. Given our research and project context, we gave priority to graphical simulation and performance analysis and developed environmental and visualization concepts to tailor and render their results.

Graphical simulation (GS)[2] is the direct execution of a model as it is virtually "brought to life." It promotes structural and behavioral assessment, providing an intuitive feel for the workings of the model and yielding quantitative results based on the simulation interval. Standard simulation ("forward simulation") may be carried out in the opposite direction ("backward simulation"). Given a forward simulation, subsequent backward simulation need *not* reproduce the markings of the forward simulation[16]; rather both, forward and backward simulation, should be reproducible in their own right and may be run in reverse order ("replay simulation").

Performance analysis (PA), on the other hand, addresses the quantitative aspect, for instance time, cost and other target functions, of a model's performance. It is carried out with analytical or simulation methods based on Markovian analysis of the model's reachability set, that is the set of all markings that can be reached from an initial mark-

[2] Note that what we call *graphical* simulation is sometimes referred to as *animation.*

ing by means of a sequence of transition firings.

In developing a system architecture for the visualization of DA, it is expedient to leverage off recent work in algorithm animation [3, 12]. One might then be tempted to try and reconcile GS and PA resolution methods by generating GS results via PA algorithm animation. This approach may work for small models; in general, however, the sheer size of models would inacceptably slow down GS and worse, because of combinatorial explosion, storage capacity would not suffice. Moreover, GS would fail in the cases where PA failed to provide a solution. Accordingly, Jensen reports that "for the moment it is, however, only possible to construct occurrence graphs for relatively small systems and for selected parts of large systems" [13], and Marsan calls for resorting to simulation techniques[18]. Hence, for practical and theoretical reasons, DA comprising both GS and PA has to cope with at least two complementary algorithms and their respective visualization.

Since DA resolution methods apply to formal models, system modelling should be considered as the preliminary stage to system analysis. Accordingly, the prerequisite for comprehensive support of DA is an adequate modelling environment. Modelling environments will be discussed in section 4; visualization support for DA will be the topic of section 5.

### 2.3 Related Work

Dozens of packages and tools making use of Petri net formalisms have been reported [6, 5]. From our investigations we conclude that none of them fully support PA and GS, nor do they further their integration. Below, we briefly discuss the five systems that have proven most influential for our work, be it for their high quality GS or their rich PA functionality

*PACE* [10, 20] is a modelling system that supports hierarchical Colored Petri nets and is integrated into the Objectworks/Smalltalk environment. It provides a graphical editor with good support for editing model hierarchies. The GS component of PACE is fairly elaborate, allowing for smooth GS, setting breakpoints based on the global clock and model topology, running GS in the background for obtaining statistical results quickly, and replaying previous simulations.

*Design/CPN* [19] makes use of Colored Petri nets and supports hierarchical models and temporal specifications. Its graphical editor allows for semantic checks and for adding comments to the models. Design/CPN supports GS as well as the detection of potentially unsafe situations. Various predefined functions are available for computing performance-related measures by means of user-defined code segments, however analytical resolution methods are not supported. In [13], an occurrence graph tool is announced. This tool would support the calculation, analysis and interactive inspection of occurrence graphs and would be synchronized with GS. To our knowledge, the tool has not, to date, been released.

*GreatSPN* [1] is based on Generalized Stochastic Petri nets and also supports deterministic timed transitions. The package includes a simple graphical editor for model specification. It supports exact analytical solutions as well as Monte Carlo simulation to derive performance results displayable on the edited model via histogram representation. GS is not supported, nor can the user define his or her own result functions.

The *SPNP* [26] package is a system for the solution of Stochastic Reward nets, i.e., Petri nets whose states and transitions may have reward attributes. SPNP allows for resolution by simulation, for logical assertion of the model, and for sensitivity analysis. A graphical user interface is not provided with the package.

*CPN/AMI* [9] is a generic modelling environment whose central underlying formalism are Colored Petri nets. *Macao*, its user interface component, supports graph editing and user interaction with the services provided by the various underlying CPN/AMI tools. GS is supported and can be carried out in different modes, such as automatic, subnet, step by step and back step. The user is allowed to define and associate an observation net to a given simulated net, and transitions can be complemented with scripts. Analysis features focus on linear invariants calculus, reduction methods, and flow analysis. PA *per se* is not supported.

There are various modelling tools based on formalisms other than Petri nets. Some of them exhibit advanced GS and analysis features, e.g., *STATEMATE* [11] and *Time-Bench* [4]. In our work, we have tried to carry over the relevant environmental and visualization concepts found in these tools.

## 3 Macrotec

The elaboration and validation of our environmental and visual concepts have been closely intertwined with the development of Macrotec, our toolset for business modelling. These concepts have been and are being implemented in Macrotec, eventually leading to a prototype modelling environment that comprehensively supports dynamic analysis. In this section, we shall first describe our modelling approach and its underlying formalism, the so-called *Macronets*.[3] We shall then provide an overview of the functional components of Macrotec and their architectural assembly.

### 3.1 Modelling Approach

Our modelling approach [2] combines several concepts that have originally been developed in separate contexts, namely, entity-relationship modelling, specialization and inheritance in the sense of object-oriented languages, event analysis, and analysis of data flow as well as resource utilization. We have integrated these concepts into a uniform modelling framework, whose semantics has been defined through the formalism of Petri nets. In what follows, we describe this framework.

The generic Macronet model is shown in Figure 1. An *action*, represented by a rectangle, is a unit of work which produces a change in the system's state. An *entity* represents a resource, for example a human actor required in the accomplishment of an action or a product created by an action. A *place*, represented by an oval, is a container of entities. In object-oriented terminology, a place defines the base class for a collection of entities, much like Colored Petri net places contain entities belonging to the same type. Furthermore, all the actions and places of a particular model may be specified as class instances (with class-specific attributes) in an overall class hierarchy of actions and places, respectively.

[3] "Macronet" is a contraction of the words *Macro*scope, our project's name, and Petri *net*.

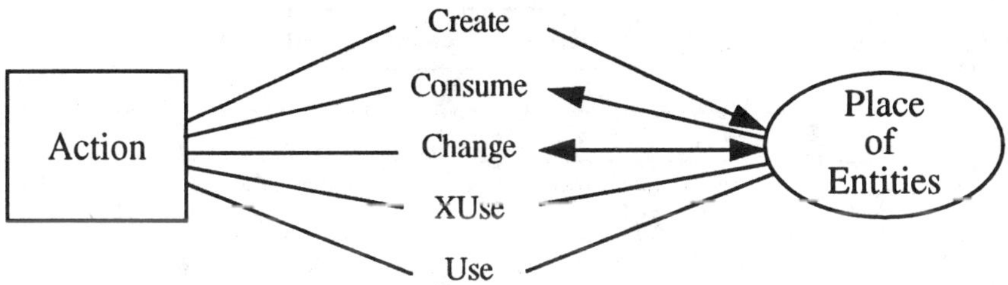


Figure 1: Generic model for Macronets.

A relation connects an action to a place and dictates the action's behavior in respect to that place. Five classes of relations are available, namely, *Create, Consume, Change, Use, and XUse*. They force, respectively, entity creation, deletion, alteration, use, and exclusive use by the action. A Use relation indicates that the given action uses a number of entities in the place without modifying their state. In case several actions are connected to the place, they may use the same entities. As long as an entity is used by one or several actions, its state cannot be changed. The XUse relation has the same semantics as the Use relation, except that one single entity may only be used by one action at a time.

Relations may be directly mapped into Petri net arcs, except for the Use relation which has no Petri net equivalent [2]. A multiplicity attribute is common to each class of relations and determines the number of entities involved in a state change. Action behavior may be further refined by constraining, via required attribute values, the entity instances that may be involved in the action. This "action protocol" enforced by relations promotes ease-of-use through its simplicity, yet is expressive enough to adequately constrain behavior.

Macronets support timed and immediate actions, i.e., actions whose time durations are zero. Timed actions can be described through random, exponentially distributed functions or fixed values. In this respect, Macronets are a combination of Generalized Stochastic Petri nets and deterministic nets. Actions may be further characterized by a priority which is taken into account in case of firing conflicts.

Macronets are hierarchical. An action or a place or a subnet thereof may be refined by a submodel, and consistency may be preserved throughout hierarchical levels.

Macronets may be evaluated by any resolution technique available to Petri nets. Note that we developed our own GS algorithm, which supports the three above-mentioned types of action durations. We use, however, the SPNP package resolution techniques for PA. This is a reasonable approach, given the fact that Macronets may be automatically translated into Petri nets without loss of important information [2]. Moreover, SPNP, which supports reward nets, gives Macrotec users access to tailorable result functions.

Macronets are a powerful formalism for addressing the specifics of business modelling. For one thing, they explicitly constrain action behavior to an expressive set of operations which correspond to atomic business activities. Furthermore, since they encompass hierarchical data and models, they allow, respectively, for the smooth mapping of business objects together with their mutual dependencies, and for the seamless rendering of business architectures that have been developed in a top-down approach.

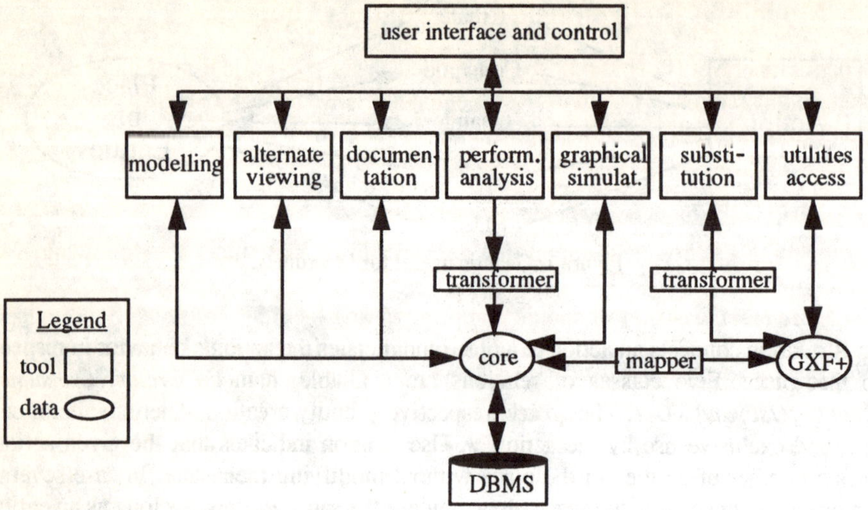


Figure 2: Macrotec toolset: architecture overview.

### 3.2 Toolset

To support and validate our approach, we have engineered the Macrotec toolset. Its overall architecture is depicted in Figure 2 (for details see 16). The toolset comprises a *modelling* tool for the graphical editing and validation of models complemented with facilities for automatic graph layout; an *alternate viewing* tool permitting user-defined iconic representations of actions and places (the building blocks of our models), as well as event and entity-flow representations; an annotation tool for hypertext-based *documentation*; a *PA* tool based on SPNP [26]; a *GS* tool; a *substitution* tool supporting multi-level modelling; and, finally, *utilities* and service components for automatic graphical layout and database access. All information to and from the tools is managed by the *core* representation, which can be mapped into the *GXF+* graphical exchange format and which interacts with the underlying *database management system.*

One typical application domain of Macrotec is the modelling of enterprise information systems for prescriptive usage. Figure 3, for example, depicts the model of a simple product delivery system. Once a product has been ordered, a shipment request may be issued, and the shipment may be prepared and scheduled. The delivery services then sends a truck with the product to the final destination, provided a truck is available. Figure 3 shows the model state after loading a user-specified marking and set of dynamic attributes. The user may edit and refine the model, adopting a top-down or bottom-up approach by respectively decomposing or abstracting parts of the model. The built-in validation component of the modelling tool ensures consistency between the different levels of the model.

## 4 Modelling Environment for Dynamic Analysis

Dynamic analysis (DA) should be embedded in an appropriate modelling environment. In this section, we shall discuss the major requirements for such an environment and detail our solution approach.

### 4.1 Requirements

A modelling environment supporting DA should at least meet the following five requirements:

*(1) Model consistency.* It should comprise editing facilities enforcing the semantic constraints between a model's building blocks. Furthermore, it should be complemented with an overall model validation scheme. Consistency checks may thus ensure input correctness to the DA resolution methods.

*(2) Selective display.* During modelling and analysis, it is desirable to have the attributes of the different model components displayed directly on the screen. If this is done without any filtering, the screen may get cluttered and the user be overwhelmed with information. Selective display mechanisms may help.

*(3) Hypertext-based documentation.* Such an environment should exhibit hypertext-based documentation facilities that allow for annotating model components with informal data as well as for creating links between components. Particularly, these facilities may serve as an informal DA documentation tool.

*(4) Model substitution.* Facilities for model decomposition and abstraction are needed. The environment should support multi-level modelling by providing appropriate substitution mechanisms. DA may then be run in a hierarchical way, be it at the various abstraction levels at once or simply at one particular level.

*(5) Coping with complexity.* The environment should provide mechanisms for coping with the complexity of models. The user should be able to reduce large models into more comprehensible ones that preserve the DA behavior of the original models.

### 4.2 Our Approach

With the Macrotec toolset, we are attempting to provide an environment meeting the above-mentioned requirements. In what follows, we describe some of the most salient Macrotec features, for details refer to [24] and [15].

To ensure *model consistency* (requirement *(1)*), the modelling tool in Macrotec supports graphical editing that enforces the semantic constraints imposed by the Macronet formalism. Figure 3 shows the base window through which the user has access to the full functionality of Macrotec. The user may select an icon from the palette in order to edit in the drawing text, actions, places, or relations. The classes of actions and places together with their attribute values may be specified via dialog boxes. The displayed icons are class-specific and thus visually indicate the class of a particular model component; the user can specify them via *class-icon maps* [24]. Through ongoing consistency checks during editing, the user is prevented from creating invalid models.

*Selective display* mechanisms (requirement *(2)*) are pervasively available in Mac-

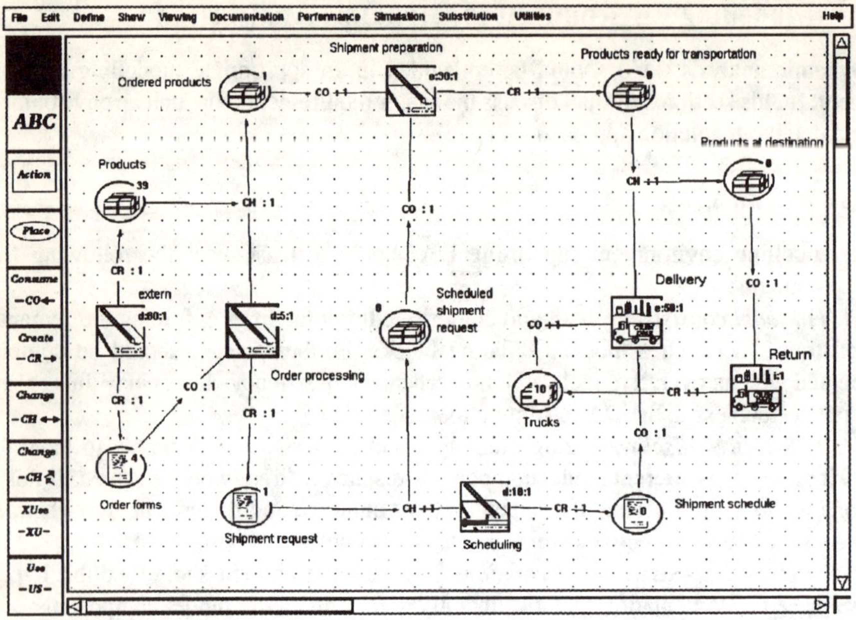


Figure 3: Sample model *Delivery System* during graphical simulation.

rotec, allowing for turning on/off the display of the attributes of the various model components, including their text labels, static attributes, attributes for DA (cf. Figure 6), results of PA (see Figure 7), class icons (turned on in Figure 3, turned off in Figure 4), and annotation icons (indicating the existence of component annotations). These mechanisms allow the user to focus on the data that are most relevant for a particular modelling activity.

*Hypertext-based documentation* (requirement *(3)*) is being addressed by *Hypertec,* Macrotec's built-in documentation tool (cf. Figure 2). Hypertec allows for the informal documentation of models via structured annotation, and for model navigation via hypertext links. The links can either be derived from the underlying model (*system links*) or be introduced by the user (*user-defined links*). Hypertec enables the user to add documentation to DA, for instance, the interpretation of DA results, informal hints for carrying out certain DA tasks, etc. To this end, the user may, for instance, link model parts that share common or correlated DA behavior and add appropriate annotations. Hypertec is described in detail in [25].

Facilitating *model substitution* (requirement *(4)*), Macrotec supports both manual and, to some extent, automatic substitution (cf. Figure 2). Users may define a set of components to be a substitution group and associate to the group a substitution model, which can be defined in a separate window and subsequently swapped with the original group. Throughout these activities, Macrotec's validation scheme ensures model consistency. Automatic substitution in its current form allows for simple decomposition of models into a set of subordinate models [2].

*Coping with complexity* (requirement *(5)*) is probably the most challenging of the requirements. It is being attacked, as for now, by Macrotec's entity flow view (cf. Fig-

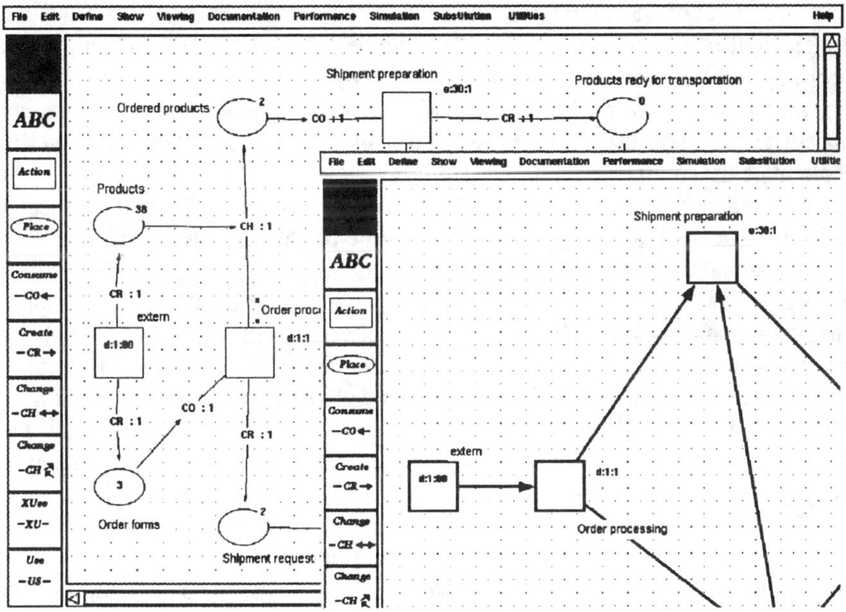


Figure 4: Macrotec's entity flow view: window at bottom right displaying entity flow view derived from model displayed in background window.

ure 2, *alternate viewing*). In this view, the flow of entities through the model is automatically highlighted, and intermediate places are eliminated. For instance, in Figure 4, the window at the bottom right displays the entity flow view which is derived from the model displayed in the background window. The algorithm for view generation is based on a set of substitution and elimination rules and is described in [24]. A generated entity flow can be interpreted as a subset of an untimed GS *trace*, showing a set of potential paths between the different model components. The analysis of entity flows may help determine, for instance, the shortest path between two model components. Via the entity flow view, Macrotec may provide information statically, which otherwise would have to be obtained by DA. In this way, the need for DA, potentially a time-consuming activity, can be reduced.

# 5 Visualization Support for Dynamic Analysis

For comprehensive support of DA, the modelling environment discussed in the previous section should be complemented with adequate visualization mechanisms. In this section, we shall discuss the requirements for DA visualization and present our solution approach.

## 5.1 Requirements

To support DA visualization, we have identified the following six requirements:

*(1) Tight integration of GS and PA.* Whereas traditional DA approaches clearly

separate the usage of GS and PA, we call for their complementary application and suggest their tight integration. In this way, DA can be carried out more efficiently. DA, comprising both GS and PA, must be supported by a comprehensive user interface with features encompassing DA access and control, as well as result definition and display. The access to the GS and PA engines should be centralized, thus favoring their alternate usage.

*(2) Customization of GS.* Users should be able to customize GS to their specific needs. For instance, it should be possible to define or choose from alternative result presentations and to specify the details of GS rendering. The prerequisite for customizability at the user level is a flexible composition scheme for GS design. For displaying basic simulation results, GS may be based on visual building blocks such as the simulation global time, the state of an action (inactive, enabled, firable), the duration of an action and the time left before it fires, as well as the representation of tokens involved in a firing. The building blocks may then be combined to produce snapshots of model behavior. In turn, consecutive snapshots may be assembled into animation scenes [3], allowing for the visualization of system behavior and GS.

*(3) Tight integration of GS modes.* The various simulation modes found in systems such as Macrotec call for their tight integration at the user interface level. The user should be able to switch easily between them and have access to context information, i.e., the history of GS.

*(4) Multiple views for GS.* Model behavior should be visualized from different perspectives. GS might thus be carried out simultaneously in multiple, complementary and synchronized views.

*(5) Flexible definition of PA results.* Visual support for PA should include facilities for result definition. The user should be able to graphically specify which results for which model elements and parts should be calculated. Furthermore, it should be possible to specify new result functions. For instance, given the basic indices automatically generated by PA (for example, the throughput of an action or the average number of entities in a place), the user may combine them into higher-level indices. The latter may more closely exhibit the performance characteristics of the model under consideration and thus facilitate PA interpretation.

*(6) Flexible presentation of PA results.* PA results should not only be available in result files, but also directly on the model under consideration. They may, at the user's discretion and depending on their nature (values, functions), be displayed either as numerical values or in graphical form. To avoid screen cluttering, selective display mechanisms are required (see section 4).

### 5.2 Our Approach

When developing the Macrotec toolset, we have tried to incorporate visualization support that responds to the discussed requirements. The discussion of our approach (for details refer to [21] and [15]) is organized into four subsections. First, the visualization of integrated DA is addressed, then more specifically, the visualization of GS and of PA are discussed, and finally, a scenario of DA in Macrotec is presented.

**Visualization of Integrated Dynamic Analysis in Macrotec.** Macrotec supports the

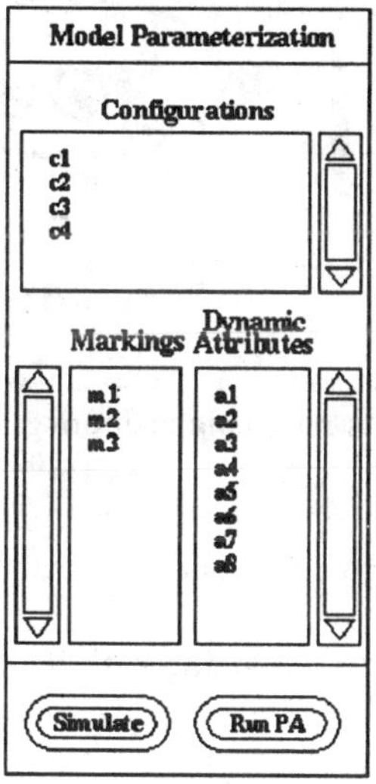


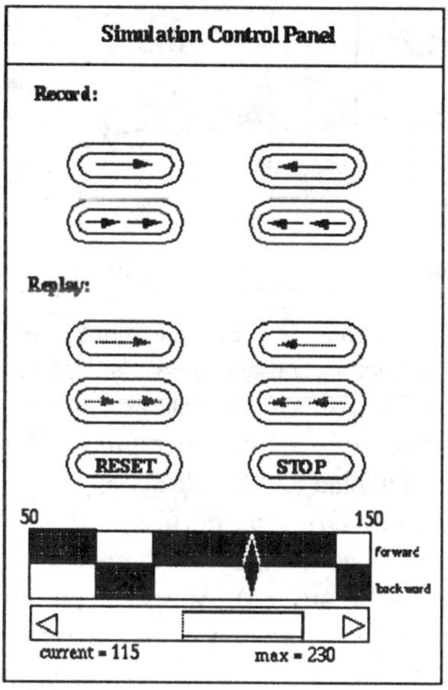


Figure 5: Model parameterization for integrated dynamic analysis (left).Compact disc player metaphor for control of graphical simulation (right).

*tight integration of GS and PA* (requirement *(1)*) by centralizing the access to GS and PA via the *Model Parameterization* dialog box, shown in Figure 5 (left). The dialog box contains three distinct areas. The top area displays the list of static configurations belonging to the model, i.e., the different versions. The bottom left area shows, for a particular configuration, the list of initial markings, and the bottom right area, for the same configuration, the list of dynamic attributes settings. These three elements of model parameterization are user-defined. Combined together, they fully describe a model and constitute the input for the DA engines.

The Model Parameterization box promotes alternation between GS and PA. Furthermore, it facilitates model experimentation, letting the user easily compose alternate parameterizations. It also supports the reuse of DA results, since it makes the markings generated and stored from previous DA readily available (provided there are adequate underlying storage facilities, which is the case in Macrotec).

**Visualization of Graphical Simulation in Macrotec.** In Macrotec, *customization of GS* (requirement *(2)*) is currently addressed at the developer level. We provide a rich visualization catalog of atomic objects and operations [21], from which GS may incrementally be composed. In the following, we shall discuss a few compositions which were assembled from the catalog of elements.

Figure 6 illustrates four compositions. Composition a) represents an action as a *rectangle* object and its priority and duration as *text* objects. The duration is composed of a function type and its parameter. In the figure, the function type is deterministic (*d*), its parameter equals *3* time units, and the action has a priority of *1*.

Visual feedback of the state of an action (inactive, enabled, or firable) is achieved

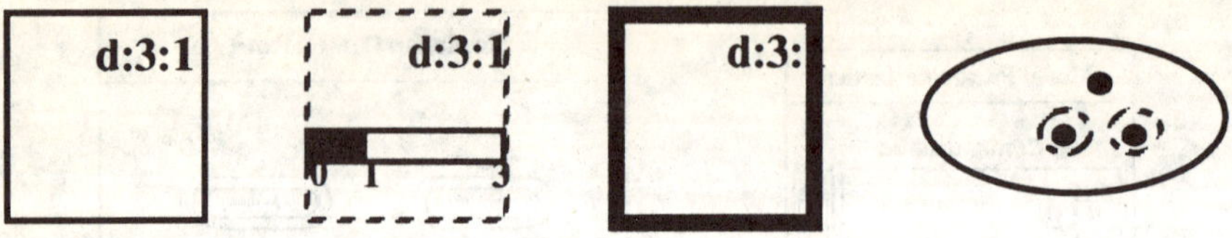


(a) (b) (c) (d)

Figure 6: Four compositions built from Macrotec's visualization catalog: a) action with deterministic duration, b) enabled action with one unit remaining until firing, c) firable action, d) two active entities.

by a number of catalog operations. These include *dotpattern* and *highlight* for expressing enabled and firable states, respectively (Figure 6 b) and c); see also Figure 3, firable action *Order processing*).

An enabled action may be characterized by the time remaining until it completes execution. In Figure 6 b), we use a continuous feedback element of the visualization catalog called *slidebar* to display the time until the action fires. During execution, the slidebar continually decreases its display value, showing the time that has elapsed from the action's total duration, until the remaining time is zero and the action becomes firable.

Entities may be represented by *circles* and places by *ovals* (Figure 6 d)). Entities involved in the execution of an action are distinguished from inactive entities. In Figure 6 d), we are using *drawhandles* to represent active entities. The actual completion of an action's execution may trigger a *path* operation, a composition of numerous *moves*, which shows the trajectory of entities from input to output places. This operation proves essential in complex models, where many firable actions may share one single input place. The actual values of entities are accessed through hypertext techniques.

The visualization catalog provides flexibility of presentation. Alternative operations and objects may be used to produce a different visual feedback of some model behavior or representation. For example, for action duration functions, we can provide quantitative and qualitative feedback by representing them, respectively, by text objects on some rectangle object, or by operations acting on a rectangle object. Such a operation is, for instance, *collapse*, which reduces the rectangle dimensions according to a specified reference function. A second example is the different representation of the number of entities, as dots in Figure 6 d), and as digits in Figure 3.

Due to the fine granularity of atomic elements of the catalog, the GS display is highly tailorable. *Show* and *hide* are available object operations which enable selective display, helping users focus on particular behavioral aspects of the model.

For portability reasons, the Macrotec catalog does not support the use of colors. The presented catalog and composition approach are aimed at the developer of GS. Furthermore, we believe that it will be an excellent platform for implementation of a *GS-builder* tool that will allow users to customize their GS.

To achieve *tight integration of GS modes* (requirement *(3)*), we have developed the compact-disc player metaphor as the basis of GS execution control. GS methods are accessible through a *Simulation Control Panel* (Figure 5 (right)) whose design was inspired by the user interface of compact-disc players.

The control panel contains two sets of buttons to trigger GS methods, the *Record* buttons and the *Replay* buttons (Figure 5 (right)).

Pushing one of the Record buttons triggers forward and backward GS respectively (represented by the direction of the arrows), and the production of new markings. Forward and backward GS may be recorded in step-by-step or in continuous fashion. Step-by-step GS terminates execution after the generation of a single marking, whereas the latter executes until a specified number of markings has been produced. Step-by-step and continuous GS are represented by single and double arrows respectively. The marking sequence produced by step-by-step GS may, in case of conflicts, be controlled via manual selection of firable actions. For continuous GS, the system resolves conflicts randomly, based on user-defined priorities.

The Replay buttons allow for the display of the model states that have previously been generated by Record methods. The history of recorded markings may be played forward in order to advance to more recent markings, or backward to return to more remote markings. Moreover, replays, in both directions, may be step-by-step or continuous, with the same effect as the corresponding recording commands.

The two buttons at the bottom of the Simulation Control Panel trigger Stop and Reset methods. The effect of the Stop method is to quit GS or to force premature termination of continuous replay or recording. The Reset method allows for the reinitialization of the model to its original state.

Other than providing execution control over GS, the Simulation Control Panel also gives context information about the current GS session and allows for direct access to recorded markings. These additional features are provided by the *history view*, the bottom part of the panel. The panel shows the progress of GS recording through incremental updates of the history view such that each marking is represented by a thin shaded bar. The width of this bar is, in the case of timed GS, proportional to the time elapsed since the preceding marking; for untimed GS, it takes on a default value. The history view is ordered by marking sequences, that is, consecutive markings are represented by adjacent bars. Markings recorded by forward GS are shown in the upper part of the view whereas backward recordings are displayed in the bottom part. The history view can show up to 100 markings at once, and its overall size is limited only by storage capacity. In Figure 5 (right), the current marking is *115*, which was produced by forward GS, and the total number of generated markings is *230*, implying that 230 markings have been recorded and that the current marking has been reached by replay GS. The cursor of the history view (black diamond in Figure 5 (right)) can be moved by the user and provides direct access to the recorded markings. This facility is suited for coarse-grain positioning ("jumps through history") as opposed to the replay functions which allow for navigation to adjacent markings. Note that, once the user starts recording, all the previously recorded markings that are more recent than the current marking are erased.

The history view is an example of a GS view that complements Macrotec's base GS view (cf. Figure 3). The two views are synchronized in that the history view's current marking always corresponds to the marking which is displayed in the base view.

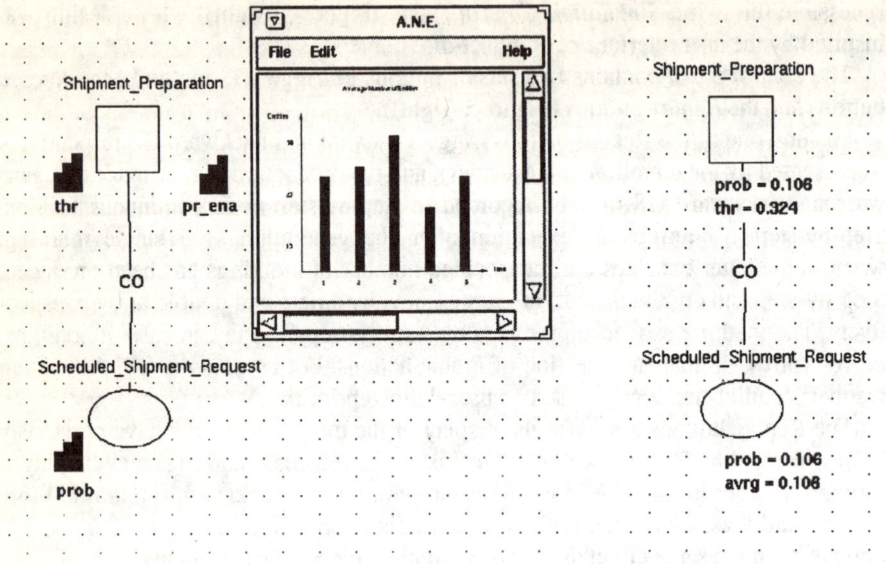


Figure 7: Example of PA result presentations: transient (left), steady-state (right).

The history/base view pair is thus a first step towards *multiple views for GS* (requirement *(4)*). Currently, further alternative GS views are not provided yet in Macrotec.

**Visualization of Performance Analysis in Macrotec.** Macrotec provides a rich set of PA results which belong to one of three classes, according to the model component to which they apply. *Flexible definition of PA results* (requirement *(5)*) is addressed in that the user can specify via direct manipulation the model components whose PA results should be displayed.

The first class applies to individual places and actions. They are the average number of entities in a place, the percentage utilization of a place, the throughput of an action, and finally, an action's probability of firing. The second class of results apply to groups of places and groups of actions. The indices computed over individual components can be computed over groups of components as well. Users may select components in the model and store the resulting group under a specific label. The third class of PA results apply to markings. They include the determination of whether or not a marking is reachable and if it is, the time required to return to it. Markings of interest are defined by entering the number of entities into the model places and storing them under a label.

Flexible PA result definition encompasses the support of complex equations. To our knowledge, all existing tools that support the equating of performance indices [26] require textual intervention by means of a programming language such as C. Visual programming techniques might support the user in combining primary indices into coherent equations. Due to lack of resources, we are only providing uncostly definition facilities, having users enter equations textually.

In coping with *flexible presentation of PA results* (requirement *(6)*), Macrotec makes PA results available in alternative representations that vary from textual displays to histograms and result icons. In Figure 7 (left), for example, the availability of results from a transient PA are indicated by icons. After selecting the *A.N.E.* (average number of entities) icon of place *Scheduled_Shipment_Request*, the icon is replaced by a pop-up window that displays the result data in histogram form. In Figure 7 (right), the results from a steady-state PA are shown in textual form. At any time, the user can hide all or part of the displayed results (selective display).

The second class of results, group results, can be displayed in pop-up windows. During the display of a pop-up window, the components of the group under consideration are highlighted. In order to identify the group, the user has to select the group label from the list of defined groups. Finally, the third class of results, results over markings, are also displayed in pop-up windows.

**Scenario of Dynamic Analysis in Macrotec.** To close this section, we present a short scenario where GS and PA come into play. The user performs a transient analysis on a model such as the one shown in Figure 3. The user wishes to test an undesirable marking, that of having 0 trucks (place *Trucks;* in the Figure, however, we do have 10 trucks) while at least one entity is present in each input place of the action *Delivery.* The PA results indicate that this marking is in fact reachable. From the reachability set, the user selects the target marking, gives it a label, and saves it. The marking can then be retrieved and used as an input to the GS engine. Running backward GS in step-by-step fashion will serve to trace the states leading to the undesirable marking. Model parameters may be modified, saved and tested through PA, followed by further backward GS runs. PA and GS may be repeated on varying models, and results may be saved and compared.

# 6 Engineering, Evaluation and Future Work

Below, we shall first report on the design and implementation of the Macrotec toolset and our experience in using it. Then, we shall discuss its evaluation in the course of ECORP, an ongoing evaluation project at CRIM, and, finally, our plans for future work.

## 6.1 Design, Implementation, and Experience

Macrotec is a Unix-based system that was mostly developed in C++ (minor parts were written in C), running under SunWindows, NeWS, and the X11 window system. Its user interface has been implemented with the *ET++* application framework [27]; as database we are using Gemstone. As its PA tool we chose and integrated the SPNP package [26]; GS was implemented from scratch. We recently completed version 2.0, its second stable version. We invested around six person years in its development.

Algorithm animation systems such as Artkit [12] inspired our development of GS and our choice of ET++. ET++ is a powerful object-oriented class library integrating user interface building blocks with high-level application framework components. In many ways, ET++'s internal schemes for managing graphical applications are similar to those of widely-used algorithm animation systems based on Smalltalk's MVC paradigm [17].

With the version 1.0 prototype, completed in March 1993, we modelled some medium-sized practical examples, including an outsourcing scenario for one of our sponsor's clients. According to our original plans, we then decided to focus on improving the environmental and visualization aspects of the prototype. The development of version 2.0 proved not only the extensibility of the original design, but also the applicability of our concepts to variants of our modelling approach such as alternate viewing (see Figure 2), and to other domains [14, 8]. Not surprisingly, we obtained at recent demonstrations and from personal communication numerous suggestions for functional improvements, for instance, code generation when modelling system designs, and integration of business rules for financial modelling.

### 6.2 Evaluating Macrotec with the ECORP Approach

The increasing number of Petri net tools calls for comparative evaluation, so that both users and developers may make more informed choices and leverage off existing work. Inspired by previous project experience at CRIM in the domain of computerized control and by the needs of the Macrotec project, our colleagues and the first author have launched the ECORP project (Etude Comparative des Outils basés sur les Réseaux de Petri).

The major result of this project is *ECORP*, a new, comprehensive approach for evaluating such tools, addressing technical, managerial as well as usage issues. At the core of ECORP is a questionnaire defining an extensive list of fine-grained yes/no questions together with comment fields. The questionnaire aims at covering all technical and commercial aspects of Petri net tools including issues such as maintenance, documentation and training. In this way, we hope to address both the research community and industry, which is, given CRIM's mission [7], a prime audience of the project. Specifically, The questionnaire consists of 72 entries, addressing tool identification, general information (covering all non-technical aspects), functionality, and performance. The leftmost column of Table 1, for instance, lists an excerpt of the criteria on dynamic analysis features as found in the part on tool functionality.

The evaluation questionnaire is complemented with two reference examples. These examples are based on our previous experience with CASE tools and CRIM projects, addressing Petri net modelling and analysis of the loan processing activities in a bank branch and the computerized control of the doors of a two car subway train. Although this choice of examples is somewhat arbitrary, we contend that they span a sufficiently broad application area of software development to yield pertinent results. ECORP evaluators filling out the questionnaire are required to implement these examples with their respective tools. In this way, they are forced to actually learn the tools and thus gain the knowledge to give informed answers. Furthermore, by using identical computing platforms at CRIM (Sun workstations) for all the tool evaluations, we are able to make some quantitative comparisons.

To date, we have evaluated two commercial tools, PACE [10] and Design/CPN [19], and two research prototypes, Voltaire [22] and Macrotec (cf. Table 1; for a complete discussion, see [5]). As a side product and service of the ECORP project, we have installed at CRIM a Petri net tool server for anonymous ftp (*petri@crim.ca*). It currently contains twenty tool descriptions, providing a new, computerized, and more comprehensive version of the tool overviews found in the literature [6]. Over the past six months, we registered more than 300 external server accesses.

| Criteria | PACE | Design/CPN | Voltairc | Macrotec |
|---|---|---|---|---|
| Forward Simul. | Yes. | Yes. | Yes. | Yes. |
| Backward Simulation. | No, but forw. sim. may be rolled back. | No. | No. | Yes, step-by-step only. |
| Control over simulation (step-by-step; automatic with/ without visualization). | Yes. Simulation may be step-by-step or continuous, with or without graphical simulation (animation). | Yes. Simulation may be step-by-step or continuous. Simulation is graphical, except when running SAC. | No. Simulation is automatic, without visualization. | Yes. Simulation may be step-by-step or continuous. Simulation is always graphical. |
| Fine-grained interactive control. | Yes. Simulation may be interrupted at any time, either manually or via breakpoints. | Yes. Simul. may be interrupted at any time except for SAC (Super Automatic Code). | Yes. Simulation may be interrupted at any time via breakpoints. | Yes. Simulation may be manually interrupted at any time. |
| Possibility of breakpoints. | Yes. either associated with places or via global time. | Yes, by giving the number of steps to be executed. | Yes, via global time. | Yes, by giving the number of steps to be executed. |
| ... | ... | ... | ... | ... |

Table 1: ECORP questionnaire, *Dynamic Analysis*: selected criteria with results [5].

The Macrotec and ECORP projects have been quite synergistic. ECORP has benefitted from preliminary evaluation work we conducted for Macrotec. In turn, the elaboration of the ECORP approach and the screening of available tools have given invaluable input to the Macrotec project team. Not surprisingly for this kind of self-evaluation, Macrotec did very well in the ECORP evaluation, especially in respect to the first reference example and environment and visualization aspects.

### 6.3 Future Plans

In the near future, we would like to make some extensions to the version 2.0 prototype. Furthermore, we intend to apply and adapt our approach to domains other than business modelling, such as organizational reengineering and process engineering. The ambitious goal of comprehensive dynamic analysis support, however, entails work that goes beyond the Macrotec 2.0 prototype. Among the activities which have yet to commence is the visualization of sensitivity analysis as for instance offered by the SPNP package. Another activity we will be pursuing is the graphical definition of result functions. This could eventually lead to a visual programming approach. We will also be promoting "static GS" in the sense of Macrotec's entity flows, by experimenting with different sets of rules and visual representations, and by leveraging off existing work, e.g., [9]. Finally, we would like to provide support for the interpretation of analysis results and for the handling of evolving models.

## 7 Conclusion

To comprehensively support the dynamic analysis of Petri net based models, we have identified a series of requirements and concepts, focusing on environmental and visualization aspects. Macrotec, a toolset for business modelling based on Macronets, a variation of the Petri net formalism, is our attempt towards meeting these requirements. Our experience with Macrotec indicates that our concepts indeed substantially facilitate DA, and that they are applicable to domains other than business modelling. It is our contention that a system such as Macrotec is an important step towards the ambitious goal of comprehensive support for dynamic analysis.

## Acknowledgement

In addition to the authors, the Macrotec system in its current form has been designed and implemented by Richard Lajoie, Fayez Saba, Tao Tao, Nathalie Rico, and Mario Marquis. We appreciate the contributions of our other colleagues in the IT Macroscope Project, Gregor v. Bochmann and Rachida Dssouli. The ECORP evaluation project is being carried out by Anne-Claire Debaque, Paul Freedman, Jean-Michel Goutal, Michel Levy, Fayez Saba, and the first author, with assistance from Pierre-Yves Gagné and Ghassan Youssef.

## References

1. Gianfranco Balbo and Giovanni Chiola. *GSPN 1.6 manual.* University of Torino, Torino, Italy, June 1993.
2. G. v. Bochmann, A. Debaque, R. Dssouli, A. Jaoua, R. Keller, N. Rico, and F. Saba. A method for architectural modelling and dynamic analysis of information systems and business processes. Technical Report CRIM-92/12/10, Centre de recherche informatique de Montréal (CRIM), Montreal, December 1992.
3. Mark H. Brown. Perspectives on algorithm animation. *Proceedings of the Conference on Human Factors in Computing Systems*, pages 33–38, 1988.
4. R.J.A. Buhr, G.M. Karam, C.M. Woodside, R. Casselman, G. Franks, H. Scott, and D. Bailey. TimeBench: A CAD tool for real-time system design. In *Proceedings of the Second International Symposium on Environments and Tools for Ada (SETA2)*, Washington, D.C., January 1992.
5. Anne-Claire Debaque, Paul Freedman, Jean-Michel Goutal, Rudolf K. Keller, Michel Levy, and Fayez Saba. The ECORP approach to Petri net tool evaluation. Technical report, Centre de recherche informatique de Montréal (CRIM), Montreal, Canada, March 1994.
6. Fritz Feldbrugge. Petri net tool overview 1992. In Grzegorz Rozenberg, editor, *Advances in Petri Nets 1993*, pages 169–209. Springer-Verlag, 1992.
7. Paul Freedman, Denis Roy, and Louise Quesnel. Crim: a new model for technology development and technology transfer. In *Canadian Conference on Electrical and Computer Engineering*, Vancouver, B.C., September 1993. IEEE.
8. Richard Furuta and P. David Stotts. Programmable browsing semantics in Trellis. In *Proceedings of the First ACM Conference on Hypertext*, pages 27–42, Pittsburgh, PA, November 1989.
9. C. Girault and R. Estraillier(contacts). *CPN/AMI environment documentation.* Universit'e Paris VI, Paris, France, November 1993.

10. Grossenbacher Elektronik, St. Gallen, Switzerland. *PACE: Tool Reference Manual*, 1992. version 1.0.0.
11. David Harel, Hagi Lachover, Amnon Naamad, Amir Pnueli, Michal Politi, Rivi Sherman, Aharon Shtull-Trauring, and Mark Trakhtenbrot. STATEMATE: A working environment for the development of complex reactive systems. *IEEE Transactions on Software Engineering*, 16(4):403–414, April 1990.
12. Scott E. Hudson and John T. Stasko. Animation support in a user interface toolkit: Flexible, robust and reusable abstractions. In *Proceedings of the Sixth Annual Symposium on User Interface Software and Technology*, pages 57–67, Atlanta, GA, November 1993. ACM.
13. Kurt Jensen. Coloured petri nets: A high level language for system design and analysis. In G. Rozenberg, editor, *Advances in Petri Nets 1990*. Springer-Verlag, 1990.
14. Rudolf K. Keller, Anurag Garg, and Tao Tao. HyperRef - on-line support for research literature assessment and documentation. In *Proceedings of the Eleventh Annual International Conference on Systems Documentation*, pages 163–175, Waterloo, Ontario, Canada, October 1993. ACM.
15. Rudolf K. Keller, Richard Lajoie, Marianne Ozkan, Fayez Saba, Xijin Shen, and Tao Tao. Macrotec version 2.0 user manual. Technical Report Tig-93-9, Centre de recherche informatique de Montréal (CRIM), Montreal, September 1993.
16. Rudolf K. Keller, Richard Lajoie, Marianne Ozkan, Fayez Saba, Xijin Shen, Tao Tao, and G. v. Bochmann. The Macrotec toolset for CASE-based business modelling. In *Proceedings of the Sixth International Workshop on Computer-Aided Software Engineering*, pages 114–118, Singapore, July 1993.
17. Glenn E. Krasner and Stephen T. Pope. A cookbook for using the model-view-controller user interface paradigm in smalltalk-80. *Journal of Object-Oriented Programming*, 1(3):26–49, August/September 1988.
18. M. Ajmone Marsan. Stochastic Petri nets: An elementary introduction. 11th International Symposium on Protocol Specification, Testing, and Verification, 1991. Published as tutorial notes.
19. Meta Software Corporation. *Design/CPN User Manual Version 2.0*, 1992.
20. H. Oswald, R. Esser, and R. Mattmann. An environment for specifying and executing hierarchical Petri nets. In *Proceedings of the Twelfth International Conference on Software Engineering*, pages 164–172, Nice, France, March 1990.
21. Marianne Ozkan. Visualization of the dynamic analysis of business models based on the petri net formalism. Master's thesis, McGill University, Montreal, PQ, Canada, July 1993.
22. Pierre Parent and Oryal Tanir. Voltaire: a discrete event simulator, December 1991. Computer Systems Laboratory, McGill University. Handout at the Tools Fair of the Fourth International Workshop on Petri Nets and Performance Models, Melbourne, Australia.
23. Lawrence Peters and Ron Schultz. The application of Petri-nets in object-oriented enterprise simulations. In *Proceedings of the Twenty-Sixth Annual Hawaii International Conference on System Sciences*, Hawaii, January 1993.
24. Xijin Shen. Environment support for business modelling: Concepts, architecture, and implementation. Master's thesis, McGill University, Montreal, PQ, Canada, January 1994.
25. Tao Tao. Applying hypertext concepts to business modelling. Master's thesis, McGill University, Montreal, PQ, Canada, July 1993.
26. K. S. Trivedi, J. K. Muppula, S. P. Woolet, and B.R. Haverkort. Composite performance and dependability analysis. *Performance Evaluation*, 14(3-1):197–215, 1992.
27. André Weinand, Erich Gamma, and Rudolf Marty. Design and implementation of ET++, a seamless object-oriented application framework. *Structured Programming*, 10(2):63–87, April-June 1989.

# Partial Order Semantics of Box Expressions

Maciej Koutny*

Department of Computing Science
University of Newcastle
Newcastle upon Tyne, NE1 7RU, U.K.

**Abstract.** We develop a partial order semantics for the process expressions underlying the Petri Box Calculus. We aim at a semantics which would be equivalent to the standard partial order semantics of the Petri nets (Boxes) corresponding to such expressions. The solution we present is a variant of step sequence semantics in which actions are annotated with an additional information about the relative position of the parts of the expression from which they were derived, as first proposed by Degano, De Nicola and Montanari. This information is then used to capture all essential causal dependencies among actions, leading to the definition of a partial order of action occurrences. To represent Petri net markings within process expressions we employ an overbarring and underbarring technique which is related to that used in the event systems due to Boudol and Castellani. The partial order operational model turns out to be consistent with that defined in the Petri net theory. More precisely, if an expression can execute a partial order then the same holds for the corresponding Petri Box. The converse holds for all guarded expressions.

**Keywords:** Causality, partial order theory of concurrency; net-based algebraic calculi; structured operational semantics.

## 1 Introduction

The Petri Box Calculus (PBC) [2, 3, 15] has been developed to provide compositional semantics of high level programming constructs in terms of a class of Petri nets with simple (entry, exit and communication) interfaces, called Petri Boxes. ([15] gives a Box semantics to occam [19], while [4] gives a Box semantics to a more indigenous Petri net based notation.) In its main concepts PBC resembles Milner's CCS [20] but distinguishes itself from CCS by being more general (allowing, for instance, non-tail-end-recursion and supporting iteration and sequence operators) and by featuring an incremental multiway synchronisation in interprocess communication based on more expressive action labels. The PBC model also supports a compositional semantics in terms of Petri nets and their associated partial order ('true concurrency') behaviour ([3] gives a basic semantics, [2] extends this to general refinement and recursion, while [5, 18] specify the partial order semantics).

* Work done within the Esprit Basic Research Working Group 6067 CALIBAN (Causal Calculi Based on Nets).

The Petri Box Calculus basically consists of two parts: (i) a syntactic domain of Box expressions; and (ii) a semantic domain of Petri Boxes. There exists a semantic homomorphism from Box expressions to Boxes. This function and its properties are described in [3]. Boxes are Petri nets equipped with an interface. They behave like black boxes extended with a communication facility and can be combined appropriately at their interfaces. Box expressions, on the other hand, are an extension and modification of CCS [20] with asynchronous multilabel communication; they are like a generalised asynchronous version of SCCS [6, 11].

In this paper we develop a partial order semantics for the Box expressions. We aim at at a semantics which would be equivalent to the standard partial order semantics of the corresponding Petri nets (Boxes). The solution we present is a variant of step sequence semantics in which actions are annotated with an additional information about the relative position of the parts of the expression from which they were derived. The structured operational semantics of Box expressions is given in the style of [23], while the annotations are similar to those in [7, 10, 22, 24]. The information conveyed by annotations is then used to capture essential causal dependencies among executed actions, from which partial order executions can be constructed.

In our approach to step sequence semantics, we use equations in addition to action rules to axiomatise what is basically the Petri net transition rule. To represent Petri net markings we employ an overbarring and underbarring technique which is related to that used in the event systems of [7]. In particular, we employ *marked expressions* which can be used to represent evolving Petri Boxes (Box expressions correspond to unmarked or initially marked Petri Boxes). The partial order semantics of marked expressions is consistent with that defined in the Petri net theory. More precisely, if an (arbitrary) marked expression $D$ can execute a partial order, $D \xrightarrow{\pi} H$, then the same holds for the corresponding Box, $Box(D) \xrightarrow{\pi} Box(H)$. The converse holds for all guarded expressions (although we conjecture that this restriction can be removed).

The paper is structured as follows. Section 2 introduces the syntax of Box expressions, their intended meaning, and the semantic model of Petri Boxes without repeating the details which can be found in [3] and [16]. Section 3 defines the notion of a marked expression, which relates to the notion of a Box expression roughly like the notion of a marked net relates to that of an unmarked net. We then introduce structural equations on marked expressions which correspond to equality on the corresponding marked Boxes. Sections 4 defines a step sequence semantics of marked expressions (in which actions are annotated with an additional information about the relative position of the parts of the expression from which they were derived) in the SOS style, by defining inference rules for marked expressions. It also presents some examples. Section 5 shows how partial orders can be derived from step sequences defined in the previous section. The consistency and completeness results—our main technical results—are then presented. Section 6 contains concluding remarks, and a brief comparison with other work. Appendix A defines marked Boxes and Appendix B shows how marked Boxes can be constructed from the corresponding expressions. Appendix C provides diagrams of marked Boxes that correspond to our running examples.

Due to space limitations, proofs are omitted. They will be available in a full version of the paper (some can be found in the report [16] on which this paper is based).

## 2 Box Expressions and Petri Boxes

The communication capabilities of process expressions will be formalised using communication labels, $\mathcal{L}$, which are finite multisets of more primitive action names.[2] The set of action names, $\mathcal{A}$, is fixed and comes equipped with a 'conjugation' bijection, $\hat{}: \mathcal{A} \rightarrow \mathcal{A}$. We assume $\hat{a} \neq a$ and $\hat{\hat{a}} = a$, for all $a \in \mathcal{A}$.

The syntax of Box expressions $E$ is defined in Table 1, where $\alpha \in \mathcal{L}$ is a communication label, $a \in \mathcal{A}$ is an action name, and $X$ is a variable.[3] When $\alpha$ is a singleton multiset, $\alpha = \{a\}$, and if unambiguity is ensured, we leave out the curly brackets. We adopt the standard notion of a free and bound variable, and of a closed expression. Table 2 shows examples of Box expressions[4].

**Table 1.** Box expressions

| | |
|---|---|
| $E ::= \alpha \mid$ | basic action |
| $X \mid$ | variable |
| $E; E \mid$ | sequence |
| $E \,\square\, E \mid$ | choice |
| $E \| E \mid$ | concurrent composition |
| $[E * E * E] \mid$ | iteration |
| $[a : E] \mid$ | scoping |
| $\mu X.E$ | recursion |

The basic semantics of a Box expression $E$ is given by a mapping that produces a labelled (possibly infinite) Petri net[5] whose labelling indicates an interface at which it can be composed – denoted by $Box(E)$. Every Box has one or more places labelled $e$ (for entry) and one or more places labelled $x$ (for exit). Other places are labelled by $\emptyset$. Transitions are labelled either by communication labels in $\mathcal{L}$, or by variable names. If the communication label is $\emptyset$ then the

[2] We use standard operations on multisets: sum, +, difference, \, and $k$–multiple of a multiset, $k \cdot \alpha$.

[3] We use $a, b, c, \ldots$ to denote action names, $\alpha, \beta, \ldots$ to denote communication labels, $X, Y, \ldots$ to denote variables, and $E, F, G, \ldots$ to denote Box expressions.

[4] [3] also defines synchronisation and restriction which here are combined into a single scoping operator.

[5] Actually, an equivalence class thereof; but for the purposes of this paper, we may ignore this distinction.

transition is called internal (and treated as $\tau$ in CCS); otherwise the transition belongs to the communication interface.

On the set of Boxes, operations corresponding to those in Table 1 are defined, as explained in the last paragraph in Appendix B. In Appendix C we show Boxes associated to the examples from Table 2.

**Table 2.** Examples of Box expressions

| | | |
|---|---|---|
| $F_1 = (a; b) \| c$ | $F_2 = (a; b) \| ((c \,\square\, d) \| e)$ | $F_3 = [a : (b; a) \| (\hat{a}; c)]$ |
| $F_4 = [a * b \| c * d]$ | $F_5 = \mu X.((a; X) \,\square\, b)$ | $F_6 = [a : (\{a, b\} \| \{\hat{a}, \hat{a}\}) \| \{a, c\}]$ |

$\mu X.E$ denotes the recursion operator [1, 20]. Sequence and choice are standard. The iterative construct $[E_1 * E_2 * E_3]$ means 'perform $E_1$ once, then perform zero or more repetitions of $E_2$, then once $E_3$'. The basic expression $E = \alpha$ means 'execute a single action with communication capabilities $\alpha$ and terminate'. The concurrent composition operator is basically a disjoint union; for instance, $a \| \hat{a}$ can perform an $\{a\}$ action and $\{\hat{a}\}$ action, but no synchronised action (in contrast to $a.nil | \hat{a}.nil$ in CCS). Synchronisation can only be achieved through the scoping operator; we explain the scoping construct $[a : E]$ in terms of its intended semantics, $Box([a : E]) = [a : Box(E)]$.

Scoping of a Box $Box(E)$ may add/remove transitions to/from it according to certain criteria applied to the labels of transitions. It consists of two consecutive steps: synchronisation and restriction: The synchronisation step is a 'repeated application' of the basic CCS synchronisation mechanism, i.e. synchronisation over pairs $(a, \hat{a})$ of conjugate action names. For instance, in $F_6 = [a : (\{a, b\} \| \{\hat{a}, \hat{a}\}) \| \{a, c\}]$, the first subexpression $\{a, b\}$ can synchronise with the $\{\hat{a}, \hat{a}\}$ (as it were, using the $a$ of $\{a, b\}$ coupled with one of the $\hat{a}$'s) and $\{\hat{a}, \hat{a}\}$ can also synchronise with $\{a, c\}$ (using the other $\hat{a}$ coupled with the $a$ of $\{a, c\}$). This yields a 3-way synchronisation. The resulting transition is labelled $\{b, c\}$ (action names participating in the synchronisation are deleted) and inherits the connectivity from the transitions corresponding to $\{a, b\}$, $\{\hat{a}, \hat{a}\}$ and $\{a, c\}$. In the restriction step all transitions that have at least one $a$ or $\hat{a}$ in their label are simply removed. For instance, the transition labelled by $\{b, c\}$ is indeed a possible action of $Box(F_6)$ because not only is it a legal synchronisation, but it survives the restriction as well.

Although the syntactic substitution $E[X \leftarrow F]$ lies outside the signature of our process algebra, it is needed to formalise the rewriting of recursive terms. To ensure that it is consistent with transition refinement in the domain of Boxes [2], i.e. that $Box(E[X \leftarrow F]) = Box(E)[X \leftarrow Box(F)]$ holds[6], we use two simple

[6] $Box(E)[X \leftarrow Box(F)]$ denotes a simultaneous transition refinement in which all transitions labelled $X$ in $Box(E)$ are replaced by $Box(F)$.

rules. The first says that any bound variable names may need to be changed to avoid clashes. The other deals with the problem caused by substitutions like $[a : a\|X][X \leftarrow \hat{a}]$ and $[b : b\|X][X \leftarrow \hat{a}]$. These should result in expressions generating the same Box, but if we were to perform a 'dumb' substitution, the resulting expressions, $[a : a\|\hat{a}]$ and $[b : b\|\hat{a}]$, would violate this. The reason is that scoping binds the name of the action name it involves. [16] defines consistent re-naming of action names, which allows changing bound action names to avoid name clashes, where $a$ and $\hat{a}$ are bound in $E$ if they occur in a scoping context $[a : E]$. For our example, we re-name $a$ to $c$, resulting in $[c : c\|X][X \leftarrow \hat{a}] = [c : c\|\hat{a}]$, which satisfies $Box([c : c\|\hat{a}]) = Box([b : b\|\hat{a}])$. We will treat as syntactically equivalent any two Box expressions which can be shown equal by changing bound variables and/or by consistent re-naming of action names.

## 3 Marked Expressions and Structural Equations

Typically, in a process algebra, the action rule $E \xrightarrow{a} F$ involves a change in the structure of $E$; e.g. $a.nil + b.nil \xrightarrow{a} nil$. However, the transition rule of Petri nets involves the modification of a net's marking, but not of its structure. There are, therefore, two natural ways of axiomatising the transition rule: (1) the occurrence of a transition in a Petri net leads to a modification of its structure (as well as of its marking); and (2) the action rule for Box expressions does not change the structure of the expression. In this work we have adopted the latter.

To capture the change of the marking in the corresponding Petri net, we employ the concept of overbarring and underbarring (subexpressions of) a Box expression $E$, yielding what we call a marked expression. The overbarred subexpression $\overline{E}$ means that $E$ is enabled (or is in its initial state) and may subsequently occur, while $\underline{E}$ means that $E$ is in its final state. For instance, $\overline{(a; b) \square c}$ means 'the expression $(a; b) \square c$ is enabled', while $(\overline{a}; b) \square c$ means 'the overbarred subexpression $a$ is enabled'. On the other hand, $(\underline{a}; b) \square c$ means 'the underbarred subexpression $a$ has just occurred and is now in its final state'.

Table 3 presents the syntax of marked expressions $D$.[7] The role of an index $i$ in the iteration operator will be explained later.[8] The notions of a bound variable, etc., are defined as for Box expressions.

We extend $Box(.)$ to a mapping which yields, for each marked expression $D$, a marked Box — $Box(D)$ — a Box with a marking. This can be done compositionally, by modifying the construction given in [3]: one only has to mark the entry (exit) places in the subnets corresponding of overbarred (underbarred) subexpressions. In Appendix A we formally define marked Boxes and in Appendix B the required extension to the $Box(.)$ mapping. Appendix C shows the marked Box $Box((\underline{a}\|\overline{b}) \square c)$.

[7] We use $D, H, \ldots$ to denote marked expressions. Note that marked expressions comprise recursively defined processes since they include $\overline{E}$, for each Box expression $E$.

[8] Basically, it is used to distinguish between odd and even executions of the body of the iteration construct.

**Table 3.** Marked expressions ($E$ is a Box expression, $a \in \mathcal{A}$ and $i = 0, 1$)

$$
\begin{array}{llll}
D ::= \overline{E} & \mid \underline{E} & \mid D; E & \mid \\
\quad\;\; E; D & \mid D \,\square\, E & \mid E \,\square\, D & \mid \\
\quad\;\; D \| D & \mid [D * E * E]_i & \mid [E * D * E]_i & \mid \\
\quad\;\; [E * E * D]_i & \mid [a : D] & &
\end{array}
$$

The operational semantics of marked expressions will be defined in the SOS style [23], but the structure of an executed marked expression will be left unchanged. Instead, the change of state will be indicated by changing the overbarring and underbarring. For example, we will have a transition leading from $(\overline{a}; b) \,\square\, c$ to $(\underline{a}; b) \,\square\, c$.

Inference rules are not sufficient to ensure that a marked expression $D$ is capable of executing exactly the same actions as $Box(D)$. For example, $Box((\overline{a; b}) \,\square\, c)$ enables a transition labelled with $a$, yet $(\overline{a; b}) \,\square\, c$ cannot mimic this using the inference rules alone. We deal with this problem by defining a relation $\equiv$ on marked expressions which identifies some of the expressions generating the same Box. For example, using the equation $\overline{E; F} \equiv \overline{E}; F$ we will be able to rewrite $(\overline{a; b}) \,\square\, c$ to $(\overline{a}; b) \,\square\, c$ and after that use inference rules to derive a transition labelled with $a$.

Tables 4 shows a set of basic equations (including one context rule, EC, where $\mathcal{C}\{.\}$ denotes an arbitrary context) that match the operators of the syntax. We then define $\equiv$ to be the least equivalence relation on marked expressions satisfying the basic equations.[9]

The following result clarifies the relationship between two equivalences on marked expressions: structural (Table 4) and induced by the $Box(.)$ mapping.

**Proposition 1 [16].** *If $D \equiv H$ then $Box(D) = Box(H)$.* $\square$

A converse result does not hold. However, if $Box(D) = Box(\overline{\lfloor D \rfloor})$ then $D \equiv \overline{\lfloor D \rfloor}$, where $\lfloor D \rfloor$ is obtained from $D$ by leaving out all overbars, underbars and indices $i$. Note that $Box(D) = Box(H)$ and $\lfloor D \rfloor = \lfloor H \rfloor$ does not, in general, imply $D \equiv H$. For example, if $D = \mathbf{stop}; (\overline{a}; (\mathbf{stop}; (a; \mathbf{stop})))$ and $H = \mathbf{stop}; (a; (\mathbf{stop}; (\overline{a}; \mathbf{stop})))$, where $\mathbf{stop} = [a : a]$.

It is also possible to obtain what might be interpreted as a restricted converse of the context rule EC. For example, if $D$ is a marked expression such that $Box(\underline{\lfloor D \rfloor}) \neq Box(D) \neq Box(\overline{\lfloor D \rfloor})$ and $D \equiv H \,\square\, E$, then $D = H_0 \,\square\, E$ and $H_0 \equiv \overline{H}$, for some $H_0$.

[9] The application of the equations in Table 4 to marked expressions does not lead out of their syntax [16].

**Table 4.** **Equations for marked Box expressions**

| *Operator* | *Equation(s)* | |
|---|---|---|
| **Context rule** | $\dfrac{D \equiv H}{\mathcal{C}\{D\} \equiv \mathcal{C}\{H\}}$ | EC |
| **Sequence** | $\overline{E;F} \equiv \overline{E};F$ | ES1 |
| | $\underline{E};F \equiv E;\overline{F}$ | ES2 |
| | $E;\underline{F} \equiv \underline{E;F}$ | ES3 |
| **Choice** | $\overline{E \,\square\, F} \equiv \overline{E} \,\square\, F$ | EC1 |
| | $\overline{E \,\square\, F} \equiv E \,\square\, \overline{F}$ | EC2 |
| | $\underline{E} \,\square\, F \equiv \underline{E \,\square\, F}$ | EC3 |
| | $E \,\square\, \underline{F} \equiv \underline{E \,\square\, F}$ | EC4 |
| **Concurrent composition** | $\overline{E \| F} \equiv \overline{E} \| \overline{F}$ | ECC1 |
| | $\underline{E} \| \underline{F} \equiv \underline{E \| F}$ | ECC2 |
| **Iteration** $(i = 0, 1)$ | $\overline{[E * F * G]} \equiv [\overline{E} * F * G]_0$ | EI1 |
| | $[\underline{E} * F * G]_i \equiv [E * \overline{F} * G]_{1-i}$<br>$\equiv [E * \underline{F} * G]_i$<br>$\equiv [E * F * \overline{G}]_{1-i}$ | EI2 |
| | $[E * F * \underline{G}]_i \equiv \underline{[E * F * G]}$ | EI3 |
| **Scoping** | $\overline{[a : E]} \equiv [a : \overline{E}]$ | ESc1 |
| | $[a : \underline{E}] \equiv \underline{[a : E]}$ | ESc2 |
| **Recursion** | $\overline{\mu X.E} \equiv \overline{E[X \leftarrow \mu X.E]}$ | ER |

## 4 Operational Rules for Step Sequences

We first define a version of the step sequence semantics for marked expressions. The basic idea is to retain a (minimal) sufficient information about the causality structure of marked expressions, to be able to determine whether two actions occurring in a step sequence are independent or causally related. For example, in the expression $(a;\underline{b})\|((c \,\square\, \overline{d})\|\underline{e})$ the action generated by $d$ will be annotated so as to indicate that it originates from the left branch of the right branch of the topmost concurrent composition operator. As we have sequential composition rather than prefixing, we ensure that in an expression such as $(a\|b);(c\|d)$, all four dependencies are recorded. A similar scheme is also devised for iteration. We encode the information about the position of an action within a marked expression using annotations. These are finite strings $\sigma$ of markers, $\sigma \in \{\|_0, \|_1, ;_0, ;_1, *_0, *_1\}^*$.
The markers $\|_0$ and $\|_1$ are used to record the concurrency structure; the markers $;_0$ and $;_1$ are used to record the dependencies resulting from sequential composition; while $*_0$ and $*_1$ serve a similar purpose for iteration. Informally, $\|_0$ should be read as 'the left branch', $\|_1$ as 'the right branch', $;_0$ should be read as 'the first component', and $;_1$ as 'the second component'. In the case of iteration, the meaning of $*_0$ and $*_1$ can be explained in the following way: A typical execution of $\overline{[E * F * G]}$ can be thought of as coming from a sequential composition $\overline{E;F;F;\ldots;F;G}$. Then $*_0$ will be used to mark actions originating from $E$, all the even copies of $F$ (and $G$, if the number of the $F$'s is odd); while $*_1$ will mark actions originating from the other parts of the expression. The $i$ index (in the syntax) is used to indicate which of the two markers, $*_0$ or $*_1$, should be used.

For instance, by annotating the action generated by $d$ in $(a;\underline{b})\|((c \,\square\, \overline{d})\|\underline{e})$ with the string $\sigma = \|_1\|_0$ we indicate that first the right branch of a concurrent composition and after that the left branch of the nested concurrent composition is taken. For the transitions resulting from synchronisation we will need sets of annotations, each annotation representing one of the constituent basic actions. The information retained by annotations will be sufficient to reconstruct, for each step sequence derived, the independence relation on the transitions in the corresponding Box.

Let $D$ and $H$ be two marked expressions and $\Gamma$ be a step ($\Gamma$ is always a set). By means of inference rules, we define a relation $D \xrightarrow{\Gamma} H$, which is informally intended to mean: '$D$ may execute the step $\Gamma$ and become $H$'. The general form of $\Gamma$ is $\Gamma = \{\alpha_1^{A_1}, \ldots, \alpha_k^{A_k}\}$, where each $\alpha_j$ is a communication label (a multiset of action names), and each $A_j$ is a nonempty finite set of annotations. Informally, $\Gamma$ denotes a concurrently enabled set of communication labels or, in Petri net terminology, a 'step' of the Box underlying $D$ and $H$. As an example, $\underline{(\overline{a};b)} \,\square\, c$ enables the step $\Gamma = \{\{a\}^{\{;_0\}}\}$ which contains a single action, and $(\overline{\{a,d\}};b)\|\overline{\{c,c\}}$ enables $\Gamma = \{\{a,d\}^{\{\|_0;_0\}}, \{c,c\}^{\{\|_1\}}\}$. The Box associated to the latter contains two concurrently enabled transitions, one labelled by $\{a,d\}$ and the other labelled by $\{c,c\}$.

Table 5 shows a set of inference rules (and a single axiom, AB) that match the operators of the syntax; $\Gamma$ and $\Delta$ are generic names for (annotated) steps.

In the rule ASc for scoping, we use the following notation: $\Gamma = \{\alpha_1^{A_1}, \ldots, \alpha_k^{A_k}\}$ is defined to belong to $syn_a$ if it determines a transition that can be executed by the scoping construct $[a : D]$. For this to be the case, the number of $a$'s must balance the number of $\hat{a}$'s and be equal to $k-1$, because otherwise there would still be something to be removed by the implicit restriction operator (see Section 2). Either both numbers are zero, in which case $k = 1$ and $\Gamma$ contains a single transition of the body $D$ which involves neither $a$ nor $\hat{a}$ (and therefore survives restriction); or $k \geq 2$ and both numbers are equal to $k-1$; then each transition $\alpha_j$ contains at least one $a$ or $\hat{a}$, and the $\alpha_j$'s can be synchronised to yield a single transition of $[a : D]$.

Formally, $\Gamma \in syn_a$ if there are exactly $k-1$ occurrences of $a$ and exactly $k-1$ occurrences of $\hat{a}$ in $\Gamma$ and, furthermore, if $k \geq 2$ then each $\alpha_i$ contains an occurrence of $a$ or $\hat{a}$. We also define

$$\Gamma \ominus a = ((\alpha_1 + \cdots + \alpha_k) \backslash (k-1) \cdot \{a, \hat{a}\})^{A_1 \cup \cdots \cup A_k}.$$

The rule ASc assumes that $D$ can execute a set of transitions which can be partitioned[10] onto a number of $\Gamma$'s which can survive scoping, as described above, $\Gamma_1, \ldots, \Gamma_m$. Since each $\Gamma_i$ gives rise to a transition $\Gamma_i \ominus a$ after scoping, the resulting step $\{\Gamma_1 \ominus a, \ldots, \Gamma_m \ominus a\}$ can be executed by $[a : D]$.

Note that the rule for scoping specifies that a synchronised action is labelled with annotations representing the positions of all of its constituent actions (by taking the union of such annotations).

In Table 5, the operation on steps, $\Gamma(\rho)$, where $\rho$ is a marker, is defined so that each annotation $\sigma$ within $\Gamma$ is changed to $\rho\sigma$. For example,

$$\{\{a\}^{\{\epsilon\}}, \{b\}^{\{;_1\|_0, \|_1\}}\}(*_1) = \{\{a\}^{\{*_1\}}, \{b\}^{\{*_1;_1\|_0, *_1\|_1\}}\}.$$

For the example $F_6$ discussed in Section 2, we have $k = 3$ and

$$\Gamma = \{\{a, b\}^{\{\|_0\|_0\}}, \{\hat{a}, \hat{a}\}^{\{\|_0\|_1\}}, \{a, c\}^{\{\|_1\}}\}.$$

Such a $\Gamma$ belongs to $syn_a$ since there are 2 each of $a$ and $\hat{a}$, and each communication label contains a least one of them; moreover, $\Gamma \ominus a = \{b, c\}^{\{\|_0\|_0, \|_0\|_1, \|_1\}}$.

In what follows, if

$$D \equiv D_1 \xrightarrow{\Gamma_1} D_2 \equiv D_3 \xrightarrow{\Gamma_2} \cdots \xrightarrow{\Gamma_n} D_{2n} \equiv H,$$

then we denote $D \xrightarrow{\omega} H$, where $\omega = \Gamma_1 \ldots \Gamma_n$, and say that $\omega$ is an annotated step sequence leading from $D$ to $H$.

A straightforward induction on the structure of marked expressions shows that applying the rules in Table 5 to marked expressions does not lead out of their syntax. Note that the rule for scoping does not affect steps which do not contain any $a, \hat{a}$-labels. Moreover, only the axiom AB changes directly the overbarring and underbarring of expressions, while the rules for sequence, choice, concurrent composition and iteration pass the result of a rewriting in a subexpression.

[10] Note that $\uplus$ denotes disjoint union.

**Table 5.** Inference rules

| *Operator* | *Inference rule(s)* | |
|---|---|---|
| Basic action | $\overline{\alpha} \xrightarrow{\{\alpha^{\{\epsilon\}}\}} \underline{\alpha}$ | AB |
| Sequence | $\dfrac{D \xrightarrow{\Gamma} H}{D;E \xrightarrow{\Gamma(;_0)} H;E}$ | AS1 |
| | $\dfrac{D \xrightarrow{\Gamma} H}{E;D \xrightarrow{\Gamma(;_1)} E;H}$ | AS2 |
| Choice | $\dfrac{D \xrightarrow{\Gamma} H}{D \Box E \xrightarrow{\Gamma} H \Box E}$ | ASC1 |
| | $\dfrac{D \xrightarrow{\Gamma} H}{E \Box D \xrightarrow{\Gamma} E \Box H}$ | ASC2 |
| Concurrent composition | $\dfrac{D \xrightarrow{\Gamma} D'\ ,\ H \xrightarrow{\Delta} H'}{D \Vert H \xrightarrow{\Gamma(\Vert_0) \cup \Delta(\Vert_1)} D' \Vert H'}$ | ACC1 |
| | $\dfrac{D \xrightarrow{\Gamma} D'}{D \Vert H \xrightarrow{\Gamma(\Vert_0)} D' \Vert H}$ | ACC2 |
| | $\dfrac{H \xrightarrow{\Gamma} H'}{D \Vert H \xrightarrow{\Gamma(\Vert_1)} D \Vert H'}$ | ACC3 |
| Iteration ($i = 0, 1$) | $\dfrac{D \xrightarrow{\Gamma} H}{[D * F * G]_i \xrightarrow{\Gamma(*_i)} [H * F * G]_i}$ | AI1 |
| | $\dfrac{D \xrightarrow{\Gamma} H}{[E * D * G]_i \xrightarrow{\Gamma(*_i)} [E * H * G]_i}$ | AI2 |
| | $\dfrac{D \xrightarrow{\Gamma} H}{[E * F * D]_i \xrightarrow{\Gamma(*_i)} [E * F * H]_i}$ | AI3 |
| Scoping | $\dfrac{D \xrightarrow{\Gamma_1 \uplus \cdots \uplus \Gamma_m} H\ ,\ \Gamma_i \in syn_a}{[a : D] \xrightarrow{\{\Gamma_1 \ominus a, \ldots, \Gamma_m \ominus a\}} [a : H]}$ | ASc |

### 4.1 Examples

We now illustrate the application of the rules of Table 5 on examples. In the first one we show that $\overline{F_1} \xrightarrow{\omega_1} \underline{F_1}$, where $\omega_1 = \{a^{\|0;0}\}\{b^{\|0;1}, c^{\|1}\}$: [11]

$$\begin{aligned} \overline{F_1} &\equiv (\overline{a}; b) \,\|\, \overline{c} \xrightarrow{\{a^{\|0;0}\}} (\underline{a}; b) \,\|\, \overline{c} \\ &\equiv (a; \overline{b}) \,\|\, \overline{c} \xrightarrow{\{b^{\|0;1}, c^{\|1}\}} (a; \underline{b}) \,\|\, \underline{c} \equiv \underline{F_1} \end{aligned}$$

The second example shows that $\overline{F_2} \xrightarrow{\omega_2} \underline{F_2}$, where $\omega_2 = \{a^{\|0;0}, d^{\|1\|0}, e^{\|1\|1}\}\{b^{\|0;1}\}$:

$$\begin{aligned} \overline{F_2} &\equiv (\overline{a}; b) \,\|\, ((c \,\Box\, \overline{d}) \,\|\, \overline{e}) \xrightarrow{\{a^{\|0;0}, d^{\|1\|0}, e^{\|1\|1}\}} (\underline{a}; b) \,\|\, ((c \,\Box\, \underline{d}) \,\|\, \underline{e}) \\ &\equiv (a; \overline{b}) \,\|\, \underline{((c \,\Box\, d) \,\|\, e)} \xrightarrow{\{b^{\|0;1}\}} (a; \underline{b}) \,\|\, \underline{((c \,\Box\, d) \,\|\, e)} \equiv \underline{F_2} \end{aligned}$$

The third example demonstrates how sets rather than single annotations arise in the process of generating steps by showing that $\overline{F_3}$ can execute an annotated step sequence $\omega_3 = \{b^{\|0;0}\}\{\emptyset^{\{\|0;1, \|1;0\}}\}\{c^{\|1;1}\}$:

$$\begin{aligned} \overline{F_3} &\equiv [a : (\overline{b}; a) \| (\overline{\hat{a}}; c)] \xrightarrow{\{b^{\|0;0}\}} [a : (\underline{b}; a) \| (\overline{\hat{a}}; c)] \\ &\equiv [a : (b; \overline{a}) \| (\overline{\hat{a}}; c)] \xrightarrow{\{\emptyset^{\{\|0;1, \|1;0\}}\}} [a : (b; \underline{a}) \| (\underline{\hat{a}}; c)] \\ &\equiv [a : \underline{(b; a)} \| (\hat{a}; \overline{c})] \xrightarrow{\{c^{\|1;1}\}} [a : \underline{(b; a)} \| (\hat{a}; \underline{c})] \equiv \underline{F_3} \end{aligned}$$

An application of the rules for iteration is exemplified thus:

$$\begin{aligned} \overline{F_4} &\equiv [\overline{a} * b \| c * d]_0 \xrightarrow{\{a^{*0}\}} [\underline{a} * b \| c * d]_0 \\ &\equiv [a * \overline{b} \| \overline{c} * d]_1 \xrightarrow{\{b^{*1\|0}, c^{*1\|1}\}} [a * \underline{b} \| \underline{c} * d]_1 \\ &\equiv [a * \overline{b} \| \overline{c} * d]_0 \xrightarrow{\{b^{*0\|0}, c^{*0\|1}\}} [a * \underline{b} \| \underline{c} * d]_0 \\ &\equiv [a * \overline{b} \| \overline{c} * d]_1 \xrightarrow{\{b^{*1\|0}, c^{*1\|1}\}} [a * \underline{b} \| \underline{c} * d]_1 \\ &\equiv [a * b \| c * \overline{d}]_0 \xrightarrow{\{d^{*0}\}} [a * b \| c * \underline{d}]_0 \equiv \underline{F_4} \end{aligned}$$

We end with an application of the rule for recursion:

$$\begin{aligned} \overline{F_5} &\equiv (\overline{a}; F_5) \,\Box\, b \xrightarrow{\{a^{;0}\}} (\underline{a}; F_5) \,\Box\, b \\ &\equiv (a; ((a; F_5) \,\Box\, \overline{b})) \,\Box\, b \xrightarrow{\{b^{;1}\}} a; ((a; F_5) \,\Box\, \underline{b})) \,\Box\, b \end{aligned}$$

[11] We simplify the notation by using $a^{\|0;0}$ to denote $\{a\}^{\{\|0;0\}}$, etc.

## 5 Deriving Partial Orders

For each annotated step sequence $\omega$ it is possible to define a unique (up to isomorphism) labelled partial order $\pi(\omega)$. The labelling is in terms of communication labels $\alpha \in \mathcal{L}$, thus allowing a direct comparison with partial orders generated by Petri Boxes.

Two annotations, $\sigma$ and $\sigma'$, are independent, $\sigma \asymp \sigma'$, if they can be decomposed as $\sigma = \sigma_0\|_i\sigma_1$ and $\sigma' = \sigma_0\|_{1-i}\sigma_2$, for some annotations $\sigma_0, \sigma_1, \sigma_2$ and $i \in \{0,1\}$. That is, if they appear in two different branches of a concurrent composition construct. The definition extends in a natural way to sets of annotations, $A$ and $B$, and annotated steps, $\Gamma = \{\alpha_1^{A_1}, \ldots, \alpha_k^{A_k}\}$ and $\Delta = \{\beta_1^{B_1}, \ldots, \beta_m^{B_m}\}$, in the following way: $A \asymp B$ if for all $\sigma \in A$ and $\sigma' \in B$, $\sigma \asymp \sigma'$; and $\Gamma \asymp \Delta$ if for all $i \leq k$ and $j \leq m$, $A_i \asymp B_j$.

We will use labelled partial orders $\pi = (Z, \prec, l)$, where $\prec \subseteq Z \times Z$ is the ordering relation (irreflexive and transitive), and $l$ is a labelling function with the domain $Z$.

We give the definition of $\pi(\omega)$ for an arbitrary sequence $\omega$ of annotated steps,

$$\omega = \{\alpha_{11}^{A_{11}}, \ldots, \alpha_{1k_1}^{A_{1k_1}}\} \cdots \{\alpha_{m1}^{A_{m1}}, \ldots, \alpha_{mk_m}^{A_{mk_m}}\}.$$

For such an $\omega$, $\pi(\omega) = (Z, \prec, l)$, where $Z$ comprises for each $\alpha_{ij}$ one distinct element $z_{ij}$, $l(z_{ij}) = \alpha_{ij}$, and $\prec$ is the transitive closure of the relation $\prec_0$ defined by:

$$z_{ij} \prec_0 z_{kl} \Leftrightarrow i < k \wedge A^{ij} \not\asymp A^{kl}.$$

The element $z_{ij}$ can be thought of as an event that corresponds to the occurrence of $\alpha_{ij}$ in the sequence $\omega$. $z_{ij} \prec_0 z_{kl}$ can be interpreted as saying that event $z_{ij}$ occurs before $z_{kl}$ if it occurs prior to it in the sequence and if their annotations do not specify them as concurrent. Note that $\pi(\omega)$ is always a partial order (since $\prec_0$ is acyclic).

For example, $\omega_3 = \{b^{\|_{0;0}}\}\{\emptyset^{\{\|_{0;1},\|_{1;0}\}}\}\{c^{\|_{1;1}}\}$ defined in the previous section generates the partial order

$$\pi(\omega_3) = (\{x, y, z\}, \prec, \{(x, b), (y, \emptyset), (z, c)\})$$

such that $x \prec y \prec z$ since $\|_{0;0} \not\asymp \|_{0;1}$ and $\|_{1;0} \not\asymp \|_{1;1}$ imply $x \prec_0 y \prec_0 z$ (but $x \not\prec_0 z$ since $\|_{0;0} \asymp \|_{1;1}$). Whenever $D \xrightarrow{\omega} H$ we will also denote $D \xrightarrow{\pi(\omega)} H$.

The result below shows that steps generated by marked expressions comprise mutually independent actions, any two consecutive independent steps can be joined together, and that each step can be split up in an arbitrary manner.

**Proposition 2 [16].** *Let $D$ and $H$ be marked expressions, and let $\Gamma$ and $\Delta$ be annotated steps, $\Gamma = \{\alpha_1^{A_1}, \ldots, \alpha_k^{A_k}\}$.*

1. *If $D \xrightarrow{\Gamma} H$ and $i \leq k$ then for all distinct $\sigma, \sigma' \in A_i$, $\sigma \asymp \sigma'$.*
2. *If $D \xrightarrow{\Gamma} H$ then for all $i \neq j$, $A_i \asymp A_j$.*
3. *If $D \xrightarrow{\Gamma \uplus \Delta} H$ then there is $J$ such that $D \xrightarrow{\Gamma} J \xrightarrow{\Delta} H$.*

*4.* *If* $D \xrightarrow{\Gamma} H \equiv J \xrightarrow{\Delta} K$ *and* $\Gamma \asymp \Delta$ *then there are* $L, M$ *such that*

$$D \equiv L \xrightarrow{\Gamma \uplus \Delta} M \equiv K.$$

□

*Remark.* The annotation scheme introduced above is minimal in the sense that we cannot leave out any of the markers and still be able to recover the partial order semantics from (annotated) step sequence semantics. We illustrate this point on examples. Let $E$ and $F$ be Box expression defined thus:

$$E = (a; b)\|(c; \mathbf{stop})$$
$$F = (a\|c); (b\|\mathbf{stop}).$$

It is easy to check that the partial orders generated by $Box(\overline{E})$ are a proper subset of those generated by $Box(\overline{E \,\square\, F})$, as $Box(\overline{E})$ cannot generate a partial order in which $a$ and $c$ are concurrent and both precede $b$. We then observe that without the markers $\|_0$ and $\|_1$ we would not have noticed any difference between the step sequences generated by $\overline{E}$ and $\overline{F}$ and involving $a, b, c$. In both cases these would look as follows:

$$\{a^{;_0}, c^{;_0}\}\{b^{;_1}\} \;\; \{a^{;_0}\}\{c^{;_0}\}\{b^{;_1}\} \;\; \{c^{;_0}\}\{a^{;_0}\}\{b^{;_1}\} \;\; \{a^{;_0}\}\{b^{;_1}\}\{c^{;_0}\} \;\; \{a^{;_0}\}\{b^{;_1}, c^{;_0}\}.$$

A similar conclusion could be drawn if we left out $;_0$ and $;_1$, as the only annotated step sequences involving $a, b, c$ would then be:

$$\{a^{\|_0}, c^{\|_1}\}\{b^{\|_0}\}\{a^{\|_0}\}\{c^{\|_1}\}\{b^{\|_0}\}\{c^{\|_1}\}\{a^{\|_0}\}\{b^{\|_0}\}\{a^{\|_0}\}\{b^{\|_0}\}\{c^{\|_1}\}\{a^{\|_0}\}\{b^{\|_0}, c^{\|_1}\}.$$

To show that we cannot do without $*_0$ and $*_1$, we consider another example. Let $E$ and $F$ be Box expressions defined in the following way:

$$E = [a\|c * b\|\mathbf{stop} * \mathbf{stop}]$$
$$F = [a * b * \mathbf{stop}] \;\|\; [c * \mathbf{stop} * \mathbf{stop}].$$

We observe that $Box(\overline{E})$ cannot generate a partial order in which $a$ precedes $b$, and both events are concurrent with $c$, whereas $Box(\overline{E \,\square\, F})$ can. Moreover, without $*_0$ and $*_1$, the annotated step sequences involving $a$, $b$, $c$ generated by $\overline{E}$ and $\overline{E \,\square\, F}$ are the same and equal to those obtained by leaving out $;_0$ and $;_1$ in our previous example. □

### 5.1 Consistency and completeness results

In this section we establish a direct relationship between the partial order semantics of marked expressions generated by annotated step sequences, and the partial orders of transition occurrences generated by the corresponding marked Boxes.

Let $\mathcal{B}$ and $\mathcal{B}'$ be marked Boxes. In what follows we use $\mathcal{B} \xrightarrow{\pi} \mathcal{B}'$ to denote that there are nets $\Sigma \in \mathcal{B}$, $\Sigma' \in \mathcal{B}'$ and a step sequence $\phi$ of transitions labelled with communication labels such that $\Sigma \xrightarrow{\phi} \Sigma'$ (according to the step sequence occurrence rule for Petri nets [21]), and $\phi$ generates (via the induced process[12]) a labelled partial order of transition occurrences $\pi$ (i.e. the labels of the partial order are the communication labels associated with transitions). The properties of marked Boxes (in particular, 1-safeness) ensure that the definition does not depend on the choice of representatives.

The first result shows that every partial order generated by a marked expression is realisable by the corresponding Box. The second result shows the converse implication, however only for guarded processes. We leave the proof of the unguarded case as an open problem.

An expression is guarded if every subexpression $\mu X.E$ is guarded in $X$, meaning that for every free occurrence of $X$ within $E$ there is a subexpression of $E$ of the form $F;G$ or $[F * G * G']$ such that the $X$ occurs within $G$ or $G'$.

**Theorem 3 (Consistency).** *Let $D$ and $H$ be closed marked expressions and $\pi$ be a labelled partial order. If $D \xrightarrow{\pi} H$ then $Box(D) \xrightarrow{\pi} Box(H)$.*

*Proof.* (Basic idea) Suppose that $\omega = \Gamma_1 \ldots \Gamma_n$, $\pi(\omega) = \pi$ and

$$D \equiv D_1 \xrightarrow{\Gamma_1} D_2 \equiv D_3 \xrightarrow{\Gamma_2} \cdots \xrightarrow{\Gamma_{n-1}} D_{2n-2} \equiv D_{2n-1} \xrightarrow{\Gamma_n} D_{2n} \equiv H.$$

The proof proceeds by induction on the maximal depth of a tableau used to generate steps in the above derivation. □

**Theorem 4 (Completeness).** *Let $D$ be a closed guarded marked expression, $\pi$ be a labelled partial order and $\mathcal{B}$ be a marked Box. If $Box(D) \xrightarrow{\pi} \mathcal{B}$ then there is a marked expression $H$ such that $Box(H) = \mathcal{B}$ and $D \xrightarrow{\pi} H$.*

*Proof.* (Basic idea) The proof proceeds by induction on the structure of the marked expression $D$. In this (as well as in the previous) proof one makes an extensive use of the results of [5] which specify what partial orders a Box can generate. For the case involving the scoping construct, Proposition 1(3,4) is essential. □

It is worth noting that for guarded marked expressions, the above two results establish a pomset bisimulation between the expressions and the corresponding marked Boxes.

## 6 Concluding Remarks

The approach presented in this paper is closely related to other existing studies of the relationship between algebraic and net-theoretic models of concurrency. Petri nets have been used to provide a non-interleaving semantics for languages based on CCS or TCSP in, for example, [7, 8, 9, 10, 12, 13, 14, 22, 24, 25]. In general, all these papers had taken as a starting point a CCS-like (or CSP-like)

[12] Which is unique since (nets belonging to) marked Boxes are 1-safe.

notation with its standard interleaving semantics and then provided a translation of process algebra expressions into Petri nets so that the standard interleaving (or a postulated non-interleaving) semantics of the expressions and the standard Petri net semantics of the corresponding nets would be consistent. Our situation in this respect was rather different as we were given at the outset both the language (Box expressions) and its translation into nets (Petri Boxes obtained via the *Box*(.) mapping), and had to find a partial order semantics of expressions consistent with the standard partial order semantics of nets.

For a detailed account of the various approaches proposed in the literature the interested reader is referred to, e.g., [7, 9]. We will only mention the relationship with the work of Boudol and Castellani, and Degano, De Nicola and Montanari. Although Boudol and Castellani [7] treat only finite (recursion-free) terms, there are two aspects of their technical development which are close to ours. The first is the notion of a 'proved transition' (itself related to an earlier proposal in [10]) which directly corresponds to our annotated communication label. Annotating actions with their syntactic contexts makes possible the definition of concurrency on proved transitions, which in turn leads to the definition of a partial order semantics. Another aspect of [7] relevant to our work are 'marked terms' which resemble our marked expression. The feature they share is that as an expressions evolves, its structure does not change. In the case of marked terms, the execution of an action marks that action within the expression as 'used'. This should be contrasted with the way marked expressions are defined, as there the only information which is kept after the execution of an action is the resulting marking. Another important issue is the treatment of recursion in [9], where recursion is performed by 'eager' unfolding which, in the same way as ER, removes the need for a separate inference rule for recursion. Both [7] and [9] are examples of models where the places and transitions of Petri nets are created syntactically (i.e. directly from an expression) which should be contrasted with the way Petri Boxes are defined in [3].

This paper shows a close correspondence between the operational semantics of marked expressions (SOS rules and equations) and the standard partial order semantics of the corresponding marked expressions. Consistency is proved to hold for arbitrary expressions, while completeness for the guarded ones.

The algebra of processes we use is richer than those usually treated elsewhere. In particular, by having an explicit iteration operator we can specify and deal smoothly (c.f. example $F_4$ in Section 4.1) with a class of infinite processes without having to resort to a general recursion. By using over- and underbarring we can define a simple rule for iteration, whereas this can cause problems in the usual treatment of iteration [17] where one 'unwinds' iteration each time it is entered:

$$(a; b; c)^* \xrightarrow{a} b; c; (a; b; c)^*.$$

Finally, [16] proves that a standard step sequence semantics of marked expressions (which can be obtained by simply ignoring anything to do with annotations) is fully consistent with the step sequence semantics of marked Boxes. However, the proof of completeness (which also includes unguarded expressions) cannot be easily adapted to deal with partial order semantics. Nevertheless, we conjecture that the completeness result for partial order semantics does hold.

**Acknowledgements**

This paper is based on the report [16] which subsumes results on the operational semantics of Box expressions developed in collaboration with Eike Best and Javier Esparza. I would like to acknowledge their fundamental contributions to the development of the semantics of Box expressions presented in this paper. I would also like to thank Raymond Devillers for his detailed comments on the earlier versions of the work reported here, and acknowledge fruitful discussions with other CALIBAN members. Finally, I would like to thank the anonymous referees for their useful comments and suggestions for improvement.

## References

1. J.C.M.Baeten and W.P.Weijland: Process Algebra. Cambridge Tracts in Theoretical Computer Science (1990).
2. E.Best, R.Devillers and J.Esparza: General Refinement and Recursion Operators in the Box Calculus. Proc. of STACS-93, Springer-Verlag Lecture Notes in Computer Science Vol. 665, 130-140 (1993).
3. E.Best, R.Devillers and J.Hall: The Petri Box Calculus: a New Causal Algebra with Multilabel Communication. Advances in Petri Nets (ed. G.Rozenberg), Springer-Verlag Lecture Notes in Computer Science Vol.609, 21-69 (1992).
4. E.Best and R.P.Hopkins: $B(PN)^2$ – a Basic Petri Net Programming Notation. Proc. of PARLE-93, Springer-Verlag Lecture Notes in Computer Science Vol. 694, 379-390 (1993).
5. E.Best and H.G.Linde-Göers: Compositional Process Semantics of Petri Boxes. Proc. of MFPS (Mathematical Foundations of Programming Semantics), Springer-Verlag Lecture Notes in Computer Science (1993).
6. G.Boudol: Notes on Algebraic Calculi of Processes. In: Logics and Models of Concurrent Systems. K.R.Apt (ed.), 261-304 (1985).
7. G.Boudol and I.Castellani: Flow Models of Distributed Computations: Event Structures and Nets. Rapport de Recherche, INRIA, Sophia Antipolis (July 1991).
8. F.de Cindio, G.De Michelis, L.Pomello and C.Simone. Milner's Communicating Systems and Petri Nets. In: Selected Papers of 3rd European Workshop on Applications and Theory of Petri Nets, IF 66 (Springer-Verlag, Heidelberg), 40-59 (1983).
9. P.Degano, R.De Nicola and U.Montanari: A Distributed Operational Semantics for CCS Based on C/E Systems. Acta Informatica 26 (1988).
10. P.Degano, R.De Nicola and U.Montanari: Partial Order Derivations for CCS. In: Proc. FCT, Lecture Notes in Computer Science Vol.199, Springer Verlag, 520-533 (1985).
11. R. de Simone: Higher-level Synchronising Devices in MEIJE-SCCS. Theoretical Computer Science Vol.37, 245-267 (1985).
12. R.J. van Glabbeek and F.V.Vaandrager: Petri Net Models for Algebraic Theories of Concurrency. Proc. PARLE'87, Lecture Notes in Computer Science Vol.259, Springer Verlag, 224-242 (1987).

13. U.Goltz: On Representing CCS Programs by Finite Petri Nets. Arbeitspapiere der GMD Nr.290 (February 1988).
14. U.Goltz and A.Mycroft: On the Relationships of CCS and Petri Nets. In: J.Paredaens (ed.), Proc. 11th ICALP, Lecture Notes in Computer Science Vol.154, Springer Verlag, 196-208 (1984).
15. J.Hall, R.P.Hopkins and O.Botti: A Petri Box Semantics of **occam**. Advances in Petri Nets (ed. G.Rozenberg), Springer-Verlag Lecture Notes in Computer Science Vol.609, 179-214 (1992).
16. M.Koutny, J.Esparza and E.Best: Operational Semantics for the Petri Box Calculus. Hildesheimer Informatik-Berichte 13/93 (October 1993).
17. W.Li and P.E. Lauer: Using the Structural Operational Approach to Express True Concurrency. Technical Report 85-01, Departmant of Computer Science and Systems, McMaster University (1985).
18. H.G.Linde-Göers: Compositional Branching Processes of Petri Boxes. Ph.D. Thesis, Universität Hildesheim (October 1993).
19. D.May: **occam**. SIGPLAN Notices, Vol.18(4), 69-79 (April).
20. R.Milner: Communication and Concurrency. Prentice Hall (1989).
21. M.Nielsen and P.S.Thiagarajan: Degrees of Nondeterminism and Concurrency. Proc. of 4th Conf. on Foundations of Software Technology and Theoretical Computer Science, Springer-Verlag Lecture Notes in Computer Science Vol.181 (eds. M.Joseph and R.Shyamasundar), 89-117 (1984).
22. E.R.Olderog: Operational Petri Net Semantics for CCSP. In: G. Rozenberg (ed.), Advances in Petri Nets 1987, Springer-Verlag Lecture Notes in Computer Science, Vol. 266, 196-223 (1987).
23. G.Plotkin: A Structural Approach to Operational Semantics. Report DAIMI FN-19, Århus University, Computer Science Department, Aarhus, Denmark (1981).
24. D.Taubner: Finite Representations of CCS and TCSP by Automata and Petri Nets. Lecture Notes in Computer Science, Vol. 369, Springer Verlag (1989).
25. G.Winskel: Petri Nets, Algebras, Morphisms and Compositionality. Info. Control 72, 197-238 (1987).

# A Compositional Construction of Marked Boxes

A labelled Petri net is a tuple $\Sigma = (S, T, W, \lambda)$ where $(S, T, W)$ is a (possibly infinite) weighted Petri net: $S$ is the set of places, $T$ is the set of transitions, and the connection mapping is given by $W : ((S \times T) \cup (T \times S)) \to \mathbf{N}$. The labelling mapping $\lambda$ has the domain $S \cup T$; for each $s \in S$, $\lambda(s) \in \{e, \emptyset, x\}$, and for each $t \in T$, $\lambda(t)$ is a communication label or a variable name. As usual, we define ${}^\bullet z = \{y \mid W(y,z) > 0\}$ and $z^\bullet = \{y \mid W(z,y) > 0\}$ for every $z \in S \cup T$, and require that ${}^\bullet t \neq \emptyset \neq t^\bullet$, for every transition $t$. We also define the entry and exit places of $\Sigma$ as respectively ${}^\bullet\Sigma = \{s \in S \mid \lambda(s) = e\}$ and $\Sigma^\bullet = \{s \in S \mid \lambda(s) = x\}$. A marked net is a tuple $\Sigma = (S, T, W, \lambda, M)$ such that $(S, T, W, \lambda)$ is a labelled Petri net and $M : S \to \mathbf{N}$ is a marking. In what follows a labelled net will sometimes be treated as a marked one with a marking $M$ such that for each $s$, $M(s) = 0$.

Two nets, $\Sigma_1$ and $\Sigma_2$, will be called re-naming equivalent if there is a sort-preserving relation $\rho \subseteq (S_1 \times S_2) \cup (T_1 \times T_2)$ such that $\rho$ is (both ways) surjective on places; $\rho$ is (both ways) surjective on transitions; $\rho$ is arc-(weight-)preserving; $\rho$ is label-preserving; $\rho$ is marking-preserving (if the net is marked); and $\rho$ is bijective on transitions labelled with variable names.

A Petri Box is a $\rho$-equivalence class $\mathcal{B} = [\Sigma]$ such that $\Sigma = (S, T, W, \lambda)$ is a labelled net satisfying: (1) ${}^\bullet\Sigma$ and $\Sigma^\bullet$ are non-empty; (2) for every $s \in {}^\bullet\Sigma$, ${}^\bullet s$ is empty, and (3) for every $s \in \Sigma^\bullet$, $s^\bullet$ is empty. Note that the definition is correct since it does not depend on the choice of representative of $[\Sigma]$. The marked Box is defined in a similar way, by taking a marked net $\Sigma = (S, T, W, \lambda, M)$ and extending the definition of $\rho$-equivalence so that $(s, r) \in \rho$ implies $M(s) = M(r)$.

# B The Box(.) mapping for marked expressions

We consider the syntax of Table 3 and refer the reader to [3] for the corresponding definitions on Box expressions. The $Box(.)$ mapping for marked expressions is defined compositionally, as follows (for brevity we omit symmetric cases):

$$\begin{array}{ll} Box(\overline{E}) & = \overline{Box(E)} \\ Box(D; E) & = Box(D); Box(E) \\ Box(E \,\square\, D) & = Box(E) \,\square\, Box(D) \\ Box(D \| H) & = Box(D) \| Box(H) \\ Box([D * E * F]_i) & = [Box(D) * Box(E) * Box(F)] \\ Box([a : D] & = [a : Box(D)] \end{array}$$

The various operators involving (marked and Petri) Boxes are defined below.

*Initial and final marking*

Let $\mathcal{B}$ be a Petri Box. Then $\overline{\mathcal{B}} = [(S, T, W, \lambda, M_e)]$, where $(S, T, W, \lambda)$ is any representative of $\mathcal{B}$, and $M_e$ is a marking defined by $M_e(s) = 1$ if $s \in^\bullet \Sigma$, and $M_e(s) = 0$ otherwise. Similarly, $\underline{\mathcal{B}} = [(S, T, W, \lambda, M_x)]$, where $M_x(s) = 1$ if $s \in \Sigma^\bullet$, and $M_x(s) = 0$ otherwise. The results of [5] ensure that both definitions are representative-independent, in the sense that no matter with what representative we start, the behaviours (both in terms of interleavings and in terms of processes) of the resulting marked nets are the same. The same is also true of the other constructs defined below.

*Sequential composition*

Let $\mathcal{B}_1$ be a marked Box and $\mathcal{B}_2$ be a Petri Box (or vice versa). Let $\Sigma_i = (S_i, T_i, W_i, \lambda_i, M_i) \in \mathcal{B}_i$ for $i = 1, 2$, be disjoint nets. Then $\mathcal{B}_1; \mathcal{B}_2 = [\Sigma]$, where

$$\Sigma = (S, T_1 \cup T_2, W_1 \cup W_2 \cup W, \lambda_1 \cup \lambda_2 \cup \lambda, M_1 \cup M_2 \cup M)$$

is a marked net satisfying the following: The domains of $W_i$, $\lambda_i$ and $M_i$ are truncated to exclude places deleted from $S_1$ and $S_2$ in forming of $S$, while $S, W, \lambda, M$ are defined as follows (below $s = (s_1, s_2) \in S \setminus (S_1 \cup S_2)$): $\lambda(s) = \emptyset$

and $M(s) = M_1(s_1) + M_2(s_2)$ and

$$S \quad = S_1 \cup S_2 \cup (\Sigma_1^\bullet \times {}^\bullet\Sigma_2) \setminus (\Sigma_1^\bullet \cup {}^\bullet\Sigma_2)$$

$$W(t,s) = \begin{cases} W_i(t,r_i) & t \in T_i,\ s = (r_1,r_2) \in S \setminus (S_1 \cup S_2)\ (i=1,2) \\ 0 & t \in T_i,\ s \in S \cap (S_1 \cup S_2) \setminus S_i \quad (i=1,2) \end{cases}$$

$$W(s,t) = \begin{cases} W_i(r_i,t) & t \in T_i,\ s = (r_1,r_2) \in S \setminus (S_1 \cup S_2)\ (i=1,2) \\ 0 & t \in T_i,\ s \in S \cap (S_1 \cup S_2) \setminus S_i \quad (i=1,2) \end{cases}$$

*Iteration (one of the forms)*

Let $\mathcal{B}_1$ and $\mathcal{B}_3$ be Petri Boxes and $\mathcal{B}_2$ be a marked Box. Moreover, assume that $\Sigma_{ij} = (S_{ij}, T_{ij}, W_{ij}, \lambda_{ij}, M_{ij}) \in \mathcal{B}_i$ for $i = 1,2,3$ and $j = 1,2$. Moreover, we re-define $M_{22}$ to be an empty marking. Then $[\mathcal{B}_1 * \mathcal{B}_2 * \mathcal{B}_3] = [\Sigma]$, where:

$$\Sigma = (S, \bigcup_{i,j} T_{ij}, W \cup \bigcup_{i,j} W_{ij}, \lambda \cup \bigcup_{i,j} \lambda_{ij}, M \cup \bigcup_{i,j} M_{ij}).$$

The set $S$ of places is given by:

$$\begin{aligned} S = {} & \textstyle\bigcup_{i,j} (S_{ij} \setminus {}^\bullet\Sigma_{ij} \setminus \Sigma_{ij}^\bullet) \\ & \cup ({}^\bullet\Sigma_{11} \times {}^\bullet\Sigma_{12}) \cup (\Sigma_{31}^\bullet \times \Sigma_{32}^\bullet) \\ & \cup (\Sigma_{11}^\bullet \times {}^\bullet\Sigma_{31} \times {}^\bullet\Sigma_{21} \times \Sigma_{22}^\bullet) \cup (\Sigma_{12}^\bullet \times {}^\bullet\Sigma_{32} \times {}^\bullet\Sigma_{22} \times \Sigma_{21}^\bullet) \end{aligned}$$

$W$ is defined similarly as before, for example, if $t \in T_{11}$ and

$$s = (s_1, s_2, s_3, s_4) \in (\Sigma_{11}^\bullet \times {}^\bullet\Sigma_{31} \times {}^\bullet\Sigma_{21} \times \Sigma_{22}^\bullet)$$

then $W(t,s) = W_1(t,s_1)$. The labelling of the newly created places assigns the $e$ to every place in ${}^\bullet\Sigma_{11} \times {}^\bullet\Sigma_{12}$, $x$ to every place in $\Sigma_{31}^\bullet \times \Sigma_{32}^\bullet$ and $\emptyset$ to every other new place.

*Scoping*

Let $\mathcal{B}$ be a marked Box and $\Sigma_0 = (S_0, T_0, W_0, \lambda_0, M_0) \in \mathcal{B}$. Then $[a : \mathcal{B}] = [\Sigma]$, where

$$\Sigma = (S_0, T, W, \lambda_0|_{S_0} \cup \lambda, M_0)$$

and $T$ is the set of newly created transitions defined in the following way: For every multiset $\Omega$ of transitions of $T_0$ satisfying $\lambda_0(\Omega) \in syn_a$ we create a unique transition $t_\Omega$ which is labelled $\lambda(t_\Omega) = \lambda_0(\Omega) \ominus a$ and has the connectivity defined by:

$$W(t_\Omega, s) = \textstyle\sum_{t \in \Omega} \Omega(t) \cdot W_0(t,s)$$
$$W(s, t_\Omega) = \textstyle\sum_{t \in \Omega} \Omega(t) \cdot W_0(s,t)$$

In the above, a multiset of communication labels $\lambda_0(\Omega) = \{\alpha_1, \ldots, \alpha_k\}$ belongs to $syn_a$ if there are exactly $k-1$ occurrences of $a$, and exactly $k-1$ occurrences of $\hat{a}$ in the $\alpha_i$'s and, furthermore, if $k \geq 2$ then each $\alpha_i$ contains $a$ or $\hat{a}$. We also define

$$\gamma \ominus a = (\alpha_1 + \cdots + \alpha_k) \setminus (k-1) \cdot \{a, \hat{a}\}$$

The original definition of the $Box(.)$ mapping for Box expressions can be recovered from those given above by simply ignoring anything which concerns markings.

# C Examples of Boxes

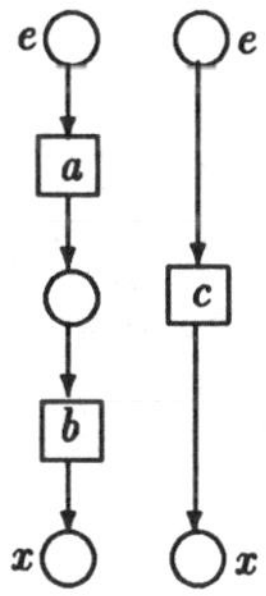


$Box((a;b)\|c)$

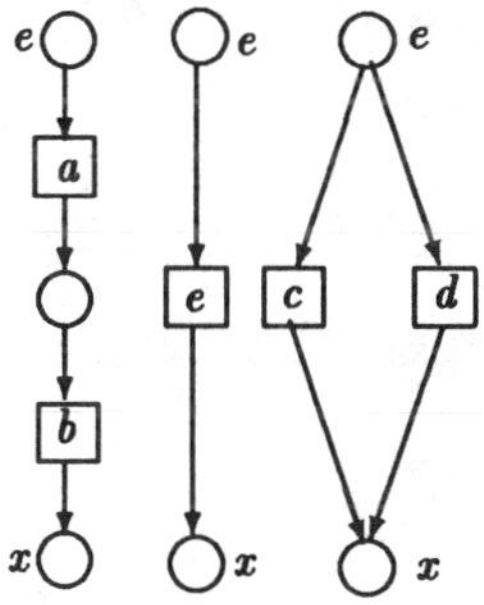


$Box((a;b)\|((c \,\square\, d)\|e))$

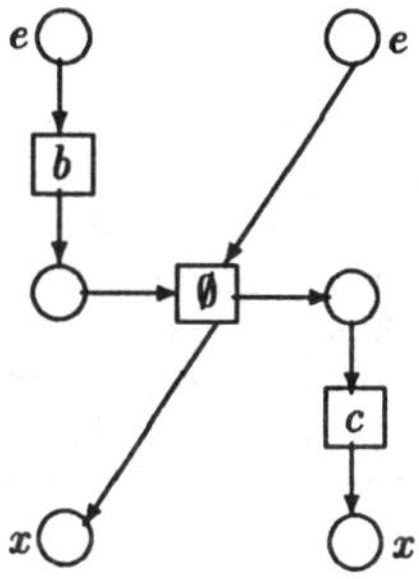


$Box([a : (b;a)\|(\hat{a};c)])$

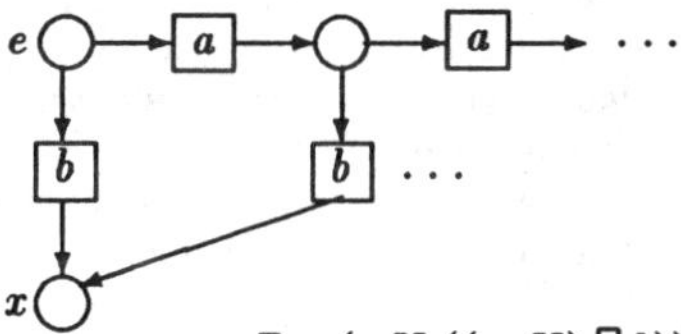


$Box(\mu X.((a;X) \,\square\, b))$

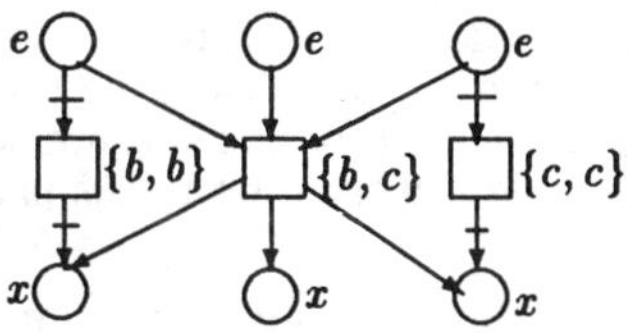


$Box([a : (\{a,b\}\|\{\hat{a},\hat{a}\})\|\{a,c\}])$

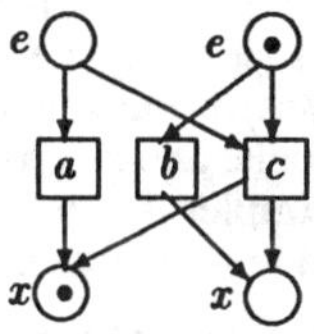


$Box((\underline{a}\|\overline{b}) \,\square\, c)$

# A General Systematic Approach to Arc Extensions for Coloured Petri Nets

Charles Lakos,
Computer Science Department,
University of Tasmania,
GPO Box 252C,
Hobart, TAS, 7001.
Australia.
Email: C.A.Lakos@cs.utas.edu.au

Søren Christensen,
Computer Science Department,
Aarhus University,
Ny Munkegade, Bldg 540,
DK-8000 Aarhus C,
Denmark.
Email: schristensen@daimi.aau.dk

**Abstract:** This paper proposes an approach to arc extensions in CP-nets which is claimed to be both general and systematic. It is general because the enabling rules cater for true concurrency as well as an interleaving semantics and because it encompasses the other proposals for arc extensions that have been made recently in the Petri Net literature, often in the context of the requirements of specific application domains. It is systematic because it proposes a set of fundamental arcs in the context of a general complementary place construction and then considers how these arcs can be combined in arbitrary ways. Because of the utility of some of these compound arcs and the minimal overhead in implementing them, it is argued that CP-net tools should provide explicit support for them.

## 1 Introduction

A number of proposals have been made for extending Petri Nets with arcs and places having modified semantics. (See for example [1], [2], [3], [4].) Some of these apply to simple net models such as PT-nets while others apply to high-level net models such as CP-nets. These proposals are usually determined by the convenience demands of modelling in specific problem domains and, as such, there is no claim to provide a set of extensions that is in any sense complete. Furthermore, as Devillers [3] shows in some detail, there are subtle variations in the semantics or enabling conditions proposed even at the level of simple net models, and this has a consequent impact on the support for true concurrency as opposed to a simpler interleaving semantics. We review some of these proposals in more detail in §6 and compare them with our own proposals.

In this paper, we present a generalised complementary place construction. On the basis of this construction, coupled with symmetry considerations, we propose a set of four fundamental arc types. Building on the work of Christensen and Hansen [2], the semantics of these arcs is defined so that all support true concurrency and do not violate the diamond rule. These fundamental arcs can be combined to form arbitrary compound arcs of which four or five are considered to be of general utility. Furthermore, by allowing linear projection functions to be associated with arc inscriptions, our proposals encompass the extensions that have been presented elsewhere in recent years.

In part, a motivation for this work has been the desire to formalise the concept of place summary functions, which is provided in the LOOPN language and simulator ([9], [10]). These functions serve to deduce some value from the multiset of tokens which is the current place marking. For example, they make it possible to determine if all tokens in a place have a given colour, or if all tokens have a colour from a subrange of possible colours, or even determining the number of tokens in the place satisfying a given condition. In practice, these functions are often exported from a module or subnet and thereby serve as module access functions, allowing other module instances to examine the state of the given module without modifying that state by the transfer of tokens. Experience from practical modelling has demonstrated the value of this particular provision for building Petri Nets with a high degree of encapsulation [11].

In order to illustrate the issues covered by the various arc extensions, we present a simple example – a refinement of the dining philosophers problem. Given in fig 1.1 is a Coloured Petri Net for this traditional problem, taken from Valmari [14].

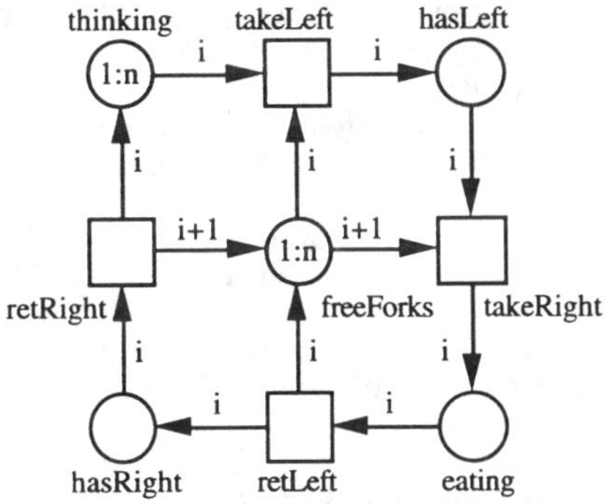


**Fig 1.1:** Dining philosophers Petri Net

We wish to extend this system to what we might call the *rowdy philosophers*. Here, when philosophers become hungry and raise their left fork, they may become rowdy and cause damage if the number of philosophers in such a state exceeds (by some predetermined margin) the number who are already eating. In order to protect against such potentially nasty situations, a hotel proprietor periodically checks for such a possibility and, if detected, will force the rowdy philsophers to recommence thinking. The use of place summary functions can be illustrated in the context of this problem by noting that it may be desirable to evaluate some or all of the following properties (which are given in mathematical notation, rather than complicating the presentation with LOOPN syntax):

- Is philosopher *i* not thinking? This condition could be given by a function on place *thinking*:
  not_thinking(i) = $(i \notin M(thinking))$
- Is any philosopher eating? This condition could be given by a function on place *eating*:
  any_eating = $(M(eating) \neq \emptyset)$
- How many philosophers have only their left fork? This condition could be given by a function on place *hasLeft*:
  num_hasLeft = $| M(hasLeft) |$

The above fanciful problem can be considered as representative of operating system or computer network management where overload situations are monitored and hopefully preempted before excessive performance degradation occurs. In seeking a solution to this problem, there are a number of constraints or consequences:

- True concurrency is to be supported. Thus, a number of philosophers must be able to pick up their left fork concurrently, and the proprietor should be able to act concurrently with other independent activities (such as a philosopher returning a right fork).

- It is necessary to be able to examine the complete marking of a place atomically. This is required in order to detect potentially nasty situations where we need to know the number of tokens in each of places *hasLeft* and *eating*. It would be possible to set up redundant places to hold these numbers as single tokens, but then the concurrency would be limited (contrary to our first constraint).
- It is necessary to be able to clear a place as an atomic action. This is necessary for fairness, to ensure that all rowdy philosophers are simultaneously ejected.

The second issue above is the one addressed by place summary functions. If atomic examination of a place marking is not possible, then a subnet is required to incrementally update the function result to reflect the place changes. This introduces extraneous intermediate states when the function result is not available, and complicates any formal proof together with the associated application of analysis techniques [12].

In considering the generalised *complementary place* construction and the arc extensions that flow from it, we assume that the reader is familiar with the notation and terminology of standard CP-nets as presented by Jensen in [7] and [8]. (These are not included here due to space constraints.) We commence in §2 by introducing the four basic arcs – input, output, test and inhibitor – and show why these are considered to be a fundamental set relative to the generalised complementary place construction. The formal definition of CP-nets extended with these arcs, together with their associated enabling rules is also given here, while the proof that these extended net models are equivalent to standard CP-nets is given elsewhere [13]. In §3 we consider the possible combinations of these arcs and show how their formal semantics can be derived from those already presented in §2. In §4 we consider a further extension to the basic arcs by associating a projection function with the arc inscriptions. The implications for tool support of these net extensions are considered in §5. In §6 we compare our arc extensions and their derivatives with others that have been proposed in the literature and we finish with the conclusions in §7.

## 2 Simple arcs and complementary places

There are four basic arcs which we consider to be fundamental. These are input arcs, output arcs, test arcs and inhibitor arcs (also known as threshold arcs). Apart from test arcs, these have all been widely discussed in the literature. Using the graphical notation of Christensen and Hansen [2] these are drawn as in fig 2.1, together with the enabling condition and occurrence effect for each one. These are given in terms of a binding element *(t,b)* which specifies the transition *t* and the binding *b* of the variables of the transition.

**Graphic**

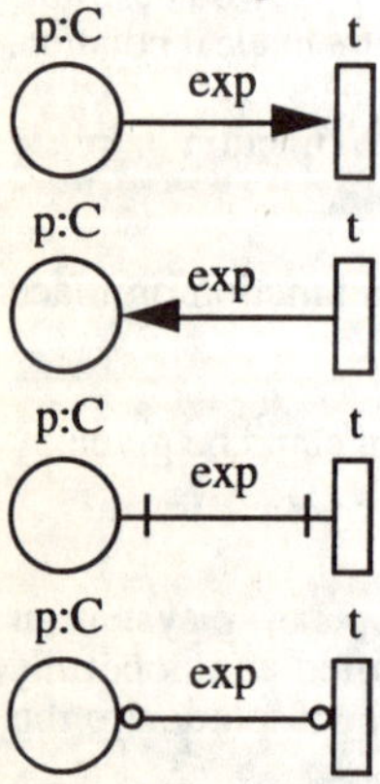


| Name | Enabling condition | Occurrence effect |
|---|---|---|
| Input | M(p) ≥ exp<b> | M'(p) = M(p) – exp<b> |
| Output | No constraint | M'(p) = M(p) + exp<b> |
| Test | M(p) ≥ exp<b> | M'(p) = M(p) |
| Inhibitor | M(p) ≤ exp<b> | M'(p) = M(p) |

**Fig 2.1:** Basic arcs and their interleaving semantics for binding element (t,b)

The enabling condition depends on the current marking *M* and the arc inscription *exp* evaluated in the binding, i.e. *exp<b>*. The occurrence effect indicates the change in marking $M$ leading to a new marking $M'$. The annotation $p{:}C$ indicates that place $p$ contains tokens of colour set $C$. Note that these conditions are specified for the effect of the arc in isolation, i.e. for an interleaving semantics.

Multiple binding elements are concurrently enabled if their respective enabling conditions (as above) are satisfied and if their cumulative occurrence effects can be applied. Furthermore, it is important to ensure that the diamond rule is satisfied. This means that binding elements $(t_1,b_1)$ and $(t_2,b_2)$ are concurrently enabled only if they can occur together or in any order. These conditions are specified formally in definitions 2.1 and 2.2.

The provision of inhibitor arcs in PT-nets and CP-nets ([1], [2], [3]) is normally restricted to arcs incident on capacity places. The capacity place is then translated into the place together with a complementary place. The complementary place holds the capacity multiset *less* the marking of the original place. This relationship is maintained by ensuring that any addition (removal) of tokens to (from) the place is matched by a corresponding operation on the complementary place. In this way, it is never possible to exceed the place capacity because it is not possible to remove sufficient tokens from the complementary place.

The problem with this approach is that some finite capacity (or at least some multiset which contains at most a finite number of appearances of each colour) needs to be associated with each place having an incident inhibitor arc. This restriction is lessened somewhat by the fact that this capacity may be imposed only on some colours or colour combinations, by the use of capacity projection functions (see [2] for example). We remove this restriction altogether by the construction of a generalised complementary place.

Our basic construction can be demonstrated informally by considering an arbitrary place $p$ with colour set $C$. We generate a new colour set $Z{\times}C$, where $Z$ is some infinite colour set such as the positive integers. Each appearance of a colour is thus paired with a unique integer, thereby providing an essentially infinite supply of colour appearances, each uniquely identified by the associated integer. The place $p$ is then translated into a new place $p'$ and a complementary place $p''$, both holding tokens of colour $Z{\times}C$. Fig 2.2 demonstrates the construction for a simple Coloured Petri Net with one place and two transitions.

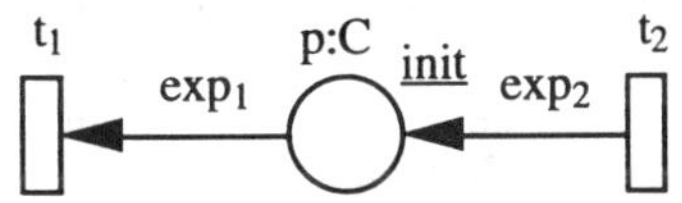


**Fig 2.2:** Simple CP-net with input and output arcs

The place $p$ holds tokens from some colour set $C$ and its initial marking is given by *init*. Transition $t_1$ removes tokens from place $p$, while transition $t_2$ deposits tokens into $p$ according to the arc inscriptions $exp_1$ and $exp_2$ respectively. The above subnet is translated into the behaviourally equivalent subnet with a complementary place, as shown in fig 2.3, and further explained in the subsequent comments.

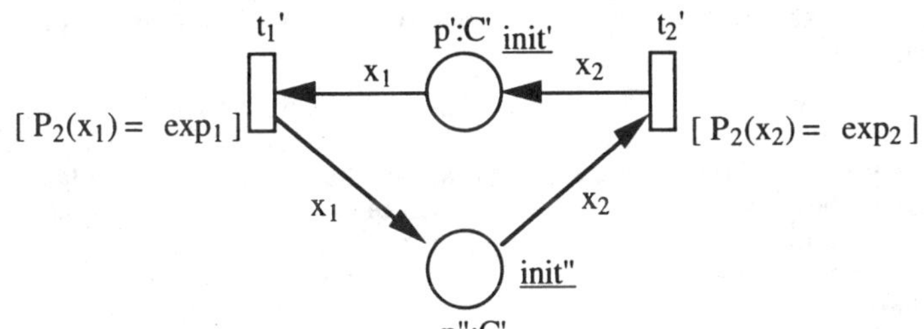


**Fig 2.3:** Translation into a CP-net with generalised complementary place

The notation of fig 2.3 is given below, followed by explanatory comments:

- each primed symbol (e.g. p') is a modified version of one from the original net
- each quoted symbol (e.g. p") relates to the generalised complementary place

- $Z$ = the set of positive non-zero integers (note the non-standard usage of this symbol)
- $C' = Z \times C$ (each element of C is paired with each integer)
- $\text{init}' = \sum_{c \in C} \sum_{i=1}^{\text{init}(c)} 1\text{`}(i,c)$ (suitable choice for initial marking – see below)
- $\text{init}'' = C' - \text{init}'$ (complementary initialisation)
- $x_1$, $x_2$ are new variables of colour $(Z \times C)_{MS}$ (subscript MS indicates a multiset type)
- $P_2$ is a function which projects a multiset of pairs onto a multiset of the second components

In this translation, the colour set *C'* is used to ensure that each appearance of a colour in a multiset over *C* is uniquely identified (by pairing it with a distinct integer). In this way, we have an essentially infinite supply of appearances of each colour from *C*, but uniquely identified by the associated integer. Since the definition of a marking implies that no marking of *p* can ever have an infinite appearance of any one colour, the interaction with the complementary place *p"* will never constrain the firing of transitions $t_1'$ and $t_2'$.

The initial marking *init'* of place *p'* is constructed by arbitrarily pairing the *init(c)* appearances of colour *c* in *init* with the integers *1..init(c)*. All other pairings of this colour with integers are assigned to the complementary place.

Subsequently, each addition of colour *c* to place *p* is mapped into the addition of a colour *(i,c)* to *p'* (for some arbitrary *i*), together with the removal of the same colour *(i,c)* from the complementary place, *p"*. Similarly for the removal of colour *c* from place *p*. The above is guaranteed by the new arcs and their inscriptions.

The new arc inscriptions make use of variables which can take multisets as values. Such variables are not explicitly mentioned in the definition of standard CP-nets [8], but they are implicitly covered by allowing arbitrarily complex data structures as colour sets.

The new arc inscriptions are related to the previous ones by an additional conjunct in each transition guard. This conjunct requires that the new arc inscription projected onto its second component matches the original. In this way, the complementary place always maintains the complement (relative to *C'*) of the marking of place *p'*, i.e.

$$M(p') + M(p'') = C'.$$

The translation to generalised complementary places can also be done for test arcs and inhibitor arcs. For example, the simple Petri Net of fig 2.4 can be translated into the Petri Net with a complementary place of fig 2.5.

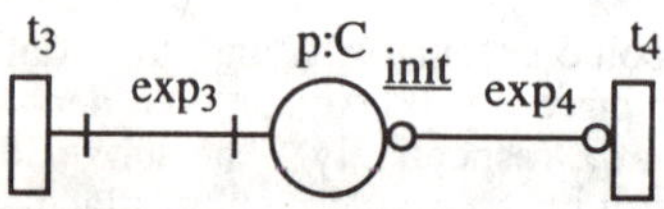


**Fig 2.4:** Simple CP-net with test and inhibitor arcs

Note that we use test arcs so that multiple transitions with inhibitor arcs to the same place can be concurrently enabled.

There is a pleasing symmetry in the above arcs and their interaction with the complementary place. The input and output arcs have a dual effect on token movement relative to the complementary place. The test arc does not affect the marking of the place and hence requires no arc to the complementary place. The dual of this is a test arc on the complementary place which does not affect the translated place, and which provides the mechanism for inhibitor arcs. In this way we see that this basic set of input, output, test, and inhibitor arcs covers the alternatives for interacting with a place and its complement and thus provides the rationale for identifying this set of arcs as fundamental.

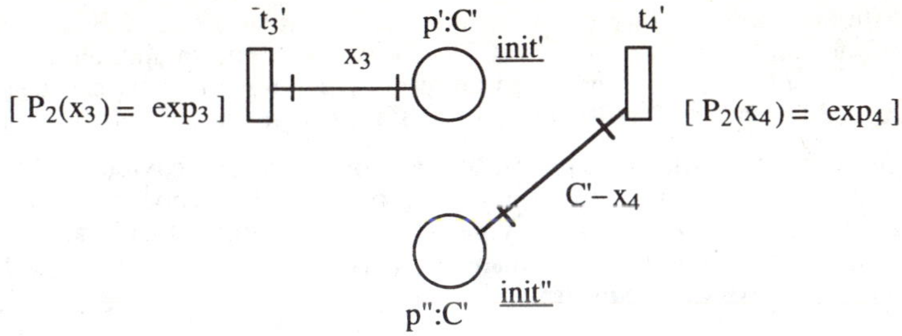


**Fig 2.5:** Translation into a CP-net with general complementary place

The formal definitions of CP-nets extended by the above test arcs and inhibitor arcs are now presented. These definitions assume the standard CP-net definitions ([7], [8]), which are also reproduced in [13].

**Definition 2.1: A CP-net with test arcs and (generalised) inhibitor arcs** is a tuple $CPN_{TI} = (CPN, TS, IS)$ where:

(i) $CPN = (\Sigma, P, T, A, N, C, G, E, I)$ is a non-hiearchical CP-net as in [7], [8].

(ii) $TS = (A_T, N_T, E_T)$ is a test arc specification with:
 (a) $A_T$ is a set of test arcs, such that: $A_T \cap (P \cup T \cup A) = \emptyset$.
 (b) $N_T : A_T \to P \times T$ is a test node function defined on $A_T$.
 (c) $E_T$ is the test expression function defined from $A_T$ into expressions such that:
 $$\forall a \in A_T: [Type(E_T(a)) = C(p(a))_{MS} \wedge Type(Var(E_T(a))) \subseteq \Sigma]$$
 where p(a) is the place of $N_T(a)$.

(iii) $IS = (A_I, N_I, E_I)$ is an inhibitor arc specification with
 (a) $A_I$ is a set of inhibitor arcs, such that: $A_I \cap (P \cup T \cup A \cup A_T) = \emptyset$.
 (b) $N_I : A_I \to P \times T$ is the inhibitor node function defined on $A_I$.
 (c) $E_I$ is the inhibitor expression function defined from $A_I$ into expressions:
 $$\forall a \in A_I : [Type(E_I(a)) = C(p(a))_{MS} \wedge Type(Var(E_I(a))) \subseteq \Sigma]$$
 where p(a) is the place of $N_I(a)$.

This definition can be compared to that of Christensen and Hansen [2]. It differs at point (iii) since the inhibitor arcs do not need to be incident on capacity places. This is possible by virtue of the generalised complementary place construction. On the other hand, it does not include projection functions on inhibitor arcs but this can be included in a more general way as syntactic sugar, as demonstrated in §4.

**Definition 2.2**: A step $Y \in \mathbb{Y}$ is **enabled** in a marking $M \in \mathbb{M}$ of the extended CP-net $CPN_{TI}$ (from definition 2.1) iff the following properties are satisfied:

(i) CP-net enabling:
$$\forall p \in P: \sum_{(t,b) \in Y} E(p,t)<b> \leq M(p).$$

(ii) Test arc enabling:
$$\forall a \in A_T: \forall (t,b) \in Y: E_T(a)<b> \leq M(p(a)) - \sum_{(t',b') \in Y} E(p(a),t')<b'>$$
where p(a) is the place of arc a.

(iii) Inhibitor arc enabling:
$$\forall a \in A_I: \forall (t,b) \in Y: E_I(a)<b> \geq M(p(a)) + \sum_{(t',b') \in Y} E(t',p(a))<b'>$$
where p(a) is the place of arc a.

The only difference between this definition and that of Christensen and Hansen [2] is the simpler inhibitor arc enabling rule. This flows directly from the simpler definition of inhibitor arcs in definition 2.1 and gives an attractive symmetry between test arc and inhibitor arc enabling rules. The firing rule is the standard one for CP-nets.

The extended CP-nets defined above can be transformed into behaviourally equivalent standard CP-nets, as is formally proved in the full paper [13]. It should be noted that the transformation maintains the concurrency properties of the original and transformed net, and is not restricted to an interleaving semantics. This could be emphasised by referring to it as a *concurrency preserving equivalence*.

## 3 Compound Arcs

Given that the four basic arcs presented in §2 are claimed to form a fundamental set, it is appropriate to consider how they can be combined to give more complex arcs. It is possible to form combinations with two, three or four constituent simple arcs and it is possible to have the same arc inscriptions on some or all of the constituents. If we restrict our attention to two-fold combinations, with the same arc inscription, we arrive at the table of fig 3.1 giving the various compound arcs together with their interleaving semantics.

| Graphic | Name | Enabling condition | Occurrence effect |
|---|---|---|---|
| 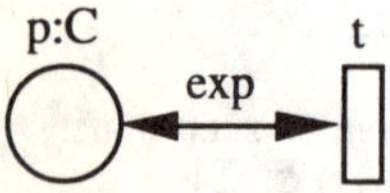  | Reserve | M(p) ≥ exp<b> | M'(p) = M(p) |
| 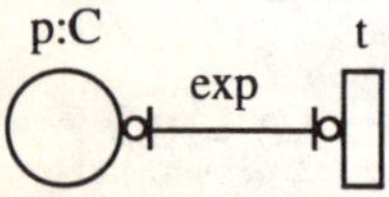  | Equal | M(p) = exp<b> | M'(p) = M(p) |
| 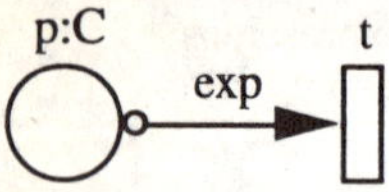  | Clear | M(p) = exp<b> | M'(p) = Ø |
| 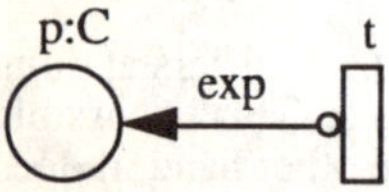  | Set | M(p) = Ø | M'(p) = exp<b> |
| 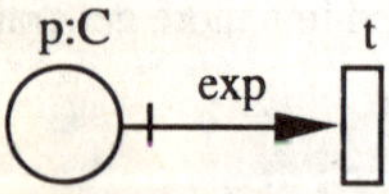  | Halve | M(p) ≥ 2*exp<b> | M'(p) = M(p) – exp<b> |
| 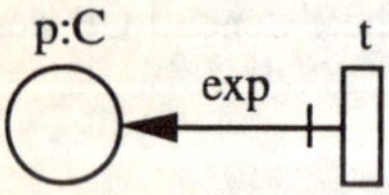  | Double | M(p) ≥ exp<b> | M'(p) = M(p) + exp<b> |

**Fig 3.1:** Table of compound arcs with interleaving semantics for binding element (t,b)

As in fig 2.1 the table gives the enabling condition for a binding element *(t,b)* in a marking *M*, with the occurrence effect leading to marking *M'*. The enabling condition and occurrence effect entries have been derived from fig 2.1 in combination with suitable concurrency assumptions (as in definition 2.2).

This table gives all possible two-fold arc combinations with the same arc inscription. These are all sensible, but those called *halve* and *double* are of doubtful utility. The two-fold arc combinations with different arc inscriptions could be named *update*, *in-range*, *max-restricted-input*, *max-restricted-output*, *min-restricted-input*, *min-restricted-output* respectively. However, we do not consider these further since it is just as easy to handle the arcs separately, both for the modeller and for the tool implementors.

It is also possible to consider more complex arc combinations. However, most of these are not even considered to be sensible since they are only enabled when both place marking and arc inscription are the empty multiset, an effect that can be achieved more simply in other ways. There is one three-fold arc combination, called *clear and set* (*reset* in terms of Billington [1]), which we consider to be sensible and useful, and which is given in fig 3.2.

| Graphic | Name | Enabling condition | Occurrence effect |
|---|---|---|---|
| 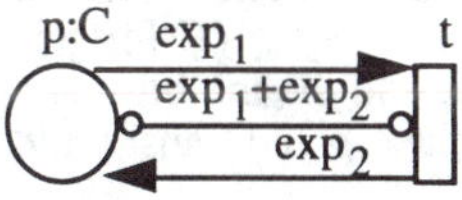 | Clear and set | $M(p) = exp_1\langle b\rangle$ | $M'(p) = exp_2\langle b\rangle$ |

**Fig 3.2:** Table entry for triple arc with interleaving semantics for binding element (t,b)

Apart from *halve* and *double*, all of the above arc extensions have been proposed elsewhere to satisfy the demands of particular application domains. It is very attractive to be able to present these more complex arc extensions as combinations of some basic set, since the combinations can serve as the definitions. Furthermore, once the behavioural equivalence of the extended CP-nets of §2 with standard CP-nets has been proved (see [13]), then no further proofs are required for the other arc combinations. We now demonstrate for three example combinations that these definitions give the appropriate enabling conditions even for concurrently-enabled binding elements.

The first arc we consider is the **equal arc**. This is defined as the combination of a test arc and an inhibitor arc with the common arc inscription, *exp* say. Suppose that these arcs are incident on place *p*. From the test arc enabling rule, we have:

$$\forall (t,b) \in Y: exp\langle b\rangle \leq M(p) - \sum_{(t',b') \in Y} E(p,t')\langle b'\rangle.$$

while, from the inhibitor arc enabling rule, we have:

$$\forall (t,b) \in Y: exp\langle b\rangle \geq M(p) + \sum_{(t',b') \in Y} E(t',p)\langle b'\rangle.$$

Clearly, these two conditions can only both be satisfied iff

$$\forall (t,b) \in Y: exp\langle b\rangle = M(p) \text{ and } \sum_{(t',b') \in Y} E(p,t')\langle b'\rangle = \sum_{(t',b') \in Y} E(t',p)\langle b'\rangle = \emptyset.$$

This condition specifies that the arc inscription (of the equal arc) must exactly match the place marking and no other transition in the step can modify that marking. It therefore follows that the place cannot be updated concurrently with an equal arc. This is exactly the intuitive meaning of an equal arc, and it demonstrates that the semantics of the compound arc can be derived from the semantics of the constituent simpler arcs.

The second arc we consider is the **clear arc**. This is defined as the combination of an input arc and an inhibitor arc with the same arc inscription, *exp* say. Suppose that these arcs are incident on place *p* and transition *t*. From the CP-net enabling rule, we have:

$$\sum_{(t,b) \in Y} E(p,t)\langle b\rangle \leq M(p).$$

For the arc under consideration, the first sum includes the term $exp\langle b\rangle$ and hence

$$exp\langle b\rangle \leq M(p)$$

From the inhibitor arc enabling rule, we have:

$$\forall (t,b) \in Y: exp\langle b\rangle \geq M(p) + \sum_{(t',b') \in Y} E(t',p)\langle b'\rangle.$$

The latter two conditions can only both be satisfied if

$$exp\langle b\rangle = M(p)$$

Furthermore, if we substitute this in the normal enabling rule and in the inhibitor arc enabling rule, we conclude that there is only one occurrence of $t$ in the step, and no other arc removes tokens from the place, i.e.

$$\forall(t,b)\in Y: exp\langle b\rangle = M(p) \wedge \sum_{(t',b')\in Y} E(p,t')\langle b'\rangle = exp\langle b\rangle \wedge \sum_{(t',b')\in Y} E(t',p)\langle b'\rangle = \emptyset.$$

Again, this is the intended semantics of the clear arc.

Thirdly, we consider the more complex **clear and set arc**. This is defined as the combination of an input, an output and an inhibitor arc – the input arc, say $a_G$ ($G$ for get) has inscription $x$, the output arc, say $a_P$ ($P$ for put) has inscription $y$, and the inhibitor arc, say $a_I$ has inscription $x+y$. Suppose that these arcs are incident on place $p$ and transition $t$. From the CP-net enabling rule, we have:

$$\sum_{(t,b)\in Y} E(p,t)\langle b\rangle \leq M(p).$$

For the input arc under consideration, the first sum includes the term $x\langle b\rangle$ for the input arc $a_G$. From the inhibitor arc enabling rule (for arc $a_I$), we have:

$$\forall(t,b)\in Y: (x+y)\langle b\rangle \geq M(p) + \sum_{(t',b')\in Y} E(t',p)\langle b'\rangle.$$

The sum includes the term $y\langle b\rangle$ for the output arc $a_P$. These conditions can only hold if

$$\forall(t,b)\in Y: x\langle b\rangle = M(p) \wedge \sum_{(t',b')\in Y} E(p,t')\langle b'\rangle = x\langle b\rangle \wedge \sum_{(t',b')\in Y} E(t',p)\langle b'\rangle = y\langle b\rangle.$$

Once again, this is the desirable semantics of the arc.

The above examples serve to demonstrate that the various arc extensions summarised in the table of fig 3.1 have the interleaving semantics proposed there. Furthermore, the formal definitions satisfy the demands of the diamond rule and thereby support true concurrency. The formal enabling conditions for the compound arcs are summarised in appendix A.

## 4 Arcs with projected inscriptions

Just as compound arcs provide an important syntactic extension to the basic arcs of §2, another important syntactic extension is to allow arc inscriptions to be modified by a linear projection function. Others ([1], [4]) have argued the utility of arcs which affect a subset or a partition of a place marking. For example, Billington [1] quotes the example of modelling the Cambridge Ring network. Here, when a broadcast message is transmitted, the source-destination pairs for stations which have received the broadcast are stored in a place of the Petri Net. When the broadcast is lost or finally discarded, these source-destination pairs need to be removed and returned to a place holding the acceptable pairs for generating new messages. Since there may be a number of broadcast messages being delivered concurrently, it is necessary to remove only the source-destination pairs relevant to a given broadcast. This can be done with an arc to purge a partition of the marking, the partition being determined by having a specified source.

In the context of this paper, this example requires a *clear* arc with a projected inscription to restrict the tokens being removed. We can consider a generic arc with a projected inscription, as in fig 4.1. The intention is that this arc should have an effect on place $p$ which, under the projection function $f$, is given by $exp$.

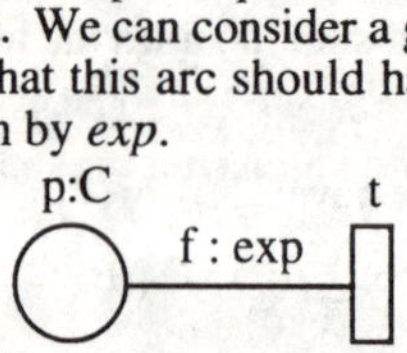


**Fig 4.1:** Generic arc with projected inscription

In the following analysis, it is convenient to identify, for an arbitrary multiset argument, that part which contributes to the function result of $f$ and that part which does not. The latter is referred to as *junk*, though the term is only relative to a particular projection function. So, we define a characteristic function $f'$ and a complementary characteristic function $f''$ as follows:

$$\forall c \in C: \; f'(c) = (\text{if } f(c)=\emptyset \text{ then } \emptyset \text{ else } c)$$

$$\forall c \in C: \; f''(c) = (\text{if } f(c)=\emptyset \text{ then } c \text{ else } \emptyset)$$

These two functions can be linearly extended to apply to multisets over C.

Because of their simplicity and their clearly identified applications, we initially restrict our attention to projected inscriptions where the projection function is a *subset function*. (A subset function maps multisets over colour set C to multisets over colour set C, restricting its argument to the elements of some specified subset of C.)

In the case of $f$ being a subset function, we have that $f'$ is identical to $f$. For such subset functions, the table of basic arcs with enabling conditions is as in fig 4.2. Comparison of this table with that of fig 2.1 shows that the projected inscription *f:exp* on an input or output arc is simply interpreted as an inscription *f'(exp)*. This guarantees that only tokens which are not junk will be added or removed by the projected arc. On the other hand, the test and inhibitor arcs do not move tokens and hence the amount of junk is of no concern. Accordingly, the projected inscriptions are interpreted as inscriptions including a term *f''(y)*, with *y* a free multiset variable, so that *f''(y)* will consist only of junk. Then, a test arc can test for as little junk as desired, while the inhibitor arc can encompass as much junk as required.

| **Graphic** | **Name** | **Enabling condition** | **Occurrence effect** |
|---|---|---|---|
| 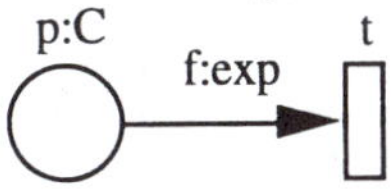  | Input | M(p) ≥ f'(exp)<b> | M'(p) = M(p) – f'(exp)<b> |
| 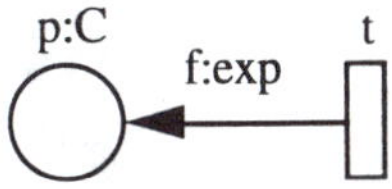  | Output | No constraint | M'(p) = M(p)+ f'(exp)<b> |
| 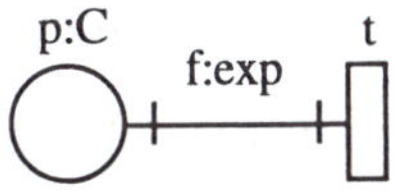  | Test | M(p) ≥ exp<b> – f''(y) | M'(p) = M(p) |
| 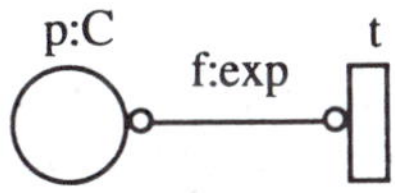  | Inhibitor | M(p) ≤ exp<b> + f''(y) | M'(p) = M(p) |

**Fig 4.2:** Arcs with subset inscriptions + interleaving semantics for binding element (t,b)

Given these interpretations, the basic arcs with projected inscriptions can be combined as in §3 to produce compound arcs with common inscriptions. For example, it is possible to have a clear arc to clear out a subset of a place marking, as shown in fig 4.3.

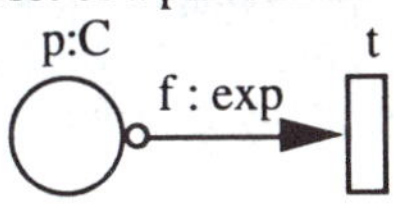


**Fig 4.3:** Subset clear arc

From the table of fig 4.2, this compound arc is interpreted as the subnet of fig 4.4.

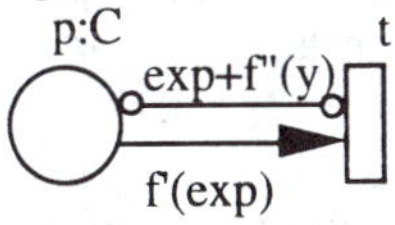


**Fig 4.4:** Expanded subset clear arc

In order to investigate whether the formal semantics of the subset clear arc corresponds to our intuition, we return to the formal enabling conditions of definition 2.2. First, however, we note the following general properties of the characteristicfunction and complementary characteristic function:

$$\forall\, m \in C_{MS}:\ m = f'(m) + f''(m)\ \wedge\ f'(f'(m)) = f'(m)\ \wedge\ f''(f''(m)) = f''(m)$$

$$\forall\, m \in C_{MS}:\ f(f''(m)) = f'(f''(m)) = f''(f'(m)) = \emptyset$$

$$\forall\, m_1, m_2 \in C_{MS}:\ m_1 \le m_2 \Leftrightarrow f'(m_1) \le f'(m_2)\ \wedge\ f''(m_1) \le f''(m_2)$$

Now, the inhibitor arc enabling rule from definition 2.2 gives us:

$$\forall (t,b) \in Y:\ exp\langle b\rangle + f''(y)\langle b\rangle\ \ge\ M(p) + \sum_{(t',b')\in Y} E(t',p)\langle b'\rangle.$$

This is true iff:

$$\forall (t,b) \in Y:\ f'(exp\langle b\rangle)\ \ge\ f'(M(p)) + f'\left(\sum_{(t',b')\in Y} E(t',p)\langle b'\rangle\right)$$

and $$\forall (t,b) \in Y:\ f''(exp\langle b\rangle) + f''(y)\langle b\rangle\ \ge\ f''(M(p)) + f''\left(\sum_{(t',b')\in Y} E(t',p)\langle b'\rangle\right).$$

Because of the free variable *y*, the latter places no restriction on the amount of junk on the right hand side of the inequality. The input arc enabling rule is:

$$\sum_{(t,b)\in Y} E(p,t)\langle b\rangle \le M(p)$$

which holds iff

$$f'\left(\sum_{(t,b)\in Y} E(p,t)\langle b\rangle\right) \le f'(M(p))$$

and $$f''\left(\sum_{(t,b)\in Y} E(p,t)\langle b\rangle\right) \le f''(M(p)).$$

The latter is independent of the clear arc, while the sum in the former includes the term *f'(exp<b>)*. Combined with the inhibitor arc rule this gives:

$$\forall (t,b) \in Y:\ f'(M(p))\ =\ f'(exp\langle b\rangle)\ =\ f'\left(\sum_{(t',b')\in Y} E(p,t')\langle b'\rangle\right)$$

and $$f'\left(\sum_{(t',b')\in Y} E(t',p)\langle b'\rangle\right)\ =\ \emptyset.$$

In other words, the arc removes exactly the non-junk elements of the marking, and no other concurrently-enabled arc can modify the non-junk elements. At the same time, there is no constraint on the modification of junk elements.

This therefore guarantees that we clear a subset of the place marking. In this way, compound arcs with projected inscriptions can encompass the proposals made in the literature for modifying a subset or a partition of a place marking. If this were all that was to be done with projected inscriptions, it could be argued that no specific tool support is required – the user could simply apply the translations given in fig 4.2 by hand. However, it is possible to generalise projected arc inscriptions to apply to arbitrary linear projection functions, and that is what we now propose to do.

We argue that such projected arc inscriptions can be defined as syntactic sugar on top of the simple and compound arcs of §2 and §3. To demonstrate this, it is simpler to consider how the above may be implemented in terms of redundant places. A *redundant place* is one whose marking can be derived from a linear combination of the markings of one or more other places in the Petri Net.

In the following, we assume the use of a linear projection function *f* on an arc incident on place *p* as in fig 4.1. We use the definitions for the characteristic function $f'$ and the complementary characteristic function $f''$, exactly as before. We set up a redundant place $p'$ and ensure the invariance of $M(p') = f(M(p))$. In doing so, we consider the various arcs incident on place $p$ and how they would be translated to interact with the redundant place $p'$.

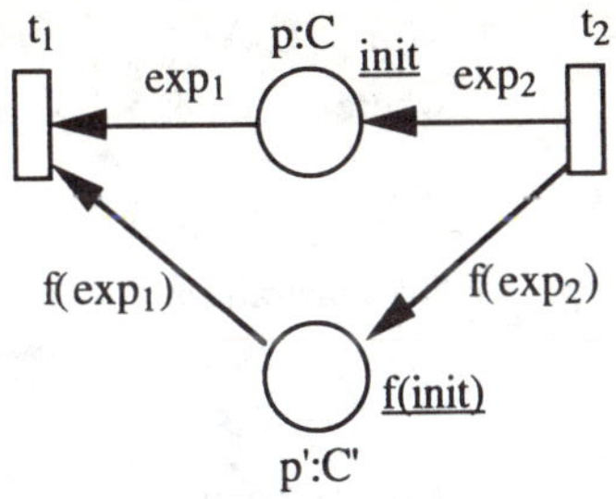


**Fig 4.5:** Translation into a net with (projected) redundant place

Firstly, consider input and output arcs with normal arc inscriptions, as in fig 2.2. This can be translated into the subnet of fig 4.5, where the effect on place $p$ is reflected by a corresponding effect on the redundant place $p'$. On the other hand, the input and output arcs could have projected arc inscriptions, as in fig 4.6, in which case the effect on place $p$ and on the redundant place $p'$ can be shown in the translation of fig 4.7.

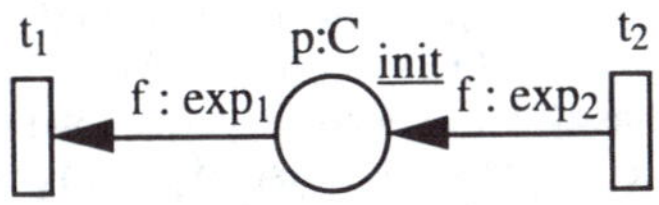


**Fig 4.6:** Simple (extended) net with projected arc inscriptions

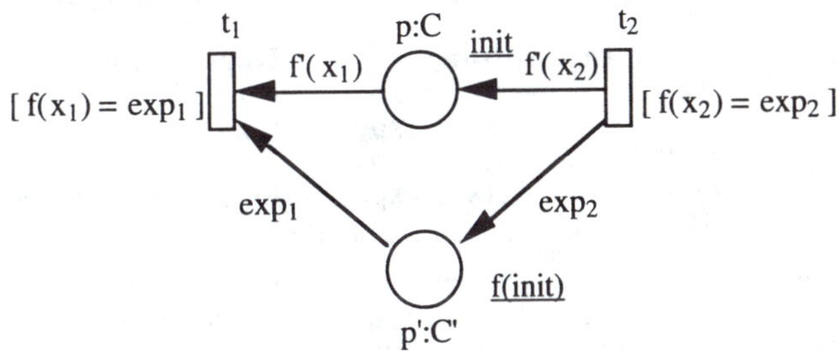


**Fig 4.7:** Translation into a net with (projected) redundant place

Since test and inhibitor arcs do not modify a place marking, they do not need to have a dual test on the original or redundant place. Thus a normal test arc can simply be translated into a test arc on the original place, while a test arc with a projected arc inscription can be translated into a test arc on the redundant place. The same applies to inhibitor arcs.

The above comments can be summarised in the table of fig 4.8, giving the simple arcs and associated enabling conditions.

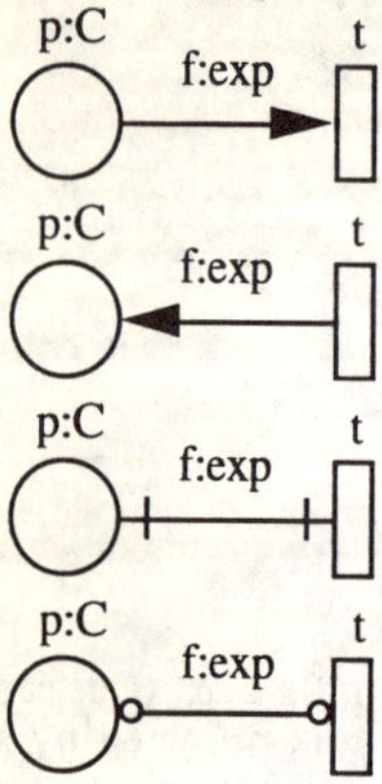


| Graphic | Name | Enabling condition | Occurrence effect |
|---|---|---|---|
| | Input | M(p) ≥ f'(x) where f(x) = exp<b> | M'(p) = M(p) – f'(x) where f(x) = exp<b> |
| | Output | No constraint | M'(p) = M(p) + f'(x) where f(x) = exp<b> |
| | Test | f(M(p)) ≥ exp<b> | M'(p) = M(p) |
| | Inhibitor | f(M(p)) ≤ exp<b> | M'(p) = M(p) |

**Fig 4.8:** Arcs with projected inscriptions + interleaving semantics for binding element (t,b)

It is important to note that the input / output arcs differ from the test / inhibitor arcs in that the former require the existence of a value *x* such that *f(x) = exp<b>*, while the latter do not. The addition of this guard has become necessary, once we move from the special case of *f* being a subset function to the more general case of *f* being an arbitrary linear function. As a result, the syntactic expansions of projected arc inscriptions into normal arc inscriptions and associated transition guards is somewhat irregular, as in the following table:

| Arc kind | Inscription | Guard | Comment |
|---|---|---|---|
| Input | f'(x) | f(x) = exp | No junk allowed |
| Output | f'(x) | f(x) = exp | No junk allowed |
| Test | x – f''(y) | f(x) ≥ exp | Inequality + junk allowed |
| Inhibitor | x + f''(y) | f(x) ≤ exp | Inequality + junk allowed |

**Fig 4.9:** Syntactic translations for projected arc inscriptions

It would clearly be desirable to make this table more regular by having the one guard *f(x)=exp<b>* in all cases. This can be done without making the enabling condition more restrictive provided that function *f* is *dense*, i.e. it satisfies the property:

$$\forall c \in C: \ |f(c)| = 1.$$

It is important to note that the above syntactic expansions are equally valid if function *f* has other parameters besides the multiset. If we were to implement projected arc inscriptions with such functions in terms of redundant places, there would need to be a redundant place for every possible combination of parameter values.

It is now possible to apply these syntactic expansions and determine precise enabling conditions for compound arcs with projected inscriptions. However, the derivations are almost identical to those given above for the special case of *f* being a subset function, and so we omit them. The results of such derivations are summarised in appendix A.

## 5 Tool implications

The arc extensions described in this paper can all be supported by tools, but there are a number of computational issues to be resolved in this regard.

Firstly, it is important to observe that the implementation does not need to perform the translation into ordinary CP-nets. It is sufficient to implement the operational semantics given by the altered enabling rules. Given that some of the compound arcs (such as the equal arc and the clear arcs) evaluate to a single value, their operational semantics will be quite trivial to implement.

Secondly, despite the generality of the construction, caution needs to be exercised in the unconstrained use of these compound arcs by the modeller. A tool should support these extensions, or at least the four basic arcs (input, output, test and inhibitor) together with the equal arc and the clear and set arc. However, if many different compound arcs are used in a single model, then it will be difficult to understand, especially for someone not familiar with the semantic details of the different arcs.

Thirdly, if a tool is to support simulation and occurrence graph analysis of the extended net models, then it will need to determine the enabled bindings in a given marking. The test is trivial for some concrete nets, while being undecidable for others. If we consider an ordinary input or test arc incident on place *p*, inscribed by a single free variable of type *C(p)*, then the number of different values for the variable is bounded by the number of distinct colours in the marking of the place. However, if the arc inscription on the input or test arc has a single free variable of type $C(p)_{MS}$, then the number of possible values is

$$\prod_{c \in C(p)} (M(p)(c) + 1).$$

On the other hand, an ordinary output arc incident on place *p*, inscribed by a single free variable of type *C(p)*, has at most $|C(p)|$ different values. This can obviously be very large (even if *C(p)* is finite), but if the arc inscription has a free variable of type $C(p)_{MS}$ then there will be infinitely many possible values. Finally, for inhibitor arcs the arc expression must exceed the marking *M(p)*, and hence with a single variable of type *C(p)* as arc inscription, the inhibitor arc is only enabled if $|M(p)| = 1$ and the variable must be bound to *c* where *c* is a colour appearing in the marking. On the other hand, if the arc inscription has a single free variable of type $C(p)_{MS}$, then we again have infinitely many values to consider.

The compound arcs impose much stronger demands on the values of the arc expressions, thus making them easier to compute. For example, the equal and clear arcs will have at most one value of the arc expression, while the set arc will be enabled only if *M(p)* is empty. The double arc restricts the enabled bindings just like a test arc. The halve arc works like test and input arcs but it demands that the value of the arc expression must be present twice, i.e. the number of different values of an inscribed multiset variable would be:

$$\prod_{c \in C(p)} (M(p)(c) \text{ div } 2 + 1)$$

In summary, it would be sufficient for many application areas to support equal, clear, and set arcs, and these are much easier to handle in a tool than even their separate test and inhibitor arc components.

## 6 Comparison with other proposed arc extensions

The Petri Net literature contains many proposals for extending nets in order to make them more suitable for modelling purposes. In this section, we review some of these proposals for extensions at the level of arc extensions and place capacities and compare them with the extensions we have considered earlier in this paper.

In an early paper, Jensen [5] proposes a number of arcs for what are now known as Condition-Event Nets, including the possibility of non-deterministic transitions. These are motivated by the desire to define the semantics of the programming language *Delta* in terms of Petri Nets. The suggestions are summarised in the table of fig 6.1.

| | | Initially unmarked | | | |
|---|---|---|---|---|---|
| | | Cannot fire | Remains unmarked | Becomes marked | Unknown marking |
| Initially marked | Cannot fire | Blocked | Test unmarked | Post | – |
| | Remains marked | Test marked | Dummy | Set marked | Increase |
| | Becomes unmarked | Pre | Set unmarked | Change | – |
| | Unknown marking | – | Decrease | – | Random |

**Fig 6.1:** Arc extensions proposed by Jensen for Condition-Event Nets

This table is read by considering the combination of actions depending on the different initial markings. For example, a transition which keeps an unmarked place unmarked, and changes a marked place to an unknown marking (either marked or unmarked) is called *decrease*. In other words, the options suggested are the traditional input and output arcs (called *pre* and *post*), arcs which test for particular markings (called *test marked* and *test unmarked*), arcs which set the place to a particular marking (called *set marked* and *set unmarked*), one which always changes the marking (called *change*), and arcs which are less predictable (called *increase*, *decrease* and *random*). It is interesting to note that all of these possibilities are covered by our own proposals. Even the rather esoteric *decrease* and *increase* could be said to correspond to our own *halve* and *double* arcs.

Billington [1] proposes a Petri Net model with place capacities, inhibitor arcs and threshold arcs. He uses complementary places relative to finite place capacities to model inhibitor and threshold arcs. (Billington's *inhibitor* and *threshold* arcs correspond to our inhibitor arcs with empty and non-empty arc inscriptions, respectively.) In considering the requirements of telecommunications systems he introduces the concepts of *reset* arcs (to reset the marking of a place), *purge* arcs (to clear the marking of a place) and the refinements of these to apply to partitions and subsets of the place colour sets. These extensions are given in terms of combinations of input, output and inhibitor arcs. The proof of equivalence of these extended nets with CP-nets is given only for an interleaving semantics. The correspondence between Billington's proposals and our own can be summarised in tabular form:

| **Billington** | **New proposals** |
|---|---|
| purge | clear |
| reset | clear and set |
| transfer a marking | one clear and one clear and set arc |
| purge a subbag | clear with a projection |
| purge a partition | clear with a projection |
| purge a subset of a partition | clear with a projection |

There are many points of contact between Billington's work and ours. For example, the combination of an input arc with an inhibitor arc with the same inscription for a clear or purge operation is the same. However, in contrast to Billington, we argue that our arc extensions form some fundamental set, we do not require finite capacities to implement inhibitor arcs, and we are careful to specify enabling rules that work with true concurrency.

Christensen and Hansen [2] have proposed net extensions for capacity places, test arcs and inhibitor arcs in the context of CP-nets. The place capacities are specified by a capacity projection function (which can restrict the capacity to various colours or colour combinations)

and a capacity multiset (which limits the number of appearances of each colour). The inhibitor arcs are defined in terms of capacity places, and specify a threshold (in terms of the capacity projection function) which is not to be exceeded by the place marking. The requirement of connecting inhibitor arcs to capacity places is a common one, but it can be an awkward constraint in practice. A significant contribution of this work (which has been used in developing the current proposals) is the specification of enabling rules which support true concurrency, as well as an interleaving semantics. We have already shown that this work generalises theirs by removing the need to define inhibitor arcs on capacity places, and by considering the possible compound arcs. It is also worth noting that their capacity place can be modelled under our proposals by a place where every input arc (of the place) is coupled with an inhibitor arc (with the place capacity as arc inscription).

Heuser and Richter [4] consider the problems of modelling information systems with Coloured Petri Nets and Predicate-Transition Nets. They propose Petri Net extensions for finite place capacities, arc inscriptions with multiset variables, exclusive and shared use of side-conditions, and arc inscriptions for the set (or subset) of all entities in a place marking. These have much in common with the other proposals above, albeit under different names. As previously discussed, the arc inscriptions with multiset variables pose no theoretical problem ( [7], [8]). The side conditions are the test arcs and inhibitor arcs of other presentations. However, they distinguish between exclusive and shared use of these side conditions. In exclusive use, the references to the side conditions remove the tokens and replace them afterwards. In shared use, the tokens need to be present, but they are not removed (and not restored afterwards) and hence can be simultaneously shared by others. They also propose (as does Billington) that arc inscriptions should be able to specify a multiset of tokens from some subset of available token colours. Heuser and Richter do not provide a formal semantics or enabling conditions for their suggested extensions. As a result, it is not possible to evaluate their concurrency properties, though the shared use of side conditions indicates that support for true concurrency was intended. The correspondence between Heuser and Richter's proposals and our own can be summarised in tabular form:

| **Heuser and Richter** | **New proposals** |
|---|---|
| capacity place | inhibitor arc combinations (discussed above) |
| restoring entry arc | reserve arc |
| restoring exit arc | inhibitor arc |
| maintaining entry arc | test arc |
| maintaining exit arc | inhibitor arc |
| total arcs | above arcs with projection functions |

Lakos and Keen ([9], [10]), in proposing object-oriented extensions to Coloured Petri Nets in the language LOOPN, define the concept of place summary functions. These functions serve to determine some value from a place marking without modifying the place. These functions can be exported from Petri Net modules (or subnets), in which case they provide a non-destructive way of inspecting the state of a module. A place summary functions can be emulated simply as an equal arc with a projected inscription. The attractive thing about this solution is that it automatically gives the appropriate concurrency limitations, namely that there may be an arbitrary number of concurrent references to the function, but *not* concurrent with any modification to the place.

To conclude this section, we present a solution to the original rowdy philosophers problem in fig 6.2 making use of our arc extensions. In that diagram, the *eject* transition uses an equal arc to determine the exact marking of the place *eating*, and simultaneously uses a *clear* arc to remove all the rowdy philosophers. The transition guard guarantees that the philosophers are in fact rowdy, by checking that the number of philosophers with left fork raised exceeds the number of eating philosophers by a specified margin (here *incr*). The output arcs ensure that these rowdy philosophers and their forks are suitably handled. The logical demands of the compound arcs ensure that they cannot be enabled concurrently with philosophers entering or leaving the places *hasLeft* or *eating*, but other concurrent activities are allowed.

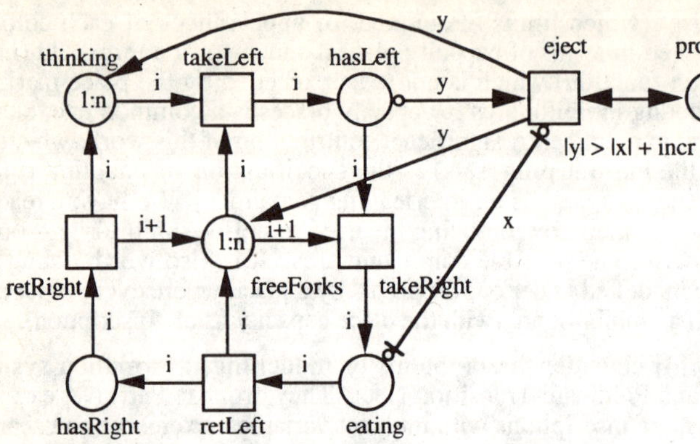


**Fig 6.2:** Solution to the rowdy philosophers problem

# 7 Conclusions

This paper has considered the implications of a generalised complementary place construction in the context of CP-nets. On the basis of this construction together with symmetry considerations, it was demonstrated that input arcs, output arcs, test arcs and inhibitor arcs supply a set of four simple, fundamental arcs. These can be combined in arbitrary ways to form useful compound arcs, of which four or five are considered to be of practical utility.

By extending the arc inscriptions to be able to specify projection functions, the arc extensions presented recently in the literature can be emulated. In particular, place purging and place summary functions can be neatly supported in this framework. Similarly, a capacity place, with the generality of Christensen and Hansen [2] can also be simply described in this formalism.

The presentation has been careful to ensure that the semantics of the various arcs is amenable to concurrently enabled and occurring steps, and not just an interleaving semantics. As Devillers [3] notes, others have not always had the same concerns. Note that our constructions can be reduced to an interleaving semantics as a special case, and they can also be reduced to the case of finite capacity bounds.

The cost of this construction has been the introduction of an infinite colour set for uniquely identifying each colour appearance in a multiset. This precludes the analysis of the Petri Net by unfolding it into a finite PT-net, since this unfolded net would have an infinite number of places, transitions and arcs. There is also a computational cost involved in supporting multiset variables as arc inscriptions, which have been introduced in some of the constructions. However, it turns out that this computational cost is much less for compound arcs than for simple, unconstrained arcs.

While it would be desirable for the implementors of individual tools to support the arc extensions in this paper, users should be selective in the extensions included in any particular diagram. It may be that some will be of particular interest in specific application domains. Unconstrained use of the arc extensions will make it hard for others to understand the diagram and remember the semantics of each.

## Acknowledements

The authors are pleased to acknowledge the detailed and helpful discussions held with Kurt Jensen and Kjeld Mortensen.

## References


[1] Billington, J. *Extensions to Coloured Petri Nets and their Application to Protocols* PhD thesis, Computer Laboratory, University of Cambridge (1988).

[2] Christensen, S. and Hansen, N.D. *Coloured Petri Nets Extended with Place Capacities, Test Arcs and Inhibitor Arcs* Proceedings of 14th International Conference on Application and Theory of Petri Nets, Chicago (1993).

[3] Devillers, R. *The Semantics of Capacities in P/T Nets* Advances in Petri Nets 1989. Lecture Notes in Computer Science vol 424, Springer-Verlag.

[4] Heuser, C.A. and Richter, G. *Constructs for Modeling Information Systems with Petri Nets* Proceedings of the 13th International Conference on Application and Theory of Petri Nets, Springer Verlag (1992).

[5] Jensen, K. *Net Models in System Description* PhD thesis, Computer Science Department, Aarhus University, Aarhus, Denmark (1980).

[6] Jensen, K. *Coloured Petri Nets*, Advances in Petri Nets 1986, Lecture Notes in Computer Science 254, Springer-Verlag.

[7] Jensen, K. *Coloured Petri nets: A high level language for system design and analysis* Advances in Petri Nets 1990, Lecture Notes in Computer Science 483, Springer Verlag.

[8] Jensen, K. *Coloured Petri Nets: Basic Concepts, Analysis Methods and Practical Use* Volume 1, EATCS Monographs in Computer Science, Springer Verlag (1992).

[9] Lakos, C.A. *LOOPN – Language for Object-Oriented Petri Nets* Technical Report 91-1, Department of Computer Science, University of Tasmania (1991).

[10] Lakos, C.A. and Keen, C.D. *Modelling Layered Protocols in LOOPN* Proceedings of 4th International Workshop on Petri Nets and Performance Models, Melbourne (1991).

[11] Lakos, C.A. and Keen, C.D. *Modelling a Door Controller Protocol in LOOPN* Proceedings of the TOOLS Europe '93 Conference, (1993).

[12] Lakos, C.A. *CP-nets Extended with Place Summary Functions* Technical Report TR93-6, Department of Computer Science, University of Tasmania (1993).

[13] Lakos, C.A. and Christensen, S. *A General Systematic Approach to Arc Extensions for Coloured Petri Nets* Technical Report TR93-8, Computer Science Department, University of Tasmania (1993).

[14] Valmari, A. *Stubborn Sets for Coloured Petri Nets* Proceedings of 12th International Conference on the Application and Theory of Petri Nets, Aarhus (1991).

## Appendix A: Enabling rules for compound and projected arcs

This appendix presents the full enabling rules for the compound and projected arcs introduced in §3 and §4. These results can be derived by combining the enabling rules of the constituent arcs, as we did for a couple of examples in each of the respective sections.

**Definition A.1**: Given a step $Y \in \mathbb{Y}$ and a marking $M \in \mathbb{M}$ of the extended CP-net $CPN_{\Pi}$ (from definition 2.1), we define (for simplicity)

$$get(p) = \sum_{(t',b') \in Y} E(p,t') < b' >$$

and

$$put(p) = \sum_{(t',b') \in Y} E(t',p) < b' >.$$

Then, step Y is **enabled** iff the following properties are satisfied:

(i) CP-net enabling:

$$\forall p \in P: \ get(p) \leq M(p).$$

(ii) Test arc enabling:

$$\forall a \in A_T: \forall (t,b) \in Y: \ E_T(a) < b > \ \leq M(p(a)) - get(p(a)).$$

(iii) Inhibitor arc enabling:

$$\forall a \in A_I: \forall (t,b) \in Y: \ E_I(a) < b > \ \geq M(p(a)) + put(p(a)).$$

(iv) Reserve arc enabling:

$$\forall a_1, a_2 \in A: N(a_1) = (p,t) \wedge N(a_2) = (t,p) \wedge E(a_1) = E(a_2) \Rightarrow$$
$$E(a_1) \leq get(p(a_1)) \leq M(p(a_1)).$$

(v) Equal arc enabling:

$$\forall a_T \in A_T: \forall a_I \in A_I: N_T(a_T) = N_I(a_I) \wedge E_T(a_T) = E_I(a_I) \Rightarrow$$
$$\forall (t,b) \in Y: \ E_T(a_T) < b > \ = M(p(a_T)) \ \wedge \ get(p(a_T)) = put(p(a_T)) = \emptyset.$$

(vi) Clear arc enabling:

$$\forall a \in A: \forall a_I \in A_I: N(a) = N_I(a_I) \wedge E(a) = E_I(a_I) \Rightarrow$$
$$\forall (t,b) \in Y: \ E(a) < b > \ = get(p(a)) = M(p(a)) \ \wedge \ put(p(a)) = \emptyset.$$

(vii) Set arc enabling:

$$\forall a \in A: \forall a_I \in A_I: N(a) = (t,p) \wedge N_I(a_I) = (p,t) \wedge E(a) = E_I(a_I) \Rightarrow$$
$$\forall (t,b) \in Y: \ get(p(a)) = M(p) = \emptyset \ \wedge \ put(p(a)) = E(a) < b >.$$

**Definition A.2**: Given a step $Y \in \mathbb{Y}$ and a marking $M \in \mathbb{M}$ of the extended CP-net $CPN_{\Pi}$ (from definition 2.1), we assume (for the sake of simplicity) that all arcs have projected inscriptions, and define:

| $\forall a \in A \cup A_T \cup A_I$: | $E(a)$ | = arc inscription according to the table of fig 4.9 |
|---|---|---|
| | $E'(a)$ | = expression from the projected inscription |
| | $F(a)$ | = projection function on arc inscription |
| | $F'(a)$ | = characteristic function for F(a) |
| | $F''(a)$ | = complementary characteristic function for F(a). |

Given these definitions, we can define get(p) and put(p) as in definition A.1.

Then, step Y is **enabled** iff the following properties are satisfied:

(i) CP-net enabling:

$$\forall p \in P: \ get(p) \leq M(p) \ \wedge$$
$$\forall a \in A: \ \forall (t,b) \in Y: \ F(a)(E(a)) < b > \ = E'(a) < b > \ \wedge \ F''(a)(E(a)) < b > \ = \emptyset.$$

(ii) Test arc enabling:

$$\forall a \in A_T: \forall (t,b) \in Y: \ E_T'(a) < b > \ \leq F(a)(M(p) - get(p)).$$

(iii) Inhibitor arc enabling:

$$\forall a \in A_I: \forall (t,b) \in Y: \ E_I'(a) < b > \ \geq F_I(a)(M(p) + put(p)).$$

(iv) Reserve arc enabling:

$$\forall a_1,a_2 \in A: N(a_1) = (p,t) \wedge N(a_2) = (t,p) \wedge E'(a_1) = E'(a_2) \wedge F(a_1) = F(a_2) \Rightarrow$$
$$\forall (t,b) \in Y: \; E'(a_1)<b> \leq F(a_1)(get(p)) \leq F(a_1)(M(p)) \wedge$$
$$F''(a_1)(E(a_1))<b> = F''(a_1)(E(a_2))<b> = \emptyset.$$

(v) Equal arc enabling:

$$\forall a_T \in A_T: \forall a_I \in A_I: N_T(a_T) = N_I(a_I) \wedge E_T'(a_T) = E_I'(a_I) \;\wedge\; F_T(a_T) = F_I(a_I) \Rightarrow$$
$$\forall (t,b) \in Y: \; E_T'(a_T)<b> = F_T(a_T)(M(p(a_T))) \wedge$$
$$F_T(a_T)(get(p(a_T))) = F_T(a_T)(put(p(a_T))) = \emptyset.$$

(vi) Clear arc enabling:

$$\forall a \in A: \forall a_I \in A_I: N(a) = N_I(a_I) \wedge E'(a) = E_I'(a_I) \;\wedge\; F(a) = F_I(a_I) \Rightarrow$$
$$\forall (t,b) \in Y: \; E'(a)<b> = F(a)(get(p(a))) = F(a)(M(p(a))) \wedge$$
$$F(a)(put(p(a))) = \emptyset = F''(a)(get(p(a))).$$

(vii) Set arc enabling:

$$\forall a \in A: \forall a_I \in A_I: N(a) = (t,p) \wedge N(a_I) = (p,t) \wedge E'(a) = E_I'(a_I) \;\wedge\; F(a) = F_I(a_I)$$
$$\Rightarrow$$
$$\forall (t,b) \in Y: \; E'(a)<b> = F(a)(put(p(a))) \wedge$$
$$F(a)(get(p(a))) = F(a)(M(p(a))) = \emptyset = F''(a)(get(p(a))).$$

# Liveness in Bounded Petri Nets which are Covered by T–Invariants

Kurt Lautenbach and Hanno Ridder

University Koblenz–Landau
Institute for Software Technology
Rheinau 1
56075 Koblenz, Germany
E–Mail: {laut, ridder}@informatik.uni-koblenz.de

**Abstract:** In this paper a criterion is introduced that is sufficient for the liveness in Petri nets which are bounded and covered by non–negative T–invariants.

**Keywords:** liveness, linear invariants, deadlocks

## 1 Introduction

Petri nets are well known models for the representation and analysis of distributed systems (see, e.g. [Rei91, Bau90]). The net itself reflects the topological structure of the system, whereas the variability of the markings represents the dynamic behaviour.

The liveness of a marking, i.e. the fact that every transition can be enabled again and again, is a property that is as important as formally hard to treat.

In contrast to marked graphs, where a marking is live iff all non–negative S–invariants are marked, nets with shared places show a certain "lack of structure" w.r.t. all known methods to investigate liveness. Of course, there is the famous liveness criterion for extended free choice nets according to which a net is live iff all its deadlocks contain a marked trap. But here the lack of structure is compensated by the fact that the structure of extended free choice nets is "relatively simple".

Continuing this way of loosely speaking about net structures, the method we introduce in this paper consists of adding structure to the nets without changing their dynamic properties.

In detail, by adding of so–called regulation circuits to a net we get a new net consisting of only one (elementary) T–invariant. This new net has two desirable properties. Firstly, by increasing the marking of the regulation circuits sufficiently, the new net becomes an arbitrarily exact approximation of the original net. Secondly, the structure of one–T–invariant nets is rich enough for liveness investigations. So, in order to test a net for liveness we examine wether it can be approximated by live one–T–invariant nets. The idea of transforming nets into one–T–invariant nets is not new (cf. [Lau77, KL82, CCS91]). What is new is to exploit consequently the possibility of approximating the original nets by the one–T–invariant nets.

The class of bounded nets which are covered by T–invariants guarantees a considerable modelling power. All systems with a system wide ability to reproduce situations and without an unlimited increase of the number of objects can be modelled by means of such nets. From the view point of net theory it is important that this class contains a large class of non–free–choice nets.

The paper is organized as follows. Section 2 contains basic definitions and notations, in particular S– and T–invariants, deadlocks and traps. In section 3 the concept of a controlled deadlock is introduced. Controlled deadlocks cannot get unmarked. The mechanism that prevents a controlled deadlock from geting unmarked is quite different from a marked trap inside the deadlock. Next, we deal with liveness in one–T–invariant nets. These nets have the property of being live iff they are weakly live, and sufficient for the weak liveness is that no deadlock can be emptied. Finally, we treat the liveness of bounded nets covered by T–invariants by approximating these nets by one–T–invariant nets.

## 2 Basic Definitions and Notations

This section contains the basic definitions and notations which will be needed in the rest of the paper.

### 2.1 Place/Transition Nets

**Definition 1.**

1. A *net* is a tripel $\mathcal{N} = (S, T, F)$ with
   (a) $S$ and $T$ are finite, nonempty and disjoint sets.
   (b) $F \subseteq (S \times T) \cup (T \times S)$.
   The elements of $S$ are called *places* and the elements of $T$ *transitions*.
2. The *preset* of a node $x \in S \cup T$ is defined as ${}^\bullet x = \{y \in S \cup T \mid (y, x) \in F\}$. The *postset* of $x \in S \cup T$ is $x^\bullet = \{y \in S \cup T \mid (x, y) \in F\}$. The preset (postset) of a set is the union of the presets (postsets) of the elements.

**Definition 2.** Let $\mathcal{N} = (S, T, F)$ be a net.

1. A *marking* of a net $\mathcal{N} = (S, T, F)$ is a mapping $M : S \to \mathbb{N}$.
2. The pair $(\mathcal{N}, M)$ is called a *net system* or *marked net*; $M$ is called the *initial marking*.
3. A transition $t \in T$ is called *enabled* under $M$, in symbols $M[t\rangle$, iff
   $\forall s \in {}^\bullet t : M(s) \geq 0.$
4. If $M[t\rangle$, the transition $t$ may *occur*, resulting in a new marking $M'$, in symbols $M[t\rangle M'$, with

$$M'(s) = \begin{cases} M(s) - 1 \text{ if } & s \in {}^\bullet t \setminus t^\bullet \\ M(s) + 1 \text{ if } & s \in t^\bullet \setminus {}^\bullet t \\ M(s) & \text{otherwise} \end{cases}$$

   for all $s \in S$.

5. The set of all *reachable markings*, in symbols $[M_0\rangle$, of a marking $M_0$ is the smallest set, such that
   $M_0 \in [M_0\rangle$
   $M \in [M_0\rangle \wedge M[t\rangle M' \Rightarrow M' \in [M_0\rangle$
   holds.
6. If $M_0[t_1\rangle M_1[t_2\rangle \ldots [t_n\rangle M_n$, then $\sigma = t_1 t_2 \ldots t_n$ is called an *occurence sequence*.

**Definition 3.** Let $(\mathcal{N}, M_0)$ be a net system and $\mathcal{N} = (S, T, F)$.

1. A transition $t \in T$ is called *live* under $M_0$, iff $\forall\, M \in [M_0\rangle\, \exists\, M' \in [M\rangle : M'[t\rangle$.
2. The net $\mathcal{N}$ is called *dead* under $M_0$, iff $\nexists\, t \in T : M_0[t\rangle$.
3. The net $\mathcal{N}$ is called *weakly live* or *deadlock-free* under $M_0$, iff $\forall\, M \in [M_0\rangle\, \exists\, t \in T : M[t\rangle$.
4. The net $\mathcal{N}$ is called *live* under $M_0$, iff $\forall\, t \in T : t$ is live under $M_0$.

**Definition 4.** Let $(\mathcal{N}, M_0)$ be a net system and $\mathcal{N} = (S, T, F)$. $(\mathcal{N}, M_0)$ is called *bounded* iff

$$\exists\, k \in \mathbb{N}\, \forall\, M \in [M_0\rangle, s \in S : M(s) \leq k.$$

## 2.2 Invariants

A net $\mathcal{N} = (S, T, F)$ is called *pure* iff $\nexists\, s \in S, t \in T : (s,t) \in F \wedge (t,s) \in F$. We assume that in the following all nets are pure.

**Definition 5.** Let $\mathcal{N} = (S, T, F)$ be a net.

1. A column vector $v : S \to \mathbb{Z}$ indexed by $S$ is called an *S–vector*.
2. A column vector $w : T \to \mathbb{Z}$ indexed by $T$ is called a *T–vector*.
3. A matrix $[\mathcal{N}] : S \times T \to \mathbb{Z}$ indexed by $S$ and $T$ such that

$$[\mathcal{N}](s,t) = \begin{cases} -1 \text{ if } & s \in {}^\bullet t \setminus t^\bullet \\ \;\;1 \text{ if } & s \in t^\bullet \setminus {}^\bullet t \\ \;\;0 \text{ otherwise} & \end{cases}$$

   for all $s \in S$ and for all $t \in T$ is called the *incidence matrix* of $\mathcal{N}$.

We denote column vectors where every component equals 0 by $\mathbf{0}$ and column vectors where every component equals 1 by $\mathbf{1}$.

**Definition 6.** Let $i$ be an S–vector and $j$ a T–vector of $\mathcal{N} = (S, T, F)$.

1. $i$ is called an *S–invariant* iff $i \neq \mathbf{0}$ and $i^t * [\mathcal{N}] = \mathbf{0}^t$.
2. $j$ is called a *T–invariant* iff $j \neq \mathbf{0}$ and $[\mathcal{N}] * j = \mathbf{0}$.
3. $\| i \| = \{s \in S \mid i(s) \neq 0\}$ ($\| j \| = \{t \in T \mid j(s) \neq 0\}$) is called the *support of* $i$ ($j$).
4. An S–invariant $i$ (T–invariant $j$) is called *non–negative* iff
   $\forall\, s \in S : i(s) \geq 0$ ($\forall\, t \in T : j(t) \geq 0$).

5. A non–negative S–invariant $i$ (T–invariant $j$) has *minimal support* iff
   $\nexists$ S–invariant $i' : \|i'\| < \|i\|$ ($\nexists$ T–invariant $j' : \|j'\| < \|j\|$).
6. A non–negative S–invariant $i$ (T–invariant $j$) is called *canonical* iff the greatest common divisor of its components is 1.
7. A non–negative S–invariant $i$ (T–invariant $j$) is called *elementary* iff it is canonical and has minimal support.
8. The *subnet* $\mathcal{N}_i = (S_i, T_i, F_i)$ generated by an S–invariant $i$ is defined by
   $S_i := \|i\|$
   $T_i := {}^\bullet S_i \cup S_i^\bullet$
   $F_i := F \cap ((S_i \times T_i) \cup (T_i \times S_i))$
   The subnet $\mathcal{N}_j = (S_j, T_j, F_j)$ generated by a T–invariant $j$ is defined analogously.
9. $\mathcal{N}$ is called *covered* by an S–invariant $i$ (T–invariant $j$) iff
   $\forall s \in S : i(s) \neq 0 \;\; (\forall t \in T : j(t) \neq 0)$.

The importance of S– and T–invariants, which have been first introduced in [Lau73], is given by the properties of the generated subnets. Subnets $\mathcal{N}_i$ generated by S–invariants $i$ are the only ones with a token load invariance $i^t * M = i^t * M_0$ for all $M \in [M_0\rangle$. Subnets $\mathcal{N}_j$ generated by T–invariants $j$ are the only ones in which the reproducibility of markings is possible.

### 2.3 Deadlocks and Traps

**Definition 7.** Let $\mathcal{N} = (S, T, F)$ be a net.

1. A nonempty set $H \subseteq S$ is called a *deadlock* iff ${}^\bullet H \subseteq H^\bullet$.
2. A nonempty set $H \subseteq S$ is called a *trap* iff $H^\bullet \subseteq {}^\bullet H$.
3. Let $H$ be a deadlock (trap). $H$ is called *minimal* iff there is no deadlock (trap) contained in $H$ as a proper subset.

We call a place $s \in S$ is *marked* by a marking $M$ iff $M(s) > 0$ and a nonempty set of places $H \subseteq S$ is called marked by a marking $M$ iff at least one element of $H$ is marked.

Once a deadlock lost all tokens it remains unmarked. Once a trap gained at least one token it remains marked.

In [Lau87][ES92] it is demonstrated that deadlocks (traps) in strongly connected nets can be calculated as special multisets of circuits. A net $\mathcal{N} = (S, T, F)$ is called *strongly connected* iff the directed graph $(S \cup T, F)$ is strongly connected. A *circuit* in a net $\mathcal{N} = (S, T, F)$ is defined as a circuit in the directed graph $(S \cup T, F)$.

**Definition 8.** Let $\mathcal{N} = (S, T, F)$ be a net. Let $C$ be a nonempty multiset of circuits of $\mathcal{N}$ and $C(f)$ denote the number of circuits of $C$ which pass through $f \in F$.

1. $C$ is called a *D–system* iff $\forall s \in S \; \exists n_s \in \mathbb{N} \; \forall t \in {}^\bullet s : C(t, s) = n_s$.
   This means, in a D–system $C$ the same number $n_s$ of circuits passes through all *input arcs* of a place $s \in S$. We want to assign $C(s) := n_s$ for all $s \in S$.

2. $C$ is called a *T-system* iff $\forall s \in S \, \exists n_s \in \mathbb{N} \, \forall t \in s^\bullet : C(s,t) = n_s$.
   This means, in a T-system $C$ the same number $n_s$ of circuits passes through all *output arcs* of a place $s \in S$. We want to assign $C(s) := n_s$ for all $s \in S$.
3. $\| C \| = \{s \in S \mid C(s) > 0\}$ is called the *support* of $C$.
4. A D-system (T-system) $C$ is called *minimal* iff
   (a) $\nexists$ D-system (T-system) $C' : \| C' \| \subset \| C \|$
   (b) the g.c.d. of the $C(s)$ with $C(s) > 0 \wedge s \in S$ is 1.

**Theorem 9.** *Let $\mathcal{N} = (S, T, F)$ be a strongly connected net.*

*1. If $C$ is a D-system (T-system), then $\| C \|$ is a deadlock (trap).*
*2. If $H \subseteq S$ is a minimal deadlock (trap), then there exists a minimal D-system (T-system) $C$ such that $\| C \| = H$.*

The proof is given in [Lau87, ES92].

# 3 The Liveness Condition

## 3.1 Controlled Deadlocks

In this subsection the concept of a controlled deadlock is introduced. Controlled deadlocks cannot get unmarked.

**Definition 10.** Let $(\mathcal{N}, M_0)$ be a net-system, let $i$ be an S-invariant and let $D \subseteq S$ be a deadlock of $\mathcal{N}$. The deadlock $D$ is called *controlled* by the S-invariant $i$ under $M_0$ iff

$$i^t * M_0 > 0 \wedge \forall s \in S \setminus D : i(s) \leq 0.$$

**Lemma 11.** *Let $(\mathcal{N}, M_0)$ be a net-system and let $D \subseteq S$ be a deadlock of $\mathcal{N}$. If $D$ is controlled by an S-invariant $i$ under $M_0$ the following holds*

$$\forall M \in [M_0\rangle : D \text{ is marked under } M.$$

**Proof**

Since $i$ is an S-invariant the following holds: $i^t * M_0 > 0 \Rightarrow \forall M \in [M_0\rangle : i^t * M > 0$. Only the places $s \in D$ have positive entries in $i$. Hence, since $i^t * M > 0$, there must be entries $M(s) > 0 \wedge s \in D$. So $D$ is marked under $M$. □

In figure 2, $D = \| d \|$ is a deadlock, $i_1$ and $i_2$ are S-invariants of the net $\mathcal{N}_1$ in figure 1. Every positive entry in $i_1$ or $i_2$ belongs to a place in $D$. $i_1^t * M = 0$ and $i_2^t * M = 1$ holds. Thus the deadlock $D$ is controlled by the S-invariant $i_2$ and remains marked.

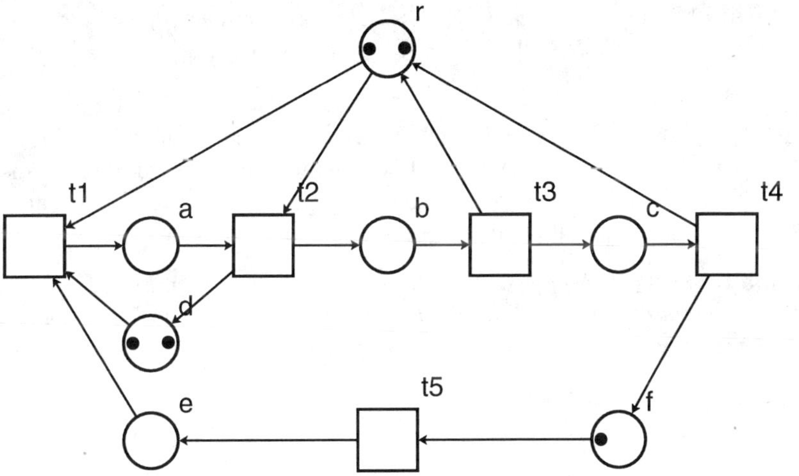


**Fig. 1.** net $\mathcal{N}_1$

| $[\mathcal{N}_1]$ | $t_1$ | $t_2$ | $t_3$ | $t_4$ | $t_5$ | $M$ | $d$ | $i_1$ | $i_2$ | $x_1$ | $x_2$ |
|---|---|---|---|---|---|---|---|---|---|---|---|
| $a$ | 1 | −1 | | | | 0 | 0 | 0 | 0 | 0 | 0 |
| $b$ | | 1 | −1 | | | 0 | 2 | 2 | 1 | 0 | 1 |
| $c$ | | | 1 | −1 | | 0 | 1 | 1 | 0 | 0 | 1 |
| $d$ | −1 | 1 | | | | 2 | 0 | −1 | 0 | 1 | 0 |
| $e$ | −1 | | | | 1 | 0 | 0 | 0 | −1 | 0 | 1 |
| $f$ | | | | 1 | −1 | 1 | 0 | 0 | −1 | 0 | 1 |
| $r$ | −1 | −1 | 1 | 1 | | 2 | 1 | 1 | 1 | 0 | 0 |
| $\Delta_d^t$ | −1 | 1 | 0 | 0 | 0 | | | | | | |

**Fig. 2.** incidence matrix, initial marking, deadlock and S–invariants of $\mathcal{N}_1$

**Definition 12.** Let $\mathcal{N} = (S, T, F)$ be a strongly connected net and $d$ the D–system of a deadlock $D = \|d\|$. The T–vector $\Delta_d$ with $\Delta_d^t = d^t * [\mathcal{N}]$ is called the *defect* of $D$.

If $\Delta_d(t) = 0$ for all $t \in T$ then $D$ is an S–invariant. The transitions $\{t \in T \mid \Delta_d(t) < 0\}$ decrease the token load on $D$ and the transitions $\{t \in T \mid \Delta_d(t) > 0\}$ increase the token load on $D$. If $D$ cannot lose all tokens these two sets of transitions must be regulated in some way:

**Lemma 13.** *Let $(\mathcal{N}, M_0)$ be a strongly connected net–system and $\mathcal{N} = (S, T, F)$, $d$ the D–system of a deadlock $D = \|d\|$ with defect $\Delta_d \neq 0$ and $x$ a non–negative S–vector. $D$ is controlled by the S–invariant $(d - x)$ if*

$$x^t * [\mathcal{N}] = \Delta_d^t \wedge (d - x)^t * M_0 > 0.$$

**Proof**

$(d - x)$ is an S–invariant, because $x^t * [\mathcal{N}] = \Delta_d^t \wedge d^t * [\mathcal{N}] = \Delta_d^t \Rightarrow (d - x)^t * [\mathcal{N}] = 0^t$. $d$ and $x$ are both non–negative S–vectors, so $\forall s \in S : (d - x)(s) > 0 \Rightarrow d(s) > 0 \Rightarrow s \in D$. This implies $\forall s \in S \setminus D : (d - x)(s) \leq 0$. □

Now we are able to formulate a suffient condition for a deadlock to remain marked.

**Theorem 14.** *Let $(\mathcal{N}, M_0)$ be a strongly connected net–system and $d$ the D–system of a deadlock $D = \|d\|$ with defect $\Delta_d$. If*

$$\exists\, S\text{–}vector\ x \geq 0 : x^t * [\mathcal{N}] = \Delta_d^t \wedge (d - x)^t * M_0 > 0$$

*then*

$$\forall M \in [M_0\rangle : D \text{ is marked under } M$$

*holds.*

**Proof**

The theorem follows immediatly from lemma 11 and lemma 13. □

This theorem shows that a deadlock $D$ remains marked if it is controlled by an S–invariant $(d - x)$ or if $d$ itself is a marked S–invariant, in which case $\Delta_d = 0$ and $x = 0$ hold.

Moreover, the theorem explains the concept of a controlling S–invariant. If $i = (d - x)$ is an S–invariant that controls the deadlock $D = \|d\|$, then $x$ is the vector form of a "path system" $\|x\|$ that leads from $\{t \in T \mid \Delta_d(t) > 0\} = \{t_2\}$

to $\{t \in T \mid \Delta_d(t) < 0\} = \{t_1\}$. The less tokens are located on $\|x\|$, the less tokens can be removed from $D$. $(d-x)^t * M_0 > 0$ indicates that if one would try to remove all tokens from $D$, one first would empty $\|x\|$. But in this case there are still tokens on $D$, and the next transition $t$ that would change the token load on $D$ $(\Delta_d(t) \neq 0)$ would necessarily increase it $(\Delta_d(t) > 0)$.

In figure 2, $i_2$ controlls $D$. The corresponding "path system" $\|x_2\| = \{b, c, e, f\}$ leads from $\{t \in T \mid \Delta_d(t) > 0\}$ to $\{t \in T \mid \Delta_d(t) < 0\}$ with $i_2 = d - x_2$. The token load of $\|x_2\|$ under $M$ equals 1, which is small enough to prevent $D$ from being emptied. $i_1$ does not control $D$ because the token load of $\|x_1\| = \{d\}$ equals 2, which is not small enough to prevent $D$ from being emptied.

The theorem provides not a necessary condition for a deadlock to remain marked, as the following two examples show.

In figure 4, $D = \|d\|$ is a deadlock and $i$ is an S–invariant of the net $\mathcal{N}_2$ in figure 3. The deadlock $D$ remains marked though there exists no S–invariant that can be positively marked *by* $M$ in order to control $D$.

In figure 6, $D = \|d\|$ is a deadlock of the net $\mathcal{N}_3$ in figure 5. The deadlock $D$ remains marked though there simply exists no S–invariant that could control $D$ ($x^t * [\mathcal{N}_3] = \Delta_d^t$ has no non–negative integer solution for $x$ with $x \neq d$).

**Lemma 15.** *Let $(\mathcal{N}, M)$ be a net–system and $\mathcal{N} = (S, T, F)$. If $\mathcal{N}$ is dead under $M$, the set of all unmarked places forms a deadlock.*

**Proof**

Let $H \subseteq S$ be the set of all places unmarked under $M$. Suppose $H$ is no deadlock. Then $\exists t \in T : t \in {}^\bullet H \wedge t \notin H^\bullet$. Since all places $s \in S \setminus H$ are marked under $M$, $t$ is enabled under $M$. This is a contradiction to the assumption that $\mathcal{N}$ is dead under $M$. So $H$ must be a deadlock. □

**Lemma 16.** *Let $(\mathcal{N}, M_0)$ be a net–system. $\mathcal{N}$ is weakly live under $M_0$ if every minimal deadlock is controlled by an S–invariant under $M_0$.*

**Proof**

Suppose $\mathcal{N}$ is dead under $M \in [M_0\rangle$. Then the set of all unmarked places forms a deadlock $D$ (Lemma 15). Thus also the minimal deadlocks contained in $D$ are not marked under $M$. This is a contradiction to the assumption that every minimal deadlock is controlled by an S–invariant (Lemma 11). So $\mathcal{N}$ must be weakly live. □

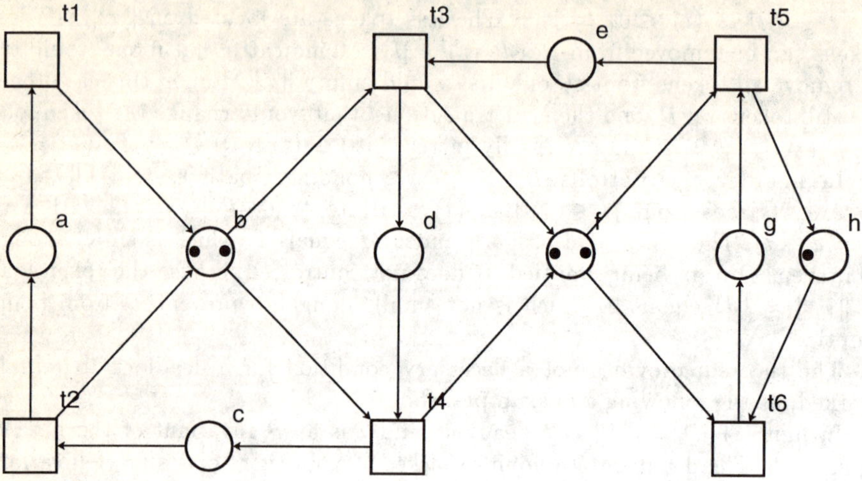


**Fig. 3.** net $\mathcal{N}_2$

| $[\mathcal{N}_2]$ | $t_1$ | $t_2$ | $t_3$ | $t_4$ | $t_5$ | $t_6$ | $M$ | $d$ | $i$ |
|---|---|---|---|---|---|---|---|---|---|
| $a$ | $-1$ | 1 | | | | | 0 | 1 | 1 |
| $b$ | 1 | 1 | $-1$ | $-1$ | | | 2 | 1 | 1 |
| $c$ | | $-1$ | | 1 | | | 0 | 2 | 2 |
| $d$ | | | 1 | $-1$ | | | 0 | 0 | 0 |
| $e$ | | | $-1$ | | 1 | | 0 | 0 | $-2$ |
| $f$ | | | 1 | 1 | $-1$ | $-1$ | 2 | 0 | $-1$ |
| $g$ | | | | | $-1$ | 1 | 0 | 0 | $-1$ |
| $h$ | | | | | 1 | $-1$ | 1 | 0 | 0 |
| $\Delta_d^t$ | 0 | 0 | $-1$ | 1 | 0 | 0 | | | |

**Fig. 4.** incidence matrix, initial marking, deadlock and S–invariant of $\mathcal{N}_2$

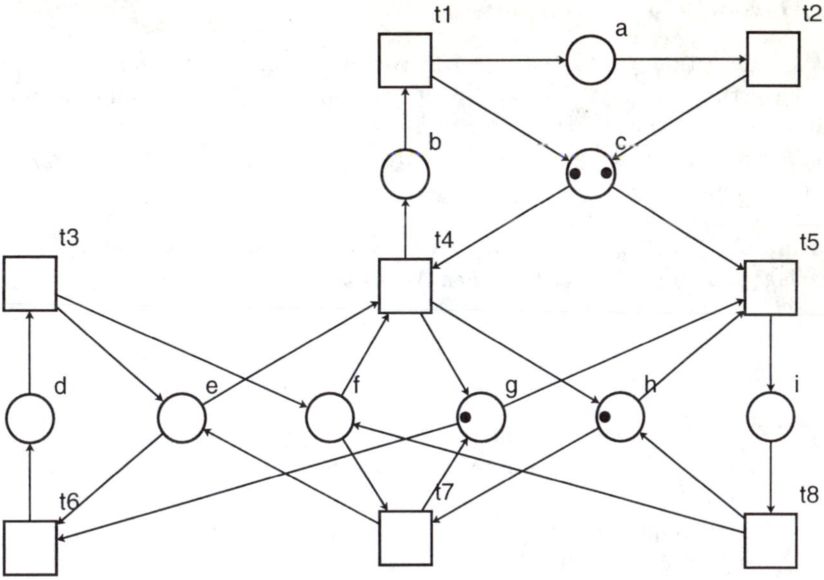


**Fig. 5.** net $\mathcal{N}_3$

| $[\mathcal{N}_3]$ | $t_1$ | $t_2$ | $t_3$ | $t_4$ | $t_5$ | $t_6$ | $t_7$ | $t_8$ | $M$ | $d$ |
|---|---|---|---|---|---|---|---|---|---|---|
| $a$ | 1 | −1 | | | | | | | 0 | 1 |
| $b$ | −1 | | | 1 | | | | | 0 | 1 |
| $c$ | 1 | 1 | | −1 | −1 | | | | 2 | 2 |
| $d$ | | | −1 | | | 1 | | | 0 | 0 |
| $e$ | | | 1 | −1 | | −1 | 1 | | 0 | 0 |
| $f$ | | | 1 | −1 | | | −1 | 1 | 0 | 0 |
| $g$ | | | | 1 | −1 | −1 | 1 | | 1 | 0 |
| $h$ | | | | 1 | −1 | | −1 | 1 | 1 | 0 |
| $i$ | | | | | 1 | | | −1 | 0 | 0 |
| $\Delta_d^t$ | 0 | 0 | 0 | 1 | −1 | 0 | 0 | 0 | | |

**Fig. 6.** incidence matrix, initial marking and deadlock of $\mathcal{N}_3$

### 3.2 Liveness in T–invariants

In this subsection we first deal with liveness in one–T–invariant nets. These nets have the property of being live iff they are weakly live. Sufficient for the weak liveness is that no deadlock can be emptied. Then we treat the liveness of bounded nets covered by T–invariants by approximating these nets by one–T–invariant nets.

**Lemma 17.** *Let $\mathcal{N}$ be a net which is bounded and covered by a elementary T–invariant $j$. If $\mathcal{N}$ is weakly live, then $\mathcal{N}$ is live.*

**Proof**

Since $\mathcal{N}$ is weakly live no occurence sequence can end in a deadlock. But in every infinite occurence sequence there must be a repetition of markings because $\mathcal{N}$ is bounded. Since $j$ is the only T–invariant markings must be reproduced by $j$. So $\mathcal{N}$ is live because $\mathcal{N}$ is covered by $j$. □

Lemma 17 holds only for T–invariants with minimal support as the example in figure 7 and figure 8 shows. $\mathcal{N}_4$ is covered by the non–negative T–invariant $j_1 + j_2$. The only deadlock $D = \| d \|$ is a marked S–invariant ($\Delta_d^t = d^t * [\mathcal{N}_4] = 0^t \wedge d^t * M_0 > 0$). $\mathcal{N}_4$ is weakly live but not live under $M_0$.

The idea is now to couple the two elementary T–invariants $j_1$ and $j_2$ by regulation circuits, such that there exists only one elementary and covering T–invariant. Then lemma 17 can be applied. By increasing the number of tokens on the regulation circuits the original net is approximated.

**Definition 18.** Let $\mathcal{N}$ be a net which is covered by a non–negative canonical T–invariant $j$. We define $\phi(\mathcal{N}, j)$ as the net $\mathcal{N}$ augmented by $j(t) - 1$ copies of every transition $t$ if $j(t) > 1$. $\phi$ maps a net and a T–invariant to a net.

**Lemma 19.** *Let $\mathcal{N}$ be a net which is covered by a non–negative canonical T–invariant $j$. The T'–vector $j' = \mathbf{1}$ is a non–negative canonical T–invariant of $\phi(\mathcal{N}, j) = (S', T', F')$.*

**Proof**

follows directly from the construction of $\phi(\mathcal{N}, j)$. □

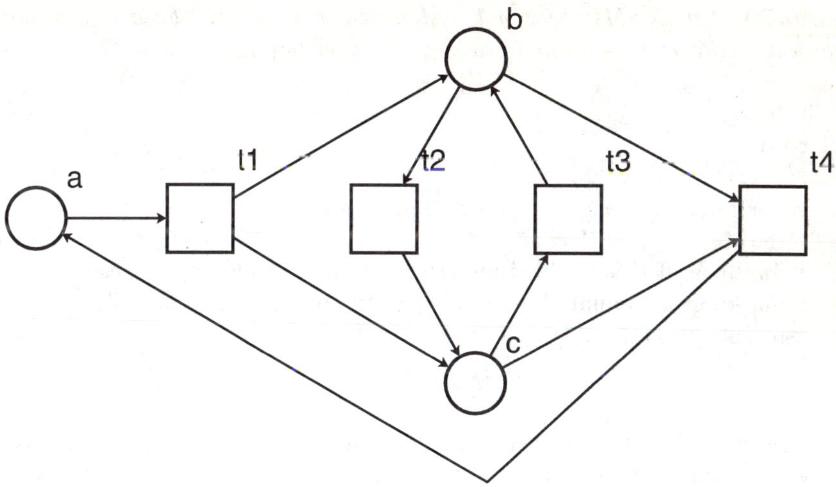


**Fig. 7.** net $\mathcal{N}_4$

| $[\mathcal{N}_4]$ | $t_1$ | $t_2$ | $t_3$ | $t_4$ | $M_0$ | $M_1$ | $d$ |
|---|---|---|---|---|---|---|---|
| $a$ | $-1$ | | | $1$ | $0$ | $1$ | $2$ |
| $b$ | $1$ | $-1$ | $1$ | $-1$ | $1$ | $0$ | $1$ |
| $c$ | $1$ | $1$ | $-1$ | $-1$ | $0$ | $0$ | $1$ |
| $\Delta_d^t$ | $0$ | $0$ | $0$ | $0$ | | | |
| $j_1^t$ | $1$ | $0$ | $0$ | $1$ | | | |
| $j_2^t$ | $0$ | $1$ | $1$ | $0$ | | | |

**Fig. 8.** incidence matrix, initial markings, deadlock and T–invariants of $\mathcal{N}_4$

**Lemma 20.** *Let $(\mathcal{N}, M_0)$ be a net–system which is covered by a non-negative canonical T–invariant $j$. $[M_0\rangle$ in $\phi(\mathcal{N}, j)$ is equal to $[M_0\rangle$ in $\mathcal{N}$.*

**Proof**

For all transitions $t$ in $N$ exists a transition $t'$ in $\phi(\mathcal{N}, j)$ such that ${}^\bullet t = {}^\bullet t'$ and $t^\bullet = t'^\bullet$ and vice versa. Hence $t$ is enabled under $M_0$ in $N$ iff $t'$ is enabled under $M_0$ in $\phi(\mathcal{N}, j)$ and the successor markings are equal. The equality of the reachability sets follows by induction. □

If a net $\mathcal{N} = (S, T, F)$ is covered by the T–invariant $j = \mathbf{1}$, $T$ can be partitioned by a set $Y$ of elementary T–invariants such that every $t \in T$ belongs to exactly one $j_{min} \in Y$.

**Definition 21.** Let $\mathcal{N} = (S, T, F)$ be a net. A *regulation circuit* between two transitions $t_1, t_2 \in T$ consists of two places $s_1, s_2 \in S$ with ${}^\bullet s_1 = \{t_1\} \wedge s_1^\bullet = \{t_2\} \wedge {}^\bullet s_2 = \{t_2\} \wedge s_2^\bullet = \{t_1\}$.

**Definition 22.** Let $\mathcal{N}$ be a net which is covered by the T–invariant $j = \mathbf{1}$. We define $\psi(\mathcal{N})$ as the net $\mathcal{N}$ augmented by regulation circuits between transitions of different elementary T–invariants, which have at least one common place in their presets. $\psi$ is a relation between nets.

**Lemma 23.** *Let $\mathcal{N}$ be a net which is covered by the T–invariant $j = \mathbf{1}$. $j$ is an* elementary *T–invariant of $\psi(\mathcal{N})$.*

**Proof**

follows from the construction of $\psi(\mathcal{N})$. □

**Definition 24.** Let $\mathcal{N} = (S, T, F)$ be a net which is covered by the T–invariant $j = \mathbf{1}$. For every marking $M$ of $\mathcal{N}$

$$M^n(s) = \begin{cases} n & \text{if } s \in S' \setminus S \\ M(s) & \text{otherwise} \end{cases}$$

is called the *parametrized marking* for $\mathcal{N}' = (S', T', F') = \psi(\mathcal{N})$ belonging to $M$.

**Lemma 25.** *Let $\mathcal{N} = (S, T, F)$ be a net which is covered by the T–invariant $j = \mathbf{1}$, $\mathcal{N}' = \psi(\mathcal{N})$ and $M_0'$ a marking of $\mathcal{N}'$. Every occurence sequence $\sigma$ with $M_0' [\sigma\rangle M'$ of $\mathcal{N}'$ is also an occurence sequence with $M_0' | S [\sigma\rangle M' | S$ of $\mathcal{N}$.*

**Proof**

We show the lemma for $\sigma = t$ with $t \in T$. The general case follows then by induction.
By construction, $^\bullet t$ in $\mathcal{N}$ is a subset of $^\bullet t$ in $\mathcal{N}'$. Hence, since $M_0'$ enables $t$ in $\mathcal{N}'$, $M_0' \mid S$ enables $t$ in $\mathcal{N}$. Since $^\bullet t$ and $t^\bullet$ in $\mathcal{N}$ are equal to $^\bullet t \cap S$ and $t^\bullet \cap S$, the successor marking in $\mathcal{N}$ is $M' \mid S$.
□

**Theorem 26.** *Let $(\mathcal{N}, M_0)$ be a net-system and $\mathcal{N} = (S, T, F)$ which is bounded and covered by the T-invariant $j = \mathbf{1}$. $(\mathcal{N}, M_0)$ is live if the following holds:*

$$\exists\, n_0 \in \mathbb{N}\, \forall\, n \in \mathbb{N}, n \geq n_0 : (\psi(\mathcal{N}), M_0^n) \text{ is weakly live.}$$

**Proof**

The T-vector $j = \mathbf{1}$ is an elementary and covering T-invariant of $\mathcal{N}' = (S', T', F') = \psi(\mathcal{N})$ (lemma 23). Hence, since lemma 17, we conclude that $(\mathcal{N}', M^n)$ is live for all $n \in \mathbb{N}, n \geq n_0$.
Suppose $(\mathcal{N}, M_0)$ is not live. Then it exists a marking $M_1 \in [M_0\rangle$ and a transition $t \in T$ such that $t$ is not live under $M_1$. Let $\sigma$ be the occurence sequence yielding from $M_0$ to $M_1$. Now choose $n_0 \in \mathbb{N}$ large enough such that $\sigma$ is also an occurence sequence in $\mathcal{N}'$ with $M_0^n[\sigma\rangle M_1'$ for all $n \geq n_0$. $M_1' \mid S$ is equal to $M_1$. $t$ is live under $M_1'$ in $\mathcal{N}'$, because $\mathcal{N}'$ is live. Hence, since every occurence sequence in $(\mathcal{N}', M_1')$ is also an occurence sequence in $(\mathcal{N}, M_1)$ (lemma 25), $t$ is live under $M_1$ in $\mathcal{N}$. This is a contradiction to the assumption that $\mathcal{N}$ is not live. □

Now we present the main theorem of this paper, which combines the previous results.

**Theorem 27.** *Let $(\mathcal{N}, M_0)$ be a net-system which is bounded and covered by a non-negative canonical T-invariant $j$. $(\mathcal{N}, M_0)$ is live if for $\mathcal{N}' = \psi(\phi(\mathcal{N}, j))$ the following holds:*

$$\begin{aligned} &\exists\, n_0 \in \mathbb{N}\, \forall\, n \in \mathbb{N}, n \geq n_0 \\ &\quad \forall \text{ minimal deadlocks } D \text{ with D-system } d \text{ and defect } \Delta_d \text{ in } \mathcal{N}' \\ &\quad\quad \exists \text{ S-vector } x \geq 0 : x^t * [\mathcal{N}'] = \Delta_d^t \wedge (d - x)^t * M_0^n > 0. \end{aligned}$$

**Proof**

Due to theorem 14 $(\mathcal{N}', M_0^n)$ is weakly live for all $n \geq n_0$. As a result of lemma 19 we get that the T'-vector $j' = \mathbf{1}$ is a non–negative T–invariant of $\phi(\mathcal{N}, j)$. Hence, since theorem 26, we conclude that $(\phi(\mathcal{N}, j), M_0)$ is live. As a result of lemma 20 we get the proposition. □

To check whether for $d$ and $x$ an $n_0$ exists with $(d-x)^t * M^n > 0$ for all $n \in \mathbb{N}, n \geq n_0$ is quite simple, because $(d-x)^t * M^n = n * \sum_{s \in S' \setminus S} d(s)$ $- n * \sum_{s \in S' \setminus S} x(s) + \sum_{s \in S} d(s) * M(s) - \sum_{s \in S} x(s) * M(s)$, where $S$ $(S')$ is the set of all places in $\mathcal{N}$ $(\mathcal{N}')$ and $S' \setminus S$ is the set of all places on regulation circuits. Since all sums are not dependent on $n$, one may write $(d-x)^t * M^n = n * c_1 - n * c_2 + c_3 - c_4 > 0$ for all $n \in \mathbb{N}, n \geq n_0$ iff either $c_1 \geq c_2 \wedge c_3 > c_4 \wedge n_0 \geq 0$ or $c_1 > c_2 \wedge c_3 \leq c_4 \wedge n_0 > (c_4 - c_3)/(c_1 - c_2)$. Since it doesn't matter how large $n_0$ is, because we are only interested if such an $n_0$ exists, the condition can be simplified to $(c_1 \geq c_2 \wedge c_3 > c_4) \vee (c_1 > c_2 \wedge c_3 \leq c_4)$.

Figure 9 and figure 10 show the transformed net $\mathcal{N}_4'$ of the net $\mathcal{N}_4$ in figure 7 and figure 8.

$\mathcal{N}_4'$ is live under the parametrized marking $M_1^n$ for all $n \in \mathbb{N}, n \geq 1$, since the minimal deadlocks $\| d_1 \|$ and $\| d_2 \|$ are respectively controlled by the S–invariants $i_1 = d_1 - x_1$ and $i_2 = d_2 - x_2$, $(\forall\, n \in \mathbb{N}, n \geq 1 : i_1^t * M_1^n = 1 \wedge i_2^t * M_1^n = 1)$. So, the net $\mathcal{N}_4$ is live under $M_1$.

Under the parametrized marking $M_0^n$ the minimal deadlock $\| d_1 \|$ is not controlled by the S–invariant $i_1$, $(\forall\, n \in \mathbb{N} : i_1^t * M_0^n = 0)$. We observe that neither the transformed system $(\mathcal{N}_4', M_0^n)$ is live nor $(\mathcal{N}_4, M_0)$ is live.

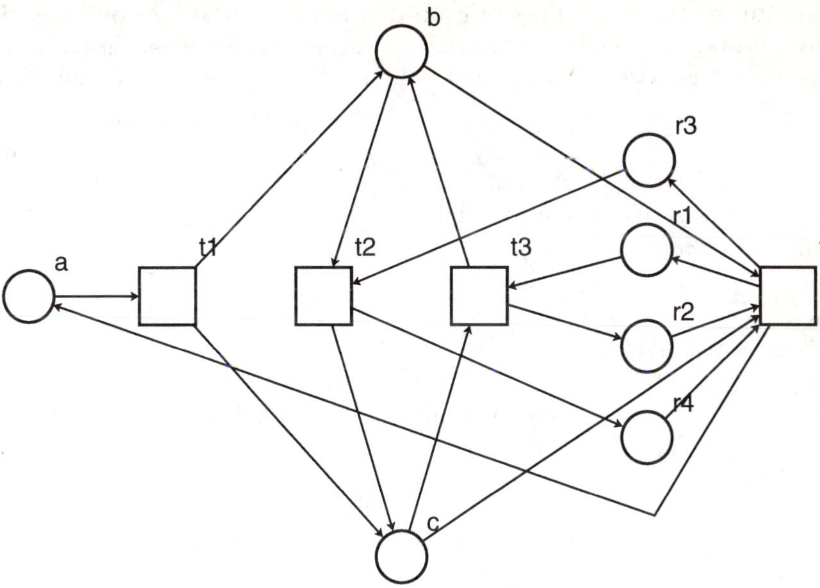


**Fig. 9.** net $\mathcal{N}_4'$

| $[\mathcal{N}_4']$ | $t_1$ | $t_2$ | $t_3$ | $t_4$ | $M_0^n$ | $M_1^n$ | $d_1$ | $d_2$ | $i_1$ | $i_2$ | $x_1$ | $x_2$ |
|---|---|---|---|---|---|---|---|---|---|---|---|---|
| $a$ | $-1$ | | | $1$ | $0$ | $1$ | $1$ | $1$ | $1$ | $1$ | $0$ | $0$ |
| $b$ | $1$ | $-1$ | $1$ | $-1$ | $1$ | $0$ | $0$ | $1$ | $0$ | $1$ | $0$ | $0$ |
| $c$ | $1$ | $1$ | $-1$ | $-1$ | $0$ | $0$ | $1$ | $0$ | $1$ | $0$ | $0$ | $0$ |
| $r_1$ | | | $-1$ | $1$ | $n$ | $n$ | $0$ | $1$ | $-1$ | $1$ | $1$ | $0$ |
| $r_2$ | | | $1$ | $-1$ | $n$ | $n$ | $0$ | $0$ | $0$ | $0$ | $0$ | $0$ |
| $r_3$ | | $-1$ | | $1$ | $n$ | $n$ | $1$ | $0$ | $1$ | $-1$ | $0$ | $1$ |
| $r_4$ | | $1$ | | $-1$ | $n$ | $n$ | $0$ | $0$ | $0$ | $0$ | $0$ | $0$ |
| $\Delta_{d_1}^t$ | $0$ | $0$ | $-1$ | $1$ | | | | | | | | |
| $\Delta_{d_2}^t$ | $0$ | $-1$ | $0$ | $1$ | | | | | | | | |

**Fig. 10.** incidence matrix, initial markings, deadlocks and S–invariants of $\mathcal{N}_4'$

**Algorithm to check liveness of bounded nets covered by non–negative T–invariants.** The algorithm returns the answer "live", if our criterion can be applied. Otherwise the algorithm returns "?", because our criterion is not necessary.

Input: A net–system $(\mathcal{N}, M)$ with $\mathcal{N} = (S, T, F)$, which is bounded and covered by a non–negative canonical T–invariant $j$.
Output: "live" or "?"

begin
  $\mathcal{N}' := \psi(\phi(\mathcal{N}, j))$;
  $(S', T', F') := \mathcal{N}'$;
  $\mathcal{D} := \{d \mid d$ is the D–system of a minimal deadlock $D = \| d \|$ of $\mathcal{N}'\}$;
  forall $d \in \mathcal{D}$ do begin
    $\Delta_d^t := d^t * [\mathcal{N}']$;
    if $\Delta_d \neq 0$ then begin (* $d$ is no S–invariant *)
      $\mathcal{X} := \{x \mid x$ is a non–negative elementary solution of $x^t * [\mathcal{N}] = \Delta_d^t\}$;
      if $\nexists x \in \mathcal{X}$ :
          $(\sum_{s \in S}(d - x)(s) * M(s) \leq 0 \wedge \sum_{s \in S' \setminus S}(d - x)(s) > 0) \vee$
          $(\sum_{s \in S}(d - x)(s) * M(s) > 0 \wedge \sum_{s \in S' \setminus S}(d - x)(s) \geq 0)$ then
        return("?");
    end;
  end;
  return("live");
end;

## 4 Conclusion

In this paper we have introduced a sufficient criterion for the liveness in Petri nets which are bounded and covered by T–invariants.

The criterion is not necessary. The nets of figures 3 and 5 have deadlocks $D$ which are not controlled even though they remain marked.

The difference between both cases is, that the deadlock $D$ in the net $\mathcal{N}_3$ of figure 5 is structurally not controllable, whereas in the net $\mathcal{N}_2$ of figure 3 it depends on the marking $M$ whether the deadlock $D$ is controlled by $i$ or not.

Scientifically, it might be interesting to generalize the method w.r.t. the second case. From a practical point of view, we doubt — at least for the moment — that this kind of generalization could really end up in some applicable criterion for liveness.

In its current form, however, our criterion allows tests for liveness also in such cases where all methods we know about cannot be used. So, for the net–systems of figures 1 and 7 neither the rank theorem [Des93] nor the decomposability method [Hac78, JV80, Sta90] can be applied.

## References

[Bau90] Bernd Baumgarten. *Petri–Netze: Grundlagen und Anwendungen.* BI-Wiss.–Verl., Mannheim; Wien; Zürich, 1990. (in German).

[CCS91] J. Campos, G. Chiola, and M. Silva. Properties and performance bounds for closed free choice synchronized monoclass queueing networks. *IEEE Transactions on Automatic Control*, 36(12), December 1991.

[Des93] Jörg Desel. Regular marked petri nets. To appear in: Graph theoretic Concepts in Computer Science, 1993.

[ES92] Javier Esparza and Manuel Silva. A polynomial-time algorithm to decide liveness of bounded free choice nets. *Theoretical Computer Science*, (102):185–205, 1992.

[Hac78] M. Hack. Extended state–machine allocatable nets (esma), an extension of free choice petri net results. Computation Structures Group Memo 78 – 1, MIT, Project MAC, 1978.

[JV80] M. Jantzen and R. Valk. Formal properties of place/transition nets. *Net Theory and Application*, LNCS 84:165–212, 1980.

[KL82] W.E. Kluge and K. Lautenbach. The orderly resolution of memory access conflicts among competing channel processes. *IEEE Transactions of Computers*, C–31(3):194–207, 1982.

[Lau73] K. Lautenbach. Exakte bedingungen der lebendigkeit für eine klasse von petri–netzen. Berichte der GMD 82, GMD, Bonn, 1973. (in German).

[Lau77] K. Lautenbach. Ein kombinatorischer ansatz zur beschreibung und erreichung von fairness in scheduling–problemen. *Applied Computer Science*, 8:228–250, 1977. (in German).

[Lau87] Kurt Lautenbach. Linear algebraic calculation of deadlocks and traps. In Voss, Genrich, and Rozenberg, editors, *Concurrency and Nets*, pages 315–336. Springer Verlag, 1987.

[Rei91] Wolfgang Reisig. *Petri nets.* Springer Verlag, 2nd edition, 1991.

[Sta90] H. Peter Starke. *Analyse von Petri-Netz-Modellen.* Teubner, 1990. (in German).

# Modelling the Work Flow of a Nuclear Waste Management Program

Kjeld H. Mortensen
Computer Science Department,
Aarhus University
Ny Munkegade, Bldg. 540
DK-8000 Aarhus C, Denmark
Phone: +45 8942 3188
Fax: +45 8942 3255
E-mail: khm@daimi.aau.dk

Valerio Pinci
Meta Software Corporation

125 Cambridge Park Drive
Cambridge, MA 02140, U.S.A.
Phone: +1 617 576 6920
Fax: +1 617 661 2008
E-mail: pinci@metasoft.com

**Abstract.** In this paper we describe a modelling project to improve a nuclear waste management program in charge of the creation of a new system for the permanent disposal of nuclear waste.

SADT (Structured Analysis and Design Technique) is used in order to provide a work-flow description of the functions to be performed by the waste management program. This description is then translated into a number of Coloured Petri Nets (CPN or CP-nets) corresponding to different program functions where additional behavioural inscriptions provide basis for simulation. Each of these CP-nets is simulated to produce timed event charts that are useful for understanding the behaviour of the program functions under different scenarios. Then all the CPN models are linked together to form a single stand-alone application that is useful for validating the interaction and co-operation between the different program functions.

A technique for linking executable CPN models is developed for supporting large modelling projects and parallel development of independent CPN models.

## 1 Introduction

A large nuclear waste program[1] in U.S.A. is responsible for permanently disposing used nuclear fuel and similar high-level nuclear waste. The objective of the program is to establish a capability to accept, transport and store nuclear waste by 1998, and to start the storage of nuclear waste in a geological repository by 2010. The program has quite unique characteristics; provide safe nuclear waste isolation for 10,000 years with unprecedented oversight and control by different affected and interested groups. Additionally the program must take into account changing conditions in its environment as, e.g., changes in the current legislation. The program director therefore decided to develop a new approach to improve the current nuclear waste systems management strategy and to carefully design the program, much like physical systems are designed. A general design methodology is used to capture the functionality of the system[2]. The methodology includes use of Structured Analysis and Design Technique (SADT) [6] and Coloured Petri nets (CP-nets or CPN) [3].

The physical waste management system is composed of groups of people, documents, and equipment. They need to co-operate and interact with each other. A group of people in charge of a specific domain needs to exchange many kinds of information between other groups often with very different domains. In order to ensure an efficient and consistent co-operation and interaction between groups, the Nuclear Waste Management System (NWMS) is modelled (with SADT) and the resulting model is analysed by translating it into CP-nets which subsequently are simulated. Major components of the system are identified which are called the programmatic

---

[1]By program we mean an organised set of activities directed toward a common purpose. A program is typically made up of, e.g., technology based activities and projects.
[2]By system we here refer to a complex organisation.

functions. The process of modelling and analysing will be referred to as the programmatic functional analysis (PFA). The modellers take the perspective of functional behaviour on the nuclear waste management system. E.g., one of the programmatic functions characterises sites for storage of nuclear waste, and another major function performs systems engineering.

The PFA of the nuclear waste management program is the effort of identifying all activities which must be performed in order to clarify interdependencies between the activities and bring the physical system into being (thus meeting the program objective). Models will provide a means for people in the program to understand their position in the overall program, thus becoming able to identify their functions, e.g., how co-operation and interaction should take place with other people in the system.

The result of the analysis is to be used for developing policies and guidelines that determine how to actually implement the system itself. The programmatic functions are the functions that bring the physical nuclear waste management system into being. The main stages of the physical system functions are simply stated: acceptance, transportation, storage, and disposal of nuclear waste. These operations comprise the stages in the nuclear waste management program.

We will look at some of the programmatic functions in this paper, namely the functions that are prepared for simulation in a CP-net design and simulation tool.

## 1.1 CP-nets, SADT, and Similarities

CP-nets are recognised as a useful modelling language for validation and simulation of complex concurrent systems (see, e.g., the book [5]). CP-nets have a hierarchy concept, which is very similar to the activity hierarchy resulting from functional decomposition in SADT. However, SADT is a top-down method while modelling with CP-nets do not need to be. See [6] for an in-depth description of SADT. There is also a direct correspondence between non-decomposed functions in SADT models and transitions in CP-nets. These similarities make it possible to do a semi-automatic translation from SADT diagrams into CP-nets, where the hierarchy and the connections between activities of the SADT diagram are preserved. Places are automatically created, but net inscriptions with behavioural information have to be added manually. Adding inscriptions to the model provides a basis for simulation, and thus also a means for validating the original intention of the SADT model. See [11] and [9] for a more detailed description of the process of going from an SADT model to a CP-net.

Above, we have described that SADT and CP-nets have a number of important similarities. But the two languages also complement each other in several interesting ways. SADT does not have a formal framework while CP-nets have. SADT cannot be simulated where CP-nets can. Furthermore, CP-nets do not have explicit guidelines for a structured and systematic approach to the model creation process – but SADT does. Both methods build on principles and concepts that are easy to learn and understand. The use of CP-nets in conjunction with SADT implies that the dynamics of an SADT model easily can be examined. Hence, we obtain a better understanding of behavioural properties.

SADT in conjunction with CP-nets is the chosen method in this project for doing the PFA. We use two tools in the project. One is for editing SADT diagrams (Design/IDEF [1]), the other for editing and simulating CP-nets (Design/CPN [2]). Design/CPN is able to load SADT diagrams and translate them into CP-nets.

## 1.2 The Top-level SADT Function

With the purpose of further introduction to SADT and the waste management system, let us take a look at the top-level SADT function from the modelled system (see figure 1).

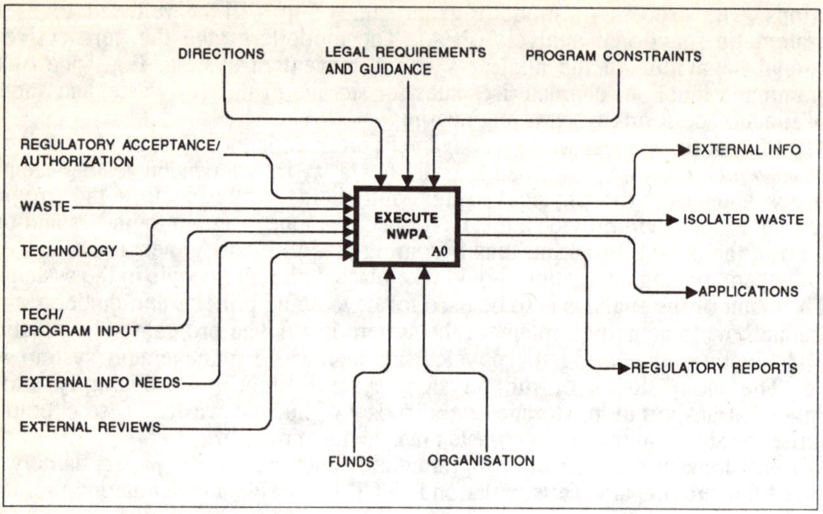


**Fig. 1.** The topmost SADT function ("Execute Nuclear Waste Policy Act").

An SADT diagram consists of three basic elements: boxes, arrows, and labels.

1) The boxes denote functions or activities. A box will always contain a short text describing what the activity stands for. A box can be decomposed, giving a more detailed description of the function by means of another SADT diagram. Each box also has a number indicating its sequence number on the page and the contents of its decomposition. E.g., "A0" means the top-level function, "A3" is the third function on the first level of decomposition, and "A31" is the first function on the second level of decomposition, of the function "A3". The notation "3.1" is a variant of "A31".

2) The arrows denote incoming and outgoing flows of functions. Incoming flows can be of three different types; controls, inputs, or mechanisms. The type of an arrow is determined by the side of the box it enters on or exits from. A function maps input (arrows coming in on the left of the box) to output (going out of the right), possible under some constraints (control arrows entering from top), and by means of mechanisms (arrows entering from below). Mechanisms are often used for modelling resources. Several arrows can join or one arrow can branch into several. (If not explicitly stated, input, control, and mechanisms will go under the common term "input".)

3) Labels are associated with arrows. A label gives the name of the kind of information flowing on the arrow.

As the function above is the topmost SADT function, it provides an overview of what kind of interaction the waste management system has with the external environment. The function takes, e.g., waste and technology as input. When the function is performing, it isolates nuclear waste in sites which are allocated by functions found on lower levels of decomposition. The waste management program, represented by the above function, operates under a number of constraints imposed by its external environment. The program gets directions (a control arrow) from a department in the U.S. government. It also operates under requirements and guidance from the legislative environment, e.g., the Congress. The resources or mechanisms necessary for the nuclear waste management system to operate are an organisation (people and machinery) and funds. On the lower level of decomposition, the resource arrows (mechanisms) are omitted which means that resources are not taken into account in the rest of the model.

The rest of the paper is organised as follows: In section 2 we provide the description of the project and look at some of the SADT diagrams from the model. Section 3 describes the work involved as the SADT diagrams are translated into a CPN model. Examples will be given. We proceed with simulation of the individual submodels in section 4, and in section 5 we explain how the submodels are merged together, resulting in one model which is then simulated. Finally, we have the conclusion where we also discuss work in progress and suggest further work – all in section 6.

# 2 Project Description

In this section we elaborate on our participation in the project. The improvement of the nuclear waste management program is still an ongoing activity, but the work described in this paper was confined to six months (in 1991).

## 2.1 Configuration Overview

Two groups of people with very different backgrounds and qualifications participated in the project. The main group performing programmatic functional analysis (which we will refer to as the PFA team) was responsible for designing the systems management strategy. The PFA team produced, among other things, functional descriptions and SADT diagrams. The size of the team varied between 15 and 25 people.

The waste management program interacts with and is restricted by a variety of groups in the world outside. The program has to work within laws (which may change) of the government, and the program has to handle and interpret many sorts of data coming from, e.g., geologists that analyse potential storage sites. Therefore, many people with knowledge and experience in various domains like geologists and lawyers are present in the team. These people do not have any SADT or CPN modelling background.

The CPN team consisted of the authors of this paper having several years of experience with SADT and CPN. Instead of training the PFA team in using CPN, we, with the CPN modelling expertise, were hired by the program management. The CPN team took the produced SADT diagrams, and based on some additional written descriptions of the intended functional behaviour, the CPN team created a corresponding CPN model which simulates some aspects of the dynamic behaviour of the system. Simulation and timed event charts gave the necessary means for validating properties and answering questions about the behaviour of the model. One interesting question is if there are any unintended blocking of the information flow in the system. If the flow is blocked and other activities are dependent on blocked information, it can under some circumstances lead to a deadlock in the model. This does not mean that the system in the real world also will deadlock. But it means that the set of activities in the model that are going to be implemented in the real world will operate less optimal, because the system will discover delays. These flaws then need to be fixed (if possible) which is more expensive than fixing the model in the first place.

The purpose of our involvement was to provide an executable model of the PFA team's SADT model, in order to give a basis for validation of the accuracy and completeness of the programmatic functional analysis they performed.

From the PFA team's point of view, the interaction and co-operation between the various submodels can be compared with a protocol. As the SADT submodels are made independently by different people it is not guaranteed that the submodels will fit together when the model is viewed as a whole. The SADT modellers made the choice of CP-nets, because CP-nets could help them to validate, by means of the simulation tool, that the models really would be able to interact and co-operate correctly. (Formal analysis of CP-nets was also of interest but never used because of lack of tool support at that time.) The modellers also wanted to be able to investigate the dynamics of the models in more detail and obtain a concrete understanding, since this cannot be accomplished by just looking at SADT diagrams.

## 2.2 The Major Components of the Nuclear Waste Program

In this project we focus on seven major programmatic system functions. Many other functions are also in the SADT model, but they do not have nearly as high priority as the seven major functions we describe in this paper and are thus not considered. Furthermore, it would not be possible to complete the CPN modelling efforts with the available resources if we did not limit our scope. The following is a short description, which explains the main purpose of the considered major functions:

- Provide Program Control (**PPC**): This function controls and provides overall management direction for the NWMS program.
- Ensure Regulatory Compliance (**ERC**): Identifies regulations which applies to the program and the physical system, and ensures they comply with these regulations.
- Perform Systems Engineering (**PSE**): The function translates the NWMS program mission requirements into a set of functions, requirements, and interfaces for the physical system.
- Design Engineered System (**DES**): The function is divided into the four phases; conceptual, preliminary, final, and as build design.
- Identify and Characterise Sites (**ICS**): Provides site information for consideration in system evaluations, and also identifies and screens potentially acceptable sites.
- Evaluate Integrated System (**EIS**): The purpose of this function is to reduce program technical performance risks.
- Perform Confirmation/Construction/Operational Testing (**PCOT**): This function plans, conducts, and documents tests to verify that the NWMS physical system conforms to, e.g., technical requirements.

The latter five functions (PSE, DES, ICS, EIS, and PCOT) are the basis for the decomposition of a function called "Configure System", shown in figure 2. PPC and ERC are located outside the Configure System function and the relationship with PSE, DES, ICS, EIS, and PCOT is not easily visualised by a simple figure.

The five functions in figure 2 interact throughout the NWMS program. Typically PSE provides input information to one of DES, ICS, EIS, or PCOT. They in turn provide a result which is either a success (a final result) or a request for more information or additional action. The result is processed by PSE, sometimes in co-operation with PPC. And so it goes on through all stages of the NWMS program.

## 2.3 Understanding the SADT Model

Throughout the project the work procedure of the CPN team was as follows. Typically the PFA team finished a first version of one of the submodels. Our task was then to translate the SADT diagram into a CPN diagram and finish the model by adding behavioural inscriptions so that a simulation could take place.

As an example of a diagram page from the SADT model, the decomposition of the function "Analyse Performance Variances" is shown in figure 3, which is a function within PPC. (We will henceforth base our examples on this function.) The purpose of the function "Analyse Performance Variances" is to identify performance variances (e.g., delays) in the waste management program. It bases its analysis on, e.g., information about the physical system provided from PSE and general information about the status of the program assembled from many origins. If there are variances, the function is also responsible for identifying the cause and issue alternative corrective actions (with associated risks) in order to reduce the variance. It also determines where the action should be taken and develop recovery plans for future actions. The analysis results from the function in the figure are used to make change requests of which approved changes are sent to many functions in ERC as a part of general program information. ERC is then responsible for realising the requests. Thus, the scope of the function really covers most of the waste management program.

To help the CPN team to understand the intended behaviour of the SADT models, these came together with clarifying function descriptions. Initially the form was only unstructured textual descriptions. The textual descriptions contained information about the relationship between input and output when a function is activated. Figure 4 shows an example of a textual description for the function "Identify Corrective Actions" from figure 3.

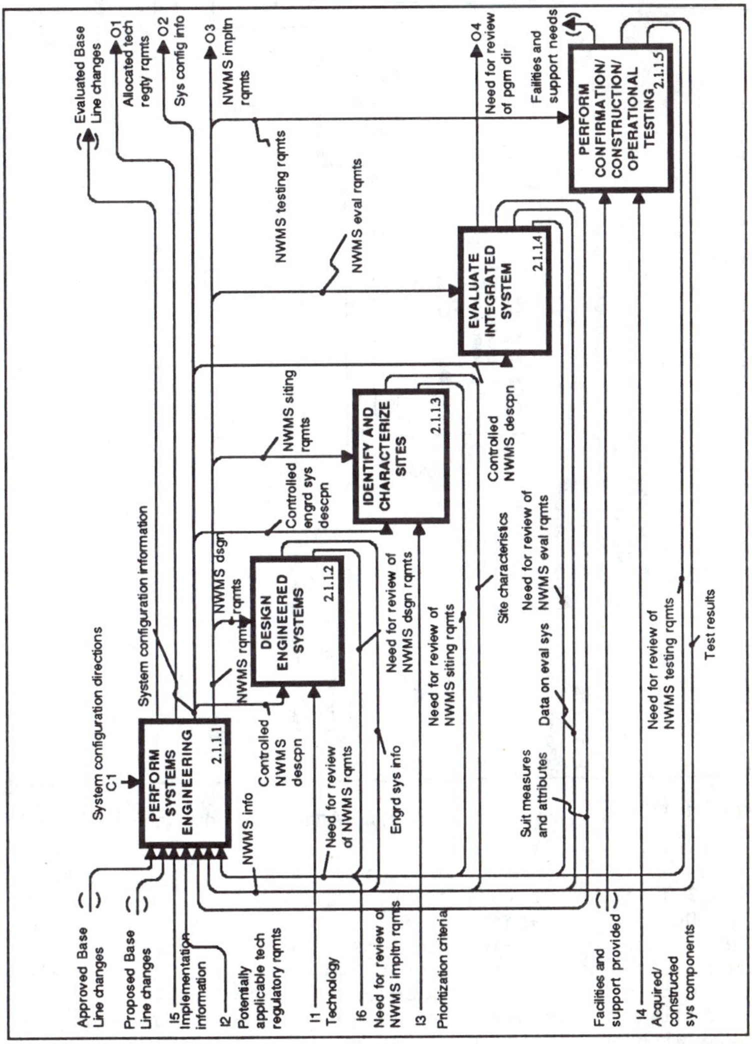


**Fig. 2.** Decomposition of the function "Configure System", which shows the relation between five of major programmatic functions; PSE, DES, ICS, EIS, and PCOT.

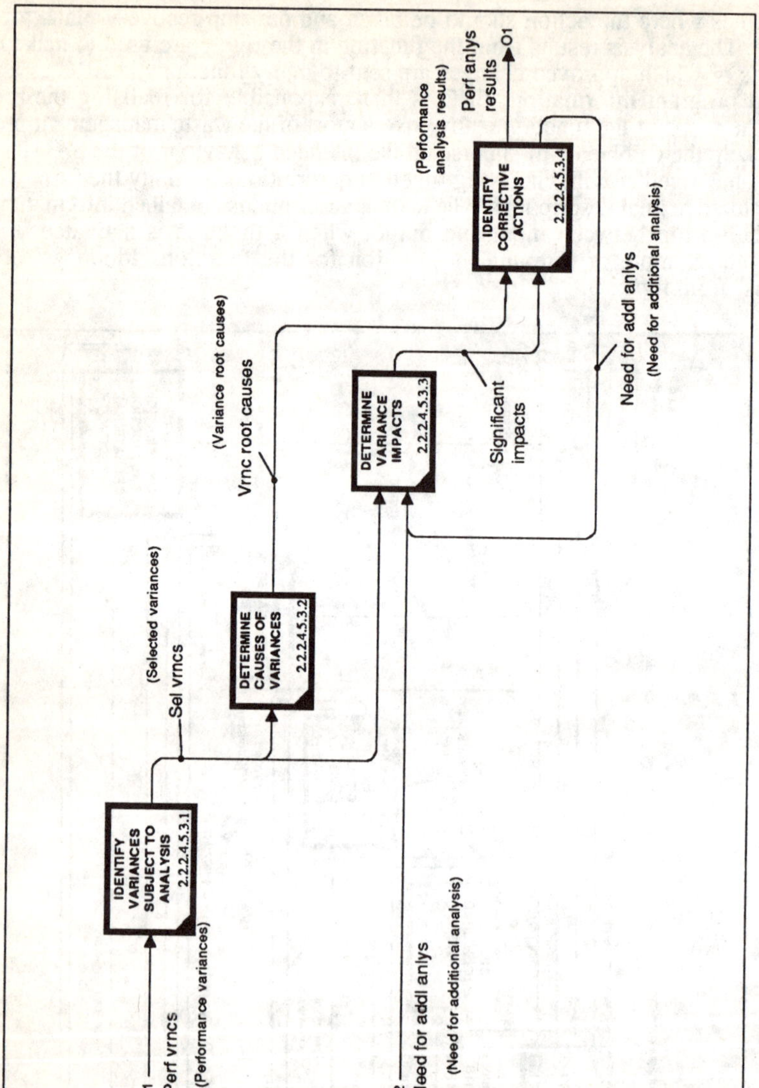


**Fig. 3.** Decomposition of the function "Analyse Performance Variances"

2.2.2.4.5.3.4 IDENTIFY CORRECTIVE ACTIONS
<u>MISSION</u>: Develop alternative actions to correct root causes and mitigate impacts of the variances affecting the program.
<u>SCOPE</u>: Analyse the variance root causes and impacts and develop alternative corrective actions to correct the cause or mitigate the impact. Analysis includes identifying and evaluating the risks associated with each corrective action.

**Fig. 4.** Description provided with the function "Identify Corrective Actions" (figure 3).

The textual description is divided into the mission and the scope of the function. The mission just explains briefly what the function itself is supposed to do in terms of input and output, and the scope explains to what extent the rest of the waste management program is involved in the actions of the function.

In the beginning of the project, this kind of textual description was the only available explanation of the model. Communicating the intended functional behaviour in the above form showed quickly to be insufficient and ambiguous, therefore error prone and a waste of time for us. To support the textual descriptions, a more formal description language for functional behaviour was introduced. The exchanged information became decision tables and descriptions which both speeded up the process of understanding the behaviour and reduced the number of misunderstandings dramatically in comprehending the intended functionality. Decision tables were useful for describing exactly when output is produced, given that some combination of inputs has changed. An example of this format can be seen in figure 5.

Arrow labels give the name of the arrow from the SADT diagram. Types of the arrow can be "in", "control", or "out", as provided by the SADT terminology. The operative cycles should be read as follows. Conditions ("x") for the first execution of the function is specified in columns labelled "1". Subsequent execution is specified by columns labelled "S". Each column in the table describes one alternative. Thus, the function above will initially produce *either* "Performance analysis results" *or* "Need for additional analysis" when both inputs are available (have changed). Similar for subsequent execution.

DECISION TABLE:

| | | Operative Cycles | | | | |
|---|---|---|---|---|---|---|
| Arrow label | Type | 1 | 1 | S | S | Notes |
| Significant impacts | In | x | x | x | x | 1,2 |
| Variance root causes | In | x | x | x | x | 1,2 |
| Performance analysis results | Out | x | | x | | |
| Need for additional analysis | Out | | x | | x | |

1) Initial execution requires both the Significant Impacts and Variance Root Causes and will produce *either* the Performance Analysis Results (which includes alternative corrective actions is applicable) *or* the Need for Additional Analysis.
2) Subsequent executions of the function behave in the same manner as the initial execution.

**Fig. 5.** Decision table for the function "Identify Corrective Actions" from figure 3.

As mentioned earlier, the decision tables provided a significant speedup in the process of understanding the SADT model. We were also, at an early stage, able to predict malfunctioning behaviour in the model. E.g., with the above decision table we can identify a blocking of the information flow on the page in figure 3. From the table one can read that both inputs are always required for the function to execute. In the case that "Identify Corrective Actions" issue "Need for additional analysis", subsequent execution of "Determine Variance Impacts" results in a change in only one of the inputs of "Identify Corrective Actions". But this function requires both inputs to change in order to execute and therefore erroneously waits instead of processing the local feedback immediately. This was reported to the PFA team and they modified the decision table so that subsequent execution of "Identify Corrective Actions" only required a change in one of its inputs. These flaws and similar errors were reported to the PFA team on a very early stage, and thus saved time later when they were not so easy to fix.

## 3 Entering the Environment of CPN

This section is about the CPN model that was built. We do not show the total model, but only a typical example to give a feeling for the general intentions.

The final SADT model has seven major submodels. A total of 116 SADT diagram pages were drawn summing up to more than 300 non-decomposed functions, which in the corresponding CPN model are transitions. Each diagram page was automatically converted into CPN and manually extended with behavioural inscriptions.

### 3.1 The CPN Submodels

In the following we describe the translation process. We take each SADT submodel and translate it automatically into a CP-net. In the translation process, places are added in the net where necessary to get a syntactically correct CP-net, but the structure of the original SADT diagram is preserved. Now, based on the textual descriptions and decision tables, we add inscriptions to the CP-net so that we end up with a complete model ready for simulation. As an example we have taken the SADT diagram from figure 3 and completed the corresponding CP-net with inscriptions. The result can be seen in figure 6. Grey places and arcs are extensions to the original SADT diagram. Inscriptions are also added manually, except from colour sets.

There are a lot of details in the figure, but we will only explain the most important ones. The basic colour set used for most places in the model is a specific record called `InfoObject`. All other colour sets are structurally equal to `InfoObject`:

```
color InfoObject = record Ver:int * Info:string * Av:bool;
color Perf_vrncs = InfoObject;
color Sel_vrncs  = InfoObject;
...
```

The `Ver` field contains a version number. It is used to identify changes when new tokens are issued. `Info` is the information field. The `Info` field of a token from an output arc in the CP-net will have the name of the corresponding arrow in the SADT diagram. `Av` denotes the availability of the token. This field is only used rarely in the model, thus not considered futher in this paper. As most places use the colour set `InfoObject` we could have chosen just to use this colour set name on these places instead of using many different names. But this would violate one of the principles of the SADT method, viz. that the type (or kind) of information flowing on arrows is associated with the arrow name, not with the implementation of the type. As it is the intention that the PFA team, who are used to read SADT diagrams, should be able to read the CPN model, it is preferable to use the arrow names from the SADT model as colour set names in the CPN model.

The grey places are modelling local information about the current version number of the tokens on the input places, except the place called "st" which has information about whether it is the first time the transition occurs or not. Initially all places contains one token with the record value `{Ver=0, Info="", Av=true}` or `first` on "st" places. With this at hand we can easily implement the decision tables by writing a guard that take advantage of the local information.

The guards in this figure determine when a new version of a token has arrived on the input places. We use the convention that a guard is placed above its transition and the guard-expression is enclosed in "`[]`". Notice that the third guard ("Determine Variance Impacts") has logic for the case when it occurs the first time and subsequently. Recall the decision table in figure 5.

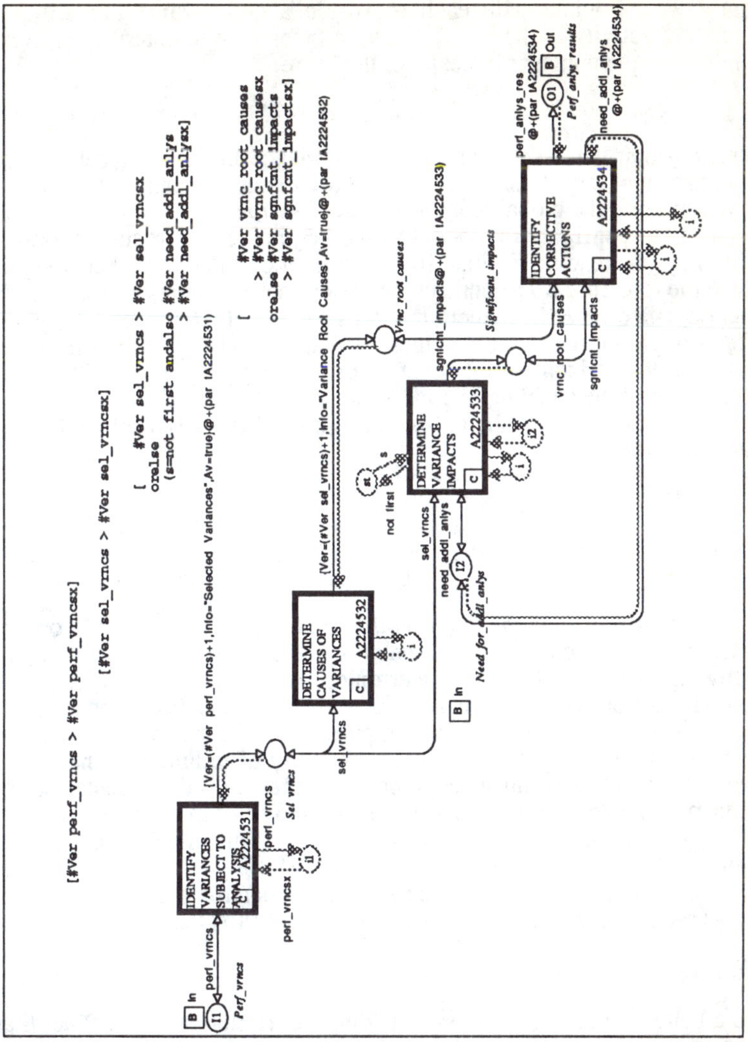


**Fig. 6.** The CP-net as a result of translating the SADT page from figure 3. Grey places and arcs are extensions to the original SADT diagram.

Then there are the code segments, which are actions called as a side effect when the transition occurs. The "C" in the lower left corner of a transition indicates that the transition has a code segment. The contents of the code segments are not visible in the figure. The code segments are responsible for creating, among other things, values used for generating the output tokens, determined by a set of configurable reference variables. These are: the maximum number of times a transition will fail (or produce abnormal output), the probability of failure, and the amount to decrement the probability of failure each time it fails. In the next paragraphs we will discuss these variables in detail. The code is also responsible for updating the graphics in the time event charts, which shows the relation between time and the occurrence of transitions. We will not show the contents of the code segments because it is just a lot of trivial code for manipulating reference values and the graphical output.

Finally we associate with each transition a delay value modelling that the corresponding activity in the SADT diagram takes time to complete. The value is a global reference to an integer typically of the form:

```
val tA2224533 = ref 1;
```

here meaning that the activity "Determine Variance Impacts" (labelled A2224533) takes one time unit to process its inputs in order to make a complete output. We have chosen to represent the global reference values like this, because it is then convenient (for the user) to configure the model in the beginning of the simulation (or at any other time) by just reading external files with a set of values and then assigning them directly to the references. The contents of the reference value is accessed by means of the function called `par`. In general, models with global references are not safe, since side effects might influence the enabling calculations as the global scope of references violate the locality principle. But we only use reference values on output arcs, which are independent of the enabling rule in CP-nets.

The four lines related to the CP-net in figure 6 from a typical configuration file have the following format:

```
1      (* A2224531 IDENTIFY VARIANCES SUBJECT TO ANALYSIS *)
1      (* A2224532 DETERMINE CAUSES OF VARIANCES          *)
1      (* A2224533 DETERMINE VARIANCE IMPACTS             *)
1 2 100 20   (* A2224534 IDENTIFY CORRECTIVE ACTIONS      *)
```

The first number is always the duration information. If there are additional numbers on the same line, in groups of three numbers like the last line above, then they have the following meaning: the first determines the maximum number of times the function will fail or generate a request for something, i.e., something unusual. E.g., in this case the abnormal or unusual output is "Need for additional analysis". The second parameter determines the initial probability of failure, and the third and last parameter determines how much the probability of failure is decremented each time the transition produces abnormal output. In the above output file we have that the transition will fail at most two times. The first time, it will fail with a 100% probability and the second time it will fail with a 80% probability. Subsequently the transition will never fail. This simple scheme of handling non-determinism in the model was satisfactory from the point of view of the PFA team.

### 3.2 Adding Inscriptions, an Example

To give a better idea of the process of adding inscriptions to the CP-net that comes from the automatic translation of SADT diagrams, we will complete the function "Identify Corrective Actions". There are many approaches on adding inscriptions to the CPN model. We chose to keep the amount of structural changes in the model to a minumum, so that it resembled the SADT model as much as possible.

Most of the efforts of adding inscriptions will appear to be rather trivial. Work is in progress to automate this process [10]. Let us first take a look at the corrected table from figure 5 (recall that it had an error that could lead to a blocking of the information flow). The corrected table is located in figure 7. If we cut out the example function "Identify Corrective Actions" from figure 6 with surrounding arcs and places we get the result as in figure 8 (with added inscriptions).

In the process of adding inscriptions we first add extra places (labelled "i"); one for each arc coming in from the left of the transition (the SADT input arrows). This is in order to be able to detect that input to the function has been updated, reflected by an increment in the version number (the `#Ver` field of the token). Therefore we store the latest information on "local" places.

DECISION TABLE:

| Arrow label | Type | Operative Cycles 1 | 1 | S | S | S | S | Notes |
|---|---|---|---|---|---|---|---|---|
| Significant impacts | In | x | x | x | | x | | |
| Variance root causes | In | x | x | | x | | x | |
| Performance analysis results | Out | x | | x | x | | | |
| Need for additional analysis | Out | | x | | | x | x | |

**Fig. 7.** The corrected decision table of "Identify Corrective Actions" from figure 5.

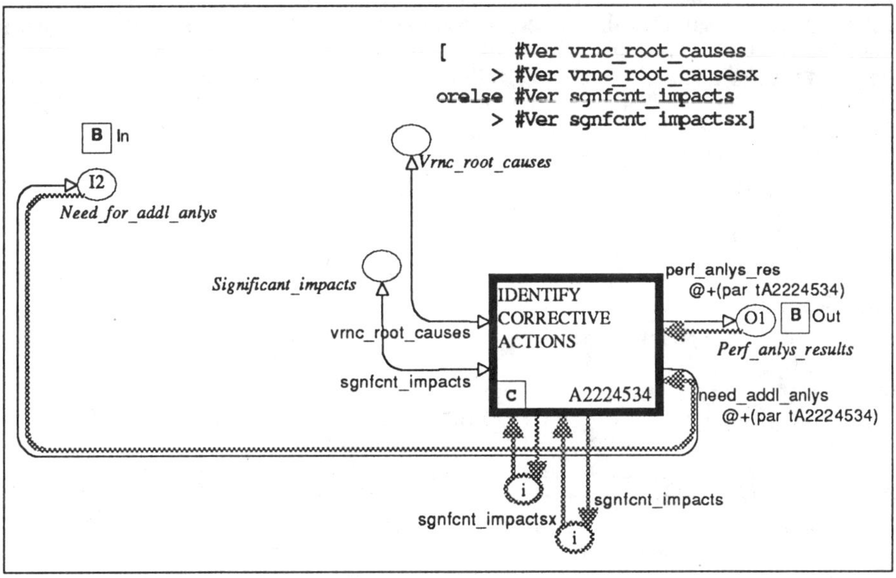


**Fig. 8.** The transition "Identify Corrective Actions" cut out of figure 6.

Next, we add initial markings, which is the multi-set 1`{Ver=0, Info="", Av=true}. On these places there will always be just one token. Thus in order not to consume the token upon occurrence of the transition, we add extra arcs (indicated in the figure with grey) in the other direction to put back the same token (input) or an updated token (output and local copies of input). Arc inscriptions are then added so we can refer to tokens from the surrounding places (on output we also add time delay expressions "@+delay"). Now we are ready to determine when the transition can occur by creating the guard. We do this by first inspecting the decision table from figure 7. From the table we get that the first time the transition occurs, we need new versions on both inputs. Subsequently the transition can occur if just one of the inputs changes. Thus, a guard that has not been optimised, can look as follows:

```
[ (<first time>     andalso <Variance Root Causes changed>
                    andalso <Significant Impacts changed>)
orelse
  (not <first time> andalso (<Variance Root Causes changed>
                      orelse <Significant Impacts changed>)) ]
```

Based on the knowledge that the two precedent transitions initially change the two inputs to this example transition at the same point in time, and that the Boolean expression "(X and Y) or (¬X and Y)" is equivalent to "Y" we get the optimised guard as seen in figure 8:

```
[        <Variance Root Causes changed>
 orelse <Significant Impacts changed> ]
```

Finally, we get to the code segment. Apart from updating the graphics in the timed event charts, it also produces values for variables used on the output arcs. This transition can issue "Need for additional analysis" to which we will associate a probability. We say that the transition (or the function) fails if it issues "Need for additional analysis" otherwise it will behave normally if it issues "Performance analysis results". The probability of failure will decrease on every failure with some user specified amount (which crudely models that activities learn from their failures). Following is a very rough sketch of the code segment:

```
input <empty>
output <perf_anlys_res, need_addl_anlys>
action
  if <graphics output is on> then
    <update timed event chart>
  else
    <do not update chart>

  if <failed less than the max # of times> then
    if <we should fail this time> then
      <decrement probability of failure>
      <increment number of times failed so far>
      <output> need_addl_anlys
    else
      <output> perf_anlys_res
  else
    <output> perf_anlys_res
```

### 3.3 Work Schedule

We quickly got experience in predicting how long it would take to complete a CP-net with inscriptions and test it. This knowledge is very valuable when making work schedules that are realistic and thus possible to keep up with.

One part of the work is to add inscriptions to the CP-net that is generated from the SADT diagrams. A minor part of the inscription work is to understand the textual descriptions and the decision tables as provided by the PFA team. The major part is to add the actual inscriptions. The amount of inscription work is somewhat proportional to the number of non-decomposed transitions. These transitions require most of the work, while adding arc inscriptions and extra places are less laborious. The time it took to test a major submodel tended mostly to depend on the complexity of the inscriptions.

Just to give an idea of the amount of work it takes to complete a CP-net, we give a few examples of how long it took to complete three of the major submodels. The numbers in the following table are based on the work of one person.

| Model | Number of Pages | Non-decomposed transitions | Time to do inscription work | Time to complete testing |
|---|---|---|---|---|

| PPC | 12 | 34 | 6 days | 5 days |
|---|---|---|---|---|
| ERC | 17 | 45 | 7 days | 6 days |
| PCOT | 11 | 18 | 3 days | 3 days |

From the table it can be seen that it takes almost as long time to test and fix a model as it takes to build it. Work is in progress with the intention to reduce the time spent doing the inscription work by automating the process. As mentioned in the previous subsection, adding the inscriptions manually tend to be more or less trivial. Most likely, adding many of the inscriptions automatically to the CPN model can significantly reduce the time spent on testing, because syntactic and some semantic errors are not introduced in the process.

### 3.4 Pushing the Technology

Because of the size of the total SADT model, it was evident from the beginning of the project that we were forced to break it down into smaller and more manageable submodels. It was fairly easy to do so, because of the inherent modularity of the SADT model. Another argument for breaking down the full model was that it would enable more people to be able to work on the model at the same time, creating a flexible work environment. However, as the technology has advanced significantly since then, it should today be possible to have the full model on one machine and even to have more than one person working on the same document at the same time. Alas, this was not possible during the period of our modelling project.

As more and more complex inscriptions were added to the CPN models, we soon realised, that they would not fit into the memory of the machines. This was simply because the machines were running Mac OS version 6.0.7, which only supports 8Mb addressable memory. We wanted to be able to simulate each of the major CPN-submodels individually, with a minimum of extra efforts in setting up an initialisation environment for each submodel. The initialisation uses the knowledge based on assumptions about how the major submodels interact with each other. The interaction is not as well defined on a lower level of decomposition. In fact, that is what we would like to get more information about from the simulation of each major submodel. Therefore it was unfavourable to break up the submodels even further, because it would not make the test of each major submodel very convincing.

Two solutions were found to the memory problem. We applied both in order to be able to proceed our modelling efforts. The first solution was to upgrade our Macintosh machines to Apple's new OS System 7, allowing us to add more physical memory. The second solution was to improve the simulation code generated for the executable models. The effort of investigating different approaches resulted in less code generation and also improved code. This reduced the size of the generated simulation code by an average factor of 40%. As a side effect, the execution time of the simulation models was also reduced significantly.

## 4 Simulating Individual CPN Submodels

After all the inscription work on the CPN model of the waste management system (as described in the previous section), we next generate simulation code from the CPN model and enter simulation mode. In this section we will look at the output from a typical simulation run.

The CPN model takes input in form of an input file, specified by the user, with parameters to initialise the model. The output of the model is a set of graphical reports showing when in time and how often a transition occurs. We will refer to these reports as the timed event charts. Whenever one of the transitions occurs, it updates the graphics in the timed event charts. It is these reports which were sent back to the PFA team for review. The intention was to give them a better and more detailed understanding of the dynamics behind the SADT models they created.

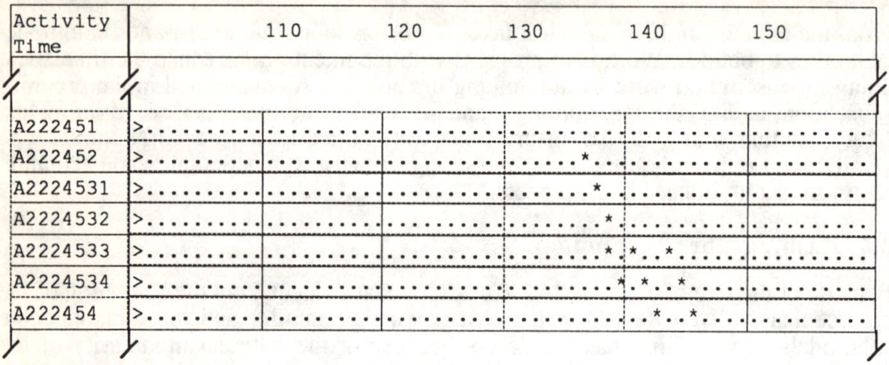


**Fig. 9.** Extract of a timed event chart illustrating the relation between time and occurring transitions. See also figure 3 with the corresponding SADT diagram page.

Each CPN submodel produces a page with a graphical report. Figure 9 shows an output from the submodel PPC (Provide Program Control). The discrete time is displayed from left to right and transitions are shown from top to bottom. A "." in the body of the report, means that the transition (or function) did not occur at that time, while an "*" means that the transition occurred once at a particular time. If the transition occurs more than once at the same point in time, a number (instead of an "*") will indicate how many times it occurred. (A "?" will show up if the transition occurs more than nine times, without the time advancing.) The timed event charts are updated during simulation and more pages with charts are created as needed.

### 4.1 Informal Validation of the SADT Models

The main purpose of our involvement is to validate the accuracy and the completeness of the programmatic functional analysis. More specifically: do the models reflect the intended behaviour and how well do each submodel interact and co-operate with other submodels?

To answer such questions, simulation is used extensively. As submodels are completed, each of them is simulated individually. In order to be able to do that, a primitive initial state is generated based on assumptions on how the environment of the model would behave. This kind of simulation is not only able to identify the worst misunderstandings and bugs, but also to give important and valuable feedback to the PFA team so that they can modify, if necessary, their SADT models accordingly.

Especially a lot of blockings in the information flow (like the one mentioned in section 2.3) have been discovered during the construction and simulation of the individual submodels. Blocking of the flow caused by the interaction of a submodel with the other models, are not investigated or discovered until the models are merged together.

The timed event charts were inspected by the team responsible for the SADT models in order to determine whether the functions, in the waste management program, will be performed satisfactory or not. Often the inspections resulted in requests for changes in the CPN model or redesign of pages in the SADT model. The reports provide important knowledge about how the waste management system can be improved.

## 5 Merge of the CPN Submodels

We have built a set of submodels that can be simulated individually. In order to validate that they actually work together it is necessary to find a way of merging them together into one large model. Once we have one CPN model containing all the submodels which can be simulated, it is possible to look at the interaction between the major functions.

The effort of having translated the SADT models into CPN, has produced seven CPN submodels each having 7–20 pages, 150–350 places, and 20–50 non-decomposed transitions. Because of the size and complexity of each submodel, the attempt of generating one single graphical CPN model was quickly abandoned[3]. Instead we developed a technique for linking the binary formats of the simulation code from the CPN models.

### 5.1 Using Standard ML

We use the functional programming language Standard ML (SML) for the purpose of generating code for the CPN simulator. In this project we use an ML compiler, that is developed at the University of Edinburgh, as the simulation engine, and for generating stand-alone executables.

ML was originally designed with the purpose to be used as a meta language for theorem proving. Robin Milner is one of the persons behind the design of Standard ML and David MacQueen introduced the module concept [7,8].

Standard ML is a strongly typed functional programming language supporting, among other things, parametric modules (functors) and type inference. Most ML-compilers have an interactive environment, where it is possible to save compiled declarations to files (e.g., functors), which then later can be loaded into a new environment. These features will show to be useful for the purpose of merging models.

### 5.2 The Interface between the Major Functions

One SADT diagram page (see figure 2) gives the necessary information about the interface between the functions PSE, DES, ICS, EIS, and PCOT. The remaining two (PPC and ERC) are on a different level of decomposition, and thus not shown here. An arrow between two major subfunctions shows that there is an interface relationship, and thus the location where the subfunctions interact and need to co-operate.

### 5.3 Generating the Stand-alone Executable

Viewed from outside, only the interface of a CPN submodel is interesting. The interface consists of a set of places from which tokens can be added or removed. The colour set (type) of these places should be specified. As a submodel has to be executed we also need ML-functions in the interface for calculating the enabling and execution of transitions.

The SML compiler (Edinburgh version) has a feature for saving individual parametric modules (called functors) as binary data in a file for later retrieval. The interface specification will therefore be captured in a functor and saved, resulting in a

[3]However, the technology today will not inhibit one from handling models of this size.

binary file with all the information about the submodel but only the possibility for manipulating markings and execution of transitions through the interface.

For all the CPN submodels we have one common environment with all the colour set declarations among other things. This common environment is a model in itself. When we need to insert each submodel into the common environment, SML provides a safe way of doing this. SML is a strongly typed programming language, and thus ensures that the types in the interface of a submodel match with any other submodel it is connected to.

### 5.4 The Principle of Model Merge

Merging models with our developed technique is similar to modular linking features used in software development environments. The principle of model merge is simple: given a set of models with a well defined interface, create a common environment for controlling the simulation of each submodel (see figure 10).

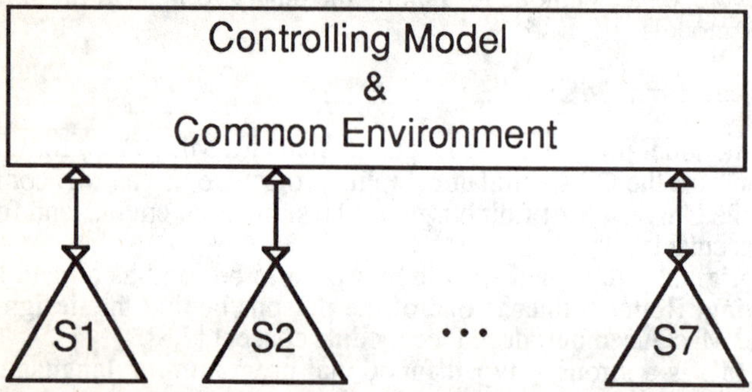


**Fig. 10.** The principle of merging submodels into a common environment.

Each submodel has an interface which is a set of places. The interface is determined by how the submodels are connected with arcs on the higher level. E.g., from figure 2 we can determine the common places (i.e., the interface) between PSE, DES, ICS, EIS, and PCOT. What do we need in order to control a submodel for the purpose of simulation? We need to be able to read and write markings in the interface places, and we need to have a facility for calculating the enabling and executing a step. The signature (type specification) of the functor in which the submodel is encapsulated and later accessed, typically has the following appearance:

```
functor SubM () :
  sig
    val SubM'calc_enab : unit -> TIME
    val SubM'exec      : unit -> unit

    val SubM'assign :
      (((P1 tms) ref) * ((P2 tms) ref) *
       ((P3 tms) ref) * ((P4 tms) ref) *
       ((P5 tms) ref) * ((P6 tms) ref)) -> bool

    val SubM'get : unit ->
      (((P1 tms) ref) * ((P2 tms) ref) *
       ((P3 tms) ref) * ((P4 tms) ref) *
       ((P5 tms) ref) * ((P6 tms) ref))
  end
= struct ... end;
```

Here we have a submodel with six places (P1, P2, ..., P6) in the interface. The `tms` denotes the type timed multi-set. Furthermore, there are four ML-functions for accessing and manipulating the places in the interface and for executing the submodel. The first `SubM'calc_enab` calculates the enabling and returns the minimal time at which something is enabled. The function `SubM'exec` executes one step in the submodel. The two last functions `SubM'assign` and `SubM'get`, respectively assigns and reads markings to and from the places in the interface. The assign function returns true if the marking assigned is different from the previous marking in the interface otherwise false. We use this functionality to determine if a recalculation of the enabling is needed (an expensive operation), i.e., when something changed in the interface marking. The four ML-functions above will be the only way of accessing the submodel, once saved to a file.

We can then save each submodel, encapsulated in the functors, into separate files. How does the controlling model look like? Submodels are passive components, so we need to implement a simple simulation engine in the controlling model. The engine should do the following during simulation:

1. Read minimal enabling time from the individual submodels. This step is necessary in a timed simulation, and it identifies what the minimal time is, when one or more transitions are enabled. If the time found is less than the current time (the global model time), we choose the current time instead. The global time is managed by the controlling model, while each submodel has its own local time. The global time is derived from the local time of the submodels; thus the global time is the model time.
2. Select an appropriate candidate model for execution. Candidates found in this step will have the smallest enabling time. We only execute models with the smallest enabling time less than or equal to the global model time (if any exists).
3. Execute one step in the selected model.
4. Transfer markings between places in the interfaces by means of the `assign` and `get` functions.
5. Recalculate the enabling and minimal time for involved models, if interface marking has changed. In this step, only submodels with changed markings on interface places, need recalculation of enabling. This step can also advance the current model time if it is needed for anything to be time-enabled.
6. If anything is enabled, restart at 1, else terminate the simulation.

The above simple algorithm is implemented as part of the controlling model.

Each submodel is then loaded into the controlling model while the SML type system ensures that there are no type conflict. The SML type system will capture human errors, typically when trying to glue places together between interfaces where the places have differing types.

We now have one single model consisting of all the submodels and we can save this model as a stand-alone executable, ready for simulation.

### 5.5 Simulating the Merged CPN Model

We are now able to simulate the whole model. The size of the model is large, as a result from merging the major submodels: there are more than 110 pages, 2000 places, and 300 transitions (not counting substitution transitions). The file size of the binary simulation image is in the order of 5MB.

As the simulation proceeds, we end up with report pages from the submodels. Typically a simulation session will produce in the order of 60 windows/pages of timed event charts, which makes it rather difficult to find and view interesting pages.

To make it easier to navigate through all this information we introduce the concept of hierarchical reports. It is simply a page where you can get a high level view of what is happening in each of the submodels. An example of a hierarchical report can be seen in figure 11.

| Activity Time | | 310 | 320 | 330 | 340 | 350 | 360 | 370 |
|---|---|---|---|---|---|---|---|---|
| PPC | >********* | ********** | ********** | *********2 | **........ | .......... | .......... | ********** |
| ERC | >....*.... | .......... | ...*...... | .........* | .......... | .......... | .......... | *......... |
| PSE | >*2**..... | .....2**2* | 2**....... | ....***22. | .......... | .......... | ..***2*2** | .......... |
| DES_CD | >......... | .......... | .......... | .........* | ****2***** | ***2****** | **........ | .......... |
| DES_PD | >......... | .......... | .......... | .......... | .......... | .......... | .......... | .......... |
| DES_FD_AB | >......... | .......... | .......... | .......... | .......... | .......... | .......... | .......... |
| ICS | >......... | .......... | .......... | .........* | .......... | .......... | .......... | .......... |
| EIS | >......... | .......... | .......... | .......... | .......... | .......... | .......... | .......... |
| PCOT | >......... | .......... | .......... | .......... | .......... | .......... | .......... | .......... |

**Fig. 11**. A hierarchical report, giving a high level view of the activity in the model.

A row has a link to the page where the timed event chart from the submodel is updated. To follow a link, the user just double clicks on a row and then the corresponding timed event chart of the submodel will show up on the screen. It is also possible to follow the link backwards, in order to return quickly to the hierarchical report.

In the above figure one can see that PPC first is active. Later the conceptual design phase (DES CD) takes over, then shortly PSE before PPC regains control. Therefore, we also find the hierarchical report useful for giving a quick overview of submodels in where transitions occur. It quickly shows if anything is wrong, e.g., if a submodel unintentionally starts at a wrong time. In the above figure one can see that ICS does something at time 339. This is not intended and needs to be investigated further.

## 6 Conclusion and Future Developments

Both the CPN team and the PFA team have learned useful lessons and gained valuable experience in this project. The PFA team obtained an improved SADT model. They were "forced" to provide a more formal description of the behaviour of the activities in the SADT model. To accomplish this level of description the PFA team needed to peruse each activity in the model more carefully. As a result they discovered many ambiguities and inconsistencies – the SADT model was essentially incomplete. Consequently, more concise SADT diagrams were created and more importantly many errors were found at a very early stage. Additionally, the ability to simulate the models gave the PFA team new insight into behaviour. Studying the SADT model in itself does not necessarily provide unequivocal information. At the stage of simulation, more errors were discovered and improvements to the model carried out.

The CPN team identified and implemented improvements in the modelling and simulation tool as a consequence of practical work with a large and non-trivial model. Hardware limits forced us to reduce the memory requirements. By rewriting the code used for simulation we improved performance. We developed a technique for merging CPN models and we also created support for parallel working on models. Both techniques performed well in practice.

When we deal with work flow models, a significant part of the translation from SADT to CPN can be automated. In this way the CP-net methodology becomes available for non-CPN experts. More importantly; the project described in this paper contributes to the opening of a path in this area of automatic translation, reducing the turn-around time. Work is in progress to create support for a more automated translation from SADT to CPN [10]. A translation tool is under development, which is specific for work flow models. The tool supports automatic generation of a CPN model, based on the SADT diagrams with additional behavioural information. Furthermore, the tool will also add simulation code for various kinds of analysis of the work flow (e.g., bottle neck analysis), and simulation code for generating textual and graphical reports. The textual output will be in a format such that advanced external analysis tools can be included as needed.

The purpose of the PFA team's work is to describe the work flow of the nuclear waste management program and ensure that their model is consistent and complete. An obvious extension to the model is to take into account mechanisms, i.e., resources like funds and people. This gives the possibility to create more sophisticated and detailed scenarios by means of simulation. An extended analysis of the resulting information can then be performed to determine more precisely the configuration of the nuclear waste management system. Another useful aspect to have in models is time. Time delays are already included in this model, but they were never exploited fully. For most simulations all time delays were set to 1. This was sufficient for validating consistency in the flow of the model. But more advanced results can be obtained by giving individual functions non-trivial delays. The model can then be used to do completion time analysis of functions with realistic simulation scenarios.

**Acknowledgements**. We would like to thank Kurt Jensen, Jens Bæk Jørgensen, Charles Lakos, and Søren Christensen for patiently reviewing draft versions of this paper, providing constructive critique and food for discussion.

## References


[1] Design/IDEF User's Manual. Meta Software Corporation. Cambridge, Mass., 1992.

[2] Design/CPN User's Manual. Meta Software Corporation. Cambridge, Mass., 1992.

[3] K. Jensen: *Coloured Petri Nets: A High-level Language for System Design and Analysis*. In: G. Rozenberg (ed.): Advances in Petri Nets 1990, Lecture Notes in Computer Science Vol. 483, Springer-Verlag 1991, 342–416. Also in [4], 44–122.

[4] K. Jensen and G. Rozenberg (eds.): *High Level Petri Nets*, Springer-Verlag, 1991.

[5] K. Jensen: *Coloured Petri Nets. Basic Concepts, Analysis Methods and Practical Use*. EATCS Monographs on Theoretical Computer Science, Springer-Verlag, 1992.

[6] D.A. Marca, C.L. McGowan: *SADT*. McGraw-Hill, New York, 1988.

[7] R. Milner, M. Tofte & R. Harper (1990). *The Definition of Standard ML*. MIT Press.

[8] L.C. Paulson: *ML for the Working Programmer*. Cambridge University Press, 1991.

[9] V.O. Pinci, R.M. Shapiro: *An Integrated Software Development Methodology Based on Hierarchical Coloured Petri Nets*. In: G. Rozenberg (ed.): Advances in Petri Nets 1991, Lecture Notes in Computer Science Vol. 524, Springer-Verlag, 1991, 227–252. Also in [4], 649–666.

[10] V.O. Pinci, R.M. Shapiro: *Work Flow Analysis*. Meta Software Corporation. Cambridge, Mass., 1993.

[11] R.M. Shapiro, V.O. Pinci, R. Mameli: *Modelling a NORAD Command Post Using SADT and Coloured Petri Nets*. In: P.E. Lauer (ed.): Functional Programming, Concurrency, Simulation and Automated Reasoning, Lecture Notes in Computer Science Vol. 693, Springer-Verlag, 1993.

# Data Structures and Algorithms for Extended State Space and Structural Level Reduction of the GSPN Model

Donna S. Nielsen and Leonard Kleinrock
University of California, Los Angeles

**Abstract: This paper extends the applicability of exact analysis of the GSPN model by providing methods to improve the time and space complexity of both state space and structural level reduction. For state space level reduction, we maximize the concurrent firing of immediate transitions. For structural level reduction, we minimize the number of generated replicas for timed transitions by using branch and bound techniques to create concurrent replicas that simulate the firing of the timed transition followed by the simultaneously firing of multiple immediate transitions.**

## 1 Introduction

State space level reduction techniques eliminate vanishing markings by concurrently firing multiple immediate transitions in a single state space transition [1, 4]. Structural level reduction techniques eliminate immediate transitions, the source of vanishing markings, by iteratively creating replicas of a timed transition to simulate the firing of the timed transition followed by the firing of an immediate transition [2, 6].

This paper provides data structures and algorithms to extend the applicability and to efficiently implement both state space and structural level reduction of the GSPN model. With respect to state state space level reduction, we rely on knowledge of the given marking, as well as the GSPN structure, to maximize the concurrent firing of immediate transitions. In addition, we develop efficient algorithms to generate the concurrent transition firing combinations and their corresponding firing probabilities. With respect to structural level reduction, we avoid the generation of redundant replicas by applying state space level reduction techniques at the structural level to generate concurrent replicas which simulate the firing of a timed transition followed by the concurrent firing of multiple immediate transitions. In addition, we use branch and bound techniques to both

**This work was sponsored by the Advanced Research Projects Agency of the U.S. Department of Defense under Contract MDA 972-91-J-1011, Advanced Networking and Distributed Systems.**

avoid the generation of infeasible replicas and permit efficient determination of the feasibility of generated replicas.

The remainder of the paper is organized as follows: Section 2 defines the data structures to represent the GSPN model. Section 3 and Section 4 develop the algorithms for state space and structural level reduction, respectively. Finally, Section 5 provides concluding remarks.

# 2 Data Structures for the GSPN Model

We assume the reader is familiar with the structural level definition and basic properties of the GSPN model. The parameters in the formal eight-tuple are standard GSPN notation as defined in [3].

$$GSPN \triangleq (P, T, \Pi(\cdot), W^{-}(\cdot), W^{+}(\cdot), W^{H}(\cdot), M_0, \Lambda(\cdot)).$$

We restrict the GSPN model to include only two priority levels such that immediate transitions have priority over timed transitions. The notation $W^{-}(p,t)$, $W^{+}(p,t)$, and $W^{H}(p,t)$ denotes the multiplicity of place $p$ in the input, output, and inhibitor functions of transition $t$. When appropriate, we employ a vector representation for a bag such that each vector component equals the corresponding bag multiplicity. Boldface type distinguishes the vector representation of a bag. If no inhibitor arc exists from place $p$ to transition $t$, then the multiplicity of place $p$ in the inhibitor bag of $t$ is infinity. An inhibitor arc with weight infinity is equivalent to the absence of an inhibitor arc. We adopt the infinite arc weight representation to facilitate the use of bag operations.

In this section, we provide data structures to represent the structural level specification of the GSPN model. These data structures support the implementation of both state space and structural level reduction. Either a directed graph or a bag represents each component of the formal GSPN tuple.

## 2.1 Representation of Directed Graphs and Bags

We represent a directed graph with an adjacency list for each node in the graph. Let $G = (V, E)$ be a directed graph with the set of nodes $V$ and set of arcs $E$. A node $w$ is in the adjacency list of a node $v$ if and only if there exists a directed arc from $v$ to $w$.

We represent a bag $B$ of a set $S$ with a variable length ordered $K$-tuple $((n_k, x_k) : 1 \leq k \leq K \leq |S|)$. The elements in $S$ are assigned an arbitrary numerical ordering and each variable $x_k$ indexes a set element. Each variable $n_k$ specifies the multiplicity in bag $B$ of the set element indexed by $x_k$. The ordering of the list requires the index $x_k$ to be less than the index $x_{k+1}$. We use the ordered tuple to represent the input, output, and inhibitor functions of a GSPN. For an inhibitor function, the absence of a place in the tuple represents a multiplicity of infinity for that place in the corresponding bag.

The power set of a bag is the set of all subbags. Both state space and structural level algorithms defined in Section 3 and Section 4 require the determination of which subbags in a given power set satisfy application specific feasibility constraints.

## 2.2 Tree Representation for a Power Set

A tree structure representation of the power set facilitates the use of branch and bound techniques to avoid enumeration and examination of infeasible subbags in a given power set. Each node of the tree represents a subbag in the power set. The root node of the tree is at level zero and represents the empty set. The path from the root node to any given node defines the subbag corresponding to that node. Specifically, the arc label from level $k$ to level $k+1$ on the path specifies the $(k+1)st$ ordered pair in the tuple that represents the subbag corresponding to the given node.

To perform a BFS generation of a power set tree for a bag $B$, a branching function defines the outgoing arcs of a generated node. Specifically, let the tuple $\{(n_1, x_1)(n_2, x_2)\ldots, (n_k, x_k)\}$ represent the subbag associated with a generated node $u$. The branching function creates a node $v$ and a directed arc with label $(n_{k+1}, x_{k+1})$ from node $u$ to $v$ if and only if $x_k < x_{k+1} \leq |S|$ and the multiplicity $n_{k+1}$ is less than or equal to the multiplicity in the bag $B$ of the set element indexed by $x_{k+1}$. This branching function both ensures each node in the tree represents a subbag in the power set and avoids the generation of multiple instances of the same subbag. For each generated node, the evaluation of application specific feasibility constraints determines the feasibility of the subbag. Likewise, the evaluation of an application specific bounding function determines if all subbags in the subtree rooted at a generated node are infeasible, thereby avoiding the generation of identified infeasible subtrees. A good bounding function should prune a substantial number of nodes in the power set tree, while maintaining an efficient evaluation at each generated node.

## 2.3 Graphical Representation of Immediate Transitions

We represent the set of immediate transitions by a directed graph which portrays a partial order among transitions with respect to their enabling conditions. We assume the trivial restriction that the GSPN contains no source immediate transitions.

### 2.3.1 Structural Enabling Cover Relation

The structural enabling cover relation provides sufficient structural conditions for the enabling of one transition to dictate the enabling of another transition. Transition $t_i$ is a structural enabling cover ($SEC$) for transition $t_j$ if and only if the input function of $t_j$ is a subbag of the input function of $t_i$, and the inhibitor function of $t_i$ is a subbag of the inhibitor function of $t_j$. For example, given the GSPN in Figure 1, $t_2$ is a structural enabling cover of $t_1$. The $SEC$ relation is a transitive, antisymmetric, reflexive relation that provides a partial ordering on a set of immediate transitions. Formally,

$$t_i \; SEC \; t_j \text{ iff } W^-(t_j) \subseteq W^-(t_i) \cap W^H(t_i) \subseteq W^H(t_j). \tag{1}$$

### 2.3.2 Enabling Graph

The enabling graph (EG) is a directed acyclic graph that depicts the partial order among the set of immediate transitions with respect to the structural enabling cover relation. Each node $v$ of an EG contains a set of transitions $T_v$ such that there exists a path of zero length or more from the node containing transition $t_i$ to the node containing transition $t_j$ if and only if $t_i$ is a structural enabling cover for $t_j$. An EG is minimal if there does not exist a subgraph of the EG which also reflects the $SEC$ partial order. Figure 1 shows a GSPN immediate subnet and its corresponding enabling graph.

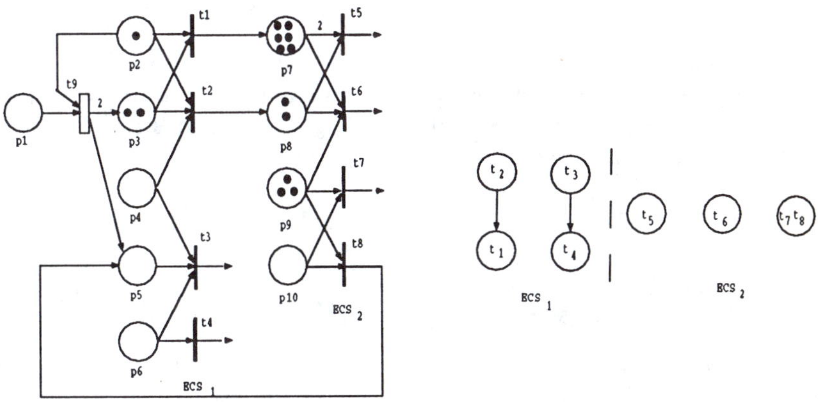


Figure 1: GSPN Subnet and Enabling Graph

## 2.4 Extended Conflict Sets

The partitioning of the set of immediate transitions into extended conflict sets (ECSs) effectively partitions the connected components of the enabling graph. In other words, transitions in the same connected components of the EG must be in the same ECS. Specifically, the $SEC$ partial order depicted in the EG dictates that the transitions contained in any two nodes connected by a directed path are in symmetric structural conflict. And, the transitive and reflexive closure of the symmetric structural conflict relation dictates that the transitions contained in any two nodes connected by an undirected path are in the same ECS. Figure 1 shows the partitioning of the connected components of the EG into ECSs.

# 3 State Space Level Reduction Methods

The concurrent firing of immediate transitions in a single state space transition eliminates the intermediate vanishing markings that derive from all possible firing order permutations. In this section, we extend the reduction methods developed in [1] to permit the concurrent firing of transitions both in different ECSs and within an ECS of a structurally confused immediate subnet.

## 3.1 Enabling and Firing Rules for Concurrent Firings

The enabling bag $EB(M)$ is a bag of transitions that specifies not only which transitions are enabled in marking $M$, but also the number of enablings $EB(t, M)$ of each transition $t$. Formally,

$$EB(t,M) = \begin{cases} 0 & t \notin E(M) \\ \max\{\, n \in \mathbb{N} \mid nW^-(t) \subseteq M \,\} & \text{otherwise} \end{cases} . \tag{2}$$

A transition firing combination $X$ is a bag of transitions such that the multiplicity of a transition in the bag specifies the number of times the transition fires. Formally, the concurrent firing of a bag of transitions $X$ in marking $M$ results in the new marking $M'$:

$$M' = M - \sum_{t \in T} X(t)W^-(t) + \sum_{t \in T} X(t)W^+(t). \tag{3}$$

## 3.2 Marking Dependent Properties

The disabled status of certain transitions in a given vanishing marking can eliminate both confusion and conflict present in the GSPN structure. In this section, we define marking dependent counterparts of structural properties to identify the absence of confusion and conflict among transitions, thereby permitting the maximum concurrent firing of immediate transitions.

### 3.2.1 Marking Confusion-Free Property

In a given vanishing marking, an immediate subnet possibly containing structural confusion is marking confusion-free if for any enabled transition $t_k$ and any disabled transition $t_j$ in the same ECS, the causally connected set $CCS_{t_k}(t_j)$ contains no enabled transitions. In other words, the firing of an enabled transition in one ECS cannot enable a disabled transition in another ECS. For example, in Figure 1, the immediate subnet containing $ECS_1$ and $ECS_2$ is marking, but not structurally, confusion-free. Formally, for marking $M$, an immediate subnet is marking confusion-free if:

$$\forall t_k \in E(M)\ \ \forall t_j \in ECS(t_k) \cap \overline{E(M)},\ CCS_{t_k}(t_j) \cap E(M) = \emptyset. \tag{4}$$

### 3.2.2 Marking Free-Choice Property

In a given vanishing marking, a non free-choice ECS exhibits the free-choice property for a specified number of transition firings if the set of enabled transitions in the ECS remains unchanged for these transition firings. Let $n$ be the minimum number of enablings over all the enabled transitions in a given ECS. If the firing of any enabled transition in the ECS does not increment the inhibitor place of any other enabled transition in the ECS, then all the enabled transitions in the ECS remain enabled for $n$ transition firings within the ECS. In addition, if the immediate subnet of the ECS is marking confusion-free then all disabled transitions in the given ECS remain disabled for these $n$ transition firings. For example, $ECS_2$ in Figure 1 is not free-choice; however, $ECS_2$ is marking free-choice for the first two transition firings. Formally, for a a given marking $M$, an ECS in a marking confusion-free subnet is marking free-choice(n) if for all $t_i, t_j \in ECS \cap EB(M)$,

$$n = \min_{t_i}\{EB(t_i, M)\} \cap \left(W^+(t_j) - W^-(t_j)\right) \cap W^H(t_i) = \emptyset. \tag{5}$$

## 3.3 Concurrent Transition Firing Combinations

In this section, we define methods to identify which transition combinations can fire concurrently, along with their associated firing probabilities.

In a marking free-choice(n) ECS, the set of enabled transitions remains unchanged throughout the firing of any $n$ transitions within the ECS. Thus, the firing probability of each of the $n$ transitions is independent of its order in the firing sequence. This independence permits the concurrent firing of the first $n$ transitions to fire within the ECS. Specifically, the set of concurrent transition firing combinations within a free-choice(n) ECS is the set of all unique transition combinations when selecting $n$ transitions to fire, with replacement and without regard to order, from the set of enabled transitions in the given ECS.

The general branch and bound generation of a power set tree as defined in Section 2 provides an efficient method to generate all concurrent transition firing combinations within a marking free-choice(n) ECS. Without loss of generality let $t_1$ through $t_m$ be the set of enabled transitions in the ECS. Let $T_n$ denote the bag of that contains a multiplicity of $n$ for each transitions $t_1$ through $t_m$. The feasibility constraint dictates that a subbag of $T_n$ is a concurrent transition firing combination if and only if the cardinality of the subbag is $n$. Likewise, the bounding function avoids the generation of all nodes that represents subbags with a cardinality that exceeds $n$. In other words the leaf nodes of the generated portion of the power set tree represent the concurrent transition firing combinations. To achieve a reduction in space requirements, the algorithm discards any infeasible node after the generation of all the node's outgoing arcs.

The firing probability associated with any concurrent transition firing combination $X$ must equal the probability of firing the transitions in $X$ one at a time in any different firing order. Specifically, the firing probability of $X$ is simply the product of the firing probabilities for each transition in $X$ times the number of permutations, defined by the multinomial coefficient, to account for all possible firing orders. Formally,

$$Prob\{X \text{ fires }\} = \frac{n!}{\prod_{t_i \in X} X(t_i)!} \prod_{t_i \in X} \left[ \frac{\Lambda(t_i)}{\sum_{t_j \in E(M) \cap ECS(t_i)} \Lambda(t_j)} \right]^{X(t_i)} \tag{6}$$

By direct extension of the results in [3], the underlying Markovian process of the GSPN is independent of the firing order of transitions in different ECSs of a marking confusion-free immediate subnet. This independence permits the concurrent firing of transitions in the different ECSs. Thus, the Cartesian product of the sets of concurrent transition firing combinations over each ECS defines the set of concurrent transition firing combinations for the immediate subnet. Likewise, the firing probability associated with any concurrent transition firing combination for the immediate subnet is simply the product over the firing probabilities of the corresponding concurrent transition combinations for each ECS within the subnet.

### 3.4 Time and Space Complexity Analysis

In [1], they discuss the reduction in vanishing markings achieved by the concurrent firing of transitions in different ECSs. In this section, we provide a theoretical analysis of the reduction in the number of vanishing markings generated through the concurrent, rather than sequential, firing of transitions within an ECS. For a given marking free-choice(n) ECS with $m$ distinct enabled transitions, the number of resulting markings from firing all feasible concurrent firing combinations is simply the number of ways to distribute $n$ non-distinct objects into $m$ distinct cells. The sequential firing of transitions generates the intermediate vanishing markings to account for all possible transitions firing combinations after the firing of each transition within the sequence. Formally, the reduction in the number of vanishing markings generated by the concurrent, rather than sequential firing of $n$ transitions is:

$$\sum_{i=1}^{n-1} \binom{m+i-1}{m-1}. \tag{7}$$

## 4 Structural Level Reduction Methods

In this section, we modify the structural level reduction algorithms proposed in [2, 5] to avoid the generation of redundant replicas that simulate different firing order permutations of the same transition firing combination. To replicate a timed transition, the proposed structural reduction algorithm first computes all feasible bags of enabled immediate transitions directly after the firing of the given timed transition. Given a feasible bag of enabled immediate transitions, the state space reduction methods defined in Section 3 provide all the feasible concurrent transition firing combinations and their corresponding firing probabilities. The direct applicability of state space level reduction techniques at the structural level is due to the dependency of the state space reduction techniques on only the bag of enabled transitions, rather than the actual markings. Each generated replica corresponds to both a feasible bag of enabled transitions and a concurrent transition firing combination within the bag. The computed enabling conditions for the feasible bag of enabled transitions dictate the input and inhibitor functions of a replica, while the input and output functions of the transitions in the concurrent transition firing combination dictate the replica's output function. Iterative replications of a timed transition result in replicas that simulate the firing of the timed transition followed by each possible firing combination of immediate transitions. Within this section we provide an outline of the modified structural reduction algorithm and a corresponding example showing the reduction of transition $t_9$ in Figure 1. For simplification, the example does not include inhibitor arcs; however, all formulas fully account for the effect of any inhibitor arcs.

## 4.1 Maximum Bag of Enabled Transitions

A timed transition's maximum enabled bag (MEB) is a bag of immediate transitions such that the multiplicity of each transition in the MEB corresponds to its maximum possible number of enablings directly after the firing of the timed transition. Since all feasible bags of enabled transitions must be a subbag of the MEB, the cardinality of the MEB's power set specifies the potential number of replicas created during a single replication step. To minimize this exponential complexity, we extend the GSPN structural properties defined in [3] to establish stringent sufficient structural conditions which restrict the maximum number of direct enablings. These structural properties must also be inclusive enough to determine when the GSPN structure prohibits any enablings of immediate transitions, thereby signifying the completion of the iterative replication of the timed transition.

### 4.1.1 Structural Enable Relation

The structural enable ($SE(n)$) relation provides necessary structural conditions for the firing of a transition to achieve $n$ direct enablings of another disabled transition. For the firing of transition $t_i$ to enable transition $t_j$, the firing of $t_i$ must either increment an input place or decrement an inhibitor place of $t_j$. If the incremented input place of $t_j$ is also an input place to $t_i$ then the arc weight from the input place to $t_i$ must be less than the arc weight from the input place to $t_j$. This condition allows the incremented input place to contain sufficient tokens to enable $t_i$ while concurrently disabling $t_j$. Analogous conditions apply to the decremented inhibitor place. If the firing of a timed transition can enable an immediate transition by decrementing an inhibitor place, then the maximum number of enablings of the immediate transition depends on the initial marking and is indeterminate with respect to the GSPN structure. Otherwise, the increased token count in the incremented input places of the immediate transition dictate an upper bound on the maximum number of enablings. For example, in Figure 1, the firing of $t_9$ structurally enables both $t_1$ and $t_2$ two times and $t_3$ one time. Formally, $t_i\ SE(n)\ t_j$ iff $\exists p \in P$,

$$((W^+(p,t_i) - W^-(p,t_i) < 0) \cap (W^H(p,t_i) > W^H(p,t_j))) \cup \left((W^-(p,t_i) < W^-(p,t_j)) \cap \left\lceil \frac{W^+(p,t_i) - W^-(p,t_i)}{W^-(p,t_j)} \right\rceil \geq n\right). \quad (8)$$

### 4.1.2 Structural Disable Relation

The structural disable ($SD(n)$) relation provides sufficient structural conditions for the firing of a transition to prohibit $n$ direct enablings of another transition. Specifically, if the firing of transition $t_i$ increments an inhibitor place of transition $t_j$ by a token count which exceeds the weight of the inhibitor arc, then the structure prohibits the firing of $t_i$ to directly enable $t_j$. Otherwise, if there exists a place which is an inhibitor place to $t_i$ and an input place to $t_j$, then the enabling and subsequent firing of the $t_i$ dictates an upper bound on the maximum number of direct enablings of $t_j$. This particular circumstance is somewhat unusual for a non-structurally reduced GSPN; however, during the iterative replication process this restriction on the enabling conditions of a replica specifies which markings map into which feasible bags of enabled transitions. Formally, $t_i\ SD(n)\ t_j$ iff $\exists p \in P$,

$$W^+(p,t_i) \geq W^H(p,t_j) \cup \left\lceil \frac{W^H(p,t_i) + W^+(p,t_i) - W^-(p,t_i)}{W^-(p,t_j)} \right\rceil < n. \quad (9)$$

Note that the structural disable relation provides sufficient conditions for the firing of a transition to guarantee the disabling of another transition; whereas, the structural conflict relation provides necessary conditions for the firing of a transition to have the potential to disable another transition.

### 4.1.3 Structurally Maximum Bag of Enabled Transitions

The structural enable and disable relations provide sufficient structural conditions to restrict the maximum number of direct enablings of an immediate transition upon firing a given timed transition. In addition, the structural enabling cover relation dictates that if $t_j\ SEC\ t_k$ then the maximum number of enablings of transition $t_j$ cannot exceed the maximum number of enablings of $t_k$. For example, in Figure 1, since $t_3$ is a structural enabling cover for $t_4$ and the firing of $t_9$ permits no enabling of $t_4$, then, even though the firing of $t_9$ increments an input place of $t_3$, $t_3$ also must have no enablings. Formally, the multiplicity of immediate transition $t_j$ in the maximum enabling bag of $t_i$ is:

$$\max\left\{n \mid t_i\ SE(n)\ t_j \cap t_i\ \overline{SD(n)}\ t_j \cap\ MEB(t_k) \geq n\ \forall t_k : t_j\ SEC\ t_k\right\}. \quad (10)$$

The MEB for timed transition $t_9$ in Figure 1 is $\{2t_1, 2t_2\}$. In [5], we provide an algorithm that generates the set of immediate transitions that can be directly enabled after the firing of a timed transition. This algorithm only requires minor modification, with no additional cost in complexity, to also determine the maximum number of enablings of each transition.

### 4.1.4 Indeterminate Transitions in an MEB

To simplify the algorithm for a transition replication step, we assume the timed transition's MEB contains no indeterminate transitions. A transition in the MEB is indeterminate with respect to the GSPN structure if the maximum number of direct or indirect enablings, after the firing of a timed transition and prior to reaching a tangible marking, is dependent on the initial marking. The structural enable relation establishes necessary conditions for a transition in the MEB to be indeterminate. Specifically, this indeterminacy occurs either if an incremented input place in the bag evaluated in the first inequality of Equation 8 participates in a cycle of immediate transitions or if, as stated previously, there is a decremented inhibitor place in the bag evaluated in the second inequality of Equation 8. In [5] we provide methods to perform structural reduction with indeterminate transitions.

### 4.1.5 Enabling Graph Representation of the MEB

An enabling graph representation of a given MEB facilitates the generation of all structurally feasible bags of enabled transitions within the MEB. This enabling graph depicts the $SEC$ partial order, not only among the transitions in the MEB, but also among each distinct number of enablings from one to a transition's multiplicity in the MEB. Specifically, each node $v$ of the EG contains a set of transitions $T_v$ such that there exists a path from the node containing $n_i$ enablings of transition $t_i$ to the node containing $n_j$ enablings of transition $t_j$ if and only if $n_i$ enablings of transition $t_i$ is a structural enabling cover for $n_j$ enablings of transition $t_j$. Figure 2a shows the augmented EG for $MEB(t_9)$ corresponding the GSPN in Figure 1. All future references to an enabling graph refer to the enabling graph which represents the MEB of the replicated timed transition.

We represent a subset of nodes in the EG of the MEB with the standard variable length ordered $K$-tuple $((x_k) : 1 \leq k \leq K \leq |V_{EG}|)$ such that each $x_k$ indexes a node in the $EG$. In turn, we represent a subbag of the MEB with a subset of the nodes in the EG such that the subbag of the MEB is simply the union of the bag of transitions contained in the subset of nodes.

## 4.2 Feasible Bags of Enabled Transitions

The GSPN structure restricts which subbags in the MEB can be simultaneously enabled, without forcing the enabling of any remaining transitions in the MEB. In this section, we develop methods to generate these structurally feasible bags of enabled transitions. We rely on branch and bound techniques to avoid the enumeration and examination of the entire power set of the MEB.

### 4.2.1 Extended Input and Inhibitor Functions

The extended input and inhibitor functions $W^{-}(\cdot)$ and $W^{H}(\cdot)$ map bags of transitions into bags of places and define necessary marking conditions to achieve the specified number of enablings of each transition in the bag. Formally, for a non-empty bag of transitions $T_B$:

$$W^{-}(T_B) = \bigcup_{t \in T_B} T_B(t) W^{-}(t); \quad W^{H}(T_B) = \bigcap_{t \in T_B} W^{H}(t). \qquad (11)$$

A bag of transitions $T_B$ is enabled in marking $M$ if and only if $\mathbf{W}^{-}(T_B) \leq \mathbf{M} < \mathbf{W}^{H}(T_B)$.

### 4.2.2 Enabling Tree

The general branch and bound generation of a power set tree as defined in Section 2 provides an efficient method to generate all structurally feasible bags of enabled transition for a given MEB. Since a subset of nodes in the EG represents a subbag of the MEB, the generated enabling tree essentially represents the power set, with pruning, of the set of nodes in the EG. Let $EB_v$ be the subbag of the MEB corresponding to node $v$ of the enabling tree.

The GSPN structure imposes two feasibility constraints on any feasible bag of enabled immediate transitions. The first structural constraint ensures the enablings of the transitions in a feasible bag do not force the enabling of any remaining transitions in the MEB. Specifically, if an enabled bag $EB$ has $n_i$ enablings of transition $t_i$ and these enablings are a structural enabling cover for $n_j$ enablings of transition $t_j$ then $EB$ must also contain at least $n_j$ enablings of $t_j$. For example, in Figure 2a, a feasible enabling bag which contains two enablings of transition $t_2$ must also contain two enablings of transition $t_1$.

The second structural feasibility constraint prohibits a feasible enabled bag from containing structurally mutual exclusive (SME) transitions. To simplify the algorithm for a transition replication step, we assume the given MEB contains no structurally mutual exclusive transitions. In [5], we provide structural reduction algorithms which incorporate this SME feasibility constraint.

The bounding function requires a topological ordering of the nodes in the EG such that if there exists a path from node $x$ to node $y$ in the EG then the index of node $x$ is greater than the index of node $y$. As shown for the EG in Figure 2a, the numbering of the nodes in increasing order as they are post-visited in a DFS traversal achieves this topological ordering. The bounding function prunes any subtree of the enabling tree that is rooted at an infeasible node. Specifically, if $EB_v$ does not satisfy the feasibility constraint then there must exist a transition $t_i$ with multiplicity $n_i$ in $EB_v$ and a transition $t_j$ with multiplicity $n_j$ in $\overline{EB_v}$ such that $n_i t_i \; SEC \; n_j t_j$. For any node $w$ in the subtree rooted at node $v$, since $EB_v$ is a subbag of $EB_w$ then $EB_w$ also contains $n_i$ enablings of transition $t_i$.

In addition, due to the topological ordering of the nodes in the EG, $EB_w$ also cannot contain $n_j$ enablings of transition $t_j$. Thus $EB_w$ is also infeasible.

Let $(x_1, x_2, \ldots x_k)$ represent the set of EG nodes associated with node $u$ of the enabling tree. The following steps define the BFS generation of all direct descendents of node $u$.

**Branching Function:** For all $x_{k+1}$ such that $x_k < x_{k+1} \leq |V_{EG}|$, generate node $v$ and create an arc from $u$ to $v$ with label $x_{k+1}$.

**Node attributes:**

$EB_v = EB_u \cup T_{x_{k+1}}$.

$W^-(EB_v) = W^-(EB_u) \cup W^-(T_{x_{k+1}})$.

$W^H(EB_v) = W^H(EB_u) \cap W^H(T_{x_{k+1}})$.

**Feasibility Constraints:** $EB_v$ not feasible if $\forall t_i, t_j \in EB_v$, $EB_v(t_i) = n_i \Rightarrow EB_v(t_j) \geq n_j$ if $n_i t_j$ $SEC$ $n_j t_j$.

**Bounding Function:** Prune subtree rooted at $v$, if $EB_v$ is infeasible.

The potential for complexity reduction in this branch and bound technique is two-fold: the bounding eliminates the generation of some subbags in the power set of the MEB and the systematic tree generation permits a worst case complexity of $O(|P| + |MEB|)$ to determine the feasibility for a generated subbag. Figure 2b shows the enabling tree associated with $MEB(t_9)$. The numbering of the nodes in the enabling tree is in accordance with the order of the BFS node generation.

### 4.2.3 Enabling Function

Given the enabling tree, the replication of the timed transition requires the determination of the enabling conditions corresponding to each structurally feasible bag. The function $Enab(\cdot)$ maps a structurally feasible enabled bag into a set of markings such that marking $M$ is an element of $Enab(T_B)$ if and only if $T_B$ is enabled in marking $M$. Formally,

$$Enab(T_B) = \{ M \mid \mathbf{W}^-(T_B) \leq \mathbf{M} < \mathbf{W}^H(T_B) \}. \qquad (12)$$

Any marking in the enabling function of a given feasible enabled bag ensures the enabling of all transitions in the given bag, but the marking does not prohibit the enabling of any remaining transitions in the MEB. Thus, a single marking may map into more than one enabling function. The complete specification of a feasible bag's enabling conditions requires more stringent marking conditions to ensure the disabling of the remaining transitions in the MEB.

## 4.3 Feasible Bags of Disabled Transitions

Corresponding to each structurally feasible bag $T_B$ of enabled transitions is a feasible bag of disabled transitions $\overline{T_B}$. In this section, we establish the marking conditions that disable the transitions in a given feasible bag of disabled transitions, without forcing the disabling of any remaining transitions in the MEB.

### 4.3.1 Disabling Input and Inhibitor Function

The disabling input and inhibitor functions $W_D^-(\cdot)$ and $W_D^H(\cdot)$ map bags of transitions into bags of places and define necessary structural conditions to disable, by restricting the token count of single place, the specified number of enablings of each transition in the bag. The subscript '$D$' emphasizes that the resulting bags of places define disabling conditions rather than enabling conditions. Formally, for a non-empty bag of transitions $T_B$:

$$W_D^-(T_B) = \bigcap_{t \in T_B} T_B(t) W^-(t); \quad W_D^H(T_B) = \bigcup_{t \in T_B} W^H(t). \tag{13}$$

The specified multiplicity of each transition in the bag $T_B$ is disabled in marking $M$ if $\mathbf{W}_D^-(T_B) \not\leq \mathbf{M} \cup \mathbf{M} \not< \mathbf{W}_D^H(T_B)$.

### 4.3.2 Disabling Tree

In this section, we identify all feasible bags of disabled transitions that satisfy necessary structural conditions to permit the disabling of the transitions in the bag, by restricting the token count in a single place, without forcing the disabling of any remaining transitions in the MEB. We refer to these bags as singularly disabled bags.

The branch and bound generation of a power set tree provides an efficient method to generate all singularly disabled bags of transitions. Analogous to the generation of the enabling tree, the generated disabling tree essentially represents the power set, with pruning, of the set of nodes in the EG. Let the disabled bag $DB_v$ be the subbag of an MEB associated with node $v$ of the disabling tree.

The structural feasibility constraint for a bag of singularly disabled transitions requires the existence of at least one place such that the multiplicity of that place in the disabling function of $DB_v$ exceeds the multiplicity of that place in the enabling function of $\overline{DB_v}$ or the multiplicity of that place in the disabling inhibitor function of $DB_v$ is less than the multiplicity of that place in the inhibitor function of $\overline{DB_v}$. For example, given $MEB(t_9)$ for the GSPN in Figure 1, any bag of transitions that contains $t_1$ and does not contain $t_2$ is not a feasible disabled bag since the disabling of $t_1$ always forces the disabling of $t_2$. Also, the bag $\{2t_1, t_2\}$ is not a feasible disabled bag since the disabling of these transition enablings forces the disabling of $t_1$.

To both retain the EG node numbering that was used to generate the enabling tree and apply the bounding function, we alter the branching function for the generation of the disabling tree such that $x_{k+1} < x_k \leq |V_{EG}|$. The bounding function prunes any subtree rooted at any node $v$ if there does not exist a complement node in the enabling tree corresponding to $\overline{DB_v}$. Specifically, using arguments analogous to those described for the enabling tree, if $\overline{DB_v}$ is not a feasible enabled bag then $DB_v$ is not a feasible disabled bag because the disabling of a transition in $DB_v$ forces the disabling of a transition in $\overline{DB_v}$. Due to the topological ordering of the nodes in the $EG$ and the specified branching function, for any node $w$ in the subtree rooted at node $v$, the bag

$DB_w$ will contain this same infeasibility. Thus, an efficient BFS generation of the disabling tree requires a parallel reverse BFS traversal of the enabling tree to determine the existence of the complement node, as well as efficiently access the complement node attributes to efficiently evaluate the feasibility constraint. A second bounding function prunes all nodes of a subtree rooted at a node that is infeasible because the corresponding bag of transitions does not share a common disabling condition.

Given the GSPN in Figure 1 and the enabling tree in Figure 2b, Figure 2c shows the disabling tree associated with $MEB(t)$. The numbering of the nodes in the tree facilitates the generation of another tree and is not in accordance with the order of the BFS generation. Let $(x_1, x_2, \ldots x_k)$ represent the set of EG nodes associated with node $u$ of the disabling tree. The following steps define the BFS generation of all direct descendents of node $u$.

**Branching Function:** For all $x_{k+1}$ such that $x_{k+1} < x_k \leq |V_{EG}|$, generate node $v$ and create an arc from $u$ to $v$ with label $x_{k+1}$.

**Node attributes:**

$DB_v = DB_u \cup T_{x_{i+1}}$

$W_D^-(DB_v) = W_D^-(DB_u) \cap W^-(T_{x_{i+1}})$.

$W_D^H(DB_v) = W_D^H(DB_u) \cup W^H(T_{x_{i+1}})$.

**Feasibility Constraints:** $DB_v$ feasible if

$W_D^-(DB_v) - W^-(\overline{DB_v}) \neq \emptyset \cup W^H(\overline{DB_v}) - W_D^H(DB_v) \neq \emptyset$

**Bounding Function:** Prune subtree rooted at $v$, if $\overline{DB_v}$ is not a feasible enabled bag or $W_D^-(DB_v) = \emptyset \cap W_D^H(DB_v)$ contains a multiplicity of infinity for each place.

### 4.3.3 Disabling Function

The function $Disab(\cdot)$ maps a bag of transitions into a set of markings such that a marking $M$ is an element of $Disab(T_B)$ if and only if the token count of a single place marking in $M$ disables $T_B$ without disabling the remaining transitions in the MEB. $Disab(T_B)$ defines the set of markings in terms of an expression that is a union of single place marking inequalities. For example, given the disabling tree in Figure 2c, $Disab(\{2t_1, 2t_2\}) = (m_1 < 2) \cup (m_2 < 2)$. Formally, $Disab(T_B) =$

$$\{ M \mid \mathbf{W}^-(\overline{T_B}) \leq \mathbf{M} \not\geq \mathbf{W}_D^-(T_B) \cup \mathbf{W}_D^H(T_B) \not< \mathbf{M} < \mathbf{W}^H(\overline{T_B})\} \qquad (14)$$

In the following section we remove the restriction that the disabling of a feasible bag of disabled transitions must occur through single place marking inequalities.

### 4.3.4 Minimal Disabling Sets

We define a minimal disabling set to be a set of singular disabled bags such that the union of the singular disabled bags over any proper subset of the minimal disabling set does not equal the union over the entire disabling set. Let the bag of transitions $T_B$ equal the union of all the singular disabled bags in a given minimal disabling set. The intersection of the respective disabling functions of all the singular disabled bags defines a set of markings which disables $T_B$ without forcing the disabling of any of the remaining transitions in the MEB. The minimality property ensures that the intersection of the disabling functions associated with the disabled bags in any proper subset of the minimal set results in the disabling of only a proper subbag of $T_B$.

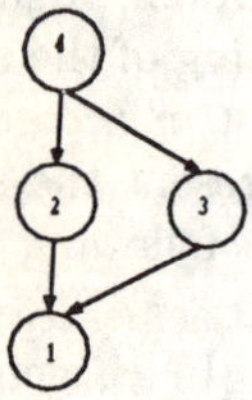

| $v$ | $T_v$ |
|---|---|
| 1 | $t_1$ |
| 2 | $t_2$ |
| 3 | $2t_1$ |
| 4 | $2t_2$ |

2a

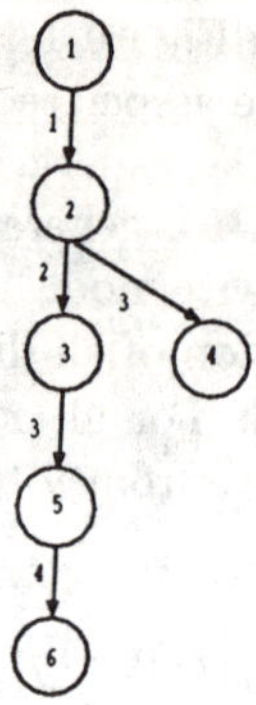

2b

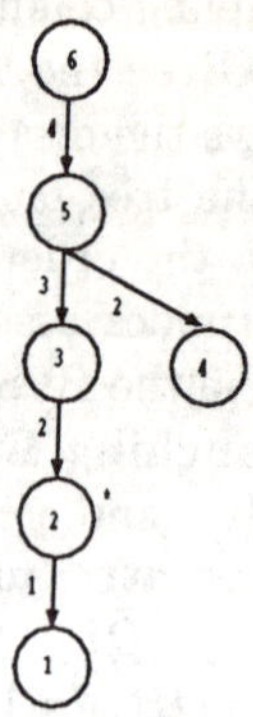

2c

*infeasible

| $v_{EB}$ | $v_{DB}$ | $EB_v$ | $DB_v$ | $W^-(EB_v)$ | $W_D^-(DB_v)$ |
|---|---|---|---|---|---|
| 1 | 1 | $\emptyset$ | $\{t_1t_2\}$ | - | $\{p_1p_2\}$ |
| 2 | 2 | $\{t_1\}$ | $\{2t_1t_2\}$ | $\{p_1p_2\}$ | $\{p_1p_2\}$ |
| 3 | 3 | $\{t_1t_2\}$ | $\{2t_12t_2\}$ | $\{p_1p_2p_3\}$ | $\{2p_12p_2\}$ |
| 4 | 4 | $\{2t_1\}$ | $\{t_2\}$ | $\{2p_12p_2\}$ | $\{p_1p_2p_3\}$ |
| 5 | 5 | $\{2t_1t_2\}$ | $\{2t_2\}$ | $\{2p_12p_2p_3\}$ | $\{2p_12p_22p_3\}$ |
| 6 | 6 | $\{2t_12t_2\}$ | $\emptyset$ | $\{2p_12p_22p_3\}$ | - |

Node Attributes

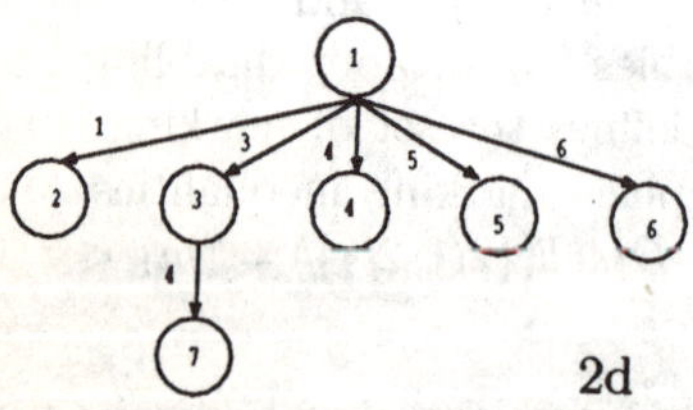

2d

| $v$ | $MS_v$ |
|---|---|
| 1 | $\emptyset$ |
| 2 | $\{\{t_1t_2\}\}$ |
| 3 | $\{\{2t_12t_2\}\}$ |
| 4 | $\{\{t_2\}\}$ |
| 5 | $\{\{2t_2\}\}$ |
| 6 | $\{\{2t_12t_2\}\{t_2\}\}$ |

Figure 2: Structural Reduction Graphs

The general branch and bound generation of a power set tree provides an efficient method to generate all minimal disabling sets. In this application, the generated minimal tree represents the power set, with pruning, of the set of feasible nodes in the disabling tree. In turn, each feasible node in the disabling tree represents a singular disabled bag.

Efficient evaluation of the feasibility constraint requires a topological ordering of the nodes in the disabling tree such that if node $x$ is less than node $y$ then the singular disabled bag $DB_x$ is not a subbag of $DB_y$. The numbering of the nodes in decreasing order as they are visited in a BFS traversal, as shown in Figure 2c, achieves the required ordering. Let $MS_v$ be the minimal set represented by node $v$ in the minimal tree and let $T(MS_v)$ be the union of the singular disabled bags in $MS_v$. The feasibility constraint ensures the minimality property by dictating that any added singular disabled bag must contain a transition not in $T(MS_v)$. Without the specified topological ordering of the disabling tree, the feasibility constraint would also have to ensure none of the singular disabled bags in $MS_v$ were a subbag of this added singular disabled bag.

Clearly the bounding function can prune all subtrees rooted at an infeasible node. Figure 2d shows the minimal tree corresponding to the disabling tree in Figure 2c. Let the $k$-tuple $(x_1, x_2, \ldots x_i)$ represent minimal set $MS_u$ of disabled bags.

**Branching Function:** For all $x_{i+1}$ such that $x_i < x_{i+1} \leq |V_{DT}|$, generate node $v$ and create an arc from $u$ to $v$ with label $x_{i+1}$.

**Node attributes:**

$MS_v = MS_u \cup \{MDS_{x_{i+1}}\}$.

$T(MS_v) = T(MS_u) \cup MDS_{x_{i+1}}$.

**Feasibility Constraints:** $MS_v$ is feasible if $DB_{x_{i+1}} \not\subseteq T(MS_u)$.

**Bounding Function:** prune subtree rooted at node $v$ if $MS_v$ is not minimal.

## 4.4 Marking Function

The marking function $M(T_B)$ maps a structurally feasible bag of enabled transitions into a set of markings. A marking is an element of $M(T_B)$ if and only if the marking enables all the transitions in $T_B$ and disables all the remaining transitions $\overline{T_B}$. Formally, $M(T_B)$ defines the set of markings in terms of a disjunctive normal form expression of inequalities on places. For example, given $MEB(t_9)$ for the GSPN in Figure 1:

$$M(t_1, t_2) = \{M \mid (1 \leq m_1 < 2 \cap 1 \leq m_2 \cap 1 \leq m_3) \cup (1 \leq m_1 \cap 1 \leq m_2 < 2 \cap 1 \leq m_3)\} \tag{15}$$

The set of markings defined by an 'and' clause of $M(T_B)$ must be a subset of $Enab(T_B)$ to ensure the enabling of $T_B$. Additional place marking inequalities of an 'and' clause guarantee the disabling of all transitions in $\overline{T_B}$. The number of 'and' clauses represent the different ways to accomplish the disabling of the transitions in $\overline{T_B}$, without the disabling of the transitions in $T_B$. The generation of minimal disabling sets permits the generation of marking functions with the minimum number of 'and' clauses. Specifically, a minimal disabling set in $MS(\overline{T_B})$ is associated with each 'and' clause such that there exists exactly one place marking inequality for each singular disabling bag that results in the disabling of all the transitions in $\overline{T_B}$. Formally,

$$M(T_B) = \bigcup_{MS \in MS(\overline{T_B})} Enab(T_B) \cap \bigcap_{T_{B'} \in MS} (Disab(T_{B'}). \tag{16}$$

In disjunctive normal form, we represent the $k_{th}$ 'and' clause of $M(T_B)$ with the two bags of places, $M^-(k,T_B)$ and $M^H(k,T_B)$ such that

$$M(T_B) = \{ M \mid \bigcup_k (\mathbf{M}^-(k,T_B) \leq \mathbf{M} \leq \mathbf{M}^H(k,T_B))\}. \tag{17}$$

The marking functions of each structurally feasible bag of enabled transitions in an MEB effectively partitions the set of all markings. In other words, each marking must result in the enabling of exactly one of the feasible enabled bags and the disabling of the remaining transitions in the MEB. However, a marking can satisfy multiple 'and' clauses of the marking function for a given feasible bag of enabled transitions.

## 4.5 Generation of Replicas

For all structurally feasible bags of transitions $T_B$ enabled directly after the firing of a given timed transition $t$, for all 'and' clauses $k$ in the marking function of $T_B$, and for all concurrent transition firing combinations $X$ given $T_B$, a single transition replication step creates the replicas $(t, X, T_B, k)$. This replica simulates the firing of the timed transition $t$ followed by the concurrent firing of the transitions in $X$ given $T_B$ is the bag of enabled immediate transitions.

Given the marking bags $M^-(k,T_B)$ and $M^H(k,T_B)$, the computations of the input and inhibitor functions for replica $(t,X,k,T_B)$ are straightforward bag operations. The input and inhibitor functions of the replica must permit the enabling of the replicated transition $t$ and upon firing $t$ produce a marking which satisfies the $k^{th}$ 'and' clause of $T_B$'s marking function. The output function of the replica simply produces the change in marking which results from the firing of both the replaced timed transition and the replica's concurrent transition firing combination. Formally, the attributes of replica $(t,X,k,T_B)$ are:

$$\begin{aligned}
W^-(t,X,k,T_B) &= W^-(t) + (M^-(k,T_B) - W^+(t)) \\
W^H(t,X,k,T_B) &= W^H(t) \cap (W^-(t) + (M^H(k,T_B) - W^+(t))) \\
W^+(t,X,k,T_B) &= (W^+(t) - W^-(X)) + W^+(X) \\
\Lambda(X,k,T_B) &= \Lambda(t)\text{Prob }\{X \text{ fires } \mid T_B \text{ enabled }\}.
\end{aligned}$$

## 4.6 GSPN Decomposition

In this section, we develop a method to decompose the GSPN into subnets, perform structural level reduction on each of these subnets, and aggregate the generated subnet replicas to construct the structurally reduced net corresponding to the original GSPN. This proposed technique is analogous to the state space level reduction method which decomposes a state into immediate submarkings, performs state space evolution of each immediate subnet, and aggregates the resulting tangible submarkings to generate the reachable tangible markings.

### 4.6.1 Augmented Immediate Subnets

The GSPN decomposition step first augments each immediate subnet to include all timed transition such that the transition's MEB contains a transition in the immediate subnet. Each included timed transition retains its corresponding input and inhibitor functions. The output function of an included timed transition equals the output function of the timed transition after the removal of all places that are not in the given immediate subnet. Each timed transition in an augmented immediate subnet has a firing rate of one.

For each timed transition, the decomposition also creates a subnet consisting solely of the single timed transition. In this subnet, the timed transition retains its input function, inhibitor function, and firing rate. The output function of the timed transition in the subnet equals the output function of the timed transition after the removal of all places that are in any immediate subnet. Figure 3b shows the decomposed GSPN subnets corresponding to the GSPN shown in Figure 3a.

### 4.6.2 Aggregation of Transition Replicas

The structural reduction of the subnets in the decomposed GSPN and the subsequent aggregation of the replicas among these subnets constructs the same reduced net created by structural reduction of the original GSPN. The Cartesian product of the replicas for a given timed transition in each structurally reduced subnet defines the concurrent replicas for this transition in the structurally reduced net of the original GSPN. In other words, each concurrent replica of a timed transition simulates the concurrent firing of one replica for this timed transition from each of the structurally reduced subnets that contains replicas corresponding to the given timed transition. The input and output functions of a concurrent replica are simply the union of the input and output functions, respectively, of the corresponding replicas in the reduced subnets. Likewise, the inhibitor function of a concurrent replica is simply the intersection of the inhibitor functions of the corresponding replicas in the subnets. The firing rate of the concurrent replica is the product of the firing rates of the corresponding subnet replicas. Figure 3c shows the structurally reduced GSPN corresponding to the GSPN, while Figure 3d shows the structurally reduced subnets corresponding to the decomposed GSPN subnets. The direct derivation of a GSPN's reachability set from the structurally reduced subnets, rather than actual construction of the aggregated structurally reduced GSPN, achieves additional improvements in time and space complexity.

## 4.7 Time and Space Complexity Analysis

Branch and bound methods avoid the generation of most structurally infeasible replicas, while concurrent replicas avoid the generation of most redundant replicas. In addition, the systematic generation of the tree structures, which represent each replica's attributes, permits the efficient determination of feasible replicas. In [5], we provide theoretical complexity analysis for the branch

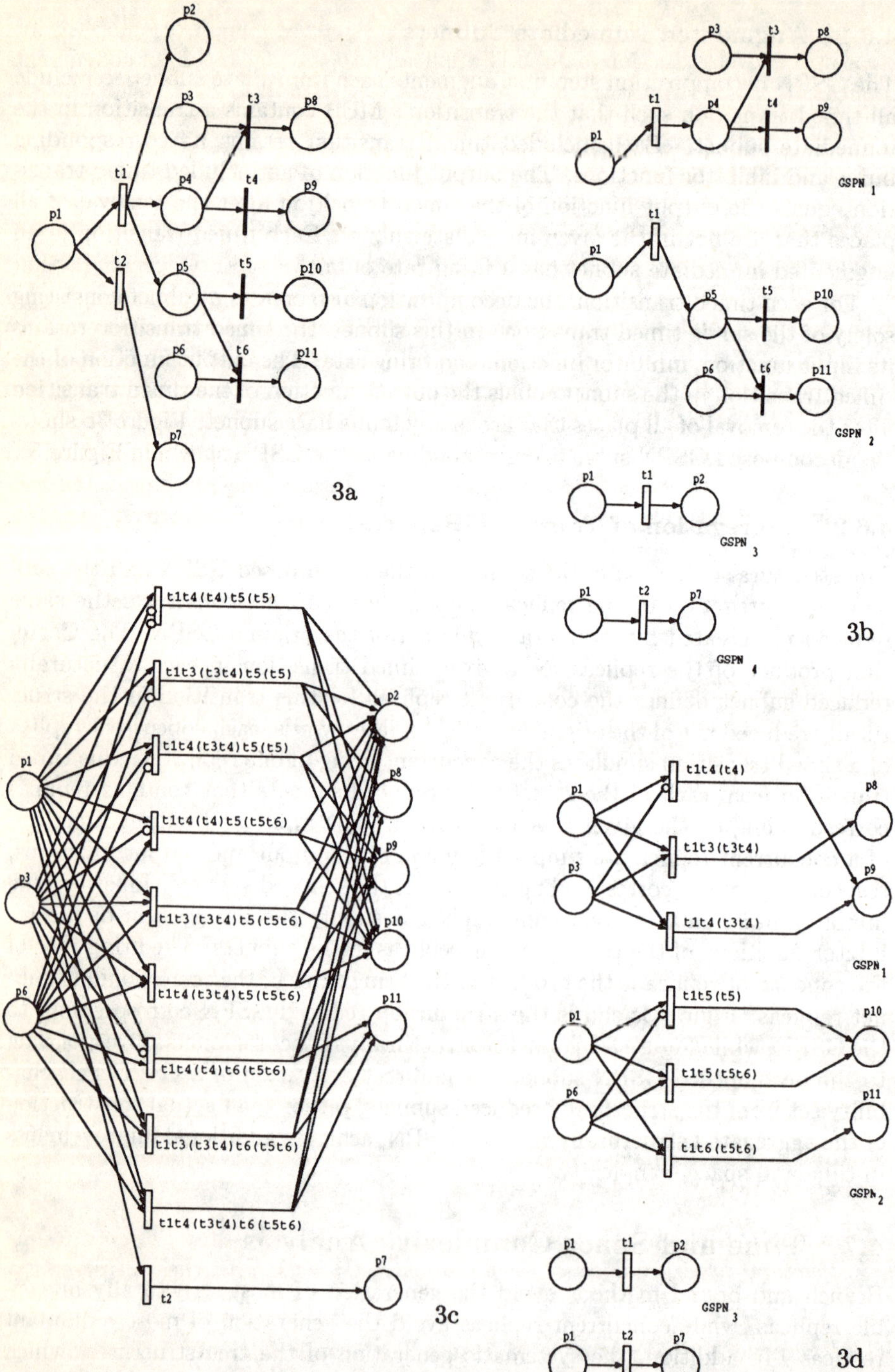


Figure 3: GSPN Decomposition

and bound generation of non-concurrent replicas and the complexity analysis for state space level reduction is directly applicable to the generation of concurrent replicas.

# 5 Conclusion

The contribution of this paper is to provide algorithms and corresponding data structures to efficiently implement both state space and structural level reduction on the GSPN model. Structural level reduction has some inherent advantages over state space level reduction. Since the GSPN structure can provide a factorization of state space transitions into significantly fewer net transitions, there is the potential for a corresponding factorization of complexity between state space and structural level reduction. In other words, the elimination of a single immediate transition can achieve a reduction equivalent to the elimination of several state space transitions and their adjacent vanishing markings. On the other hand, structural level reduction may generate replicas that are not enabled by any marking in the reachability set.

# References

[1] G. Balbo, G. Chiola, G. Franceschinis, and G. Roet. On the Efficient Construction of the Tangible Reachability Graph of Generalized Stochastic Petri Nets. In *Proceedings of the International Workshop on Petri Nets and Performance Models*, pages 136–145, August 1987.

[2] Giovanni Chiola, Susanna Donatelli, and Giuliana Franceschinis. GSPNs versus SPNs: What is the Actual Role of Immediate Transitions? In *Proceedings of the Fourth International Workshop on Petri Nets and Performance Models*, pages 20–31, December 1991.

[3] Giovanni Chiola, Marco Marsan, Gianfranco Balbo, and Gianni Conte. Generalized Stochastic Petri Nets: A Definition at the Net Level and its Implications. *IEEE Transactions on Software Engineering*, 19(2):89–107, February 1993.

[4] Mark A. Holliday and Mary K. Vernon. A Generalized Timed Petri Net Model for Performance Analysis. *IEEE Transactions on Software Engineering*, SE-13(12):1297–1310, December 1987.

[5] Donna S. Nielsen. *Performance Evaluation of Distributed Systems Using Structural Level Analysis of Performance Petri Nets.* PhD thesis, University of California, Los Angeles, June 1994.

[6] C. Simone and M. Ajmone Marsan. The Application of EB-Equivalence Rules to the Structural Reduction of GSPN Models. *Journal of Parallel and Distributed Computing*, 15(3):296–302, July 1992.

# Petri Net Analysis Using Boolean Manipulation

Enric Pastor, Oriol Roig, Jordi Cortadella, and Rosa M. Badia

Department of Computer Architecture
Universitat Politècnica de Catalunya
08071 Barcelona, Spain

**Abstract.** This paper presents a novel analysis approach for bounded Petri nets. The net behavior is modeled by boolean functions, thus reducing reasoning about Petri nets to boolean calculation. The state explosion problem is managed by using *Binary Decision Diagrams* (BDDs), which are capable to represent large sets of markings in small data structures. The ability of Petri nets to model systems, the flexibility and generality of boolean algebras, and the efficient implementation of BDDs, provide a general environment to handle a large variety of problems. Examples are presented that show how all the reachable states ($10^{18}$) of a Petri net can be efficiently calculated and represented with a small BDD ($10^3$ nodes). Properties requiring an exhaustive analysis of the state space can be verified in polynomial time in the size of the BDD.

## 1 Introduction

*Petri nets* were initially proposed by C.A. Petri in 1962 for describing information processing systems, characterized as being concurrent, asynchronous, distributed, parallel, nondeterministic, and/or stochastic. Many different application areas have considered Petri nets for the modeling and analysis of their systems. Among them, we could mention operating systems, communication protocols, distributed systems, multiprocessor systems, etc.

Several methods for Petri net analysis have been proposed in the literature. They can be classified into three categories [11]: the reachability tree method, the matrix-equation method and reduction or decomposition techniques. While the first method is only applicable to small nets due to the explosion of the number of states, the second and third methods are restricted to special classes of nets.

In this paper, a novel analysis approach applicable to any type of bounded Petri net is presented. It is based on the description of the net behavior by means of *boolean functions*, thus reducing *reasoning* to *calculation* [2]. Questions like *"is there any marking with a deadlock ?"* or *"can transitions $t_1$ and $t_2$ be fired concurrently ?"* or properties like *liveness*, *safeness* and *persistence* can be answered and verified by properly manipulating the functions that describe the system.

Supported by CYCIT TIC 91-1036 and Dept. d'Ensenyament de la Generalitat de Catalunya

The exponential complexity involved in the enumeration of the markings of a net is managed by using *Binary Decision Diagrams* (BDD) [3]. BDDs have been widely and successfully used in the areas of logic synthesis and verification of digital circuits, and their appeal comes from the capability of representing large sets of coded data with small data structures.

One of the most interesting applications for this novel technique comes from the area of logic synthesis and verification of asynchronous circuits. Rosenblum and Yakovlev [12] and Chu [5] proposed the use of *Signal Transition Graphs* (STGs) to describe the behavior of asynchronous sequential circuits. An STG is an interpreted Petri net where transitions correspond to rising or falling transitions of digital signals. Previous methods based on the explicit enumeration of the reachable states for logic synthesis [7] suffer the state explosion problem, due to the arbitrary interleaving of concurrent transitions, while unfolding methods for verification [10] suffer a lack of flexibility and generality. With boolean manipulation techniques, both logic synthesis and verification of asynchronous circuits can be comprised in a unique and fairly general environment, which is also computationally capable of dealing with large systems, due to the efficient data representation and manipulation provided by BDDs. Although the main interest of the authors comes from the area of asynchronous circuits, the underlying theory of this technique is applicable to any kind of Petri net. Boundedness is the only restriction imposed by the approach.

The paper is organized as follows. In Sect. 2 we review the definition and some basic properties of Petri nets. Section 3 sketches the fundamental concepts on boolean algebras and algebras of classes. Logic functions, Boole's expansion theorem and logic abstractions are presented in Sect. 4. BDDs are described in Sect. 5. The main result of this paper, the isomorphism between boolean algebras and bounded Petri nets, is presented in Sect. 6. The reachability analysis algorithm is outlined in Sect. 7, and some reduction techniques to improve the efficiency of the algorithms are described in Sect. 8. Algorithms for the verification of properties such as safeness, liveness, and persistence are presented in Sect. 9. Section 10 sketches the extension to $k$-bounded nets. Some experimental results are analyzed in Sect. 11. Finally, the paper concludes in Sect. 12 with a discussion of the scope of this paper and future work.

## 2 Petri Nets: Definitions and Basic Properties

A *Petri net* [11] is a 4-tuple, $N = \langle P, T, F, m_0 \rangle$, where $P = \{p_1, p_2, \ldots, p_n\}$ is a finite set of places, $T = \{t_1, t_2, \ldots, t_m\}$ is a finite set of transitions, satisfying $P \cap T = \emptyset$ and $P \cup T \neq \emptyset$, $F \subseteq (P \times T) \cup (T \times P)$ is a set of arcs (flow relation), and $m_0 : P \rightarrow \mathbb{N}$ is the initial marking. The symbols ${}^\bullet t$, $t^\bullet$, ${}^\bullet p$ and $p^\bullet$ define, respectively, the pre-set and post-set of every place $p$ or transition $t$.

A *marking* of a Petri net is an assignment of a nonnegative integer to each place. If $k$ is assigned to place $p$, we will say that $p$ is marked with $k$ tokens. The structure of a Petri net defines a set of firing rules that determine the behavior of the net. A transition $t$ is enabled when each $p \in {}^\bullet t$ has at least one token.

The Petri net moves from one marking to another by firing one of the enabled transitions. When a transition $t$ fires, one token is removed from each place $p \in {}^\bullet t$ and one token is added to each place $p \in t^\bullet$. If $m_1$ and $m_2$ are markings, we will denote by $m_1[t\rangle m_2$ the fact that $m_2$ is reached from $m_1$ after transition $t$ being fired. A marking $m'$ is said to be *reachable* from a marking $m$ if there exists a sequence of transition firings that transforms $m$ into $m'$. The set of reachable markings from $m$ is denoted by $[m\rangle$.

We denote by $m(p)$ the number of tokens in place $p$ for the marking $m$. Thus, a marking can be represented by a vector of integers, $m = (m(p_1), \ldots, m(p_n))$.

**Definition 1.** A Petri net $N = \langle P, T, F, m_0\rangle$ is said to be *bounded* if $[m_0\rangle$ is a finite set.

**Definition 2.** A Petri net $N = \langle P, T, F, m_0\rangle$ is said to be *k-bounded* if for any $m \in [m_0\rangle$ and for any place $p \in P$, $m(p) \leq k$.

**Definition 3.** A Petri net is said to be *safe* if it is 1-bounded.

As starting point, we will restrict the proposed approach to *safe Petri nets*. Extensions to $k$-bounded nets will be presented in Sect. 10.

## 3 Boolean Algebras

In this section we briefly sketch some basic theory on boolean algebras. Most of the fundamental concepts presented here have been extracted from [2].

### 3.1 Boolean Algebras

A *boolean algebra* is a quintuple

$$(\mathrm{B}, +, \cdot, 0, 1) \;, \tag{1}$$

where B is a set called the carrier, $+$ and $\cdot$ are binary operations on B, and 0 and 1 are elements of B, such that $\forall a, b, c \in \mathrm{B}$ the following postulates are satisfied:

1. *Commutative Laws*: $a + b = b + a$; $a \cdot b = b \cdot a$
2. *Distributive Laws*: $a + (b \cdot c) = (a + b) \cdot (a + c)$; $a \cdot (b + c) = (a \cdot b) + (a \cdot c)$
3. *Identities*: $a + 0 = a$; $a \cdot 1 = a$
4. *Complement.* $\forall a \in \mathrm{B}$, $\exists a' \in \mathrm{B}$ such that: $a + a' = 1$; $a \cdot a' = 0$

As it is well known, the system $(\{0, 1\}, +, \cdot, 0, 1)$, with $+$ and $\cdot$ defined as the *logic OR* and *logic AND* operations respectively, is a boolean algebra (also known as the *switching algebra*). From now on, and since we will limit our scope to *logic functions*, we will always assume that $\mathrm{B} = \{0, 1\}$.

### 3.2 Logic Functions and Boolean Algebras of Logic Functions

An $n$-variable *logic function* (also called *switching function*) is a mapping

$$f : \mathrm{B}^n \longrightarrow \mathrm{B} \ . \tag{2}$$

Let $F_n(\mathrm{B})$ be the set of $n$-variable logic functions on B. Then the system

$$(F_n(\mathrm{B}), +, \cdot, 0, 1) \ , \tag{3}$$

is also a boolean algebra, in which $+$ and $\cdot$ signify addition and multiplication of logic functions, and 0 and 1 signify the zero- and one-functions ($f(x_1, \ldots, x_n) = 0$ and $f(x_1, \ldots, x_n) = 1$). The cardinality of $F_n(\mathrm{B})$ (number of different $n$-variable logic functions) is $2^{2^n}$.

### 3.3 Algebra of Classes (Subsets of a Set)

The *algebra of classes* of a set $S$ consists of the set $2^S$ (the set of subsets of S) and two operations on $2^S$: $\cup$ (union) and $\cap$ (intersection). This algebra satisfies the postulates for a boolean algebra and the system $(2^S, \cup, \cap, \emptyset, S)$ is a boolean algebra.

Next, the *Representation Theorem* (Stone, 1936) establishes the basis of the approach presented in this paper:

**Theorem 4.** *Every finite boolean algebra is isomorphic to the boolean algebra of subsets of some finite set $S$.*

Consequently, Stone's theorem states that reasoning in terms of concepts such as *union*, *intersection*, *empty set*, etc ..., in a finite set of elements is isomorphic to performing logic operations ($+$, $\cdot$) with logic functions. Furthermore, from Stones's theorem it can be easily deduced that the cardinality of the carrier of any boolean algebra must be a power of two. In particular, the algebra of classes of a set $S$ ($|S| = 2^n$) is isomorphic to the boolean algebra of $n$-variable logic functions.

## 4 Logic Functions

In this section, we present some fundamental concepts on logic functions used along the paper.

Given the boolean algebra of $n$-variable logic functions, we call a *vertex* each element of $\mathrm{B}^n$. The on-set (off-set) of a function $f$ is the set of vertices where the function evaluates to 1 (0). Each vertex of the on-set is also called a *minterm*. A *literal* is either a variable or its complement. A *cube* $c$ is a set of literals, such that if $a \in c$ then $a' \notin c$ and vice versa. A cube is interpreted as the boolean product of its elements. The cubes with $n$ literals are in one-to-one correspondence with the vertices of $\mathrm{B}^n$.

### 4.1 Boole's Expansion

The functions

$$f_{x_i} = f(x_1, \ldots, x_{i-1}, 1, x_{i+1}, \ldots, x_n) \tag{4}$$

and

$$f_{x'_i} = f(x_1, \ldots, x_{i-1}, 0, x_{i+1}, \ldots, x_n) \tag{5}$$

are called the *cofactor* of $f$ with respect to $x_i$ and $x'_i$ respectively. The definition of cofactor can also be extended to cubes. If $c = x_1 c_1$, $x_1$ being a literal and $c_1$ another cube, then:

$$f_c = (f_{x_1})_{c_1} \quad . \tag{6}$$

**Theorem 5** *Boole's expansion. If $f : \mathrm{B}^n \rightarrow \mathrm{B}$ is a boolean function, for all $(x_1, x_2, \ldots, x_n) \in \mathrm{B}^n$:*

$$f(x_1, x_2, \ldots, x_n) = x_i \cdot f_{x_i} + x'_i \cdot f_{x'_i} \quad .$$

### 4.2 Abstractions

Abstractions are of fundamental use in our framework. They have a direct correspondence to the existential and universal quantifiers applied to predicates in boolean reasoning. The *existential* and *universal abstractions* of $f$ with respect to $x_i$ are defined as:

$$\exists_{x_i} f = f_{x_i} + f_{x'_i} \quad ; \qquad \forall_{x_i} f = f_{x_i} \cdot f_{x'_i} \quad . \tag{7}$$

As an example, let us consider the function: $f = bc + ab'c' + a'c$ . The cofactor with respect to $a$ and $a'$ are: $f_a = bc + b'c'$ and $f_{a'} = c$ , and the abstractions with respect to $a$ are $\exists_a f = f_a + f_{a'} = b' + c$ and $\forall_a f = f_a \cdot f_{a'} = bc$ . $\exists_a f$ is the function that evaluates to 1 for all those values of $b$ and $c$ such that there is a value of $a$ for which $f$ evaluates to 1. $\forall_a f$ is the function that evaluates to 1 for all those values of $b$ and $c$ such that $f$ evaluates to 1 for any value of $a$.

## 5 Binary Decision Diagrams

A logic function can be represented in many ways, such as *truth tables*, *Karnaugh maps* or *minterm canonical forms*. Another form that can be much more compact is the *sum of products*, where the logic function is represented by means of an equation, i.e.,

$$f = bc + ab'c' + a'c \quad . \tag{8}$$

These techniques are inefficient for fairly large functions. However, all these forms can be canonical [2]. A form is canonical, if the representation of any function in that form is unique. Canonical forms are useful for verification techniques, because equivalence test between functions is easily computable.

Recently, Binary Decision Diagrams (BDDs) have emerged as an efficient canonical form to manipulate large logic functions. The introduction of BDDs is relatively old [8], but only after the recent work of Bryant [3] they transformed into a useful tool. For a good review on BDDs we refer to [1, 3, 13].

We will present BDDs by means of an example. Given (8), its BDD is shown in Fig. 1(a). A BDD is a *Directed Acyclic Graph* with one root and two leaf nodes (0 and 1). Each node has two outgoing arcs labeled T (*then*) and E (*else*). To evaluate $f$ for the assignment of variables $a = 1$, $b = 0$, and $c = 1$, we only have to follow the corresponding directed arcs from the root node. The first node we encounter is labeled with variable $a$, whose value is 1. Given this assignment, the T arc must be taken. Next, a node labeled with variable $b$ is found. Since $b = 0$ the E arc must be taken now. Finally the T arc for variable $c$ reaches the 0 leave node.

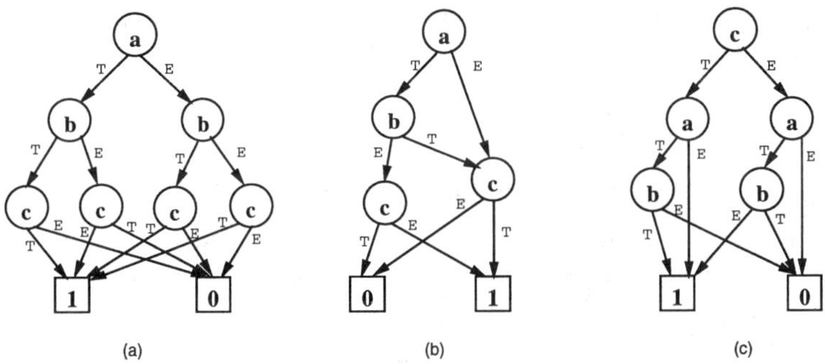


**Fig. 1.** BDD example

### 5.1 Reduced and Ordered BDDs

The representation of a function by means of a BDD is not unique. Figures 1(a), 1(b) and 1(c) depict different BDDs representing (8). The BDD in 1(b) can be obtained from 1(a) by successively applying reduction rules that eliminate *isomorphic subgraphs* from the representation [3]. The BDD in 1(c) has a different variable ordering.

All BDDs shown in Fig. 1 are *ordered* BDDs. In an ordered BDD, all variables appear in the same order along all paths from the root to the leaves. If a BDD is *ordered* and *reduced* (no further reductions can be applied) then we have a Reduced Ordered BDD (ROBDD). Given a total ordering of variables, an ROBDD is a canonical form [3]. Figures 1(b) and 1(c) are ROBDDs with variable ordering $a < b < c$ and $c < a < b$ respectively. The shape and size of an ROBDD depend on the ordering of its variables.

Some important properties of ROBDDs are:

- The size of the BDD can be exponential in the number of variables [9]; however BDDs are a compact representation for many functions.
- Boolean binary operations can be calculated in polynomial time in the size of the BDDs.

– Some interesting problems like satisfiability, tautology and complementation are solved in constant time using BDDs.

Henceforth, we will implicitly assume that BDDs are reduced and ordered. Note that each BDD node represents at the same time a function whose root is the node itself. This property allows the implementation of BDD packages managing all BDDs using the same set of variables in only one multi-rooted graph [1].

## 5.2 Boolean Operations with ROBDDs

Let us see, first, how to calculate the BDD for (8) given the ordering $a < b < c$. We will use $(v, T, E)$ to denote a node labeled with variable $v$, and $T$ and $E$ as "Then" and "Else" BDDs respectively. Applying Boole's theorem to expand $f$ with variable $a$ we have:

$$f \ = \ a \ f_a \ + \ a' \ f_{a'} \ , \tag{9}$$

with $f_a = bc + b'c'$, and $f_{a'} = c$. Expanding variable $b$ in $f_a$ and $f_{a'}$ yields to

$$f = a \ (b \ f_{ab} \ + \ b' \ f_{ab'}) \ + \ a' \ (b \ f_{a'b} \ + \ b' \ f_{a'b'}) \tag{10}$$

with $f_{ab} = c$, $f_{ab'} = c'$, $f_{a'b} = c$, and $f_{a'b'} = c$. Thus the BDD for (8) is

$$f \ = \ (a, (b, c, c'), (b, c, c)) \ . \tag{11}$$

Note that the logic functions $f_{ab} = f_{a'b} = c$ and $f_{ab'} = f_{a'b'} = c'$ are isomorphic and must be represented with the same node if we want to preserve canonicity.

BDDs can be created by combining existing BDDs by means of boolean operations like AND, OR, and XOR. This approach is implemented using the *if-then-else* operator (ITE), defined as follows:

$$\text{ite}(F, G, H) = F \cdot G + F' \cdot H \ , \tag{12}$$

where $F$, $G$, $H$ are logic functions represented by BDDs. The interesting property of the ITE operator is that it can directly implement all two-operand logic functions. For example:

$$\text{AND}(F, G) = \text{ite}(F, G, 0) \ , \ \ \text{XOR}(F, G) = \text{ite}(F, G', G) \ . \tag{13}$$

Let $Z = \text{ite}(F, G, H)$, and let $v$ be the *top* variable of $F$, $G$, $H$. Then the BDD for $Z$ is recursively computed as follows [3]:

$$Z = (v, \text{ite}(F_v, G_v, H_v), \text{ite}(F_{v'}, G_{v'}, H_{v'})) \ , \tag{14}$$

where the terminal cases are:

$$\text{ite}(1, F, G) = \text{ite}(0, G, F) = \text{ite}(F, 1, 0) = \text{ite}(G, F, F) = F \ . \tag{15}$$

The code for the ITE algorithm is shown in Fig. 2. Note that the algorithm keeps the BDD reduced by checking if $T$ equals $E$, and checking in a *unique-table* if the produced node already exists in the graph. In this way, all isomorphic subgraphs are always eliminated.

Unless there is a terminal case, every call to the procedure generates two other calls, so the total number of ITE calls would be exponential in the number of variables. To avoid this exponentiality, ITE uses a table of pre-computed operations (*computed table*). The *computed table* acts as a cache memory, in such a way that the most recently used results are stored in this table. The effect of this computed table is to cause ITE to be called at most once for each possible combination of the nodes in $F$, $G$, $H$. So the complexity of the algorithm, under the assumptions of infinite memory and constant access time (hash) tables, is reduced to $O(|F| \cdot |G| \cdot |H|)$.

```
ite (F,G,H) {
    if ( terminal case ) return result for terminal case;
    else if ( {F, G, H} is in computed-table )
        return pre-computed result;
    else {
        let v be the top variable of { F, G, H };
        T = ite (F_v,G_v,H_v);
        E = ite (F_v',G_v',H_v');
        if T equals E return T;
        R = find_or_add_unique_table (v,T,E);
        insert_computed_table ({ F, G, H }, R);
        return R;
}   }
```

**Fig. 2.** The *ITE* algorithm

An important consequence of representing all BDDs in the same graph is that checking the equivalence between two BDDs can be done in constant time (two BDDs representing the same function have the same root node). Counting the number of vertices represented by a BDD can be done in linear time in the size of the BDD.

## 6 Modeling Safe Petri Nets with Boolean Algebras

Let $N = \langle P, T, F, m_0 \rangle$ be a safe Petri net. A marking in $[m_0\rangle$ can be represented by a set of places $m$, where $p_i \in m$ denotes the fact that there is a token in $p_i$. Therefore, any set of markings in $[m_0\rangle$ can be represented by a set $M$ of subsets of $P$. Let $M_P$ be the set of all markings of a safe Petri net with $|P|$ places ($|M_P| = 2^{|P|}$). The the system

$$(2^{M_P}, \cup, \cap, \emptyset, M_P) \tag{16}$$

is the boolean algebra of sets of markings. This system is isomorphic with the boolean algebra of $n$-variable logic functions, where $n = |P|$.

We will indistinctively use $p_i$ to denote a place in $P$, or a variable in the boolean algebra of $n$-variable logic functions. Therefore, there is a one-to-one correspondence between markings of $M_P$ and vertices of $\mathrm{B}^n$. A marking $m \in M_P$ is represented by means of an *encoding function* that provides a binary mapping from $M_P$ into $B^n$, that is, $\mathcal{E} : M_P \rightarrow B^n$, where the image of a marking $m \in M_P$ is encoded into an element $(p_1, \ldots, p_n) \in B^n$, such that:

$$p_i = \begin{cases} 1 \text{ if } p_i \in m \\ 0 \text{ if } p_i \notin m \end{cases} . \tag{17}$$

As an example, both the vertex $(1, 0, 1, 0) \in B^4$ and the cube $p_1 p_2' p_3 p_4'$ represent the marking in which $p_1$ and $p_3$ are marked and $p_2$ and $p_4$ are not marked.

## 6.1 Characteristic Functions and Binary Relations

The *characteristic function* $\chi_V$ of a set of vertices $V \subseteq B^n$ is defined as the logic function that evaluates to 1 for those vertices of $B^n$ that are in $V$, i.e.,

$$\forall v \in B^n \ , \ v \in V \Leftrightarrow \chi_V(v) = 1 \ . \tag{18}$$

Extending the use of the *encoding function* $\mathcal{E}$, each set of markings $M \in 2^{M_P}$ has a corresponding *characteristic function* $\chi_M^{\mathcal{E}} : B^n \rightarrow B$, that evaluates to 1 for those vertices that correspond to markings belonging to $M$. The image of $M \subseteq 2^{M_P}$ according to $\mathcal{E}$ is the set $V \subseteq B^n$, defined by:

$$V = \{\mathcal{E}(m) : m \in M_P\} \ . \tag{19}$$

From now on, given the *encoding function* $\mathcal{E}$, we will define the *characteristic function* of $M$ as the characteristic function of the set $V$, that is, $\chi_M = \chi_V$. For example, given the Petri net depicted in Fig. 3(a), the characteristic function of the set $M = \{\{p_2, p_5\}, \{p_2, p_3, p_5\}, \{p_1, p_2, p_5\}, \{p_1, p_2, p_3, p_5\}, \{p_1, p_2, p_3, p_4, p_5\}\}$ is calculated as the disjunction of each boolean code $\mathcal{E}(m), m \in M$. The resulting function $\chi_M = p_1 p_2 p_3 p_5 + p_2 p_4' p_5$ , represents the set of markings in which $p_1$, $p_2$, $p_3$, and $p_5$ are marked or $p_2$ and $p_5$ are marked and $p_4$ is not marked.

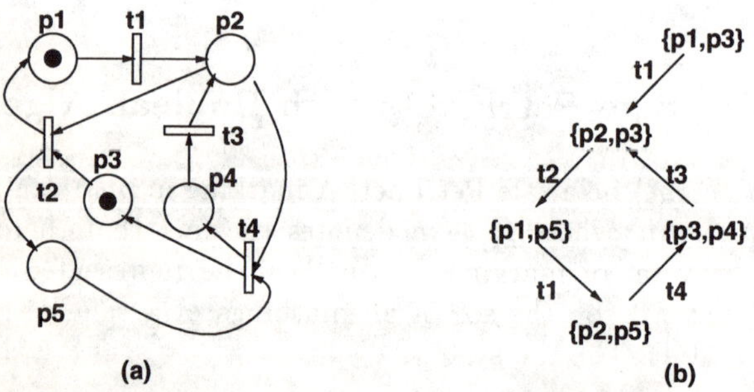


**Fig. 3.** (a)Petri net, (b)reachable markings

Hence and for sake of simplicity we will indistinctively use $M$ and $\chi_M$ to denote the characteristic function of the set of safe markings $M$.

All set manipulations can by applied directly to the characteristic functions. For example, given the sets of safe markings $M_1, M_2 \in M_P$:

$$\chi_{M_1 \cup M_2} = \chi_{M_1} + \chi_{M_2} \ ; \quad \chi_{M_1 \cap M_2} = \chi_{M_1} \cdot \chi_{M_2} \ ; \quad \chi_{\overline{M_1}} = \chi'_{M_1} \cdot \chi_{M_P} \ . \tag{20}$$

When implemented with BDDs, characteristic functions provide, in general, compact and efficient representations.

Characteristic functions can also be used to represent *binary relations*, that is, subsets of a cartesian product between two sets. To represent the binary relation $R \subseteq M_1 \times M_2$, it is necessary to use different sets of variables to identify the elements of $M_1$ and $M_2$. Given the binary relation $R$ between sets $M_1$ and $M_2$, the elements of $M_1$ that are in relation with some element of $M_2$, are the set:

$$V = \{m_1 \in M_1 : \exists m_2 \in M_2, (m_1, m_2) \in R\} \ , \tag{21}$$

and using the characteristic function of $R$, the characteristic function of $V$ is computed by:

$$\chi_V(x_1, \ldots, x_n) = \exists_{y_1, \ldots, y_n} \chi_R(x_1, \ldots, x_n, y_1, \ldots, y_n) \ . \tag{22}$$

## 6.2 Transition Firing

We define the *transition function* of a Petri net as a function

$$\delta : 2^{M_P} \times T \to 2^{M_P} \ , \tag{23}$$

that transforms, for each transition, a set of markings $M_1$ into a new set of markings $M_2$ as follows:

$$\delta(M_1, t) = M_2 = \{m_2 \in M_P : \exists m_1 \in M_1, \ m_1[t\rangle m_2\} \ . \tag{24}$$

This concept is equivalent to the one-step reachability in Petri nets.

Equation (23) can be generalized to be the *transition function* of a Petri net:

$$\Delta : 2^{M_P} \to 2^{M_P} \ , \tag{25}$$

where all the transitions are processed in the same function. $\Delta$ transforms a set of markings $M_1$ into the set of markings $M_2$ that can be reached from $M_1$ in one step (one transition firing). Equation (25) can be obtained by computing:

$$\Delta(M) = \bigcup_{\forall t \in T} \delta(M, t) \ . \tag{26}$$

Note that (25) calculates the image of several markings simultaneously. Using the terminology for verification of sequential machines [6], $\Delta$ performs the *constrained image computation* of the net.

There are three different techniques to implement the *constrained image computation* for transitions using BDDs: by *topological image computation*, by the *transition function* $\delta$ and by the *transition relation* associated to $\delta$. In the remainder of this section we will study the topological image computation. We refer the reader to [6] for the other techniques.

### 6.3 Topological Image Computation

*Constrained image computation* for transitions can be efficiently implemented by using the topological information of the Petri net and the characteristic function of sets of markings. First of all, we will present the characteristic function of some important sets related to a transition $t \in T$:

$$\begin{aligned}
\mathrm{E}_t &= \prod_{p_i \in {}^\bullet t} p_i && (t \text{ enabled}), \\
\mathrm{NPM}_t &= \prod_{p_i \in {}^\bullet t} p_i' && (\text{no predecessor of } t \text{ is marked}), \\
\mathrm{ASM}_t &= \prod_{p_i \in t^\bullet} p_i && (\text{all successors of } t \text{ are marked}), \\
\mathrm{NSM}_t &= \prod_{p_i \in t^\bullet} p_i' && (\text{no successor of } t \text{ is marked}).
\end{aligned}$$

Given these characteristic functions, the *constrained image computation* for transitions is reduced to calculate:

$$\delta(M,t) = (M_{\mathrm{E}_t} \cdot \mathrm{NPM}_t)_{\mathrm{NSM}_t} \cdot \mathrm{ASM}_t \ . \tag{27}$$

We will show with an example how this formula "simulates" transition firing. In the example of Fig. 3(a), given the set of markings

$$M = p_1p_2'p_3p_4'p_5' + p_1'p_2p_3p_4'p_5' + p_1p_2'p_3'p_4'p_5 \tag{28}$$

we will calculate $M' = \delta(M,t_1)$. First, $M_{\mathrm{E}_{t_1}}$ (the cofactor of $M$ with respect to $\mathrm{E}_{t_1} = p_1$) selects those markings in which $t_1$ is enabled and removes its predecessor places from the characteristic function:

$$M_{\mathrm{E}_{t_1}} = p_2'p_3p_4'p_5' + p_2'p_3'p_4'p_5 \ . \tag{29}$$

Then the product with $\mathrm{NPM}_{t_1} = p_1'$ simulates the elimination of the tokens in the predecessor places:

$$M_{\mathrm{E}_{t_1}} \cdot \mathrm{NPM}_{t_1} = p_1'p_2'p_3p_4'p_5' + p_1'p_2'p_3'p_4'p_5 \ . \tag{30}$$

Next, taking the cofactor with respect to $\mathrm{NSM}_{t_1} = p_2'$ removes all successor places from the characteristic function:

$$(M_{\mathrm{E}_t} \cdot \mathrm{NPM}_t)\left(M_{\mathrm{E}_{t_1}} \cdot \mathrm{NPM}_{t_1}\right)_{\mathrm{NSM}_{t_1}} = p_1'p_3p_4'p_5' + p_1'p_3'p_4'p_5 \ . \tag{31}$$

Finally, the product with $ASM_{t_1} = p_2$ adds a token in all the successor places of $t_1$:

$$M' = p_1'p_2p_3p_4'p_5' + p_1'p_2p_3'p_4'p_5 \ . \tag{32}$$

Note that (23) is correctly defined only for safe Petri nets. However, safeness can be also verified by using $\delta$, as it will be shown in Sect. 9.

```
traverse_Petri_net (N = ⟨P,T,F,m0⟩) {
/* Let Δ be the transition function of N */
    Reached = From = {m0};
    repeat {
        To = Δ(From);
        New = To − Reached;
        From = New;
        Reached = Reached ∪ New;
    } until (New = ∅);
    return Reached; /* The set of all reached markings from m0 */
}
```

**Fig. 4.** Algorithm for *symbolic traversal*

## 7 Net Traversal and Reachable Markings

Once the constrained image computation has been defined, the set $[m_0\rangle$ can be calculated by *symbolic traversal.* We will use an approach similar to *symbolic breadth-first* traversal for Finite State Machines [6]. This method allows to process several markings simultaneously by using their characteristic function and the constrained image computation.

The algorithm presented in Fig. 4 traverses the Petri net and calculates $[m_0\rangle$. The union and difference of sets are performed by manipulating their characteristic functions.

Each iteration of the traversal obtains all the markings reachable from the set *"From"* in one step. Only those markings that are *"New"* in the set of reachable markings are considered for the next iteration. The algorithm iterates until no new markings are generated. The number of iterations performed by the algorithm is determined by the maximum number of firings from the initial marking to the first occurrence of any of the reachable markings, and its called the *sequential depth* of the Petri net.

The final set of reachable markings are shown in Fig. 3(b), where the nodes represent markings and the edges the firing transitions. Note that the *sequential depth* of this Petri net is four.

## 8 Petri Net Reductions

Petri nets can be reduced to simpler ones by using transformation rules that preserve the properties of the system being modeled. By using these rules, the complexity inherent to the reachability analysis can be effectively reduced.

In [11], a set of six transformations that preserve the properties of liveness, safeness, and boundedness were proposed. Here we illustrate how these transformations can be used to simplify the breadth-first traversal analysis. Fig. 5 depicts the set of transformations actually used.

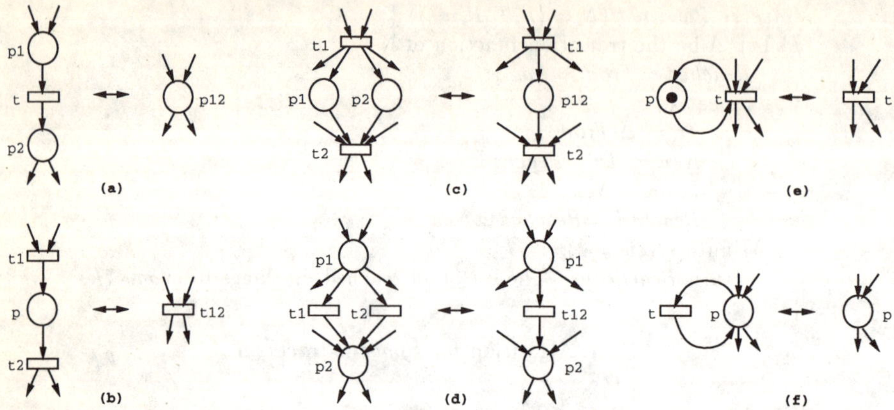


**Fig. 5.** Transformations preserving liveness, safeness and boundedness

The original Petri net $N$ is reduced into a new net $N'$ by applying these transformations. Then, the reachability analysis technique presented in Sect. 7 can be used more efficiently with $N'$ due to the reduction in both, the number of places and the *sequential depth* of the net. Given the set of reachable markings $[m'_0\rangle$ of $N'$, the set of reachable markings $[m_0\rangle$ of the original net $N$ is derived using an inverted transformation on $[m'_0\rangle$. The inverted transformations are shown in Tab. 1.

**Table 1.** Petri net reductions and their inverse transformations

| Forward Transformations | Backward Transformations |
|---|---|
| (a) series places fusion | $R = R'_{p_{12}} \cdot (p_1 \oplus p_2) + R'_{p'_{12}} \cdot (p'_1 p'_2)$ |
| (b) series transitions fusion | $R = R'_{\mathrm{E}_{t_{12}}} \cdot (\mathrm{E}_{t_1} \oplus \mathrm{E}_{t_2}) + R'_{\mathrm{E}'_{t_{12}}} \cdot (\mathrm{E}'_{t_1} \cdot \mathrm{E}_{t_2})$ |
| (c) parallel places fusion | $R = R'_{p_{12}} \cdot (p_1 p_2) + R'_{p'_{12}} \cdot (p'_1 p'_2)$ |
| (d) parallel transitions fusion | $R = R'$ |
| (e) self-loop place | $R = R' \cdot p$ |
| (f) self-loop transition | $R = R'$ |

For example, Fig. 5(a) depicts how a net can be transformed into another by fusing places $p_1$ and $p_2$ into place $p_{12}$. If $R'$ is the set of reachable markings of the resulting net, the set of markings in the original net can be derived as follows:

$$R = R'_{p_{12}} \cdot (p_1 \oplus p_2) + R'_{p'_{12}} \cdot (p'_1 p'_2) \ , \tag{33}$$

denoting that a token in $p_{12}$ implies that either $p_1$ or (exclusive or) $p_2$ were marked and no token in $p_{12}$ implies that neither $p_1$ nor $p_2$ were marked in the original net. Similar substitutions can be applied for other types of transformations.

# 9 Verification of Properties

In this section we show how different Petri net properties can be verified by boolean manipulation on the set of reachable markings. From the wide range of properties that can be verified with this approach we have chosen three of them as examples: *safeness*, *liveness* and *persistence*. Some properties can be easily specified with a boolean equation, thus not requiring any traversal to be verified. Others require partial or complete traversals of the net. However, symbolic traversing by means of BDDs makes their computation affordable even for large nets.

## 9.1 Safeness

The calculation of $[m_0\rangle$ by means of constrained image computation is done under the assumption that the Petri net is safe. This calculation is erroneous if some of the markings is unsafe [2], since unsafe markings are not representable by encoding each place with one variable of the boolean algebra. A similar reasoning can be done for $k$-bounded nets.

According to (27), unsafe markings are removed from the set of reachable markings. However, detecting if some unsafe marking is reachable from $[m_0\rangle$ can be done by identifying a marking $m$ in which a transition $t$ is enabled, and some successor place $p$ of $t$, and not predecessor of $t$, is already marked. In that situation, after firing transition $t$, place $p$ will have two tokens. Formally:

$$\text{N is not safe} \Leftrightarrow \exists(m \in [m_0\rangle, t \in T, p \in P) \text{ such that}$$
$$t \text{ is enabled in } m,\ p \in t^\bullet,\ p \notin {}^\bullet t \text{ and } m(p) = 1.$$

Given the set of reachable markings $[m_0\rangle$, the algorithm depicted in Fig. 6 detects whether a Petri net is safe or not by checking one equation for each transition.

```
is_safe (N = ⟨P,T,F,m0⟩ , [m0⟩) {
    foreach t ∈ T do {
        Succ_p = 0;
        Enabled = [m0⟩ · E_t;
        foreach (p_i ∈ t• ∧ p_i ∉ •t) do { Succ_p = Succ_p + p_i }
        if (Enabled · Succ_p ≠ 0) return false;
    }
    return true;
}
```

**Fig. 6.** Algorithm for *safeness checking*

## 9.2 Liveness

A Petri net is said to have a *deadlock* if there is a marking where no transition can be fired. A transition is said to be *dead* (L0-live) if it can never be fired

[2] In this context, unsafe markings are those with more than one token is some place.

in any firing sequence from $m_0$. A transition that can be fired at least once in some firing sequence from $m_0$ is said to be *potentially fireable* (L1-live). All these properties can be verified with simple equations.

The set of markings where a *deadlock* occurs is calculated:

$$\mathit{Deadlock} \equiv ([m_0\rangle \cdot \prod_{t\in T} \mathrm{E}'_t) \neq 0) \ . \tag{34}$$

The set of markings where a transition is *potentially fireable* is calculated as:

$$\mathrm{Fireable}_t = [m_0\rangle \cdot \mathrm{E}_t \ . \tag{35}$$

If $\mathrm{Fireable}_t = 0$, then transition $t$ is L0-live, otherwise it is L1-live.

To verify if a transition can be fired an infinite number of times (L3-liveness), or if transition can be fired an infinite number of times from any reachable marking of $[m_0\rangle$ (L4-liveness), requires more elaborate techniques. Both problems can be reduced to the calculation of the *Strongly Connected Components* of $[m_0\rangle$.

**Definition 6.** A Strongly Connected Component (SCC) $U$ of a directed graph $G = (V, E)$, is a maximal set of vertices $U \subseteq V$, such that for every pair of vertices $u$ and $v$ in $U$ we have both $u \rightsquigarrow v$ and $v \rightsquigarrow u$; that is, vertices $u$ and $v$ are reachable from each other.

**Definition 7.** A Strongly Connected Component $U$ of a directed graph $G = (V, E)$ is *terminal* (TSCC) if from the vertices in $U$ it is not possible to reach any vertex in $V \backslash U$.

A transition $t$ enabled in all the TSCCs markings of the Petri net is L4-live, because from any marking of $[m_0\rangle$ we will reach some $\mathrm{TSCC}_i$ where $t$ can be fired an infinite number of times. L4-liveness of transition $t$ can be computed as follows:

$$t \text{ is L4-live} \iff \bigwedge_{\forall i} (\mathrm{TSCC}_i \cdot \mathrm{E}_t \neq 0) \ . \tag{36}$$

If there is some $\mathrm{SCC}_i$ where transition $t$ is enabled, then $t$ is L3-live because there is at least a firing sequence from $[m_0\rangle$ that leads to $\mathrm{TSCC}_i$ where $t$ can be fired an infinite number of times. L3-liveness for transition $t$ can be calculated as follows:

$$t \text{ is L3-live} \iff \bigvee_{\forall i} (\mathrm{SCC}_i \cdot \mathrm{E}_t \neq 0) \ . \tag{37}$$

The algorithm to compute the TSCCs and SCCs of $[m_0\rangle$ is shown in Fig. 7. First, the *Transitive Closure* ($C_T$) of the *Transition Relation* is computed, where $C_T(x, y) = 1$ if there is a firing sequence from $x$ that leads to $y$ ($x \rightsquigarrow y$) [4]. The following steps compute the sets of markings that are in any SCC (InSCC) or in any TSCC (InTSCC). Finally, each individual SCC (TSCC) is obtained from InSCC (InTSCC).

```
Let T_R be the Transition Relation of N.

compute_SCC_TSCC (N = ⟨P,T,F,m_0⟩ , [m_0⟩) {
    C_T = compute_Transitive_Closure ( T_R );
    C_Y = C_T(x,y) · C_T(y,x); C_NY = C_T(x,y) · C_T(y,x)';
    InSCC = ∃_y C_Y(x,y);
    InTSCC = (∃_y C_NY(x,y))';
    SCC_1...m = extract_Strongly_Connected_Components ( InSCC );
    TSCC_1...m = extract_Strongly_Connected_Components ( InTSCC );
}
```

**Fig. 7.** Algorithm to compute the SCC and TSCC sets of $[m_0\rangle$

### 9.3 Persistence

A Petri net is said to be persistent if, for any two enabled transitions, the firing of one transition will not disable the other.

The algorithm depicted in Fig. 8 verifies persistence for a Petri net. For each transition $t_1$, the set of markings with $t_1$ enabled are calculated. Next, the sets of markings reachable in one step by firing any transition different from $t_1$ are obtained. If $t_1$ is not enabled in any of those markings, then the net is not persistent.

```
is_persistent (N = ⟨P,T,F,m_0⟩ , [m_0⟩) {
    foreach t_1 ∈ T do {
        Enabled = [m_0⟩ · E_t1;
        foreach t_2 ∈ T, t_2 ≠ t_1 do {
            To = δ(Enabled, t_2);
            Not_enabled = To · E'_t1;
            if (Not_enabled ≠ 0) return false;
    } }
    return true;
}
```

**Fig. 8.** Algorithm to verify persistence

## 10 Extension to k-Bounded Petri Nets

This section presents the modifications needed to extend the boolean manipulation techniques to $k$-bounded Petri nets.

A $k$-bounded place $p \in P$ can be represented with a set of boolean variables, $v_1, \ldots, v_q$ to encode the up-to-$k$ possible number of tokens. The number of required variables depends on the type of encoding. If an *one-hot* encoding is used, $k$ variables are needed. For example, in a 3-bounded Petri net the number of tokens in place $p$ could be represented by three variables. With a *binary encoding* $\lceil \log_2(k+1) \rceil$ variables would be required (see Tab. 10).

The one-hot encoding can be implemented using a *transition function* simpler than the binary encoding, however the number of variables, which is a critical

**Table 2.** Encoding of $k$-bounded places ($k = 3$)

| # tokens | one-hot encoding | binary encoding |
|---|---|---|
| 0 | $v_3'v_2'v_1'$ | $v_2'v_1'$ |
| 1 | $v_3'v_2'v_1$ | $v_2'v_1$ |
| 2 | $v_3'v_2v_1'$ | $v_2v_1'$ |
| 3 | $v_3v_2'v_1'$ | $v_2v_1$ |

parameter in the efficiency of BDD algorithms, is larger than for the one-hot encoding. Comparative studies, analyzing the size of the BDDs and the performance of the algorithms, are necessary to decide which is the practical limit for each type of encoding.

## 11 Experimental Results

In this section we illustrate the power of using boolean reasoning and BDDs for the analysis of Petri nets. We have chosen two simple and scalable examples to show how the approach can generate all the states for fairly large nets. We present the results corresponding to the calculation of the set of reachable markings, which dominates the complexity of the analysis. Most properties can then be verified in a straightforward manner from $[m_0\rangle$, as shown in Sect. 9.

### 11.1 The Dining Philosophers

The first example is the well-known *dining philosophers* paradigm represented by the Petri net shown in Fig. 9. The net has $7n$ places and $5n$ transitions, $n$ being the number of philosophers sitting at the table. By successively applying the reductions depicted in Fig. 5, the complexity of the net can be reduced down to $6n$ places and $4n$ transitions.

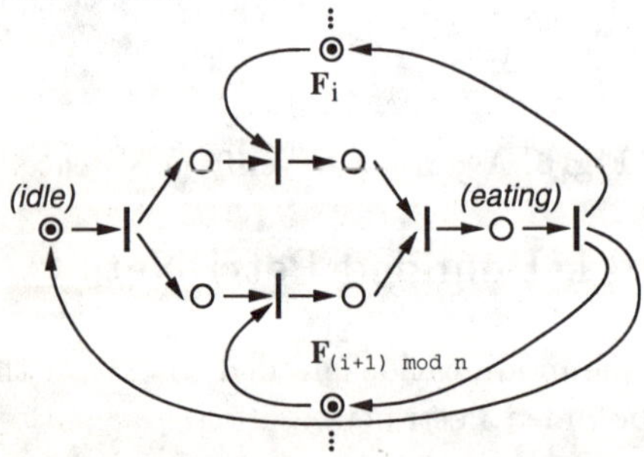


**Fig. 9.** Petri net for a dining philosopher

Table 3 shows the number of states of the original and the reduced Petri net, the size of the BDDs representing the reachable markings and the number of iterations and CPU time spent by the traversal algorithm. CPU times have been obtained by executing the algorithms on a Sun SPARC 10 workstation, with a 64Mbyte main memory.

It is worthwhile to point out how a small BDD (1347 nodes $\approx$ 21 Kbyte memory) can represent the complete set of markings of the Petri net for 28 philosophers ($4.8 \times 10^{18}$). The BDD representing $[m_0\rangle$ has been calculated by using the traversal algorithm presented in Fig. 4. The number of executed iterations corresponds to the sequential depth of the reduced net.

**Table 3.** Results for the *dining philosophers* example

| # of philos. | states original | states reduced | BDD size orig. | BDD size red. | BDD size peak (red.) | # of iters. | CPU (secs.) |
|---|---|---|---|---|---|---|---|
| 8 | $2.2 \times 10^5$ | $1.0 \times 10^5$ | 429 | 347 | 1354 | 17 | 15 |
| 12 | $1.0 \times 10^8$ | $3.3 \times 10^7$ | 677 | 547 | 3230 | 25 | 137 |
| 16 | $4.7 \times 10^{10}$ | $1.1 \times 10^{10}$ | 925 | 747 | 5906 | 33 | 731 |
| 20 | $2.2 \times 10^{13}$ | $3.5 \times 10^{12}$ | 1173 | 947 | 9382 | 41 | 1952 |
| 24 | $1.0 \times 10^{16}$ | $1.1 \times 10^{15}$ | 1421 | 1147 | 13658 | 49 | 4208 |
| 28 | $4.8 \times 10^{18}$ | $3.6 \times 10^{17}$ | 1669 | 1347 | 18734 | 57 | 8274 |

Figure 10 depicts the number of states represented by the BDD *"Reached"* at each iteration for the reduced net. The slope between iterations 27 and 43 illustrates the ability of the approach to process large sets of markings in parallel. It is important to notice that, although the number of reached states is lower, the size of the BDD *"Reached"* at intermediate iterations can be larger than the final BDD. This is a usual phenomenon in the traversal of sequential machines using BDDs. The peak BDD size achieved during the traversal is also shown in Tab. 3, and the evolution of the BDD size during the traversal is depicted in Fig. 11.

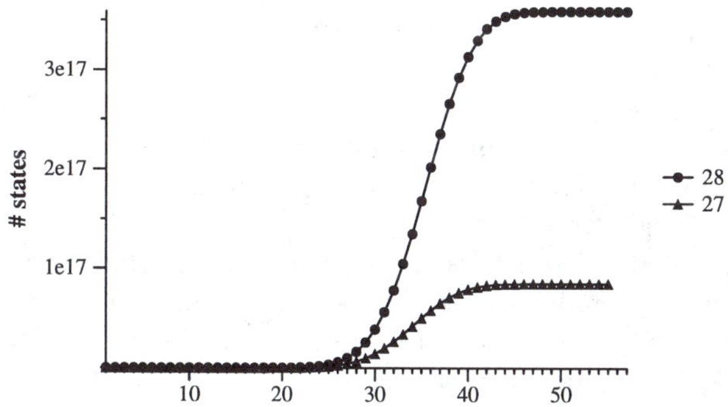


**Fig. 10.** Number of states reached at each iteration

## 11.2 Slotted Ring

The second example models a protocol for Local Area Networks called *slotted ring*. The Petri net is depicted in Fig. 12. The example is scalable for any number of nodes in the network. The results corresponding to the traversal of the net are presented in Tab. 4.

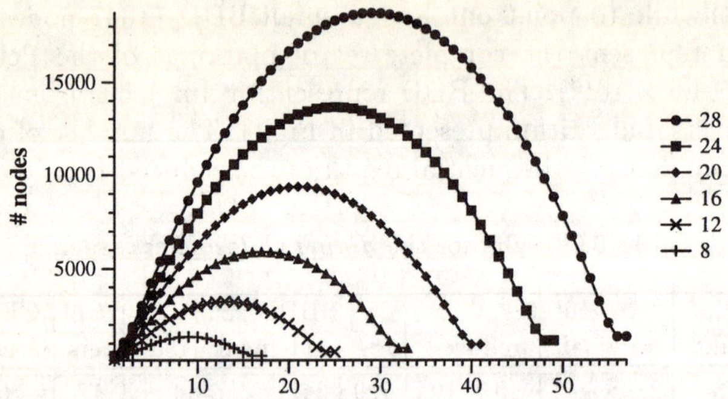


**Fig. 11.** Size of the BDD "Reached" at each iteration of the traversal

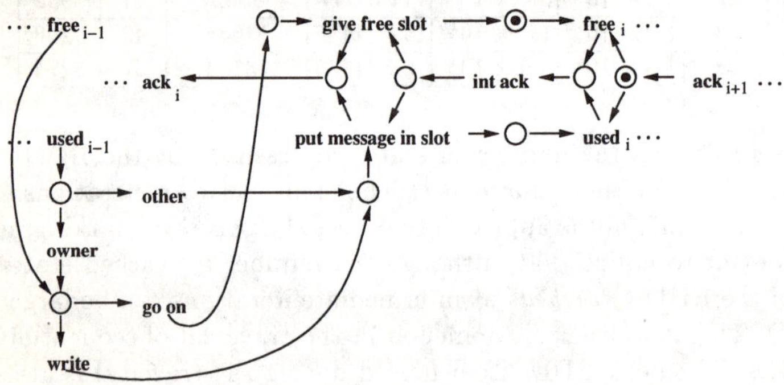


**Fig. 12.** Slotted ring protocol for one node

## 12 Conclusions and Future Work

This paper proposes the combination of boolean reasoning and BDD algorithms to manage the state explosion produced in Petri net analysis. This technique has been successfully used for the analysis and verification of sequential machines and synthesis of logic circuits.

It has been shown that BDDs can represent large sets of markings ($10^{18}$ in the example) with a small number of nodes ($10^3$). Once the reachable markings have been generated, many properties can be verified in a straightforward manner. Therefore, BDDs are proposed as an alternative to the reachability tree, providing a compact representation of the markings of a bounded net.

Many issues are still under research to increase the applicability of the approach. The ordering of variables is a topic of major interest that must be studied in order to reduce even more the size of the BDDs, thus speeding-up BDD operations. As mentioned in Sect. 10, encoding methods for $k$-bounded nets must also be explored. The combination of further reduction techniques and analysis with BDDs is another area for future research. Finally, the representation of unbounded nets by means of BDDs is a challenge not discarded by the authors yet.

**Table 4.** Results for the slotted ring example

| # of nodes | states original | states reduced | BDD size orig. | BDD size red. | BDD size peak (red.) | # of iters. | CPU (secs.) |
|---|---|---|---|---|---|---|---|
| 2 | $2.1 \times 10^2$ | 52 | 158 | 56 | 70 | 11 | 1 |
| 3 | $4.0 \times 10^3$ | $5.0 \times 10^2$ | 210 | 91 | 177 | 19 | 2 |
| 4 | $8.2 \times 10^4$ | $5.1 \times 10^3$ | 356 | 151 | 488 | 28 | 7 |
| 5 | $1.7 \times 10^6$ | $5.4 \times 10^4$ | 540 | 223 | 1024 | 39 | 27 |
| 6 | $3.7 \times 10^7$ | $5.8 \times 10^5$ | 758 | 311 | 2156 | 51 | 105 |
| 7 | $8.0 \times 10^8$ | $6.2 \times 10^6$ | 1014 | 411 | 4150 | 65 | 453 |
| 8 | $1.7 \times 10^{10}$ | $6.8 \times 10^7$ | 1304 | 527 | 7280 | 80 | 1600 |
| 9 | $3.8 \times 10^{11}$ | $7.5 \times 10^8$ | 1632 | 655 | 12259 | 97 | 4080 |

## References


1. K. S. Brace, R. L. Rudell, and R. E. Bryant. Efficient implementation of a BDD package. In *Proc. of the 27th DAC*, pages 40–45, June 1990.
2. F. M. Brown. *Boolean Reasoning: The Logic of Boolean Equations*. Kluwer Academic Publishers, 1990.
3. R. E. Bryant. Graph-based algorithms for boolean function manipulation. *IEEE Transactions on Computers*, C-35(8):677–691, August 1986.
4. J. R. Burch, E. M. Clarke, K. L. McMillan, D. L. Dill, and L. J. Hwang. Symbolic model checking: $10^{20}$ states and beyond. In *Proc. of the Fifth Annual Symposium on Logic in Computer Science*, June 1990.
5. Tam-Anh Chu. *Synthesis of Self-timed VLSI Circuits from Graph-theoretic Specifications*. Ph.D. thesis, MIT, June 1987.
6. O. Coudert, C. Berthet, and J. C. Madre. Verification of sequential machines using boolean functional vectors. In L. Claesen, editor, *Proc. IFIP International Workshop on Applied Formal Methods for Correct VLSI Design*, pages 111–128, Leuven, Belgium, November 1989.
7. L. Lavagno, K. Keutzer, and A. Sangiovanni-Vincentelli. Algorithms for synthesis of hazard-free asynchronous circuits. In *Proc. of the 28th. DAC*, pages 302–308, June 1991.
8. C. Y. Lee. Binary decision programs. *Bell System Technical Journal*, 38(4):985–999, July 1959.
9. H-T. Liaw and C-S. Lin. On the OBDD representation of generalized boolean functions. *IEEE Transactions on Computers*, 41(6):661–664, June 1992.
10. K. L. McMillan. Using Unfoldings to Avoid the State Explosion Problem in Verification of Asynchronous Circuits. In *Proc. of the 4th Workshop on Computer-Aided Verification*, June 1992.
11. T. Murata. Petri Nets: Properties, analysis and applications. *Proc. of the IEEE*, Vol. 77(4):541–574, April 1989.
12. L. Ya. Rosenblum and A. V. Yakovlev. Signal graphs: From self-timed to timed ones. In *International Workshop on Timed Petri Nets*, pages 199–206, 1985.
13. H. Touati, H. Savoj, B. Lin, R. K. Brayton, and A. Sangiovanni-Vincentelli. Implicit enumeration of finite state machines using BDD's. In *Proc. of the ICCAD*, pages 130–133, November 1990.

# Modelling of Pilot Behaviour Using Petri Nets

W. Ruckdeschel, R. Onken

Institut für Systemdynamik und Flugmechanik
Universität der Bundeswehr München
Werner-Heisenberg-Weg 39
D-85577 Neubiberg

**Abstract.** This paper focusses on a petri-net based pilot behaviour model as part of the knowledge based Cockpit Assistant System CASSY. The model represents rule-based knowledge of commercial aviation pilots regarding plan execution.

The modelling method and the applied modular constructs are presented. An overview is given to analysis goals and results and the used tools. It is shown that the net model was successfully integrated in the CASSY data processing by use of a real-time net interpreter. The presented system is an example for an industrial size petri net application.

## 1 Introduction

Increasing commercial air traffic, economical constraints and the complexity of modern aircraft put great demands on crew performance. Investigations show that about 75 percent of civil aviation aircraft accidents can at least partly be attributed to human overcharge or error [1]. In the past this problem was mainly treated by increasing cockpit automation. However, this approach changed the types of occuring errors, but did not solve the problem itself [2]. This is explicable by findings on human cognitive behaviour which differentiate between skill-, rule- und knowledge-based levels [3]. Today's auto-pilot and flight management systems support only the skill-based and partly the rule-based level of crew activity.

Knowledge-based assistant systems represent one particular approach to support human operators both within the rule- und knowledge-based level [4]. Such systems should primarily support the crew's situation assessment and, secondarily, in case the situation overcharges the crew, reduce the crew's workload by technical means [5][6]. This approach aims at supplementing human capabilities by functions which comply with pilot's needs.

For the development of such crew assistant as much knowledge as possible has to be considered concerning human factors, the pilot's normative and individual behaviour in situation assessment, planning and decision making and plan execution. This paper focusses on the petri-net based modelling, implementation and analysis of rule-based pilot behaviour in plan execution.

The paper is organized as follows. Section 2 gives a brief overview to our cockpit assistant system. Section 3 presents the knowledge base of the pilot model, section 4 the modelling process. Section 5 shows the structure and an example of the realized net model. In section 6 first analysis results and further analysis steps are presented. Section 7 summarizes the used net tools. Further work - mainly concerning the application of high-level nets - is discussed in section 8.

## 2 Cockpit Assistant System CASSY

The **C**ockpit **A**ssistant **Sy**stem CASSY is being developed at the University of the German Armed Forces, Munich, in cooperation with the DASA-Dornier company, Friedrichshafen. Like a third electronic crew member CASSY is based on a complete understanding of the current situation. Therefore, interfaces to the aircraft, to the crew and to air trafic control (ATC) are needed. The main components are shown in figure 1 and are briefly described in the following.

The **Automatic Flight Planner** (AFP) [7] provides a detailled flight plan as basis for the flight. This module considers crew and ATC constraints, weather conditions and system failures within planning and conflict detection tasks. If a new situation requires flight plan changes, the replanning is done autonomously. The planning results are presented to the crew as recommendations. If not rejected or modified by the crew, agreement for the new plan is achieved.

The actual flight plan is the reference for plan execution by the crew as well as by the **Piloting Expert** (PE) [8]. The PE is construed as a model of pilot crew and covers normative and individual crew behaviour. On the basis of this knowledge, expected pilot actions are determined, considering flight plan, local ATC instructions, aircraft and environmental constraints. Modelling of rule-based pilot behaviour by petri nets is focused in the following sections.

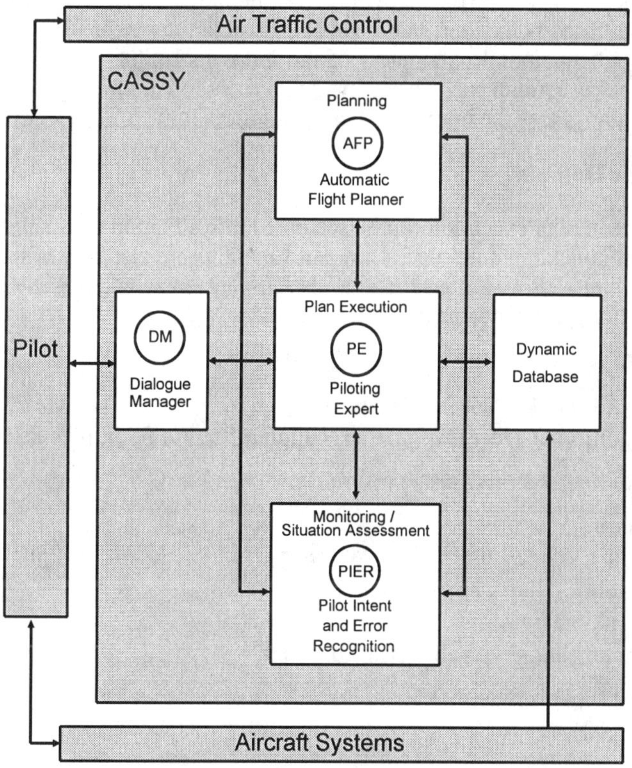


Fig. 1: Structure of Cockpit Assistant System CASSY

The module **Pilot Intent and Error Recognition** (PIER) [9] compares this reference behaviour to the actual crew actions. The interpretation of deviations is based on expert knowledge about possible situation dependent pilot intentions and results in a classification as pilot error or intent. If error is assumed, warnings to the crew are performed by speech output. In case of a verified intent hypothesis this information is reported to the other CASSY modules and leads - e.g. by replanning - to CASSY's adaption to the changed situation.

The **Dialogue Manager** (DM) [10] comprises all components for information transfer between CASSY and the pilot crew, including the information flow management. Extensive use of speech communication is made in either direction. Speech input is performed by a speaker independent speech recognition system, speech output by synthetic voice.

CASSY has been implemented in two flight simulators in Munich and Friedrichshafen. The hardware platform is a Silicon Graphics INDIGO Unix workstation (85 MIPS), the software is realized in C.

# 3 Knowledge Base

Modelling of pilot behaviour within the CASSY module Piloting Expert is done in two ways: The *normative model* describes deterministic pilot behaviour as documented in pilot handbooks and air traffic regulations. Modelling is done primarily within the domain of rule-based behaviour, but covers admissible tolerances also. The *individual model* contains behavioural parameters of the individual pilot and is developed as a real-time adaptive component. This paper will focus on knowledge domain, representation and analysis of the normative behaviour model.

## 3.1 Knowledge Domain

Pilot behaviour in plan execution can be separated into situation assessment and action processing components. Some special support functions are also part of the knowledge base. Behaviour modelling is done for all flight segments (taxi, takeoff, departure, cruise, approach, landing) and concerns the following tasks:

a) modelling of situation assessment:

- recognition of actual flight segment
- recognition of process of plan execution related to flight plan and procedures

b) modelling of pilot actions / pilot performance:

- primary flight guidance
  (altitude, course, airspeed, power setting, climb/descent rate, pitch attitude)
- operation of flaps, landing gear, speed brakes
- radio navigation
- communication with air traffic control

c) model-based support functions:

- callouts (of important checkpoints, e.g. altitudes)
- checklist processing (normal, abnormal, emergency)

### 3.2 Analysis of Knowledge Base

To choose an adequate modelling formalism, the pilot tasks were analysed with regard to causal, temporal and structural relations. This analysis gave the following characteristics:

- Piloting tasks are *strongly concurrent*. This can be stated in the domain of situation assessment as well as in the parallel processing of several tasks (e.g. maintaining altitude, reducing airspeed, radial tracking, ATC communication).
- Processing of pilot tasks (e.g. radio navigation) is driven by situation-dependent choices of different rule domains (e.g. cruise navigation or approach navigation), this is a *choice between (excluding) alternatives*.
- The basic element within the considered tasks is always a causal relation, which can be formulated by a *production rule* (if ..., then ...).
- The situation space as well as the pilot's action space can by described by *discrete states* (e.g. flight segments, flaps setting) and *state transitions* (e.g. flight segment transition, flaps setting transition).
- State transitions are driven by *discrete events* (e.g. "passing station X", "reaching altitude Y", "system Z breakdown").
- Pilot behaviour can be broken down into several *levels of abstraction*, like flight segments and their decomposition into sub-segments in the domain of situation assessment as well as a holding procedure and its decomposition into single actions in the domain of pilot actions.

### 3.3 Representation of Knowledge Base

One of the most important objectives of this modelling activity was to obtain a homogenous representation of the considered pilot behaviour. Homogenity should be required with respect to low expense for software tools and - if available - to enable formal analysis methods. It is obvious that the knowledge representation method to be chosen must be adequate to the system characteristics named above.

In former systems knowledge was often represented by production rules and so called production systems. However, production systems become difficult to control if the number of rules increases. Reasons must be seen in the lack of methods for structuring and decomposition. Finally, concurrency cannot be represented explicitly by production rules. Thus, modelling of pilot behaviour solely by production systems is no longer adequate.

Another alternative was the use of finite automata. However, the number of states which had to be modelled explicitly is enormous in view of the concurrencies to be considered.

These considerations led to the choice of petri nets by making use of different petri net classes, adequately matched to the particular properties of the knowledge domains considered:

- A considerable part of the knowledge is well representable by condition/event-nets.
- Another part of knowledge, even for modelling of multiple resources, requires at least use of place/transition-nets.

- Finally, a further part can in principle also be formulized by place/transition-nets, however for multiple identical model structures individual tokens are demanded and thus the application of high-level nets is suggested.

To bound the model complexity and the expense for net tools (primarily of the real-time tools, see 7.) at first the class of place/transition-nets was chosen and used extensively for modelling. The application of high-level nets is discussed in section 8.

Recently several petri net applications in the domains of civil aviation and aerospace arised, e.g. [11][12][13]. Modelling is done for simulation as well as for analysis purposes, partly by high-level nets. Regarding the criticality of aviation / aerospace software and with respect to safety and the resulting certification processes, further - even industrial - applications should be expected in future.

# 4 Model Design Process

When applying petri net theory to a concrete technical process the problem encounters of missing general rules concerning the formulation of application knowledge into petri net constructs. In general, the question is: How does the transformation *real world* → *model* look like and which rules have to be applied ?

Typical questions arising in the modelling process are:

- Which real world components have to / may / must not be formulated as places / transitions ?
- Which levels of net modularization are suitable ?
- How can local testability of a large net construct be secured ?

In the following, some characteristics of our petri net application are summarized. Especially we try to illustrate the design process, beginning with single production rules and rounding up with a hierarchically structured net system.

## 4.1 Semantic of Net Primitives

### Places

Discrete states in the field of situation assessment and during pilot action procedures are represented by places (C/E-nets). Examples are flight segments ("final approach"), conditions for subsequent actions ("turn right after passing altitude A") and states of discrete aircraft systems ("flaps 20 degrees"). Multiple resources, e.g. redundant navigation devices, are represented by multiple marked places (S/T-nets). Within the scope of modelling pilot workload, limited pilot resources are also modelled by multiple marked places.

### Transitions

Transitions are used to represent situation state transitions, e.g. between flight segments ("final approach → landing") and discrete aircraft systems ("landing gear up → down"). In the domain of pilot actions transitions represent for instance changes between basic tasks ("cruise → descent"), navigation instrument settings and callouts of checklist items ("landing gear down ?").

Because transitions are typically used to model *state transitions*, their firing time is assumed to be zero. In case of a model-relevant state transition time the transition is decomposed into a place and timeless firing transitions.

### 4.2 Knowledge Transfer Production Rules → Net Construct

The knowledge base to start with consisted of production rules which had to be transformed into net constructs. In the following example the transfer of two simple rules is shown. The rules are:

- IF (flight segment = "Final Approach") AND (altitude < 50 feet over ground) THEN new flight segment: "Landing"
- IF (flight segment = "Final Approach") AND (recognized crew intent = "Missed Approach") THEN new flight segment: "Missed Approach"

Either rule can be represented as place-transition-place construct. The transitions are attributed by external conditions (altitude and crew intent criteria), see fig. 2a. It is evident that the identical net pre-conditions of both transitions ("Final Approach") lead to a joined net construct (see fig. 2b). Analoguously, if pre- and post-states of different rules are identical, they are connected sequentially.

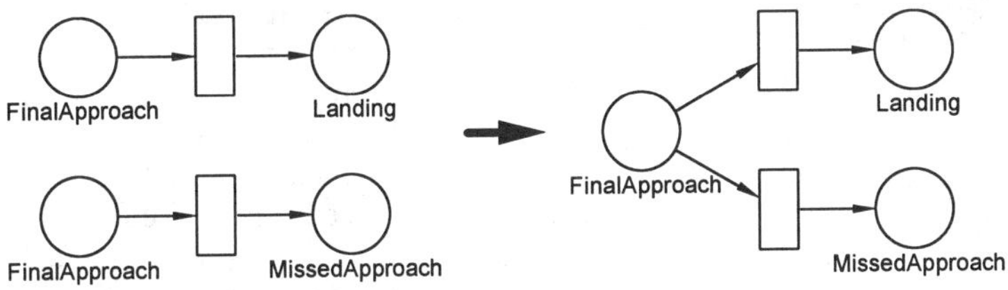


Fig. 2a Fig. 2b

### 4.3 Modular Construction

Size and complexity of the knowledge to be represented by petri nets for the application domain considered require extensive use of modularization techniques. Several requirements have to be satisfied:

1) Subsystems must be testable and analyzable on the local level. For this reason, activation and deactivation of coupling mechanisms is needed.
2) Because no tokens may be inserted or removed dynamically, all subsystems have to be strongly connected and marked.
3) Token flow between subsystems is not allowed. Nevertheless, implicit token flow is realized by the activation of subsystems (see 4.4).
4) Requirement 3) implies that access to places of other subsystems is permitted read-only.

The literature names different kinds of modular net construction, e.g. place and transition refinements, place and transition fusion sets, invocation transitions etc. [14][15]. According to the mentioned requirements we choose the place and transition fusion sets for our application. This method allows to give a place or a transition multiple graphical representations, even in different nets.

Modular construction often requires access to state or state transition information established in other subsystems. This information must be accessed in a read-only way. Thus we use two coupling mechanisms, both construed of place and transition fusion sets.

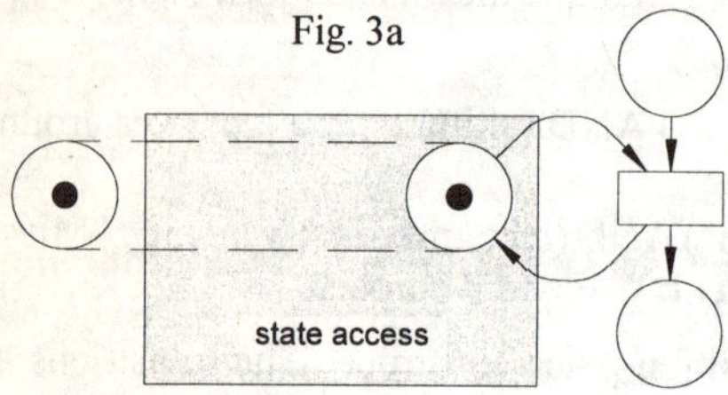


- *read-only access to state*
  The required state information is imported into the client net by place fusion. The place can now be accessed by test (double) arcs. Thus no token flow between the nets is allowed (see fig. 3a).

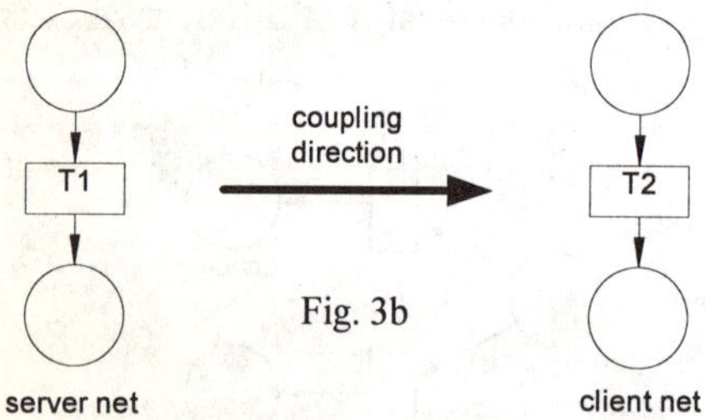


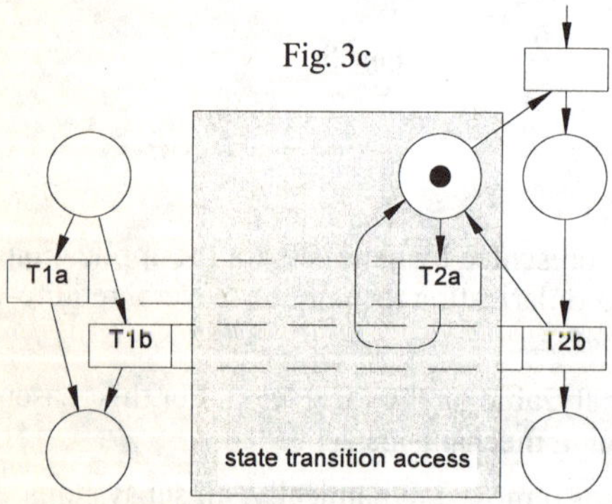


- *read-only access to state transition*
  This coupling mechanism is used to model an unidirectional dependence between two transitions (see fig. 3b): Firing of $T_1$ should be a precondition for firing of $T_2$, but $T_1$ should fire independently.
  The construct is realized by splitting transition $T_1$ into $T_{1a}$ and $T_{1b}$ and by importing them into the client net via transition fusion ($T_{2a}$, $T_{2b}$). A place complement is needed to guarantee firing of either $T_{1a}$ or $T_{1b}$. Fig. 3c shows that the state transition of the server net (firing of $T_{1a}$ or $T_{1b}$) is *not* restricted by the state of the client net, while firing of $T_{2b}$ is coupled with the state transition of the server net (firing of $T_{1b}$). Such coupling ist often applied for reset purposes.

### 4.4 Hierarchical Construction

The design of the net model is done in a top-down way. In many cases already modelled behaviour aspects have to be expanded by a more detailed model. Of course, net models cannot be extended boundless (graphical representation, testability etc.). Thus it has to be decided which parts of the net model are suited to be located in a subsystem and which coupling mechanisms are applied. In many cases it is desired to refine a state which is represented in the coarse net by a single place. A direct replacement of the coarse state by a subsystem does not comply with the modularization requirements mentioned above (primarily 1)). Since the coarse state carries semantical information (e.g. accessed by other nets), it is essential *not to substitute* the coarse state (as done by place refinement). For these reasons we refine states by duplicating their "interface" transitions into a subsystem, where the coarse state is modelled in more detail. In case the coupling mechanisms are deactivated, this construction preserves the behavioural properties of the coarse net. To fulfil requirements 1) and 2), a marked complement place is added to the subnet (see fig. 4). Coupling of the two nets is done by transition fusion sets of the interface transitions. This construction is an extention of pure place refinement (see [14]).

Figure 4 illustrates this technique using the example of fig. 2: We consider the place "FinalApproach", i.e. the decision between "Landing" and "MissedApproach" has to be modelled in more detail.

The upper path of the subnet in fig. 4 represents a lateral decomposition of the flight segment "Final Approach". This is done to pay regard to the outer marker (OM) beacon. A recognized crew intent "MissedApproach" is told to the net by a message from the CASSY-module PIER. This message has to be received concurrently to the described flight segement decomposition. This is modelled by the places "WaitForCrewIntent", "CrewIntent" and the transition "recv message" (lower path of the subnet). In case no crew intent message is received, the flight segment "FinalApproach - BehindOM" terminates regulary by firing of transition $T_2$ (altitude condition, see section 4.2, rule 1). In case a crew intent occurs, the subnet terminates via transition $T_3$. Different actions are performed dependent on the actual flight segment (transitions $T_4$, $T_5$).

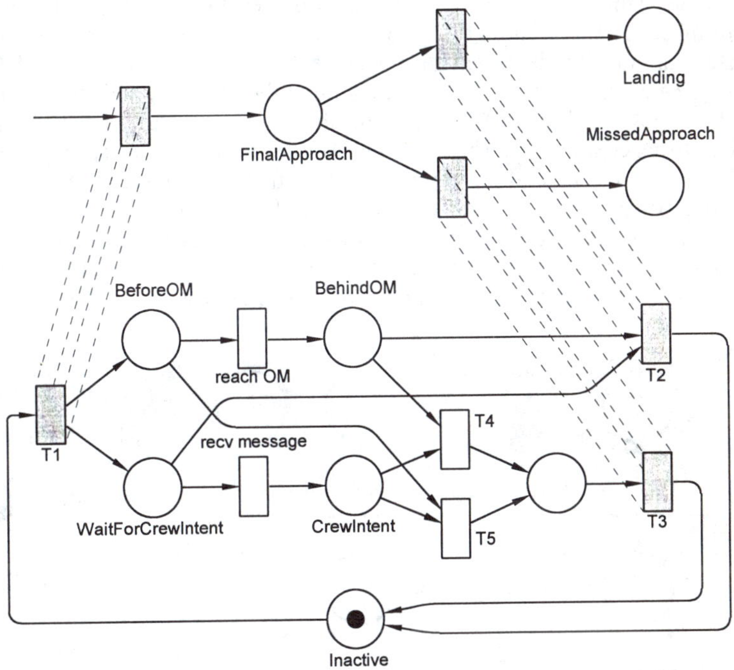


Fig. 4: Refinement of Place "Final Approach"

## 4.5 Process Interface

Modelling of pilot behaviour in plan execution can be separated into situation assessment and action components (see 3.1). As a basis for a rule-based situation assessment model a discrete situation space with well-defined state transitions has to be established.

The rule-based situation assessment process is characterized by a permanent considera-tion of all possible state transitions with regard to the actual situation state vector. These state transitions are typically defined as discrete limits within the - more or less - continuous state space of aircraft and environment (e.g. "passing station X", "reaching altitude Y").

For the assessment of the actual situation the state transition itself suffices. Nevertheless, the processes leading to this state transition influence the dynamics of situation assess-ment. Obvious questions like "what is earlier reached - station X or altitude Y ?" show that the causal structur of the underlying processes have essential effects on the assess-ment results.

Relative to real-time situation assessment this "why ?" of state transitions can be neglected. But for an overall investigation of the dynamics of situation assessment, the causal structure of aircraft and environment has to be made accessible to analysis methods. This means these systems have to be modelled by petri nets, including qualities like "x happens before y". This extended modelling has to be done in future.

For real-time situation assessment state transitions within the net model have to be executed dependent on external (aircraft / environment) conditions, in the following called "firing conditions". These firing conditions can be understood as states within a - not realized - aircraft / environment net model (see fig. 5a). These two net models can be connected by a common transition. The marking of the places $P_1$ and 'firing condition' enables the firing of transition $T_1$. With regard to the firing of $T_1$ the net structure containing 'firing condition' can be neglected. This leads to a compressed representation, see fig. 5b. A disadvantage of this representation is that a token has to be inserted when the condition occurs. Besides, if the condition is left without having fired $T_1$, the token has to be removed to avoid subsequent, faulty firing. For this reason we formulate this firing condition briefly as transition attribute (guard), see fig. 5c. A back-transformation to the other representations, e.g. for analysis purposes, can easily be done.

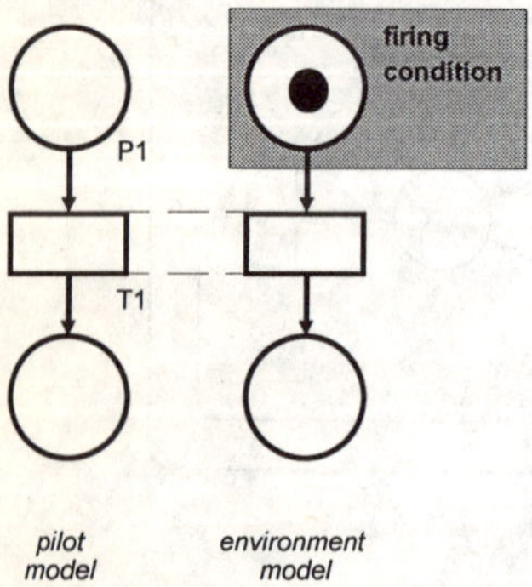


Fig. 5a

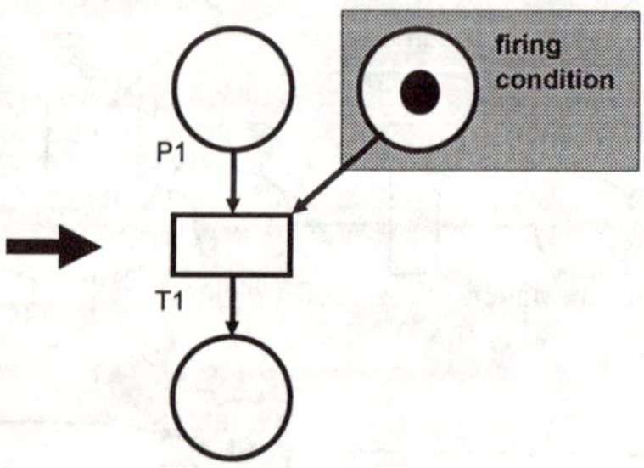


Fig. 5b

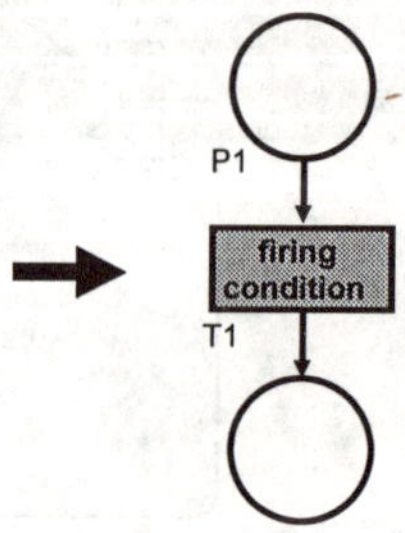


Fig. 5c

## 5 Model Structure

Figure 6 gives an overview of the (strongly simplified) model structure. The structurization of the net model was done according to knowledge structures as far as possible. The primary structurization characteristic are the pilot tasks within plan execution: recognition of flight segment, primary flight guidance (altitude, course, airspeed), system operation (gear, flaps, radio navigation) etc., see section 3.1 for details. These are typical examples for concurrent tasks.

To come up with subnets of handy size and complexity (not more than 10-15 places, reset constructs excluded), most of the tasks need further subdivision. For this purpose subclasses of behaviour within the main tasks had to be identified. An efficient structurization was done by separating behaviour with regard to different situation characteristics. The resulting behaviour classes are always related to excluding situation elements, therefore they are exclusive alternatives. The situation characteristics can mainly be attached to two groups: flight segment subsets (e.g. *departure airspeed* behaviour) and orders derived from the flight plan (e.g. *course* behaviour for "*proceed* to station X").

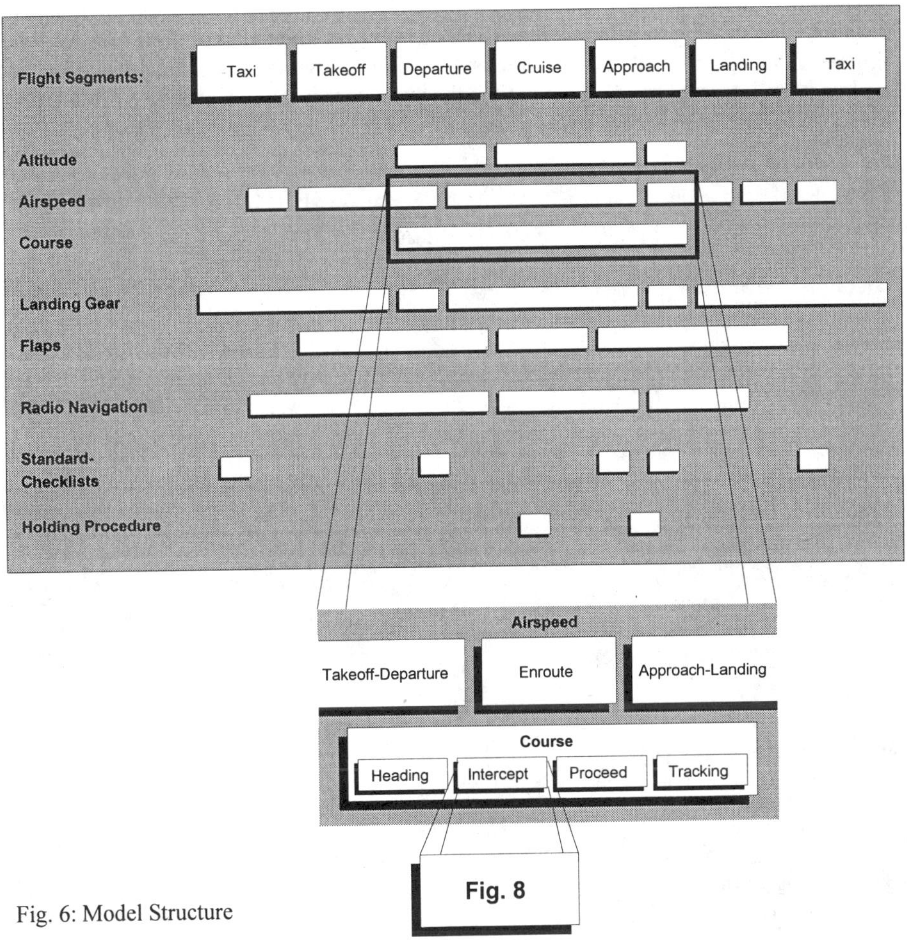


Fig. 6: Model Structure

Figure 6 shows the concurrent task models in vertical direction. Alternative (excluding) submodels related to flight segments are drawn in horizontal direction. Submodels are represented by white boxes; their size, complexity and hierarchical depth (i.e. number of subnets) differs widely. Nets for coordination purposes and their couplings with task model nets are also neglected in this illustration.

A small part of the model structure is zoomed out and discussed in more detail. Fig. 6 shows the course model and a part of the airspeed submodels. Subdivision of airspeed behaviour is done related to flight segments (takeoff-departure, enroute, approach-landing subnets). The actual course selection behaviour class is derived from the flight plan and is a choice between "turn to *heading* H", "*intercept* course C of station S", "*proceed* to station S", and "*tracking* from station A to station B". A simplified "Intercept" subnet is described in the following (see fig. 7 / 8).

**Example**

An *interception* is carried out to reach a given (magnetic) course to a given station (e.g. a radio navaid). This can be required within published departure or approach procedures or can be ordered by air traffic control. In the general case, an interception covers 4 sections (see fig. 7): turning to a special intercept heading ($S_1$), maintaining on intercept heading ($S_2$), turning to given course ($S_3$), tracking on given course ($S_4$). Sections are skipped if the aircraft fulfils the characteristics of a following section, e.g. if the aircraft is already on intercept heading at the time the procedure is started, section $S_1$ is skipped.

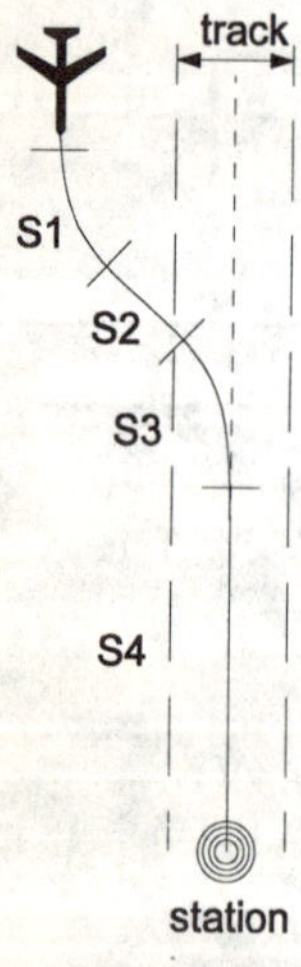


Fig. 7: Intercept Procedure

S1 turning to intercept heading
S2 maintaining on intercept heading
S3 turning to station
S4 tracking to station

t1 intercept heading is reached
t2 turn to station should be started
t3 heading to station is reached

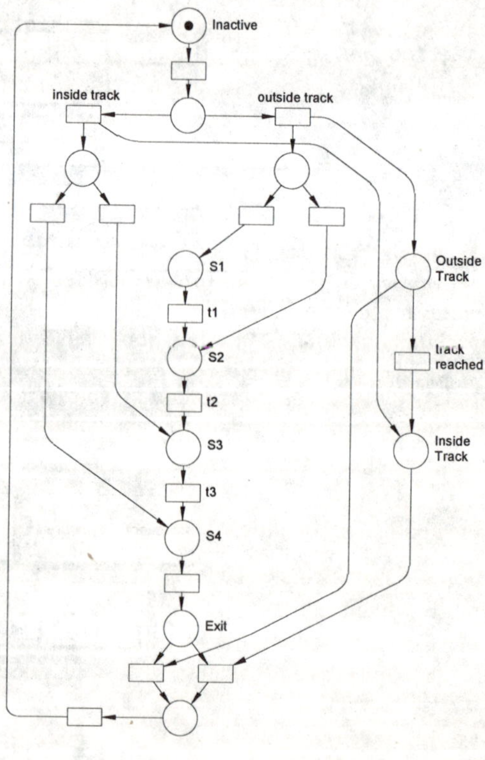


Fig. 8: Intercept Net

In addition to this *heading* behaviour, the given course to the station and an admitted course deviation specify a lateral *track*. After having reached this track, the aircraft should not leave the track until a new lateral procedure is started (see fig. 7). In this example it is assumed that the transition point between section $S_2$ and $S_3$ (start of second turn) is always placed outside the track.

The *intercept net* is mainly characterized by two concurrent constructs:

- A sequence of 4 places ("$S_1$" to "$S_4$") represents the 4 subsequent behaviour sections described above. The transitions connecting these states are attributed with heading conditions or other lateral conditions.
- The places "OutsideTrack", "InsideTrack" and the transition "track reached" are used for the modelling of the tracking performance mentioned.

After the net became active several conditions have to be considered within the initial section of the intercept procedure (not discussed here in detail). A further concurrent construct is needed to enable a net reset from all (stable) states, for instance in case of a changed flight plan (not shown in fig. 8).

As final result of the modelling process we expect at least 250 subnets with about 2500 places and 4000 transitions. At the time being, the model covers 1800 places and 2800 transitions in 170 subnets.

In the following, the size of this effort is summarized. This petri net activity was started at the end of 1991. Net modelling was done by the authors and in part by engineer students (5 persons, together 22 man-months). A part of the students had knowledge within the application domain (e.g. helicopter pilots). The students did the net modelling of small knowledge domains with great interest. The tools were mainly developed by computer science students (5 persons, together 24 man-months).

## 6 Analysis - Goals and Results

When the development was started, model testing was done mainly by simulation. Only few simple properties like connectedness were checked automatically after parsing the net declaration. Simulation tests have to be carried out anyway, even to check numerical results and the interfacing of the pilot model. However, the main problem of testing only by simulation runs is that it is impossible to reach all (critical) net states and state transitions within the test run. With respect to reversibility all net states are critical, because if a reset condition occurs and the net is unable to perform the reset successfully, this must lead to deadlock or at least to malfunction of the net and the parent nets.

For this reason formal analysis methods are applied to the net model in the meantime. The analysis strategy is "bottom-up", thus in a first step all subnets are checked to satisfy some obligatory qualities. These are at least:

- strong connectedness
- boundedness
- reversibility
- liveness and safeness

Analysis was done for all 170 subnets by the analysis tool INA [16][17]. Net analysis proved about one essential defect in every tenth net, mainly incomplete reset structures (remark: generally successful simulation runs are done since one year !).

The next analysis step is to combine (already checked) subnets into more complex net systems, step by step, and to prove the required properties again for the more complex net. This has not been done yet.

Besides these general properties there are other special properties which can be derived from the pilot model specification. Examples are:

- exclusive states
  (e.g. exclusive flight segments, exclusive aircraft system states)
- state sequences
  "states $S_1, S_2, ..., S_n$ have to occur / may never occur subsequently"
  (e.g. flight segments)
- state refinements
  "activity of refined state $S_r$ requires activity of coarse state $S_c$"
- predetermined reset procedures
  "firing of transition $T_r$ carries over every net marking to the initial marking"
  (e.g. subnet reset)

These properties can be proved using facts, invariants or special evaluations of the reachability graph (critical with large models). Some of the listed properties have already been investigated on the basis of the reachability graph generated by INA.

For extensive verification a checking tool enabling the formulation of logical terms (numerous and/or-operations, negations) and avoiding the calculation of the reachability graph is desirable.

## 7 Tools

As a main requirement, the net model, as presented, is to be used not only in the design phase but also as final implementation of the CASSY-module Piloting Expert. For this purpose, a real-time petri net interpreter was needed. This central role of highly integrated real-time net interpretation is an atypical aspect of this application and has some unfavourable effects on the suitability of commercial petri net tools. In the following, the main tool requirements are summarized:

- availability of real-time tools (interpreter and monitor) on the CASSY hardware platform (Silicon Graphics workstation)
- interpreter interface to program language C for integration of transition guard / action functions (process interface)
- strict separation of interpreter (simulator) and graphical user interface
- graphical *and* textual net declaration (especially important to large nets and declaration of transition attributes)

Because of these requirements only a fraction of the commercial tools could be applied.

A description language for place/transition-nets was defined enabling the declaration of modular constructs and process interfacing by transition attributes (guard and action functions).

By use of the commercial tool Design/OA [18] a graphical editor was developed which supports the required coupling and refinement methods and the local treatment of subsystems.

The net interpreter is integrated in the real-time data processing of the assistant system and does *not* have any graphical interfaces. The central requirement for interpreter development was to gain response times not dependent on net size and nearly independent on the number of actual active transitions. The process couplings are achieved by use of an open interface to progamming language C. Transition guard and action functions implemented in C are automatically linked to the net data structures and to the interpreter kernel.

Debugging of net simulations is supported by a graphical monitor system using OSF/Motif. The monitor receives actual marking information from the interpreter. Transition firing (overwriting of transition guard functions) can interactively be done by the monitor. Special attention was paid to a net activity dependent choice of presented information. Because of system size this information reduction is indispensable.

As presented in the last section, the commercial tool INA is used for net analysis.

## 8 Further Work

### 8.1 Use of High-level Nets

In section 3 it was stated that parts of the model gained greater compactness and expressiveness by use of high-level nets. Individual tokens can reduce net-external data flow and enable folding of identical net structures. By use of different interpretations (algebras) of the same net structure very compact model can be construed. An example for the re-modelling of a pilot behaviour aspect by use of high-level nets is presented in the following.

Checklist processing of pilots was modelled by place/transition-nets. The nets have been simplified for better clearness of description (see fig. 9):
Checklists can be disabled, for instance within inadequate flight segments. A checklist is enabled as soon as the checklist processing can be allowed regarding the flight segment and further situation elements. Checklist processing itself is started when the pilot requests the check or when a timeout condition (e.g. the latest point within flight progress to do the check) occurs, see fig. 9a. After the check has been started, a sequence of $n$ checklist items is treated. This is done for each item in the same way and modelled in $n$ structurally identical subnets ("Item N"). These subnets hold the complementary places "ItemSet" and "ItemNotSet" with "ItemNotSet" as initial state (the reset of these states is not described here). After subnet activation, the concrete check is done or skipped dependent on the state of the item. If the item has to be checked and the pilot confirms the check (e.g. by speech input), it is made sure that the check has been done correctly (e.g. by checking the aircraft system state). If the check was successful, a state transition to "ItemSet" is performed. After this, the subnet "Item $_i$" terminates and the subnet "Item $_{i+1}$" is activated simultaneously.

It is essential to know whether all checklist items have been checked successfully or if there are items remaining. These states are represented by the complementary places

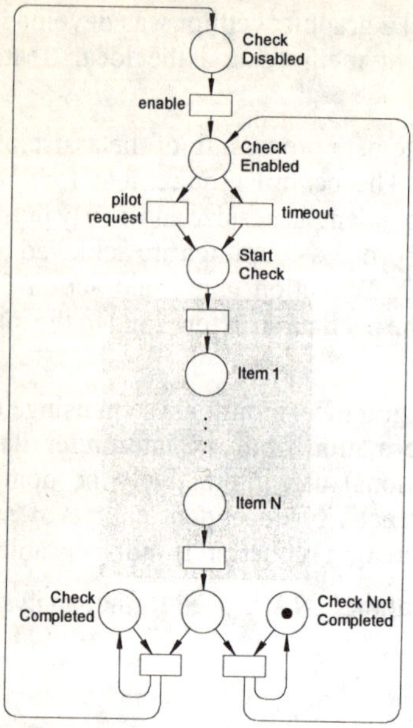


Fig. 9a: Checklist Subnet "Frame"

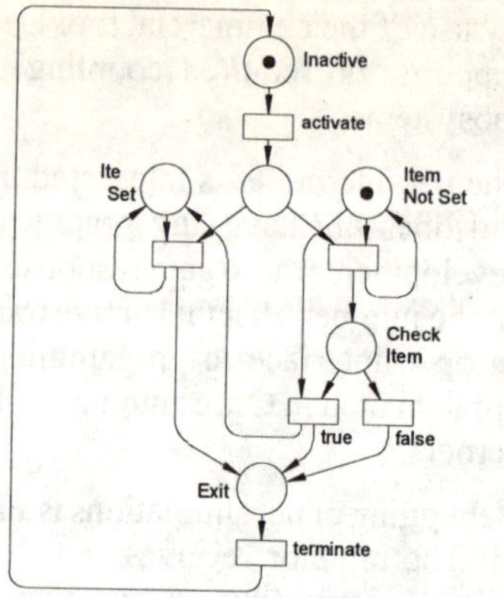


Fig. 9b: Checklist Subnet "Item N"

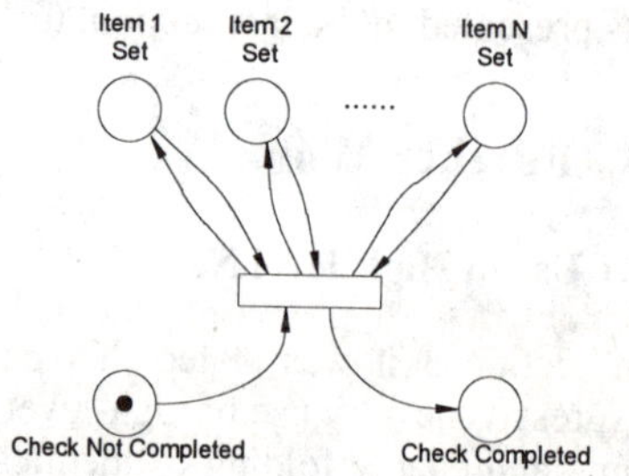


Fig. 9c: Checklist Subnet "ItemState"

"Check Completed" and "CheckNotCompleted" and are evaluated in the "ItemState" subnet. They are made accessible to the "Frame" net by place fusion.

After the last item has been processed, the checklist is enabled again (for later processing of the remaining items) or finally disabled (if the check is complete).

As presented, checklist models contain identical net structures for the treatment of each checklist item. By use of a high-level net class, a first folding step reduces *n* item-subnets to one item-subnet used for *n* items. Fig. 10 shows the high-level net construct which replaces the highlighted area of the S/T-net in fig. 10. The sequence of *n* items is replaced by a cyclical structure; variable tokens ($item_1 \ldots item_n$) represent different checklist items. The item sequence is realized by use of the successor function $SUC\ (item_k) = item_{k+1}$.

After this step all checklist items are modelled by the same subnet, i.e. different transition attributes representing concrete aircraft systems are folded into a single transition. Thus a transition interpretation (i.e. choice of transition guard and action functions) dependent on firing modes is needed.

Having done this first folding step, all checklist models are structurally identical; they still differ in the number of items and in the transition attributes. Thus, a second folding

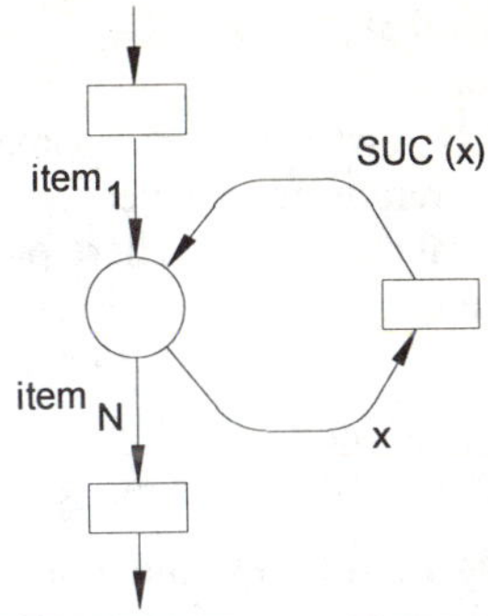


Fig. 10: High-level Net Construct for Checklist Modelling

step can be done to reduce *m* checklist-nets to one checklist-net used for *m* checklists. To perform this step, different interpretations of one net structure are required.

To allow these folding steps, the high level net class must enable variable cyclical structures (e.g., *n* subsequent items of checklist) and variable semantics of a single net structure (e.g., *m* different checklists).

Another example for the use of high-level nets is the modelling of pilot behavior in case of navigation instrument breakdowns: This requires a distinction between multiple, redundant devices. Use of individual tokens enables a folding of identical net structures and thus improves model compactness.

The choice of net class and specification algebra was done in cooperation with Prof. W. Reisig, Humboldt-Universität Berlin [19]. The extension of the net tools is in work.

**8.2 Further Analysis**

As stated in section 4.5, the net model is coupled with the environment (primarily flight plan and aircraft data pool) by transition guard functions. These couplings are often used to choose between exclusive alternatives (e. g. to select the required course behaviour model dependent on the flight plan). Such couplings gain external (in software modules realized) conflict solutions not accessible for net analysis methods.

To enable further analysis, it is intended to generate net moduls dynamically from the actual flight plan. These net models are related to a special flight plan and thus require less external conflict solutions. Of course, validity of analysis results will be restricted to the special scenario.

In section 4.5 the problem of so far neglected causal structures of aircraft and environment was mentioned. To analyse deadlock situations, e. g. resulting from unusual flight plan processing by the crew, at least exemplary net modelling of aircraft and environment has to be done. In control theory this step corresponds to the use of a process model to evaluate the qualities of a controller in the closed loop.

By use of such models it is possible to use the net quality of (state) reachability. The success of a given flight mission can be investigated and questions can be answered like "which events must / must not occur in different flight segments to guarantee mission success ?".

## 9 Conclusion

The pilot crew in the environment of increasing commercial air traffic and complex modern aircrafts needs extensive assistance to meet the requirements of flight safety and economical constraints. Therefore, electronic crew assistance can offer great benefits in monitoring, planning and decision making for complex situations. CASSY is a development effort in this direction.

To judge the actual crew behaviour as part of an overall situation assessment process a reference is needed. This reference is provided by a model of normative crew behaviour.

Resulting from an examination of crew behaviour, petri nets are considered to be an adequate knowledge representation method. By use of different - in part high level - net classes model expressiveness can be maximized. Application of high-level nets to parts of the knowledge base leads to a considerable reduction of net extent compared to pure place/transition-net modelling.

The great size of the rule-based knowledge base requires a modular and hierarchical model structure. Some frequently applicable standard coupling and refinement constructs could be established.

First analysis steps detected model faults which could not be located in many simulation runs before. Analysis of connected subsystems is expected to enable further model improvements.

The net knowledge base is used not only in the design phase but also as part of the final implementation. Net data processing is done by a package of - partially commercial - tools consisting of editor, parser, real-time interpreter, monitor and analyzer. Extensions for high-level net processing are in work.

CASSY is being developed and tested in a flight simulator since 1991, an evaluation with 14 professional pilots at the end of 1993 yielded good acceptance and priorities for the further development of CASSY. In 1994 CASSY will be tested in an experimental aircraft.

## Acknowledgements

The authors are very grateful to W. Reisig for his cooperation, for valuable comments on our work and for reviewing a draft version of the paper. We thank four anonymous referees for their sharp comments.

## References

[1] Rüegger, B.: *Human error in the cockpit*, Swiss Reinsurance Company, Aviation Department, 1990.

[2] Chambers, A.B.; Nagel, D.C.: *Pilots of the Future: Human or Computer?*, Communications of the ACM,Vol. 28, No. 11, November 1985.

[3] Rasmussen, J.: *Skills, Rules and Knowledge; Signals, Signs and Symbols, and Other Distinctions in Human Performance Models*, IEEE-SMC-13, No. 3, 1983.

[4] Dudek, H.-L.: *Wissensbasierte Pilotenunterstützung im Ein-Mann-Cockpit bei Instrumentenflug*, Dissertation, Universität der Bundeswehr München, 1990.

[5] Onken, R.: *New Developments in Aerospace Guidance and Control: Knowledge-based Pilot Assistance,* IFAC Symposium on Automatic Control in Aerospace, München, 1992.

[6] Onken, R.: *Funktionsverteilung Pilot-Maschine: Umsetzung von Grundforderungen im Cockpitassistenzsystem CASSY*, DGLR-Fachausschuß Anthropotechnik, Berlin, 1993.

[7] Prevot, T., Onken, R.: *On-board Interactive Flight Planning and Decision Making With the Cockpit Assistant System CASSY*, Human Machine Interaction and Artificial Intelligence in Aerospace, Toulouse, 1993.

[8] Ruckdeschel, W., Onken, R.: *Petrinetz-basierte Pilotenmodellierung*, in: Scheschonk, G., Reisig, W. (eds.), *Petri-Netze im Einsatz für Entwurf und Entwicklung von Informationssystemen*, Springer, Berlin, Heidelberg, NewYork, 1993.

[9] Wittig, T., Onken, R.: *Inferring pilot intent and error as a basis for electronic crew assistance*, Human Computer Interaction 1993, Orlando, 1993.

[10] Gerlach, M., Onken, R.: *A Dialogue Manager as Interface between Aircraft Pilots and a Pilot Assistant System*, Human Computer Interaction 1993, Orlando, 1993.

[11] Huck, V.: *Petri-Netze im Flughafenverkehr - Möglichkeiten, Grenzen, Perspektiven*, DLR-Mitteilung 91-17, Braunschweig, 1991.

[12] Kreher, H.-J., Stöhr, P., Schroer, W.: *Simulationsmodelle zur Analyse des Bodenrollverkehrs auf Flugplätzen*, in: *Entwurf komplexer Automatisierungssysteme*, TU Braunschweig, 1993.

[13] Lloret, J.C., Roux, J.L., Algayres, B., Chamontin, M.: *Modelling and Evaluation of a Satellite System Using EVAL, a Petri Net Based Industrial Tool*, in: Jensen, K. (ed.), *Application and Theory of Petri Nets 1992*, Lecture Notes in Computer Science, vol. 616, Springer, Berlin, Heidelberg, New York, 1992, 379-383.

[14] Vogler, W.: *Modular Construction and Partial Order Semantics of Petri Nets*, Lecture Notes in Computer Science, vol. 625, Springer, Berlin, Heidelberg, New York, 1992.

[15] Huber, P.; Jensen, K.; Shapiro, R.M.: *Hierarchies is Coloured Petri Nets*, in: Rozenberg, G. (ed.): *Advances in Petri nets 1990*, Lecture Notes in Computer Science, vol. 483., Springer, Berlin, Heidelberg, New York, 1990, 313-341.

[16] Starke, P.: *INA Integrated Net Analyzer*, Version 1.3, Berlin, 1993.

[17] Starke, P.: *Analyse von Petri-Netz-Modellen*, B.G. Teubner , Stuttgart, 1990.

[18] Meta Software Corporation: *Design/OA*, Version 3.0, Cambridge, 1991.

[19] Reisig, W.: *Petri Nets and Algebraic Specifications*, in: Theoretical Computer Science 80 (1991), 1-34.

[20] Abel, D.: *Petri-Netze für Ingenieure*, Springer, Berlin, Heidelberg, New York, 1990.

[21] Hanisch, H.-M.: *Petri-Netze in der Verfahrenstechnik*, Oldenbourg, München, Wien, 1992.

[22] Reisig, W.: *A Primer in Petri Net Design*, Springer, Berlin, Heidelberg, New York, 1992.

[23] Jensen, K., Rozenberg, G. (eds.): *High-level Petri Nets*, Springer, Berlin, Heidelberg, New York, 1991.

# Simulation and Analysis of a Document Storage System

Gert Scheschonk and Michael Timpe

C.I.T. Communication and Information Technology GmbH
Ackerstr. 71-76 D-13355 Berlin
Tel.: [+49](30) 4 63 60 77
Fax: [+49](30) 4 64 16 10
e-mail: sheshonk@tfh-berlin.de

**Abstract.** The paper discusses the modeling and simulation of a Document Storage system using colored Petri nets and the Design/CPN tool. The system modeled is part of a system which is capable of managing and delivering documents in the range of 20 million up to 30 million documents. This corresponds to a storage capacity of 10 TB up to 15 TB. The 660 concurrent users working with the system request about 10,000 documents within one hour. The goal of this project was to develop a model of the critical parts of the system in order to identify bottlenecks and to find the right scaled hardware configuration. Special modeling techniques are needed in order to process such a large number of tokens i.e. user requests.

## 1 Introduction

The system described in the following is part of a system which must be capable of providing its users with about 20 million documents. The documents are stored as image files. This is corresponding to an estimated storage system of 10 Tera bytes. The number of documents will increase from 20 million in the year 1991 up to 30 million in the year 2001. The number of users working with the storage system will increase from 1200 up to 2000. Nevertheless, the number of concurrent users has the constant value of 660.

The purpose of this paper is to show how Petri nets can advance alternative approaches in evaluating the performance of complex computer systems. Many different methods and tools are used to evaluate the performance of complex systems (see [5]). Most of them are text based systems that do not aid human computer interaction for system modeling. Modern tools support graphical representation and allow the user to use native descriptions of complex, concurrent systems.

The project described in this paper was done for the Bull AG in order to model and to evaluate the performance of an essential part of the system. This part was critical in terms of performance and therefore in forcing a bottleneck analysis. It was required that the requested documents be delivered within a certain time interval.

An essential aim of the simulation was to verify whether or not it was possible to fulfill the required response time. One of the requirements was that the response time of the servers for a request of documents ensured an average response time for the first document. The subsequent documents should then be delivered in their requested sequence within a fixed time interval.

The documents were randomly distributed over the servers. This results in a request for a number of documents being processed by several servers, whereby it could occur that a document requested first could arrive last. In order to ensure that the document first requested would arrive first within the required response time, it was necessary to check when, on average, the last document of a request would become available.

# 2 Application Modeling

## 2.1 General Description of the System

The storage system has three different sites (locations). The core of the system is installed in only one of the three sites. The others will have access to the system via remote access or via a local implementation.

The documents managed by the system are classified according to their importance (i.e. frequency of the requests) into four service levels which reflect four hardware levels. The service levels differ in the specified response time for a request on a document.

- Service Level 1 (SL1) holds the most frequently used documents. 80% of all requests refer to a document stored in this service level. The user must be able to read these documents within seconds, i.e. a time period that can be ignored in the context of the remainder of the requests.
- Service Level 2 (SL2): 10% of all requests are served by SL2. The specified response time for a request of a document on this level is 5 minutes.
- Service Level 3 (SL3): 9% of all requests are served by SL3. The specified response time for a request of a document on this level is 10 minutes.
- Service Level 4 (SL4): 1% of all requests are served by SL4. For requests at this level, no response time is specified. This level is also used for backups of documents from all other service levels.

For handling the four service levels, different hardware configurations have been chosen according to the specified response time: The documents of SL1 are stored on Optical Disks (OD). These ODs are accessible to all users at their local workstations. The documents from the other service levels are stored in a global storage system. The storage media are ODs in SL2 and WORMs (Write Once - Read Many) in SL3 and SL4 and are organized in juke boxes (JB) (see Fig. 1). The documents are distributed at random over the media. An Index Server stores the location of all documents at its site.

Every user has access to the stored data via a workstation which is connected to other workstations via a token ring and, if necessary, also via FDDI (Fiber

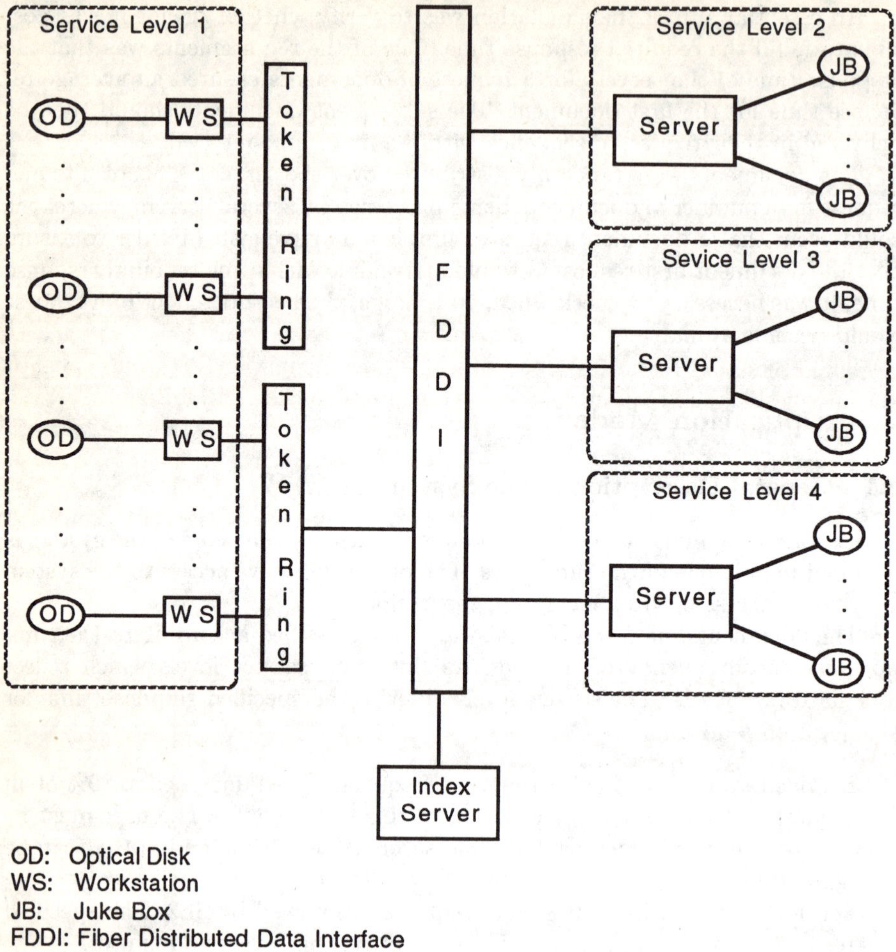


**Fig. 1.** Global architecture

Distributed Data Interface). The token rings are connected to the server via a backbone ring (FDDI). Each server handles a number of juke boxes with the media that stores the image files of the documents. Site 1 will have the complete equipment. Site 2 is restricted to SL1 through SL3 and Site 3 to SL1 through SL2.

The document retrieval appears as follows:

1. The user has to request the locations of the documents to be read via the Index Server. This process is called "request for location".
2. The Index Server returns the location of all requested documents.
3. The user has to request the documents from the server at the document's service level. This process is called "request for delivery".

4. The server loads the document from its juke boxes and sends it to the user's workstation.

## 2.2 Focussing on Critical Parts

To form a more concrete idea of the system, the following assumptions were made. These assumptions were drawn from skills of the client:

- There are 660 concurrent users who request 6000 single documents from Service Level 2 and 4600 single documents from Service Sevel 3 within 1 hour.
- The users only request a group of single documents. Such a group is called "Request"; each request on a single document is called "SubRequest". "Requests" are classified by short and long "Requests". Short "Requests" contain from 1 to 3 "SubRequests"; long Requests contain 4 to 12. The total number of "SubRequests" is 10,600 (= 6000 on SL2 + 4600 on SL3).
- The "Requests" are equally distributed over the time interval (1 hour) as required from the client. It is also possible to implement other distributions like Poisson in the simulation model.

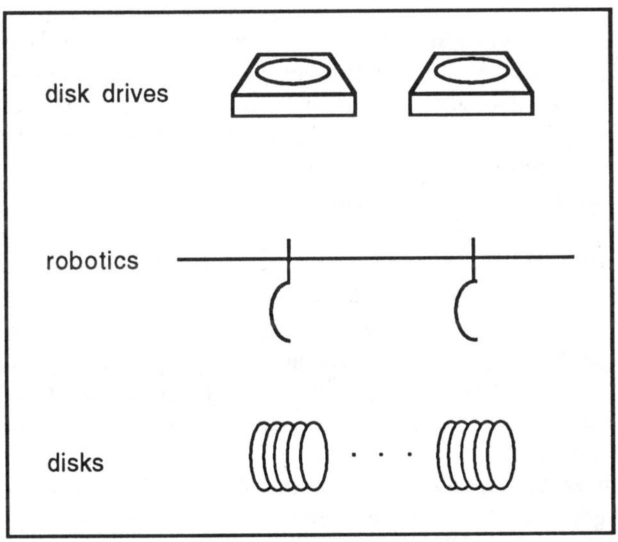


**Fig. 2.** Schema of a juke box

The modeling and simulation of the whole system is very expensive in terms of person power. Thus it is not necessary to simulate those parts of the system that are not critical, e.g. there are no performance problems expected at the index server. Requests on documents of Service Level 1 do not cause network traffic at the servers because the users access these documents directly via their local optical disk drive. Requests on documents of Service Level 4 have no specified response time; therefore a simulation is not necessary either. As documents of this service level are rarely requested, they do not affect the response time of other requests.

The purpose of the simulation is to evaluate the performance of the servers' juke boxes. It should be determined if the juke boxes are able to read the requested documents from the storage media within the specified response time. Another question is the topology of the juke boxes, i.e. the number of juke boxes in each server; and the number of disk drives and disks in each juke box (see Fig. 2).

Another reason for modeling only the critical part of the system was to ensure meeting the deadline set down by the client. As parts of the system are not critical in terms of performance, a bottleneck analysis of them is not necessary.

The critical parts of the system are the juke boxes. The mechanical process of putting the disk into the disk drive via a robotic is very time consuming, i.e. pick the disk, load it, spin it up, and read the documents. Therefore, the simulation model should represent only this critical part in detail.

## 3 Petri Net Modeling

There are two main advantages of Petri nets over traditional modeling and simulation tools. First, the graphical representation of nets gives people who are not involved in the modeling process an easy, achievable insight into the model. These nets are much easier to understand than the models of traditional tools, which have only textual representations of the system.

The second advantage is the straightforward design process using Petri nets. The whole design process, from specification to analysis, could be done with Petri nets. Traditional tools give the developer only the possibility to simulate the system. They are rarely built according to specification.

The decision of using Petri nets and the Design/CPN tool for modeling and simulating this system is based on these advantages and the capability of the tool in managing complex systems by using hierarchical structuring. In addition, the client learned how to read, create, and simulate models with Design/CPN in a one week tutorial.

Simulating nets with many thousands of tokens with Design/CPN is not a trivial task. A good trade-off between clear and easy specification model and an efficient simulation model is essential. The following section will show why Design/CPN has performance problems in simulating nets with thousands of tokens and what solution can be found for this particular system.

### 3.1 Sizing down the Number of Tokens

The main drawback of Design/CPN's simulation component is the time needed for simulating industrial-sized models. The model described in this paper needs about 15,000 tokens to represent all requests and all disk queues. The long simulation time is caused by the necessary calculations whether a transition is enabled or not. The performance of the simulation is strongly influenced by the structure of the Petri net tokens (see [1]). Three facts work together in order to determine whether the transition is enabled or not: the multiset of tokens of the input place, the input arc inscription and the transition guard. That means that each combination of variable bindings have to be checked against these conditions. For a large number of tokens, this is a very time consuming process. The solution is to hold one token queue on a place instead of individual tokens (see Fig. 3). This is done without losing information and time behavior. This transformation of token yields to token with a FIFO property.

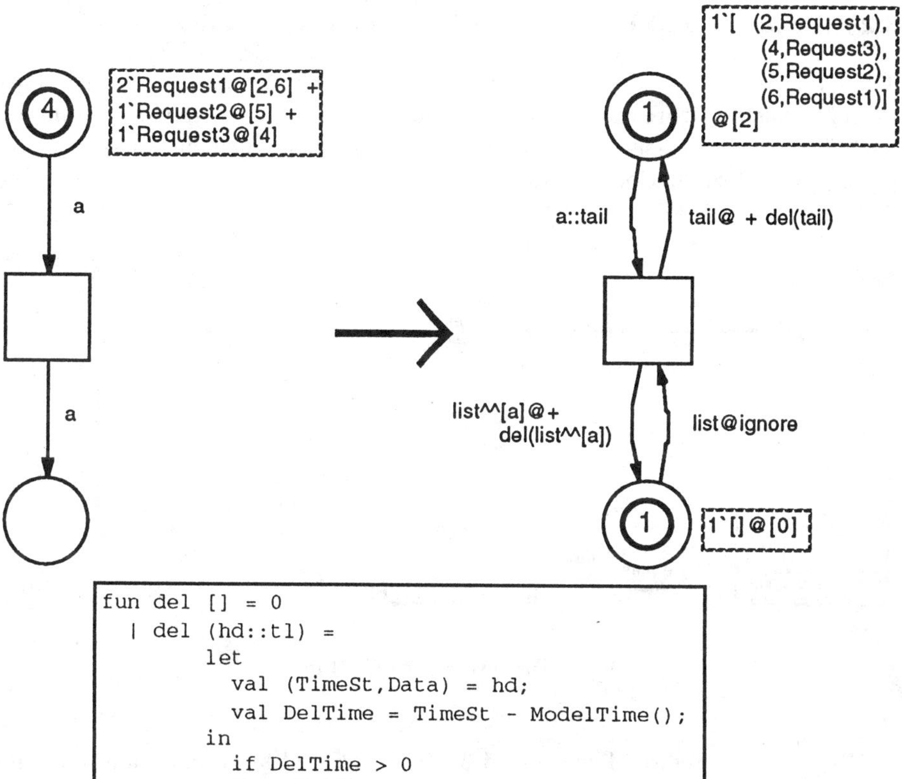


```
fun del [] = 0
  | del (hd::tl) =
        let
          val (TimeSt,Data) = hd;
          val DelTime = TimeSt - ModelTime();
        in
          if DelTime > 0
          then DelTime
          else 0
        end;
```

**Fig. 3.** Translation to structured tokens

The time stamp of the token queue must always be equal to its first element.

This is ensured by a delay function caused by the removal of the head of the token queue. The removed head is appended to the token queue of the output place. The FIFO behavior of the system guarantees that the "internal" time stamp of the appended element is greater than all "internal" time stamps of all other tokens in the token queue. A transition, whenever it occurs, must be able to put tokens on output places and the time stamp of the queue must therefore be ignored.

This transformation of the net leads to fast simulation times but in terms of a specification model, the readability of the net might be reduced. Most net inscriptions do not describe the functionality of the modeled system; they are only necessary to handle the specialized token structure. Furthermore this method of translating cannot be used in general as it implies a loss of concurrency.

### 3.2 Structure and Description of the Petri Net Model

The Colored Petri Net (CPN) in Fig. 4 shows the top level of the model. This model consists of three main parts: the initialization of the model, the model itself (i.e. the processing of requests for the delivery of documents) and the reports on the simulation results.

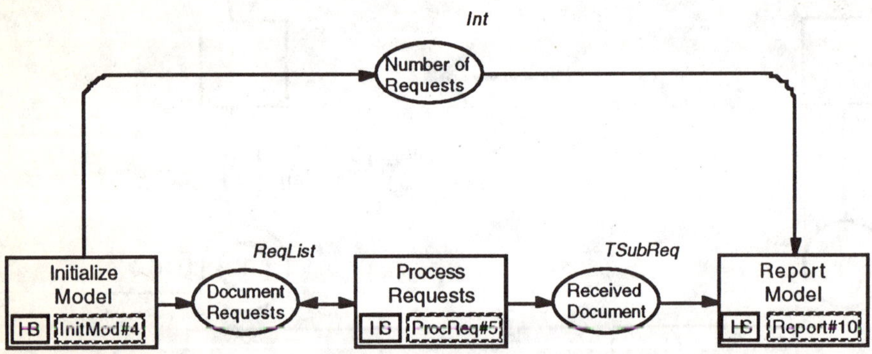


**Fig. 4.** Top level of the CPN model

Requests for Service Level 2 and 3 will be created in order to initialize the model. The generated requests are equally distributed over the simulation interval. This is a skill drawn by the client. The distribution is done by an SML function which gives the possibility to implement other distribution strategies like Poisson distribution. The number of the generated requests can be specified in a parameter file. Each request itself contains one to twelve subrequests which refer to different documents. All the created requests will be distributed over a specified time interval. This results in a time stamp for each subrequest. A random number generator defines the target disks of these subrequests.

The user wants to receive the documents in the same order as requested. At the request time, the first subrequest of a request is the one that is most urgent because the corresponding document must be delivered first. A priority mechanism takes into account the order number of a subrequest and ensures that the user is able to read the documents in the requested sequence.

Fig. 5 shows the hierarchical structure of the model. All net inscriptions and SML functions used in the model are defined on additional pages. This paper focusses on some refinements of the model, i.e. the juke box process.

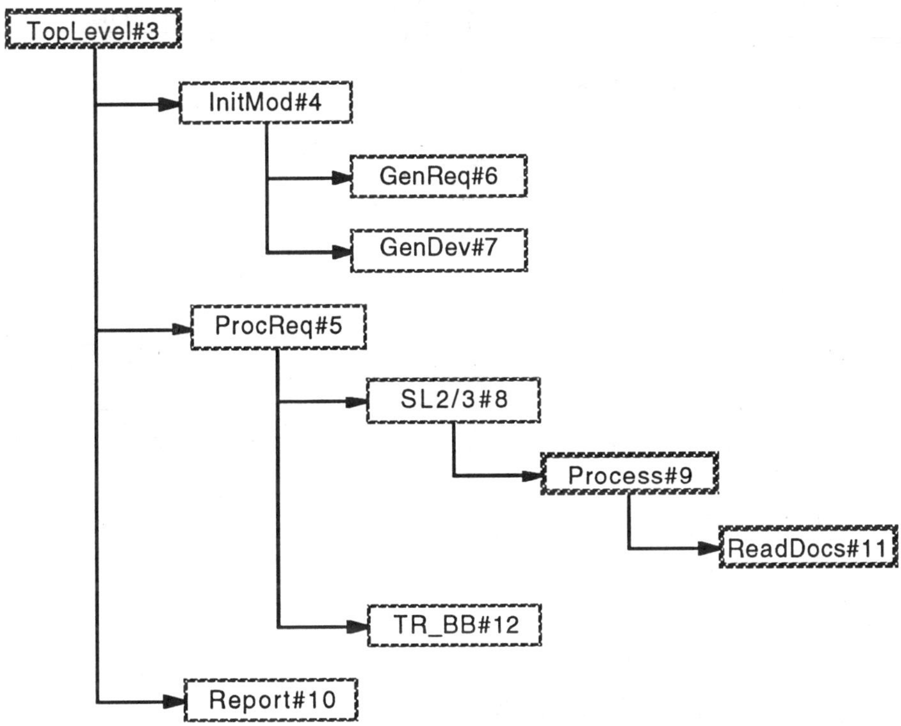


**Fig. 5.** Hierarchical structure of the model

The process of the requests starts with enqueueing the subrequests into the appropriate disk queue. Each disk queue is assigned to one disk. The disk queue priority is composed of both the current priority of the disk queue and the priority of subrequests at enqueueing. Before selecting a disk queue for reading from the disk, the priorities of all disk queues will be updated in order to consider the idle times of the subrequests in the disk queues.

The selection of the disk queue, the subrequests of which are next to be read from the disk, depends on different factors (see Fig. 6). First, a disk drive and a corresponding robotic must be available in order to load a new disk. From all disk queues of this juke box, the one with the highest priority is selected, once

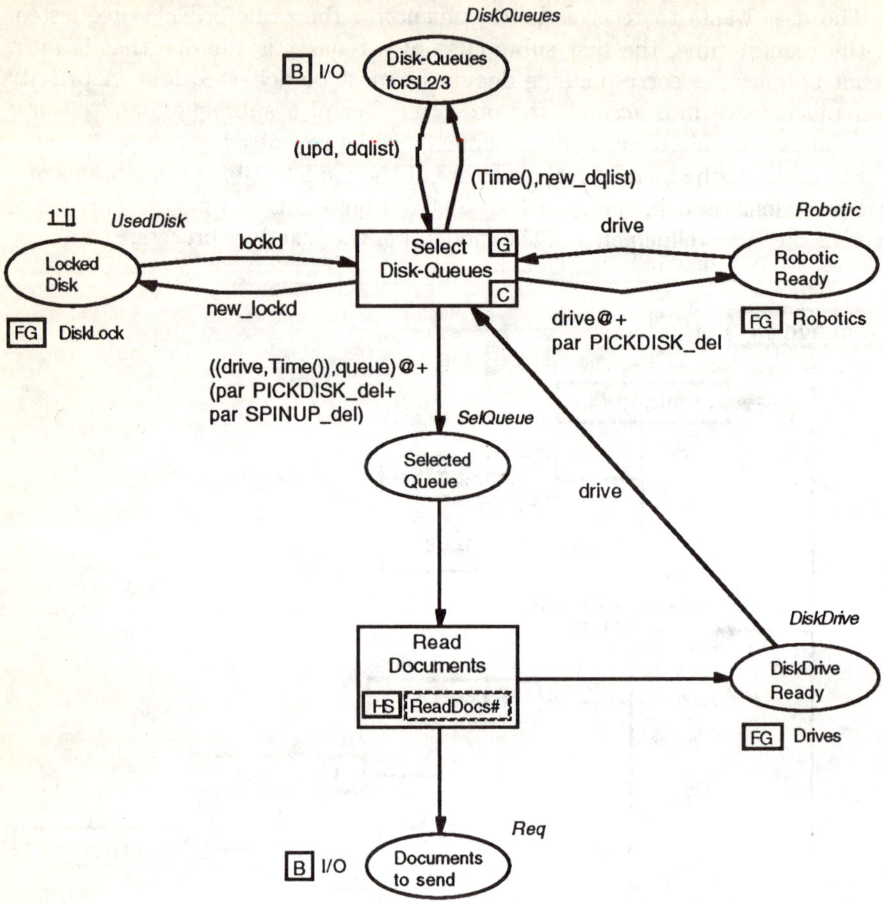


**Fig. 6.** Petri net model of the juke box (Process#9)

the disk drive and the robotic is free. If a disk has been chosen, it will be put into a list that contains all disks in use. The only disks that can be chosen are those not currently included in that list. This ensures that a disk cannot be read by different disk drives at the same time.

The robotic inserts the disk into the disk drive once a disk queue has been selected. After the disk is spinned up, all documents for this subrequest are read consecutively from the disk (see Fig. 7).

The read documents are temporarily stored in the server's memory. If the size of the stored documents exceed the memory size, the server has to swap some documents to the virtual memory. This swapping process causes a significant delay and is recorded. If all requested documents are read from the disk, the disk drive and the disk are released in order to serve other requests.

The server sends the documents back to the user's workstation via the FDDI

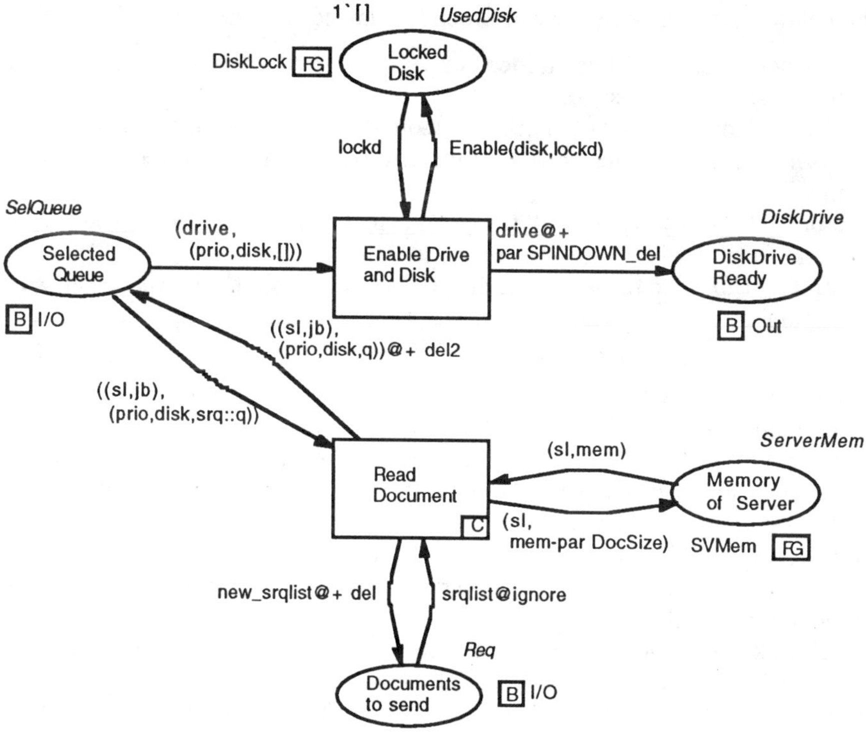


**Fig. 7.** Petri net model of reading the documents (ReadDocs#11)

backbone ring and the token ring and clears them from the memory. The time the documents have to wait until they have been sent is also significant and is recorded.

When the document reaches the workstation, all recorded delays and all other data are written into the report file for further analysis.

### 3.3 Measuring the Process Time of the Requests

To measure the process time of one subrequest, the net is divided into different parts in order to get the time a subrequest needs to pass this part of the model. If the total process time exceeds the specified maximum response time, it is able to identify which part of the net is responsible for the lack in performance. The arrival and departure time of a subrequest is recorded for each part.

The following data is necessary in order to analyze the system:

- transfer time: the time needed for the transfer of a subrequest from the workstation to the server
- waiting time: the time that the subrequest waits in the disk queue
- reading time: the time that the juke box needs to read the requested document
- storing time: the time that the document has to be stored in the server's RAM until it will be sent to the workstation
- sending time: the transfer time of the document from the server to the workstation

In order to measure this data, we have to specify measuring points in the system. At each measuring point (see Fig. 8), the model time will be added to the subrequest's time list.

The first point required is the creation time of the subrequest t0. The first delay is the transfer of the request from the workstation to the server, but because the requests are very brief in comparison to the documents, this transfer delay can be ignored. Nevertheless, in order to get a complete list of measuring points, the model time t1, i.e. the time the server receives the request, is added to the subrequest's time list.

The next delay needed for the evaluation of the model is the waiting time of the subrequest in the disk queue. All incoming subrequests are added by the server to the corresponding disk queues. If no disk drive or robotic is available, or the disk storing the requested document is in use, the subrequest cannot be processed and has to wait in the disk queue. This waiting time is calculated by subtracting t1 from t2, the time of adding the subrequest to the disk queue and the time of selecting a free disk drive respectively.

What renders this waiting time so important is the fact that it signifies a lack of devices such as disk drives and robotics.

After loading the disk into the disk drive, the juke box starts to read the requested documents. The requests are processed in the same order as they are added to the disk queue which means that a request has to wait until all previous requests are processed. This waiting time (loading the disk and reading the preceding documents from the selected disk queue) is expressed by the difference between t3 and t2. The time the disk queue is selected in order to process next is t2. The time the juke box starts to read the requested document is t3.

Reading a document results in a time delay expressed by t4-t3 which can sometimes be influenced by a memory swap of the server. If the size of the read documents exceeds the server's memory, it swaps some documents on its internal hard disk. This "swapping" results in a longer reading time.

The time the server starts to send the document to the user's workstation is t5, while t6 is the time the document becomes available to the user on his workstation. By simply subtracting t5 from t6, the transfer time can be calculated.

Many of the delays described above are constant. They are added to the system to produce a more flexible model. It is possible that the model has to simulate documents with variable length, so that transfer time is variable, too.

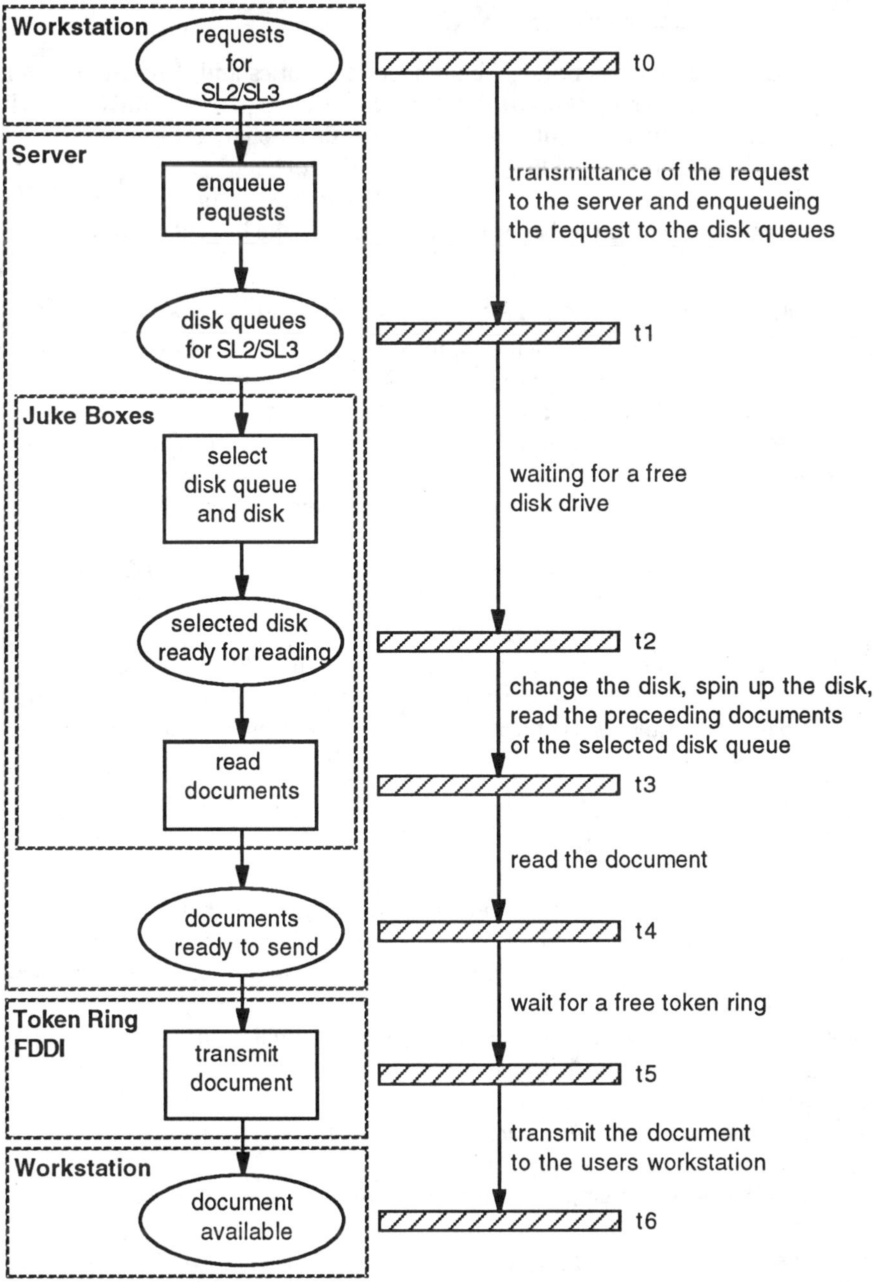


**Fig. 8.** Measuring points

### 3.4 Simulation Parameter

The fact that the dimension, i.e. the number of robotics and disks in each juke box, of the system was not known exactly forces the use of a parameter file. Its main advantage is the possibility to simulate various scenarios without changing the net structure and recompiling the complete model.

At the beginning of the simulation, the simulation parameters are read from a parameter file specified by the user. The parameter file has the following structure:

- Number of subrequests for service level 2 (SL2) and Service Level 3 (SL3);
- maximum length of short requests; maximum length of long requests; occurrence rate of short requests (e.g. every second request is a short one);
- initial value for the subrequests; document feed rate;
- possible delays (in time units): time of transmission of a document from server to workstation; spin up time of disk drives; spin down time of disk drives; reading time of document; time of disk pick up; swap time in the case of memory overflow;
- configuration of the system: number of juke boxes in SL2/SL3; number of disk drives per juke box in SL2/SL3; number of robotics per juke box in SL2/SL3; number of disks per juke box in SL2/SL3;
- size of server's memory and size of documents in KB;
- period of time, in seconds, over which the requests are to be distributed;
- number of documents that can be transmitted concurrently by the token ring;

## 4 Simulation Results

Many simulation runs have to be made to find the right system. There are many possibilities to change the model in order to get shorter response times or to get a more cost effective system. One possibility is to change the number of devices used in the system.
The following parameter file was used for the first simulation:

```
(* Number of Requests on SL2 and SL3 *)
6400 4800
(* max. length of Short Req; Long Req; rate of Short Req *)
3 12 2
(* initial value constant; document feed rate *)
144 12
(* Delays in s: Token Ring; SpinUp; SpinDown; ReadDoc; PickDisk; Swap *)
2 2.5 2 1 8 0.5
(* JB; DD; Rob; Disk *)
6 5 2 2 2 2 540 410
(* ServerMem Size; DocumentSize *)
16000 165
(* time interval in seconds *)
```

```
3600
(* concurrency of Token Ring *)
7
```

The important fact shown by the simulation data is the response time the system needs to respond the user requests. The diagram in Fig. 9 shows what percentage of all subrequests are responded within which time.

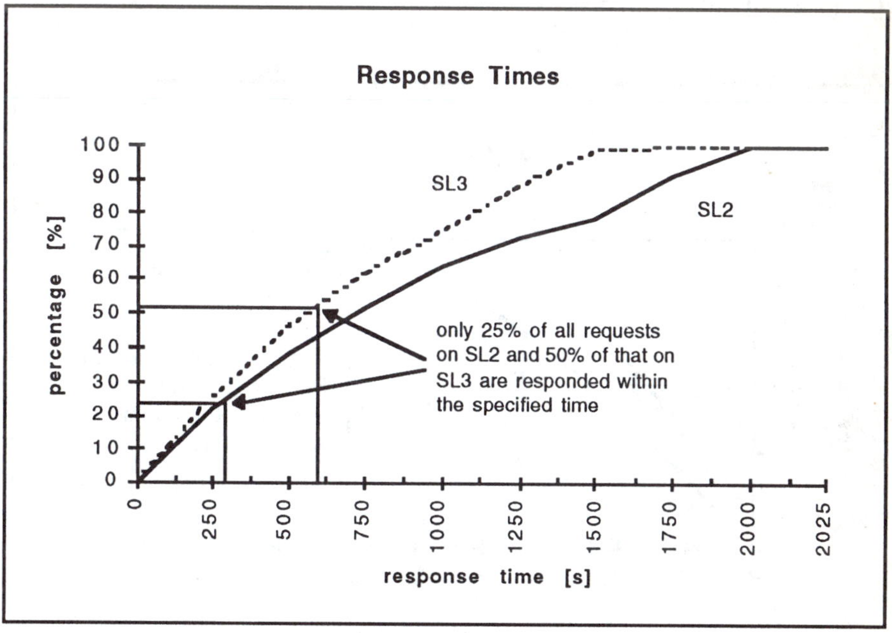


**Fig. 9.** Requests processed within 300s for SL2 and 600s for SL3

Fig. 9 first shows that only 25% of all subrequests on Service Level 2 and only 50% of all subrequests on Service Level 3 have complied with the specified response time. The system processes requests on SL3 faster than on SL2.

Which parts of the system are responsible for the long response times? The longest delay is caused by waiting for a free disk drive and a free robotic (t2-t1), i.e. waiting in the queues. The delay while inserting the disk, spinning it up, and reading the preceding documents is defined by (t3-t2). Subrequests on documents from the Service Level 2 have to wait much longer than from Service Level 3.

The length of the processed disk queues indicates that only an average of 1.4 documents are read from the disk. After reading these documents, another disk must be loaded. Thus many loading processes are necessary. During the loading process, the disk drive cannot read other documents.

The disk drives do not work to full capacity so they cannot be responsible for the long waiting time in the disk queues of both service levels.

Adding juke boxes could solve the aforementioned problem leading to a higher number of disk drives which can read documents. The following simulation uses juke boxes with two disk drives and only one robotic. The juke boxes use disks as media with an extremely high storing capacity. The question is whether this change results in shorter response times (see Fig. 10).

A different simulation run was based on a juke box with one robotic and disks with a much higher capacity. This increases the number of juke boxes for SL 2 to 8 and for SL3 to 7.

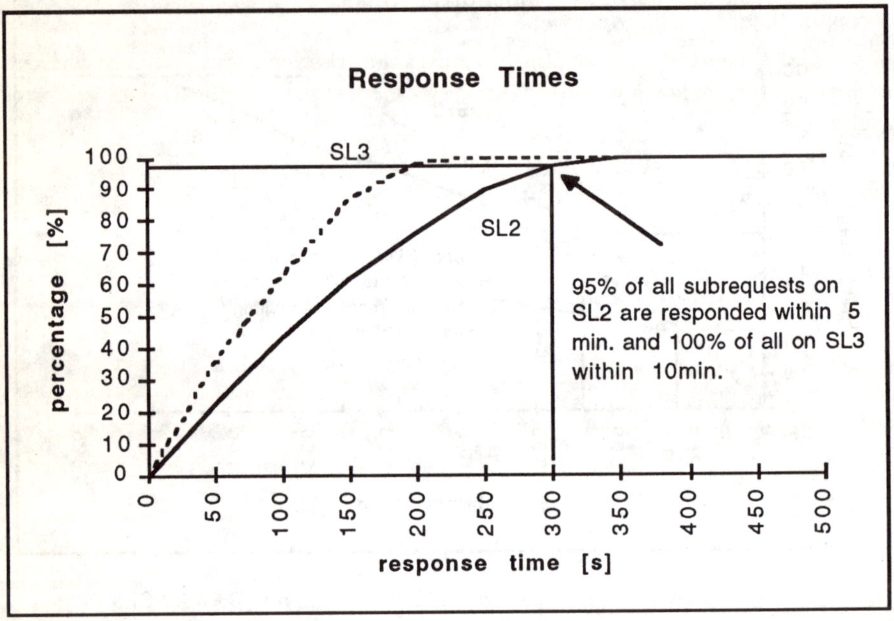


**Fig. 10.** Response times on adding of juke boxes

The disk drives are working at the same capacity as in the earlier simulation. The low capacity is a result of the short disk queues. They contain only an average of 1.6 requests when they are processed and so many disk loading processes are necessary. Therefore, there is no need for larger disks to store more documents. Reducing the number of disks does not result in longer disk queues.

## 5 Project Schedule and Conclusion

The project started with a one-week-training course that was not only based on introducing Petri net modeling, but also on handling the tool Design/CPN itself. Although all participants were experienced system analysts or programmers, none of them however had previous knowledge of Petri nets. Two employees from the client's group attending the seminar were chosen for the project. Two experts who had worked before with Design/CPN and with Petri nets joined the team. Due to a very tight schedule, the modeling and the simulation of the whole system was not possible. Therefore, the part of the system to be examined more closely was to be agreed within a two-day-meeting. The result focussed on the juke box, the behavior of which was not clearly predictable under the required workload. The different number of juke boxes available and the different hardware configurations with a different time behavior added to this unpredictability.

The project took place over a period of approx. 4 weeks, 3 of which were dedicated to implementing (i.e. designing the model and implementing all functions needed for the net inscription and verification of the function). The last week remained for the evaluation. Verification always meant the verification of the whole system as the net was developed, completed or changed step by step. This disregards smaller functions that could also be tested on their own. Another week was needed to simulate different scenarios provided by the client. The spread sheet tool Excel was used for evaluating the data exported by Design/CPN. The success of the project is reflected in the following benefits to the client:

- After the project, the client was capable of stipulating other scenarios, simulating, and evaluating on his own.
- Under certain hardware configurations of the juke box bottlenecks were encountered. These could be avoided by a change in the configuration or another type of juke box.
- The client recognized Colored Petri Nets as a useful method which could also solve other problems arising within their company.

Design/CPN uses directly executable, hierarchical colored Petri nets to model and simulate system behavior. In this project version 1.9.1 was used. This version already contains fundamental statistical and chart facilities. The statistical variables' facility provides the ability to calculate and access frequently needed statistical measurements such as standard deviation, variance, average, minimum, maximum, and many others.
The main advantages shown by the Design/CPN tool can be summarized as follows:

- Possibility of implementing detailed functionalities by programming with a functional language (Standard ML) instead of addional net structures. Using SML instead of nets causes a faster simulation.
- The interpretative language SML allows the fast verification of small functions without translating the complete model.
- Use of extremely high structured tokens (e.g. one token contains more than 10,000 subrequests each with its own internal structure).
- The simulation time was reduced to the lowest possible level by high structured tokens.
- By using text files to specify input data, different scenarios were evaluated without changing the model.

- High flexibility of Design/CPN allows modeling and evaluation of huge industrial-sized systems.

However, the project also revealed several aspects to be improved:

- The time stamps for each subrequest had to be controlled by SML programming in such high structured tokens.
- The use of statistical variables provided by Design/CPN raised the simulation time by factor five. This decrimental effect occurs because the statistical variables are updated whenever the user adds a new value by evaluating a code region during the execution of the CPN model.
- The chart facilities of Design/CPN are not flexible enough. It is better to further analyze the simulation data with an external spread sheet tool.
- It is in the nature of interpretative languages that they are not as powerful concerning execution time than compiler languages. The development of an SML compiler could speed up the simulation times of Design/CPN.

## Acknowledgement

We would like to thank U. Störmer and Degener from Bull AG whose constructive discussion and suggestions were helpful to get a proper simulation model, and A.Vogt from C.I.T. GmbH who supported us in analyzing the simulation data. Their impressive overview of the system was a great help.

## References

1. Cherkasova, L.; Kotov, V.; Rokicki, T.:
*On Net Modeling of Industrial Size Concurrent Systems.*
LNCS 691, Proceedings of the 14th International Conference
Chicago, Illinois, USA, June 21-25, Springer-Verlag 1993
2. Jensen, K.:
*An Introduction to High-Level Petri Nets*
Proceedings of the 1985 International Symposium on Circuits and Systems,
Kyoto 1985, IEEE, 723-726
3. Jensen, K.:
*Coloured Petri Nets*
In: EATCS Monographs on Theoretical Computer Science, Volume 1,
Springer Verlag, 1992
4. Meta Software Corporation:
*Design/CPN Manual*
Cambridge, USA, 1991
5. Randy, D. Brent; James R. Gross.:
*A Demonstration of Alternative Approaches for an Inventory Simulation Model*
Oshkosh, Wisconsin, 1990

# COOPERATIVE NETS

C. Sibertin-Blanc
CERISS-IRIT, Université Toulouse 1
Place A. France, 31042 Toulouse Cedex, France
E-mail: sibertin@irit.fr

**Abstract:**

The behavior of some kinds of systems features a high rate of dynamic evolution. The system running causes the introduction of new components whereas some others disappear, and links between components are dynamically set: a component sometimes interacts with given components and sometimes with others. It is uneasy to capture such evolution inside Petri nets whose structure is fixed; but it is necessary to respect the Petri net semantics and keep the possibility to apply the structural analysis technics.

The paper introduces two extensions of Petri nets dealing with this problem, Communicative Nets and Cooperative Nets. They enable to model a system as a collection of nets which encapsulate their behavior, while interacting by means of message sending or a client/ server protocol; a net may instantiate another net, and the links between nets are dynamic. An algorithm is given which captures this dynamicity by building a single fixed net whose behavior is equivalent to a whole system.

## I. INTRODUCTION

This paper presents yet another Petri net based formalism for the modeling, analysis and simulation of systems. The objective is to provide designers with a formalism which on one hand has a high expressive power and on the other hand has a strong theoretical basis.

By a high expressive power we intend the possibility to model complex distributed systems made up of a number of entities which have their own internal structure and behavior, and also interact the ones with the others. The state of such entities may be accurately modeled only by means of complex data structures, which are also required to implement their data processing capabilities. The behavior of some kinds of systems -for instance the set of windows of a user interface [Palanque...93] or a Workflow System [Ellis 93]- causes the creation of entities which are dynamically introduced into the system, whereas other entities vanish. In addition, we want a formalism featuring mechanisms to structure the model of a system in a hierarchical or Object-Oriented way, to ease the reuse of components and their successive modifications, and to support incremental design.

By a strong theoretical basis we intend a well defined semantics and, in our case, the possibility to apply results of the Petri nets theory on the analysis of systems. In addition, the formalism must enable an incremental analysis by means of a net algebra which deduces the properties of a system from the properties of its components and the way they are composed.

The main difficulty results from the contradiction between the structural evolution of systems and the rigidity of Petri nets. The system dynamicity requires to consider entities which appear and disappear, which sometimes interact with given entities, sometimes with others. And the rigidity of Petri nets enforces to use predefined components whose structure and relationships are fixed; indeed, although it is possible to make simulation of nets whose structure evolves while they run, it is not possible to analyse such nets and to prove any behavioural property.

Several extensions of Petri nets based on Abstract Data Types have already been proposed. Some of them cope only with the data processing aspect and do not worry about the interactions between components and thus do not provide supports for structuring the model of a system

[Sibertin 85, Vauterin 87, Heuser...91, van Hee...91]; others allow for net composition but without encapsulation of their behavior (nets are related by transition fusion) nor dynamic liking [Battiston...87, Buchs...91] .
Among the extensions of Petri nets based on the Object-Oriented approach, some of them have a semantics which is not well defined or so complex that it is not tractable [Bruno 86, Di Giovanni 91, Sonnenchein 93], and the others do not allow for dynamic interactions between nets [Valette...88, Bachatène...93].
Loose coupling between entities is necessary to support the dynamic binding in the framework of a static structure: if a net needs to be structurally related (by places or transitions) to another net in order to interact with it, either it can interact only with some defined nets or its structure must evolve. Communicative and Cooperative Nets achieve this loose coupling by linking the nets by arcs, which is the exact implementation of communication by message sending. This is the loosest coupling, provided that the tokens received by a net are collected in interface places whose only purpose is to store them. Coupling by arcs respects for the encapsulation of behavior since the enabling of a transition in a net is decided locally from its inner state and the received tokens; on the contrary, a net related to other nets by transition or place fusion can't decide the enabling of its transitions without referring to the structure and the state of its associates. The dynamicity of interactions also requires that the partner of a communication is selected at the time when a transition occurs; thus each entity must be able to refer to other ones by their name, these names being recorded into tokens [Valk 86].

With regard to the expressive power, Communicative and Cooperative Nets support the following features:
• an Object is an entity which has its own name and state, and belongs to an object class defining its structure; this structure is a Communicative (or a Cooperative) Net.
• a token of an Object is a t-uple of items which are either data structures defined according to some Data Types or Object names.
• while a transition occurs, the data structures included in its input tokens may be processed by functions defined upon their types.
• Communicative Objects interact by message sending. To this end, a Communicative Net features accept-places which are spring places where tokens may be put by any entity but the Net; it also includes send-transitions whose occurrences put a token in such a place of the Net of an Object whose name is recorded in an input token.
• Cooperative Objects interact trough a client/ server protocol; as a server, a Cooperative Net features services having an input and a result signature; as a client it includes "invocation patterns" whose occurrences request a given service of a specified Object by sending an input token, and get back a result token when the server has completed the service rendering.
• an Object may cause the dynamic instantiation of another Object.
• there is a true concurrency between Objects: if two transitions of distinct Objects are enabled, they may occur simultaneously.
• the set of Communicative (and Cooperative) Nets is structured according to an inheritance hierarchy allowing the polymorphism: an interaction launched by an Object is typed by an object class, but the actual partner of an occurrence of this interaction may belong to a more specialised class[1].

From a theoretical point of view, Communicative and Cooperative Nets feature the following characteristics:
• their semantics is well defined in terms of the Petri nets theory, so that all results on ordinary Petri nets may be applied to Communicative and Cooperative Nets;
• there exists a net algebra enabling to deduce the properties of a system from the properties of the nets of its components, so that both the incremental design and the incremental analysis are feasible;
• there exists an algorithm which, from the Nets of the initial Objects of a System of Objects, builds a single Synthetic Net which on one hand is static and includes no interaction and on the other hand has a behavior equivalent to the whole System behavior. This algorithm captures the System's dynamicity into an ordinary Object. The Synthetic Net enables to treat a whole System as a single Object, and its analysis is a global way to prove properties about a System.

[1] Let's notice that polymorphism requires dynamic binding, that is the possibility for successive occurrences of an interaction to involve different partners.

The remainder of the paper is organised as follows.
Section II is an informal introduction to Communicative Objects illustrated with a new version of the dining philosophers: they may leave the table or be invited to join it ! This example neglects the data aspect because this aspect is not the focus of the paper, but it highlights the dynamic interactions between nets.
Section III defines Communicative Objects, Communicative Nets and their semantics, and section IV gives a synthesising algorithm which maps a dynamic System of Objects to an equivalent lower-level static net.
Section V provides the definition of Cooperative Objects, which stand at a higher abstraction level and allow for more structured models than Communicative Nets, while section VI gives some indications about their use for system modeling and analysing.

# II. Informal introduction to Communicative Objects

Like in most Object-Oriented formalisms, a Communicative Object is an instance of its object type (or object class). An *object type* has a domain, which is the set of names of all its potential instances, and it is associated with a Communicative Net which is the common structure of its instances. Each *Object* owns two properties: its name which is fixed and is the only one in the set of all object names, and its actual state which is a marking of the Communicative Net of its type.
A *System of Objects* is a set of Objects which refer the ones to the others; starting from an initial state, a System of Objects will evolve either by changes in the state of its Objects when then fire enabled transitions, or by the introduction of new Objects into the System.
The definition of any Communicative Net rests upon a *system of data types* which provides what is required to process data: a set of data types, a domain for each data type, and a set of functions defined on these types.
Each *place* of a Communicative Net is typed by a data type, an object type, or a tuple of them, and it may contain only tokens of this type. Thus a *token* is a tuple of data values and/or Object names. *Arcs* are labelled by tuples of variables which are typed according to the type of places. Some places are declared as *accept-place*, and any net has permission to put tokens in such places. For each transition it may be defined an *action* which is performed at each occurrence; it consists either in calling a function of the system of data type, or in requesting the creation of a new Object to be included in the System of Objects, or in sending a token in an accept-place of an Object of the System.

### An example

As an example, we will consider the very well known table of philosophers, with a slight sophistication: a philosopher may leave the table as he likes it, and introduce a new guest. This example includes no data type, but it illustrates how Objects interact and how new Objects are introduced into a System.
`Fork` is a data type which has only a domain associated to it.
The Communicative Net of the object type Philosopher is shown in Figure 1, where accept-places are indicated by a small double arrow. When he has his Left and Right Forks (in places `LFork` and `RFork`), a philosopher can `starteating` and become `eating`, and when he is eating he can `stopeating` and make his forks to be available again. When the accept-place `GiveLFork` contains a token and the `LeftFork` is available, the transition whose action is `neighb.TakeRFork(f)` may occur: a token made up of the fork bound to the variable `f` is put into the accept-place `TakeRFork` of the philosopher bound to the variable `neighb`. When he has `NoFork`, a philosopher may send the message `GiveRFork` to his left neighbour whose name is stored in the place `LNeighb`; then he waits until his accept-place `TakeLFork` receives a fork (from this neighbour if he behaves as expected). The adventure of a fork at the philosopher's right side is symmetrical, since the behavior of the left side of a philosopher is intended to agree with the behavior of the right side of his left neighbour, and conversely. The accept-places `NewLNeighb` and `NewRNeighb` enable a philosopher to know who are his neighbours.
Since the philosophers may leave and join the table, the forks must be stored in a repository. This is the role of the Object `Heap` whose place `FreeForks` contains the available forks (Figure 2). Any Object may give a fork to `Heap` by means of the accept-place `Take`, and ask `Heap` to send a fork by the place `Give`. Remark that `Heap` is the proper name of a given Object, and not an Object name variable.
A philosopher may quit the table when he has no right fork and a left fork. To this end, he has to give his fork to Heap and to introduce his left neighbour to his right neighbour and conversely his

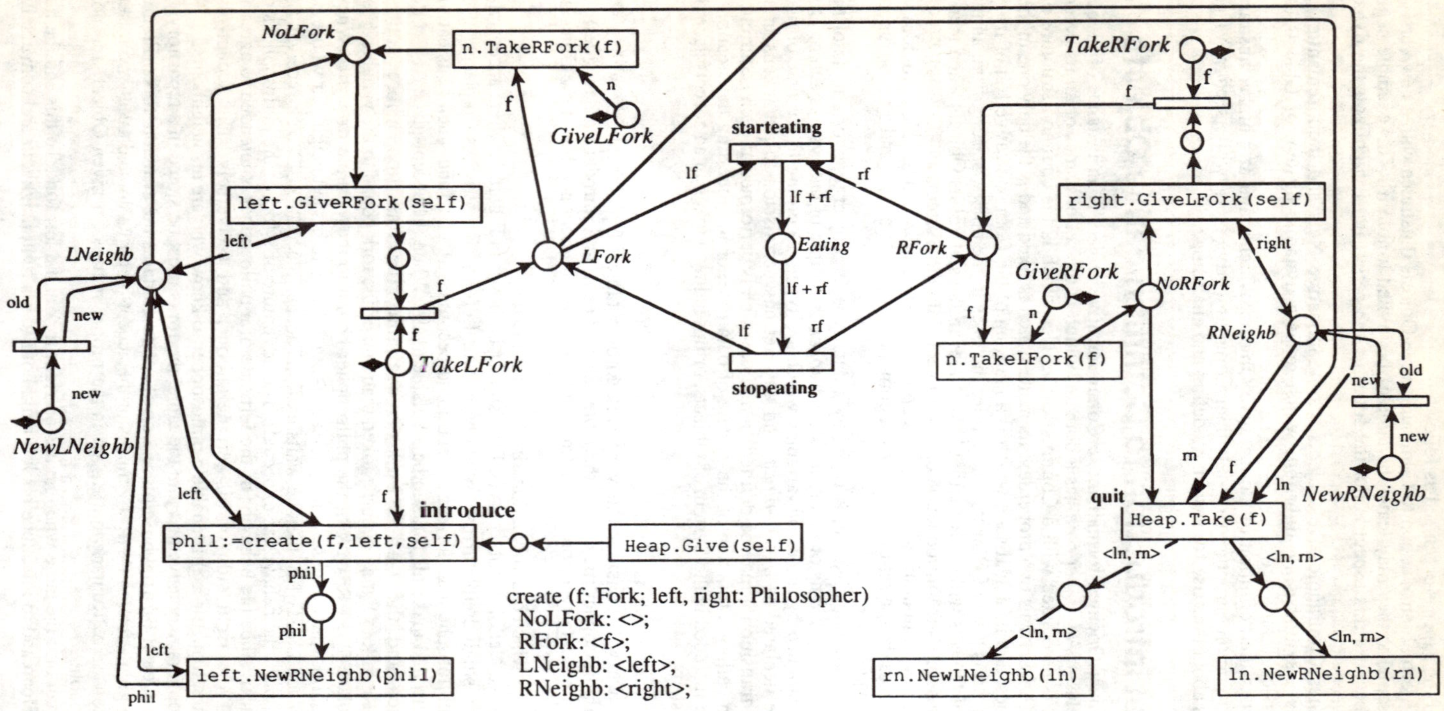


**Figure 1:** The Communicative Net of the object type Philosopher

right neighbour to his left neighbour. Then he is no longer known by any other Object and his state is a dead marking: he can vanish. A philosopher could also leave when he has a fork on his right and no fork on his left, provided that his neighbours continue to share a single fork.
A philosopher may introduce another philosopher on his left when he has no left fork. First he has to ask for a fork to `Heap`; then he requests the creation of a new philosopher which receives as initial marking this fork at his right side, the left neighbour of the introducing philosopher at his left, and the introducing philosopher himself at his right. Finally, it introduces this new guest to his old left neighbour and changes his own left neighbour.
The initial marking of a philosopher is given in Figure 1 by the create primitive. The place `NoLFork` contains one raw token. Places `LNeigh`, `RNeigh` and `RFork` contain a token whose value is set according to the parameters of the *create* primitive. Object creation could be shown by a transition having an input accept-place of type <Fork, Philosopher, Philosopher> and an arc towards each place containing a token at the initial marking. This transition appears in the synthetic Net of Figure 3, but it would needlessly overload the diagram of Figure 1.
The initial state of a table of philosophers is a System of Objects including the heap of forks and at least two philosophers. Philosophers exchange their forks as usually; they introduce new philosophers provided that they can get an extra fork so that the number of guests is bounded by the number of forks; they also may give their fork back and leave the table. The last philosopher would be both his left and right neighbour and he could get a left fork only by giving to himself his right fork and vice versa ...; but this case never happens because the two last philosophers can not leave the table: they have the same Object as their left and right neighbours and the transition `Quit` is not enabled (distinct input variables of a transition must be bound to distinct object names).

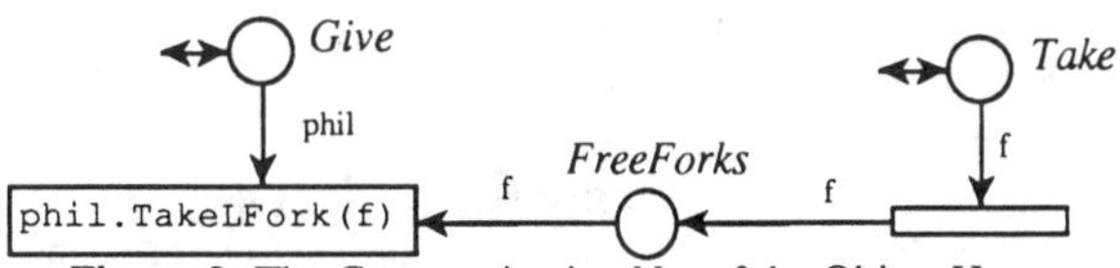


**Figure 2:** The Communicative Net of the Object Heap

# III. Definition of Communicative Objects

First, we give some mathematical notations and definitions.

**Definition 1** **- Tuples**

Let E be a set and **P**(E) its power set.

1. An element of $E^* = \cup_{n \in \mathbb{N}} E^n$ is called a *tuple* and it is noted $x = <e_1, \dots, e_n>$. $<>$ is the unique element of $E^0$.
2. The function *supp** : $E^* \longrightarrow$ **P**(E) is such that $supp^*(<x_1, \dots, x_n>) = \{x_1, \dots, x_n\}$. For any $e \in E$, we say that e *appears* in x, and we note $e \in x$, if $e \in supp^*(x)$.
3. Any order relation $\leq$ in E defines a partial order relation $\leq^*$ in E* such that $<x_1, \dots, x_n> \leq^* <y_1, \dots, y_m>$ iff $n = m$ and $x_i \leq y_i$ for $i = 1 \dots n$.
4. Let F be a set and $f : E \longrightarrow F$ be a function. f is extended to a homomorphism $f : E^* \longrightarrow F^*$ such that $f(<x_1, \dots, x_n>) = <f(x_1), \dots, f(x_n)>$.

**Definition 2** **- Multisets**

Let E be a set.

1. For any function $x : E \longrightarrow \mathbb{N}$, $supp_{\mathbb{N}}(x) = x^{-1}(\mathbb{N} \setminus \{0\}) = \{e \in E;\ x(e) \neq 0\}$ is the support of x. We say that e *appears* in x and note $e \in x$ if $e \in supp_{\mathbb{N}}(x)$.
2. A function $x : E \longrightarrow \mathbb{N}$ having a finite support is called a *multiset on E*, and it is noted $x = \Sigma_{e \in E}\ x(e).e$. We note *N(E)* the set of multisets on E.
3. $\mathbb{N}(E)$ is provided with an addition $x + y = \Sigma_{e \in E} (x(e) + y(e)).e$, a subtraction $x - y = \Sigma_{e \in E} (x(e) - y(e)).e$ and a partial order relation defined by $x \leq y$ iff $\forall\ e \in E$, $x(e) \leq y(e)$.
4. Let F be a set and $f : E \longrightarrow F$ be a function. f is extended into a homomorphism $f : \mathbb{N}(E) \longrightarrow \mathbb{N}(F)$ defined by $f(x) = \Sigma_{e \in E}\ x(e).f(e)$, and according to Definition 1.4 to a homomorphism $f : \mathbb{N}(E^*) \longrightarrow \mathbb{N}(F^*)$.
5. We define *supp* : $\mathbb{N}(E^*) \longrightarrow$ **P**(E) by $supp(x) = \cup_{e^* \in supp_{\mathbb{N}}(x)} supp^*(e^*)$.

## Types, Objects and Systems of Objects

A system of data types is a set of types whose domains are sets of predefined constants, together with a set of functions on these domains. Example of data types are Integer, Boolean, Character, Date or any enumerated domains. The syntax of their definition is out the scope of this paper, and a much more comprehensive definition of systems of data types is given in [Sibertin 92] [2].

A system of types includes a system of data types together with a set of object types having their own domains. Each object type is defined by:

- a domain which is the set of names of its instances,
- a Communicative Net which is the structure of its instances,
- a marking of this net defining the initial value of its instances.

Unlike the elements of data types domains, elements of object types domains may not be processed: they are objects names which allow only to refer to Objects for message sending, Object creation and testing the identity relation between Objects.

**Definition 3** **- System of data types**

A *system of data types* $\Sigma_d$ is a tuple ($C_d$, Dom, DF) defined as follows:

1. $C_d$ is a set whose elements are called data types.
2. For any $t \in C_d$, *Dom (t)* is a set called its *domain*, and for $t \neq t'$, either $Dom(t) \cap Dom(t') = Ø$, or $Dom(t) \subset Dom(t')$; in this latter case we say that *t specialises t'* (or t' *generalises* t).
   The elements of Dom(t) are called *constants* and $CONST = \cup_{t \in C_d} Dom(t)$.
   The function *Type* : CONST —> $C_d$ is defined by:
   $$Type (e) = t \text{ iff } e \in Dom(t) \text{ and } \forall t' \in C_d \, [e \in Dom(t') \Rightarrow t \text{ specialises } t'].$$
3. DF is a set of *data functions* of the kind
   $$f : Dom (Sig_{in} (f)) \longrightarrow Dom (Sig_{out} (f))$$
   where $Sig_{in} (f) \in C_d^*$ and $Sig_{out} (f) \in C_d^*$.

**Definition 4** **- System of object types**

A *system of object types* $\Sigma_o$ is a tuple ($C_o$, Dom, Net, Init) defined as follows:

1. $C_o$ is a set whose elements are called object types.
2. For any $t \in C_o$, *Dom (t)* is a set called the *domain* of t, the elements of Dom(t) are called *object names*, and $NAME = \cup_{t \in C_o} Dom(t)$.
   Like for data types,
   $t \neq t'$ implies $Dom(t) \cap Dom(t') = Ø$ or $Dom(t) \subset Dom(t')$, in which case *t specialises t'*, and the function *Type* : NAME —> $C_o$ is defined by:
   $$Type (e) = t \text{ iff } e \in Dom(t) \text{ and } \forall t' \in C \, [e \in Dom(t') \Rightarrow t \text{ specialise } t'].$$
3. For any $t \in C_o$, *Net (t)* is a Communicative Net (cf. definition 7 below); if t specialises t' then any accept-place of Net(t') is also an accept-place of Net(t) having a specialised type.
4. For any $t \in C_o$, *Init (t)* is a parameterized marking of Net(t), that is a marking whose tokens include constant values and parameters of any data or object type.

**Definition 5** **- System of types**

Let $\Sigma_d = (C_d, Dom_d, DF)$ be a system of data types and $\Sigma_o = (C_o, Dom_o, Net, Init)$ be a system of object types. $\Sigma = (C_o, C_d, Dom, DF, Net, Init)$ is a system of types if $C_d \cap C_o = Ø$, $Dom = Dom_d \cup Dom_o$, and $CONST \cap NAME = Ø$.
Then we define $C = C_d \cup C_o$ and $U$ = CONST $\cup$ NAME the *Universe* of $\Sigma$, so that *Dom* : $C^* \longrightarrow P(U)^*$, *specialise* is a partial order relation on $C^*$, and *Type* : $U^* \longrightarrow C^*$.

The association of a system of data types and a system object types into a system of types is quite loose; this allows for generic systems of types parameterized by the data types.

An Object is a marked Communicative Net identified by a proper name.
Objects are not designed to evolve alone but to interact ones with others inside a System of Objects. The cohesion of a System of Objects is kept by the fact that each member has a different name and only refers to objects which belong to this System of Objects.

---

[2] A system of types is not an Algebra in the sense of Abstract Data Type [Ehrig…85] since data types and functions are provided with a concrete interpretation; using this concept would carry no technical difficulty, but it is not the focus of the paper.

**Definition 6** **- Object, System of Objects**

Let $\Sigma$ be a system of types, and $t \in C_o$ be an object type.

1. $o = (id, m)$ is an *instance of t*, or an *Object* of type t, if $id \in Dom(t)$ and m is a marking of Net(t).
   We define *name(o)* = id, *Type(o)* = Type(id), *Net(o)* = Net (Type (id)) and *ref(o)* = $(\cup_{p \in P}$ supp $(m(p))) \cap$ NAME, where P is the set of places of Net(t).
2. A set of objects $O = \{(id_1, m_1), \ldots, (id_n, m_n)\}$ is called a *System of Objects* on $\Sigma$ if:
   i) $\forall o \in O$, o is an instance of an object type of $\Sigma$;
   ii) $\forall i, j \in \{1, \ldots, n\}$ $[i \neq j \Rightarrow id_i \neq id_j]$;
   iii) $\forall o \in O$ [ref (o) $\subset \{id_1, \ldots, id_n\}$ = *name(O)*].

## Communicative Nets: syntax and semantics

A Communicative Net is defined for a given system of types; its tokens are tuples of constants or object names of types of this system, and its transitions may applied for the data functions. Communicative Nets communicate by message sending. As a receiver a Net is provided with accept-places where other Nets may deposit message tokens. As a sender a Net includes message sending transitions whose each occurrence put a token in such a place of the addressee Net.

**Definition 7** **- Communicative Net**

Let $\Sigma = (C_o, C_d, Dom, DF, Net, Init)$ be a system of types. A *Communicative Net* on $\Sigma$ is a 8-uple N = <P, T, V, Pre, Post, Precond, Action, Accept> defined in the following way.

1. *P* is a finite set of places, provided with a function *Type* : P —> C*.
   *Accept* is a subset of P whose elements, called accept-places, have no input transition: $\forall p \in$ Accept, $[{}^\bullet p = \emptyset]$.
2. *T* is a finite set of transitions.
3. $(V_t)_{t \in T}$ is a family of sets of variables on C indexed by T, and $V = \cup_{t \in T} V_t$.
   The typing function is $Type_t : V_t \longrightarrow C$
   (the subscript will be omited where there is no ambiguity).
4. The forward incidence function *Pre* : P x T —> $\mathbb{N}(V^*)$
   associates to a (place, transition) couple a multiset of variable tuples such that:
   i) Pre (p, t) $\in \mathbb{N}(V_t^*)$;
   ii) $\forall v^* \in$ Pre (p, t) $[Type_t(v^*) = Type(p)]$.
5. The backward incidence function *Post* : P x T —> $\mathbb{N}(V^*)$
   associates to a couple (place, transition) a multiset of variable tuples such that:
   i) Post (p, t) $\in \mathbb{N}(V_t^*)$;
   ii) $\forall v^* \in$ Post (p, t) $[Type_t(v^*)$ specialises Type(p)].
   moreover we must have $V_t = V_{in}(t) \cup V_{out}(t)$, where $V_{in}(t) = \cup_{p \in P}$ supp (Pre (p, t)) and $V_{out}(t) = \cup_{p \in P}$ supp (Post (p, t)) are respectively the set of input variables and the set of output variables of t. We also note $V_{in \backslash out}(t) = V_{in}(t) \setminus V_{out}(t)$, $V_{out \backslash in}(t) = V_{out}(t) \setminus V_{in}(t)$, $V_{in,d}(t) = \{v \in V_{in}(t); Type(v) \in C_d\}$, $V_{in,o}(t) = \{v \in V_{in}(t); Type(v) \in C_o\}$, and the like for $V_{out,d}$, $V_{out,o}$, $V_{in \backslash out,d}$ etc.
6. For each transition $t \in T$, *Precond (t)* is an expression $f(v_1, \ldots, v_h)$,
   where f is a boolean function of DF, $v_i \in V_{in,d}(t)$ for i = 1 … h, and Type $(v_1, \ldots, v_h)$ specializes $Sig_{in}(f)$.
7. For each transition $t \in T$, *Action (t)* is an expresion of one of the three following kinds:
   i) a data *function call*: $\langle r_1, \ldots, r_k \rangle := f(v_1, \ldots, v_h)$
      where f is a function of DF,
      $v_i \in V_{in,d}(t)$ for i = 1 … h and Type $(v_1, \ldots, v_h)$ specializes $Sig_{in}(f)$,
      $V_{out \backslash in}(t) \subset \{r_1, \ldots, r_k\}$, $r_i \neq r_j$ if $i \neq j$, and $Sig_{out}(f)$ specializes Type $(r_1, \ldots, r_k)$;
      if f is the null function (with h = k= 0) it is said that t has no action.
   ii) a *message sending*: v.mes $(v_1, \ldots, v_h)$
      where $v \in V_{in,o}(t)$, mes is an accept-place of Net (Type (v)),
      $v_i \in V_{in}(t) \cup \{self\}$[3] for i = 1 … h, and Type $(v_1, \ldots, v_h)$ specializes Type (mes);

---

[3] `self` is a predefined variable which is allways bound to the name of the object which evaluates the expression in which it occurs; in message sending, the self variable enables objects to pass their identity on.

in this case, we must have $V_{out\backslash in}(t) = \emptyset$ and t is called a *send-transition.*
iii) an *object creation*: v.create $(v_1, \ldots, v_h)$
where $\{v\} = V_{out\backslash in}(t)$ is an object type variable, $v_i \in V_{in}(t)$ for $i = 1 \ldots h$,
and Type$(v_1, \ldots, v_h)$ specializes the parameter of Init(Type(v));
t is then called a *create-transition.*

8. A *marking* of N is a function $M : P \longrightarrow \mathbb{N}(U^*)$, where U is the universe of $\Sigma$,
such that $\forall\, p \in P, \forall\, j \in M(p)$ [Type(j) specializes Type(p)]

The enabling of a transition requires a substitution which binds its input variables tuples to tokens; a constant or object name which is also a variable is always bound to itself, and an object name cannot be bound to two different variables (different variable names correspond to different roles which cannot be mixed unless this fusion is specified by the use of a single variable name; this rule avoids troubles which also occur in programming languages when a procedure call give the same actual value to two formal reference parameters).
The enabling of a transition of the net of an Object in a System relies only on the marking of this Object and not on the state of other Objects. To fire a transition of an Object is a local decision depending just on this Object.

**Definition 8 - Transition enabling**

Let $\Sigma$ be a system of types, N a Communicative Net on $\Sigma$, M a marking of N, and t a transition of N.

1. A *substitution for t* is a function $S : V_{in}(t) \longrightarrow U$
such that $\forall\, v \in V_{in}(t)$ [Type (S(v)) specialises Type(v)], $S(e) = e$ if $e \in U$ and the restriction of S to $V_{in,o}$ is an injection.
2. t is *enabled by S from M*, or (t, S) is enabled from M, which is noted $M \xrightarrow{(t,S)}$ iff:
i) for any place p of N, S (Pre (p, t)) $\leq$ M(p);
ii) S (Precond(t)), the proposition obtained by substituting $S(v_i)$ for $v_i$ in Precond(t), is true.

A transition occurrence transforms a System of Objects into another one.The occurrence of a transition concerns the object which fires this transition, but also the System of Objects it belongs to: the occurrence of a send-transition alters the marking of the addressee Object, and the occurrence of a create-transition introduces a new Object into the System.

**Definition 9 - Transition occurrence**

Let $\Sigma$ be a system of types, O a System of Objects on $\Sigma$, $o = (id, m) \in O$, t a transition of Net(o), and S a substitution such that (t, S) is enabled from m. The occurrence of (t, S) in o produces a new system of objects O', and we note $O \xrightarrow{(o,t,S)} O'$, which is so defined:

1) t has no action:
let m' be the marking of Net(o) defined by m'(p) = m(p) - S (Pre (p, t)) + S (Post (p, t));
then O' is the System of Objects resulting from the substitution of (id, m') for (id, m) in O.
2) the action of t is a data function call $<r_1, \ldots, r_k> := f(v_1, \ldots, v_h)$:
let S' be the substitution for t such that
$S'(v) = S(v)$ if $v \notin \{r_1, \ldots, r_k\}$,
$S'(v)$ = the $i^{th}$ component of $f(S(v_1), \ldots, S(v_h))$ if $v = r_i \in \{r_1, \ldots, r_k\}$,
and m' the marking of Net (o) defined by m'(p) = m(p) - S (Pre (p, t)) + S' (Post (p, t));
then O' is the system of objects resulting from the substitution of (id, m') for (id, m) in O.
3) the action of t is a message sending $v_o$.mes $(v_1, \ldots, v_h)$:
Let m' be the marking of Net(o) defined by m'(p) = m(p) - S (Pre (p,t)) + S (Post(p, t)),
$o_o = (id_o, m_o) \in O$ such that $S(v_o) = id_o$,
and $m_o'$ the marking of Net($o_o$) defined by $m_o'(mes) = m_o(mes) + <S(v_1), \ldots, S(v_h)>$
and $m_o'(p) = m_o(p)$ for the other places;
then O' is the system of objects resulting from the substitution of (id, m') and ($id_o$, $m_o'$) for (id, m) and ($id_o$, $m_o$) in O; In the case where $o = o_o$, $<S(v_1), \ldots, S(v_h)>$ is added to m'(mes) and (id, m') is substituted for (id, m) in O; thus an Object can send a message to itelf.
4) the action of t is an object creation $v_0$.create $(v_1, \ldots, v_h)$:
let S' be the substitution for t such that
$S'(v_0)$ = some element $id_0 \in$ Dom (Type $(v_0)$) \ name(O),
$S'(v) = S(v)$ if $v \neq v_0$,
and m' the marking of Net(o) defined by m'(p) = m(p) - S (Pre (p, t)) + S' (Post (p, t));

then O' is the system of objects resulting from the substitution of (id, m') for (id, m) and the addition of ($id_0$, Init(Type($v_0$))(S ($v_1$), ..., S($v_h$))) in O.

The dynamicity of the structure of a System results from the creation of new Objects while the system is running, the set of Nets included in a System being therefore not fixed. The dynamicity of communications in a System results from the links between Nets: Nets are not acceded by their name but trough references to their name, so that the addressee of the occurrence of a send-transition is determined by the substitution enabling the transition; moreover the polymorphism allows the very class of the addressee to vary according to the inheritance hierarchy.

Let's consider two transitions $t_1$ and $t_2$ of a net N, substitutions $S_1 : V_{in}(t_1) \longrightarrow U$ and $S_2 : V_{in}(t_2) \longrightarrow U$, and M a marking of N such that ($t_1$, $S_1$) and ($t_2$, $S_2$) are both enabled. They are *concurrently enabled* if for any place p of N, $S_1(Pre(p, t_1)) + S_2(Pre(p, t_2)) \leq M(p)$. In this case there is no conflict on the tokens required for their respective occurrence and the occurrence of one of them does not put back into question the enabling of the other; they may occur in the same step, and their *concurrent occurrence* is the natural extension of definition 9.
As a consequence of the locality of transition enabling, enabled transitions of different Objects of a System are always concurrently enabled and the "diamond property" holds: the result of their occurrence is the same, they occur concurrently or in whatever order. Thus the Communicative Objects formalism encapsulates the behavior of each Object; a System of Objects may really be thought as a collection of Objects which evolve concurrently on their own without any global control.

**Theorem 1** **- concurrent enabling of transitions**

Let $\Sigma$ be a system of types, O a System of Objects on $\Sigma$, and for i = 1...n, $o_i = (id_i, m_i) \in O$, $o_i \neq o_j$ if $i \neq j$, $t_i$ a transition of Net($o_i$), and $S_i$ a substitution such that ($t_i$, $S_i$) is enabled from $m_i$.
Then ($t_1$, $S_1$), ..., ($t_n$, $S_n$) are concurrently enabled,
and there exists a System of Objects O' such that for any permutation $\sigma$ of {1, ..., n} we have
$O \xrightarrow{(o_{\sigma(1)},\, t_{\sigma(1)},\, S_{\sigma(1)})} O_{\sigma,1} \xrightarrow{(o_{\sigma(2)},\, t_{\sigma(2)},\, S_{\sigma(2)})} O_{\sigma,2} \cdots\cdots \xrightarrow{(o_{\sigma(n)},\, t_{\sigma(n)},\, S_{\sigma(n)})} O'$.

We omit the proof which is without difficulty. To be more precise, O' is defined except for the name of the Objects created by the occurrences of ($t_1$, $S_1$). ... .($t_n$, $S_n$) since these names are not determined by the semantics of Object creation in Definition 9.4. It could be by ordering the Domains of object types and providing each new Object with the smallest unused name of its type.

# IV. The synthesis of a System of Objects

We will show in this section that a System of Objects is equivalent to a single Object whose transitions actions include only data functions calls and neither message sending nor object creation.
More precisely, let O be a System of Objects; an algorithm is given to build a Communicative Net *SyntNet (O)* and its marking *SyntInit (O)*, such that a transition sequence is enabled in O if and only if it is in SyntNet (O) from SyntInit (O). This algorithm solves the contradiction mentioned in the introduction of the paper by capturing into a single fixed Net all the dynamicity of Communicative Objects. It also provides a way to define the semantics of Communicative Nets which avoids the message sending and object creation transitions: a transition t is enabled in a System O if and only if it is in SyntNet (O) from SyntInit (O)
This algorithm also gives an additional support for both the incremental design and the analysis of systems. Indeed, a System of Objects turns out to have the same nature than its components and to be a Communicative Net. Many facilities for structuring a model result from this recursive property which is a basis of the Structured Design, namely the possibility to treat a sub-system as a single Object. Moreover, a synthetic Net includes no communication, and it may be analysed like any Petri net while disregarding its data structure [Sibertin 85], or as a Pr/T Net [Genrich 86]. The results of this analysis may be interpreted in terms of the components Nets because the synthesis algorithm preserves their structure.

The synthetic Net of a System is built in several steps which first enrich the Nets of the object types taking part in the System, and then gather these Nets into a single one by fusion of places.

### Step 1: determining the involved Nets

The first step of the process consists in defining the set of Nets which must be taken into account to build SyntNet(O). This set includes the Nets of the initial Objects of O plus the Nets of Objects which may be dynamically created during the running of O; it is the transitive closure of the reference relationship between object types, starting from the members of O.

### Step 2: provisions for setting up communications

The purpose of this step is to extend the Nets in order to support the communications established by send- and create-transitions.

1° Each send-transition receives an additional output place, called its *send-place*, to hold the issued tokens. If the transition's action is "$v_0$.mes ($v_1$, ..., $v_h$)", the type of this place is Type($v_1$, ..., $v_h$, $v_0$), and the arc from the transition toward the place is labelled by $<v_1, ..., v_h, v_0>$.

2° For each create-transition whose action is "$v_0$.create($v_1$, ..., $v_h$)", two places are added. The *unbornObject-place* is an input place holding the names of the Objects whose creation will be requested by occurrences of this transition; its type is Type($v_0$) and the arc toward the transition is labelled by $v_0$. The *send-place* is an output place which gets the creation requests issued by the transition; its type is Type($v_1$, ..., vh, $v_0$), and the arc from the transition toward the place is labelled by $<v_1, ..., v_h, v_0>$.

### Step 3: folding the instances of each class into a single net

The purpose of this step is to deal with both the object creation and the dynamic setting of message receivers. For each object type t, a Communicative Net *ClassNet (t)* is built which will operate the creation of all its instances and hold their marking. This ClassNet is obtained by applying the following transformations to the net resulting from step 2.

1° For each transition which has no input place, or whose only input place is an accept-place, an *identity-place* is added; it is both an input and an output place of the transition, its type is <> and its initial marking is one zero-arity token.

2° A *creation transition* is added to the net, together with a *create-place*. This place is the only input place of the creation transition, and its type is the one of the parameter of Init(t). The creation transition has an arc towards each place having a non-empty initial marking (that is such that Init(t) (p) $\neq \emptyset$) and towards each identity-place. These arcs are labelled by tuples of variables in such a way that the transition occurrence results in an initial marking Init(t).

3° The type of the class is appended to the type of each place, except for the send- and unbornObject-places introduced during step 2, but including the identity- and create-places. Thus the place type <> becomes <t>. Note that in consequence, the send- and accept-place of each message sending have the very same type, so that it will be possible to merge them. The *self* variable is accordingly appended to each arc label.

Whereas ClassNet(t) only depends on Net(t) and Init(t), its initial marking depends also on the markings of the initial Objects of O; it will be noted *ClassInit (t, O)*. ClassInit (t, O) is an actual marking, not a parameterized one, which stamps tokens of the marking of each Object of type t with the name of that Object, and then adds the tokens of each place of ClassNet (t).

For a place p of Net(t), let $O_t(p) = \{(id_i, m_i) \in O \; ; \; Type(id_i) = t \text{ and } m_i(p) \neq \emptyset\}$ be the set of Objects of type t having at least one token in place p; then ClassInit (t, O)(p) = $\Sigma_{o\in O_t(p)}$ append ($m_i$(p), $id_i$). If p is a send- or create-place, its marking is empty. The marking of unbornObject-places will be defined in the next step. Finely, the marking of each identity-place is $\Sigma_{id\in name(O) \,\&\, Type(id)=t} <id>$.

### Step 4: the synthetic net and its marking

The synthetic Net results from the fusion of the send- and accept-places of the ClassNets. But a send- place can't be merged with a single accept-place because of polymorphism: when a message sending "$v_0$.mes ($v_1$, ..., $v_h$)" occurs the variable $v_0$ may be bound to any object whose type specialises Type ($v_0$). SyntNet (O) is obtained by the application of the following operations to the ClassNets.

1° In each tree of the inheritance hierarchy, all accept-places having the same name are merged. According to 3 of Definition 4, they have the same type except for the last component which is the object type itself, and for which the most general type is kept. After this operation, all occurrences of a given send- or create-transition put its tokens in the same places and the links between Nets are now static.

2° For each send-transition, its send-place is merged with the accept-place corresponding to the message. The last component of the type of the send-place may be more specialised than the one

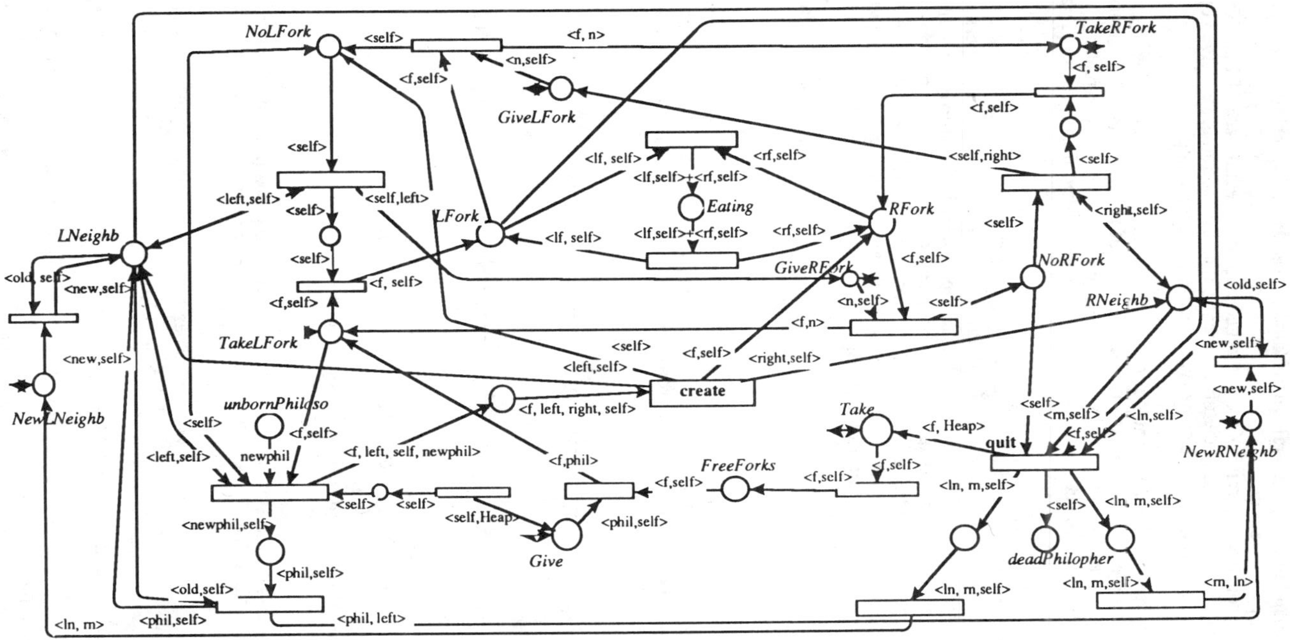


**Figure 3**: the synthetic Net of the table of Philosophers

of the accept-place and the most general type is kept. Moreover, the transition's action (consisting in a message sending) is given up.
3° For each create-transition, its send-place is merged with the create-place of ClassNet (t), where t is the type of the variable supporting the object creation, and its action is also given up.
4° All unbornObject-places having the same type are merged into a single one.

The initial marking *SyntInit* (O) is the one defined by the ClassInit markings. The places resulting from the merging of several places get as marking the sum of the markings of the merged places, while the other places keep their previous marking. Each unbornObject-place p receives as marking the set of unused Object names which have its very type, that is (Dom (Type (p)) \ $\cup_{t \text{ specialises Type(p)}}$ Dom (t)) \ Name (O).

The synthetic Net generated for our philosopher table is shown Figure 3. It features an extra `unbornPhilosophers` place and a `creation` transition for the Philosophers, but not for `Heap` which is a single-copy Object. Although the accept-places have turned into ordinary places, they are shown with their special shape for clarity. The Net also features an extra arc for each send- and create- transition, and a place `deadPhilosophers` has been added. Although it is drawn to be as clear as possible, it is much more complicated than the Nets of Figures 1 and 2; this difference measures the additional expressive power brought by the semantics of Communicative Nets. Analysing this net reveals for instance a place invariant corresponding to the circuit of forks, place invariants for the philosophers which are either unborn, or in activity or dead, and also that this net is live if the places `unbornPhilosophers` and `deadPhilosophers` are fusioned.

The relationships between an Object System and its synthetic net as well as the very nature of synthetic nets is given by the following theorems. Theorem 2 states that any transition sequence enabled in O is enabled in SyntNet(O) from SyntInit(O), and it may be illustrated by the following diagram:

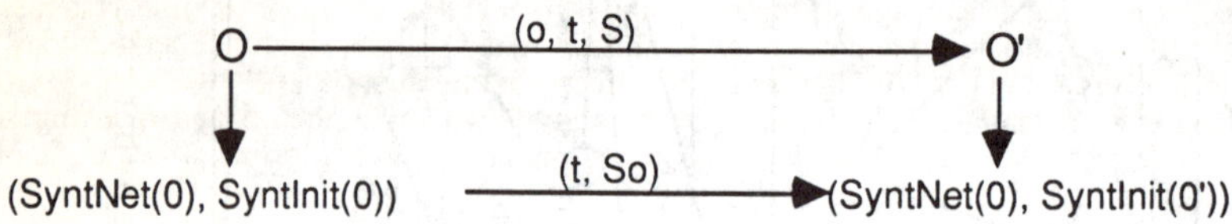


**Theorem 2**

Let O be a System of Objects on a system of types $\Sigma$, o = (id, m) $\in$ O, t a transition of Net(o), S a substitution such that (t, S) is enabled from m, and O $\xrightarrow{(o,\, t,\, S)}$ O'.
1. SyntNet (O) = SyntNet (O');
2. Let $S_o$ be the extension of S such that $S_o$ (self) = id; then in SyntNet (O) we have SyntInit(O) $\xrightarrow{(t,\, S_o)}$ SyntInit (O').[4]

Although the synthesis of a System of Objects is not reversible, we have the following converse of theorem 2 which establishes the behavioural equivalence of O and (SyntNet(O), SyntInit(O)).

**Theorem 3**

Let O be a System of Objects on a system of types $\Sigma$, t a transition of SyntNet(O), S a substitution such that (t, S) is enabled from SyntInit(O), SyntInit(O) $\xrightarrow{(t,\, S)}$ M', and o = (id, m) $\in$ O such that S(self) = id.
Then O $\xrightarrow{(o,\, t,\, S)}$ O', and SyntInit (O') = M'.

These two theorems are proved by reasoning on the synthesis algorithm which provides all transitions with the input variable `self`.
They state to what extent SyntNet(O) is equivalent to O. Any transition sequence enabled in O is enabled from SyntInit, and conversely. Thus O and <SyntNet(O), SyntInit(O)> have the same behavior, and any behavioural property about O may be decided by the study of SyntNet(O) or

---

[4] As for Theorem 1, this equality holds except for the name of the Object created by t since the selection of a token in the unbornObject-places is not deterministic.

<SyntNet(O), SyntInit(O)>. Moreover, SyntNet(O) is a definitive equivalent of O since it is the synthetic net of any Object System O' reachable from O.

SyntNet is a Communicative Net since the transformations made by the above algorithm do not introduce new material. Its accept-places are the accept-places of the initial Nets which receive no message. But it includes neither send- nor create-transition and it can run alone. In fact, it is a data Net, that is a net defined on a system of data types. It sets no communications with other Nets and runs alone since the actions of its transitions are only Data Function calls. The algorithm shows that the semantics of Communicative Objects could be defined with only data types and without the concept of Object. To be more precise, the following theorem result from Definitions 3, 4, 5 and 7:

**Theorem 4**

Let $\Sigma = (C_O, C_S, \text{Dom}, \text{DF}, \text{Net}, \text{Init})$ be a system of types and O a System of Objects on $\Sigma$. Then SyntNet (O) is a Communicative Net on the system of data types $\Sigma' = (C_O \cup C_S, \text{Dom}, \text{DF})$.

# V. Cooperative Objects

We will now introduce the Cooperative Objects formalism which stands at a higher abstraction level and allows for designing more structured models.

When two entities have to communicate, one initiates the interaction and the other accepts (or not) to engage in it. But the completion of most of the interactions requires information to be transmitted both from the initiator toward the target entity and from the target toward the initiator. It is because the initiator pursues some objective while initiating the interaction and, in order to continue its work, it needs to be aware of the work done by the target. For instance, a philosopher sends his neighbour a message `GiveFork` only because it wants to receive a fork. To deal with this fact, Cooperative Objects interact through a client/server protocol. The difference between Communicative and Cooperative Nets is similar to the difference between Actor languages [Agha 86] and Object-Oriented languages [Meyer 92]; the formers allow for more flexible systems, and the latters for more structured and reliable systems designed using a rigorous methodology (for instance [Rumbaugh...91] or [DeChampeaux 91] among many others).

Our philosopher example as Cooperative Objects is shown in Figures 5 and 6, using some graphical conventions which will be explained in section VI. After an informal presentation of the client/ server communication protocol, we will define Cooperative Nets and Objects, and then sketch the synthesis process of a System of Cooperative Objects.

## The client/ server protocol

As a server, a Cooperative Object offers some services at the other Nets' disposal; rendering a service consists in doing some work and returning a value. As a client, a Cooperative Object applies to other Objects for requesting services. The invocation by a client of a service offered by a server encapsulates a complex interaction which is processed according to the four-steps following protocol:

1. the client fires a request-transition, which sends into a place of the server a token which is the calling parameter of the request;
2. the server accepts the request and consumes this parameter token;
3. the server processes the service request and produces into one of its places a token as the result of the request;
4. the client fires a retrieve-transition, which takes this result token.

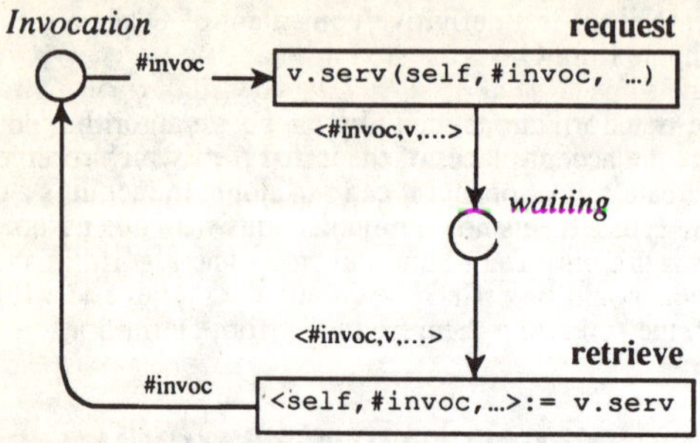


**Figure 4:** Pattern of a safe invocation

### Definition of Cooperative Nets

All the aspects of Communicative Objects concerning data types, object types, Object instances and Systems of Objects (cf. definitions 3, 4, 5 and 6 of section III) remain the same in Cooperative Objects; Communicative Nets are just replaced by Cooperative Net. From Communicative Nets to Cooperative Nets, only the interaction mechanism is changed. It is no longer the message sending paradigm, but the client/ server paradigm, and it requires the following changes:

- accept-places are replaced by a couple of places for each service, one accept-place for receiving the parameter of requests and one return-place for the produced results;
- services are requested by request-transitions, which act as send-transitions;
- retrieve-transitions are needed, whose an occurrence takes a token from a return-place;
- a Cooperative Net inherits from another if it offers the same services, plus some other ones (we will see in the next section that Cooperative Nets support a more comprehensive notion of inheritance).

**Definition 10** **- Cooperative Net**

Let $\Sigma = (C_o, C_d, \text{Dom}, \text{DF}, \text{Net}, \text{Init})$ be a system of types. A *Cooperative Net* on $\Sigma$ is a 8-uple $N = \langle P, T, V, \text{Pre}, \text{Post}, \text{Precond}, \text{Action}, \text{Service} \rangle$ where:

1. P, T, V, Pre, Post and Precond are as in Communicative Nets (cf. definition 6).
2. *Service* is a bijective relation between two disjoint subsets of P, that is there exist $P_a \subset P$ and $P_r \subset P$ such that $P_a \cap P_r = \varnothing$,
   $P_a = \{p_a^s ; s \in \text{Service}\}$ is the set of *accept-places*,
   $P_r = \{p_r^s ; s \in \text{Service}\}$ is the set of *return-places*,
   and $\forall p \in P_a\ [{}^\bullet p = \varnothing]$, $\forall p \in P_r\ [p^\bullet = \varnothing]$.
3. for each service s, there exists at least a minimal place invariant $I^s$ such that $I^s(p_a^s) = I^s(p_r^s) = 1$ and $\forall p \in P\ [[I^s(p) \neq 0 \text{ and } s \neq s'] \Rightarrow I^{s'}(p) = 0]$;
   moreover, for each place p appearing in the support of such an invariant, the two first components of its type are ANY[5] and INTEGER, that is Type(p) = <ANY, INTEGER, ...>, and the variable tuples labeling arcs adjacent to those places have the form <caller, #invocation, ... >.
4. For each transition $t \in T$, *Action (t)* is an expresion of one of the four following kinds:
   i) a data *function call*: $\langle r_1, \ldots, r_k \rangle := f\,(v_1, \ldots, v_h)$,
   ii) an *object creation*: $v.\text{create}\,(v_1, \ldots, v_h)$
   both defined as in Communicative Nets;
   iii) a *service request*: $v.\text{serv}\,(v_1, \ldots, v_h)$
   where $v \in V_{in,o}(t)$, serv is a service of Net(Type(v)),
   $v_i \in V_{in}(t) \cup \{\text{self}\}$ for i = 1 ... h, Type $(v_1, \ldots, v_h)$ specializes Type $(p_a^{serv})$), and $V_{out \setminus in}(t) = \varnothing$;
   such a transition, called a *request-transition,* has the same syntax and semantics than a send-transition of Communicative Nets.
   iv) a *service retrieve*: $\langle r_1, \ldots, r_k \rangle := v.\text{serv}$
   where $v \in V_{in,o}(t)$, serv is a service of Net(Type(v)),
   Type $(r_1, \ldots, r_k)$ generalizes Type $(p_r^{serv})$), $V_{out \setminus in}(t) \subset \{r_1, \ldots, r_k\}$, and $r_i \neq r_j$ if $i \neq j$;
   in this case t is called a *retrieve-transition.*

[5] ANY is the union of the domains of all object types, so that it matches with the type of any Object.

The item 3 of the above definition enables a Cooperative Net not to confuse the requests it is being processing. Indeed a service of an Object may be requested at the same time by several Objects, by several request-transitions of the same Object, or by several occurrences of the same transition, and it is necessary to distinguish the tokens taking part to the processing of these different requests, namely the result tokens. The idea is to avoid a global mechanism, and to identify a request by the name of the calling Object and an invocation number provided by it. According to item 3, this request stamp is recorded into the two first components of parameter tokens and it is transmitted along the places of the service's minimal invariant, so that the result and parameter tokens of a request bear the same stamp. The existence of services' minimal invariants also contributes to enforce that each request causes one and only one result [Sibertin 93].

A client must obey some rules while invoking a service in order for the invocations to occur as intended. First, it must issue request parameters having a right stamp; then it must seek for the result of a service just as it has requested this service and take the only one token intended for it. Obeying these rules results in safe invocations, whose pattern is shown Figure 4. In the following, only Cooperative Nets with safe invocations will be considered.

**Definition 11** **- safe invocations**

Let N and N' be two Cooperative Nets, serv a service of N', and `Invocations` a predefined place of N whose type is INTEGER and whose initial marking is any number of distinct integer values.

A sub-net of N is a *safe invocation* of serv if it is made up of:

1. a request-transition $t_1$ with action v.serv (self, #invocation, $v_1, \dots, v_h$), and Pre (Invocations, $t_1$) = <#invocation>;
2. a retrieve-transition $t_2$ with action <self, #invocation, $r_1, \dots, r_k$> := v.serv, and Post (Invocations, $t_2$) = <#invocation>;
3. a waiting place $p_w$ between $t_1$ and $t_2$ such that $p_w \in t_1{}^\bullet$, ${}^\bullet p_w = \{t_1\}$, ${}^\bullet t_2 = \{p_w\}$ and $p_w{}^\bullet = \{t_2\}$, and Post ($p_w$, $t_1$) = Pre ($p_w$, $t_2$) = <#invocation, v, $w_1, \dots, w_n$>;
4. $t_2$ has no Precondition and $\{r_1, \dots, r_k\} \cap \{w_1, \dots, w_n\} = \varnothing$, so that the only restriction on a token removed from $p_r^{serv}$ by an occurrence of $t_2$ bears on the stamp of this token.

## Transition enabling and occurrence

The semantics of Cooperative Nets rests upon the one of Communicative Nets, so that the enabling and occurrence rules have to be defined only for retrieve-transitions.

The intuitive semantics of a retrieve-transition with action `<r1, ..., rk> := v.serv` is that when it occurs, a token bound to <r1, ..., rk> is removed from the return-place of the service `serv` in the net of the Object bound to `v`.

**Definition 12** **- retrieve-transition enabling**

Let $\Sigma$ be a system of types, O a System of Objects on $\Sigma$, o = (id, m) $\in$ O, t a retrieve-transition of Net(o) whose action is <$r_1, \dots, r_k$> := $v_o$.serv.

1. A *substitution for t* is a function $S : V_{in}(t) \cup \{r_1, \dots, r_k\} \cup \{v_o\} \longrightarrow U$ such that $\forall\, v \in V_{in}(t) \cup \{r_1, \dots, r_k, v_o\}$ [Type (S(v)) specialises Type (v)].
2. Let $o_o = (id_o, m_o) \in O$ such that $S(v_o) = id_o$, and $p_r^{serv}$ be the return place of the service serv of Net($o_o$); then t is *enabled by S from O* iff:
   i) for any place p of Net(o), S (Pre (p, t)) $\leq$ m (p);
   ii) <$S(r_1), \dots, S(r_k)$> $\leq m_o(p_r^{serv})$.

**Definition 13** **- retrieve-transition occurrence**

Let $\Sigma$, O, o = (id, m) and t be as above, and S a substitution such that (t, S) is enabled from O.

The occurrence of (t, S) in o produces a new System of Objects O' which is so defined:

Let m' be the marking of Net(o) defined by m'(p) = m(p) - S (Pre (p, t)) + S (Post (p, t)),
$o_o = (id_o, m_o) \in O$ such that $S(v_o) = id_o$,
and $m_o$' the marking of Net($o_o$) defined by $m_o'(p_r^{ser}) = m_o(p_r^{ser})$ - <$S(r_1), \dots, S(r_k)$>
and $m_o'(p) = m_o(p)$ for the other places;

Then O' is the System of Objects resulting from the substitution of (id, m') and ($id_o$, $m_o$') for (id, m) and ($id_o$, $m_o$) in O.

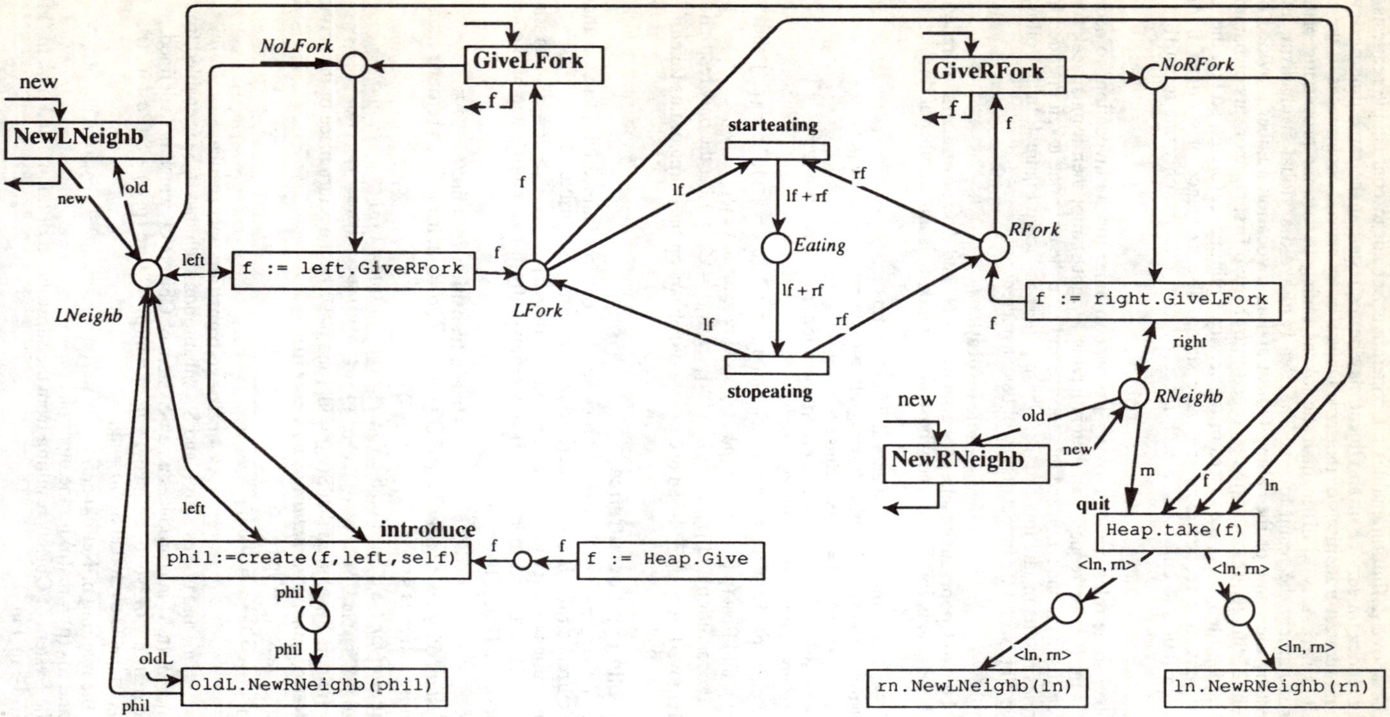


**Figure 5:** The Cooperative Net of the object type Philosopher

One very valuable property of Communicative Nets is the concurrent enabling of transitions in different objects (Theorem 1). This property still holds for Cooperative Nets, although services' return-places may be shared by several retrieve-transitions. In fact, safe invocations warrant the Object encapsulation; there no conflict for result tokens, since the request stamp determines the only Object, retrieve-transition and token staying in its waiting place by which this result may be taken.

**Theorem 5** **- concurrent enabling of transitions**

Let $\Sigma$ be a system of types, O a System of Objects on $\Sigma$, and for $i = 1\ldots n$, $o_i = (id_i, m_i) \in O$, $t_i$ a transition of Net($o_i$), and $S_i$ a substitution such that $(t_i, S_i)$ is enabled from O.
Then $(t_1, S_1), \ldots, (t_n, S_n)$ are concurrently enabled, and the result of their occurrence is the same, they occur concurrently or in whatever order.

There is an alternative way to define Cooperative Nets: a service may be identified by a couple including an accept-place for receiving the request parameters and a return-transition for sending the results to the callers. This would avoid to consider retrieve-transitions: definitions 12 and 13 above become useless and Cooperative Nets become just a special case of Communicative Nets. Then the semantics of Cooperative Nets is simpler, but their syntax is more complicated (in each Net, all invocation patterns which request a given service must share the place where the return-transition put the result). From an Engineering point of view, it is an important drawback and this is the reason why we pick the definition given here. From a theoretical point of view, the two solutions are equivalent since their synthesis process results in the same Net.

## The synthesis of a System of Cooperative Objects

A System of Cooperative Objects may be synthetised into a single data Net, like a System of Communicative Objects does (another version of this algorithm may be found in [Bastide 92]). Since retrieve-transitions are the only extra material it is enough to indicate how they are treated.
At the step 2, each retrieve-transition of the Net of the object type TO receives an additional input place, called a retrieve-place. If the transition action is <self, #invocation, $r_1, \ldots, r_k$> := $v_o$.serv, the type of this place is <TO, INTEGER>.Type($r_1, \ldots, r_k, v_o$), and the arc from this place to the transition is labelled by <self, #invocation, $r_1, \ldots, r_k, v_o$>.
At the step 3, it is useless to append the object type to the type of retrieve-places, since the safe invocation mechanism already ensures its occurrence.
At the fourth step, the accept-places of services are merged according to the inheritance hierarchy and also the return-places. Then send-places are merged with accept-places, and return-places with retrieve-places.

Now we see why the two means to define Cooperative Nets are equivalent: merging a return-place with a retrieve-place is exactly the same operation than merging a send-place with an accept-place.
Cooperative Objects give the opportunity to treat the Object creation like an ordinary service, with an accept-place for the parameters of the initial marking of the new Object and a return-place for its name. In this case, the unbornObject places are no longer in the Net requesting the Object creation and they are gathered into the ClassNet of the created Object, in input of the creation transition. This compels to request an Object creation by a safe invocation, but brings a better cohesion of ClassNets.
The synthetic Net of a System of Cooperative Objects has the same properties than the one of a System of Communicative Objects, and Theorems 2, 3 and 4 still hold.

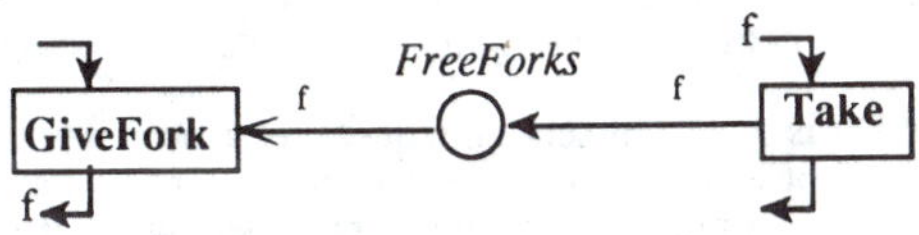


**Figure 6:** the Heap as a Cooperative Net

# VI. Modeling with Cooperative Objects

Cooperative Objects have a wide expressive power enabling to make concise and accurate models of real systems. They also feature theoretical properties which ease both the design process and the validation of models. We will sketch some of their properties.

## The refinement of a System

Communicative and Cooperative Nets both enable to structure the model of a system by means of quite independent components, and thus favour re-use and incremental design.
Cooperative Nets in addition enable to refine a model, and thus allow for hierarchical top-down (and bottom-up) design, by modeling a system at different abstraction levels. A net $N_1$ is refined by introducing a new net $N_2$ and linking them through the client/ server relationship. To this end, it is enough to transform some transitions of $N_1$ into invocation patterns for services offered by $N_2$, so that $N_1$ becomes a client of the server $N_2$. On the other hand, Communicative Nets do not allow for refinement of a net $N_1$ with another net $N_2$; if there are send-transitions only in $N_1$, $N_1$ has no feed-back on the effect of the sent messages and thus it can't control $N_2$, and if there are send-transitions both in $N_1$ toward $N_2$ and in $N_2$ toward $N_1$ it could be very difficult to decide if $N_2$ actually refines $N_1$ or if $N_1$ refines $N_2$.
Moreover, Cooperative Nets bring the possibility to distinguish the specification of a Net -how it behaves from the point of view of its clients- from its actual implementation.

## The analysis of a System

The composition of Petri nets by places fusion or by arcs has been studied (for instance [Baumgarten 88, Vogler 91 or Souissi...90]), and some of the obtained results may be applied to Communicative Nets. But there are fewer results than for nets composed by transitions fusion, or the hypothesises do not match the requirements of real systems modeling [Souissi...90].
Cooperative Nets offer many possibilities, mainly because the client/ server protocol allows for both asynchronous and synchronous communications; indeed, the amount of tokens at the initial marking in the place `Invocations` determines the rate of asynchronism, and a single token corresponds to almost synchronous communications. These possibilities have been investigated within the framework of ordinary Petri nets in [Sibertin 93]. The client/ server protocol is defined in this paper in a static way, that is there is no dynamic instantiation of nets neither dynamic determination of the addressee of an invocation. The gained results may not be applied to the Net of individual Objects, but to the ClassNets which gathers the marking of all Objects of a given class; in this case, the number of Nets to consider is fixed and the links between these Nets are static, since all occurrences of a request-transition put a token in the same place and all occurrences of a retrieve-transition take a token from the same place. We briefly overview these results.

In order for the client/ server protocol properly works, all servers must actually return a result for each accepted request. In this case, the system is *correct* and any token staying in a waiting place may leave after a finite delay. This is a first property which has to be checked. The second point is the availability of a Net Algebra enabling to deduce the properties of a system from the properties of its components, so that the incremental design of a system is supported by its incremental analysis. The theory developed in [Sibertin 93] addresses both questions, and the trends of the results is towards easing the designers' work.
It turns out that a Cooperative Net has four essential characteristics, two as client and two as server:

- the *demand* of a client is the set of service request sequences it may issue,
- the *confidence degree* of a client is the maximum number of tokens which may be in its waiting places, and it measures how tightly it controls the processing of its requests by its servers,
- the *supply* of a server is the set of service request sequences it may accept,
- the *reliability degree* of a server measures to what extent it is able to actually accept the next service requests of a request sequence which belongs to its supply, whatever are its state and the order in which it has accepted previous requests.

The proper use of Cooperative Nets consists in relating a client Net C and a server Net S in such a way that

$$\text{demand}(C) \subset \text{supply}(S) \text{ and } \text{confidence}(C) \leq \text{reliability}(S).$$

In this case, it is shown that the System is correct, and that the following properties may be (recursively) deduced from the ones of its components: the places invariants, the sets of enabled

transition sequences and reachable markings, and the liveness of transitions.
Two servers may be considered as equivalent if they offer the same services and are observationally equivalent for any client. Then two servers are equivalent if and only if they have the same supply and the same reliability. This provides the formal basis to define the behavior inheritance between Nets which actually ensures the net polymorphism.

**Definition 14** **- behavior inheritance**

Let $\Sigma$ be a system of types, N1 and N2 two Cooperative Nets on $\Sigma$ such that all the services of N1 are also offered by N2.
Then N2 inherits the behavior of N1 if and only if:
i) supply (N1) $\subset$ supply (N2),
ii) reliability (N1) $\leq$ reliability (N2).

If N2 inherits the behavior of N1, Then any client may use N2 in the same way as N1.

### The graphical representation of Cooperative Nets

We will now give some rules for simplifying the graphical representation of Coo-Net and get concise diagrams. This is an important point with regard to engineering concern, since the Nets become easier to edit and to understand. These rules are applied in Figures 5 and 6 showing the philosophers and the heap of forks modeled as Cooperative Objects.
The first simplification is to hide the accept and return places of services, and instead to show the transitions whose occurrences start and complete the services' execution. The accept transitions, which are after accept-places and consume their tokens, are drawn with a pending incoming arc labelled by the service parameter, the name of the service being inscribed in bold characters; and the return transitions, which provide their tokens to return-places, are drawn with a pending outgoing arc labelled by the service result. Of course a transition may be both an accept and a return one. This representation highlights that a service is mainly a work to be done, whereas accept and return places only support the token transfer, and it alleviates the diagram by avoiding arcs when a service has several accept transitions.
A second simplification is to hide the request stamps which are included into the parameter tokens of services' requests and propagated along the place invariants of services as far as the result tokens. Indeed, the components ANY and INTEGER may be automatically added to the type of all concerned places, and the variables `caller` and `#invocation` added to the valuation of their surrounding arcs. As a consequence, the references to these stamps may be omitted in the clients' invocation patterns, more precisely in the action of the request and retrieve transitions and in the type of the waiting places.
A third simplification is to hide the `Invocations` place of a client, if its initial marking is infinite or includes so many tokens than it never prevents a request-transition to occur. In this case, showing this place and its adjacent arcs brings no information upon the behavior of the Net.
The last simplification consists in gathering the request-transition, the waiting place and the retrieve-transition of an invocation pattern into a single *invoke-transition* whose action is of the kind

$$<r_1, \ldots, r_k> := v.serv\ (v_1, \ldots, v_h).$$

Indeed, it is possible to recover the invocation pattern from the invoke-transition, namely the arc valuations by keeping all the input variables of the request-transition in Post $(p_w, t_{request})$ and Pre $(p_w, t_{retrieve})$. The justification for this simplification is that Cooperative Nets are intended to be used to design correct systems of objects (just as programming languages are, and we have seen above that the means to do so exist). In this case, no token is infinitely locked in a waiting place; invoke-transitions have the same semantics as transitions, except that some delay may take place between the removal of tokens from the input places and their deposit into the output places. Thus the occurrence of an invoke-transition has to be considered as an activity which requires some amount of time.
Figure 5 has to be compared with Figure 1. It is easier to understand, namely because the causal relationship between requesting and getting a fork is syntactically shown. It may also be analysed as an ordinary Petri net, whereas the presence of accept-places in the Communicative Net of Figure 1 is cumbersome.

# Conclusion

Cooperative Objects have already proved their suitability for modeling different kinds of systems such as manufacturing systems [Bastide...91], information systems [Sibertin 91], user-driven interfaces [Palanque...93] or Workflow systems. They are in fact both a formalism suitable for system modelling at an abstract level and an Object-Oriented programming language. In the implementation which is in progress, they are provided with mechanisms enabling data flow between Objects, that is synchronous communications which do not modify the Object control flows. These mechanisms have not been introduced in this paper because they are outside the scope of the Petri nets semantics, but they are essential to model many systems.

# References


[Agha 86] G. AGHA
An overview of Actor Languages. SIGPLAN Notices, Vo 21, n°10, October 1986.

[Bachatène...93] H. BACHATENE, J-M. COUVREUR
A reference Model for Modular-Colored Petri Nets; Proceedings of the IEEE Intern. Conf. on Systems, Man and Cybernetics, session on Petri Nets and the O-O Approach, Le Touquet France, October 1993.

[Bastide...91] R. BASTIDE, C. SIBERTIN-BLANC.
Modelling flexible manufacturing systems by means of CoOperative Objects. Computer Applications in Production and Engineering CAPE 91, IFIP (G. Doumeingts, J. Browne, M. Tomjanovich Editors, North-Holland); Bordeaux (F), Sept. 1991

[Bastide 92] R. BASTIDE
Cooperative Objects : a formalism for modelling concurrent systems. Thesis of the University Toulouse III, February 1992. (in french).

[Battiston...87] E. BATTISTON, F. DE CINDIO, G. MAURI.
OBJSA net: a class of high-level nets having objects as domains. ATPN 87, LNCS 340, 1988.

[Baumgarten] B. BAUMGARTEN
On internal and external characterization of PT-net building bloc behaviour ; Advances in Petri Nets 88, LNCS 340.

[Bruno...86] G. BRUNO, M. BALSAMO
Petri Net based object-oriented modelling of distributed systems. Proceedings OOPSALA' 86, Sept. 1986.

[Buchs...91] D. BUCHS, N. GERFI.
C0-OPN: a concurrent object oriented approach. ATPN 91, LNCS 524, 1991.

[De Champeaux 91] D. De CHAMPEAUX
Object-Oriented analysis and Top-Down software development. ECOOP'91, LNCS 512, Springer-Verlag, 1991.

[Di Gionani 91] R. DI GIOVANNI
HOOD Nets; ATPN 91, LNCS 524, 1991.

[Ehrig...85] H. EHRIG, B. MAHR
Fundamentals of algebraic specification. Springer Verlag, 1985.

[Ellis...93] C. A. ELLIS, G. J. NUTT
Modeling and Enactment of Workflow Systems. ATPN'93, Chicago (Il.), LNCS 691, 1993.

[Genrich 86] GENRICH H. J.
Predicate/Transition Nets; in Petri Nets : Applications and relationships to other models of concurrency (W. Brauer, W. Reisig, G. Rosenberg editor), LNCS 254, Springer

[Heuser...91] C.A. HEUSER, E. MEIRA PERES.
ER-T Diagrams: An approach to specifying database Transactions. 8th Conf. on Entity-Relationship Approach, 1991.

[Huber...89] HUBER P., JENSEN K., SHAPIRO R. M.
Hierarchies in Coloured Petri nets. APN 1990, LNCS 483, Springer.

[Meyer 92] B. MEYER
Object-Oriented Sogtware Construction, seceond ed. Prentice-Hall, 1993.

[Palanque...93] P. PALANQUE, R. BASTIDE, L. DOURTE, C. SIBERTIN-BLANC
Design of User-Driven Interfaces Using Petri Nets and Objects. Proc. CAISE'93, Conf. on Advanced Information System Engineering, LNCS 685, Paris (F), June 1993.

[Rumbaugh...91] J. RUMBAUGH, M.BLAHA, W. PREMERLANI, F. EDDY, W. LORENSEN
Object-Oriented Modeling and Design Prentice-Hall, 1991.

[Sibertin 85] C. SIBERTIN-BLANC
High level Petri nets with data structure. 6th European workshop on Theory and applications of Petri Nets. Espoo (Finland) june 1985.

[Sibertin 91] C. SIBERTIN-BLANC Cooperative Objects for the conceptual modelling of organizational Information Systems. The Object-Oriented approach in Information Systems. Quebec 28-31 October 1991. Elsevier Science Publisher B.V. © IFIP.

[Sibertin 92] C. SIBERTIN-BLANC
A functional semantics of Petri Nets with Objects. CERISS Internal report, dec. 1992.

[Sibertin 93] C. SIBERTIN-BLANC
A client-server protocol for the Composition of Petri Nets. ATPN'93, LNCS 691, Springer-Verlag, 1993.

[Sonnenschein 93] M. SONNENSCHEIN
An Introduction to GINA. Proceedings of the IEEE Intern. Conf. on Systems, Man and Cybernetics, session on Petri Nets and the O-O Approach, Le Touquet (F), October 1993.

[Souissi 90] Y. SOUISSI
On Liveness Preservation by composition of Nets via a set of places ; 11th nternational Conference on applications and theory of Petri Nets, Paris, June 90.

[Valette...88] R. VALETTE, M. PALUDETTO, B. P. LABREUILLE, P. FARAIL
Object-Oriented approach HOOD and Petri nets for real time system design. Proc. Intern. Conf. Software Engineering & its Applications, Toulouse (F), Dec. 1988.

[Valk 86] R. VALK
Nets in Computer Organization. in Petri nets: Applications and relationships to other models of concurrency. W. Brauer, W. Reisig, G. Rosenberg editor, LNCS 255, Springer Verlag 19986.

[Van Hee...91] K.M. VAN HEE, P.A.C. VERKOULEN
Integration of a Data Model and High-Level Petri Nets. ATPN 91, LNCS 524, 1991.

[Vautherin 87] J. VAUTHERIN
Parallel system specification with coloured Petri Nets and algebraic specification. APN 87, LNCS 266.

[Vogler 91] W. VOGLER
Modular Construction and Partial Order Semantics of Petri Nets; Report Techn. Univ. München, 1991.

# Well-formedness of Equal Conflict Systems *

Enrique Teruel and Manuel Silva

Departamento de Ingeniería Eléctrica e Informática
Centro Politécnico Superior de Ingenieros de la Universidad de Zaragoza
María de Luna 3, E-50015 Zaragoza (Spain)
Tel: + 34 76 517274 Fax: + 34 76 512932 e-mail: eteruel@cc.unizar.es

**Abstract.** The aim of this work is to deepen into the structure theory of Place/Transition net systems. In particular, it considers well-formedness (i.e. structural liveness and boundedness) of Equal Conflict net systems, a class which generalises the well-known (Extended) Free Choice allowing weights but keeping the total autonomy of conflicts. This contribution, together with [18, 19, 20, 21], shows that many results that were known for some subclasses of ordinary Petri nets can be extended to non ordinary nets. For this purpose, some important classical concepts are re-defined, namely components and allocatability.

*Keywords:* Place/Transition net systems; Structure theory; Rank theorem; Well-formedness; Decomposition.

*Topics:* Analysis and synthesis, structure and behaviour of nets; System design and verification using nets.

## 1 Introduction and Related Work

*Structure theory* investigates the relationship between the behaviour of a net system and its *structure*, i.e. the linear algebraic and graph theoretic objects and properties associated to the net and the initial marking. The ultimate goals of structure theory are usually phrased as (see for instance [1, Sect. 3.1]):

- The *analysis* problem: the avoidance of the state space explosion problem, developing analysis methods which do not require the construction of the state space
- The *synthesis* problem: the design of refinement and composition operators which are known to preserve properties of interest.

It is a common trend in structure theory to concentrate on restricted classes of systems and particular properties, in order to obtain powerful results. It has often been the case that the classes under consideration were ordinary (i.e. no multiple arcs), partially due to the intensive use of certain graph theoretic techniques which are not always well suited for weighted nets. Nevertheless, multiple

* This work has been partially supported by the projects CONAI (DGA) P IT-6/91, Esprit BRA Project 7269 (QMIPS), and Esprit W.G. 6067 (CALIBAN).

arcs are convenient to properly model systems with *bulk* services and arrivals, and it is not always possible to implement them by means of ordinary subnets without spoiling the applicability of interesting structural techniques. This posed a challenging question: do the "nice" results obtained for some ordinary subclasses *require* ordinarity? The interest of a negative answer is twofold: both deeper understanding of the causes of these results, and broader applicability of structural techniques would be gained. The present work is part of this undertaking. More precisely, it deals with the class of *Equal Conflict* net systems introduced in [20], which is a weighted generalisation of the renowned *(Extended) Free Choice* subclass [3, 5, 6, 7, 22, ...]. In [20] a conceptually simple algebraic characterisation of liveness, and the existence of home states in live and bounded Equal Conflict systems were presented. We concentrate here on the structural characterisation of well-behavedness, meaning liveness and boundedness, and well-formedness, meaning structural (str.) liveness and boundedness. The obtained characterisation of well-formedness, which generalises the rank theorem for Free Choice nets, has polynomial time complexity and allows to deduce the duality theorem as a corollary. It has also much to do with decomposition issues, but these involve some further developments that are out of the scope of this paper [21], although we announce the principal decomposition result at the conclusion.

The major effort has been the setting of some concepts and results for general Place/Transition net systems. This includes a non-trivial necessary condition for well-formedness, contained in Sect. 3, and the generalisation of two classical concepts: *components* (and *decomposability*) and *allocations* (and *allocatability*), which are contained in Sect. 4 and 5, respectively. Once this is done, it is relatively easy to obtain a simple characterisation of well-formedness for Equal Conflict nets (Sect. 6). The proofs in this work are rather independent of the classical theory, which is generally taken as a starting point, and they are fortunately shorter than the existing ones for the ordinary case, so we believe that they contribute to the understanding of the structure theory of Free Choice models, not to mention that they apply to a more general subclass. Although they do use some graph theoretic arguments, most of the results are stated in linear algebraic terms. This was somehow expected since linear algebraic properties and objects are rather insensible to the weighting.

## 2 Preliminaries and Notation

Although the reader is assumed to be familiar with Petri nets' theory, in this section we recall the basic concepts and introduce the notation to be used. For the sake of readability, whenever a net or system is defined it "inherits" the definition of all the characteristic sets, functions, parameters... with names conveniently marked to identify whose is which. More comprehensive presentations of Petri net concepts are, for instance, [2, 13, 17].

**Place/Transition net and related concepts.** A P/T *net* is a triple $\mathcal{N} = (P, T, W)$ where $P$ and $T$ are disjoint finite sets of *places* and *transitions* ($|P| = n$,

$|T| = m$), and $W: (P \times T) \cup (T \times P) \mapsto \mathbb{N}$ defines the *weighted flow relation*: if $W(u,v) > 0$, then we say that there is an *arc* from $u$ to $v$, with *weight* or *multiplicity* $W(u,v)$. *Ordinary* nets are those where $W: (P \times T) \cup (T \times P) \mapsto \{0,1\}$. Since a P/T net can be seen, and drawn, as a bipartite weighted directed graph, several graph concepts, like paths, circuits, connectedness (without loss of generality, nets are assumed to be connected), strong connectedness, etc. can be extended to nets. In particular, let $v \in P \cup T$; its *preset* and *postset* are given by: ${}^\bullet v = \{u \mid W(u,v) > 0\}$, and $v^\bullet = \{u \mid W(v,u) > 0\}$. The preset (postset) of a set of nodes is the union of presets (postsets) of its elements. A place (transition) $v$ such that $|v^\bullet| > 1$ / $|{}^\bullet v| > 1$ is a *choice* / *attribution (fork / join)*. Any of these constructs is said to be *balanced* when all the corresponding arcs have the same weight. The weighted flow relation can be alternatively defined by: $Pre(p,t) = W(p,t)$, $Post(p,t) = W(t,p)$. These functions can be represented by matrices[2]. If $\mathcal{N}$ is *pure* (i.e. $\forall p \in P, \forall t \in T: Pre[p,t] \cdot Post[p,t] = 0$; without loss of generality, nets are assumed to be pure), then the weighted flow relation is represented by the *incidence matrix* $C = Post - Pre$. By reversing arcs or interchanging places and transitions we get the *reverse*, $\mathcal{N}^r$, or the *dual*, $\mathcal{N}^d$, of $\mathcal{N}$, with incidence matrices $-C$ and $-C^T$, respectively. Both transformations together lead to the *reverse-dual*, $\mathcal{N}^{rd}$, with incidence matrix $C^T$. A net $\mathcal{N}'$ is *subnet* of $\mathcal{N}$ ($\mathcal{N}' \subseteq \mathcal{N}$) iff $P' \subseteq P$, $T' \subseteq T$ and $W'$ is the restriction of $W$ to $P'$ and $T'$. Subnets are *generated by* subsets of nodes of both kinds. A subnet generated by a subset $V$ of nodes of a single kind is assumed to be that generated by $V \cup {}^\bullet V \cup V^\bullet$. Subnets generated by a subset of places (transitions) are called *P-(T-)subnets*, or *open (closed)* subnets [7].

**Place/Transition system and related concepts.** A function $M: P \mapsto \mathbb{N}$ is called *marking*, and can be represented by a vector. A P/T *system* is a pair $(\mathcal{N}, M_0)$ where $\mathcal{N}$ is a P/T net and $M_0$ is the *initial* marking. A transition $t$ is *enabled* at $M$ iff $M \geq Pre[P,t]$. Being enabled, $t$ may *occur* (or *fire*) yielding a new marking $M' = M + C[P,t]$, and this is denoted by $M \xrightarrow{t} M'$. An *occurrence sequence* from $M$ is a sequence $\sigma = t_1 \cdots t_k \cdots \in T^*$ such that $M \xrightarrow{t_1} M_1 \cdots M_{k-1} \xrightarrow{t_k} \cdots$. If the firing of sequence $\sigma$ yields the marking $M'$, this is denoted by $M \xrightarrow{\sigma} M'$. The *firing count vector* of a sequence $\sigma$ is defined as $\vec{\sigma}[t] = \#(t,\sigma)$, where $\#(t,\sigma)$ denotes the number of occurrences of $t$ in $\sigma$, a notation that is extended to sets in the natural way. Therefore, if $M \xrightarrow{\sigma} M'$, then $M' = M + C \cdot \vec{\sigma}$. The set of all the occurrence sequences from $M_0$, the *language*, is denoted by $L(\mathcal{N}, M_0)$, and the set of all the markings reachable from $M_0$, the *reachability set*, is denoted by $R(\mathcal{N}, M_0)$.

---

[2] Places and transitions are supposed to be arbitrarily, but fixedly, ordered. Therefore rows and columns can be indexed by the sets $P$ and $T$. The submatrix of $A$ corresponding to rows in $\pi \subseteq P$ and columns in $\tau \subseteq T$ is denoted by $A[\pi,\tau]$. (Braces are omitted in singletons in this context.) The usual multiplication of scalars, vectors and/or matrices $A$ and $B$ is denoted by $A \cdot B$. The componentwise comparison of $A$ and $B$ is denoted by $A \geq B$, while $A > B$ denotes $A \geq B$ but $A \neq B$.

**Flows and semiflows.** Let $(\mathcal{N}, M_0)$ be a P/T system with incidence matrix $C$. *Flows* (*semiflows*) are integer (natural) annullers of $C$. Right and left annullers are called T- and P-(semi)flows, respectively. A semiflow is called *minimal* when its support[3] is not a proper superset of the support of any other, and the greatest common divisor of its elements is one. Unless explicitly stated, we shall not consider the trivial flow, i.e. vector 0. A couple of straightforward properties of semiflows are:

**Proposition 1.** *Let $\mathcal{N}$ be a P/T net and let $X$ be a T-semiflow of $\mathcal{N}$.*

1. *If $t \in \|X\|$, then for all $p \in t^\bullet$, $p^\bullet \cap \|X\| \neq \emptyset$, and for all $p' \in {}^\bullet t$, ${}^\bullet p' \cap \|X\| \neq \emptyset$*
2. *If $X$ is minimal, then there is no other $X'$ minimal T-semiflow of $\mathcal{N}$ such that $\|X\| = \|X'\|$.*

Flows are important because they induce certain invariant relations which are useful for reasoning on the behaviour. Actually, several structural properties are defined in terms of flows, or similar vectors:

$$\mathcal{N} \text{ is consistent (str. repetitive)} \Leftrightarrow \exists X \geq \mathbb{1} \text{ such that } C \cdot X = (\geq)\, 0$$
$$\mathcal{N} \text{ is conservative (str. bounded)} \Leftrightarrow \exists Y \geq \mathbb{1} \text{ such that } Y \cdot C = (\leq)\, 0,$$

where $\mathbb{1}$ denotes the vector with all entries equal to one.

Regarding flows and subnets, the next property follows immediately (as an illustration, we state both the direct and the "reverse-dual" formulations):

**Proposition 2.** *Let $\mathcal{N}$ be a P/T net.*

1. *Let $\mathcal{N}' \subseteq \mathcal{N}$ be a P-subnet. If $X$ is a T-flow of $\mathcal{N}$, then $X[T']$ is a T-flow of $\mathcal{N}'$. If $Y'$ is a P-flow of $\mathcal{N}'$, then $Y$ such that $Y[P'] = Y'$ and $Y[P - P'] = 0$ is a P-flow of $\mathcal{N}$.*
2. *Let $\mathcal{N}' \subseteq \mathcal{N}$ be a T-subnet. If $Y$ is a P-flow of $\mathcal{N}$, then $Y[P']$ is a P-flow of $\mathcal{N}'$. If $X'$ is a T-flow of $\mathcal{N}'$, then $X$ such that $X[T'] = X'$ and $X[T - T'] = 0$ is a T-flow of $\mathcal{N}$.*

**Boundedness, liveness, and other properties.** A P/T system is *bounded* when every place is bounded, i.e. its token content is less than some bound at every reachable marking, it is *live* when every transition is live, i.e. it can ultimately occur from every reachable marking, and it is *deadlock-free* when at least one transition is enabled at every reachable marking. Boundedness is necessary whenever the system is to be implemented, while liveness is often required, specially in reactive systems. They are so important that the name *well-behaved* has been coined for live and bounded systems. A net $\mathcal{N}$ is *str. bounded* when $(\mathcal{N}, M_0)$ is bounded for *every* $M_0$, and it is *str. live* when *there exists* an $M_0$ such that $(\mathcal{N}, M_0)$ is live. Consequently, if a net $\mathcal{N}$ is str. bounded and str. live there exists some marking $M_0$ such that $(\mathcal{N}, M_0)$ is well-behaved. In

[3] The set $\|X\|$ of the non-zero components of vector $X$.

such case, non well-behavedness is exclusively imputable to the marking, and we say that the net is *well-formed.* (Note that with this definition, in general, well-formedness is *not* necessary for well-behavedness. However, in [5, 22] well-formedness is *defined* as the existence of a live and bounded marking. It will be shown that both definitions coincide in the case of Equal Conflict net systems.) A well-known polynomial time necessary condition for well-formedness, based solely on purely structural properties (i.e. properties that can be defined without any reference to the behaviour) is str. boundedness and str. repetitiveness, which, for convenience sake, will be called *well-structuredness.* Some relations between these concepts are summarised in the following statement:

**Theorem 3** [12, 16]. *Let* $(\mathcal{N}, M_0)$ *be a* P/T *system.*

1. $(\mathcal{N}, M_0)$ *well-behaved* $\Rightarrow$ $\mathcal{N}$ *strongly connected and consistent*
2. $\mathcal{N}$ *well-formed* $\Rightarrow$ $\mathcal{N}$ *well-structured*
3. $\mathcal{N}$ *well-structured* $\Leftrightarrow$ $\mathcal{N}$ *consistent and conservative*
4. $\mathcal{N}$ *well-structured* $\Rightarrow$ $\mathcal{N}$ *strongly connected.*

**Syntactical subclasses.** For the purpose of this work, let us define here some syntactical subclasses of P/T nets, that have already been studied in some detail in [19, 20].

**Definition 4** [19, 20]. Let $\mathcal{N}$ be a P/T net.

1. $\mathcal{N}$ is *Choice-free* (CF) iff $\forall p \in P : |p^\bullet| \leq 1$
2. $\mathcal{N}$ is *Join-free* (JF) iff $\forall t \in T : |^\bullet t| \leq 1$
3. $\mathcal{N}$ is *Equal Conflict* (EC) iff $^\bullet t \cap {}^\bullet t' \neq \emptyset \Rightarrow Pre[P,t] = Pre[P,t']$.

These subclasses generalise the ordinary Marked Graphs, State Machines and (Extended) Free Choice, respectively. Other generalisations are possible, of course, but there are arguments to support the claim that these ones are quite adequate to allow the extension of major results of the ordinary structure theory. As an example, *weighted T-systems* [18], which are a weaker generalisation of Marked Graphs than CF nets, are not adequate to study decomposition issues, as will be illustrated. (We want to mention that CF and JF nets were defined long ago, in [8, 11], and they have been studied, from many and quite diverse points of view, in the literature, receiving several different names. For instance, our CF nets are often named *structurally persistent,* a name due to this property they have that they are persistent for any initial marking. We rather liked having a more syntax-oriented naming, though.) Let us recall here a few of the basic properties of these subclasses that will be used later:

**Theorem 5** [19]. *Let* $\mathcal{N}$ *be a strongly connected* CF *net.*

1. *If* $\mathcal{N}$ *has a T-semiflow then it is consistent*
2. *If* $\mathcal{N}$ *is consistent then it has a unique minimal T-semiflow.*

**Theorem 6.** *Let $\mathcal{N}$ be a conservative* CF *net. For every T-flow $X$ of $\mathcal{N}$, there exists some T-semiflow $X'$ such that $\|X'\| \subseteq \|X\|$.*

*Proof.* Without loss of generality, the flow $X$ can be decomposed as follows: $X = X' - X''$, where $X' > 0$, $X'' \geq 0$, and $\|X'\| \cap \|X''\| = \emptyset$. We claim that $X'$ is indeed a T-semiflow. Since $C \cdot X = 0$, it follows that $C \cdot X' = C \cdot X''$. We prove first that $C \cdot X' \geq 0$. Assume contrary: let $C[p, T] \cdot X' = C[p, T] \cdot X'' < 0$. Since all entries in $X'$ and $X''$ are non-negative and, by the CF assumption, only one entry in row $C[p, T]$, is negative, namely $C[p, p^\bullet]$, it follows that the only element in $p^\bullet$ is in $\|X'\| \cap \|X''\|$, which was assumed to be empty, contradiction. Therefore $C \cdot X' \geq 0$. If $C \cdot X' > 0$, $\mathcal{N}$ would not be conservative against the hypothesis, so necessarily $C \cdot X' = 0$. To see this, apply the *alternatives theorem* [14]:

$$\exists X' \geq 0 : C \cdot X' > 0 \Rightarrow \nexists\, Y \geq \mathbb{1} : Y \cdot C \leq 0$$

□

**Theorem 7** [20]. *A bounded strongly connected* EC *system is live if and only if it is deadlock-free.*

## 3 Conflicts, Arbiters, and Well-formedness

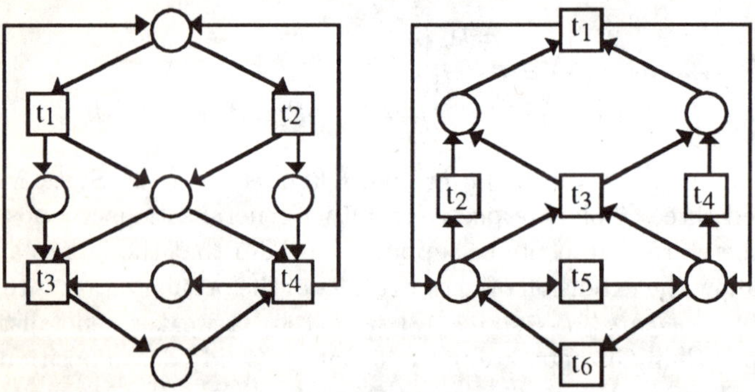


rank(C) = 3 $\qquad\qquad$ rank(C) = 3

$\mathcal{E} = \{\{t_1, t_2\}, \{t_3\}, \{t_4\}\}$ $\qquad$ $\mathcal{E} = \{\{t_1\}, \{t_2, t_5\}, \{t_3\}, \{t_4, t_6\}\}$

**Fig. 1.** Two well-structured but non well-formed nets.

Well-structuredness is necessary for well-formedness, although not sufficient, as shown by the nets in Fig. 1. In [4] a stronger, and also of polynomial time complexity, necessary condition for well-formedness of general P/T nets was introduced. We recall here that important result, together with a revised proof. We require some further concepts:

**Definition 8.** Let $\mathcal{N}$ be a P/T net, and let $t, t' \in T$.

1. $t$ and $t'$, are in *Choice (or Structural Conflict) Relation* iff $t = t'$ or ${}^\bullet t \cap {}^\bullet t' \neq \emptyset$. This relation is not transitive
2. $t$ and $t'$, are in *Coupled Conflict Relation* iff there exist $t_0, \ldots, t_k \in T$ such that $t = t_0$, $t' = t_k$ and, for $1 \leq i \leq k$, $t_{i-1}$ and $t_i$ are in Choice Relation. This is an equivalence relation on the set of transitions, and each equivalence class is a *Coupled Conflict Set*. We denote by $\mathcal{C}$ its quotient set, i.e. the set of Coupled Conflict Sets, and by $\bar{t}$ the equivalence class of $t$. This notation is extended to sets: $\overline{\mathcal{T}} = \bigcup_{t \in \mathcal{T}} \bar{t}$.
3. $t$ and $t'$, are in *Equal Conflict Relation* iff $t = t'$ or $Pre[P, t] = Pre[P, t'] \neq 0$. This is also an equivalence relation on the set of transitions, and each equivalence class is an *Equal Conflict Set*. We denote by $\mathcal{E}$ its quotient set, i.e. the set of Equal Conflict Sets.

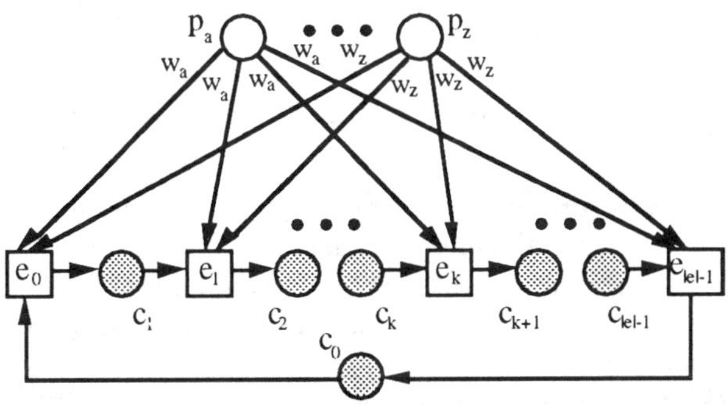

**Fig. 2.** A Circuit Arbiter merged on an Equal Conflict Set.

Choices (places with more than one output transition, these transitions being consequently in Choice Relation) are the "topological construct" making possible the existence of conflicts, because (it is obvious that) two transitions must be in Choice Relation if they are in Conflict Relation at some marking, although the converse is not true. The Coupled Conflict Relation is the transitive closure of the Choice Relation. An Equal Conflict Set is such that whenever any transition belonging to it is enabled, then all of them are. Note that whenever two transitions are in Equal Conflict Relation they are also in Choice Relation, and whenever two transitions are in Choice Relation they are also in Coupled Conflict Relation, but not the other way round, and observe that EC nets are those where $\mathcal{E} = \mathcal{C}$. The transitions in Fig. 2 are in Coupled Conflict Relation. If the shaded places are removed, they are in Equal Conflict Relation.

Some relations analogous to these can be easily defined between places, in a reverse-dual way, that is, exchanging places and transitions and reversing the flow relation in the statements. We do not intend to pursue every single possibility given by reverse-duality. Anyhow, the following concepts will be used: The

*Join Relation* is the reverse-dual of the Choice Relation, i.e. two places are in Join Relation iff they are the same place or they have a common output transition. The *Coupled Synchronisation Relation* is its transitive closure, which is an equivalence relation, where each equivalence class is a *Coupled Synchronisation Set*, and $\mathcal{S}$ denotes the set of Coupled Synchronisation Sets. Observe that the set of input places of a Coupled Conflict Set is a Coupled Synchronisation Set or it is empty, and that the set of output transitions of a Coupled Synchronisation Set is a Coupled Conflict Set or it is empty[4], so in nets without sink places or source transitions, e.g. in strongly connected nets, $|\mathcal{C}| = |\mathcal{S}|$. As an example, the places in Fig. 2 form a Coupled Synchronisation Set.

One possible way of reducing undeterminacy of non-trivial conflicts (those involving more than one transition) in a net system is merging another net system sharing the transitions in conflict (see [10]). Our mere interest on such "arbiters" is to use them in the proof of the necessary condition of the rank theorem, and for this purpose a very particular class is enough:

**Definition 9.** Let $\mathcal{N}$ be a P/T net, and let $e \in \mathcal{E}$ such that $|e| > 1$. A net $\mathcal{A}_e = (P_e, e, W_e)$ is an *(ordinary) Circuit Arbiter* for the Equal Conflict Set $e$ iff $\mathcal{A}_e$ is an ordinary net such that $P_e \cap P = \emptyset$ and its underlying graph is an elementary circuit.

Let us enumerate some straightforward properties of these arbiters: being circuits, they have the same number of places and transitions, i.e. $|e|$; the set of places of a Circuit Arbiter in a net is the support of a minimal P-semiflow; with every non empty initial marking, a Circuit Arbiter is well-behaved and *reversible*, i.e. the initial marking can be reached from whichever other reachable marking. Figure 2 represents a Circuit Arbiter merged on an Equal Conflict Set.

We are ready now to state and prove the main result in this section. It improves the necessary condition for well-formedness given by Th. 3.2 additionally requiring that the rank of the incidence matrix be less than the number of Equal Conflict Sets of the net.

**Theorem 10.** *Let $\mathcal{N}$ be a* P/T *net.*
*If $\mathcal{N}$ is well-formed then it is well-structured and* $\text{rank}(C) < |\mathcal{E}|$.

**Lemma 11.** *Let $\mathcal{N}$ be a* P/T *net, and let $e \in \mathcal{E}$ such that $|e| > 1$. Let $\mathcal{A}_e = (P_e, e, W_e)$ be a Circuit Arbiter for $e$, and let $\mathcal{N}'$ be the net $\mathcal{N}$ merged with the Circuit Arbiter $\mathcal{A}_e$ sharing the transitions in $e$. If $\mathcal{N}$ is well-formed then*

1. *$\mathcal{N}'$ is well-formed*
2. $m - 1 \geq \text{rank}(C') = \text{rank}(C) + |e| - 1$

*Proof of Lemma.* For Part 1, conservativeness of $\mathcal{N}$ and $\mathcal{A}_e$, which are P-subnets of $\mathcal{N}'$, guarantee conservativeness of $\mathcal{N}'$ (by Prop. 2.1). Regarding str. liveness, we will construct a marking $M_0'$ such that $(\mathcal{N}', M_0')$ is live. Let $M_0$ be a marking

[4] In [5] the union of a Coupled Conflict Set and its corresponding Coupled Synchronisation Set is called a *cluster*.

such that $(\mathcal{N}, M_0)$ is well-behaved. Since $(\mathcal{N}, M_0)$ is live and bounded, then the number $r_e = \max\{\min\{\#(e,\sigma) \mid \sigma t \in L(\mathcal{N}, M)\} \mid t \in T, M \in R(\mathcal{N}, M_0)\}$ is well-defined. This is a bound for the number of firings of transitions in $e$ that are *required* to enable an arbitrary transition from an arbitrary reachable marking. We define $M_0{}'$:

$$M_0{}'[p] = \textbf{if } p \in P \textbf{ then } M_0[p] \textbf{ else } (\text{i.e.}\, p \in P_e)\ r_e$$

Let $M' \in R(\mathcal{N}', M_0{}')$ and $t \in T$. We shall prove that $t$ can ultimately be enabled from $M'$. We claim that there exists a marking $M'' \in R(\mathcal{N}', M')$ such that $M''[P_e] = M_0{}'[P_e]$. In that case, since (1) $(\mathcal{N}, M_0)$ is live, (2) $M''[P] \in R(\mathcal{N}, M_0)$, and (3) $M_0{}'[P_e]$ has been defined in a way that it does not interfere when firing a sequence to enable an arbitrary $t$ from an arbitrary reachable marking, then we can fire in $(\mathcal{N}', M'')$ the same sequence that we could fire in $(\mathcal{N}, M''[P])$ in order to enable $t$. To prove the claim, let $\sigma_e = e_{i_1} e_{i_2} \cdots e_{i_k} \in L(\mathcal{A}_e, M'[P_e])$ be such that $M'[P_e] \xrightarrow{\sigma_e} M_0{}'[P_e]$, i.e. a sequence in the Circuit Arbiter returning to the initial marking. It is easy to see that a sequence such that its projection on $e$ is $\sigma_e$ can be fired in $(\mathcal{N}', M')$. The idea is firing transitions not in $e$, which does not affect the marking of places in $P_e$, until $e$ are $P$-enabled (their input places in $P$ have enough tokens, no matter how many tokens are there in other places), which will eventually happen thanks to liveness of $(\mathcal{N}, M'[P])$, then firing $e_{i_1}$ which is also $P_e$-enabled according to our definition of $\sigma_e$, then firing more transitions not in $e$ until $e$ are $P$-enabled again, then firing $e_{i_2}$ which is also $P_e$-enabled, etc.

To prove Part 2, since $\mathcal{N}'$ is well-formed (Part 1) then it is well-structured, so, in particular, it has a right annuler, hence $\text{rank}(C') \leq m' - 1$, where $m' = m$ after the way we defined $\mathcal{N}'$. For $\text{rank}(C') = \text{rank}(C) + |e| - 1$, we shall prove that $|e| - 1$ out of the $|e|$ rows corresponding to the places of the Circuit Arbiter are linearly independent. Let us fix a notation for the Equal Conflict Set and the Circuit Arbiter (see Fig. 2):

- $e = \{e_0, e_1, \ldots, e_k, \ldots, e_{|e|-1}\}$
- $P_e = \{c_0, c_1, \ldots, c_k, \ldots, c_{|e|-1}\}$
- $W(c_i, e_j) =$ **if** $i = j$ **then** $1$ **else** $0$
- $W(e_i, c_j) =$ **if** $j = i \oplus 1$ **then** $1$ **else** $0$, where $\oplus$ is the sum modulo $|e|$.

It is clear that there is one row being a linear combination of the rest, for instance $C[c_0, T] = -\sum_{p \in P_e - \{c_0\}} C[p, T]$, so we remove it and then we prove that the rows corresponding to places in $P_e - \{c_0\}$ are all linearly independent. Assume, on the contrary, that $c_k$, where $1 \leq k \leq |e| - 1$, is a linear combination of the other places (let *the other places* be denoted by $OP = P \cup P_e - \{c_0, c_k\}$):

$$C[c_k, T] = \sum_{p \in OP} \lambda[p] \cdot C[p, T] = \lambda \cdot C[OP, T] \tag{1}$$

Thus, the marking increment produced by a sequence $\sigma$ should also be a linear combination of the marking increment of the other places:

$$\Delta M[c_k] = C[c_k, T] \cdot \vec{\sigma} \overset{by\ (1)}{=} \lambda \cdot C[OP, T] \cdot \vec{\sigma} = \lambda \cdot \Delta M[OP] \tag{2}$$

Let $M_0$ be a marking such that $(\mathcal{N}, M_0)$ is well-behaved. Clearly, it is possible to fire a sequence $\sigma$ such that $\#(e_i, \sigma) = \textbf{if } i < k \textbf{ then } \omega \textbf{ else } 0$, where $\omega$ is arbitrarily large. In that case $\Delta M[c_k] = C[c_k, T] \cdot \vec{\sigma} = \omega$ is arbitrarily large, while all the entries in $\Delta M[OP]$ are finite, what contradicts (2). □

*Proof of Theorem.* Only the rank condition needs to be proven. Let $\mathcal{N}'$ be the net $\mathcal{N}$ together with Circuit Arbiters merged to *every* non-trivial Equal Conflict Set. Applying Lemma 11.2 repeatedly after each Circuit Arbiter is merged, what can be done thanks to Lemma 11.1, it follows that:

$$m - 1 \geq \text{rank}(C') = \text{rank}(C) + \sum_{e \in \mathcal{E}} (|e| - 1)$$

Rearranging the above inequality we obtain a bound for the rank:

$$\text{rank}(C) \leq m - \sum_{e \in \mathcal{E}} (|e| - 1) - 1$$

Since $\sum_{e \in \mathcal{E}} |e| = m$, this bound is $|\mathcal{E}| - 1$, so the result follows. □

Using this better condition we can, for instance, detect non well-formedness of the leftmost net in Fig. 1. We shall show later that the condition is also sufficient in the case of EC nets.

## 4 Components of a Net

T- and P-components have been shown to be important structural objects in the study of (subclasses of) ordinary P/T nets. This section is devoted to the generalisation of their definition to the weighted case, and to the presentation of a few basic results about them. We concentrate on P-components, because reverse-duality leads easily to the corresponding definitions and results regarding T-components.

Ordinary P-components are strongly connected P-graphs (i.e. every transition has one input and one output place) being P-subnets of a given net. In weighted nets, the analogue of P-graph subnets are JF subnets. (We shall discuss later why weighted P-graphs, i.e. underlying net like a P-graph but allowing weights, are paradoxically not suitable to generalise P-graphs in this context.) When a JF subnet is a P-subnet *and conservative*, then it is a P-component:

**Definition 12.** A subnet $\mathcal{N}' \subseteq \mathcal{N}$ is a *Join-free subnet* of $\mathcal{N}$ iff $\mathcal{N}'$ is JF. It is *maximal* iff there does not exist $\mathcal{N}'' \supset \mathcal{N}'$ that is also a JF subnet of $\mathcal{N}$. A strongly connected JF subnet $\mathcal{N}' \subseteq \mathcal{N}$ is a *P-component* of $\mathcal{N}$ iff $\mathcal{N}'$ is a conservative P-subnet of $\mathcal{N}$. A P/T net $\mathcal{N}$ is *P-decomposable* iff it is covered by a set, called *cover*, of its P-components.

(It follows easily that a subnet $\mathcal{N}'$ is a T-component of $\mathcal{N}$ iff $(\mathcal{N}')^{rd}$ is a P-component of $\mathcal{N}^{rd}$, and that $\mathcal{N}$ is T-decomposable iff $\mathcal{N}^{rd}$ is P-decomposable.)

The claim that this generalisation of P-components is adequate, and also the explanation of the necessity of the different requisites, are supported by several facts. Firstly, with the above definition, the components of ordinary nets are the usual ones, because ordinary strongly connected JF (CF) subnets are conservative (consistent) iff they are P-graphs (T-graphs) [11, 19]. The concepts of P- and T-decomposability correspond to State Machine and Marked Graph decomposability in the ordinary case, and the so defined components allow the generalisation of important decomposition results from the theory of ordinary nets (some of them appear in this paper, and some more in [21]). But also, and perhaps more convincingly, the following result says that a P-component is a maximal strongly connected JF subnet, and it has an associated minimal P-semiflow. Therefore, it can be seen as an individual subsystem (it has its own permanent content of tokens: on one hand, they are not mixed with the tokens moving within other subsets of places, because it is a P-subnet; on the other, they — actually their weighted sum — remain constant, because it is conservative), which has no proper synchronisations (it has no joins). It communicates with the rest via shared transitions.

**Theorem 13.** *Let $\mathcal{N}$ be a* P/T *net. If $\mathcal{N}'$ is a P-component of $\mathcal{N}$ then*

1. *$P'$ is the support of a minimal P-semiflow of $\mathcal{N}$ (i.e. P-components have an associated minimal P-semiflow)*
2. *There exists no strongly connected* JF *subnet of $\mathcal{N}$, $\mathcal{N}''$, such that $\mathcal{N}'' \supset \mathcal{N}'$ subnet of $\mathcal{N}$ (i.e. P-components are maximal strongly connected* JF *subnets)*

*Proof.* For Part 1 observe that the set of places of a P-component is the support of a P-semiflow (immediate from Def. 12 and Prop. 2.1). Since strongly connected conservative JF nets have a unique minimal P-semiflow (by the reverse-dual of Th. 5), it is minimal, because otherwise the smaller P-semiflow would be a second P-semiflow of $\mathcal{N}'$ (by Prop. 2.1 again). For Part 2, assume contrary. Take $p_a \in P'' - P'$. By strong connectedness of $\mathcal{N}''$ there exists a path from $p_a$ to any place in $P'$. Let $p_b$ be the first place in such path belonging to $P'$. Two cases are possible: either the input transition of $p_b$ in the path is in $T'$, and then $\mathcal{N}''$ is not JF, or it is not, and then $\mathcal{N}'$ was not a P-subnet. □

In general, the converses of these statements are both false, as the nets in Fig. 3 show. The whole set of places of the leftmost net is the support of a minimal P-semiflow but it does not generate a P-component because the net is not JF; its shaded subnet is a conservative maximal strongly connected JF subnet but is not a P-component because it is not a P-subnet; the rightmost net is a maximal strongly connected JF subnet of itself, but it is a P-component only when conservative, i.e. when $w = 2$. Nevertheless both converses hold in some subclasses [21].

Obviously, if a P/T net $\mathcal{N}$ is P-(T-)decomposable then it is conservative (consistent), although the converse is not true, as shown by the leftmost net

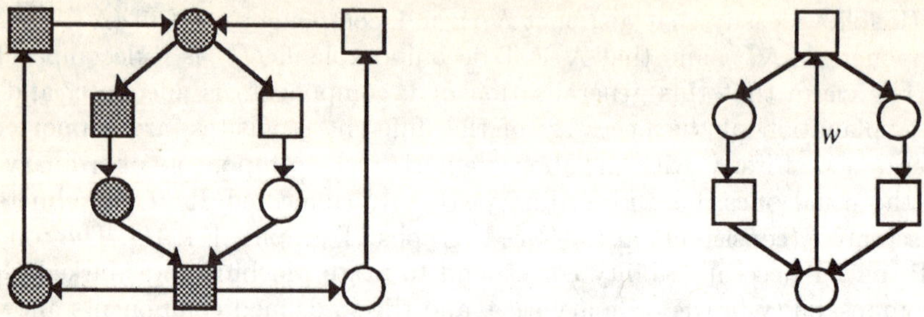


**Fig. 3.** Comparing P-components, JF subnets, and P-semiflows.

in Fig. 3. Nevertheless well-formed EC nets, among others, *are* decomposable (Th. 24 below), although decomposability is *not* enough for well-formedness, as shown by the decomposable but non str. live net in Fig. 4.

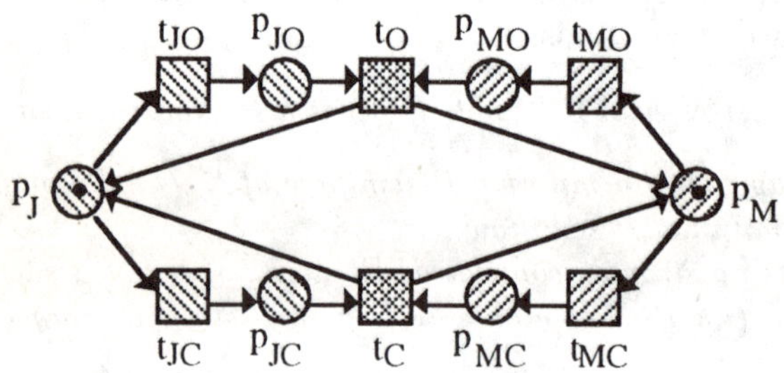


**Fig. 4.** John (the P-component generated by places with a $J$, which represents his activity) and Mary (places with an $M$) are a couple of millionaires who spend their lives going to the cinema. There are two cinemas for millionaires in their city, the Odeon and the Capitol. They both want to decide privately which cinema will they go to, but they must go together according to Society rules. Obviously, they will eventually reach a deadlock, if they happen to choose different cinemas. (Adapted from [7]. The funny interpretation is taken from [6].)

At first sight, it may seem that weighted P-graphs and T-graphs should have played the roles of their ordinary counterparts. Nevertheless, working out some examples shows that it is not so. Consider the rightmost net of Fig. 3, with $w = 2$. Since it is a well-formed EC net (actually it is even CF and JF), one would expect that it was decomposable. But the strongly connected P-graphs of this net, which are its circuits, are neither P-subnets, nor conservative, so they cannot be P-components by any means. Nevertheless our definition of a P-component is fulfilled by the whole net, which preserves its (weighted) content of tokens while it does not have any *proper* synchronisation. (It has *a sort of* synchronisation, due to the weight two which forces to assemble two tokens in front of the transition at the top, but *these tokens belong to the same individual* because they assemble in the same place, and they could even be produced by the same input transition of this place.)

## 5 Allocations over Nets. Allocatable Nets

Allocations were defined for Free Choice nets in [7], and the definition has been extended diversely [5, 22]. We introduce here a more general concept, which naturally coincides with the previous ones when restricted to the corresponding subclasses. We shall concentrate mainly on T-allocations, and the reverse-dual formulation of purely structural concepts and results goes without saying.

**Definition 14.** Let $\mathcal{C}$ be the set of Coupled Conflict Sets of a P/T net, $\mathcal{N}$. A mapping $\alpha : \mathcal{C} \mapsto 2^T$ is a *T-allocation* over $\mathcal{N}$ iff for every $c \in \mathcal{C}$, $\alpha(c)$ is a maximal subset of $c$ such that no pair of transitions are in Choice Relation. The notation is extended to sets: $\alpha(\gamma)$ denotes $\bigcup_{c\in\gamma} \alpha(c)$.

A T-allocation is a function that selects transitions not in Choice Relation out of the transitions in each Coupled Conflict Set. In the case of EC nets, $\alpha(c)$ is always a single transition. Paraphrasing [22], a T-allocation can be interpreted as a "control function". According to this control, whenever (part of) a Coupled Conflict Set $c$ is enabled, then only transitions in $\alpha(c)$ can be chosen to be fired. Thus, unallocated transitions will never be fired. To be precise, our concept of T-allocation generalises the concept of *strict* T-allocation of [22], because the allocated transitions cannot be selected arbitrarily, they must be not in Choice Relation. Therefore, it can be seen as a particular control function that statically solves conflicts. For instance, in the net of Fig. 4 the allocation $\alpha_O(\mathcal{C}) = \{t_{JO}, t_{MO}, t_O, t_C\}$ is interpreted as a control function which forces both John and Mary to choose the Odeon.

In [7], P- and T-allocations for Free Choice nets are called *State Machine* and *Marked Graph Allocations*, respectively. There, a *reduction* algorithm is given to verify whether there are components of the original net in the subnet generated by the image of a given allocation. In case there are such subnets for every allocation, and the union of them covers the net, the net is said to be (*State Machine* or *Marked Graph*) *Allocatable*. In order to re-define allocatability, we base on the following ideas:

- Hack's algorithm starts removing the unallocated nodes and then proceeds stepwise removing certain nodes according to some graph theoretic arguments. In fact, this removal policy can be interpreted in linear algebraic terms, because it is easy to see that the removed nodes are those that cannot be involved in a semiflow (see Prop. 1.1). Therefore, this part of the definition could be re-stated in linear algebraic terms, by asking for existence of semiflows in the subnet generated by allocated nodes. We shall show that these semiflows effectively correspond to components (see Th. 17 below)

– Regarding the requisite that the obtained components cover the net, we rather think that it is a property that *some* allocatable nets enjoy (see Th. 18 below), but it should not be part of the definition.

**Definition 15.** Let $\mathcal{N}$ be a P/T net. $\mathcal{N}$ is *T-allocatable* iff for every T-allocation over $\mathcal{N}$ the T-subnet generated by the allocated nodes has at least a T-semiflow.

(It follows easily that a P/T net $\mathcal{N}$ is P-allocatable iff $\mathcal{N}^{rd}$ is T-allocatable.) Going back to the interpretation of T-allocations as control functions, being T-allocatable can be informally interpreted as being capable of infinite activity whichever control function we apply. This is why we require maximality in the definition of a T-allocation. If it was not required, the net in Fig. 5 would not be T-allocatable (consider the eventual T-allocation image $\{t_1, t_3, t_5\}$), although its interpretation clearly indicates that any static policy for solving the conflict (corresponding to the actual T-allocation images, $\{t_1, t_3, t_4, t_5\}$ or $\{t_1, t_2, t_3\}$) allows infinite activity.

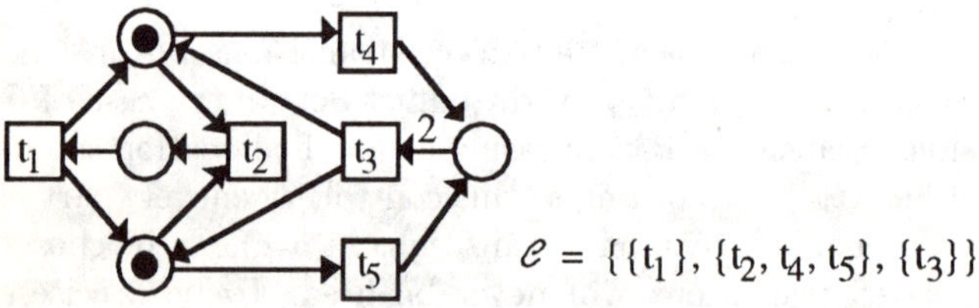


**Fig. 5.** Interpretation of T-allocations and T-allocatability.

Checking allocatability using the definition is complex: for every possible allocation (there are, in general, an exponential number of them wrt. the number of nodes) the existence of a semiflow must be checked (polynomial time problem). Next we describe an efficient sufficient condition (we wonder whether it is also necessary):

**Theorem 16.** *Let* $\mathcal{N}$ *be a conservative* P/T *net. If* $\text{rank}(C) < |\mathcal{C}|$ *then* $\mathcal{N}$ *is T-allocatable.*

*Proof.* Let $\alpha$ be an arbitrary T-allocation, and let $T' = \alpha(\mathcal{C})$ be the set of allocated transitions, which generate a conservative (by Prop. 2.2) CF (by Def. 14) T-subnet of $\mathcal{N}$, $\mathcal{N}'$. Its incidence matrix is $C' = C[P, T']$, so $\text{rank}(C') \leq \text{rank}(C)$. Since we select at least one transition from every Coupled Conflict Set, clearly $m' \geq |\mathcal{C}|$. By assumption, $\text{rank}(C) < |\mathcal{C}|$, hence $\text{rank}(C') < m'$, which means that there exists at least a T-flow of $\mathcal{N}'$, which implies the existence of a T-semiflow (by Th. 6). □

Allocatability is an important structural property. It informs on the structure of the minimal semiflows, and it can be used to study decomposability issues [21]. In first place, we prove that the semiflows of the subnet generated by an allocation effectively correspond to components, which implies that they are minimal, according to Th. 13. Then we show that allocatability together with strong connectedness implies the covering by semiflows in EC nets.

**Theorem 17.** *Let $\mathcal{N}$ be a conservative* P/T *net. If the T-subnet generated by the image of some T-allocation over $\mathcal{N}$ has a T-semiflow, then its support generates a T-component of $\mathcal{N}$.*

*Proof.* After Def. 14, the subnet generated by the allocated transitions, which is a T-subnet, is CF. The support of a T-semiflow in that subnet generates a consistent, and CF, T-subnet of the original net, and being $\mathcal{N}$ conservative, such T-subnet is also conservative (by Prop. 2.2), hence strongly connected (by Th. 3). □

**Theorem 18.** *If $\mathcal{N}$ is a T-allocatable strongly connected* EC *net, then it is consistent.*

*Proof.* Let $t_s \in T$ be an arbitrary transition. We construct $X \in \mathbb{N}^m$ such that $t_s \in \|X\|$ and $C \cdot X = 0$. To do so we find a subset of transitions, $T'$, which is the image of some T-allocation, and the sought $X$ is a (minimal) semiflow of the net $\mathcal{N}'$ generated by $T'$, which exists by the T-allocatability assumption. Such $X$ is also a minimal T-semiflow of $\mathcal{N}$ according to Prop. 2.2.

**begin**
  $T'_0 := \{t_s\}$; $i := 0$
  **while** $\overline{T'_i} \subset T$ **do**
    $T'_{i+1} := T'_i \cup \{t\}$, where $\bar{t} \cap \overline{T'_i} = \emptyset$ (1) and $t^{\bullet\bullet} \cap T'_i \neq \emptyset$ (2)
    $i := i + 1$
  **od**; $T' := T'_i$
**end**

Clearly $T'$ is the image of some T-allocation if the algorithm terminates properly, since we selected one transition per Coupled (actually Equal) Conflict Set.

To prove that $t_s \in \|X\|$, let $t_X \in \|X\|$. By the way transitions are selected at each step, namely by condition (2), there exist paths in $\mathcal{N}'$ from $t_X$ to $t_s$. By condition (1) all the places in these paths have only one output transition in $\mathcal{N}'$, so it follows from Prop. 1.1 that all of the transitions in these paths, in particular $t_s$, must be in $\|X\|$.

To prove proper termination, we show that while $\overline{T'_i} \subset T$ there always exists a transition fulfilling both conditions (1) and (2). Let $t_a \in T - \overline{T'_i}$ be a transition in a Coupled Conflict Set which has not been visited yet. By strong connectedness of $\mathcal{N}$ there exists a path from $t_a$ to some transition in $\overline{T'_i}$. Let $t_b$ be the first transition in that path which belongs to $\overline{T'_i}$, and let $t_c$ be the transition in $\overline{t_b}$ belonging to $T'_i$. Since ${}^\bullet t_b = {}^\bullet t_c$ (because $\mathcal{N}$ is EC), there is a path from $t_a$ to

$t_c$, identical to the path from $t_a$ to $t_b$ except for the last transition. Let $t$ be the only transition in ${}^{\bullet\bullet}t_c$ that is in such a path. Clearly $t^{\bullet\bullet} \cap T_i' \neq \emptyset$, because $t_c \in t^{\bullet\bullet} \cap T_i'$, and $\bar{t} \cap \overline{T_i'} = \emptyset$, because otherwise $t_b$ would not have been the first transition in the path belonging to $\overline{T_i'}$. □

Allocatability has also certain importance regarding the behaviour: in the net of Fig. 4, the T-subnet generated by $\alpha_O(\mathcal{C}) = \{t_{JO}, t_{MO}, t_O, t_C\}$ has a T-semiflow, corresponding to the T-component generated by $\{t_{JO}, t_{MO}, t_O\}$ (both John and Mary decide going to the Odeon, and they do), but the T-subnet generated by $\alpha_\dagger(\mathcal{C}) = \{t_{JO}, t_{MC}, t_O, t_C\}$ does not have any, so the net is not T-allocatable. In case $\mathcal{N}$ is an EC net, and assuming boundedness, this fact indicates the impossibility of infinite activity if conflicts are solved according to $\alpha_\dagger$. Since "choices are free", hence *any* conflict resolution is possible, this intuitively implies non str. liveness. This will be formalised later, in Th. 20. Another important consequence of being allocatable is that liveness of every P-component guarantees liveness of the whole system, that we state together with the property that the P-components of a live system are also live:

**Theorem 19.** *Let $(\mathcal{N}, M_0)$ be an EC system.*

1. *If $(\mathcal{N}, M_0)$ is live, then the system $(\mathcal{N}', M_0[P'])$ is live for every $\mathcal{N}'$ P-component of $\mathcal{N}$*
2. *If $\mathcal{N}$ is strongly connected and P-allocatable, and the system $(\mathcal{N}', M_0[P'])$ is live for every $\mathcal{N}'$ P-component of $\mathcal{N}$, then $(\mathcal{N}, M_0)$ is live.*

*Proof.* For Part 1, assume $(\mathcal{N}', M_0[P'])$ is not live. Then, it can reach a deadlock marking $M_d'$ (by Th. 7), because it is strongly connected and bounded (by Def. 12). Let $\vec{\sigma}' \in \mathbb{N}^{|T'|}$ be a firing count vector with minimum "length" $\sum_{t \in T'} \vec{\sigma}'[t]$ (it could even be zero) among those for which there exists $M \in R(\mathcal{N}, M_0)$ such that $M_d' = M[P'] + C[P', T'] \cdot \vec{\sigma}'$.

Define $\mu = \{M \in R(\mathcal{N}, M_0) \mid M_d' = M[P'] + C[P', T'] \cdot \vec{\sigma}'\}$, which is not empty according to our definiton of $\vec{\sigma}'$. A marking $M_a \in \mu$ does not enable any transition in $\overline{\|\vec{\sigma}'\|}$, because otherwise a transition in $\|\vec{\sigma}'\|$ could be fired too, by the EC property, yielding a marking contradicting minimality of $\sum_{t \in T'} \vec{\sigma}'[t]$. The marking $M_a$ does not enable either any other transition in $T'$, because otherwise this would also be enabled at $M_d'$ since the "potential sequence" $\sigma'$ does not take tokens from its input places. Since the firing of a transition not in $T'$ yields trivially another marking in $\mu$, the transitions in $T'$ are all dead in $(\mathcal{N}, M_a)$, hence $(\mathcal{N}, M_0)$ was not live, contradiction.

For Part 2, strongly connected and P-allocatable imply conservative (by the reverse-dual of Th. 18). Therefore $(\mathcal{N}, M_0)$ is strongly connected and bounded, so if $(\mathcal{N}, M_0)$ is not live, then it deadlocks (by Th. 7). Let $M_d$ be a deadlock marking, such that $M_0 \xrightarrow{\sigma_d} M_d$. Define the following P-allocation: $\forall s \in \mathcal{S} : \alpha_d(s) = \{p\} \subseteq s$ such that $M_d[p] < W(p, t)$, where $t$ in $s^\bullet$ (that is, places preventing their output transitions from being enabled are taken). Any P-component $\mathcal{N}'$ of $\mathcal{N}$ being a P-subnet of the P-subnet generated by $\alpha_d(\mathcal{S})$ is dead at $M_d[P']$, and obviously

$M_0[P']\stackrel{\sigma_d\downarrow_{T'}}{\longrightarrow}M_d[P']$, where $\sigma_d \downarrow_{T'}$ denotes the subsequence of $\sigma_d$ obtained removing all the transitions not in $T'$. Since $\mathcal{N}$ is P-allocatable there is at least one such P-component, so we are done. □

# 6 The Rank Theorem for Equal Conflict Nets, and Some Consequences

A remarkable property of (Extended) Free Choice nets is the existence of structural characterisations of well-formedness, some of which happen to have polynomial time complexity. As a consequence of the above results, one of these polynomial time characterisations can be proven valid for EC nets, generalising [3, Th. 5.4], [6, Th. 2.3], and [5, Th. 9]. This is known as the *rank theorem.* The importance of this result has been argued frequently in the recent Free Choice literature. A proof for Extended Free Choice nets was published by Desel in [5], using a couple of well-known results from the classical Free Choice theory, namely Commoner's and the decomposition theorems [7]. Our proof for EC systems is independent of the classical theory, and all the results that we require have been presented and proven in this paper.

**Theorem 20.** *Let $\mathcal{N}$ be an* EC *net. The following are equivalent:*

1. *$\mathcal{N}$ is well-formed*
2. *$\mathcal{N}$ is well-structured and* $\text{rank}(C) = |\mathcal{C}| - 1$
3. *$\mathcal{N}$ is strongly connected and P-allocatable*
4. *$\mathcal{N}$ is strongly connected and T-allocatable.*

*Proof.* That (1) implies (2) is a particular case of Th. 10, taking into account that in EC nets $\mathcal{E} = \mathcal{C}$, and also that a well-formed EC net where Circuit Arbiters have been merged to all Equal Conflict Sets *has a unique minimal T-semiflow.* For existence of this T-semiflow, notice that the "arbitered" net is consistent. For unicity, let $X$ be a T-semiflow of the arbitered net, and let $t \in \|X\|$. All the transitions in $\bar{t}$ are also in $\|X\|$, according to Prop. 1.1, because the places in the Circuit Arbiters have only one output transition. Every output place of the transitions in $\bar{t}$ has at least one output transition in $\|X\|$ according to Prop. 1.1 again, so we can apply repeatedly the same argument and, by strong connectedness, all the transitions are shown to be in $\|X\|$. Therefore, as far as all the T-semiflows have the same support, only one of them can be minimal, according to Prop. 1.2.

That (2) implies (3) and (4) is Th. 3.4 and Th. 16.

For (3) implies (1), strongly connected and P-allocatable imply conservative (reverse-dual of Th. 18), so only str. liveness needs to be proven. According to Th. 19.2, if every P-component is made live the whole net will be live. It is not difficult to see that the following marking is enough for this purpose: $M_0[p] = W(p,t)$, where $t \in p^\bullet$.

For (4) implies (3), if $\mathcal{N}$ is strongly connected and T-allocatable then $\mathcal{N}^{rd}$ is strongly connected and P-allocatable. Therefore, given that (3) implies (2),

$\mathcal{N}^{rd}$ is well-structured and $\text{rank}(C^T) = |C^{rd}| - 1$ . Since $\text{rank}(C) = \text{rank}(C^T)$ and $|S| = |C^{rd}|$, we have $\mathcal{N}$ well-structured and $\text{rank}(C) = |S| - 1$, hence $\mathcal{N}$ is P-allocatable (reverse-dual of Th. 16). □

The above formulation of the rank theorem deals with well-formedness. Next we show that, in the case of EC net systems, well-formedness is necessary for well-behavedness, that is, *well-formedness characterises the possibility of marking lively and boundedly a given* EC *net.* This makes it specially interesting having such an efficient characterisation of well-formedness as the rank theorem.

**Theorem 21.** *If* $(\mathcal{N}, M_0)$ *is a well-behaved* EC *system then* $\mathcal{N}$ *is well-formed.*

*Proof.* If $\mathcal{N}$ is not well-formed then either it is not strongly connected or it is not T-allocatable (Th. 20). If it is not strongly connected it is not well-behaved by Th. 3.1, so we concentrate on the case it is strongly connected but not T-allocatable. There exists a T-allocation such that the T-subnet generated by the T-allocated transitions has no T-semiflow. Therefore, if conflicts are solved according to such allocation, it is impossible to reach the same marking twice. If $(\mathcal{N}, M_0)$ is live this fact implies unboundedness, while if $(\mathcal{N}, M_0)$ is bounded it implies non liveness. □

Combining Th. 19, Th. 20, and Th. 21 we obtain the following structural characterisation of well-behavedness, generalising [6, Th. 2.6] and [5, Th. 25]:

**Theorem 22.** *Let* $(\mathcal{N}, M_0)$ *be an* EC *system.* $(\mathcal{N}, M_0)$ *is well-behaved iff* $\mathcal{N}$ *is well-structured,* $\text{rank}(C) = |C| - 1$, *and for every P-component* $\mathcal{N}'$ *of* $\mathcal{N}$ *the system* $(\mathcal{N}', M_0[P'])$ *is live.*

*Proof.* After Th. 21, an EC system is well-behaved iff the net is well-formed, which is characterised by Th. 20, and the marking is "adequate", which is characterised by Th. 19. □

In the case of ordinary EC (i.e. Extended Free Choice) systems the condition *every P-component is live* is reduced to *no P-component is unmarked*, what can be verified in polynomial time ($\nexists\, Y > 0 : Y \cdot C = 0 \wedge Y \cdot M_0 = 0$).

The well-known *duality theorem* of [7, part of Th. 9] appears as a corollary after the rank theorem. Obviously, EC nets are not closed under the reverse-dual transformation, since joins need not be balanced, while choices do. It is also obvious, however, that the subclass of EC nets such that their joins are balanced (e.g. the Extended Free Choice subclass) is self-reverse-dual. Every EC system is easily implementable by means of a *balanced-joins* EC system (see [21]).

**Corollary 23 (The Duality Theorem).**
*$\mathcal{N}$ is a well-formed balanced-joins* EC *net iff* $\mathcal{N}^{rd}$ *is.*

*Proof.* Follows easily from Th. 20. Observe that $\mathcal{N}$ is well-structured iff $\mathcal{N}^{rd}$ is, that $\text{rank}(C^T) = \text{rank}(C)$, where $C^T$ is the incidence matrix of $\mathcal{N}^{rd}$, that $|C^{rd}| = |S|$, and that $|S| = |C|$ in strongly connected nets. □

## 7 Conclusion

The aim of this work was to deepen into the structure theory of P/T net systems. In particular, some topics concerning well-formedness and well-behavedness have been dealt with. This contribution, together with [18, 19, 20, 21], shows that many results that were known for some subclasses of ordinary P/T net systems can be extended to non ordinary nets.

To do so, important concepts had to be re-defined because they were well suited for ordinary nets only. This was the case of *components* and *allocations*, introduced in [7], that have been considered in Sect. 4 and 5, respectively. Also the proofs had to be written almost from scratch, because the existing ones were heavily based upon ordinary concepts, namely *siphons* and *traps*. As expected, in the new material, linear algebraic arguments and formulations are conspicuous, since they are rather insensible to the weighting.

The major results contained in this paper are:

- Polynomial time characterisation of well-formedness for Equal Conflict nets (Th. 20). This generalises [3, Th. 5.4], [6, Th. 2.3] and [5, Th. 9]. Also we want to highlight that the necessary part is valid for general Place/Transition nets (Th. 10)
- Structural characterisation of well-behavedness (Th. 22). This generalises [6, Th. 2.6] and [5, Th. 25], based on a compositional liveness characterisation (Th. 19)
- Duality theorem for Equal Conflict nets (Cor. 23), which generalises [7, part of Th. 9], and which is deduced as a corollary of the rank theorem.

Our intention was not to offer an alternative proof of the rank theorem, not even a mere generalisation, but to lay some foundations to build the structure theory of Equal Conflict systems. (In fact, a complete proof of the rank theorem, apart from a couple of basic results which have been simply stated in Sect. 2, can be built out of the following theorems: Th. 6, Th. 10, Th. 16, Th. 18, Th. 19.2, and Th. 20.) In particular, with some additional effort, decomposition results for well-formed Equal Conflict nets can be obtained, the most relevant being:

**Theorem 24 (The Decomposition Theorem)** [21].
*If $\mathcal{N}$ is a well-formed* EC *net, then it is T- and P-decomposable,*

which generalises [7, part of Th. 9].

These results, together with the equivalence of liveness and deadlock-freeness in bounded strongly connected Equal Conflict systems — which leads to a simple algebraic characterisation of liveness for a given $M_0$ — and with the existence of home states in well-behaved Equal Conflict systems [20], support the claim that not only Equal Conflict net systems are an adequate generalisation of Free Choice net systems, but also that many of the nice properties of the latter are rooted on the former. Moreover, we hope that our study smooths out the way to the solution of other open problems, mainly synthesis ones and the extension to larger subclasses.

## References


1. E. Best, editor. *Final Report of Esprit BRA 3148 (DEMON)*, Feb. 1993. GMD-Studien Nr.217.
2. G. W. BRAMS. *Réseaux de Petri: Théorie et Pratique.* Masson, 1983.
3. J. Campos, G. Chiola, and M. Silva. Properties and performance bounds for closed Free Choice synchronized monoclass queueing networks. *IEEE Trans. on Automatic Control*, 36(12):1368–1382, 1991.
4. J. M. Colom, J. Campos, and M. Silva. On liveness analysis through linear algebraic techniques. In *Procs. of the AGM of Esprit BRA 3148 (DEMON)*, 1990.
5. J. Desel. A proof of the rank theorem for Extended Free Choice nets. In Jensen [9], pages 134–153.
6. J. Esparza and M. Silva. On the analysis and synthesis of Free Choice systems. In Rozenberg [15], pages 243–286.
7. M. H. T. Hack. Analysis of production schemata by Petri nets. Master's thesis, MIT, 1972. (Corrections in *Computation Structures Note* 17, 1974).
8. A. W. Holt and F. Commoner. Events and conditions. *Appl. Data Research*, 1970.
9. K. Jensen, editor. *Application and Theory of Petri Nets 1992*, volume 616 of *Lecture Notes in Computer Science.* Springer Verlag, 1992.
10. W. E. Kluge and K. Lautenbach. The orderly resolution of memory access conflicts among competing channel processes. *IEEE Trans. on Computers*, 31(3):194–207, 1982.
11. Y. E. Lien. Termination properties of generalized Petri nets. *Siam Journal on Computing*, 5(2):251–265, 1976.
12. G. Memmi and G. Roucairol. Linear algebra in net theory. In W. Brauer, editor, *Net Theory and Applications*, volume 84 of *Lecture Notes in Computer Science*, pages 213–223. Springer Verlag, 1979.
13. T. Murata. Petri nets: Properties, analysis and applications. *Proceedings of the IEEE*, 77(4):541–580, 1989.
14. K. G. Murty. *Linear Programming.* Wiley and Sons, 1983.
15. G. Rozenberg, editor. *Advances in Petri Nets 1990*, volume 483 of *Lecture Notes in Computer Science.* Springer Verlag, 1991.
16. M. W. Shields. *An Introduction to Automata Theory.* Blackwell Scientific Publications, 1987.
17. M. Silva. *Las Redes de Petri: en la Automática y la Informática.* AC, 1985.
18. E. Teruel, P. Chrzastowski, J. M. Colom, and M. Silva. On Weighted T-systems. In Jensen [9], pages 348–367.
19. E. Teruel, J. M. Colom, and M. Silva. Modelling and analysis of deteministic concurrent systems with bulk services and arrivals. In M. Cosnard and R. Puigjaner, editors, *Decentralized and Distributed Systems*, pages 213–224. IFIP Transactions A-39, Elsevier, 1994.
20. E. Teruel and M. Silva. Liveness and home states in Equal Conflict systems. In M. Ajmone Marsan, editor, *Application and Theory of Petri Nets 1993*, volume 691 of *Lecture Notes in Computer Science*, pages 415–432. Springer Verlag, 1993.
21. E. Teruel and M. Silva. Structure theory of Equal Conflict systems. Research Report GISI-RR-93-22, DIEI. Univ. Zaragoza, Nov. 1993. (Revised in Mar. 1994, 30 pages).
22. P. S. Thiagarajan and K. Voss. A fresh look at Free Choice nets. *Information and Control*, 61(2):85–113, 1984.

# Symbolic, Symmetry, and Stubborn Set Searches

Mikko Tiusanen

Helsinki University of Technology
Otakaari 1A, SF–02150 Espoo, FINLAND

**Abstract.** The *state space explosion problem* is the proliferation of states to be considered during the verification of a finite state system. This paper proposes ways to combine methods that have successfully been used to alleviate this problem during the reachability analysis of safe Petri nets (or ones with known bounds for all places): *symbolic model checking* employing data structures for *binary-decision diagrams* (BDDs), the symmetry equivalence method of Jensen et al, and the *stubborn set method* of Valmari.
The use of *Petri nets* as the basis extends the scope of the BDD-based analysis beyond hardware verification. The reachability graph of Petri nets is slightly generalized to allow for multiple initial markings, in order to correspond to the symbolic state space search using BDDs. The generalization is useful in itself if the initial marking of the system is not fully defined, e.g. for fault-tolerant systems. A stubborn set selection algorithm of Valmari is generalized to apply to sets of current markings instead of one marking at a time. This will provide the combination of the stubborn set method and symbolic state space generation using BDDs. The combination of the symmetry equivalence method with the stubborn set method is discussed: this turns out to be as straightforward as claimed by Valmari. Using BDDs to benefit in the symmetry equivalence method is also discussed.

## 1 Introduction

The *state space explosion problem* refers to the effect of having to inspect a large—computationally infeasible—number of states during the verification of a finite state system, be it hardware, software, parallel, concurrent, distributed, real-time, or some combination of these.

We shall propose a way to combine successful methods to alleviate the state space explosion problem in *reachability analysis*, that is, explicit generation of the state space of the system, namely, the *stubborn set method* by Valmari [23, 24, 22], and the *symbolic model checking* methods employing graph data structures for *binary-decision diagrams* (BDDs) [6, 7, 4, 5]. The BDDs have been successfully used to verify hardware systems with more than $10^{20}$ states. The stubborn sets, on the other hand, offer in some systems a reduction from a number of states exponential in a number of independent agents to polynomial, even linear, by reducing the number of interleavings of actions or events that need to be considered. The method has so far been restricted to generating the state space one global state at a time.

One of the first methods of doing high-level analysis of high-level Petri nets was the so called *equivalent marking* or *symmetry equivalence method* proposed in [11] for Coloured Petri nets. It uses *symmetries* to reduce the number of markings considered during the reachability graph generation. We shall follow the terminology of [19] rather than that of [11]. We will discuss how to combine this with the stubborn set method. After a thorough discussion of the symmetry method, this turns out to be as straightforward as Valmari has claimed it to be [24]. We will also mention some ways of improving the symmetry method through the use of BDDs.

We shall use *safe Petri nets* [16, 2, 3] (or more generally Petri nets with known bounds on all places) as the system model that the methods are applied on. The use of Petri nets extends the applicability of BDDs to non-hardware systems. The combination with stubborn sets potentially reduces the number of redundant interleavings considered, which has still been a problem with applications of BDDs to asynchronous hardware [6]. We shall define a simple generalization of the reachability graph of Petri nets to multiple initial markings that is potentially useful for fault-tolerant systems as such.

In the sequel we shall first review some of the relevant notation and definitions. We shall present a symbolic search algorithm using Dijkstra's *weakest precondition* or *℘-calculus*, and discuss the combinations of the methods before conclusions and considering future work.

## 2 Notations and definitions

In this section we shall present some notation on Petri nets used in the sequel when presenting the algorithms. We will also briefly review the relevant points of binary-decision diagrams and stubborn sets. We shall use Dijkstra's *language of guarded commands* and his *weakest precondition* or *℘-calculus*, see e.g. [10], in presenting the algorithms. We will use the notation $(\forall x : p(x) : q(x))$ to denote the universal quantification, more commonly given as $\forall x : (p(x)) \Rightarrow (q(x))$. Similarly, $(\exists x : p(x) : q(x))$ denotes the existential quantification $\exists x : (p(x)) \wedge (q(x))$.

The proofs we will present will be calculations of the form

$$\begin{array}{ll} & A \\ = & \{\!\!\{\text{hint as to why } A \Leftrightarrow B\}\!\!\} \\ & B \\ = & \{\!\!\{\text{hint as to why } B \Leftrightarrow C\}\!\!\} \\ & C, \end{array}$$

the conclusion being reached being that $A \Leftrightarrow C$, since $(A \Leftrightarrow B) \wedge (B \Leftrightarrow C)$. Note that this proof establishes more than $A \Leftrightarrow B \Leftrightarrow C$, since the latter is satisfied if $A = B = \textbf{false}$, $C = \textbf{true}$. Implications uniformly in one direction can also be used to conclude an implication between the top and bottom predicates, based on the tautology $((A \Rightarrow B) \wedge (B \Rightarrow C)) \Rightarrow (A \Rightarrow C)$. Keyword "splitting" refers to doing a case analysis on the elements of a set or the range of a quantification.

We shall use the notation $R^*$ to denote the reflexive and transitive closure of a binary relation $R \subseteq X^2$ over the set $X$. We will also use it as the Kleene star, denoting the set of finite strings over a set, depending on the context.

### 2.1 Petri nets

There are many closely related formal definitions of Petri nets. We shall use the following, see e.g. [15]. Let $\mathsf{N}$ be the set of natural numbers, with 0 as the least element.

**Definition 1.** A *Petri net* $N$ is a 5-tuple $N = \langle S, T; F, W, m_0\rangle$ where

- $S$ is a finite set of *places*,
- $T$ is a finite set of *transitions*,
- $F \subseteq (S \times T) \cup (T \times S)$ is a set of arcs, the *flow* relation,
- $W : F \longrightarrow (\mathsf{N} \setminus \{0\})$ is the *weight function* on the arcs, and
- $m_0 : S \longrightarrow \mathsf{N}$ is the *initial marking*, giving the initial distribution of tokens to places,

such that the sets of places and transitions are disjoint, $S \cap T = \emptyset$, and there is at least one net element, $S \cup T \neq \emptyset$. The domain of the weight function $W$ is usually extended to $(S \times T) \cup (T \times S)$ by setting $W(\langle x, x'\rangle) = 0$ iff $\langle x, x'\rangle \notin F$. Usually, the value of either of these arc weight functions on $\langle x, x'\rangle$ is denoted simply as $W(x, x')$. Any function $m : S \longrightarrow \mathsf{N}$, giving the distribution of the tokens onto places is called a *marking* of the net in question. The set of markings of a net with the places $S$ is denoted by $\mathsf{M}_S$. For any net element $x \in S \cup T$, the *preset* (input elements) of $x$ is ${}^\bullet x = \{x' \in S \cup T \mid x' \; F \; x\}$ and the *postset* (output elements) of $x$ is $x^\bullet = \{x' \in S \cup T \mid x \; F \; x'\}$. ◇

The definition of a post- or a preset of an element can obviously be extended to apply to a set of elements by taking the union of the post- or presets of the elements of the set, respectively. The behavior of a Petri net is defined as follows, see e.g. [15]:

**Definition 2.** A transition $t$ of a Petri net $N = \langle S, T; F, W, m_0\rangle$ is *enabled at a marking* $m : S \longrightarrow \mathsf{N}$ iff there are enough tokens on the input places:

$$\forall s : s \in {}^\bullet t \Rightarrow m(s) \geq W(s, t).$$

This is denoted by $m\ [t\rangle$ . An enabled transition $t \in T$ may, but need not *occur* (fire); if it does, it transforms the marking $m$ to another $m'$ that satisfies

$$\forall s : s \in S \Rightarrow m'(s) = m(s) + W(t, s) - W(s, t).$$

The relation that $m'$ is obtained from $m$ by firing the transition $t$ is denoted by $m\ [t\rangle\ m'$; this implies $m\ [t\rangle$ . We shall use $m\ [\rangle\ m'$ to denote that there is a transition $t$ such that $m\ [t\rangle\ m'$. ◇

In effect, the firing of a transition removes tokens from the input and deposits tokens on the output places of the transition, as many as the arc weights require.

**Definition 3.** The *reachability graph* of a Petri net $\langle S, T; F, W, m_0\rangle$ is the rooted, arc-labeled graph $\langle V, A; r\rangle$, where $V = \{m \in \mathsf{M}_S \mid m_0\ [\rangle^*\ m\}$ is the set of markings (finitely) reachable from the initial marking, $A = \{\langle m, t, m'\rangle \mid m \in V \wedge m\ [t\rangle\ m'\}$ are the arcs labelled with the transition involved, and $r = m_0$. A Petri net is called *k-bounded* iff there are at most $k$ tokens in any place of the net in any reachable marking. A 1-bounded net is called *safe*. ◇

We will generalize the concept of a reachability graph to better suit our purposes. This generalized reachability graph is potentially useful when discussing systems that do not have a fully defined initial (or representative) state. Fault-tolerant systems that contain a model of the faulty components (an explicit fault model) can be such, since the initial state of a faulty component can be convenient to leave unspecified. The usefulness stems from the fact that the generalized reachability graph is the union of the vertex and arc sets of all the reachability graphs of any marking in a given finite set of initial markings. Instead of doing separate reachability analyses for each of the initial markings, parts of the graphs can be shared during the analysis. We shall be abusing the notation of Petri nets by replacing the last component of the tuple comprising the Petri net by the *set* of initial markings.

**Definition 4.** The *generalized reachability graph* of a Petri net $N = \langle S, T; F, W, M \rangle$ with a set of initial markings $M$ is the directed graph $\langle V, A \rangle$ rooted at $M$, where

$$V = \{m \in \mathsf{M}_S \mid (\exists m_0 : m_0 \in M : m_0 \,[\rangle^*\, m)\}$$

is the set of markings (finitely) reachable from an initial marking, and

$$A = \{\langle m, t, m' \rangle \mid m \in V \wedge m \,[t\rangle\, m'\}$$

are the arcs labelled with the transition. ◇

Obviously, this is a very slight generalization: by adding one place and $|M|$ transitions, assuming this is finite, one can from any finite Petri net construct another with exactly one extra state, namely the initial one, and excluding that marking, an isomorphic (generalized) reachability graph (whether finite or not).

### 2.2 Stubborn sets

We shall recall what is called by Valmari [24] the *strong* theory of stubborn sets. It aims at using knowledge about the concurrency, or, more generally, independence or non-interference of transitions, to reduce the number of elements of $X^\bullet$ actually considered further during the symbolic search, say the cardinality of the set $Y$. This will bring down the *branching factor* of the search even if it does not minimize the number of markings generated.

What is characteristic of the behavior of concurrently enabled transitions apparent in a reachability graph of a Petri net is the *diamond substructure*: if transition $t$ and $t'$ are concurrently enabled at a marking $m$ they will span a square

$$\begin{array}{ccc} m & \xrightarrow{t} & m' \\ t' \downarrow & & \downarrow t' \\ m'' & \xrightarrow{t} & m''' \end{array}$$

in the reachability graph. Any set of $n > 0$ concurrently enabled transitions at a marking will span an $n$-dimensional hypercube. But not only do concurrently enabled transitions span these squares: some seemingly conflicting transitions will span these, too. These squares are, in fact, commutative diagrams. The important thing to note is that these transitions form a "well-behaved" part of the nondeterminism included in the interleaved interpretation of concurrency, which the reachability graph is based on. Concurrently enabled or—more generally—commuting transitions offer a possibility of avoiding the generation of at least one of the four markings in a square: final state of the square is guaranteed to be the same whichever way is taken to reach it. In a hypercube of higher dimension, say $n$, any path through it can be chosen to dissect it, resulting in a number of markings linear in $n$ to be generated, instead of a number exponential in $n$. Naturally, there might be states of interest within the hypercube, say, because they enable yet another transition in exactly that state. Since such a transition might be on the only path to a deadlock, it must be considered and the state in the middle must be generated while generating the reachability graph. The challenge is to formulate concepts that allow making use of commutativity among transitions to reduce the size of $Y$ while preserving the completeness of the analysis in the sense of finding all the deadlocks.

Consider a marking, say $m$, whether reachable or not, that has at least one enabled transition. Now, try to construct a partition of the the set of transitions $T$ into two sets, say $T_s$ and $\overline{T_s}$, so that the following diagram commutes for any $t \in T_s$ and $\sigma \in (\overline{T_s})^*$ such that $m\ [t\rangle \wedge m\ [\sigma\rangle$ :

$$\begin{array}{ccc} m & \xrightarrow{\sigma} & m' \\ t\downarrow & & \downarrow t \\ m'' & \xrightarrow{\sigma} & m''' \end{array}$$

That the diagram commutes can be guaranteed if firing a transition in $T_s$ cannot disable a transition in $\overline{T_s}$ nor vice versa. If one furthermore can guarantee that firing any $\sigma \in (\overline{T_s})^*$ cannot enable any transition in $T_s$, there is no need to generate any of the internal markings on the path $m\ [\sigma\rangle\ m'\ [t\rangle\ m'''$: any deadlock that can be found by generating such a path, can be found by firing some enabled transition $t \in T_s$, if a deadlock exists. If one of the blocks contains no enabled transitions, we have gained nothing. But if both contain enabled transitions, we can ignore any enabled transition in $\overline{T_s}$ when choosing $Y$, since firing any of them will not affect the enabledness of any of the stubborn transitions (the stubborn transitions will stay enabled/disabled); firing a stubborn transition will at most enable transitions that were not enabled at $m$. The only thing that remains is to ensure that there is at least one transition in $T_s$ enabled at marking $m$ to make $T_s$ a *stubborn* set. Valmari has given many alternative definitions for stubborn sets of Petri nets. We shall use the following [24]:

$$\begin{aligned} \mathit{stubborn}(T_s, m) = T_s \subseteq T \wedge (\exists t : t \in T_s : m\ [t\rangle\ ) \\ \wedge\, (\forall t : t \in T_s : [m\ [t\rangle \Rightarrow ({}^\bullet t)^\bullet \subseteq T_s] \\ \wedge\, [\neg m\ [t\rangle \Rightarrow (\exists s : m(s) < W(s,t) : {}^\bullet s \subseteq T_s)]) \end{aligned}$$

Obviously, $T$ is stubborn at any marking with at least one enabled transition. Note that this definition mixes structural elements, e.g. $({}^\bullet t)^\bullet \subseteq T_s$, with elements depending on the marking, e.g. $m(s) < W(s,t)$. So the goal is is to compute the set $Z$ of enabled transitions of a stubborn set:

$$R : \mathit{stubborn}(T_s, m) \wedge Z = T_s \cap \{t \in T \mid m\ [t\rangle\ \}.$$

The definition is suggestive of an algorithm to compute a stubborn set at a marking: given a method to pick a place $s$ in the definition, the *scapegoat*, any marking $m$ determines a binary relation $t\ R_m\ t'$: "if $t$ is in a stubborn set, so is $t'$, immediately due to definition". Starting from an enabled transition $t$ and computing the set of enabled transitions in the first-found *fragment* (maximal strongly connected component) of the graph $(T, R_m)$ that really has an enabled transition using Tarjan's algorithm [20], we have the set of enabled transitions of a stubborn set. The algorithm can be stopped after having found the first fragment.

The correctness of the algorithm relies on the fact that Tarjan's fragment search algorithm will find a fragment only after all the fragments that can be reached from it have been found. If the algorithm is stopped as soon as it finds a fragment with at least one enabled transition, the fragment *together with* all the fragments that can be reached from it forms a stubborn set. However, we are only interested in the enabled transitions in the stubborn set, and the other fragments reachable from the found one do *not* contain enabled transitions since these were found earlier, so the others can be ignored. One of the interesting aspects of this algorithm is that it inspects at most once the enabledness of any transition, making it at most linear in the number of transitions. We give an iterative version of this algorithm in Appendix A.

In the sequel, we shall use $stub(m)$ to denote a stubborn set computed by the given algorithm at the marking $m$, $m\ [t\rangle_s$ to denote $m\ [t\rangle \wedge t \in stub(m)$, $m\ [t\rangle_s\ m'$ to denote $m\ [t\rangle\ m' \wedge t \in stub(m)$, and $m\ [\rangle_s\ m'$ to denote $(\exists t : t \in stub(m) : m\ [t\rangle\ m')$. The notation $m\ [\sigma\rangle_s\ m'$ is used to denote that marking $m'$ can be reached from marking $m$ with a sequence of transitions $\sigma$ that has each of its transitions in the chosen stubborn set of the marking it is fired from in the sequence (a *stubborn transition sequence*), and the notation $m\ [A\rangle_s^*\ m'$ is used to denote that marking $m'$ can be reached from marking $m$ by a stubborn transition sequence through markings in the set $A$ only, excluding the first marking $m$, however. Finally, $A^\circ$ is used to denote the stubborn postset of the set $A$, $\{y \mid (\exists x, t : x \in A \wedge t \in stub(x) : x\ [t\rangle\ y)\}$.

## 2.3 Binary-decision diagrams

Binary-Decision Diagrams (BDDs) [12, 1] are full binary trees that represent the result of a sequence of two-way (binary) choices. Bryant [4] introduced a graph representation for these that can be used to represent Boolean functions of Boolean variables. The representation imposes (or assumes) a linear order on the argument variables; the resulting representation is unique, providing a *canonical* representation of functions. Given any canonical representation of Boolean functions of Boolean variables, this can also be seen to be a representation of sets of binary vectors, a set being represented by the *characteristic function* of the set. The characteristic function is a function that has the value 1 on those binary vectors that are in the set under consideration, and 0 on all other vectors.

The representation proposed by Bryant consists of imposing a total order on the Boolean variables appearing as arguments to a function and then systematically applying two transformations to any BDD. The order on variables determines the order of the decisions made in the BDD, starting from the root. First of the transformations makes the vertices of the resulting graph *share* any subgraph representing identical subtrees of the original full binary graph. That is, all of the identical subtrees of the original BDD are represented by a unique subgraph of the resulting graph, but the subgraph may be the immediate descendant of many vertices. Since the order on variables (decisions) is total, this results in a rooted directed, acyclic graph (DAG) where all the vertices at the same level (distance from the root) represent a choice made based on the same variable. The second transformation replaces any vertex, both the descendants of which represent the same subtree, by the representation of the common subtree. The replaced vertices are redundant, since the choice made does not affect the value of the function represented: the value is independent of the variable. Since after this transformation the distance to the root is no longer sufficient to determine the variable based on which the choice is made, this must somehow be attached to a vertex explicitly. These two transformations are conceptually repeated until they can no longer be applied. An example of the transformation for the Boolean function $(x \vee y) \wedge \overline{z}$ is given in Figure 1.

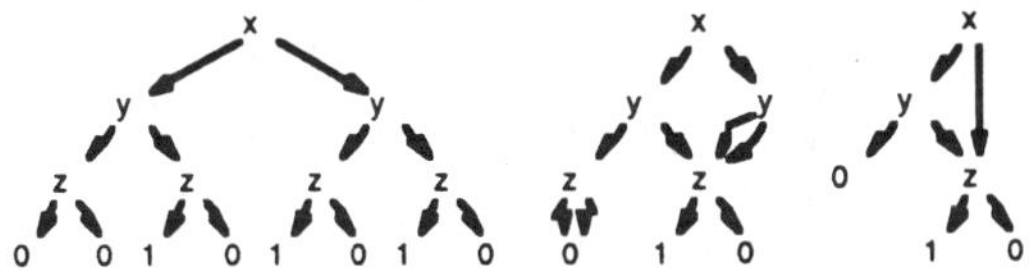

**Fig. 1.** The transformation of the BDD for Boolean function $(x \vee y) \wedge \overline{z}$. Branches to the left are taken when the variable of the node is 0, right if 1.

The interest in the representation stems from the fact that it can be used to efficiently perform operations on a sets of binary vectors, such as testing for emptiness, union, intersection, and set difference. The functional composition of functions can also be performed as an operation on the representations of these functions. Since the state of a digital circuit can on a particular level of abstraction be represented as a binary vector, the data structure has been used with success for hardware verification [5, 7, 6]. One of the keys for this success has been the realization that the next-state relation of a digital circuit can also be represented using the data structure, and that the application of this relation to a set of current states can be used to produce the set of successor states of any current state. This provides a way of doing a set-parallel, breadth-first search of the state space of a digital circuit. The efficiency of the search depends heavily on the choice of the ordering of variables, however.

## 3 Symbolic search

Consider performing a search of the generalized reachability graph of a Petri net. We shall apply the well-known white-grey-black coloring invariant of Dijkstra's, the invariants $P_0$ to $P_2$ below. Let $Q$ be the predicate stating that $V$ is the set of vertices of the generalized reachability graph of the net $N = \langle S, T; F, W, M \rangle$:

$$Q : V = \{m \in \mathsf{M}_S \mid (\exists m_0 : m_0 \in M : m_0 \,[\rangle^*\, m)\}.$$

We shall not consider the computation of the arcs of the generalized reachability graph: this is a straightforward extension of the resulting algorithm. The objective is then to compute the set of vertices of the graph to a variable, say $B$:

$$R : B = V.$$

To get the graph we must somehow generate the vertices, which calls for a repetition. Following Dijkstra, we shall choose the invariant of the repetition to consist of two parts. First, the (still to be determined) vertices of the graph are partitioned into three disjoint blocks: the black, the grey and the white vertices, with the result being computed as the set of black vertices,

$$P_0 : V = B \uplus G \uplus W.$$

The intuition is to have the black ones be the processed, the grey the still possibly unprocessed, and the white the unknown vertices of $V$: the algorithm is to "color" all the vertices black. Again following Dijkstra, we shall proceed with the idea that the grey vertices shall be a "buffer zone" between the other two blocks: $\boxed{W \boxed{G \boxed{B}}}$.

To formulate this, we shall require that for any white (unknown) vertex there be a white path from some grey (unprocessed) vertex to it.

$$P_1 : (\forall w : w \in W : (\exists g : g \in G : g\ [W\rangle^*\ w)).$$

Note that now $P_1 \wedge G = \emptyset \Rightarrow W = \emptyset$, since empty $G$ can contain no elements from which one in $W$ can be reached. As the termination function for the repetition, $|V \setminus B|$ is an obvious choice: $|V|$ should, therefore, be finite for the repetition to terminate, and the body of the repetition should color black at least one vertex that was not black previously. Based on the intuition about $P_1$, consider then coloring some non-empty subset of vertices $\emptyset \neq X \subseteq G$ black: this might result in some of the white neighbours of $X$, $X^\bullet \cap W$, not to have a grey vertex from which these can be reached. To avoid this, color all the white neighbours of $X$ grey. Setting $P : P_0 \wedge P_1$, assuming there is a function *pick* satisfying

$$G' = G \neq \emptyset \Rightarrow \wp(X := \mathit{pick}(G), \emptyset \neq X \subseteq G = G')$$

for any $G'$, and collecting the results of the analysis, we have

$$\begin{array}{l}
\{\!\{Q\}\!\}\, B, G, W := \emptyset, M, V \setminus M \{\!\{P\}\!\} \\
;\ \textbf{do}\ G \neq \emptyset \rightarrow \\
\quad X := \mathit{pick}(G) \{\!\{P \wedge \emptyset \neq X \subseteq G\}\!\} \\
\quad ; B, G, W := B \cup X, (G \setminus X) \cup (X^\bullet \cap W), W \setminus X^\bullet \\
\quad \textbf{od} \\
\quad \{\!\{P \wedge G = \emptyset\}\!\}\ \{\!\{R\}\!\}
\end{array}$$

For more details on the derivation, the reader is referred to [21]. (The significance of $G'$ in specifying the behavior of *pick* is to exclude any trivial solutions that modify $G$, an often used technicality.)

Note that this program subsumes many search strategies, since it does not restrict the implementation of *pick* to any great extent. Most common implementations of the algorithm make $M$ contain a single element and *pick* return a single element. Moreover, any implementation for the necessary set operations is equally valid, but will naturally have an effect on the performance of the search. In particular, the data structures for binary-decision diagrams (BDDs) can be used to implement the sets of markings $B$ and $G$ for safe Petri nets. Of course, since $V$ is not known prior to the search, the set $W$ will have to be implemented implicitly as the set of reachable markings not already colored, that is, not in $B$ or $G$. One of the powerful features of BDDs is that the transition relation among global states can be represented as a BDD, and, furthermore, to compute the *set* of followers of a given *set* of current global states. For Petri nets, this involves having a BDD representation of the set of binary vectors of the relation $\{\langle m, t, m' \rangle \mid m, m' \in \mathsf{M}_S \wedge t \in T \wedge m\,[t\rangle\, m'\}$. Each element of the relation can be represented as a binary vector by concatenating suitable representations of its components. The BDD represetation can be used to compute the followers of a given set of markings also represented as a BDD by using operations on the two BDDs. For a Petri net with known bounds on the marking of any place, the marking of the places can be suitably encoded as binary variables to achieve the same result. In effect, we shall assume the use of the graph representation of BDDs as an efficient implementation of operations on sets of markings and the operation $X^\bullet$ for a set $X$ of markings. We shall return to some of the aspects of the use of BDDs in Section 6. Note that the program also maintains the invariant $P_2 : B^\bullet \subseteq B \cup G$, that is, that black vertices have only coloured immediate neighbors.

## 4 Symbolic stubborn set search

We shall present an algorithm extending the stubborn set method for the computation of the generalized reachability graph of safe Petri nets. The algorithm given in Section 3 will be used as a basis. As can easily be checked, this provides also a program to do the stubborn set search, after replacing $[\ldots\rangle$ by $[\ldots\rangle_s$ and $(\ldots)^\bullet$ by $(\ldots)^\circ$ in the program. Intuitively, the stubborn set search merely generates the subgraph of the full reachability graph defined by the stubborn sets chosen at each marking. The remaining problem is to compute

$$R' : (Y = X^\circ \cap W) \wedge P,$$

starting from $Q' : \emptyset \neq X \subseteq G \wedge P$, that is, $(Y = X^\circ \setminus (G \uplus B)) \wedge P$.

The algorithm we shall give is based on the idea of running an algorithm very much like the one described above to compute a stubborn set at *one* marking with the *whole* set $X$ instead. First, all the dead markings are removed from the set $X$. This can be done, since the subset of markings $\mathsf{M}_S$ that do not enable any transitions, say $A$, can be precomputed for a safe Petri net. The intersection of $A$ and $X$ gives the dead markings in $X$. If the rest of $X$ is empty, the algorithm is not applied. When the original algorithm to compute a stubborn set at a marking is to determine whether a transition is enabled at the given marking, the set under consideration (originally $X \setminus A$) is split into those that enable the transition, and to those that do not. The splitting guarantees that the enabledness of all the transitions that have caused a split to occur is well-defined for the set under consideration. Moreover, a split occurs only if the search for a stubborn set requires it.

If one of the parts resulting from the split is empty, it is ignored, since the empty set of markings has no followers and does not contribute to the result; both parts cannot be empty, since $X \setminus A$ was non-empty originally and empty sets are ignored. If both sets are non-empty, one is set aside with sufficient information on the state of the search to be able to resume the search later, and the processing of the other part is continued. Once a search terminates, a set is chosen from those set aside, and the state of the search associated with the set is restored, and the search resumed. This is continued until there are no more sets set aside. The set of follower markings of a set is produced by firing the enabled transitions in the stubborn sets found for the set. The set $Y$ to be computed is then the union of the sets of white follower markings of all the split parts for which a stubborn set was found. There will be a stubborn set for all the non-empty parts of $X \setminus A$, since if there is no stubborn set for some part, then this has no enabled transitions since $T$ is stubborn otherwise. This contradicts the construction of $A$.

There are some complications involved, however. The enabledness of the transitions is not the only marking-dependent aspect of the computation of the stubborn sets using the given algorithm: also the choice of the scapegoat is marking dependent. Any transition with at least two input places can have two markings, both of which disable the transition, and the place that has too few tokens is unique but different for both, making the scapegoat different for these, see Figure 2 where the scapegoat depends whether the marking with the black or the striped tokens only is considered. A simple solution to the problem is to split the set under consideration further, setting aside parts in the exactly same manner if necessary. The other problem is that since the search for a stubborn set has not started from a transition that is enabled for all the markings in the set $X \setminus A$, there is no guarantee that the search will find a stubborn set, i.e. one with an enabled transition. Since all the markings enable at least one transition, any transition that has not been assigned a depth-first number in the search already can be used to continue the search.

**Fig. 2.** Ambiguity of the scapegoat.

We give the algorithm for this in Appendix B. Whenever the algorithm of Appendix A would compute the set $(t, m)^\circ$ of transitions necessarily in the stubborn set for some transition $t$ at the marking $m$, the set of markings needs potentially to be split, sometimes to many pieces due to the selection of the scapegoat. The fact that the algorithm of Appendix A is iterative is very useful here, since the state of the program is laid out in the form of explicit variables. The information to be set aside is the state of the search, basically all of the variables of the algorithm of Appendix A, with the notable exception of $Y$. There will not be a unique marking $m$, of course. Naturally, the variables that need not be implemented explicitly, like the set $W$, need not actually be set aside. The program will then repeat until the data structure used to set aside is empty.

It should be noted that the given algorithm only produces the set of reachable markings. It is fairly simple to also maintain a set of arcs between the markings produced.

## 5 Symmetry equivalence search

A symmetry is an *automorphism* on a Petri net: it is a bijection $\sigma$ that maps places to places and transitions to transitions, and respects the weight function:

$$W(x, x') = W(\sigma(x), \sigma(x'))$$

for any $\langle x, x' \rangle \in (S \times T) \cup (T \times S)$ in the unfolding. Obviously, a symmetry $\sigma : S \rightarrow S$ can be extended to a function on markings $\sigma : (S \rightarrow \mathsf{N}) \rightarrow (S \rightarrow \mathsf{N})$ by setting $(\sigma(m))(s) = m(\sigma^{-1}(s))$ (making $(\sigma(m))(\sigma(s)) = m(s)$). The crucial property of any symmetry $\sigma$ is that $m\ [t\rangle\ m' \Rightarrow \sigma(m)\ [\sigma(t)\rangle\ \sigma(m')$ for any $m, m' \in \mathsf{M}$, and $t \in T$. Since the symmetries are bijective, they are invertible, making the implication above in fact an equivalence.

We shall consider a given set of symmetries $\Sigma$ that includes the identity function and the inverse function of any of its elements, and which is closed with respect to functional composition. Such a set forms a *group* with the identity function as the identity element and functional composition as the binary operation, as can be easily verified. Based on the set $\Sigma$ being a group, we can define an equivalence relation $\simeq$ on markings by setting $m \simeq m' = (\exists \sigma : \sigma \in \Sigma : m = \sigma(m'))$. A marking $m$ is called *symmetric* iff $(\forall \sigma : \sigma \in \Sigma : m = \sigma(m))$. A symmetric marking forms a block of the equivalence relation $\simeq$ by itself, as any marking equivalent to it will also be equal to it. These definitions have followed the presentation in [19]. Given a set of initial markings $M$, a $\simeq$-reachablity graph is then a graph $\langle \Gamma, A \rangle$ such that it satisfies the predicate

$$\begin{aligned}\tilde{S}(\langle \Gamma, A \rangle) : M \subseteq \Gamma \subseteq V \wedge (\forall m : m \in \Gamma : (\exists m_0 : m_0 \in M : m_0\ [\Gamma\rangle^*\ m))\\ \wedge (\forall m, m' : m, m' \in \Gamma \wedge m \simeq m' \wedge (m \notin M \vee m' \notin M) : m = m')\\ \wedge (\forall m, m', t : m, m' \in \Gamma \wedge t \in T \wedge m\ [t\rangle\ m' : (\exists m'' : m'' \in \Gamma : m'' \simeq m'))\\ \wedge A = \{\langle m, t, m' \rangle \mid m, m' \in \Gamma \wedge t \in T \wedge (\exists m'' : m\ [t\rangle\ m'' : m' \simeq m'')\}.\end{aligned}$$

Basically, this is a graph of connected representatives of the blocks of the equivalence relation $\simeq$: there is exactly one for each block unless the initial markings contain equivalent markings, all are reachable from some initial marking, and the arcs are the ones induced between the blocks.

We shall extend the framework of Section 3 to compute a $\simeq$-reachability graph. The goal is $R : \tilde{S}(\langle B, E \rangle)$. We shall ignore the computation of $E$, however, for sake of conciseness. Let us refine the partition of the set of reachable markings to

$$P_0 : V = B \uplus G \uplus W \uplus C,$$

where $C$ is the set of "clear" or ignored markings. We shall also use the invariant

$$P_1 : (\forall w : w \in W : (\exists g : g \in G : g\ [W\rangle^*\ w)).$$

Let us weaken the goal: we shall consider it in smaller conjuncts for the sake of clarity.

$$P_2 : M \subseteq B \cup G \subseteq V,$$

$$P_3 : (\forall m : m \in B \cup G : (\exists m_0 : m_0 \in M : m_0\ [B \cup G\rangle^*\ m)),$$

$$P_4 : (\forall m, m' : m, m' \in B \cup G \wedge m \simeq m' \wedge (m \notin M \vee m' \notin M) : m = m').$$

We shall also need to describe the function of the clear block of the partition:

$$P_5 : (\forall m : m \in C : (\exists b : b \in B \cup G : b \simeq m)).$$

Let us call the whole invariant

$$P : P_0 \wedge P_1 \wedge P_2 \wedge P_3 \wedge P_4 \wedge P_5.$$

Now, since we have $P$ and that the set of grey vertices is empty in the end, the vertices will be either black or clear at the end, and we have that any successor of a black marking $b$ either black or clear, $b^\bullet \subseteq V = B \uplus C$, so by $P_5$ the missing conjunct of $R$ is implied. The initialization is $B, G, W, C := \emptyset, M, V \setminus M, \emptyset$, as can easily be checked [21].

For the progress we consider first proceeding the same way as in Section 3. The invariant $P_1$ has already been considered there. For the rest, we compute

$$\begin{array}{l}
\wp(\text{“}B, G, W := B \cup X, (G \setminus X) \cup (X^\bullet \cap W), W \setminus X^\bullet\text{”}, P_0) \\
= V = (B \cup X) \uplus (G \setminus X) \cup (X^\bullet \cap W) \uplus W \setminus X^\bullet \uplus C \\
\Leftarrow P_0 \wedge \emptyset \neq X \subseteq G,
\end{array}$$

$$\begin{array}{l}
\wp(\text{“}B, G, W := B \cup X, (G \setminus X) \cup (X^\bullet \cap W), W \setminus X^\bullet\text{”}, P_2) \\
= M \subseteq B \cup X \cup (G \setminus X) \cup (X^\bullet \cap W) \subseteq V \\
\Leftarrow \{\!\{X \subseteq G \Rightarrow G = X \cup (G \setminus X)\}\!\} \\
\quad X \subseteq G \wedge M \subseteq B \cup G \cup (X^\bullet \cap W) \subseteq V \\
\Leftarrow \{\!\{P_0 \Rightarrow W \subseteq V\}\!\} \\
\quad P_0 \wedge P_2 \wedge X \subseteq G,
\end{array}$$

and

$$\begin{array}{l}
\wp(\text{“}B, G, W := B \cup X, (G \setminus X) \cup (X^\bullet \cap W), W \setminus X^\bullet\text{”}, P_3) \\
= (\forall m : m \in B \cup X \cup (G \setminus X) \cup (X^\bullet \cap W) : \\
\qquad (\exists m_0 : m_0 \in M : m_0\ [B \cup X \cup (G \setminus X) \cup (X^\bullet \cap W)\rangle^*\ m)) \\
\Leftarrow \{\!\{X \subseteq G \Rightarrow G = X \cup (G \setminus X)\}\!\} \\
\quad X \subseteq G \\
\quad \wedge (\forall m : m \in B \cup G \cup (X^\bullet \cap W) : \\
\qquad (\exists m_0 : m_0 \in M : m_0\ [B \cup G \cup (X^\bullet \cap W)\rangle^*\ m)) \\
\Leftarrow \{\!\{\text{splitting}\}\!\} \\
\quad X \subseteq G \\
\quad \wedge (\forall m : m \in B \cup G : (\exists m_0 : m_0 \in M : m_0\ [B \cup G \cup (X^\bullet \cap W)\rangle^*\ m)) \\
\quad \wedge (\forall m : m \in X^\bullet \cap W : (\exists m_0 : m_0 \in M : m_0\ [B \cup G \cup (X^\bullet \cap W)\rangle^*\ m)) \\
\Leftarrow \{\!\{(*)\}\!\} \\
\quad P_3 \wedge X \subseteq G,
\end{array}$$

where (*) is proven by noting that $P_3 \wedge B \cup G \subseteq B \cup G \cup (\ldots)$ implies the first universal quantification, and that the second follows from the first and $X \subseteq G$ since there is always a path of length one from any $x$ to any $y \in x^\bullet$. Further, we get

$$\begin{aligned}
&\wp(\text{“}B, G, W := B \cup X, (G \setminus X) \cup (X^\bullet \cap W), W \setminus X^\bullet\text{”}, P_4)\\
= &(\forall m, m' : m, m' \in B \cup X \cup (G \setminus X) \cup (X^\bullet \cap W)\\
&\quad \wedge m \simeq m' \wedge (m \notin M \vee m' \notin M) : m = m')\\
\Leftarrow &\{\!\!\{X \subseteq G \Rightarrow G = X \cup (G \setminus X)\}\!\!\}\\
&X \subseteq G \wedge (\forall m, m' : m, m' \in B \cup G \cup (X^\bullet \cap W)\\
&\quad \wedge m \simeq m' \wedge (m \notin M \vee m' \notin M) : m = m')\\
\Leftarrow &\{\!\!\{\text{see [21]}\}\!\!\}\\
&P_0 \wedge P_4 \wedge X \subseteq G \wedge R_0 \wedge R_1
\end{aligned}$$

where

$$R_0 : (\forall m, m' : m \in B \cup G \wedge m' \in X^\bullet \cap W : m \not\simeq m')$$

and

$$R_1 : (\forall m, m' : m, m' \in X^\bullet \cap W \wedge m \simeq m' : m = m').$$

The predicates $R_0$ and $R_1$ are clearly not implied by $P$, but they depend only on the elements of known sets $X^\bullet$, $B$, and $G$.

Let us then treat $P_5$:

$$\begin{aligned}
&\wp(\text{“}B, G, W := B \cup X, (G \setminus X) \cup (X^\bullet \cap W), W \setminus X^\bullet\text{”}, P_5)\\
= &(\forall m : m \in C : (\exists b : b \in B \cup X \cup (G \setminus X) \cup (X^\bullet \cap W) : b \simeq m))\\
\Leftarrow &\{\!\!\{X \subseteq G \Rightarrow G = X \cup (G \setminus X)\}\!\!\}\\
&X \subseteq G \wedge (\forall m : m \in C : (\exists b : b \in B \cup G \cup (X^\bullet \cap W) : b \simeq m))\\
\Leftarrow &X \subseteq G \wedge (\forall m : m \in C : (\exists b : b \in B \cup G : b \simeq m))\\
= &X \subseteq G \wedge P_6.
\end{aligned}$$

The remaining problem is, how to establish the predicates $R_0$ and $R_1$ given $\emptyset \neq X \subseteq G \wedge P$. Perhaps the simplest strategy is to look at each marking $m' \in X^\bullet$, one at a time, and decide its equivalence: if $m'$ is equivalent to a black or grey marking, or a white one in $X^\bullet$ treated prior to it, $m'$ will be colored clear. This will also maintain $P_6$: all clear markings will, after making $X$ black and $X^\bullet \cap W$ grey, have an equivalent black or grey marking. Since the identity function is a symmetry, $m = m' \Rightarrow m \simeq m'$, or equivalently, $m \not\simeq m' \Rightarrow m \neq m'$, it is not necessary to separately check whether $m'$ is grey or black.

Let the symbol $\perp$ denote the smallest transition in a given but arbitrary linear order on the transitions; we have assumed there to be at least one transition in the net under consideration. Let the function *nextmark*$(m, t)$ give the marking $m'$ such that $m\ [t\rangle\ m'$ and the function *nexttrans*$(t)$ the next transition in the assumed linear order on transitions, the value $\infty \notin T$ for the largest $t \in T$. Collecting the results of the discussion, we get the program in Figure 3, where *pick* now returns any one element of its argument set.

This strategy has the drawback of making the generation of the $\simeq$-reachability graph a marking-at-a-time process. This denies, at least to a degree, the benefits of the use of efficient representations for sets of states, such as the BDDs. The *compatible projection operator* proposed by Lin and Newton [14, 13] offers a chance to improve upon this. The operator can be used to compute a function that maps each marking to a unique representative of its equivalence class in an equivalence

```
{Q} B, G, W, C := ∅, M, V \ M, ∅{P}
; do G ≠ ∅ → x := pick(G){P ∧ x ∈ G}
  ; t, S := ⊥, ∅
  ; do t ≠ ∞ →
      if x [t⟩  → y := nextmark(x, t)
        ; W := W \ {y}
        ; if [y] ∩ (B ∪ G ∪ S) ≠ ∅ → C := C ∪ {y}
          [] [y] ∩ (B ∪ G ∪ S) = ∅ → S := S ∪ {y}
          fi
      [] ¬x [t⟩  → skip
      fi
    ; t := nexttrans(t)
    od
  ; B, G := B ∪ {x}, (G \ {x}) ∪ S
  od
{P ∧ G = ∅}
```

**Fig. 3.** Symmetry equivalence search.

relation, such as the symmetry relation $\simeq$ on markings is, given in a BDD form. Since this can be precomputed for the symmetry relation of a safe Petri net (or one with known bounds on markings of all the places), the unique representative function can be precomputed and represented in BDD form, and can be applied to achieve $R_0$ and $R_1$. Obviously, the set of all reachable markings—and not only those found during the symmetry search—can also be computed given a BDD representation of the symmetry relation and the $\simeq$-reachability graph, the relation can be applied to the set of all markings in the graph, which can, if wanted, be also maintained in BDD form.

To sum up, we have presented a concise derivation of the algorithm to compute the $\simeq$-reachability graph with no restrictions placed on the symmetries. [11] considers only symmetries that are generated by maps of colors of a Coloured Petri net: any symmetry is supposed to map a place-color pair $\langle s, c\rangle$ or a transition-binding pair $\langle t, b\rangle$ to another only if the first components are equal. The approach taken in [19, 18] and here is slightly more general, since this restriction is not placed on the symmetries. Though here we have assumed that the symmetries are given, [18] considers an algorithm to compute the set of symmetries of a Petri net.

The $\simeq$-reachability graph provides only a non-reachability test if the initial marking considered is not symmetric: otherwise, given a sequence $\tilde{\sigma}$ in the $\simeq$-reachability graph there will be a corresponding sequence $\sigma$ in the ordinary reachability graph, but it is only guaranteed that the initial marking of $\sigma$ is *equivalent* to that of $\tilde{\sigma}$. For a symmetric initial marking this implies equality, otherwise not necessarily so. If the set of initial markings $M$ is closed with respect to all the symmetries $\sigma \in \Sigma$, then the reachability test is again valid, since then there is an initial marking from which the marking is reachable.

Combining the symmetry and stubborn set methods turns out to be trivial: the only modification caused by the stubborn set method to the above is the use of the stubborn set related notation, that is, $(\ldots)^{\circ}$ instead of $(\ldots)^{\bullet}$, $m\ [\langle t\rangle\rangle_s$ instead of $m\ [t\rangle$ , and, say, $nexttrans_s$ instead of $nexttrans$, and $nextmark_s$ instead of $nextmark$.

Finally, it should be noted that dropping $P_4$ and $R_1$ in order to be able to use the BDD representation of states to its fullest will make the use of symmetries superfluous. The point of using the symmetries is to avoid having multiple representatives of the equivalence classes in the $\simeq$-reachability graph. We have discussed using the *compatible projection operator* of Lin and Newton and an explicit representation of the symmetry equivalence relation in the form of a BDD to benefit in the symmetry equivalence search, however.

## 6 Conclusion

We have defined the concept of a generalized reachability graph of a Petri net that can potentially be used to benefit in analyzing the behavior of fault-tolerant systems, and discussed a symbolic, set-at-a-time algorithm for the generation of such a graph. We have also discussed the generalization of one of Valmari's stubborn set selection algorithms to operate on a set of markings at a time. The stubborn sets offer a chance of reducing the number of interleavings of transitions considered during the symbolic that still is a problem for the symbolic search for asynchronous systems [6]. The use of BDDs in these algorithms extends the field of application of this important data structure, used to much benefit in hardware verification, to software verification, among others. This extension is achieved through the use of Petri nets as the basic system model, since Petri nets have been widely applied, and not only within computer science. Combining the two methods offers then a chance for increased applicability of reachability based analysis methods by potentially increasing the size of the models that can be analysed.

More recently, Valmari has also presented an algorithm that verifies any pregiven stuttering-invariant linear time temporal logic formula [22], and then an on-the-fly verification algorithm [25], both using stubborn sets. Relating the other definitions of stubborns sets and these to the symbolic search is left for future work. A specific problem that remains for future work also is how to coordinate the nondeterministic choices appearing in the algorithms to achieve a significant extension of the applicability of the methods. These choices are picking the ordering the variables, picking $X$, picking transitions and scapegoats during the stubborn set computation, and picking which part to shelve. For choosing the ordering of the variables, there has been work done to develop heuristics based on the application to hardware verification. These can form a basis for the considerations involved.

We have also discussed in more detail than previously combining the symmetry equivalence search with the stubborn set method. The methods turned out to be compatible in that they operate on "orthogonal" aspects of the problem. Moreover, we have discussed the use of BDDs in the symmetry equivalence search. In the context of temporal logic the same ideas have been put forward by Clarke et al [8]. The use of the compatible projection operator of Lin and Newton has not been exploited so far in the symmetry equivalence search.

*Acknowledgements:* This work has been supported by National Science Foundation, Finnish Academy, the Finnish Cultural Foundation (Kalle and Dagmar Välimaa fund), the Emil Aaltonen Foundation, the Alfred Kordelin Foundation, the Information Processing Research Foundation (Tietotekniikan tutkimussäätiö), and by the Helsinki University of Technology directly, through Ernst Wirtzen fund, and through the Foundation for Financial Support of HUT. The use of the compatible projection operator of Lin and Newton was suggested to the author by Licentiate of Technology Kimmo Varpaaniemi. The comments of the anonymous referees helped improve the presentation and were greatly appreciated. Ps 127.

## References

1. S. B. Akers. Binary decision diagrams. *IEEE Transactions on Computers*, C–27:509–516, June 1978.
2. W. Brauer, W. Reisig, and G. Rozenberg, editors. *Petri Nets: Applications and Relationships to Other Models of Concurrency, Part I*, volume 254 of *Lecture Notes in Computer Science*, Berlin, Germany, 1987. Springer-Verlag. Advances in Petri Nets 1986, Part I, Proceedings of an Advanced Course, Bad Honnef, Germany, September 1986.
3. W. Brauer, W. Reisig, and G. Rozenberg, editors. *Petri Nets: Applications and Relationships to Other Models of Concurrency, Part II*, volume 255 of *Lecture Notes in Computer Science*, Berlin, Germany, 1987. Springer-Verlag. Advances in Petri Nets 1986, Part II, Proceedings of an Advanced Course, Bad Honnef, Germany, September 1986.
4. R. E. Bryant. Graph-based algorithms for Boolean function manipulation. *IEEE Transactions on Computers*, C–35(6):677–691, 1986.
5. R. E. Bryant. Symbolic Boolean manipulation with ordered binary-decision diagrams. *ACM Computing Surveys*, 24(3):293–318, 1992.
6. J. R. Burch, E. M. Clarke, and D. E. Long. Symbolic model checking with partitioned transition relations. In A. Halaas and P. B. Denyer, editors, *Proceedings of the 1991 International Conference on Very Large Scale Integration*, Aug. 1991.
7. J. R. Burch, E. M. Clarke, K. L. McMillan, D. L. Dill, and J. Hwang. Symbolic model checking: $10^{20}$ states and beyond. In *Fifth Annual IEEE Symposium on Logic in Computer Science*. IEEE, June 1990.
8. E. M. Clarke, T. Filkorn, and S. Jha. Exploiting symmetry in temporal logic model checking. In Courcoubetis [9], pages 450–462.
9. C. Courcoubetis, editor. *Computer Aided Verification*, volume 697 of *Lecture Notes in Computer Science*. Springer-Verlag, 1993.
10. E. W. D. Dijkstra. Guarded commands, nondeterminacy and the formal derivation of programs. *Communications of the ACM*, 18:453–457, Aug. 1975.
11. P. Huber, A. M. Jensen, L. O. Jepsen, and K. Jensen. *Towards Reachability Trees for High-Level Petri Nets*, volume 188 of *Lecture Notes in Computer Science*, pages 215–233. Springer-Verlag, Berlin, Germany, 1985.
12. C. Y. Lee. Representation of switching circuits by binary-decision programs. *Bell Systems Technical Journal*, 38:985–999, July 1959.
13. B. Lin and A. R. Newton. Efficient symbolic manipulation of equivalence relations and classes. In *Proceedings of Formal Methods in VLSI Design Workshop, Miami, FL, January 1991*, 1991.

14. B. Lin and A. R. Newton. Implicit manipulation of equivalence classes using binary decision diagrams. In *Proceedings of the IEEE International Conference on Computer Design: VLSI in Computers and Processors, Cambridge, MA, October 14–16, 1991. IEEE Computer Society Press, Los Alamitos, CA 1991, 654 p.*, page 6, 1991.
15. T. Murata. Petri nets: Properties, analysis and applications. *Proceedings of the IEEE*, 77(4):541–580, 1989.
16. C. A. Petri. *Kommunikation mit Automaten*. PhD thesis, Institut für Instrumentelle Mathematik, Bonn, Germany, 1962. Schriften des IIM Nr. 2. English translation: [17].
17. C. A. Petri. Communication with automata. Technical Report RADC–TR–65–377, Griffith Air Force Base, New York, 1966.
18. K. Schmidt and P. H. Starke. An algorithm to compute the symmetries of Petri nets. *Petri Net Newsletter*, 40:25–30, Dec. 1991.
19. P. H. Starke. Reachability analysis of Petri nets using symmetries. *Systeme: Analysis, Modellierung, Simulation*, 8(4/5):293–303, 1991.
20. R. E. Tarjan. Depth-first search and linear graph algorithms. *SIAM Journal of Computing*, 1(2):146–160, 1972.
21. M. Tiusanen. *Static Analysis of Ada Tasking Programs: Models and Algorithms*. PhD thesis, University of Illinois at Chicago, July 1993.
22. A. Valmari. Stubborn attack on state explosion. *Formal Methods in System Design*, 1:297–322, 1991.
23. A. Valmari. *Stubborn Sets for Reduced State Space Generation*, volume 483 of *Lecture Notes in Computer Science*, pages 491–515. Springer-Verlag, Berlin, Germany, 1991.
24. A. Valmari. Stubborn sets of coloured Petri nets. In *Papers Presented at the 12th International Conference on Application and Theory of Petri Nets*, pages 102–121, Gjern, Denmark, June 1991. Århus University.
25. A. Valmari. On-the-fly verification with stubborn sets. In Courcoubetis [9], pages 397–408.

## A An iterative stubborn set algorithm

The algorithm in Figures 4 and 5 computes the stubborn postset of a single marking $m$, $Y = \{m\}^\circ$. It will not search for a transition that is enabled at $m$, but will, if necessary go through them all. The variable $e$ is used to terminate the search once a suitable strongly connected component (SCC) has been found. The symbol $\perp$ denotes the smallest transition in a given but arbitrary linear order on the transitions; we have assumed there to be at least one transition in the net under consideration. The function $nextmark(m, t)$ gives the marking $m'$ such that $m\ [t\rangle\ m'$ and the function $nexttrans(t)$ the next transition in the assumed linear order on transitions, the value $\infty \notin T$ for the largest $t \in T$.

Note that $s(q)$ is maintained as the *set* of followers of the transition according to the definition of stubborn sets. We have taken the liberty of using the notation $(t, m)^\circ$ of this set of transitions. This makes it easier to relate the algorithm to an iterative version of Tarjan's algorithm to find the fragments of a graph.

## B A symbolic stubborn set algorithm

The algorithm in Figures 6, 7, 8, and 9 computes the stubborn postset $Y$ of a set of markings $X$, $Y = X^\circ$. The variable $e$ is used to terminate the search once a suitable

```
B, t, e := ∅, ⊥, false
; do ¬e ∧ t ≠ ∞ →
    if t ∈ B → skip
    ▯t ∉ B → B, G, W, d(t), q, s(0), s(1), s(2), f(t), r, h(0), S :=
        B ∪ {t}, (t, m)° \ {t}, T \ ((t, m)° ∪ B ∪ {t}),
        |B|, 2, ∅, t, (t, m)° \ {t}, |B|, 1, t, {t}
    ; do q ≠ 0 → EXTEND
      ; do q ≠ 0 cand s(q + 1) ∈ B →
          if d(s(q − 1)) > d(s(q + 1) ∧ s(q + 1) ∈ S →
              f(s(q − 1)) := f(s(q − 1)) min f(s(q + 1))
          ▯d(s(q − 1)) ≤ d(s(q + 1) ∨ s(q + 1) ∉ S → skip
          fi
        ; EXTEND
        od
      ; if q = 0 → skip
        ▯q ≠ 0 → B, G, W, d(s(q − 1)), q, s(q + 2), f(s(q − 1)), r, h(r), S :=
            B ∪ {s(q − 1)}, (G \ {s(q − 1)}) ∪ ((s(q − 1), m)° ∩ W),
            W \ (s(q − 1), m)°, |B|, q + 2, (s(q − 1), m)° \ {s(q − 1)},
            |B|, r + 1, s(q + 1), S ∪ {s(q + 1)}
        fi
      od
    fi
  ; t := nexttrans(t)
  od
```

**Fig. 4.** Iterative stubborn set algorithm.

```
do q ≠ 0 cand s(q) = ∅ →
  if f(s(q − 1)) = d(s(q − 1)) → {found root of an SCC}
    do h(r) ≠ s(q − 1) →
      if m [h(r)⟩ → e, Y := true, Y ∪ {nextmark(m, h(r))}
      ▯¬m [h(r)⟩ → skip
      fi
      ; f(h(r)), r, S := d(s(q − 1)), r − 1, S \ {h(r)}
    od
    ; r, S := r − 1, S \ {h(r)}
  ▯f(s(q − 1)) ≠ d(s(q − 1)) → skip
  fi
  ; if e → q := 0 ▯ ¬e → q := q − 2 fi
od
; if q = 0 → skip ▯ q ≠ 0 → s(q + 1) := pick(s(q)); s(q) := s(q) \ {s(q + 1)} fi
```

**Fig. 5.** The part *EXTEND*.

strongly connected component (SCC) has been found. The symbol $\perp$ denotes the smallest transition in a given but arbitrary linear order on the transitions; we have assumed there to be at least one transition in the net under consideration. We assume $A \subseteq \mathsf{M}_S$ is the set of markings that enable no transition of the net. The function $nextmark(m, t)$ gives the marking $m'$ such that $m\ [t\rangle\ m'$ and the function $nexttrans(t)$ the next transition in the assumed linear order on transitions, the value $\infty \notin T$ for the largest $t \in T$.

```
D, X, Y := X ∩ A, X \ A, ∅
; if D = ∅ → skip ▯ D ≠ ∅ → "report D as deadlocks" fi
; if X = ∅ → skip
  ▯ X ≠ ∅ → B, t, e := ∅, ⊥, false
  ; do ¬e ∧ t ≠ ∞ →
      if t ∈ B → skip
      ▯ t ∉ B → SPLIT1
      ; do q ≠ 0 → EXTEND
        ; do q ≠ 0 cand s(q + 1) ∈ B →
            if d(s(q − 1)) > d(s(q + 1)) ∧ s(q + 1) ∈ S →
                f(s(q − 1)) := f(s(q − 1)) min f(s(q + 1))
            ▯ d(s(q − 1)) ≤ d(s(q + 1)) ∨ s(q + 1) ∉ S → skip
            fi
          ; EXTEND
            od
        ; if q = 0 → skip ▯ q ≠ 0 → SPLIT2 fi
          od
        fi
    ; if e → UNSHELVE(X, p, B, G, W, t, e, d, q, s, f, r, h, S, E)
      ▯ ¬e → t := nexttrans(t)
      fi
      od
  fi
```

**Fig. 6.** A symbolic stubborn set search algorithm.

Note that $s(q)$ is maintained as the *set* of followers of the transition according to the definition of stubborn sets. The programs *SPLIT1* and *SPLIT2* perform as much as possible of the assignments to the program variables as possible before actually splitting the set $X$. The programs *SHELVE* and *UNSHELVE* that have not been explicitly given are assumed to save the given values as a unit and restore them to the given variables, respectively. After a restoration, the unit used to restore the values of the variables is discarded. No assumption is made as to the order of the restoration of the saved units; it is assumed, however, that if there are no more units to use for restoration at the time *UNSHELVE* is called, this will return the values of **true** and 0 for $e$ and $q$, respectively, to terminate the program.

```
do q ≠ 0 cand s(q) = ∅ →
  if f(s(q − 1)) = d(s(q − 1)) → {found root of an SCC}
    do h(r) ≠ s(q − 1) →
    if h(r) ∈ E → e, Y := true, Y ∪ {nextmark(m, h(r)) | m ∈ X}
    ▯h(r) ∉ E → skip
    fi
   ; f(h(r)), r, S := d(s(q − 1)), r − 1, S \ {h(r)}
    od
  ; r, S := r − 1, S \ {h(r)}
   ▯f(s(q − 1)) ≠ d(s(q − 1)) → skip
   fi
 ; if e → q := 0▯¬e → q := q − 2 fi
 od
; if q = 0 → skip▯q ≠ 0 → s(q + 1) := pick(s(q)); s(q) := s(q) \ {s(q + 1)} fi
```

**Fig. 7.** The part *EXTEND*.

```
B, G, W, d(t), q, s(0), s(1), f(t), r, h(0), S, E :=
   B ∪ {t}, ∅, T \ (B ∪ {t}), |B|, 2, ∅, t, |B|, 1, t, {t}, ∅
; X, X', H := X ∩ {m ∈ M_S | m [t⟩ }, X \ {m ∈ M_S | m [t⟩ }, •t
;  do X' ≠ ∅ → g := pick(H); H := H \ {g}
   X', X'' := X' \ {m ∈ M_S | m(g) < W(g, t)}, X' ∩ {m ∈ M_S | m(g) < W(g, t)}
 ; s(q) := •g \ {t}; SHELVE(X'', B, •g \ {t}, W \ •g, t, e, d, q, s, f, r, h, S, E)
   od
; if X = ∅ →
   UNSHELVE(X, B, G, W, t, e, d, q, s, f, r, h, S, E)
 ▯X ≠ ∅ →
   G, W, s(q), E := (•t)• \ {t}, W \ (•t)•, (•t)• \ {t}, E ∪ {t}
 fi
```

**Fig. 8.** The part *SPLIT1*.

```
B, G, d(s(q − 1)), q, f(s(q − 1)), r, h(r), S :=
   B ∪ {s(q − 1)}, G \ {s(q − 1)}, |B|, q + 2, |B|, r + 1, s(q + 1), S ∪ {s(q + 1)}
; X, X', H := X ∩ {m ∈ M_S | m [s(q − 1)⟩ }, X \ {m ∈ M_S | m [s(q − 1)⟩ }, •s(q − 1)
;  do X' ≠ ∅ → g := pick(H); H := H \ {g}
 ; X', X'' := X' \ {m ∈ M_S | m(g) < W(g, s(q − 1))},
    X' ∩ {m ∈ M_S | m(g) < W(g, s(q − 1))}
 ; SHELVE(X'', B, •g \ {s(q − 1)}, W \ •g, t, e, d, q, s, f, r, h, S, E)
   od
; if X = ∅ → UNSHELVE(X, B, G, W, t, e, d, q, s, f, r, h, S, E)
 ▯X ≠ ∅ → G, W, s(q), E := (•t)• \ {t}, W \ (•t)•, (•t)• \ {t}, E ∪ {t}
 fi
```

**Fig. 9.** The part *SPLIT2*.

# Compositional Analysis with Place-Bordered Subnets

Antti Valmari

Tampere University of Technology
Software Systems Laboratory
PO Box 553, FIN-33101 Tampere
FINLAND
ava@cs.tut.fi

**Abstract.** A compositional verification method for Petri nets composed of place-bordered subnets is presented. The method assumes that the Petri net can be divided into an *interesting* and an *environment* component, and it facilitates the verification of all properties which can be stated in terms of the projections of the executions of the net onto its interesting component. For instance, one can check what is the lowest upper bound of the marking of any place in the interesting component, or whether some transition of the interesting component may ever occur. Also deadlocks and a certain class of livelocks can be detected. The method is based on the process-algebraic compositional approach, but is novel in that it can be used to produce state-oriented information, and it works in a framework with asynchronous communication. In the example used for demonstrating the method, an infinite number of systems of different size is analysed with a small finite amount of effort.

## 1 Introduction

Reachability graphs of other than the smallest Petri nets are usually far too big for computer processing. Several techniques have been suggested for alleviating this problem; examples are stubborn sets [11], equivalent markings [7], parameterized markings [10], and minimal coverability graphs [4]. In [12] we suggested the application of process algebraic compositional analysis techniques to Petri nets. There we used transition-bordered subnets (i.e. subnets whose interfaces consist of transitions), because process algebraic ideas apply almost immediately to them. It is, however, often natural to use places as the interface entities instead of transitions. Therefore, in this article we tackle the significantly more difficult task of developing a compositional analysis theory for place-bordered subnets.

The basic idea of the compositional analysis method developed in this paper is to divide a Petri net into two parts or *components*: the *interesting* component and the *environment* component. The components share some places called *border places*, but otherwise they are disjoint. The environment component is taken apart and *condensed*. Condensation destroys details of the internal operation of the environment component, but does not change its interface behaviour. The result of the condensation is a behaviourally simpler net component which, nevertheless, is equivalent to the original environment component as far as the interesting component can see. It is connected to the interesting component, and the behaviour of the resulting net is then analysed using any suitable analysis technique (reachability graph construction, place or transition invariants, fact transitions, place boundedness, ...). The results of the analysis tell about the behaviour of the *original* net as seen from the point of view of the *interesting* component. However, the analysis is computationally easier than the analysis of the original net, because the environment component has been replaced by a component with almost no internal behaviour.

What kinds of properties can be analysed using our method? The basic restriction is that the properties have to be stated in terms of the places and transitions of the interesting component; they cannot refer to the places and transitions of the environment component. It is possible, for instance, to check whether the marking of some place of the interesting component may ever exceed some limit, or whether some transition of the interesting component may ever occur. The issue is not this simple, however. Can one check whether some transition of the interesting component is guaranteed to occur? The answer to this question depends on what exactly constitutes the "interface behaviour" preserved during the condensation of the environment component. The occurrence of a transition in the interesting component may depend on the willingness of the environment component to produce sufficiently many tokens for the border places. Knowing that the environment component *may* produce sufficiently many tokens is not enough, if it also may deadlock or livelock immediately before producing the tokens.

Therefore, what kinds of properties can be analysed using our method depends on the notion of "interface behaviour". The more properties are preserved in the interface behaviour, the more properties can be analysed. On the other hand, the more properties are preserved, the less the environment component can be condensed, and the less savings of effort are obtained. A trade-off has to be made. Experience from essentially the same trade-off situation in process algebra has shown that there is no natural "right" notion of interface behaviour. Instead, literally hundreds of different behavioural models have been suggested. (A discussion of the trade-offs involved in compositional analysis methods is found in [12].)

In this article the so-called *Chaos-Free Failures Divergences model* (*CFFD-model*) [13, 16] is used as the model of interface behaviour. The CFFD-model is a modification of the well-known failures-divergences model of CSP [2, 6], but, unlike the latter, it does not involve a notion of "chaos". The use of the CFFD-model is motivated by the fact that it is exactly the weakest possible compositional semantic model preserving traces, deadlocks and livelocks of finite-state systems in a typical process algebraic setting [8]. In the current framework this implies that the *projection* of every execution of the original net onto the interesting component is preserved, together with the information whether the net may deadlock or livelock at any stage of the projection. This is sufficient for most Petri net analysis tasks. (The reader is warned that there are, however, classical net properties which are *not* preserved: although liveness relative to the initial marking $IM$ (that is, is there $M \in [IM\rangle$ such that $M\ [t\rangle$?) is preserved, liveness in the classical Petri net sense of the word (for every $M \in [IM\rangle$, is there $M' \in [M\rangle$ such that $M'\ [t\rangle$?) is not preserved.) The result that the CFFD-model is the *weakest* suitable model does not necessarily carry over to the current framework, because the proof of the result relies on synchronisation patterns which are not possible with place-bordered subnets. However, it is certainly the weakest currently known suitable model.

We have not yet said anything about how the environment component is condensed. Here we take advantage of the theory and tools available in the process algebra world. In order to introduce more savings, the environment component is decomposed into several smaller place-bordered net components. The behaviour of each of them is represented as a *labelled transition system* (*LTS*). A LTS is a reachability graph -like structure, and it can be constructed similarly to a reachability graph. The LTS of each component is condensed using any CFFD-semantics -preserving condensation algorithm. A group of condensed LTSs is combined into a compound LTS, which again is condensed, and so on until a single LTS representing the environment component is obtained. This LTS can be transformed into a net component simply by interpreting its states as places and edges as transitions, and putting one token into the place corresponding to the initial state. In the practical experiments reported in this article we used the "ARA" tool [15] for LTS condensation and composition. ARA uses a condensation algorithm presented in [13].

Although our approach relies heavily on known results from process algebra, it is novel in two important ways: it applies ordinary process algebraic methods to obtain extensive *state-oriented* information in a framework with *asynchronous* communication. This is made possible by the division of the net into "interesting" and "environment" components. Our approach can be contrasted to [9], where state-oriented information was obtained by adding it directly to the semantic model.

In Section 2 we present the notions of net components and projections of executions onto them. We use the classic Dining Philosophers' net to demonstrate the ideas. Section 3 presents the CFFD-semantics of net components, gives the key theorems underlying our compositional method, and shows the transformations needed to use the ARA tool for the condensation of net components. An analysis of the Dining Philosophers' net is shown in Section 4. It turns out that in this particular example, a *fixed-point* is reached: the behaviours of environment components consisting of four and five philosophers are equivalent. Consequently, they are equivalent to any larger environment component. Therefore, the results obtained by analysing a net consisting of one interesting philosopher and four environment philosophers are valid also for all philosopher nets with more environment philosophers. (That the Dining Philosophers' system has a fixed-point which can be beneficially used in compositional analysis was observed already in [13]. Actually, there the fixed point was reached one step earlier, because the system was divided into components in a different way.)

## 2 Net Components

We first recall the formal definitions of Petri nets and their executions. For the sake of concreteness, we have chosen to use place/transition nets without capacity constraints for the development of our theory. However, the ideas of our method can be easily applied to other net formalisms as well.

Let $N$ denote the set of natural numbers $\{0,1,2,\ldots\}$.

**Definition 2.1** A *Petri net* is the quadruple $PN = (P,T,W,IM)$, where

- $P$ is a finite set of *places* and $T$ is a finite set of *transitions* such that $P \cap T = \emptyset$,
- $W$: $(P \times T) \cup (T \times P) \to N$ is the *arc weight function*, and
- $IM$: $P \to N$ is the *initial marking*. ■

**Definition 2.2** Let $PN = (P,T,W,IM)$ be a Petri net, $p \in P$, $t \in T$, and $X \subseteq P \cup T$.

- $\bullet p = \{ t \in T \mid W(t,p) > 0 \}$, $p\bullet = \{ t \in T \mid W(p,t) > 0 \}$, and $\bullet t$ and $t\bullet$ similarly.
- $\bullet X = \cup_{x \in X} \bullet x$ and $X\bullet = \cup_{x \in X} x\bullet$. ■

**Definition 2.3** Let $PN = (P,T,W,IM)$ be a Petri net.

- A *marking* of $PN$ is any function $M: P \to N$.
- Transition $t \in T$ is *enabled* at marking $M$, denoted by $M\,[t\rangle$, iff
  $\forall\, p \in P: M(p) \geq W(p,t)$.
- If $t$ is enabled at $M$ it may *occur* and produce a new marking $M'$ such that
  $\forall\, p \in P: M'(p) = M(p) - W(p,t) + W(t,p)$.
  This is denoted by $M\,[t\rangle\, M'$.
- Let $M$ and $M'$ be markings of $PN$, $n \in N$, and $t_1, t_2, \ldots, t_n \in T$. We define
  $M\,[t_1t_2\ldots t_n\rangle\, M' \;\Leftrightarrow\; \exists\, M_0,M_1,\ldots,M_n: M_0 = M \wedge M_n = M' \wedge$
  $\forall\, i \in \{1,\ldots,n\}: M_{i-1}\,[t_i\rangle\, M_i$.
  $M\,[t_1t_2\ldots t_n\rangle \;\Leftrightarrow\; \exists\, M': M\,[t_1t_2\ldots t_n\rangle\, M'$.
- The set of markings *reachable* from a marking $M$ is defined as
  $[M\rangle = \{\, M' \mid \exists\, t_1,t_2,\ldots,t_n \in T: M\,[t_1t_2\ldots t_n\rangle\, M' \,\}$. ■

We are interested in detecting *deadlocks* and *livelocks* of nets with our compositional method. A deadlock is a marking where no transition is enabled. The concept of livelock is defined relative to some set $F$ of transitions of the net. A marking $M$ is an $F$-*livelock*, if it is possible that the net executes forever starting at $M$ such that no transition in $F$ ever occurs. In other words, if a marking is not an $F$-livelock, then every infinite execution starting at it is guaranteed to contain at least one occurrence of at least one transition belonging to $F$.

**Definition 2.4** Let $PN = (P,T,W,IM)$ be a net, $M$ its marking, and $F \subseteq T$.

- $M$ is a *deadlock*, iff $\forall\, t \in T: \neg M\,[t\rangle$.
- $M$ is a $F$-*livelock*, iff $PN$ has an infinite execution $M_0\,[t_1\rangle\, M_1\,[t_2\rangle \ldots$ such that $M_0 = M$ and $\forall\, i \in \{1,2,\ldots\}: t_i \notin F$. ■

In order to be able to compose a large Petri net from smaller pieces, we define *net components*. Net components interface with each other via common places, called *border places*. Because the evolution of the marking of the border places of a net component during an execution of the net depends on the behaviours of neighbouring net components, we leave the initial marking of the border places unspecified at this stage. It will be specified later when net components are put together into an ordinary Petri net.

**Definition 2.5** A *net component* is the 5-tuple $NC = (P,T,W,IM,B)$, where

- $P$, $T$, and $W$ are as in a Petri net,
- $B \subseteq P$ is the set of *border places*, and
- $IM$: $(P - B) \to N$ is the *initial marking*.

A *marking* of $NC$ is any function $M$: $(P - B) \to N$. ■

Petri nets can be thought of as net components with empty sets of border places. Therefore, from now on we equate the Petri net $(P,T,W,IM)$ to the net component $(P,T,W,IM,\emptyset)$.

In order to simplify talking about the markings of net components, we introduce an operator for restricting a marking to a smaller set of places, and another for combining the markings of two disjoint sets of places.

**Definition 2.6**

- If $D \subseteq P$ and $M$ is a function $M: P \rightarrow N$, then $M{\downarrow}D$ is the function $(M{\downarrow}D): D \rightarrow N$ such that $\forall\, p \in D: (M{\downarrow}D)(p) = M(p)$.
- If $P_1 \cap P_2 = \emptyset$ and $M_1$ and $M_2$ are functions $M_1: P_1 \rightarrow N$ and $M_2: P_2 \rightarrow N$, then $M_1{^\wedge}M_2$ is the function $(M_1{^\wedge}M_2): (P_1 \cup P_2) \rightarrow N$ such that $(M_1{^\wedge}M_2)(p) = M_1(p)$, if $p \in P_1$, and $(M_1{^\wedge}M_2)(p) = M_2(p)$, if $p \in P_2$. ■

Two operators are used for combining net components into a net: *parallel composition* and *border place initialization.* Parallel composition requires that the net components have nothing in common except some border places. It joins the components by fusing their common border places.

**Definition 2.7** Let $NC_1 = (P_1,T_1,W_1,IM_1,B_1)$ and $NC_2 = (P_2,T_2,W_2,IM_2,B_2)$ be net components. They are *par-composable*, if $T_1 \cap T_2 = P_1 \cap T_2 = T_1 \cap P_2 = \emptyset$ and $P_1 \cap P_2 \subseteq B_1 \cap B_2$. If $NC_1$ and $NC_2$ are par-composable, then their *parallel composition* $NC_1 \mathbin{|||} NC_2$ is defined as the net component $NC = (P,T,W,IM,B)$, where

- $P = P_1 \cup P_2$, $T = T_1 \cup T_2$, $IM = IM_1{^\wedge}IM_2$, and $B = B_1 \cup B_2$.
- $W(x,y) = \begin{cases} W_1(x,y), & \text{if } x, y \in P_1 \cup T_1 \\ W_2(x,y), & \text{if } x, y \in P_2 \cup T_2 \\ 0, & \text{otherwise.} \end{cases}$ ■

Because $B_1 \subseteq P_1$ and $B_2 \subseteq P_2$ by Definition 2.5, we have $B_1 \cap B_2 \subseteq P_1 \cap P_2$. Therefore, in a parallel composition $P_1 \cap P_2 = B_1 \cap B_2$. From now on, whenever we write the parallel composition operator "|||", we assume that its operands are par-composable. This does not imply loss of generality, because we can always give problematic places and transitions new names.

Border place initialization assigns an initial marking to selected border places, and removes those places from the set of border places.

**Definition 2.8** Let $NC = (P,T,W,IM_1,B_1)$ be a net component, $D \subseteq B_1$, and let $IM_D$ be a function $IM_D: D \rightarrow N$. The *initialization* of $NC$ by $IM_D$ is denoted by $NC[IM_D]$, and it is the net component $NC = (P,T,W,IM,B)$, where $B = B_1 - D$ and $IM = IM_1{^\wedge}IM_D$. ■

The compositional method developed in this article assumes that the net $PN$ under analysis is divided into two components $NC_I$ (the "interesting" component) and $NC_E$ (the "environment") with a common set $B$ of border places. That is, $PN = (NC_I \mathbin{|||} NC_E)[IM_B]$, where $IM_B$ initializes all places in $B$.

Our last task in this section is to define the *projections* of executions of the Petri net $PN$ onto its interesting component $NC_I$. In essence, a projection shows how $NC_I$ sees the corresponding execution. In the projection, the successive markings of $PN$ are restricted to the places of $NC_I$ (this time the markings of the border places are included). Similarly, occurrences of only the transitions belonging to $NC_I$ are shown, with one exception. If an occurrence of a transition of $NC_E$ modifies the marking of some place in $B$, then it is shown, because we want the change of the marking be shown. However, the identity of the transition is not shown; it is replaced by the "anonymous transition" symbol "ε". These considerations are formalized in the following definition.

**Definition 2.9** Let $PN = (P,T,W,IM) = (P,T,W,IM,\emptyset)$ be a Petri net. Let $NC_I = (P_I,T_I,W_I,IM_I,B_I)$ and $NC_E = (P_E,T_E,W_E,IM_E,B_E)$ be net components, $B = B_I \cup B_E$, and $IM_B$ a function $IM_B: B \to N$ such that $PN = (NC_I \;|||\; NC_E)[IM_B]$. The operator $\downarrow NC_I$ on a finite execution $M_0\,[t_1\rangle\, M_1\,[t_2\rangle \ldots [t_n\rangle\, M_n$ of $PN$ is defined recursively as follows:

$$
(\, M_0\,[t_1\rangle\, M_1\,[t_2\rangle \ldots [t_n\rangle\, M_n\,)\downarrow NC_I =
\begin{cases}
M_0{\downarrow}P_I, & \text{if } n = 0,\\
(M_0{\downarrow}P_I)\,[t_1\rangle\,(\, M_1\,[t_2\rangle \ldots [t_n\rangle\, M_n\,)\downarrow NC_I, & \text{if } n > 0 \wedge t_1 \in T_I,\\
(M_0{\downarrow}P_I)\,[\varepsilon\rangle\,(\, M_1\,[t_2\rangle \ldots [t_n\rangle\, M_n\,)\downarrow NC_I, & \text{if } n > 0 \wedge t_1 \notin T_I \wedge M_1{\downarrow}P_I \neq M_0{\downarrow}P_I,\\
(\, M_1\,[t_2\rangle \ldots [t_n\rangle\, M_n\,)\downarrow NC_I, & \text{if } n > 0 \wedge t_1 \notin T_I \wedge M_1{\downarrow}P_I = M_0{\downarrow}P_I.
\end{cases}
$$

The operator $\downarrow NC_I$ on infinite executions $M_0\,[t_1\rangle\, M_1\,[t_2\rangle\, M_2\,[t_3\rangle \ldots$ of $PN$ is defined as the limit of the finite case. ■

Note that the result of the application of $\downarrow NC_I$ to an infinite execution may be finite.

Projections can now be defined simply as the results of the application of $\downarrow NC_I$ to the executions of $PN$ starting at the initial marking. In order to analyse deadlock and livelock properties of $PN$, projections of executions leading to a deadlock or livelock are called *deadlocking* and *livelocking* projections, respectively.

**Definition 2.10** Let $PN$, $NC_I$ and $NC_E$ be as in Definition 2.9. Let $k \in N$, $m_i: P_I \to N$ for $0 \le i \le k$, and $f_i \in T_I \cup \{\varepsilon\}$ for $1 \le i \le k$.

- $m_0\,[f_1\rangle\, m_1\,[f_2\rangle \ldots [f_k\rangle\, m_k$ is a *projection* of $PN$ onto $NC_I$, if $PN$ has an execution $M_0\,[t_1\rangle\, M_1\,[t_2\rangle \ldots [t_n\rangle\, M_n$ such that $M_0 = IM$, and $m_0\,[f_1\rangle\, m_1\,[f_2\rangle \ldots [f_k\rangle\, m_k = (\, M_0\,[t_1\rangle\, M_1\,[t_2\rangle \ldots [t_n\rangle\, M_n\,)\downarrow NC_I$.
- $m_0\,[f_1\rangle\, m_1\,[f_2\rangle\, m_2\,[f_3\rangle \ldots$ is an *infinite projection* of $PN$ onto $NC_I$, if $PN$ has an infinite execution $M_0\,[t_1\rangle\, M_1\,[t_2\rangle\, M_2\,[t_3\rangle \ldots$ such that $M_0 = IM$, and $m_0\,[f_1\rangle\, m_1\,[f_2\rangle\, m_2\,[f_3\rangle \ldots = (\, M_0\,[t_1\rangle\, M_1\,[t_2\rangle\, M_2\,[t_3\rangle \ldots\,)\downarrow NC_I$.
- $m_0\,[f_1\rangle\, m_1\,[f_2\rangle \ldots [f_k\rangle\, m_k$ is a *deadlocking projection* of $PN$ onto $NC_I$, if $PN$ has an execution $M_0\,[t_1\rangle\, M_1\,[t_2\rangle \ldots [t_n\rangle\, M_n$ such that $M_0 = IM$, $M_n$ is a deadlock, and $m_0\,[f_1\rangle\, m_1\,[f_2\rangle \ldots [f_k\rangle\, m_k = (\, M_0\,[t_1\rangle\, M_1\,[t_2\rangle \ldots [t_n\rangle\, M_n\,)\downarrow NC_I$.
- $m_0\,[f_1\rangle\, m_1\,[f_2\rangle \ldots [f_k\rangle\, m_k$ is a *livelocking projection* of $PN$ onto $NC_I$, if $PN$ has an execution $M_0\,[t_1\rangle\, M_1\,[t_2\rangle \ldots [t_n\rangle\, M_n$ such that $M_0 = IM$, $M_n$ is a $T_I$-livelock, and $m_0\,[f_1\rangle\, m_1\,[f_2\rangle \ldots [f_k\rangle\, m_k = (\, M_0\,[t_1\rangle\, M_1\,[t_2\rangle \ldots [t_n\rangle\, M_n\,)\downarrow NC_I$. ■

Many different executions of $PN$ may have a common projection $m_0\,[f_1\rangle\, m_1\,[f_2\rangle \ldots [f_k\rangle\, m_k$ onto $NC_I$. Therefore, $m_0\,[f_1\rangle\, m_1\,[f_2\rangle \ldots [f_k\rangle\, m_k$ may be simultaneously a deadlocking and a livelocking projection. The projection of an infinite execution may be finite. If this is the case, then it is a livelocking projection.

**Example 2.11** Figure 1 shows a part of the well-known $n$ Dining Philosophers' net $DP(n)$ divided into components. Each component corresponds to one philosopher, and the border places of the component correspond to forks. For instance, $Phil_1 = (P_1,T_1,W_1,IM_1,B_1)$, where $P_1 = \{p_{11}, p_{12}, p_{13}, p_{14}, b_1, b_2\}$, $T_1 = \{t_{11}, t_{12}, t_{13}, t_{14}\}$, and $B_1 = \{b_1, b_2\}$. Furthermore, $IM_1(p_{11}) = 1$, $IM_1(p_{12}) = IM_1(p_{13}) = IM_1(p_{14}) = 0$, and $IM_1$ is not defined for $b_1$ and $b_2$.

Let the number of philosophers be 5. The net has a deadlock $M_d$ which may be reached from the initial marking by the execution $M_0\,[t_{11}\rangle\, M_1\,[t_{21}\rangle\, M_2\,[t_{31}\rangle\, M_3\,[t_{41}\rangle\, M_4\,[t_{51}\rangle\, M_d$. The projection of this execution onto $Phil_1$ is

$$\chi = \langle 1,0,0,0,1,1\rangle\,[t_{11}\rangle\,\langle 0,1,0,0,0,1\rangle\,[\varepsilon\rangle\,\langle 0,1,0,0,0,0\rangle,$$

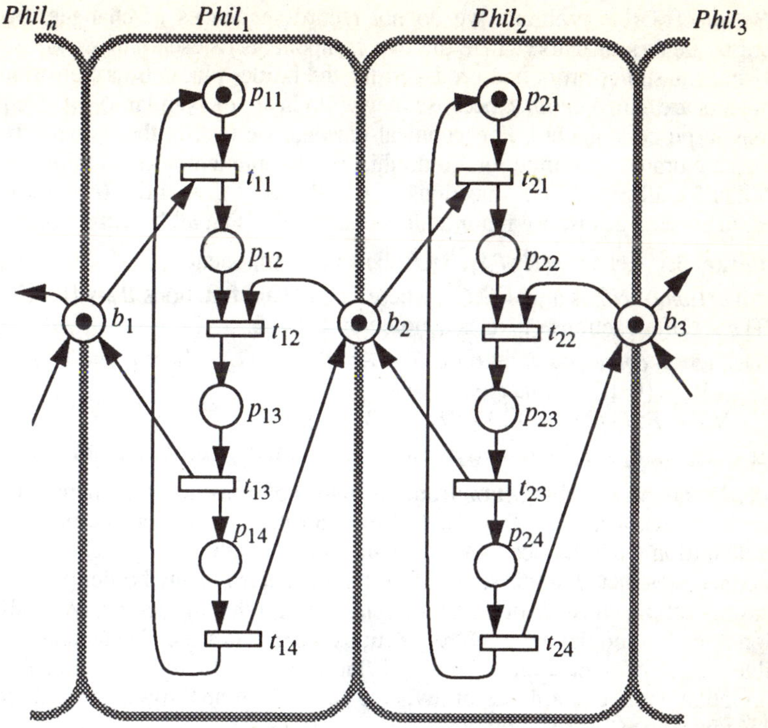


**Fig. 1.** A part of the *n* Dining Philosophers' net $DP(n)$

where each $M{\downarrow}P_1$ is shown in the form $\langle M(p_{11}), M(p_{12}), \ldots, M(b_2)\rangle$. Because $M_d$ is a deadlock, $\chi$ is a deadlocking projection.

The net has also an infinite execution $M_0\ [t_{11}\rangle\ M_1\ [t_{21}\rangle\ M_2\ [t_{31}t_{32}t_{33}t_{34}\rangle\ M_2\ [t_{31}t_{32}t_{33}t_{34}\rangle \ldots$ . The fact that $\{t_{31}, t_{32}, t_{33}, t_{34}\} \cap T_1 = \emptyset$ implies that $M_2$ is a $T_1$-livelock of $DP(5)$. Consequently, $(\ M_0\ [t_{11}\rangle\ M_1\ [t_{21}\rangle\ M_2\ ){\downarrow}Phil_1$ is a livelocking projection. But $(\ M_0\ [t_{11}\rangle\ M_1\ [t_{21}\rangle\ M_2\ ){\downarrow}Phil_1 = \chi$. Therefore, $\chi$ is simultaneously a deadlocking and a livelocking projection of $DP(5)$ onto $Phil_1$. Knowing that $\chi$ has these properties is sufficient for concluding that the five Dining Philosophers' system may deadlock, and the first philosopher in it may starve. ■

## 3 Abstraction of Net Components

It is not necessary to know $NC_E$ in full detail in order to find out the ordinary, infinite, deadlocking and livelocking projections of $PN = (NC_I \,|||\, NC_E)[IM_B]$ onto $NC_I$. Instead, suitably chosen information about the effect of $NC_E$ on border places suffices. Apparently, this information should contain at least the sequences of $NC_E$-caused changes to the markings of the border places (cf. the use of ε-symbols in Definition 2.9). However, this is not sufficient, because an individual transition $t_E$ of $NC_E$ may both consume tokens from and produce tokens for the same border place. It is not possible to say whether the change caused by $t_E$ may actually occur in a given marking of border places, unless we know also how many tokens $t_E$ wants to "borrow" from the border places during its occurrence.

Because of this problem, we do not record sequences of changes. Instead, we record sequences of pairs, where the two components of each pair show the numbers of tokens consumed from and produced for the border places by a transition of $NC_E$. Such pairs are called *actions*, because they will play a role similar to the role played by actions in process algebra. For technical reasons, we restrict the domain from which actions are drawn to a finite set. To do this, we assume that *max* is an integer constant such that for all places and transitions of *PN*, $W(p,t) \le max$ and $W(t,p) \le max$. Such a constant exists, because we required in Definition 2.1 that all nets are finite.

**Definition 3.1** Let $NC = (P,T,W,IM,B)$ be a net component.

- An *action* of *NC* is a pair $(I,O)$, where $I$ and $O$ are functions $B \to \{0,1,2,\ldots,max\}$. The set of all actions of *NC* is denoted by $\Sigma_\tau(NC)$.
- The action *corresponding* to $t \in T$ is $Act(t) = (I_t,O_t)$, where $I_t$ and $O_t$ are the functions $B \to \{0,1,2,\ldots,max\}$ such that
  $\forall p \in B$: $I_t(p) = W(p,t) \wedge O_t(p) = W(t,p)$.
- If $t_1, t_2, \ldots, t_n \in T$, then $Act(t_1t_2\ldots t_n) = Act(t_1)Act(t_2)\ldots Act(t_n)$. ■

Actions represent abstraction from the names of transitions; the name of each transition is thrown away, but the effect of the transition on border places is kept. In the next definition we introduce "$-Act(t)\rightarrow$"-notation for talking about executions of a net component abstracted in this sense. If $t$ is not connected to any border places, its corresponding action consists of constant functions producing zeroes. We call such an action *invisible* and denote it by "$\tau$". Actions other than $\tau$ are *visible*, and the set of all visible actions of a net component is called its *alphabet*. It will turn out that most of the information about occurrences of invisible actions can be thrown away. In order to do that we introduce "$=\sigma\Rightarrow$"-notation.

**Definition 3.2** Let $NC = (P,T,W,IM,B)$ be a net component and $M$ and $M'$ its markings.

- If $a \in \Sigma_\tau(NC)$, we define
  $M -a\rightarrow M' \Leftrightarrow \exists\, t \in T$: $a = Act(t) \wedge \forall p \in (P-B)$:
  $M(p) \ge W(p,t) \wedge M'(p) = M(p) - W(p,t) + W(t,p)$.
- Let $a_1, a_2, \ldots, a_n \in \Sigma_\tau(NC)$. The notations $M -a_1a_2\ldots a_n\rightarrow M'$ and $M -a_1a_2\ldots a_n\rightarrow$ are defined as generalizations of $M -a\rightarrow M'$ similarly to the definition of $M\,[t_1t_2\ldots t_n\rangle\, M'$ and $M\,[t_1t_2\ldots t_n\rangle$ from $M\,[t\rangle\, M'$.
- The particular action $(I,O)$ such that $\forall p \in B$: $I(p) = O(p) = 0$ is called the *invisible action*, and it is denoted by $\tau$.
- The *alphabet* of *NC* is $\Sigma(NC) = \Sigma_\tau(NC) - \{\tau\}$.
- Let $X$ and $Y$ be sets and $\sigma \in X^*$ (i.e. $\sigma$ is a finite string of elements of $X$). We denote the *restriction* of $\sigma$ to $Y$ (i.e. the result of removing all elements not in $Y$ from $\sigma$) by $restr(\sigma,Y)$.
- Let $\sigma \in \Sigma(NC)^*$. We write $M =\sigma\Rightarrow M'$, iff there is $\rho \in \Sigma_\tau(NC)^*$ such that $M -\rho\rightarrow M'$ and $\sigma = restr(\rho, \Sigma(NC))$.
- $M =\sigma\Rightarrow\ \Leftrightarrow \exists M'$: $M =\sigma\Rightarrow M'$. ■

As was mentioned in the introduction, it is not known what is the minimum amount of information of a net component *NC* which has to be preserved in its abstraction, but it is known (and shown soon) that the CFFD-model contains sufficient information. Any finite sequence of visible actions generated by an execution of *NC* starting at its initial marking is called a *trace*. The CFFD-model preserves all traces of *NC*. *Infinite traces* are defined in a similar way, and the CFFD-model preserves also them. In order to be able to check whether $(NC_I \mid\mid\mid NC_E)[IM_B]$ may deadlock after $NC_E$ has executed some trace, also the so-called *stable failures* are preserved. A stable failure consists of a trace and a set of visible actions such that after the trace, *NC* may be in a marking where it can execute neither any action from the set, nor the invisible action. Finally, to facilitate the analysis of livelocks, traces after which *NC* may execute invisible actions without limit are included to the CFFD-model. Such traces are called *divergence traces.*

**Definition 3.3** Let $NC = (P,T,W,IM,B)$ be a net component.

- The set of *traces* of *NC* is $tr(NC) = \{ \sigma \in \Sigma(NC)^* \mid IM = \sigma \Rightarrow \}$.
- The set of *infinite traces* of *NC* is $inftr(NC) = \{ \omega \in \Sigma(NC)^\infty \mid IM = \omega \Rightarrow \}$.
- The set of *stable failures* of *NC* is $sfail(NC) =$
  $\{ (\sigma,A) \in \Sigma(NC)^* \times 2^{\Sigma(NC)} \mid \exists M: IM = \sigma \Rightarrow M \wedge \forall a \in A \cup \{\tau\}: M \not{-}a\not\rightarrow \}$.
- The set of *divergence traces* of *NC* is $divtr(NC) =$
  $\{ \sigma \in \Sigma(NC)^* \mid \exists M_0, M_1, \ldots: IM = \sigma \Rightarrow M_0 \wedge M_0 - \tau \rightarrow M_1 - \tau \rightarrow M_2 - \tau \rightarrow \ldots \}$.
- The *CFFD-model* of *NC* is $(B, sfail(NC), divtr(NC), inftr(NC))$.
- Two net components $NC_1 = (P_1,T_1,W_1,IM_1,B_1)$ and $NC_2 = (P_2,T_2,W_2,IM_2,B_2)$ are *CFFD-equivalent*, denoted by $NC_1 \approx NC_2$, iff $B_1 = B_2$, $sfail(NC_1) = sfail(NC_2)$, $divtr(NC_1) = divtr(NC_2)$, and $inftr(NC_1) = inftr(NC_2)$. ■

(The CFFD-model given in [13, 16] contains an extra component recording the so-called *stability* of the initial marking. This component was needed to ensure compositionality with respect to the "choice" operator found in several process algebras. However, our current framework lacks the choice operator and indeed the power to mimic it, so the stability component is not needed.)

Remembering the introduction to Definition 3.3, it may be surprising to see that the CFFD-model of *NC* does not explicitly contain *tr*(*NC*). This is because, as is shown by the next theorem, *tr*(*NC*) can be uniquely determined from the other components of the model. So it is not needed as a separate component. Furthermore, with finite-state systems also *inftr*(*NC*) is unnecessary. A proof of the theorem can be found in [16].

**Theorem 3.4** Let *NC* be a net component.

(a) $tr(NC) = divtr(NC) \cup \{ \sigma \mid (\sigma,\emptyset) \in sfail(NC) \}$

(b) If $\{ M \mid \exists \rho \in \Sigma_\tau(NC)^*: IM - \rho \rightarrow M \}$ is finite, then
$inftr(NC) = \{ a_1 a_2 a_3 \ldots \mid \forall n: a_1 a_2 \ldots a_n \in tr(NC) \}$. ■

The following is the first of the two main theorems underlying the compositional method developed in this paper. It says that the CFFD-model preserves enough information for our purposes.

**Theorem 3.5** Let $PN_1$ and $PN_2$ be Petri nets of the form $PN_i = (NC_I \mid\mid\mid NC_{Ei})[IM_B]$, where $NC_I$, $NC_{E1}$, and $NC_{E2}$ have a common set *B* of border places. If $NC_{E1} \approx NC_{E2}$, then

(a) $m_0 [f_1\rangle m_1 [f_2\rangle \ldots [f_k\rangle m_k$ is a projection of $PN_1$ onto $NC_I$ if and only if it is a projection of $PN_2$ onto $NC_I$.

(b) $m_0 [f_1\rangle m_1 [f_2\rangle \ldots$ is an infinite projection of $PN_1$ onto $NC_I$ if and only if it is an infinite projection of $PN_2$ onto $NC_I$.

(c) $m_0 [f_1\rangle m_1 [f_2\rangle \ldots [f_k\rangle m_k$ is a deadlocking projection of $PN_1$ onto $NC_I$ if and only if it is a deadlocking projection of $PN_2$ onto $NC_I$.

(d) $m_0\,[f_1\rangle\, m_1\,[f_2\rangle \ldots [f_k\rangle\, m_k$ is a livelocking projection of $PN_1$ onto $NC_I$ if and only if it is a livelocking projection of $PN_2$ onto $NC_I$.

**Proof** The proof of (a) is based on taking an execution of $PN_1$ with a given projection, constructing the corresponding trace of $NC_{E1}$, showing that it is also a trace of $NC_{E2}$, and interleaving the trace and the original projection into an execution of $PN_2$. Claims (b) and (d) are proven by investigating infinite executions in the same way. To prove (c), the properties of the last states of the executions of $PN_1$ and $PN_2$ in the proof of (a) are investigated. Although the idea of the proof is simple, the proof is long, because lots of formal details are required.

We first introduce some auxiliary notation. Let $PN_i = (P_i,T_i,W_i,IM_i)$ and $NC_{Ei} = (P_{Ei},T_{Ei},W_{Ei},IM_{Ei},B)$ for $i \in \{1,2\}$, and $NC_I = (P_I,T_I,W_I,IM_I,B)$. Because both $NC_{E1}$ and $NC_{E2}$ have $B$ as their set of border places, we have $\Sigma(NC_{E1}) = \Sigma(NC_{E2})$, and we may write simply $\Sigma = \Sigma(NC_{Ei})$. We define $Act_E(t) = Act(t)$, if $t \in T_{Ei}$, and $Act_E(t) = t$, if $t \in T_I$. Furthermore, let $Act_E(t_1t_2\ldots t_n) = Act_E(t_1)Act_E(t_2)\ldots Act_E(t_n)$. Let $M_I, M'_I$: $(P_I - B) \to N$ (i.e. $M_I$ and $M'_I$ are markings of $NC_I$), $M_B, M'_B$: $B \to N$ and $M'_{Ei}$: $(P_{Ei} - B) \to N$. We write $M_I\wedge M_B =\pi\Rightarrow_i M'_I\wedge M'_B$, if and only if there are $M'_{Ei}$: $(P_{Ei} - B) \to N$ and $t_{i1}, t_{i2}, \ldots, t_{in} \in T_i$ (note that $T_i = T_I \cup T_{Ei}$) such that $M_I\wedge M_B\wedge M_{Ei}\,[t_{i1}t_{i2}\ldots t_{in}\rangle\, M'_I\wedge M'_B\wedge M'_{Ei}$ in $PN_i$ and $\pi = restr(Act_E(t_{i1}t_{i2}\ldots t_{in}), T_I \cup \Sigma)$.

Assume that $IM_I\wedge IM_B =\pi\Rightarrow_1 M'_I\wedge M'_B$ in the initial marking of $PN_1$. Let $M'_{E1}$: $(P_{E1} - B) \to N$ and $t_{11}, t_{12}, \ldots, t_{1n} \in T_1$ be those whose existence was declared in the definition of "$=\pi\Rightarrow_i$". By Definition 3.2, $IM_{E1} -\rho\rightarrow M'_{E1}$, where $\rho = restr(Act_E(t_{11}t_{12}\ldots t_{1n}), \Sigma \cup \{\tau\})$. Furthermore, $IM_{E1} =\sigma\Rightarrow M'_{E1}$, where $\sigma = restr(\rho,\Sigma) = restr(\pi, \Sigma)$. So $\sigma \in tr(NC_{E1})$. By Theorem 3.4 and $NC_{E1} \approx NC_{E2}$ we conclude that $\sigma \in tr(NC_{E2})$. There are thus $M'_{E2}$ and $t'_1, t'_2, \ldots, t'_h \in T_{E2}$ such that $IM_{E2} -Act(t'_1t'_2\ldots t'_h)\rightarrow M'_{E2}$ and $restr(Act(t'_1t'_2\ldots t'_h), \Sigma) = \sigma$.

We now construct from $\pi$ and $t'_1t'_2\ldots t'_h$ a sequence of transitions $t_{21}t_{22}\ldots t_{2m} \in T_2^*$ such that $\pi = restr(Act_E(t_{21}t_{22}\ldots t_{2m}), T_I \cup \Sigma)$ and $IM_I\wedge IM_B\wedge IM_{E2}\,[t_{21}t_{22}\ldots t_{2m}\rangle\, M'_I\wedge M'_B\wedge M'_{E2}$ in $PN_2$. If $t'_1t'_2\ldots t'_h$ is not empty, and $Act(t'_1) = \tau$ or $Act(t'_1)$ is the first symbol in $\pi$, then let $t_{21} = t'_1$; otherwise, let $t_{21}$ be the first symbol in $\pi$. In the latter case, $t_{21} \in T_I$, because $restr(Act(t'_1t'_2\ldots t'_h), \Sigma) = \sigma = restr(\pi, \Sigma)$. We delete $t_{21}$ from the head of $t'_1t'_2\ldots t'_h$ if it is its first element, and delete $Act_E(t_{21})$ from the head of $\pi$ if it is its first element (it is possible that we have to do both deletions). We produce $t_{22}$, $t_{23}$ etc. in a similar way until both $t'_1t'_2\ldots t'_h$ and $\pi$ are exhausted. By construction, $restr(t_{21}t_{22}\ldots t_{2m}, T_{E2}) = t'_1t'_2\ldots t'_h$. Also, each $Act_E(t_{2x})$ is either $\tau$ or the first symbol in the remaining part of $\pi$, so $restr(Act_E(t_{21}t_{22}\ldots t_{2m}), T_I \cup \Sigma) = \pi$. This and the fact that $\pi$ was originally obtained from an executable sequence of transitions guarantee that $t_{21}t_{22}\ldots t_{2m}$ is executable in $PN_2$ as far as the marking of the places in $P_I$ is concerned. That $restr(t_{21}t_{22}\ldots t_{2m}, T_{E2}) = t'_1t'_2\ldots t'_h$ and $IM_{E2} -Act(t'_1t'_2\ldots t'_h)\rightarrow M'_{E2}$ guarantee that the marking of the places in $P_{E2} - B$ causes no problem either. In conclusion, $IM_I\wedge IM_B\wedge IM_{E2}\,[t_{21}t_{22}\ldots t_{2m}\rangle\, M'_I\wedge M'_B\wedge M'_{E2}$ in $PN_2$.

We have now shown that $IM_I\wedge IM_B =\pi\Rightarrow_1 M'_I\wedge M'_B$ in the initial marking of $PN_1$ implies $IM_I\wedge IM_B =\pi\Rightarrow_2 M'_I\wedge M'_B$ in the initial marking of $PN_2$. Claim (a) follows from this, because all projections of $PN_i$ onto $NC_I$ can be obtained by introducing all intermediate markings to $IM_I\wedge IM_B =\pi\Rightarrow_i M'_I\wedge M'_B$ ($\pi$ determines them in a unique way), removing those $\Sigma$-steps which do not modify the marking of $P_I$, and replacing $\varepsilon$ for the action names in the remaining $\Sigma$-steps.

Similarly to the above (using $inftr(NC_{E1}) = inftr(NC_{E2})$ and $divtr(NC_{E1}) = divtr(NC_{E2})$ if $\rho$ is infinite), it can be shown that if $IM_I\wedge IM_B =\pi\Rightarrow_1$ where $\pi$ has been obtained from an infinite execution of $PN_1$, then $PN_2$ has an infinite execution from which $IM_I$^$IM_B$ $=\pi\Rightarrow_2$ may be obtained. This gives (d), because livelocking projections correspond to infinite executions such that $\pi$ contains only a finite number of instances of elements of $T_I$; and (b), because infinite projections correspond to infinite executions such that $\pi$ contains an infinite number of instances of elements of $T_I$ and / or marking-changing elements of $\Sigma$.

To prove (c), assume that $M'_I$^$M'_B$^$M'_{E1}$ is a deadlock of $PN_1$, and $\sigma$ is as in the proof of (a). Let $A$ be the set of actions $a = (I,O)$ of $NC_{E1}$ such that $\forall\ p \in B$: $I(p) \leq M'_B(p)$. We have $\forall\ a \in A$: $M'_{E1}$ $-a\nrightarrow$, because otherwise $M'_I$^$M'_B$^$M'_{E1}$ would not be a deadlock. So $(\sigma, A-\{\tau\}) \in sfail(NC_{E1})$. Because $sfail(NC_{E1}) = sfail(NC_{E2})$, the marking $M'_{E2}$ in the proof of (a) can be chosen such that $\forall\ a \in A$: $M'_{E2}$ $-a\nrightarrow$. So, if $t \in T_{E2}$ and $Act(t) \in A$, then $t$ is not enabled at $M'_I$^$M'_B$^$M'_{E2}$. If $t \in T_{E2}$ and $Act(t) \notin A$, then $t$ is not enabled at $M'_I$^$M'_B$^$M'_{E2}$, because $W(p,t) > M'_B(p)$ for some $p \in B$ by the definition of $A$. Finally, if $t \in T_I$, then $\bullet t \subseteq P_I$, and $t$ is disabled at $M'_I$^$M'_B$^$M'_{E2}$ because it is disabled at $M'_I$^$M'_B$^$M'_{E1}$. Therefore, $M'_I$^$M'_B$^$M'_{E2}$ is a deadlock of $PN_2$. This completes the proof of the theorem. ■

The second necessary theorem says that the CFFD-model is *compositional* in the sense that its corresponding equivalence is a congruence with respect to all net composition operators we use. In other words, composing equivalent net components leads to equivalent results. This theorem gives us the permission to compute the CFFD-semantics of $NC_E$ in a compositional way.

**Theorem 3.6** Let $NC$, $NC'$, $NC_1$, $NC'_1$, $NC_2$, and $NC'_2$ be net components.

- If $NC_1 \approx NC'_1$, $NC_2 \approx NC'_2$, $NC_1$ and $NC_2$ are par-composable, and $NC'_1$ and $NC'_2$ are par-composable, then $NC_1$ ||| $NC_2 \approx NC'_1$ ||| $NC'_2$.
- If $NC \approx NC'$, then $NC[IM_D] \approx NC'[IM_D]$. ■

The proof of the theorem is rather tedious, so we omit it. It is based on showing that "|||" is the same as the interleaving operator of Basic LOTOS, and "$[IM_D]$" can be built from the synchronization, renaming and hiding operators of Basic LOTOS. Then the result follows immediately from the fact that CFFD-equivalence is a congruence with respect to Basic LOTOS interleaving, synchronization, renaming and hiding, shown in [13, 16]. Alternatively, one can prove the result directly by expressing the CFFD-models of $NC_1$ ||| $NC_2$ and $NC[IM_D]$ as functions of the CFFD-models of $NC_1$, $NC_2$, and $NC$. This was the proof strategy used in [13, 16].

To implement the compositional method based on Theorems 3.5 and 3.6, we need an algorithm which, given a net component, produces a behaviourally simpler but CFFD-equivalent net component. As was mentioned in the introduction, such an algorithm and its implementation are already available in the process algebra world. The algorithm manipulates so-called *labelled transition systems* (*LTS*). We can use this algorithm if we first introduce CFFD-semantics preserving transformations from net components to LTSs and back.

In process algebra, the LTS of a system is defined as a function of its states and the "$-a\rightarrow$"-relation in almost the same way as the reachability graph of a Petri net is defined as a function of its reachable markings and the "$[t\rangle$"-relation. The only difference is that the edges of the LTS are labelled by action names. The LTS defined in this way is CFFD-equivalent to the system. Therefore, the very definition of LTSs can be used as the forward transformation.

**Definition 3.7** Let $NC = (P,T,W,IM,B)$ be a net component. The *labelled transition system* of $NC$ is the quadruple $(S,\Sigma_\tau,\Delta,is)$, where

- $S = \{ M \mid \exists \rho \in \Sigma_\tau^*: IM -\rho\rightarrow M \}$
- $\Sigma_\tau = \Sigma_\tau(NC)$
- $\Delta = \{ (M,a,M') \in S \times \Sigma_\tau \times S \mid M -a\rightarrow M' \}$
- $is = IM$ ■

The backward transformation is performed simply by interpreting the states of the LTS as places, edges as transitions, and the initial state as the only initially marked place.

**Definition 3.8** Let $(S,\Sigma_\tau,\Delta,is)$ be a LTS, where $\Sigma_\tau$ is the set of pairs of functions of the form $B \rightarrow \{0,1,2,\ldots,max\}$ for some set $B$. Its *canonical net component* is $(P,T,W,IM,B)$, where

- $P = S \cup B$ and $T = \Delta$,
- $IM(is) = 1$, and $IM(p) = 0$ when $p \in P - \{is\}$,
- if $t = (s,a,s')$ where $a = (I,O)$, then $W(s,t) = W(t,s') = 1$, and $W(p,t) = I(p)$ and $W(t,p) = O(p)$ for every $p \in B$; and $W(t,p) = W(p,t) = 0$ in all other cases. ■

That this transformation preserves CFFD-semantics can be checked easily by constructing the LTS of the result, and noticing that it is the same as the original LTS except names of states.

## 4 Analysis of the Dining Philosophers' Net

Consider again the Dining Philosophers' net shown in Figure 1. The LTS constructed from $Phil_2$ according to Definition 3.7 is shown in Figure 2. Actions are shown by giving two rows, the upper showing $I(b_2)$ and $I(b_3)$, and the lower showing $O(b_2)$ and $O(b_3)$.

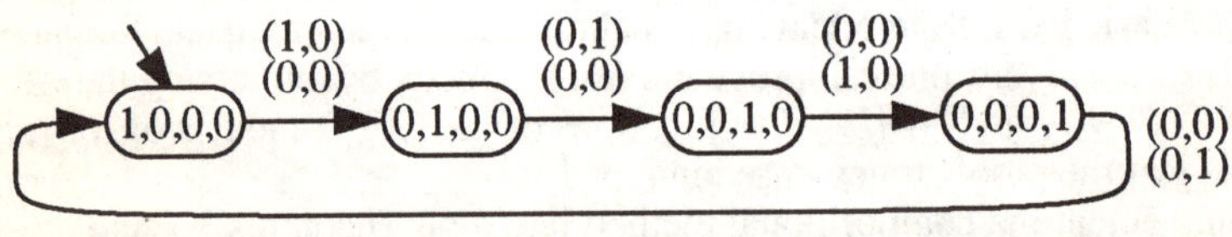


**Fig. 2.** LTS of $Phil_2$

The condensation algorithm of the ARA tool does not modify $LTS(Phil_2)$ except that it destroys the names of the states. Correspondingly, the canonical net component $cond_2$ of the condensed LTS is the same as $Phil_2$, except that the names of all transitions and the names of the four places $p_{21}, p_{22}, p_{23}$ and $p_{24}$ have been changed.

The condensation algorithm becomes useful when two philosophers are put together. The LTS corresponding to $Phil_{23} = (Phil_2 \mid\mid\mid Phil_3)[IM(b_3)=1]$ consists of 12 states and 20 edges. The ARA tool condenses it to an LTS with 7 states and 11 edges. The resulting canonical net component $cond_{23}$ is shown in Figure 3. Although it has more transitions and almost the same number of places as $Phil_{23}$, its behaviour is much simpler, because it contains always exactly one token.

Next we compute a canonical net component $cond_{234}$ corresponding to a condensed LTS of $Phil_{234} = (Phil_2 \mid\mid\mid Phil_3 \mid\mid\mid Phil_4)[IM(b_3)=1, IM(b_4)=1]$. We can either compute the LTS of $Phil_{234}$ directly (36 states and 84 edges), condense it (7 states and

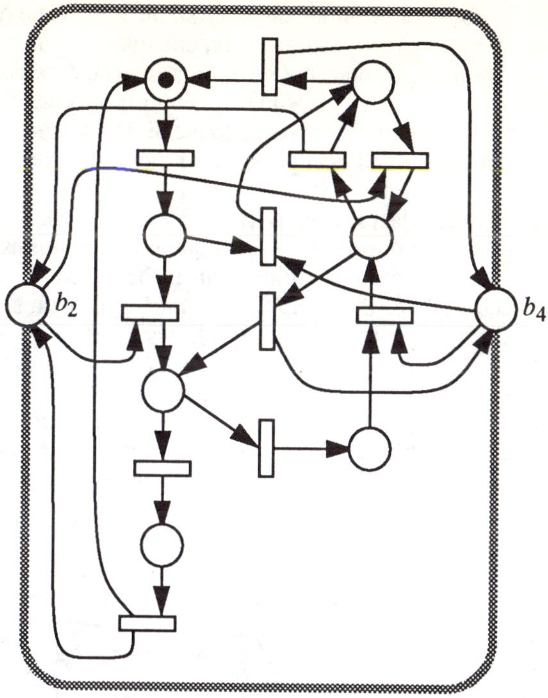


**Fig. 3.** A condensed version of $(Phil_2 \mathbin{|||} Phil_3)[IM(b_3)=1]$

15 edges), and transform the result into a net component; or we can use the result of the previous computation by computing the LTS of $(cond_{23} \mathbin{|||} Phil_4)[IM(b_4)=1]$ (20 states and 44 edges), condense it (7 states and 15 edges), and transform the result into a net component. Theorem 3.6 guarantees that the results are CFFD-equivalent.

Let us keep on putting more philosophers together and condensing the results. Table 1 shows the sizes of the LTSs of $Phil_{23\ldots n+1}$ computed directly, and compositionally as suggested by the formula $Phil_{23\ldots n+1} \approx (cond_{23\ldots n} \mathbin{|||} Phil_{n+1})[IM(b_{n+1})=1]$. It shows also the size of $cond_{23\ldots n+1}$.

**Table 1**

| | direct | | compositional | | condensed | |
|---|---|---|---|---|---|---|
| $n$ | states | edges | states | edges | states | edges |
| 1 | 4 | 4 | | | 4 | 4 |
| 2 | 12 | 20 | 12 | 20 | 7 | 11 |
| 3 | 36 | 84 | 20 | 44 | 7 | 15 |
| 4 | 108 | 324 | 20 | 56 | 6 | 14 |
| 5 | 324 | 1188 | 20 | 56 | 6 | 14 |
| 6 | 972 | 4212 | 20 | 56 | 6 | 14 |

Two things can be observed from the table. First, the size of the directly computed LTS grows rapidly in $n$ (it is known to grow exponentially), whereas the size of the compositionally constructed LTS does not seem to grow fast. As a consequence, when $n$ is large, the construction of $cond_{23...n}$ with the compositional method saves a significant amount of computational effort compared to the direct method.

Second, because $cond_{23...5}$ and $cond_{23...6}$ are of the same size, it is worthwhile to check whether they are equivalent. Strictly speaking, they cannot be, because they have different sets of border places (namely $\{b_2,b_6\}$ vs. $\{b_2,b_7\}$). But this is a superficial difference which can be removed by renaming the border places. ARA has a tool for verifying CFFD-equivalence. If we hide from it the difference in the naming of border places, it indeed declares that $cond_{23...5} \approx cond_{23...6}$. With appropriate renaming of the border places, we can now reason as follows: $cond_{23...n+2} \approx (cond_{23...n+1} \parallel\!\mid Phil_{n+2})[IM(b_{n+2})=1] \approx (cond_{23...n} \parallel\!\mid Phil_{n+1})[IM(b_{n+1})=1] \approx cond_{23...n+1}$ whenever $n \geq 5$. As a consequence, $cond_{23...n} \approx cond_{23...5}$ for every $n \geq 5$.

The above result explains why the size of $cond_{23...n}$ stops growing. But it can be used for more. Due to it, results obtained from an analysis of the behaviour of $(Phil_1 \parallel\!\mid cond_{23...5})\ [IM(b_1)=1, IM(b_2)=1]$ are valid for all Philosophers' nets of size 5 or more. This way a limited analysis effort produces results which are valid for all except the few smallest Philosophers' nets.

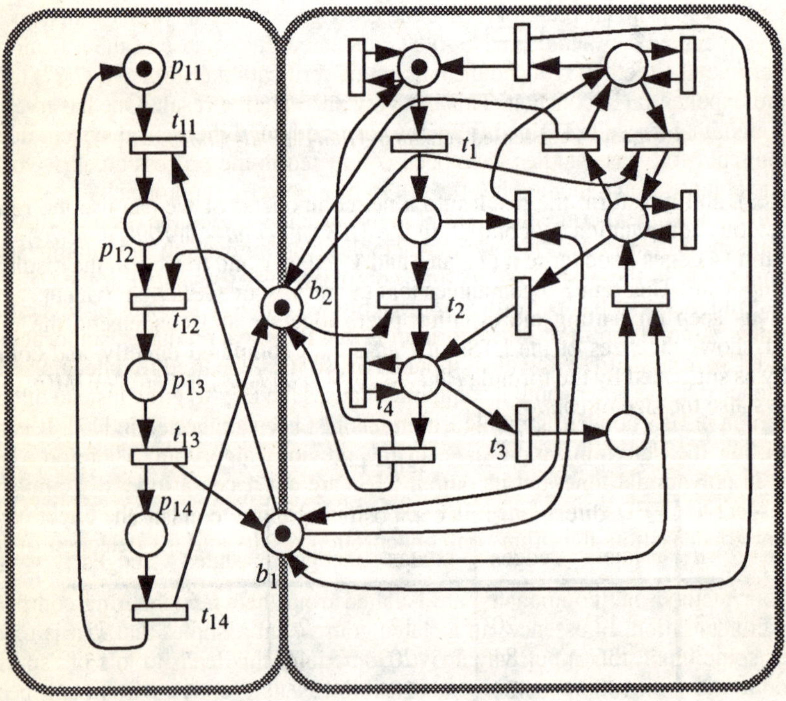


**Fig. 4.** The net $(Phil_1 \parallel\!\mid cond_{23...5})\ [IM(b_1)=1, IM(b_2)=1]$

The net $(Phil_1 \parallel\!\mid cond_{23...5})[IM(b_1)=1, IM(b_2)=1]$ is shown in Figure 4. It can reach a deadlock by executing the transition sequence $t_{11}t_1t_2t_3$. The projection of this execution onto $Phil_1$ is thus deadlocking. The projection is the same as the $\chi$ in Example 2.11. Projecting the infinite execution $t_{11}t_1t_2t_4t_4t_4$... onto $Phil_1$ produces again $\chi$, so $\chi$ is also a livelocking projection. These results are in agreement with the results of Example 2.11, but this time they are valid for any number of philosophers greater than 4.

## 5 Discussion

We developed a compositional analysis method suitable for nets consisting of subnets interfacing through shared places. The novelty of our method is in that it can produce state-oriented information (i.e. information about markings) and that it is developed in a framework where communication is not synchronous. The method is much more natural from the Petri net point of view than our earlier compositional method developed in [12], because it is more directly applicable to typical Petri net analysis questions.

The condensation of environment components in our method is in some sense similar to net reductions [1]. However, condensation modifies nets in a much more tangled way than typical net reductions do. It might even be possible to apply the condensation theory for developing and checking new net reductions: a net reduction preserves deadlocks etc. if the original and reduced net fragment are CFFD-equivalent, and CFFD-equivalence can be checked automatically.

A key question posed to every new effort-saving analysis method is: how much savings does it yield in practice? Of course, to give a fair answer to this question, extensive practical case studies have to be conducted. In the case of the $n$ Dining Philosophers' net the method reduced the amount of verification effort from $O(2^n)$ to $O(1)$, i.e. from exponential to constant. This is a very encouraging result. One has to remember, though, that a result obtained from one case study of a theoretical system does not prove much. More case studies have been conducted in the process algebra world. In them the compositional method has proven to be a powerful approach.

In the practical experiments reported in this article we used the condensation algorithm presented in [13]. This algorithm resembles determinization and minimization of finite automata. In the worst case its time complexity is exponential in the size of the LTS, but in practice it has performed reasonably well up to LTS sizes of several thousand states. A good aspect of this algorithm is that it always produces isomorphic condensed LTSs from CFFD-equivalent input LTSs, so it can be used for checking CFFD-equivalence. On the other hand, it sometimes causes the LTS to grow instead of shrink. An alternative LTS condensation algorithm has just been suggested in [17]. It is based on omitting the determinization stage in the previous algorithm. Therefore, it runs always in polynomial time, but its output LTSs are not necessarily easily comparable with respect to CFFD-equivalence. It would be interesting to check the effect of these different condensation algorithms on condensation results and the detection of fixed-points.

In our method, net components are isolated from their neighbouring components during condensation. In essence, an isolated component assumes that if it is ready to process some input, that input may arrive. Sometimes this leads to lots of "spurious" behaviour which the component does not exhibit in its proper environment, because the environment does not give it all possible inputs. There are practical examples where the behaviours of isolated components are larger than the behaviour of the complete system. Of course, in such a situation our compositional method is useless.

Apparently, to solve the spurious behaviour problem, some aspects of the behaviour of the neighbourhood of a net component should be taken into account during condensation. A nice way of doing this in the process algebra world was given in [5], and a variant of it was developed in [3]. An interesting topic for future research is how the techniques in these articles could be transported into the net component framework.

The main reason why our compositional method saves effort is that it breaks the computation of the behaviour of the parallel composition of all net components of a net into several steps. However, it is still necessary to compute behaviours of parallel compositions of few net components, and this can sometimes be costly. The article [14] transports the *stubborn set* theory [11] into the process algebra world. It develops a stubborn set -based technique for computing a reduced version of the parallel composition of labelled transition systems without first computing the full version. The reduced version is CFFD-equivalent with the full parallel composition, but it may be much smaller. This technique transports immediately into our present framework. It can be used for speeding up the computation of parallel compositions during compositional analysis.

## Acknowledgements

This work was partially funded by the Technical Research Centre of Finland (VTT) and the Technology Development Centre of Finland (TEKES), in connection with the European Community ESPRIT BRA Project REACT (6021).

## References

1. Berthelot, G.: *Transformations and Decompositions of Nets.* In Petri Nets: Central Models and Their Properties, Lecture Notes in Computer Science 254, Springer-Verlag 1987, pp. 359–376.
2. Brookes, S. D. & Roscoe, A. W.: *An Improved Failures Model for Communicating Sequential Processes.* In: Proceedings of the NSF-SERC Seminar on Concurrency, Lecture Notes in Computer Science 197, Springer-Verlag 1985, pp. 281–305.
3. Cheung, S. C. & Kramer, J.: *Enhancing Compositional Reachability Analysis with Context Constraints.* Proceedings of ACM SIGSOFT '93: Symposium on the Foundations of Software Engineering, Los Angeles, USA, December 1993, ACM Software Engineering Notes Vol. 18 Nr 5, 1993, pp. 115–125.
4. Finkel, A.: *The Minimal Coverability Graph for Petri Nets.* Advances in Petri Nets 1993, Lecture Notes in Computer Science 674, Springer-Verlag 1993, pp. 210–243. (Earlier version in Proceedings of the 11th International Conference on Application and Theory of Petri Nets, Paris, France 1990, pp. 1–21.)
5. Graf, S. & Steffen, B.: *Compositional Minimization of Finite State Processes.* In: Computer-Aided Verification '90 (Proceedings of a workshop), AMS-ACM DIMACS Series in Discrete Mathematics and Theoretical Computer Science, Vol. 3, American Mathematical Society 1991, pp. 57–73.
6. Hoare, C. A. R.: *Communicating Sequential Processes.* Prentice-Hall 1985, 256 p.
7. Jensen, K.: *Coloured Petri Nets.* In: Petri Nets: Central Models and Their Properties, Lecture Notes in Computer Science 254, Springer-Verlag 1987, pp. 248–299.

8. Kaivola, R. & Valmari, A.: *The Weakest Compositional Semantic Equivalence Preserving Nexttime-less Linear Temporal Logic*. Proceedings of CONCUR '92, Lecture Notes in Computer Science 630, Springer-Verlag 1992, pp. 207–221.

9. Kaivola, R. *Compositional Linear Temporal Logic Model-Checking for Concurrent Systems*. Licentiate Thesis, University of Helsinki, Department of Computer Science, Report C-1993-1, Helsinki, Finland, 1993, 58 + 42 pp.

10. Lindqvist, M.: *Parameterized Reachability Trees for Predicate/Transition Nets*. Advances in Petri Nets 1993, Lecture Notes in Computer Science 674, Springer-Verlag 1993, pp. 301–324. (Earlier version in Proceedings of the 11th International Conference on Application and Theory of Petri Nets, Paris, France 1990, pp. 22–42.)

11. Valmari, A.: *Stubborn Sets for Reduced State Space Generation*. Advances in Petri Nets 1990, Lecture Notes in Computer Science 483, Springer-Verlag 1991, pp. 491–515. (Earlier version in Proceedings of the 10th International Conference on Application and Theory of Petri Nets, Bonn, West Germany 1989, Vol II, pp. 1–22.)

12. Valmari, A.: *Compositional State Space Generation*. Advances in Petri Nets 1993, Lecture Notes in Computer Science 674, Springer-Verlag 1993, pp. 427–457. (Earlier version in Proceedings of the 11th International Conference on Application and Theory of Petri Nets, Paris, France 1990, pp. 43–62.)

13. Valmari, A. & Tienari, M.: *An Improved Failures Equivalence for Finite-State Systems with a Reduction Algorithm*. In: Protocol Specification, Testing and Verification XI, North-Holland 1991, pp. 3–18.

14. Valmari, A.: *Alleviating State Explosion during Verification of Behavioural Equivalence*. Department of Computer Science, University of Helsinki, Report A-1992-4, Helsinki, Finland 1992, 57 p.

15. Valmari, A., Kemppainen, J., Clegg, M. & Levanto, M.: *Putting Advanced Reachability Analysis Techniques Together: the "ARA" Tool*. Proceedings of Formal Methods Europe '93, Lecture Notes in Computer Science 670, Springer-Verlag 1993, pp. 597–616.

16. Valmari, A. & Tienari, M.: *Compositional Failure-Based Semantic Models for Basic LOTOS*. Tampere University of Technology, Software Systems Laboratory, Report 16, Tampere, Finland, July 1993, 25 p.

17. Valmari, A. & Setälä, M.: *A Fast Reduction Algorithm for Failure-Based Semantics*. Submitted for publication, 13 p.

# On Combining the Stubborn Set Method with the Sleep Set Method

Kimmo Varpaaniemi

Helsinki University of Technology, Digital Systems Laboratory
Otakaari 1, SF-02150 Espoo, Finland
Kimmo.Varpaaniemi@hut.fi

**Abstract.** *Reachability analysis* is a powerful formal method for analysis of concurrent and distributed finite state systems. It suffers from the *state space explosion problem*, however: the state space of a system can be far too large to be completely generated. This paper considers two promising methods, Valmari's *stubborn set method* and Godefroid's *sleep set method*, to avoid generating all of the state space when searching for *undesirable reachable terminal states*, also called *deadlocks*. These methods have been *combined* by Godefroid, Pirottin, and Wolper to further reduce the number of inspected states. However, the combination presented by them places assumptions on the stubborn sets used. This paper shows that at least in *place/transition nets*, the stubborn set method can be combined with the sleep set method in such a way that all reachable terminal states are found, without having to place any assumption on the stubborn sets used. This result is shown by showing a more general result which gives a sufficient condition for a method to be compatible with the sleep set method in the detection of reachable terminal states in place/transition nets.
**Topics:** system verification using nets, analysis and behaviour of nets

## 1 Introduction

*Reachability analysis*, also known as *exhaustive simulation* or *state space generation*, is a powerful formal method for detecting errors in such concurrent and distributed systems that have a finite state space. It suffers from the so called *state space explosion problem*, however: the state space of the system can be far too large with respect to the time and other resources needed to inspect all states in the space. Fortunately, errors such as *undesirable reachable terminal states*, also called *deadlocks*, can be detected in a variety of cases without inspecting all reachable states of the system. What makes deadlocks especially interesting is the fact that the verification of a *safety property* can often be reduced to the detection of deadlocks, as shown by Godefroid and Wolper [8] among others.

This paper concentrates on the problem of detecting *reachable terminal states* in *place/transition nets* [15], a class of Petri nets. Two promising methods are studied: Valmari's *stubborn set method* [16, 17, 18, 19, 20] and Godefroid's *sleep set method* [4, 5, 6, 7, 8, 24, 25]. Both methods utilize the *independence of transitions* to cut down on the number of states inspected during the search. These methods have also been *combined* by Wolper and Godefroid [24], Godefroid and Pirottin [7], and Wolper, Godefroid, and Pirottin [25] to further reduce the number of inspected states. All of these methods guarantee that all reachable terminal states are found if the complete state space is finite.

The application of the stubborn set method and the sleep set method is not limited to Petri nets. The methods have been applied to several models of concurrency by Valmari [17, 20], Godefroid and Pirottin [7], Wolper and Godefroid [24], and Peled [13] among others. Valmari [16], and Godefroid and Kabanza [6] have presented how the stubborn set method and the sleep set method can be used to improve the *graph search methods* used for artificial intelligence. The independence of rules in production systems of artificial intelligence resembles the independence of actions in concurrent and distributed systems in many senses. The essential point in the stubborn set method and the sleep set method is that they utilize the independence of actions or rules. *Refined independence relations* are important since the more refined is the independence relation the less states usually have to be inspected. Best and Lengauer [2], and Katz and Peled [10] among others have studied refined independence relations and developed general concepts of independence.

The application of the stubborn set method and the sleep set method is not limited to the detection of reachable terminal states either. Both methods can be extended to verify properties expressed as linear temporal logic formulae as shown by Valmari [18, 20, 21], Wolper and Godefroid [24], and Peled [13].

The stubborn set method is closely related to, though not necessarily based on Overman's algorithms [12]. These, according to Valmari [16, 20], are somewhat limited and not so efficient as the stubborn set algorithms. The stubborn set method can also be considered a dynamic priority method in contrary to the static priority method mentioned by Valmari and Tiusanen [22], Rauhamaa [14], and Valmari [20] among others. The static priority method is in turn a generalization of the virtual coarsening of atomic actions presented by Ashcroft and Manna [1].

The sleep set method was originally inspired by Mazurkiewicz's *trace theory* [11]. An early version of the sleep set method [4] was essentially faithful to Mazurkiewicz's trace semantics. Later, inspired by Katz's and Peled's work [10], the method has been refined to take into account *conditional independence* [7]. As suggested in [24, 25] and seen in this work, representing traces is sometimes not necessary at all.

Katz and Peled have, independently of Valmari, developed verification algorithms that use *faithful decompositions* [9]. Peled [13] states that faithful decompositions are similar to stubborn sets. Peled has recently improved and extended [13] Valmari's and Godefroid's linear temporal logic verification algorithms.

The contribution of this paper can be described as follows: The combination of the stubborn set method and the sleep set method presented by Godefroid, Pirottin, and Wolper [7, 24, 25] places assumptions on the stubborn sets used. This paper shows that at least in place/transition nets, the stubborn set method can be combined with the sleep set method in such a way that all reachable terminal states are found, without having to place any assumption on the stubborn sets used. This result is shown by showing a more general result which gives a sufficient condition for a method to be compatible with the sleep set method in the detection of reachable terminal states in place/transition nets.

The rest of this paper has been organized as follows: in Section 2, we introduce place/transition nets. The presentation does not go beyond what is necessary for the remaining sections. In Section 3, dynamically stubborn sets [14, 19] are shown to be a useful generalization of stubborn sets. Section 4 considers the sleep set method and its combination with the stubborn set method. We conclude in Section 5 by summarizing the results obtained and briefly discussing possible directions for future research.

## 2 Place/Transition Nets

In this section we give definitions of *place/transition nets* [15] that will be used in later sections.

We shall use "iff" to denote "if and only if". The *power set* (the set of subsets) of a set $A$ is denoted by $2^A$. The set of *(total) functions* from a set $A$ to a set $B$ is denoted by $(A \to B)$. The set of natural numbers, including 0, is denoted by $N$. We shall use $\omega$ to denote a formal infinite number, and $N_\omega$ to denote $N \cup \{\omega\}$. Relation $\leq$ over $N$ is extended to $N_\omega$ by defining

$$\forall n \in N_\omega \; n \leq \omega.$$

Addition and subtraction are extended similarly by defining

$$\forall n \in N \; \omega + n = \omega \wedge \omega - n = \omega.$$

Clearly, $\omega \notin N$ since no natural number can be substituted for $\omega$ in these conditions in such a way that the conditions would hold.

**Definition 1.** A *place/transition net* is a 6-tuple $\langle S, T, F, K, W, M_0 \rangle$ such that

- $S$ is the set of *places*,
- $T$ is the set of *transitions*, $S \cap T = \emptyset$,
- $F$ is the set of *arcs*, $F \subseteq (S \times T) \cup (T \times S)$,
- $K$ is the *capacity function*, $K \in (S \to N_\omega)$,
- $W$ is the *arc weight function*, $W \in (F \to (N \setminus \{0\}))$, and
- $M_0$ is the *initial marking (initial state)*, $M_0 \in \mathcal{M}$ where $\mathcal{M}$ is the set of *markings (states)*, $\mathcal{M} = \{M \in (S \to N) \mid \forall s \in S \; M(s) \leq K(s)\}$.

If $x \in S \cup T$, then the set of *input elements* of $x$ is

$$^\bullet x = \{y \mid \langle y, x \rangle \in F\},$$

the set of *output elements* of $x$ is

$$x^\bullet = \{y \mid \langle x, y \rangle \in F\},$$

and the set of *adjacent elements* of $x$ is $x^\bullet \cup {}^\bullet x$. The function $W$ is extended to a function in $(((S \times T) \cup (T \times S)) \to N)$ by defining $W(x, y) = 0$ iff $\langle x, y \rangle \notin F$. The net is *finite* iff $S \cup T$ is finite. □

Unlike Reisig [15], we do not accept $M(s) = \omega$. Such markings would be redundant in finite place/transition nets.

**Definition 2.** Let $\langle S, T, F, K, W, M_0 \rangle$ be a place/transition net. A transition $t$ is *enabled at a marking* $M$ iff

$$\forall s \in {}^\bullet t \; M(s) \geq W(s, t)$$

and

$$\forall s \in t^{\bullet}\ M(s) - W(s,t) + W(t,s) \leq K(s).$$

A transition $t$ *leads (can be fired) from a marking* $M$ *to a marking* $M'$ ($M[t\rangle M'$ for short) iff $t$ is enabled at $M$ and

$$\forall s \in S\ M'(s) = M(s) - W(s,t) + W(t,s).$$

A transition $t$ is *disabled at a marking* $M$ iff $t$ is not enabled at $M$. A marking $M$ is *terminal* iff no transition is enabled at $M$. A marking $M$ is *nonterminal* iff $M$ is not terminal. □

Our enabledness condition is weaker than Reisig's enabledness condition [15] that requires $M(s) + W(t,s) \leq K(s)$ instead of $M(s) - W(s,t) + W(t,s) \leq K(s)$.

Finite transition sequences and reachability are introduced in Definition 3. We shall use $\varepsilon$ to denote the empty sequence.

**Definition 3.** Let $\langle S, T, F, K, W, M_0\rangle$ be a place/transition net. For any $T_s \subseteq T$,

$$\begin{aligned} T_s^0 &= \{\varepsilon\}, \\ (\forall n \in N\ T_s^{n+1} &= \{\sigma t \mid \sigma \in T_s^n \land t \in T_s\}), \text{ and} \\ T_s^* &= \{\sigma \mid \exists n \in N\ \sigma \in T_s^n\}. \end{aligned}$$

The set $T_s^*$ is called the *set of finite sequences of transitions in* $T_s$, and the set $T^*$ is called the *set of finite transition sequences of the net.* A finite transition sequence $\sigma'$ is a *prefix* of a finite transition sequence $\sigma$ iff there exists a finite transition sequence $\sigma''$ such that $\sigma = \sigma'\sigma''$. A finite transition sequence $\sigma$ *leads (can be fired) from a marking* $M$ *to a marking* $M'$ iff $M[\sigma\rangle M'$ where

$$\begin{aligned} &\forall M \in \mathcal{M}\ M[\varepsilon\rangle M, \text{ and} \\ &\forall M \in \mathcal{M}\ \forall M' \in \mathcal{M}\ \forall \delta \in T^*\ \forall t \in T \\ &M[\delta t\rangle M' \Leftrightarrow (\exists M'' \in \mathcal{M}\ M[\delta\rangle M'' \land M''[t\rangle M'). \end{aligned}$$

A finite transition sequence $\sigma$ is *enabled at a marking* $M$ ($M[\sigma\rangle$ for short) iff $\sigma$ leads from $M$ to some marking. A finite transition sequence $\sigma$ is *disabled at a marking* $M$ iff $\sigma$ is not enabled at $M$. A marking $M'$ is *reachable from a marking* $M$ iff some finite transition sequence leads from $M$ to $M'$. A marking $M'$ is a *reachable marking* iff $M'$ is reachable from $M_0$. A marking $M'$ is *globally unreachable* iff $M'$ is not reachable from any other marking in $\mathcal{M}$ than $M'$. The *(full) reachability graph* of the net is the pair $\langle V, A\rangle$ such that the set of vertices $V$ is the set of reachable markings, and the set of edges $A$ is

$$\{\langle M, t, M'\rangle \mid M \in V \land M' \in V \land t \in T \land M[t\rangle M'\}. \quad \square$$

A finite transition sequence is merely a string. It can be thought of as occurring as a path in the full reachability graph iff it is enabled at some reachable marking.

**Definition 4.** Let $\langle S, T, F, K, W, M_0\rangle$ be a place/transition net. Let $f$ be a function from $\mathcal{M}$ to $2^T$. A finite transition sequence $\sigma$ *$f$-leads (can be $f$-fired) from a marking $M$ to a marking $M'$* iff $M[\sigma\rangle_f M'$, where

$$\forall M \in \mathcal{M}\ M[\varepsilon\rangle_f M, \text{ and}$$
$$\forall M \in \mathcal{M}\ \forall M' \in \mathcal{M}\ \forall \delta \in T^*\ \forall t \in T$$
$$M[\delta t\rangle_f M' \Leftrightarrow (\exists M'' \in \mathcal{M}\ M[\delta\rangle_f M'' \wedge t \in f(M) \wedge M''[t\rangle M').$$

A finite transition sequence $\sigma$ is *$f$-enabled at a marking $M$* ($M[\sigma\rangle_f$ for short) iff $\sigma$ $f$-leads from $M$ to some marking. A marking $M'$ is *$f$-reachable from a marking $M$* iff some finite transition sequence $f$-leads from $M$ to $M'$. A marking $M'$ is an *$f$-reachable marking* iff $M'$ is $f$-reachable from $M_0$. The *$f$-reachability graph* of the net is the pair $\langle V, A\rangle$ such that the set of vertices $V$ is the set of $f$-reachable markings, and the set of edges $A$ is

$$\{\langle M, t, M'\rangle \mid M \in V \wedge M' \in V \wedge t \in f(M) \wedge M[t\rangle M'\}. \quad \square$$

Definition 4 is like a part of Definition 3 except that a transition selection function $f$ determines which transitions are fired. If $f$ is clear from the context or is implicitly assumed to exist and be of a kind that is clear from the context, then the $f$-reachability graph of the net is called the *reduced reachability graph* of the net. Note that the reduced reachability graph of the net can even be the full reachability graph of the net, e.g. in the case where $f(M) = T$ for each $M \in \mathcal{M}$.

**Definition 5.** Let $\langle S, T, F, K, W, M_0\rangle$ be a place/transition net. A transition sequence $\delta$ is an *alternative sequence of a finite transition sequence $\sigma$ at a marking $M$* iff $\delta$ is a finite transition sequence, $\sigma$ is enabled at $M$, and $\delta$ leads from $M$ to the same marking as $\sigma$. A transition sequence $\delta$ is a *length-secure alternative sequence of a finite transition sequence $\sigma$ at a marking $M$* iff $\delta$ is an alternative sequence of $\sigma$ at $M$ and not longer than $\sigma$. The functions $\eta$ and $\vartheta$ from $T^* \times \mathcal{M}$ to $2^{(T^*)}$ are defined as follows: for each finite transition sequence $\sigma$ and marking $M$, $\eta(\sigma, M)$ is the *set of alternative sequences of $\sigma$ at $M$*, and $\vartheta(\sigma, M)$ is the *set of length-secure alternative sequences of $\sigma$ at $M$*. $\square$

Clearly, for each finite transition sequence $\sigma$ and marking $M$, $\vartheta(\sigma, M) \subseteq \eta(\sigma, M)$. Also, $\eta(\sigma, M)$ is empty iff $\sigma$ is not enabled at $M$.

**Definition 6.** Let $\langle S, T, F, K, W, M_0\rangle$ be a place/transition net. A transition sequence $\delta$ is a *permutation of a finite transition sequence $\sigma$* iff $\delta$ is a finite transition sequence and for each transition $t$, the number of $t$'s in $\delta$ is equal to the number of $t$'s in $\sigma$. A transition sequence $\delta$ is an *enabled permutation of a finite transition sequence $\sigma$ at a marking $M$* iff $\delta$ is a permutation of $\sigma$ and enabled at $M$. The function $\pi$ from $T^* \times \mathcal{M}$ to $2^{(T^*)}$ is defined as follows: for each finite transition sequence $\sigma$ and marking $M$, $\pi(\sigma, M)$ is the set of enabled permutations of $\sigma$ at $M$. $\square$

Clearly, if finite transition sequences are enabled permutations of each other at a marking $M$, they lead to the same marking from $M$. So, if a finite transition sequence $\sigma$ is enabled at a marking $M$, then $\pi(\sigma, M) \subseteq \vartheta(\sigma, M)$. The set $\pi(\sigma, M)$ can be nonempty even if $\sigma$ is not enabled at $M$ since some permutation of $\sigma$ can be enabled at $M$. The set of length-secure alternative sequences, as well as the set of alternative sequences, of an enabled finite transition sequence $\sigma$ at a marking can always be partitioned into sets of enabled permutations of sequences at the marking. Of course, only one of those sets is the set of enabled permutations of $\sigma$.

Figure 1 presents the functions $\eta$, $\vartheta$, and $\pi$, in a nutshell.

| | |
|---|---|
| $\eta(\sigma, M)$ | the set of alternative sequences of a finite transition sequence $\sigma$ at $M$ |
| $\vartheta(\sigma, M)$ | the set of length-secure alternative sequences of a finite transition sequence $\sigma$ at $M$ |
| $\pi(\sigma, M)$ | the set of enabled permutations of a finite transition sequence $\sigma$ at $M$ |

**Fig. 1.** The functions $\eta$, $\vartheta$, and $\pi$.

**Definition 7.** Let $\langle S, T, F, K, W, M_0 \rangle$ be a place/transition net. Let $f$ be a function from $\mathcal{M}$ to $2^T$. Then we say that $f$ *represents all sets of alternative sequences to terminal markings* iff

$$\forall \sigma \in T^* \ \forall M \in \mathcal{M} \ (M[\sigma\rangle \wedge \forall t \in T \ \neg M[\sigma t\rangle) \Rightarrow (\exists \delta \in \eta(\sigma, M) \ M[\delta\rangle_f).$$

Correspondingly, $f$ *represents all sets of length-secure alternative sequences to terminal markings* iff

$$\forall \sigma \in T^* \ \forall M \in \mathcal{M} \ (M[\sigma\rangle \wedge \forall t \in T \ \neg M[\sigma t\rangle) \Rightarrow (\exists \delta \in \vartheta(\sigma, M) \ M[\delta\rangle_f).$$

Respectively, $f$ *represents all sets of enabled permutations to terminal markings* iff

$$\forall \sigma \in T^* \ \forall M \in \mathcal{M} \ (M[\sigma\rangle \wedge \forall t \in T \ \neg M[\sigma t\rangle) \Rightarrow (\exists \delta \in \pi(\sigma, M) \ M[\delta\rangle_f). \ \square$$

The following can clearly be seen from the above.

- A function representing all sets of enabled permutations to terminal markings represents all sets of length-secure alternative sequences to terminal markings.
- A function representing all sets of length-secure alternative sequences to terminal markings represents all sets of alternative sequences to terminal markings.

**Definition 8.** Let $\langle S, T, F, K, W, M_0 \rangle$ be a place/transition net. Transitions $t$ and $t'$ *commute at a marking M* iff $M[tt'\rangle$ and $M[t't\rangle$. Transitions $t$ and $t'$ are *independent at a marking M* iff

$$(M[tt'\rangle \wedge M[t't\rangle) \vee ((\neg M[t\rangle) \wedge (\neg M[t'\rangle)) \vee$$
$$(M[t\rangle \wedge (\neg M[t'\rangle) \wedge (\neg M[tt'\rangle)) \vee (M[t'\rangle \wedge (\neg M[t\rangle) \wedge (\neg M[t't\rangle)). \ \square$$

Our definition of independence corresponds to Godefroid's and Pirottin's [7] definition of conditional independence which in turn is based on Katz's and Peled's [10] corresponding definition. Our definition of independence can be obtained from Godefroid's and Pirottin's definition of valid conditional dependency relations, Definition 5 in [7], by taking the necessary conditions for a triple of two transitions and one state to be in the complement of a valid dependency relation, and substituting terms of place/transition nets for the terms of the model of concurrency in [7] in an obvious way.

The following can clearly be seen from the above.

- Different transitions are independent at a marking iff neither of them can be fired at the marking making the other transition turn from enabled to disabled or from disabled to enabled.
- A transition $t$ commutes with itself at a marking iff $tt$ is enabled at the marking.
- A transition $t$ is independent of itself at a marking iff $tt$ is enabled or $t$ is disabled at the marking.
- Transitions commute at a marking iff they are enabled and independent at the marking.

## 3 Dynamically Stubborn Sets

This section is concentrated on *dynamically stubborn* sets [14, 19]. All the stubborn sets that have been defined in the literature are known to be dynamically stubborn. Dynamically stubborn sets seem to have all the nice properties of (statically) stubborn sets except that the definition of dynamic stubbornness does not seem to imply a practical algorithm for computing dynamically stubborn sets. We define dynamic stubbornness on the basis of Rauhamaa's principles [14].

**Definition 9.** Let $\langle S, T, F, K, W, M_0 \rangle$ be a place/transition net. Let $M$ be a marking of the net. A set $T_s \subseteq T$ *fulfils the first principle of dynamic stubbornness* (*D1* for short) *at* $M$ iff

$$\forall \sigma \in (T \setminus T_s)^* \; \forall t \in T_s \; M[\sigma t\rangle \Rightarrow M[t\sigma\rangle.$$

A transition $t$ is a *key transition of a set* $T_s \subseteq T$ *at* $M$ iff $t \in T_s$ and

$$\forall \sigma \in (T \setminus T_s)^* \; M[\sigma\rangle \Rightarrow M[\sigma t\rangle.$$

A set $T_s \subseteq T$ *fulfils the second principle of dynamic stubbornness* (*D2* for short) *at* $M$ iff $T_s$ has a key transition at $M$. A set $T_s \subseteq T$ *fulfils the first principle of strong dynamic stubbornness* (*SD1* for short) *at* $M$ iff

$$\forall \sigma \in (T \setminus T_s)^* \; \forall t \in T_s \; M[\sigma t\rangle \Rightarrow M[t\rangle.$$

A set $T_s \subseteq T$ *fulfils the second principle of strong dynamic stubbornness* (*SD2* for short) *at* $M$ iff

$$\forall \sigma \in (T \setminus T_s)^* \; \forall t \in T_s \; (M[t\rangle \wedge M[\sigma\rangle) \Rightarrow (M[\sigma t\rangle \wedge M[t\sigma\rangle).$$

A set $T_s \subseteq T$ is *dynamically stubborn at* $M$ iff $T_s$ fulfils D1 and D2 at $M$. A set $T_s \subseteq T$ is *strongly dynamically stubborn at* $M$ iff $T_s$ fulfils SD1 and SD2 at $M$ and $\exists t \in T_s \; M[t\rangle$. □

The principles D1, D2, SD1, and SD2 are illustrated in Figure 2. The principles D1, D2, SD1, and SD2 are Rauhamaa's Principles 1*, 2*, 1, and 2, respectively [14]. Clearly, a key transition of a set at a marking is enabled at the marking. Our key transitions are similar to Valmari's key transitions [17]. The difference is that Valmari's key transitions satisfy a condition that can be checked easily and is sufficient but not necessary for a transition to be a key transition in the sense of our definition. As shown in [23], our strong dynamic stubbornness is weaker than Valmari's strong dynamic stubbornness [19].

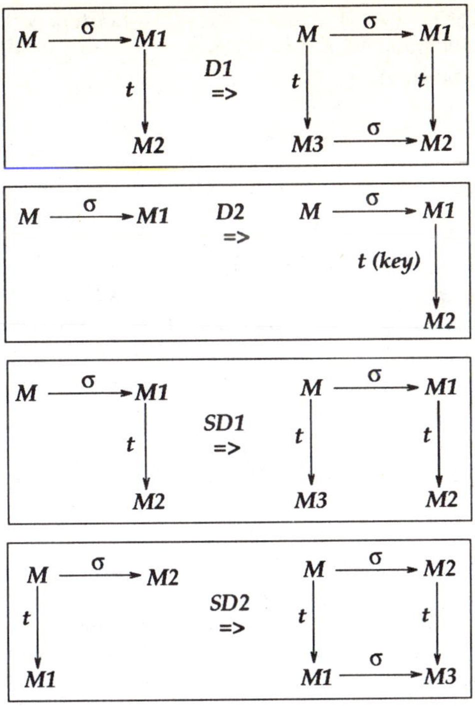


**Fig. 2.** The principles of dynamic and strong dynamic stubbornness.

*Lemma 10. A set is strongly dynamically stubborn at a marking iff the set is dynamically stubborn at the marking and each enabled transition in the set is a key transition of the set at the marking.*

*Proof.* The result follows trivially from Definition 9. □

The result in Lemma 10 is due to Valmari [17] but has missed explicit treatment. We shall soon see that dynamic stubbornness alone is sufficient as far as the detection of reachable terminal markings is concerned. While all stubborn sets are known to be dynamically stubborn, quite efficient stubborn set computation algorithms exist that can compute both strongly dynamically stubborn sets and such sets that are not strongly dynamically stubborn [16, 17, 23]. Strongly dynamically stubborn sets are useful when one wants to eliminate the *ignoring phenomenon* [17, 23]. A transition is ignored at a marking iff the transition is enabled at the marking but not fired at any marking that is reachable from the marking. The existence of ignored transitions is called the ignoring phenomenon.

**Definition 11.** Let $\langle S, T, F, K, W, M_0 \rangle$ be a place/transition net. Let $f$ be a function from $\mathcal{M}$ to $2^T$. Then we say that $f$ is *dynamically stubborn* iff for each nonterminal marking $M$, $f(M)$ is dynamically stubborn. Correspondingly, $f$ is *strongly dynamically stubborn* iff for each nonterminal marking $M$, $f(M)$ is strongly dynamically stubborn. □

***Theorem 12.*** *Let $\langle S, T, F, K, W, M_0\rangle$ be a place/transition net. Let $f$ be a dynamically stubborn function from $\mathcal{M}$ to $2^T$. Then $f$ represents all sets of enabled permutations to terminal markings.*

*Proof.* We show that

$$\forall \sigma'' \in T^* \; \forall M \in \mathcal{M} \; (M[\sigma''\rangle \land \forall t \in T \; \neg M[\sigma'' t\rangle) \Rightarrow (\exists \delta \in \pi(\sigma'', M) \; M[\delta\rangle_f).$$

We use induction on the length of $\sigma''$. The claim holds trivially when restricted to $\sigma'' = \varepsilon$. Our induction hypothesis is that the claim holds when restricted to any $\sigma''$ of length $n \geq 0$. Let $\sigma \in T^*$, $M \in \mathcal{M}$, and $M' \in \mathcal{M}$ be such that $\sigma$ is of length $n+1$, $M[\sigma\rangle M'$, and $\forall t \in T \; \neg M'[t\rangle$. The set $f(M)$ is dynamically stubborn at $M$ since some transition is enabled at $M$. The sequence $\sigma$ must contain a transition in $f(M)$ since otherwise some transition in $f(M)$ would be enabled at $M'$ by D2. Let $\delta \in (T \setminus f(M))^*$, $t \in f(M)$, and $\delta' \in T^*$ be such that $\sigma = \delta t \delta'$. By D1 we have $M[t\delta\rangle$, so $M[t\delta\delta'\rangle$. Let $M'' \in \mathcal{M}$ be such that $M[t\rangle M''$. Now $M[t\rangle_f M''$. By the induction hypothesis, $\exists \sigma' \in \pi(\delta\delta', M'') \; M''[\sigma'\rangle_f$. We thus have $t\sigma' \in \pi(\sigma, M)$ and $M[t\sigma'\rangle_f$. □

The result in Theorem 12 is due to Valmari [16, 17] but has missed explicit treatment. Theorem 12 has the consequence that if a finite transition sequence leads from a marking $M$ to a terminal marking, and $M$ occurs in the reduced reachability graph, then an enabled permutation of the sequence occurs in the graph. A dynamically stubborn set selective search thus certainly finds all reachable terminal markings if the net and the set of reachable markings are finite. If the set of reachable markings is infinite but the net is finite and a dynamically stubborn set selective search is performed in a breadth-first order for some time, then reachable terminal markings "near the initial marking" can be found. As we shall see in Section 4, the permutation preserving property makes the stubborn set method compatible with the sleep set method in the detection of reachable terminal markings though a weaker property would suffice.

We define persistence and conditional stubbornness in such a way that the definitions correspond to the definitions given by Godefroid and Pirottin [7]. Our definitions can be obtained from Godefroid's and Pirottin's Definitions 7 and 8 in [7] by substituting terms of place/transition nets for the terms of the model of concurrency in [7] in an obvious way.

**Definition 13.** Let $\langle S, T, F, K, W, M_0\rangle$ be a place/transition net. Let $M \in \mathcal{M}$. A set $T_s \subseteq T$ *fulfils the principle of persistence and conditional stubbornness* (*PE* for short) *at* $M$ iff

$$\forall \sigma \in (T \setminus T_s)^* \; \forall t \in T_s \; \forall t' \in T \setminus T_s \; \forall M' \in \mathcal{M}$$
$$(M[t\rangle \land M[\sigma\rangle M' \land M'[t'\rangle) \Rightarrow (t \text{ and } t' \text{ are independent at } M').$$

A set $T_s \subseteq T$ is *persistent at* $M$ iff $T_s$ fulfils PE at $M$ and $\forall t \in T_s \; M[t\rangle$. A set $T_s \subseteq T$ is *conditionally stubborn at* $M$ iff $T_s$ fulfils SD1 and PE at $M$ and $\exists t \in T_s \; M[t\rangle$. □

Clearly, the "$M[t\rangle \land$" in PE is redundant in the definition of persistence since all transitions in persistent sets are enabled. Our goal in the rest of this section is to re-express persistence and conditional stubbornness in terms of dynamic stubbornness.

***Lemma 14.*** *Let $\langle S, T, F, K, W, M_0\rangle$ be a place/transition net. Let $M$ be a marking of the net. A set $T_s \subseteq T$ fulfils SD2 at $M$ iff*

$$\forall \sigma \in (T \setminus T_s)^* \ \forall \delta \in (T \setminus T_s)^* \ \forall t \in T_s \ (M[t\rangle \wedge M[\sigma\delta\rangle) \Rightarrow M[\sigma t \delta\rangle.$$

*Proof.* The "if"-part is obvious. Let's prove the "only if" -part. Let a set $T_s \subseteq T$ fulfil SD2 at $M$. Let $\sigma \in (T \setminus T_s)^*$, $\delta \in (T \setminus T_s)^*$, $t \in T_s$, $M[t\rangle$, and $M[\sigma\delta\rangle$. Using SD2 for both $\sigma\delta$ and $\sigma$, we get $M[t\sigma\delta\rangle$ and $M[\sigma t\rangle$. As $\sigma t$ and $t\sigma$ lead to the same marking, we have $M[\sigma t\delta\rangle$. □

The result in Lemma 14 is due to Valmari [17] but has missed explicit treatment.

***Lemma 15.*** *Let $\langle S, T, F, K, W, M_0\rangle$ be a place/transition net. Let $M$ be a marking of the net, and $T_s$ and $T_e$ subsets of $T$ such that*

$$\{t \in T_s \mid M[t\rangle\} \subseteq T_e, \text{ and } T_e \subseteq T_s.$$

*If $T_s$ is dynamically stubborn at $M$, $T_e$ is dynamically stubborn at $M$. If $T_s$ is strongly dynamically stubborn at $M$, $T_e$ is strongly dynamically stubborn at $M$.*

*Proof.* (i) Let $T_s$ be dynamically stubborn at $M$. We show that

$$\forall \sigma \in (T \setminus T_e)^* \ M[\sigma\rangle \Rightarrow \sigma \in (T \setminus T_s)^*.$$

Let $\sigma \in (T \setminus T_e)^*$ and $\delta \in (T \setminus T_s)^*$ be such that $M[\sigma\rangle$ and $\delta$ is the longest prefix of $\sigma$ not containing any transition in $T_s$. If $\delta \neq \sigma$, the first transition after $\delta$ in $\sigma$ is enabled at $M$ by D1 for $T_s$. Since no transition in $T_s \setminus T_e$ is enabled at $M$, we conclude that $\delta = \sigma$, so $\sigma \in (T \setminus T_s)^*$.

(ii) Since strongly dynamically stubborn sets are dynamically stubborn by Lemma 10, the result of part (i) holds for them, too. Then D1 for $T_s$ implies D1 for $T_e$, D2 for $T_s$ implies D2 for $T_e$, SD1 for $T_s$ implies SD1 for $T_e$, and SD2 for $T_s$, implies SD2 for $T_e$. □

The result in Lemma 15 is new though inspired by Godefroid and Pirottin [7]. Lemma 15 states that if we remove disabled transitions from a dynamically stubborn (strongly dynamically stubborn) set, the remaining set is dynamically stubborn (strongly dynamically stubborn). For example, if a dynamically stubborn set is minimal with respect to set inclusion, by Lemma 15 the set consists of enabled transitions only.

***Lemma 16.*** *A set fulfils PE at a marking iff the set fulfils SD2 at the marking. A set is conditionally stubborn at a marking iff the set is strongly dynamically stubborn at the marking.*

*Proof.* We show that PE is equivalent to SD2. The second statement then follows directly from Definitions 9 and 13. Let $\langle S, T, F, K, W, M_0\rangle$ be a place/transition net. Let $M \in \mathcal{M}$ and $T_s \subseteq T$.

(i) We prove that SD2 implies PE. Let $T_s$ fulfil SD2 at $M$. Let $\sigma \in (T \setminus T_s)^*$, $t \in T_s$, $t' \in T \setminus T_s$, $M' \in \mathcal{M}$, $M[t\rangle$, $M[\sigma\rangle M'$, and $M'[t'\rangle$. By Lemma 14 we have both $M'[t't\rangle$ and $M'[tt'\rangle$. The transitions $t$ and $t'$ are thus independent at $M'$.

(ii) We prove that PE implies SD2. Let $T_s$ fulfil PE at $M$. We use induction on the length of finite transition sequences to show that $T_s$ fulfils SD2 at $M$. The principle SD2 is fulfilled trivially when restricted to $\varepsilon$. Our induction hypothesis is that SD2 is fulfilled when restricted to finite transition sequences of length $n \geq 0$. We show that SD2 is then fulfilled when restricted to finite transition sequences of length $n+1$. Let $\delta \in (T \setminus T_s)^*$, $t' \in T \setminus T_s$, and $t \in T_s$ be such that $M[t\rangle$, $M[\delta t'\rangle$, and $\delta$ is of length $n$. Let $M' \in \mathcal{M}$ be such that $M[\delta\rangle M'$. The transition $t'$ is then enabled at $M'$. By the induction hypothesis we have $M[\delta t\rangle$ and $M[t\delta\rangle$. The transition $t$ is thus enabled at $M'$. The principle PE then implies that $t$ and $t'$ are independent at $M'$. Transitions commute at a marking iff they are enabled and independent at the marking. So $t$ and $t'$ commute at $M'$. Thus $M'[tt'\rangle$ and $M'[t't\rangle$, and consequently $M[\delta tt'\rangle$ and $M[\delta t't\rangle$. As already mentioned, we have $M[t\delta\rangle$, so $t\delta$ leads from $M$ to the same marking as $\delta t$. We thus have $M[t\delta t'\rangle$. □

The result in Lemma 16 is new.

*Lemma 17. A set is a nonempty persistent set at a marking iff the set is a conditionally stubborn set at the marking and does not contain any transition that is disabled at the marking. The set of enabled transitions of any conditionally stubborn set is a nonempty persistent set.*

*Proof.* The first statement follows from the fact that a persistent set fulfils SD1 trivially since all its transitions are enabled. The second statement follows trivially from the first statement and Lemmata 15 and 16. □

The result in Lemma 17 is due to Godefroid and Pirottin [7] but the proof is new.

## 4 Sleep Set Method

In this section we present Godefroid's *sleep set method* [4, 5, 6, 7, 8, 24, 25]. The plain sleep set method preserves at least one sequence from each *conditional trace* [7, 10, 23] leading from the initial state to a terminal state. To prevent a transition from firing, it is put into a so called sleep set.

Wolper and Godefroid [24], Godefroid and Pirottin [7], and Wolper, Godefroid, and Pirottin [25] have combined the sleep set method with the stubborn set method. The combination is justified by the fact that the stubborn set method alone is sometimes bound to fire independent transitions at a state. The combination presented in [7, 24, 25] is such that at each encountered nonterminal state a nonempty persistent set is computed. Let us recall from Lemmata 16 and 17 that a set is a nonempty persistent set iff the set is a strongly dynamically stubborn set consisting of enabled transitions only. In [24, 25], persistence is defined on the basis of global independence but since global independence implies independence at each reachable state, the persistent sets in [24, 25] are persistent in the sense defined by Godefroid and Pirottin [7]. As mentioned immediately above Definition 13, our definition of persistence in Definition 13 corresponds to Godefroid's and Pirottin's definition [7]. The plain sleep set method can be thought of as a special case of the combined method: a simple heuristic for computing a persistent set is used. We shall not consider the plain sleep set method further.

We concentrate on a generalized version of Wolper's and Godefroid's terminal state detection algorithm [24]. The generalized version is in Figure 3. The intuitive idea of the algorithm is to eliminate such redundant *interleavings of transitions* that are not eliminated by the transition selection function $f$. We show that the algorithm is guaranteed to find all reachable terminal markings of any finite place/transition net with a finite set of reachable markings. From the finiteness of the net, from the finiteness of the set of reachable markings, and from the fact that no transition is fired twice at one marking it follows that the execution of the algorithm takes a finite time only. Any dynamically stubborn function is valid for $f$, but the algorithm is not limited to dynamically stubborn sets. It suffices that $f$ represents all sets of length-secure alternative sequences to terminal markings. The set $T_0$ can be any subset of transitions that are disabled at the initial marking. The $\varphi$ in Figure 3 can be any truth-valued function on $\mathcal{M} \times T \times T \times 2^T$ that satisfies: if $\varphi(M, t, t', T_s)$, then either $t$ and $t'$ commute at $M$ and $t' \in T_s$, or $tt'$ is disabled at $M$. For example, $\varphi(M, t, t', T_s)$ could be

- "either $t$ and $t'$ commute at $M$ and $t' \in T_s$, or $tt'$ is disabled at $M$",
- "$t$ and $t'$ are independent at $M$ and $t' \in T_s$",
- "$t$ and $t'$ commute at $M$ and $t' \in T_s$", or simply
- "false".

Note that if $M[t\rangle M'$, then $tt'$ is disabled at $M$ iff $t'$ is disabled at $M'$. So the first alternative in the above list has the effect that if $t$ is fired from $M$ to $M'$ in the algorithm in Figure 3, then the sleep set pushed onto the stack with $M'$ contains all those transitions that are disabled at $M'$. We shall consider the practicalities related to $\varphi$ and $T_0$ later in this section.

The algorithm in Figure 3 is similar to Wolper's and Godefroid's algorithm [24]. The only essential differences are that Wolper and Godefroid assume that the set corresponding to $f(M)$ is persistent, the set corresponding to $T_0$ is empty, and the condition corresponding to $\varphi(M, t, t', T_s)$ is "$t$ and $t'$ are globally independent and $t' \in T_s$".

***Theorem 18.*** *Let $\langle S, T, F, K, W, M_0 \rangle$ be a finite place/transition net such that the set of markings reachable from $M_0$ is finite. Let $f$ be a function from $\mathcal{M}$ to $2^T$ such that $f$ represents all sets of length-secure alternative sequences to terminal markings. Let $T_0$ be a subset of transitions that are disabled at $M_0$. Let $\varphi$ be a truth-valued function on $\mathcal{M} \times T \times T \times 2^T$ such that for each marking $M$, for all transitions $t$ and $t'$, and for each $T_s \subseteq T$, if $\varphi(M, t, t', T_s)$, then either $t$ and $t'$ commute at $M$ and $t' \in T_s$, or $tt'$ is disabled at $M$. Then the algorithm in Figure 3 finds all terminal markings that are reachable from $M_0$.*

*Proof.* Let $M_d$ be a terminal marking that is reachable from $M_0$.

(i) We first prove that if $X \subseteq T$, a finite transition sequence $\sigma$ leads from a marking $M$ to $M_d$, and for each $\delta$ in $\vartheta(\sigma, M)$, the first transition of $\delta$ is not in $X$, then, if $\langle M, X \rangle$ is pushed onto the stack, some element having $M_d$ as the first component will be or has already been popped from the stack.

The proof proceeds by induction on the length of $\sigma$. For $\sigma = \varepsilon$, the result is immediate. Now, assume the proposition holds for finite transition sequences of length less than or equal to $n$, where $n \geq 0$, and let us prove that it holds for a

```
make Stack empty; make H empty;
push ⟨M0,T0⟩ onto Stack;
while Stack is not empty do {
    pop ⟨M, Sleep⟩ from Stack;
    if M is not in H then {
        Fire = {t ∈ f(M)\ Sleep | M[t⟩};
        if Fire and {t ∈ Sleep | M[t⟩} are both empty then print "Terminal state!";
        enter ⟨M, a copy of Sleep⟩ in H;
    }
    else {
        let hSleep be the set associated with M in H;
        Fire = {t ∈ hSleep \ Sleep | M[t⟩};
        Sleep = hSleep ∩ Sleep;
        substitute a copy of Sleep for the set associated with M in H;
    }
    for each t in Fire do {
        let M[t⟩M';
        tSleep = {t' ∈ T | φ(M, t, t', Sleep)};
        push ⟨M', a copy of tSleep⟩ onto Stack;
        Sleep = {t}∪ Sleep;
    }
}
```

**Fig. 3.** A terminal marking detection algorithm.

finite transition sequence $\sigma$ of length $n+1$. Let $X$ be a subset of $T$, $\sigma$ lead from a a marking $M$ to $M_d$, and $\langle M, X\rangle$ have been pushed onto the stack. Let it also be the case that for each $\delta$ in $\vartheta(\sigma, M)$, the first transition of $\delta$ is not in $X$. Let us consider the actions immediately following the popping of $\langle M, X\rangle$ from the stack.

We first consider the case where $M$ is not already in $H$. Since $M$ is a non-terminal marking and $f$ represents all sets of length-secure alternative sequences to terminal markings, at least one transition in $f(M)$ is the first transition of some sequence in $\vartheta(\sigma, M)$. Moreover, every such transition is in $T \setminus X$ and is thus fired at $M$. Let $t_1$ be the first of such transitions in the firing order. Then there exists a finite transition sequence $\sigma'$ such that $t_1\sigma'$ is in $\vartheta(\sigma, M)$. From the definition of $\vartheta$ it follows that $M[t_1\sigma'\rangle M_d$ and $t_1\sigma'$ is not longer than $\sigma$. The length of $\sigma'$ is thus less than or equal to $n$. Let $t_1$ lead from $M$ to a marking $M'$. Then $M'[\sigma'\rangle M_d$. Let $\langle M', X'\rangle$ be pushed onto the stack when firing $t_1$ at $M$. We show that for each $\delta$ in $\vartheta(\sigma', M')$, the first transition of $\delta$ is not in $X'$.

Indeed, assume the opposite, i.e., there exists some transition $t'$ in $X'$ such that for some finite transition sequence $\delta'$, $t'\delta'$ is in $\vartheta(\sigma', M')$. Clearly, then $t_1t'\delta'$ is in $\vartheta(\sigma, M)$. From the condition satisfied by $\varphi$ it follows that $t_1$ and $t'$ commute at $M$ and $t'$ is in Sleep at the time of pushing $\langle M', X'\rangle$ onto the stack. The transition $t'$ cannot be equal to $t_1$ since $t_1$ is not in Sleep at the time of the pushing. Since $t_1$ and $t'$ commute at $M$, $t_1t'$ and $t't_1$ lead to the same marking from $M$. Consequently, $t't_1\delta'$ is in $\vartheta(\sigma, M)$. From the condition satisfied by $X$ it thus follows that $t'$ is not in $X$. There is then only the possibility that $t'$ has been inserted into Sleep in the "for-loop" before firing $t_1$. We have obtained a contradiction since if a transition in $f(M) \setminus X$ is the first transition of some sequence in $\vartheta(\sigma, M)$, the transition is either $t_1$ itself or fired after $t_1$. The inductive hypothesis can thus be used to establish that some element having $M_d$ as the first component will be or has already been popped from the stack.

We now consider the case where $M$ already appears in $H$. Let $Y \subseteq T$ be such that $\langle M, Y\rangle$ is in $H$. All those transitions in $Y \setminus X$ that are enabled at $s$ are fired. There are two situations: either some transition in $Y$ is the first transition of some sequence in $\vartheta(\sigma, M)$, or no such transition exists. In the first situation, we can choose a transition analogous to the above $t_1$ and proceed as above.

Let us now turn to the second situation in which no transition in $Y$ is the first transition of any sequence in $\vartheta(\sigma, M)$. This can be the case either because no transition in $Y_0$ is the first transition of any sequence in $\vartheta(\sigma, M)$ where $Y_0$ is the sleep set entered in $H$ with $M$ when $M$ was inserted into $H$, or because there are some $Y'$ and $Z$ such that $\langle M, Z\rangle$ was popped from the stack before popping $\langle M, X\rangle$ from the stack, $\langle M, Y'\rangle$ was in $H$ at the time of the popping of $\langle M, Z\rangle$ from the stack, some transition in $Y'$ is the first transition of some sequence in $\vartheta(\sigma, M)$, and no transition in $Y' \cap Z$ is the first transition of any sequence in $\vartheta(\sigma, M)$. In the former case, we can proceed as above with $Y_0$ in the place of $X$. In the latter case, we can proceed as above with $Z$ in the place of $X$, taking into account the fact that Sleep$= Y' \cap Z$ when the "for-loop" is entered.

(ii) The algorithm in Figure 3 starts by pushing $\langle M_0, T_0\rangle$ onto an empty stack. Every transition in $T_0$ is disabled at $M_0$. From the result shown in part (i) it thus follows that some element having $M_d$ as the first component will be popped from the stack. □

The result in Theorem 18 is new though inspired by Wolper and Godefroid [24], Godefroid and Pirottin [7], and Wolper, Godefroid, and Pirottin [25]. Let us recall from Theorem 12 that dynamically stubborn functions represent all sets of enabled permutations to terminal markings. So they represent all sets of length-secure alternative sequences to terminal markings, too. From Theorem 18 it thus follows that the algorithm in Figure 3 is compatible with all dynamically stubborn sets.

Let's consider an example which shows that the statement obtained from Theorem 18 by removing the word "length-secure" is not valid. Let $\varphi(M, t, t', T_s)$ iff $t$ and $t'$ commute at $M$ and $t' \in T_s$. Let $T_0 = \emptyset$. Let $M_0[a\rangle M_1$, $M_1[d\rangle M_2$, and $M_2[bc\rangle M_3$ in the net in Figure 4. Let $f$ be defined by $f(M_0) = \{a\}$ and $f(M) = T$ when $M \neq M_0$. For each $M \in \mathcal{M}$, if $M_0$ is reachable from $M$, then $M = M_0$ or $M = M_1$. The marking $M_3$ is the only terminal marking that is reachable from $M_0$. The function $f$ thus represents all sets of alternative sequences to terminal markings. However, $f$ does not represent all sets of length-secure alternative sequences to terminal markings since $M_0[cd\rangle M_3$ but for each $\sigma \in T^*$ of length less than or equal to 2, $\neg M_0[\sigma\rangle_f M_3$.

During the first visit to $M_0$, the algorithm in Figure 3 inserts $\langle M_0, \emptyset\rangle$ into $H$ and pushes $\langle M_1, \emptyset\rangle$ onto the stack. The algorithm then visits $M_1$. The transitions $b$ and $d$ are the enabled transitions in $f(M_1) = T$ at $M_1$. Let $b$ be fired before $d$ at $M_1$. The algorithm pushes $\langle M_0, \emptyset\rangle$ and $\langle M_2, \{b\}\rangle$ onto the stack since $b$ and $d$ commute at $M_1$. The algorithm then visits $M_2$ but does not fire the sleeping $b$ which is the

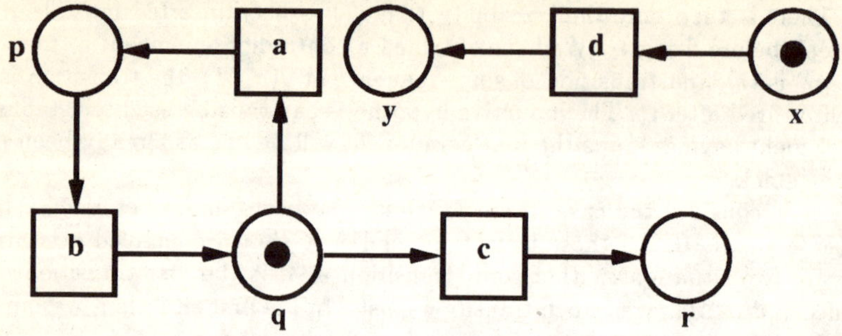


**Fig. 4.** A deceptive $f$ makes the algorithm in Figure 3 fail.

only enabled transition at $M_2$. No transition is fired during the second visit to $M_0$ since the sleep set associated with $M_0$ in $H$ is empty. The execution of the algorithm is then over. No terminal marking was found though $M_3$ is a terminal marking that is reachable from $M_0$.

***Lemma 19.*** *Let $\langle S, T, F, K, W, M_0 \rangle$ be a finite place/transition net. Let $\varphi$ be a truth-valued function on $\mathcal{M} \times T \times T \times 2^T$ such that for each marking $M$, for all transitions $t$ and $t'$, and for each $T_s \subseteq T$, if $\varphi(M, t, t', T_s)$, then $t$ and $t'$ are independent at $M$ and $t' \in T_s$. Let $T_0 = \emptyset$. Then in the algorithm in Figure 3, each sleep set associated with a marking contains only transitions that are enabled at the marking.*

*Proof.* If a transition $t'$ is in the sleep set pushed onto the stack with a marking $M'$ when a transition $t$ is fired from a marking $M$ to $M'$, then $t' \in$ Sleep at the time of the push, and each transition in Fire is enabled at $M$. The sleep set associated with the initial marking at the beginning of the execution of the algorithm is empty. If transitions $t$ and $t''$ are enabled and independent at a marking $M$, and $M[t\rangle M''$, then $t''$ is enabled at $M''$. The result thus follows by a trivial induction. □

The result in Lemma 19 is due to Wolper and Godefroid [24] despite the differences between the algorithm in Figure 3 and their terminal state detection algorithm.

In Figure 5, an implementation of the algorithm in Figure 3 with respect to $\varphi$ and $T_0$ is presented. The algorithm in Figure 5 can be obtained from the algorithm in Figure 3 by making $T_0$ empty, removing the checking of enabledness of transitions in sleep sets, and defining: $\varphi(M, t, t', T_s)$ iff $t$ and $t'$ commute at $M$ and $t' \in T_s$. We know that transitions commute at a marking iff they are enabled and independent at the marking. Checking commutation should be easier than checking independence. Lemma 19 implies that the algorithm in Figure 3 is equivalent to the algorithm in Figure 5 when $T_0$ is empty and $\varphi$ is defined: $\varphi(M, t, t', T_s)$ iff $t$ and $t'$ commute at $M$ and $t' \in T_s$. Lemma 5 thus also implies that in the algorithm in Figure 5, each sleep set associated with a marking contains only transitions that are enabled at the marking.

***Lemma 20.*** *Let $\langle S, T, F, K, W, M_0 \rangle$ be a finite place/transition net such that the set of markings reachable from $M_0$ is finite. Let $T_0$ be a subset of transitions that are disabled at $M_0$. Let $\varphi$ be a truth-valued function on $\mathcal{M} \times T \times T \times 2^T$ such that for*

```
make Stack empty; make H empty;
push ⟨M₀, ∅⟩ onto Stack;
while Stack is not empty do {
    pop ⟨M, Sleep⟩ from Stack;
    if M is not in H then {
        Fire = {t ∈ f(M) \ Sleep | M[t⟩};
        if Fire and Sleep are both empty then print "Terminal state!";
        enter ⟨M, a copy of Sleep⟩ in H;
    }
    else {
        let hSleep be the set associated with M in H;
        Fire = hSleep \ Sleep;
        Sleep = hSleep ∩ Sleep;
        substitute a copy of Sleep for the set associated with M in H;
    }
    for each t in Fire do {
        let M[t⟩M';
        tSleep = {t' ∈ Sleep | t and t' commute at M};
        push ⟨M', a copy of tSleep⟩ onto Stack;
        Sleep = {t} ∪ Sleep;
    }
}
```

**Fig. 5.** A practical implementation of the algorithm in Figure 3 with respect to $\varphi$ and $T_0$.

*each marking $M$, for all transitions $t$ and $t'$, and for each $T_s \subseteq T$, if $\varphi(M,t,t',T_s)$, then either $t$ and $t'$ commute at $M$ and $t' \in T_s$, or $tt'$ is disabled at $M$. Let's further require that for each marking $M$, for all transitions $t$ and $t'$, and for each $T_s \subseteq T$, if $t$ and $t'$ commute at $M$ and $t' \in T_s$, then $\varphi(M,t,t',T_s)$. Let's assume that the sets, the set operations (insertion, union, intersection, and difference), the stack, the stack operations, the "for-loop", and the computation of $f(M)$ in the algorithms in Figure 3 and 5 are implemented exactly in the same way. Then the algorithms visit exactly the same markings and fire exactly the same transitions in exactly the same order.*

*Proof.* Let's assume that a transition $t$ is being fired from a marking $M$ to a marking $M'$ in the algorithm in Figure 3. If a transition $t'$ is enabled at $M'$ and is in the sleep set pushed onto the stack with $M'$ when $t$ is fired at $M$, then $\varphi(M,t,t',\text{Sleep})$ holds at the time of the push but $tt'$ is enabled at $M$, so $t$ and $t'$ commute at $M$, $t' \in \text{Sleep}$ at the time of the push, and $t'$ is enabled at $M$. The result now follows by a trivial induction. □

The result in Lemma 20 is new though inspired by Wolper and Godefroid [24].

Lemma 20 states that there is no more refined implementation of the algorithm in Figure 3 with respect to $\varphi$ and $T_0$ than the algorithm in Figure 5 if $\varphi$ and $T_0$ are required to satisfy the assumptions in Theorem 18. Lemmata 19 and 20 suggest the heuristic that each sleep set associated with a marking should only contain transitions that are enabled at the marking.

Let's consider the complexity of the algorithm in Figure 5. The time taken by a check of whether two transitions commute at a marking is at most proportional to $\nu$, where $\nu$ is the maximum number of adjacent places of a transition. The cumulative time per marking spent in the "for-loop" is at most proportional to $\nu\rho^2$, where $\rho$ is the maximum number of enabled transitions of a marking, and all visits to the marking are counted. This is based on the fact that each sleep set associated with a marking contains only transitions that are enabled at the marking. The time per visit to a marking spent in the operations related to $H$ is the time of the search for the marking plus a time that is at most proportional to $\rho$. The searches in $H$ are something that cannot be avoided easily whether or not we use sleep sets at all. It depends much on the net how many times a marking is visited and how many simultaneous occurrences of a marking there are in the stack. One stack element requires space for the marking and at most $\rho$ transitions. It is not necessary to store copies of markings and transitions since pointers suffice. More clever ways to cut down on space consumption in sleep set algorithms have been presented by Godefroid, Holzmann, and Pirottin [5].

The combination of the sleep set method and the stubborn set method can really be better than the plain stubborn set method as far as the number of inspected markings is concerned. More precisely, there can be a dynamically stubborn function $f$ such that $f(M)$ can be computed by using a feasible algorithm such as the *incremental algorithm* [16], and for each dynamically stubborn function $g$, the number of vertices in the $g$-reachability graph is greater than the number of markings that are inspected by the algorithm in Figure 5 that uses $f$. The net in Figure 6 is a simple example showing this. The example is essentially the same as can be found in [25].

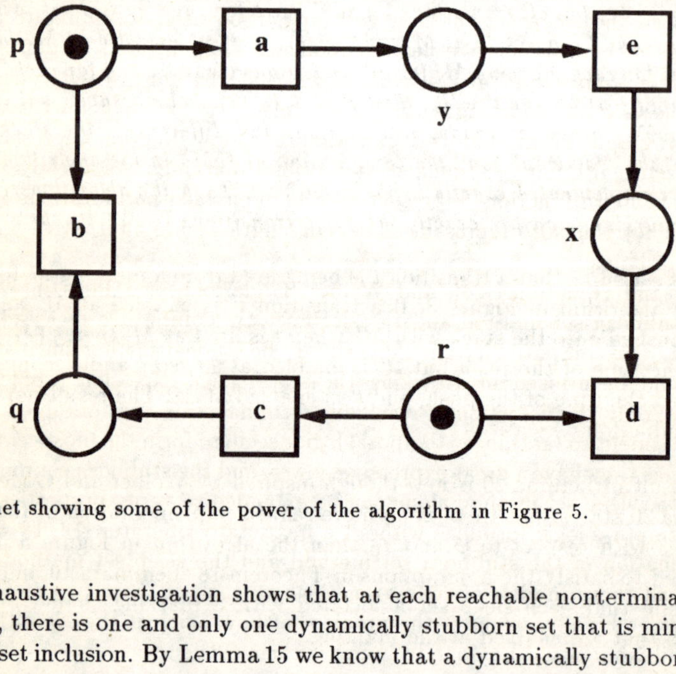


**Fig. 6.** A net showing some of the power of the algorithm in Figure 5.

An exhaustive investigation shows that at each reachable nonterminal marking of this net, there is one and only one dynamically stubborn set that is minimal with respect to set inclusion. By Lemma 15 we know that a dynamically stubborn set that is minimal with respect to set inclusion only contains enabled transitions. Another exhaustive investigation shows that at each reachable nonterminal marking of this net, the set of enabled transitions of any stubborn set computed by the incremental algorithm, using any of the definitions of stubbornness in [16, 17, 19], is a dynamically stubborn set that is minimal with respect to set inclusion. If $M$ is a nonterminal marking, let $f(M)$ be the set of enabled transitions of a stubborn set computed by the incremental algorithm. Then $f(M)$ is the only dynamically stubborn set at $M$ that is minimal with respect to set inclusion. Thus, for each dynamically stubborn function $g$, the $f$-reachability graph is a subgraph of the $g$-reachability graph.

We have $f(M_0) = \{a, c\}$. Let $a$ be fired before $c$ at $M_0$ in the algorithm in Figure 5. Let $M_0[c\rangle M'$. Since $a$ and $c$ commute at $M_0$ and $a$ is fired before $c$, $\langle M', \{a\}\rangle$ is pushed onto the stack. Let's consider the visit to $M'$ where $\langle M', \{a\}\rangle$ is popped from the stack. We have $f(M') = \{a, b\}$, but the sleeping $a$ is not fired. Let $M'[a\rangle M''$. By executing the algorithm in Figure 5 completely, we see that $M''$ is never encountered, and no nonterminal marking is visited more than once. The latter observation is important since it guarantees that all transitions that are fired at a marking $M$ are in $f(M)$. The set of inspected markings is thus a proper subset of the markings of the $f$-reachability graph.

The statistics in [7, 24, 25] concerning some analyzed protocols do not give direct information for comparing the stubborn set method with the combination of the sleep set method and the stubborn set method.

## 5 Conclusions

The stubborn set method alone is sometimes bound to fire independent transitions at a state, so the sleep set method can further be used to eliminate redundant interleavings of transitions. We have generalized Wolper's and Godefroid's terminal state detection algorithm [24] and shown that the generalized version detects all reachable terminal markings of any finite place/transition net with a finite full reachability graph, given that the transition selection function represents all sets of length-secure alternative sequences to terminal markings. As already known, dynamically stubborn functions represent all sets of enabled permutations to terminal markings. They thus also represent all sets of length-secure alternative sequences to terminal markings.

Wolper and Godefroid [24], and Wolper, Godefroid, and Pirottin [25] suggest that the stubborn set method and the sleep set are compatible in a broad area of verification. The compatibility should certainly be studied further since all available means should be utilized in attacking the state space explosion problem, and in our opinion, only the detection of reachable terminal states has obtained more than cursory treatment so far. Linear temporal logics seem to form the most central area of research since they have a great expressive power, and the stubborn set method alone as well as the sleep set method alone can be extended to verify properties expressed as linear temporal logic formulae without a next state -operator [13, 18, 20, 21, 24]. The combination of the stubborn set method and the sleep set method should be studied in all those models of concurrency where each of these two methods alone are applicable. Finally, we have the problem of how efficient the algorithms are in practice and what could be done to improve their efficiency.

## Acknowledgements

This work has been carried out in the Digital Systems Laboratory of Helsinki University of Technology. This paper is essentially an abridged version of a part of the research report [23]. I am grateful to Professor Leo Ojala for his continuous support, and Acting Associate Professor Mikko Tiusanen and Associate Professor Antti Valmari for their helpful comments. In addition, I would like to thank Lic.Tech. Marko Rauhamaa for fruitful discussions on the subject of this research.

The financial support received from the Emil Aaltonen Foundation is also gratefully acknowledged.

## References

1. Ashcroft, E. and Manna, Z.: *Formalization of Properties of Parallel Programs.* Meltzer, B. and Michie, D. (Eds.), Machine Intelligence 6. Edinburgh University Press, Edinburgh 1971, pp. 17–42.
2. Best, E. and Lengauer, C.: *Semantic Independence.* Arbeitspapiere der GMD 250, Sankt Augustin 1987, 32 p.
3. Courcoubetis, C. (Ed.): Proceedings of the 5th International Conference on Computer-Aided Verification, Elounda, Greece, June/July 1993. Lecture Notes in Computer Science 697, Springer-Verlag, Berlin 1993, 504 p.
4. Godefroid, P.: *Using Partial Orders to Improve Automatic Verification Methods.* Clarke, E.M. and Kurshan, R.P. (Eds.), Proceedings of the 2nd International Workshop on Computer-Aided Verification, New Brunswick NJ, June 1990. Lecture Notes in Computer Science 531, Springer-Verlag, Berlin 1991, pp. 176–185.
5. Godefroid, P., Holzmann, G.J., and Pirottin, D.: *State Space Caching Revisited.* von Bochmann, G. and Probst, D.K. (Eds.), Proceedings of the 4th International Workshop on Computer-Aided Verification, Montreal, June 1992. Lecture Notes in Computer Science 663, Springer-Verlag, Berlin 1993, pp. 178–191.
6. Godefroid, P. and Kabanza, F.: *An Efficient Reactive Planner for Synthesizing Reactive Plans.* Proceedings of AAAI-91, Anaheim CA, July 1991, Vol. 2, pp. 640–645.
7. Godefroid, P. and Pirottin, D.: *Refining Dependencies Improves Partial-Order Verification Methods.* In [3], pp. 438–449.
8. Godefroid, P. and Wolper, P.: *Using Partial Orders for the Efficient Verification of Deadlock Freedom and Safety Properties.* Formal Methods in System Design 2 (1993) 2, pp. 149–164.
9. Katz, S. and Peled, D.: *Verification of Distributed Programs Using Representative Interleaving Sequences.* Distributed Computing 6 (1992) 2, pp. 107–120.
10. Katz, S. and Peled, D.: *Defining Conditional Independence Using Collapses.* Theoretical Computer Science 101 (1992) 2, pp. 337–359.
11. Mazurkiewicz, A.: *Trace Theory.* Brauer, W., Reisig, W., and Rozenberg, G. (Eds.), *Petri Nets: Applications and Relationships to Other Models of Concurrency. Advances in Petri Nets 1986, Part II, Proceedings of an Advanced Course, Bad Honnef, September 1986.* Lecture Notes in Computer Science 255, Springer-Verlag, Berlin 1987, pp. 279–324.
12. Overman, W.T.: *Verification of Concurrent Systems: Function and Timing.* PhD thesis, University of California Los Angeles, Los Angeles CA 1981, 174 p.
13. Peled, D.: *All from One, One for All: on Model Checking Using Representatives.* In [3], pp. 409–423.

14. Rauhamaa, M.: *A Comparative Study of Methods for Efficient Reachability Analysis.* Helsinki University of Technology, Digital Systems Laboratory Report A 14, Espoo, September 1990, 61 p.
15. Reisig, W.: *Petri Nets: An Introduction.* EATCS Monographs on Theoretical Computer Science 4, Springer-Verlag, Berlin 1985, 161 p.
16. Valmari, A.: *State Space Generation: Efficiency and Practicality.* Doctoral thesis, Tampere University of Technology Publications 55, Tampere 1988, 170 p.
17. Valmari, A.: *Stubborn Sets for Reduced State Space Generation.* Rozenberg, G. (Ed.), Advances in Petri Nets 1990. Lecture Notes in Computer Science 483, Springer-Verlag, Berlin 1991, pp. 491–515.
18. Valmari, A.: *A Stubborn Attack on State Explosion.* Formal Methods in System Design 1 (1992) 4, pp. 297–322.
19. Valmari, A.: *Stubborn Sets of Coloured Petri Nets.* Proceedings of the 12th International Conference on Application and Theory of Petri Nets, Gjern, Denmark, June 1991, pp. 102–121.
20. Valmari, A.: *Alleviating State Explosion during Verification of Behavioural Equivalence.* University of Helsinki, Department of Computer Science, Report A-1992-4, Helsinki 1992, 57 p.
21. Valmari, A.: *On-the-Fly Verification with Stubborn Sets.* In [3], pp. 397–408.
22. Valmari, A. and Tiusanen, M.: *A Graph Model for Efficient Reachability Analysis of Description Languages.* Proceedings of the 8th European Workshop on Application and Theory of Petri Nets, Zaragoza, June 1987, pp. 349–366.
23. Varpaaniemi, K.: *Efficient Detection of Deadlocks in Petri Nets.* Helsinki University of Technology, Digital Systems Laboratory Report A 26, Espoo, October 1993, 56 p.
24. Wolper, P. and Godefroid, P.: *Partial-Order Methods for Temporal Verification.* Best, E. (Ed.), Proceedings of the 4th International Conference on Concurrency Theory, Hildesheim, August 1993. Lecture Notes in Computer Science 715, Springer-Verlag, Berlin 1993, pp. 233–246.
25. Wolper, P., Godefroid, P., and Pirottin, D.: *A Tutorial on Partial-Order Methods for the Verification of Concurrent Systems.* Tutorial material of the 5th International Conference on Computer-Aided Verification, Elounda, Greece, June/July 1993, 85 p.

# OR Causality: Modelling and Hardware Implementation

Alex Yakovlev[1*], Michael Kishinevsky[2**],
Alex Kondratyev[3], and Luciano Lavagno[4]

[1] Department of Computing Science, University of Newcastle upon Tyne, England
[2] Department of Computer Science, Technical University of Denmark, Denmark
[3] Department of Computer Science, University of Aizu, Japan
[4] Dipartimento di Elettronica, Politecnico di Torino, Italy

**Abstract.** *Asynchronous circuits behave like concurrent programs implemented in hardware logic. The processes in such circuits are synchronised in accordance with the dynamic logical and causal conditions between switching events. In this paper we investigate a paradigm called OR causality. Petri nets and Change Diagrams provide adequate modelling and circuit synthesis tools for the various OR causality types, yet they do not always bring the specifier to a unique decision about which modelling construct must be used for which type. We present a unified descriptive tool, called Causal Logic Net, which is graphically based on Petri net but has an explicit logic causality annotation for transitions. The signal-transition interpretation of this tool is analogous to, but more powerful than, the well-known Signal Transition Graph. A number of examples demonstrate the usefulness of this model in the synthesis of asynchronous control circuits.*

## 1 Introduction

Asynchronous circuits can be seen as hardwired versions of concurrent programs. Such a circuit is an interconnection of primitive components, which can either be single-output logical gates or multi-output elements, such as mutual exclusion "gates". The switching events occurring on the circuit inputs and the outputs of the gates when some Boolean conditions in the circuit are satisfied are those atomic computational actions that characterise the behaviour of any concurrent system, in which processes interact and communicate.

Event- or causality-based models have proved useful for the verification and synthesis of asynchronous, speed-independent or delay-insensitive circuits [10, 1, 6, 5]. Informally, speed-independent circuits are asynchronous circuits whose correct operation does not depend on *output delays* of the constituting gates. Delay-insensitive circuits are asynchronous circuits whose correct operation does not depend on *interconnection delays* among those gates as well. The correct

* This work has been partly supported by SERC GR/J52327.
** This work has been supported by The Danish Technical Research Council.

operation of a generic asynchronous circuit, on the other hand, may depend on specific information about gate and wire delays.

Most of the above-cited models are based on the inherently causal framework of Petri nets and their interpreted version called Signal Transition Graph (STG), in which Petri net transitions are associated with circuit signal transitions. The capability of modelling causality explicitly is crucial in designing control-dominated asynchronous circuits, traditionally modelled with timing diagrams. A number of analysis and synthesis techniques have been developed and automated. Such techniques generate hazard-free circuits from STGs under certain restrictions imposed on the structural and behavioural subclasses of STGs, logical element basis and delay models. It was proved in [15] that STGs are sufficient in their modelling power to represent speed-independent (more precisely, semi-modular as defined in [7]) circuit behaviour. As was shown in [13, 15, 4] such behaviour is characterised by two major forms of causality between signal transitions, strong (AND) and weak (OR) causality. Assume for example that event $a$ has two causes, events $b$ and $c$. The strong form of causality assumes that both $b$ and $c$ must have occurred before $a$ may occur. Therefore, in the "strong" case, every cause strongly precedes its effect. In the case of weak causality, $a$ may occur if at least one of $b$ and $c$ have occurred. I.e., in the "weak" case, an event in question may be caused by any cause belonging to the set of weak causes for the event, provided that at least one such cause has occurred.

Certain limitations inherent in STGs prevent them from efficient modelling of weak causality. The problem is with the underlying Petri nets, whose event dynamics is "biased" towards the strong form of causality (a Petri net transition fires only if *all* its input places contain tokens). Weak causality can only be represented in Petri nets *indirectly*, by using complicated place/transition interconnecting schemes (see, e.g., Fig. 4). For this, the Petri net *must* be made unsafe (allow more than one token in a place). Most of the existing STG-based methods and tools require the underlying Petri net be safe. The non-safety should generally be no problem per se, but the problem arises because there is no precise way of saying how the second token arriving in the place that models the OR causality has to be "removed" before the place can be marked again. Hence providing the designer with a simple "recipe" to specify weak causality is rather difficult.

Such problems appear to be resolved in a slightly different model, called Change Diagram [13, 5]. The Change Diagram, whose original target was circuit behaviour modelling, gives "equal priority" to the two major causality types because events have two mechanisms of enabling. To avoid the need for unsafeness when representing OR causality, Change Diagrams have a "token borrow" mechanism, which allows a place [5] to be marked with a *negative* value. It has been shown [15] that certain behaviours that are representable with finite but unbounded (i.e., with places with unlimited marking) Change Diagrams cannot be modelled using a finite STG representation.

[5] To be more precise, an arc, since a Change Diagram is a directed graph in which marking is attributed to the arcs between vertices.

Change Diagrams also have a number of shortcomings. It is possible to model some conflict-free (corresponding to semi-modularity in circuits) behaviour with a finite and bounded Petri net, while the corresponding Change Diagram would be negatively unbounded [15]. Furthermore, Change Diagrams, in their present form cannot directly represent processes with conflicts or choice.

In this paper we tackle the discrepancy between these otherwise closely related models. We demonstrate the ways in which both STGs and Change Diagrams model OR causality. With the aid of examples, we show the problems of each of the two languages in modelling some "difficult" cases. We then present a unifying model, Causal Logic Net, which is based on a Petri net graphical notation augmented with the causal logic attribute for transitions. The latter is in the form of Boolean enabling functions for net transitions. The Causal Logic Net model is thus viewed as the main theoretical contribution of this paper.

Finally, we present an example of synthesising circuits from the Causal Logic Nets, from which we elicit two major types of OR causality. This part of our study results in a crucial practical result, which is the precise recommendation for the specifier as to which modelling construct of the Causal Logic Net has to be used to represent a particular type of OR causality.

## 2 Motivating Example: Low Latency Arbiter

The following example (other examples can be found in [14]) shows the usefulness of the weak form of causality between events in designing asynchronous systems.

Consider an arbiter cell that arbitrates between two users [3]. Multi-way arbitration is organised by cascading such cells to form a tree or a chain. Each cell propagates the request in the direction from the lower level to the upper level, while the grants are generated in the opposite direction. Figure 1.(a) shows one such cell with its three request-grant handshake links $(R1, G1), (R2, G2)$ and $(R, G)$, where $(R1, G1), (R2, G2)$ stand for the links with lower levels, producing competing requests $R1$ and $R2$, and the $(R, G)$ pair is the link with the upper level. Figure 1.(b) illustrates, with the help of a timing diagram, the handshaking protocol between the links. After the first request by $R1$ is granted and the resource is released, two simultaneous requests are made by $R1$ and $R2$ and granted in turn.

Asynchronous arbiters of this type are usually implemented using an SR flip-flop and an analogue mutual exclusion element that is aimed at resolving the metastability and oscillation anomalies occurring in the flip-flop. The time it takes to resolve the arbitration can be much longer than an ordinary switching delay. The implementation of the arbiter described in [16] is advantageous over the one in [3] for it allows *not to wait* for the arbitration to be resolved by the local mutual exclusion element before propagating the request (signal $R$) to the upper level. This circuit uses a weak form of causality to produce the request on $R$ from the arrival of $R1$ or $R2$ (the first of them causes $R$ to be set), thereby allowing the arbitration resolution process to be executed in parallel with the process run in the upper level to generate the grant on $G$. Normally, of course,

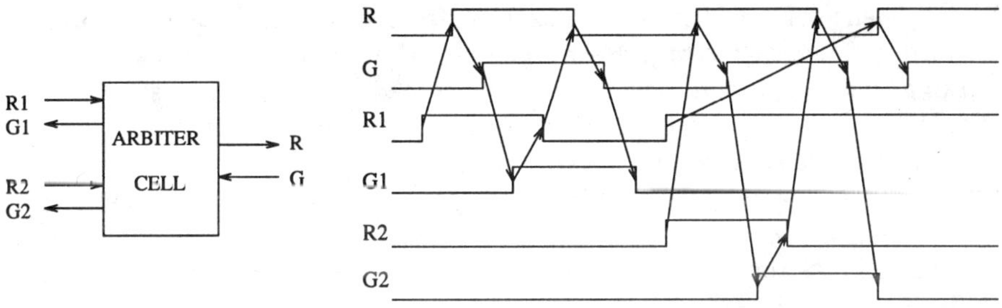


(a) (b)

**Fig. 1.** The arbiter example

when the signal on $G$ arrives the cell is ready to generate an appropriate grant either on $G1$ or $G2$ depending on which of the grants has been chosen by the mutual exclusion element.

The above example shows that using OR causality seems quite a natural way of organising the dynamic behaviour in asynchronous systems. However, as will be shown further, the descriptive tools of even such event-oriented models as Petri nets are not very well suited for the modelling of this form of causality in a universal way. The following section more formally proves the need for the modelling enhancement of Petri nets.

# 3 Modelling OR Causality in Signal Transition Graphs and Change Diagrams

## 3.1 Petri Nets and Change Diagrams

When talking about Signal Transition Graphs and their ability to model the forms of causality, in this paper we often address the properties of their underlying model, Petri nets. On the other hand, there has been no separate, uninterpreted, notation defined for Change Diagrams in the literature ([5]), so we shall use the same name to refer to the more abstract unlabelled version of such a model.

**Petri Nets.** A *Petri net* (PN) [8] is a triple $\mathcal{P} = \langle T, P, F \rangle$ where $T$ and $P$ are non-empty finite sets of transitions and places, and $F \subseteq (T \times P) \cup (P \times T)$ is the flow relation between transitions and places. A PN can be represented as a directed bipartite graph, where the arcs represent elements of the flow relation.

A PN marking is a function $m : P \rightarrow \{0, 1, 2, \ldots\}$, where $m(p)$ denotes the number of *tokens* in $p$ under marking $m$. A *marked* PN is a quadruple $\mathcal{P} = \langle T, P, F, m_0 \rangle$, where $m_0$ denotes its initial marking. A transition $t \in T$ is *enabled* at a marking $m$ if all its predecessor places are marked. An enabled transition $t$ *may fire*, producing a new marking $m'$ with one less token in each predecessor place and one more in each successor place (denoted by $m[t > m')$.

A sequence of transitions and intermediate markings $m[t_1 > m_1[t_2 > \dots m'$ is called a *firing sequence from* $m$. The set of markings $m'$ reachable from a marking $m$ through a firing sequence is denoted by $[m >$. The set $[m_0 >$ is called the *reachability set* of a marked PN with initial marking $m_0$, and a marking $m \in [m_0 >$ is called a reachable marking.

A PN marking $m$ is *live* if for each $m' \in [m >$ for each transition $t$ there exists a marking $m'' \in [m' >$ that enables $t$. Similarly, a transition $t$ is *live* if for each $m' \in [m >$ there exists a marking $m'' \in [m' >$ that enables $t$. A marked PN is live if its initial marking is live.

A marked PN is *k-bounded* (or simply "bounded") if there exists an integer $k$ such that for each place $p$, for each reachable marking $m$ we have $m(p) \leq k$. A marked PN is *safe* if it is 1-bounded.

A transition $t_1$ *disables* another transition $t_2$ at a marking $m$ if both $t_1$ and $t_2$ are enabled at $m$ and $t_2$ is not enabled at $m'$ where $m[t_1 > m'$. A marked PN is *persistent* if no transition can ever be disabled at any reachable marking.

A PN is a Marked Graph if every place has exactly one predecessor and one successor. A PN is *free-choice* if for any two transitions $t_1$ and $t_2$ that share a predecessor place $t_1$ and $t_2$ have only one predecessor (i.e. any two transitions with a common predecessor place have only one predecessor).

A marked PN $\mathcal{P} = \langle T, P, F, m_0 \rangle$ generates a state graph, called Reachability Graph, $\langle [m_0 >, E, T, \delta \rangle$, where for each edge $(m_1, m_2) \in E$, such that $m_1, m_2 \in [m_0 >$ and $m_1[t > m_2$, we have $\delta(m_1, m_2) = t$.

Note that PNs used for the representation of Signal Transition Graphs have traditionally been depicted in a "shorthand" form, which seems convenient to the circuit designer and which is adopted in this paper unless it creates confusion. In this form, PN transitions are denoted by their corresponding labels (instead of bars or boxes) and PN places are explicitly denoted by circles only if such place has more than one predecessor or successor transitions. If a place has only one predecessor and one successor, the corresponding circle is omitted and the token marking is associated with the arc (a similar notation is often used to represent Marked Graphs [2]).

**Signal Transition Graphs.** Interpreted Petri nets, where transitions represent changes in the values of circuit signals, were proposed independently as specification models for Asynchronous Logic Circuits by [10] (where they were called Signal Graphs) and [1] (where they were called Signal Transition Graphs, STGs). Both papers proposed to interpret a PN as the specification of a circuit defined on a set of signals $Y$, by labelling each transition with an element of $Y \times \{+, -\}$. A label $y_i^+$ means that signal $y_i \in Y$ changes from 0 to 1, and $y_i^-$ means that $y_i$ changes from 1 to 0, while $y_i^*$ denotes either $y_i^+$ or $y_i^-$.

An STG is a quadruple $\mathcal{G} = \langle \mathcal{P}, X, Z, \Delta \rangle$ where $\mathcal{P} = \langle T, P, F, m_0 \rangle$ is a *marked* PN, $X$ and $Z$ are (disjoint) sets of input and output signals respectively ($Y = X \cup Z$), and $\Delta : T \to (X \cup Z) \times \{+, -\}$ labels each transition of $\mathcal{P}$ with a signal transition. An STG is *autonomous* if it has no input signals (i.e. $X = \emptyset$).

Both [10] and [1] gave also synthesis methods to translate the PN into a State Transition Diagram (called Transition Diagram in [10] and State Graph in [1]) and hence into a circuit implementation of the specified behaviour.

Given an STG $\mathcal{G} = \langle \mathcal{P}, X, Z, \Delta \rangle$ and the Reachability Graph $([m_0 >, E, T, \delta)$ corresponding to its PN $\mathcal{P}$ (where $\delta$ labels each edge in $E$ with a transition in $T$), we define the associated State Transition Diagram (STD) $\mathcal{S} = \langle [m_0 >, E, \lambda \rangle$ as follows. For each state (marking) $m \in [m_0 >$, $\lambda(m)$ is a vector of signal values (also denoted $s^m$ for simplicity). We say that vector $s^m$ is a *state label* corresponding to the marking $m$. Obviously the STD labeling must be *consistent* with the interpretation of the STG signal transitions. Let $s_i^m$ denote the value of signal $y_i$ in $s^m$. Each arc $e = (m, m') \in E$ in the STD must obey the following *consistency condition*: if $\Delta(\delta(e)) = y_i^+$, then $s_i^m = 0$ and $s_i^{m'} = 1$; if $\Delta(\delta(e)) = y_i^-$, then $s_i^m = 1$ and $s_i^{m'} = 0$; otherwise $s_i^m = s_i^{m'}$.

A *finite, bounded* STG is defined as *valid* if its underlying reachability graph has a *consistent labeling* as defined above. An STD is *semi-modular* if for every state $m \in [m_0 >$ with two successor states $m', m''$ such that $(m, m'), (m, m'') \in E$ there exists $m'''$ such that $(m', m'''), (m'', m''') \in E$. Note that, by labeling consistency, if $\Delta(\delta(m, m')) = y_i^*$ and $\Delta(\delta(m, m'')) = y_j^*$, we must have $\Delta(\delta(m', m''')) = y_j^*$ and $\Delta(\delta(m'', m''')) = y_i^*$.

Figure 2 shows an example of an STG and the corresponding STD. Following the convention of [7], a signal value is denoted by 1∗ in a label if it is currently 1 and a falling transition for it is enabled in the corresponding state. Similarly 0∗ denotes a signal that is currently at 0 but enabled to rise. 0 and 1 denote stable values for the corresponding signal. The initial marking of the PN in Fig. 2.(a) (corresponding to the leftmost state in Fig. 2.(b)) appears on the edge between $y^-$ and $x^+$.

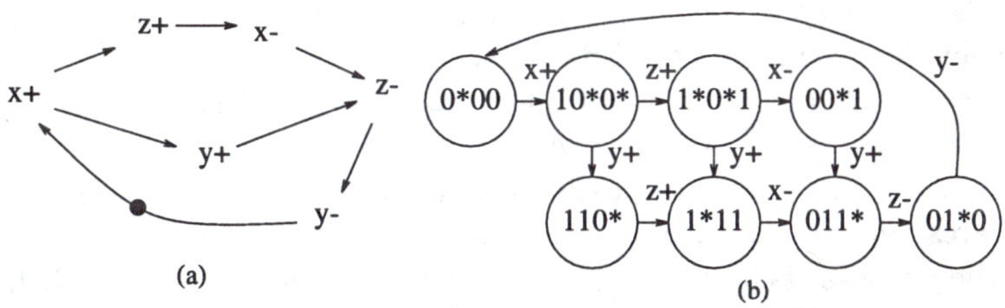


**Fig. 2.** A Signal Transition Graph and its State Transition Diagram

Another example of STG is shown in Fig. 3. This is a model of the low latency arbiter described in Sect. 2. The STG clearly demonstrates that after the arrival of either $R1$ or $R2$, the handshake $R/G$ is set in parallel with resolving the mutual exclusion inside the arbiter cell (by means of an internal mutual exclusion element whose output signals are $A1$ and $A2$ in Fig. 3).

**Change Diagrams.** The definition of Change Diagrams, described more in detail in [13, 5], is based on two explicit types of precedence relations between transitions in Asynchronous Logic Circuits:

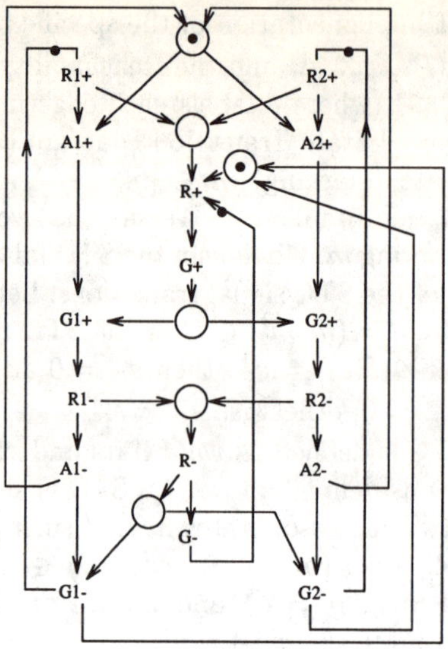


**Fig. 3.** Signal Transition Graph for the arbiter in Fig. 1

(1) the *strong precedence* relation between transitions $a^*$ and $b^*$, depicted by a solid arc in the graphical representation of Change Diagrams, means that that $b^*$ *cannot occur without the occurrence of* $a^*$.

(2) the *weak precedence* relation between transitions $a^*$ and $b^*$, usually depicted by a dashed arc, means that $b^*$ *may occur after an occurrence of* $a^*$. But $b^*$ may also occur *after some other transition* $c^*$, which is also weakly preceding $b^*$, without the need for $a^*$ to occur.

A Change Diagram (CD) is a tuple $\mathcal{D} = \langle A, \rightarrow, \vdash, M, O \rangle$, where: $A$ is a set of *transitions* or *events*, $\rightarrow \subseteq (A \times A)$ is the *strong* precedence relation between transitions, $\vdash \subseteq (A \times A)$ is the *weak* precedence relation, $M$ is a set of initially *marked* arcs, and $O$ is a set of *disengageable* arcs.

For the purpose of circuit specification all transitions from $A$ are labelled with signal transitions of a set of signals $Y$ similarly to the signal transition labelling of STGs. The relations $\rightarrow$ and $\vdash$ are mutually exclusive (i.e. $(a^*, b^*) \in \rightarrow$ implies that $(a^*, b^*) \notin \vdash$ and vice-versa), and all the predecessors of a transition $a^*$ must be either of the *strong* type or of the *weak* type. Hence the set of transitions $A$ is partitioned into *AND-type* transitions (with strong predecessors) and *OR-type* transitions (with weak predecessors).

The firing rule of CDs is similar to that of PNs, with arcs playing the role of places and flow relation elements at the same time. Each arc is assigned an integer *marking* which, unlike PN marking, can be *negative*. Initially each arc in $M$ has marking 1, and each arc not in $M$ has marking 0. An *AND-type* transition is enabled if *all* its predecessor arcs have marking greater than 0. An *OR-type* transition is enabled if *at least one* predecessor arc has marking greater than 0. When an enabled transition fires, the marking of each predecessor arc is decremented, and the marking of each successor arc is incremented. A CD is *bounded* if the marking on each arc is bounded (both above and below) in all possible firing sequences.

*Disengageable* arcs are "removed" from the CD after the *first firing* of their successor transition. They are used to represent the *initialisation sequence* of a circuit, and we will not enter into details concerning their usage.

Following [5], a State Transition Diagram $\mathcal{S} = \langle S, E, \lambda \rangle$ can be associated with a CD in a way analogous to the above defined for STGs.

A CD is *correct* if it satisfies the following conditions, ensuring that the labelling of the states in the STD is consistent:

- for all firing sequences, the signs of the transitions of each signal alternate;
- no two transitions of the same signal can be concurrently enabled in any reachable marking vector; and
- the CD is connected and *bounded* (i.e. the reachability set is *finite*).

The main theoretical result concerning CD correspondence to semi-modular STD claims that

1. each semi-modular STD without transient cycles has a corresponding correct CD, and
2. each correct CD has a corresponding semi-modular STD [5].

A *transient cycle* in an STD is defined as a cycle where at least one variable is continuously excited with the same value. See, for example, the cycle of states labeled: $0{*}0{*}0.000 \rightarrow 0{*}10.0{*}00 \rightarrow 0{*}1{*}0.100 \rightarrow 0{*}00{*}.100 \rightarrow 0{*}01.10{*}0 \rightarrow 0{*}01{*}.110 \rightarrow 0{*}00.110{*} \rightarrow 0{*}00.1{*}1{*}1 \rightarrow 0{*}00.01{*}1 \rightarrow 0{*}00.001{*} \rightarrow 0{*}0{*}0.000$ in Fig. 11. In all states of this cycle signal $b$ has value $0{*}$.

CDs are useful in practice because of the availability of low-complexity polynomial time *analysis* algorithms to decide, e.g., whether a given CD is correct, and hence it can be used as a valid specification of a semi-modular circuit. Note that this analysis can be performed by direct construction of the CD, without going through the exponential size STD ([5]). Furthermore synthesis algorithms from CDs to circuits in various technologies are outlined in [5].

The main limitation of CDs however is their inability to describe *choice* among alternative behaviours, as modeled by places with more than one successor in PNs. So a designer faced with the description, for example, of a self-timed memory element, must describe the various possible read/write cycles of each data value as an *alternation* rather than a *choice* between them.

### 3.2 Problems in Modelling OR Causality

Although both Petri nets and Change Diagrams are capable of modelling the causality paradigms of semi-modular circuits, there are some problems, and even discrepancy, in the way these formalisms represent OR causality.

**Problems with Using Petri Nets.** To show formally the difficulties of modelling OR causality by PNs let us define the notion of *observation equivalence* between PNs and CDs. We will think of such an equivalence as the similarity of the partial behaviours generated by a PN and a CD for a given subset of their transitions.

The notion of behaviour in both PNs and CDs can be defined in terms of the accepted languages. Suppose that transitions of a CD $\mathcal{D}$ and of a PN $\mathcal{P}$ are labelled by the elements from some set $B$ using partial labelling functions.

Given a labelled CD $\mathcal{D}$ or a labelled PN $\mathcal{P}$ the *language* accepted by $\mathcal{D}$ or $\mathcal{P}$, denoted as $L(\mathcal{D})$ or $L(\mathcal{P})$ , is a set of *feasible* sequences of transition labels that the model generates from its initial marking.

Denote by $\mathcal{Q}$ a CD or a PN under consideration. We now define the *projection* operator, denoted by $\downarrow$ for the labelled CDs and PNs with respect to a subset of the labels $B_1 \subset B$ as follows. If $p$ is a feasible sequence for $\mathcal{Q}$, then its projection $p \downarrow B_1$ is sequence $p$ with all labels from $B - B_1$ deleted. If $L(\mathcal{Q})$ is a language accepted by $\mathcal{Q}$, then its projection $L(\mathcal{Q}) \downarrow B_1$ is a set of sequences $\{p \downarrow B_1 \ : \ p \in L(\mathcal{Q})\}$

Two models, a CD $\mathcal{D}$ and a PN $\mathcal{P}$, are *observation equivalent* with respect to the subset of labels $B_1$ if $L(\mathcal{D}) \downarrow B_1 \ = \ L(\mathcal{P}) \downarrow B_1$.

This equivalence relation is not very restrictive, because it compares behaviour only up to the "observable points" of specifications. Moreover, we have no restrictions on the corresponding number of transitions labelled by the same symbol in the models. Therefore one can model behaviour of some particular labelled transition from the CD by a set of transitions from the PN labelled by the same symbol, and vice versa.

A stronger type, *strong observation equivalence*, disallows several PN transitions to be associated with one CD event and vice versa. This notion is useful due to the following two reasons. First, to save compactness and locality of the specification and, second, which is more important, to represent an OR-causal event directly, as a single transition in the PN model, rather than through its "simulation" by the firing of several transitions, each occurring after its own OR-cause.

Two models, a CD $\mathcal{D}$ and a PN $\mathcal{P}$, are *strongly observation equivalent* with respect to the subset of labels $B_1$ if they are observation equivalent with respect to $B_1$ and each $b \in B_1$ labels the same number of transitions both in $\mathcal{D}$ and in $\mathcal{P}$.

Figure 4.(a) shows a simple CD $\mathcal{D}$ with an OR-transition $c$, while Fig. 4.(b) and (c) show two different PNs $\mathcal{P}1$ and $\mathcal{P}2$ that are strongly observation equivalent to $\mathcal{D}$ with respect to transitions $\{a, b, c\}$ [6].

In $\mathcal{P}1$ the transition $c$ will fire after $a$ or $b$ have put a token into the place $p$. If both $a$ and $b$ have fired, two tokens will be placed in $p$. To prevent $c$ from repeatedly firing in the latter case, we introduce the additional "hidden" transition $t$ to remove the second token from $p$. Thus, in the functioning of $\mathcal{P}1$ the firings of $c$ and $t$ alternate; $c$ fires from the "first' token in $p$ and $t$ from the

[6] In case of unique labels, we do not distinguish between a transition and its label.

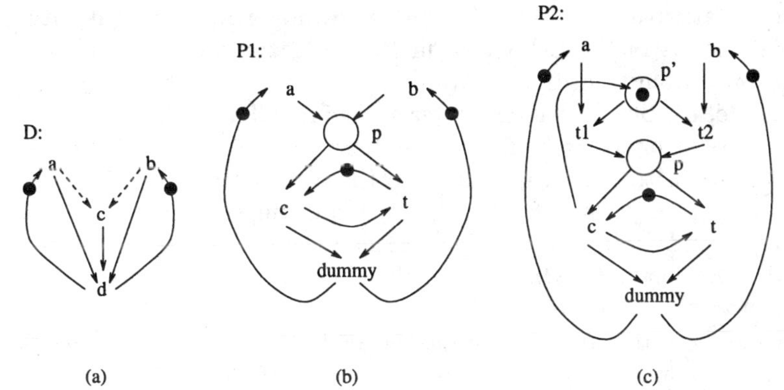


**Fig. 4.** A Change Diagram with strongly observation equivalent Petri Nets

"second". $\mathcal{P}2$ operates in a similar way, with the exception that in $\mathcal{P}2$ the second token is not allowed to pass into $p$, but is "delayed" on transitions $t1$ and $t2$ by the choice place $p'$.

It is easy to see that $\mathcal{P}1$ is unsafe (for place $p$) and $\mathcal{P}2$ is non-persistent (for place $p'$), moreover both PNs are non-free-choice (e.g., for places $p$). This is not just a shortcoming of this particular example since it is intrinsic to the OR-causality modelling in PNs as the following Proposition shows.

***Proposition 1.*** *If a safe CD $\mathcal{D}$ is strongly observation equivalent to a PN $\mathcal{P}$ with respect to the set of live* [7] *transitions $\{a, b, c\}$, where $a \vdash c$, $b \vdash c$, then the PN $P$ is either* non-persistent and non-free-choice *or* unsafe and non-free-choice.

The proof of the proposition can be found in [14].

This proposition shows the difficulties of using PNs for the design of circuits with OR-causal behaviour. Indeed, the equivalence notion introduced here is aimed at establishing the equivalence only with respect to some subset of signals. Usually these are *input* and *output* signals of the circuit, and we are looking for different implementations that have the same input-output behaviour. In this case the signals that are not participating in the equivalence relation are the *internal* signals of the circuit. If any signal is non-persistent, *no speed-independent logic circuit can implement the specified behaviour* [11]. The non-safety and non-free-choice features of specification may also be a problem, since most known STG synthesis methods work only with free-choice and safe descriptions.

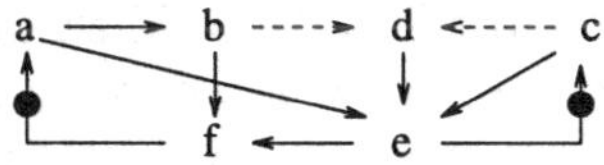


**Fig. 5.** Unsafe Change Diagram with OR causality

[7] A CD transition is live if it can be enabled infinitely many times in the CD operation.

The situation may however become worse if we have to deal with unsafe OR relations, as in the CD shown in Fig. 5. In this CD the OR-transition $d$ can fire twice from transition $c$ without any occurrence of $b$ (e.g., the sequence $a$, $c$, $d$, $e$, $c$, $d$). But if $b$ happens after $a$, then it will not trigger $d$ and the firing of $b$ will be "ignored". In ordinary PNs there is no means to distinguish from which transition the token is coming to the common place $p$. But due to the operation rules of the unsafe OR, we have to distinguish the token that comes from the second firing of $c$ (it affects $d$) and the token that comes from the $b$ firing (it does not affect $d$). This is why there is no PN strongly observation equivalent to a CD with unsafe OR.

The only way to represent the unsafe OR in PNs is to use the fact that any unsafe but $k$-bounded CD can be reduced to a safe form by *unfolding* it into $k$ periods ([5]). Such a solution, however, is not strongly observation equivalent because it implies to construct a PN that is equivalent not to the initial CD but to its $k$-period unfolding. This may significantly increase the size of the specification and force the designer to think in terms of the unfolded behaviour.

**"Unbounded" Cases.** The above analysis was applied to the case in which both PN and CD were bounded and could represent the same behaviour. Let us look at two "unbounded" cases.

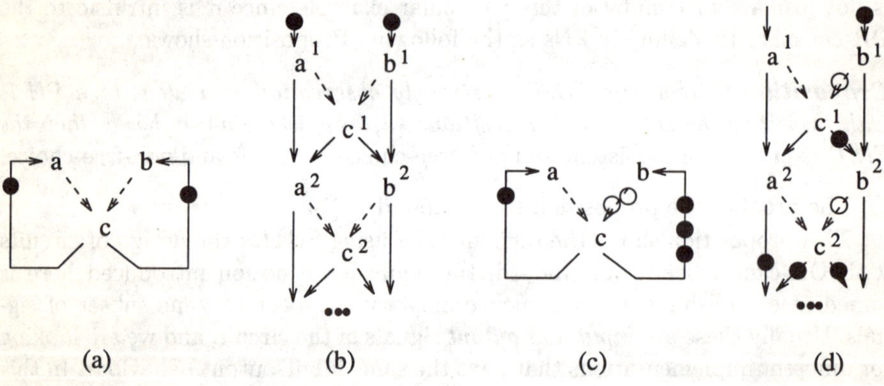


**Fig. 6.** A Change Diagram without an equivalent finite Petri net

The first example is depicted by a simple CD in Fig. 6.(a) with unbounded arc marking. Such a behaviour, in which the $i$-th ($i = 1, 2, ...$) occurrence of transition $c$ is caused either by the $i$-th occurrence of $a$ or by the $i$-th occurrence of $b$, is represented in Fig. 6.(b) as a CD unfolding ([5]). In the unfolding each transition $a^i$ represents a *unique occurrence* of the corresponding transition $a$ in a firing sequence of the CD (similarly for $b^i$ with respect to $b$ and $c^i$ with respect to $c$).

It was first noted in [15], and later formally proved in [9], that there exists no finite PN representing such a behaviour. The seemingly equivalent PN shown

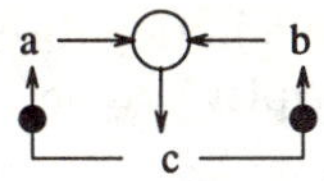


**Fig. 7.** A Petri net only seemingly equivalent to Fig. 6.(a)

in Fig. 7 describes in effect a different behaviour. The $i$-th occurrence of its transition $c$ can be caused by any combination of pairs of the form $a^k$ and $b^{i-k}$ where $k$ can be any value between 0 and $i$. The difference between these behaviours is obvious.

The CD is able to remember the number of occurrences of transitions $a$ and $b$, using the negative marking mechanism. So if $a$ fires twice, as represented in Fig. 6.(c) and 6.(d) (empty circles represent negative marking on the arc between $b$ and $c$), and then it stops firing, $c$ can fire again only after $b$ has fired *three* times, in order to "re-absorb" the negative marking. On the other hand in Fig. 7.(a), if $a$ fires twice and then stops, $c$ can begin firing again as soon as $b$ fires, because there is no way to remember an *unbounded* "debt" of tokens.

Another example is a PN shown in Fig. 8.(a). It models an initially one-place buffer that becomes two-place when transition $b$ occurs. The behaviour is semi-modular, because no transition is disabled. Yet there is no connected equivalent CD, because connected CDs can represent only semi-modular behaviours without *transient cycles*. In this example transition $b$ is continuously enabled during the cyclic firing of $a$ and $c$.

The behaviour of the CD shown in Fig. 8.(b) is only a subset (in terms of the corresponding languages build on the set of firing sequences) of that of the PN in Fig. 8.(a). It has an *unbounded* negative marking on the arc between $b$ and $c$ (Fig. 8.(c) shows such negative marking after the occurrence of $b$, followed by four occurrences of $a$ and $c$). The difference in their behaviours begins after the occurrence of transition $b$.

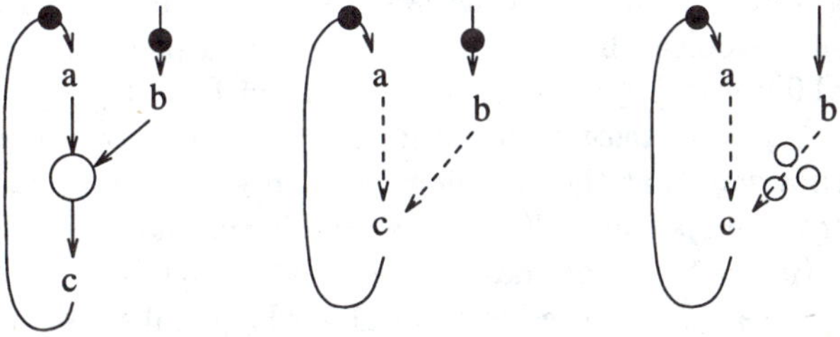


(a) (b) (c)

**Fig. 8.** A bounded Petri net with a Change Diagram that is not equivalent to it

In order to fully model this behaviour in the CD language, we have to represent the modes of one-place and two-place buffer as separate CDs. Their composition would then require some additional selection mechanism, which would however lead us outside the descriptive domain of the original CDs notation.

## 4 Causal Logic Nets

In this section we propose a new model that is based on the PN graph but has more general rules defining its dynamics. The model, called Causal Logic Net, inherits the power of both PNs and CDs. We show that both these models are only special cases of the Causal Logic Net. On the other hand, this model is a sort of "least upper bound" of PNs and CDs, so we hope that its analysis will not be drastically more complex than analysing its prototypes. The topic of analysis of Causal Logic Nets is outside the scope of the present paper.

A *Causal Logic Net* (CLN) is a quadruple $\mathcal{N} = \langle T, P, F, \beta \rangle$ where $T$, $P$ and $F \subseteq (T \times P) \cup (P \times T)$ have the same meaning as for PNs, and $\beta : T \rightarrow \mathcal{F}$ is a function which assigns each transition a Boolean function from the set $\mathcal{F}$ of Boolean functions defined on subsets of $P$, i.e. $\mathcal{F} = \{f | \exists P' \subseteq P \wedge f : \{0,1\}^{|P'|} \rightarrow \{0,1\}\}$, in such a way that $\forall t \in T : \beta(t) : \{0,1\}^{|{}^\bullet t|} \rightarrow \{0,1\}$. Note that here and further on we use standard notation for the sets of input and output places of a transition: ${}^\bullet t = \{p | (p,t) \in F\}$ and $t^\bullet = \{p | (t,p) \in F\}$. Analogous notation will be used for the sets of input and output transitions of a place: ${}^\bullet p = \{t | (t,p) \in F\}$ and $p^\bullet = \{t | (p,t) \in F\}$.

It is clear that a CLN can be represented as a PN-like (bipartite) graph in which each transition is associated with a Boolean function defined on its input places. This function is called the *enabling function* of the transition. It can be written as an expression (with brackets) using the standard mathematical notation accepted for Boolean functions, in which each place $p \in {}^\bullet t$ is associated with a literal (for simplicity, we shall use the same name $p$). For each transition $t$ the enabling function $\beta(t)$ is evaluated according to the marking of the net. A CLN marking is defined similar to a PN marking $m : P \rightarrow \{\ldots, -2, -1, 0, 1, 2, \ldots\}$ except that the range of values is the full integer range. Thus, a *marked* CLN is a quintuple $\mathcal{N} = \langle T, P, F, \beta, m_0 \rangle$, where $m_0$ denotes its initial marking.

We say that a Boolean literal $p$ associated with a place $p \in P$ is evaluated with the logical 0 if $m_0(p) \leq 0$ and it is evaluated with the logical 1 if $m_0(p) > 0$. A transition $t \in T$ is *enabled* at a marking $m$ if all its predecessor places are marked in such a way that the enabling function evaluates to 1. For example, if for $t$ with ${}^\bullet t = \{p_1, p_2\}$ and $\beta(t) = p_1 p_2$ the initial marking $m_0$ is such that $m_0(p_1) = 0, m_0(p_2) = 1$ $t$ is not enabled because $\beta(t) = 0$. On the other hand, if we define $\beta(t) = p_1 + p_2$, then the same initial marking makes $t$ enabled since now $\beta(t) = 1$. The first situation corresponds to the case of strong (AND) causality for $t$, while the second is an example of weak (OR) causality.

The remaining task is now to define the transition firing rule. First of all, like in PNs, we say that any transition $t \in T$ *may fire* under $m$ (initially, $m = m_0$), producing a new marking $m'$, if it is enabled under $m$.

**Firing Rule.** The new marking $m'$ is defined in the same way as in ordinary PNs: $\forall p \in P : m'(p) = m(p) - 1$ if $p \in {}^\bullet t,\ p \notin t^\bullet$, $m'(p) = m(p) + 1$ if

$p \in t^\bullet$, $p \notin {}^\bullet t$, and $m'(p) = m(p)$ otherwise. Thus, according to this rule some of the places, that may originally be marked by $m$ with nonnegative number of tokens, can be marked negatively ("tokens are borrowed") in $m'$.

With the effect of this firing rule we say that $m'$ is directly reachable from $m$ through the firing of $t$, using the ordinary PN notation $m[t > m'$.

We define all other notions standard for PNs, such as firing sequence, reachability, reachability set and reachability graph in an analogous way.

The following two propositions show the relationship between CLN and the previously used models.

***Proposition 2.*** *A CLN $\mathcal{N}$ is a PN iff*

1. *for each transition $t \in T$ its enabling function $\beta(t)$ is a positive unate conjunction of literals associated with all input places from ${}^\bullet t$ ($\beta$ assigns only AND expressions with positive literals), and*
2. *the initial marking of each place is nonnegative.*

An analogous proposition holds for the reduction of CLN to CD.

***Proposition 3.*** *A CLN $\mathcal{N}$ together with the signal transition interpretation $\Delta$ is a CD without disengageable arcs iff*

1. *for each transition $t \in T$ the enabling function $\beta(t)$ is either a positive unate conjunction or a positive unate disjunction of literals associated with all places from ${}^\bullet t$ ($\beta(t)$ can either be an AND or OR expression with positive literals),*
2. *for each place $p \in P$ the sets of predecessor and successor transitions contain at most one transition, i.e. $|p^\bullet| \leq 1$ and $|{}^\bullet p| \leq 1$, and*
3. *the initial marking of a place is either 0 or 1.*

The proof of these propositions is trivial. Furthermore, we can demonstrate that the CLN with the qualities given by item 1 of this proposition is also capable of adequate modelling of disengageable arcs. As stated in [5], in well-formed CDs, every disengageable arc connects a non-repeated transition with a cyclic transition. In this case, the modelling is very simple. It is shown in Fig. 9(a). For the general case when the disengageable arc is outgoing from a transition which can have other outgoing arcs that are not disengageable we have to refer to Fig. 9(b). It shows a CD fragment with such an arc together with its associated PN fragment, that is strongly observation equivalent with respect to the set of transitions in the CD.

Based on the notion of CLN we can define an updated version of an STG, by analogy with the original version of STG defined on a PN. It would however be more appropriate to give it a separate name, Causal Logic Signal Transition Graph (CL-STG). Thus, a CL-STG is a tuple $\mathcal{G} = \langle \mathcal{N}, X, Z, \Delta \rangle$ where $\mathcal{N}$ is a *marked* CLN, $X$ and $Z$ are disjoint sets of input and output signals respectively and $\Delta : T \to (X \cup Z) \times \{+, -\}$ labels each transition of $\mathcal{N}$ with a signal change symbol. Obviously $\mathcal{N}$ is an STG iff its underlying CLN satisfies the conditions of Proposition 2.

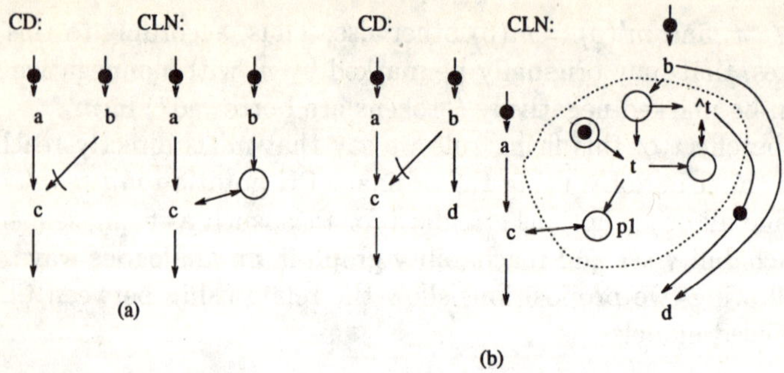


**Fig. 9.** Modelling disengageable arcs in Causal Logic Nets

We redefine the properties of a CL-STG by analogy with STGs. The most important of them is *validity* which allows the construction of the STD model. Validity (finiteness, boundedness and consistent labeling) remains unchanged for CL-STGs, since the underlying reachability graph for a CLN is defined in the same way as for ordinary PNs.

In this paper we do not attempt to investigate all the problems of analysis of properties and classes of CLNs. Instead, we have only demonstrated that this model allows the representation of any type of causality describable by means of Boolean logic.

In a more detailed presentation [14], we show how a slight extension of the firing rule, making the resulting marking conditional upon the enabling marking, can allow the CLN model to have the descriptive power of a Turing Machine. The same extension also allows to represent a non-deterministic behaviour with non-commutativity.

## 5 Circuit Synthesis Examples

In this section we show two examples of synthesis of speed-independent circuits from initial specifications using CL-STGs. These two examples also illustrate the two major types of OR causality.

The first example is an event-based *inclusive OR* element. The CL-STG description of this element is shown in Fig. 10.(a). The element, having two inputs $x_1$ and $x_2$ and two outputs $y_1$ and $y_2$, behaves in the following way. Starting from the initial states where all the signals are at 0, $y_1$ changes its output from 0 to 1 whenever *either* of its inputs changes from 0 to 1. The inputs cannot change until the other output $y_2$ has also been set to 1, which happens when both are at logical 1. Later, inputs can change back to 0 in any order, and the output $y_1$ follows the first of these, while $y_2$ again ensures "safe" operation by checking that both inputs are at logical 0.

We assume that in the CL-STG in Fig. 10.(a) the $\beta$ function for the transitions labelled with the changes of $y_1$ is a simple disjunction between the two input

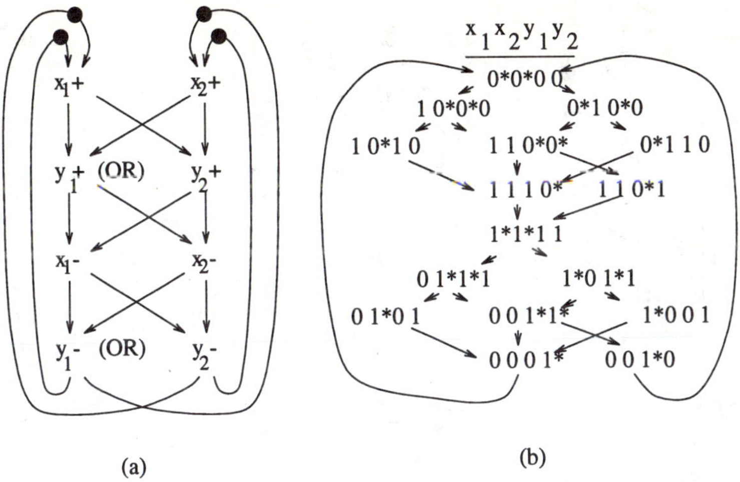


**Fig. 10.** A Causal Logic State Transition Graph and its State Transition Diagram for the inclusive OR example

arcs [8]. The $\beta$ function of all the remaining transitions is AND.

Figure 10.(b) shows an STD, which is semi-modular with respect to all signals. We can thus derive the circuit implementation using any of the existing techniques (e.g., [5, 6, 10, 12]). A set of Boolean functions for the circuit is:

$$y_1 = x_1 x_2 + (x_1 + x_2)\overline{y_2}$$
$$y_2 = x_1 x_2 + (x_1 + x_2) y_2$$

The second example is a *variable capacity* buffer, which operates initially as a one-place buffer, but upon the arrival of a request signal becomes a two-place buffer. Note that we use the term "buffer" meaning its main control flow functionality, without concerns about data path functions.

The buffering property is therefore characterised here by the number of times one handshake pair (the input of the buffer) can change its value while the other handshake pair remains in the same state. A purely abstract model, where each handshake was denoted simply by one symbol, was shown in Fig. 8.(a). We shall draw upon this model in constructing a corresponding "signalling expansion", which is shown in Fig. 11.(a). Here signals $a$ and $c$ stand for the outputs of the handshakes. The inputs are "hidden" because they are simply delayed versions of the outputs. The environment is assumed to be acknowledging the requests of the circuit, where the semantics of request is "the circuit is ready".

Note that the CL-STG in Fig. 11.(a) is actually an STG because, in this example, we do not want to use the causal logic model for transition $c^+$. This decision is due to the idea of keeping an extra token, arriving as a result of

[8] We resort to the usual "shorthand" notation style, in which places are explicitly shown only where they have more than one incoming or outgoing arc.

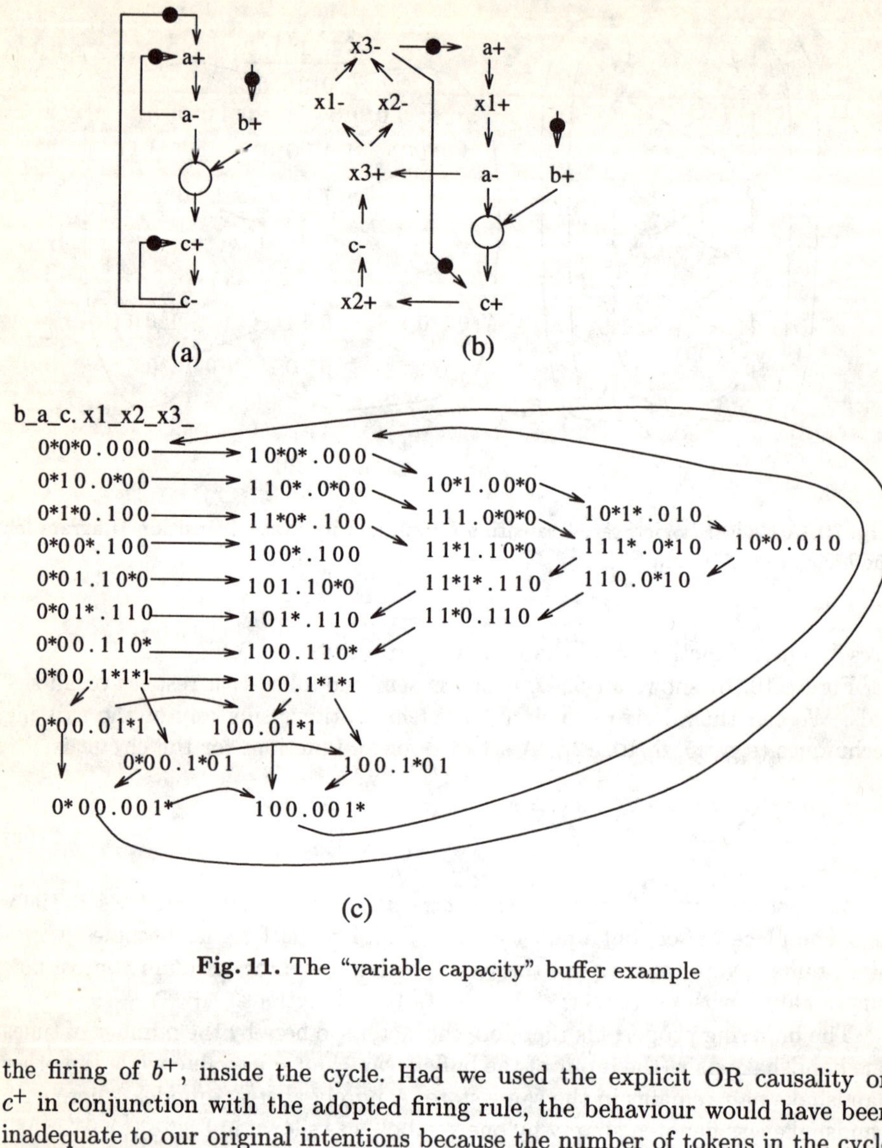


**Fig. 11.** The "variable capacity" buffer example

the firing of $b^+$, inside the cycle. Had we used the explicit OR causality on $c^+$ in conjunction with the adopted firing rule, the behaviour would have been inadequate to our original intentions because the number of tokens in the cycle would not have increased, so we would not have got the effect of a two-place buffer. The purpose of this example is to illustrate a second type of OR causality, as described in the next section.

The initial STG model is valid, but cannot be implemented directly in logic, because it first requires the introduction of hidden *state signals*. For this purpose, we introduce three internal signals, $x_1, x_2$ and $x_3$, in such a way that their transitions do not change the original ordering of the initial specification. This modified STG is shown in Fig. 11.(b). The STD generated by this STG is shown in Fig. 11.(c). From this STD it is now possible to derive the Boolean functions of the circuit implementation of the buffer (both CAD tools, SIS [6] and Forcage [5], produced the same circuit):

$$
\begin{aligned}
a &= \overline{x_1}\,\overline{x_3} \\
x_1 &= a + \overline{x_3}x_1 \\
x_2 &= a + \overline{x_3}x_2 \\
x_3 &= \overline{a}\,\overline{c}x_1x_2 + (x_1 + x_2)x_3 \\
c &= b\overline{x_2}\,\overline{x_3} + \overline{a}x_1\overline{x_2}\,\overline{x_3}
\end{aligned}
$$

## 6 The Two Types of OR Causality

The above two examples demonstrate the two major paradigms in which OR causality can be distinguished. The first is the case when the actions, say $a$ and $b$, that weakly cause another action, $c$, *do not insist* on the effect of their completion to be applied to $c$ independently of each other. This means that for every possible occurrence of $c$ after only one of the causes, say $a$, the other cause, say $b$, has no effect over the same occurrence of $a$. The unbounded case of such a paradigm was shown in Figure 6.(a), for which we could not build a finite PN representation. Now, using CLNs, the required model would be a trivial redrawing of the CD in Figure 6.(a) in such way that the transition labelled with $c$ must have its $\beta$ function equal to the simple disjunction of the input arcs.

Let us call this type of OR causality *joint* OR causality.

The other type, called *disjoint* OR causality, happens to be in the example of the variable capacity buffer. It also took place in the example of the low latency arbiter, whose STG was shown in Figure 3. This type of OR causality, inherited intact from PNs, is called disjoint because we do not allow the tokens independently arriving in one place $p$ from several cause actions to be removed or annihilated. This *additive* effect of the PN marking mechanism is adequate for the purposes of modelling.

A more transparent illustration of the fact that a joint OR causality construct *will not* be able to represent this effect is shown in Figure 12. If we analyse the behaviour of the CLN shown in Figure 12.(a), with seemingly equivalent joint OR causality, we will see that it is negatively unbounded with respect to either of the dashed arcs. Furthermore, we cannot satisfy the requirement that in every execution sequence the action labelled with $c$ has to occur as many times as the *sum* of the occurrences of actions $a_1$ and $a_2$. This requirement would be necessary to guarantee, for example, that *none of the requests to an arbiter are lost*. A satisfactory CLN model of this causality (in this case an ordinary PN) is shown in Figure 12.(b).

To summarise, the above differentiation of the OR causality resolves the uncertainty as to how the designer has to specify OR causality. Although the formal relationship between the class of semi-modular behaviours and CDs has been formally shown (as well as the corresponding classes of PNs and STGs), there has been no clear recommendation for the designer as to what construct had to be employed for modelling OR causality. It is now clear that the demonstrated modelling mismatches were due to:

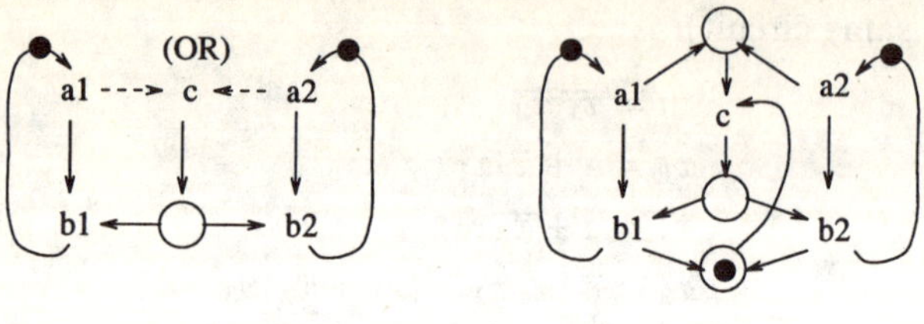


(a) (b)

**Fig. 12.** Illustration of disjoint OR causality

- the inadequacy of PNs to model *joint* OR causality and
- the inadequacy of CDs to model *disjoint* OR causality.

On the other hand, using the unified model of CLNs as an underlying description tool for CL-STG, the designer can use:

- a disjunctive transition enabling function for joint OR causality, or
- a single place with multiple predecessors for disjoint OR causality.

## 7 Conclusions

In this paper, we have demonstrated the role of OR causality in modelling asynchronous circuits. We have shown that within the framework of only the PN or the CD model the designer has certain limitations which can be crucial in capturing some useful semantic details. We have thus justified the introduction of a unified formalism, Causal Logic Nets, which is based on the classical graphical background of Petri nets but augmented with the notion of a causal enabling function defined for each transition of the net. We have been able to identify a class of Causal Logic Nets which is a unification of PN and CD. Such a class, as follows from Propositions 2 and 3, is formed by CLNs whose transitions have enabling functions that are either nonnegative simple conjunctions or nonnegative simple disjunctions of literals associated with input places. This class of CLNs enables the designer to model two types of OR causality inherent in circuit behaviour. The joint OR causality can be represented through the enabling function mechanism of transitions, while the disjoint OR causality can be modelled by net places and their input transitions.

The following table summarises our model classification in terms of the essential features of the new CLN formalism.

| *model* | *CLN features* | |
|---|---|---|
| | *enabling type* | *arc incidence on places* |
| PN (STG) | AND | Multiple input/output arcs |
| CD | AND+OR | Single input/output arcs |
| CLN | AND+OR | Multiple input/output arcs |

**Acknowledgements**

We are grateful to Leonid Rosenblum and Alexander Taubin who co-authored our work on Signal Graphs and Change Diagrams. Thanks to Michael Yoeli, Eike Best, Maciej Koutny and Marta Pietkiewicz-Koutny for helpful discussions.

## References

1. T.-A. Chu. On the models for designing vlsi asynchronous digital systems. *Integration: the VLSI journal*, 4:99–113, 1986.
2. F Commoner, A.W. Holt, S. Even, and A. Pnueli. Marked directed graphs. *Journal of Computer and Systems Sciences*, 5:511–523, 1971.
3. H.J. Genrich and R.M. Shapiro. Formal verification of an arbiter cascade. In *Proceedings of 13th Int. Conferenece on Application and Theory of Petri Nets*, Lecture Notes in Computer Science, Springer-Verlag, Berlin, 1992.
4. J. Gunawardena. Causal automata. *Theoretical Computer Science*, 101(2):265–288, 1992.
5. M. Kishinevsky, A. Kondratyev, A. Taubin, and V. Varshavsky. *Concurrent Hardware: The Theory and Practice of Self-Timed Design.* John Wiley and Sons, London, 1993.
6. L. Lavagno and A. Sangiovanni-Vincentelli. *Algorithms for Synthesis and Testing of Asynchronous Circuits.* Kluwer Academic Publishers, Boston, 1993.
7. D. Muller and W. Bartky. A theory of asynchronous circuits. In *Annals of Computation Laboratory*, pages 204–243. Harvard University, 1959.
8. J.L. Peterson. *Petri Net Theory and Modeling of Systems.* Prentice-Hall, Englewood Cliffs, N.J., 1981.
9. Marta Pietkiewicż-Koutny. Proof of the conjecture that the language generated by a certain Change Diagram cannot be generated by Petri nets. Manuscript, Feb. 1994.
10. L.Ya. Rosenblum and A.V. Yakovlev. Signal graphs: from self-timed to timed ones. In *Proceedings of International Workshop on Timed Petri Nets, Torino, Italy, July 1985*, pages 199–207. IEEE Computer Society, 1985.
11. Patil S.S. and J.B. Dennis. *The description and realization of digital systems. In: Innovative Architectures.* IEEE, N.Y., 1972.
12. V. Varshavsky, M. Kishinevsky, V. Marakhovsky, V. Peschansky, L. Rosenblum, A. Taubin, and B. Tzirlin. *Self-Timed Control of Concurrent Processes.* Kluwer AP, Dordrecht, 1990.
13. V.I. Varshavsky, M.A. Kishinevsky, A.Y. Kondratyev, L.Y. Rosenblyum, and A.R. Taubin. Models for specification and analysis of processes in asynchronous circuits. *Soviet Journal of Computer and System Sciences*, 26(2):61–76, 1989. Russian Edition - 1988.
14. A. Yakovlev, M. Kishinevsky, A. Kondratyev, and L. Lavagno. On the models for asynchronous circuit behaviour with OR causality. TR-463, University of Newcastle upon Tyne, November 1993.
15. A. Yakovlev, L. Lavagno, and A. Sangiovanni-Vincentelli. A unified STG model for asynchronous control circuit synthesis. In *Proceedings of ICCAD'92*, Santa Clara, CA, November 1992.
16. A Yakovlev, A. Petrov, and L. Lavagno. High speed asynchronous arbiter. TR-427, University of Newcastle upon Tyne Computing Science, May 1993.

# Lecture Notes in Computer Science

For information about Vols. 1–739
please contact your bookseller or Springer-Verlag

Vol. 740: E. F. Brickell (Ed.), Advances in Cryptology – CRYPTO '92. Proceedings, 1992. X, 593 pages. 1993.

Vol. 741: B. Preneel, R. Govaerts, J. Vandewalle (Eds.), Computer Security and Industrial Cryptography. Proceedings, 1991. VIII, 275 pages. 1993.

Vol. 742: S. Nishio, A. Yonezawa (Eds.), Object Technologies for Advanced Software. Proceedings, 1993. X, 543 pages. 1993.

Vol. 743: S. Doshita, K. Furukawa, K. P. Jantke, T. Nishida (Eds.), Algorithmic Learning Theory. Proceedings, 1992. X, 260 pages. 1993. (Subseries LNAI)

Vol. 744: K. P. Jantke, T. Yokomori, S. Kobayashi, E. Tomita (Eds.), Algorithmic Learning Theory. Proceedings, 1993. XI, 423 pages. 1993. (Subseries LNAI)

Vol. 745: V. Roberto (Ed.), Intelligent Perceptual Systems. VIII, 378 pages. 1993. (Subseries LNAI)

Vol. 746: A. S. Tanguiane, Artificial Perception and Music Recognition. XV, 210 pages. 1993. (Subseries LNAI).

Vol. 747: M. Clarke, R. Kruse, S. Moral (Eds.), Symbolic and Quantitative Approaches to Reasoning and Uncertainty. Proceedings, 1993. X, 390 pages. 1993.

Vol. 748: R. H. Halstead Jr., T. Ito (Eds.), Parallel Symbolic Computing: Languages, Systems, and Applications. Proceedings, 1992. X, 419 pages. 1993.

Vol. 749: P. A. Fritzson (Ed.), Automated and Algorithmic Debugging. Proceedings, 1993. VIII, 369 pages. 1993.

Vol. 750: J. L. Díaz-Herrera (Ed.), Software Engineering Education. Proceedings, 1994. XII, 601 pages. 1994.

Vol. 751: B. Jähne, Spatio-Temporal Image Processing. XII, 208 pages. 1993.

Vol. 752: T. W. Finin, C. K. Nicholas, Y. Yesha (Eds.), Information and Knowledge Management. Proceedings, 1992. VII, 142 pages. 1993.

Vol. 753: L. J. Bass, J. Gornostaev, C. Unger (Eds.), Human-Computer Interaction. Proceedings, 1993. X, 388 pages. 1993.

Vol. 754: H. D. Pfeiffer, T. E. Nagle (Eds.), Conceptual Structures: Theory and Implementation. Proceedings, 1992. IX, 327 pages. 1993. (Subseries LNAI).

Vol. 755: B. Möller, H. Partsch, S. Schuman (Eds.), Formal Program Development. Proceedings. VII, 371 pages. 1993.

Vol. 756: J. Pieprzyk, B. Sadeghiyan, Design of Hashing Algorithms. XV, 194 pages. 1993.

Vol. 757: U. Banerjee, D. Gelernter, A. Nicolau, D. Padua (Eds.), Languages and Compilers for Parallel Computing. Proceedings, 1992. X, 576 pages. 1993.

Vol. 758: M. Teillaud, Towards Dynamic Randomized Algorithms in Computational Geometry. IX, 157 pages. 1993.

Vol. 759: N. R. Adam, B. K. Bhargava (Eds.), Advanced Database Systems. XV, 451 pages. 1993.

Vol. 760: S. Ceri, K. Tanaka, S. Tsur (Eds.), Deductive and Object-Oriented Databases. Proceedings, 1993. XII, 488 pages. 1993.

Vol. 761: R. K. Shyamasundar (Ed.), Foundations of Software Technology and Theoretical Computer Science. Proceedings, 1993. XIV, 456 pages. 1993.

Vol. 762: K. W. Ng, P. Raghavan, N. V. Balasubramanian, F. Y. L. Chin (Eds.), Algorithms and Computation. Proceedings, 1993. XIII, 542 pages. 1993.

Vol. 763: F. Pichler, R. Moreno Díaz (Eds.), Computer Aided Systems Theory – EUROCAST '93. Proceedings, 1993. IX, 451 pages. 1994.

Vol. 764: G. Wagner, Vivid Logic. XII, 148 pages. 1994. (Subseries LNAI).

Vol. 765: T. Helleseth (Ed.), Advances in Cryptology – EUROCRYPT '93. Proceedings, 1993. X, 467 pages. 1994.

Vol. 766: P. R. Van Loocke, The Dynamics of Concepts. XI, 340 pages. 1994. (Subseries LNAI).

Vol. 767: M. Gogolla, An Extended Entity-Relationship Model. X, 136 pages. 1994.

Vol. 768: U. Banerjee, D. Gelernter, A. Nicolau, D. Padua (Eds.), Languages and Compilers for Parallel Computing. Proceedings, 1993. XI, 655 pages. 1994.

Vol. 769: J. L. Nazareth, The Newton-Cauchy Framework. XII, 101 pages. 1994.

Vol. 770: P. Haddawy (Representing Plans Under Uncertainty. X, 129 pages. 1994. (Subseries LNAI).

Vol. 771: G. Tomas, C. W. Ueberhuber, Visualization of Scientific Parallel Programs. XI, 310 pages. 1994.

Vol. 772: B. C. Warboys (Ed.),Software Process Technology. Proceedings, 1994. IX, 275 pages. 1994.

Vol. 773: D. R. Stinson (Ed.), Advances in Cryptology – CRYPTO '93. Proceedings, 1993. X, 492 pages. 1994.

Vol. 774: M. Banâtre, P. A. Lee (Eds.), Hardware and Software Architectures for Fault Tolerance. XIII, 311 pages. 1994.

Vol. 775: P. Enjalbert, E. W. Mayr, K. W. Wagner (Eds.), STACS 94. Proceedings, 1994. XIV, 782 pages. 1994.

Vol. 776: H. J. Schneider, H. Ehrig (Eds.), Graph Transformations in Computer Science. Proceedings, 1993. VIII, 395 pages. 1994.

Vol. 777: K. von Luck, H. Marburger (Eds.), Management and Processing of Complex Data Structures. Proceedings, 1994. VII, 220 pages. 1994.

Vol. 778: M. Bonuccelli, P. Crescenzi, R. Petreschi (Eds.), Algorithms and Complexity. Proceedings, 1994. VIII, 222 pages. 1994.

Vol. 779: M. Jarke, J. Bubenko, K. Jeffery (Eds.), Advances in Database Technology — EDBT '94. Proceedings, 1994. XII, 406 pages. 1994.

Vol. 780: J. J. Joyce, C.-J. H. Seger (Eds.), Higher Order Logic Theorem Proving and Its Applications. Proceedings, 1993. X, 518 pages. 1994.

Vol. 781: G. Cohen, S. Litsyn, A. Lobstein, G. Zémor (Eds.), Algebraic Coding. Proceedings, 1993. XII, 326 pages. 1994.

Vol. 782: J. Gutknecht (Ed.), Programming Languages and System Architectures. Proceedings, 1994. X, 344 pages. 1994.

Vol. 783: C. G. Günther (Ed.), Mobile Communications. Proceedings, 1994. XVI, 564 pages. 1994.

Vol. 784: F. Bergadano, L. De Raedt (Eds.), Machine Learning: ECML-94. Proceedings, 1994. XI, 439 pages. 1994. (Subseries LNAI).

Vol. 785: H. Ehrig, F. Orejas (Eds.), Recent Trends in Data Type Specification. Proceedings, 1992. VIII, 350 pages. 1994.

Vol. 786: P. A. Fritzson (Ed.), Compiler Construction. Proceedings, 1994. XI, 451 pages. 1994.

Vol. 787: S. Tison (Ed.), Trees in Algebra and Programming – CAAP '94. Proceedings, 1994. X, 351 pages. 1994.

Vol. 788: D. Sannella (Ed.), Programming Languages and Systems – ESOP '94. Proceedings, 1994. VIII, 516 pages. 1994.

Vol. 789: M. Hagiya, J. C. Mitchell (Eds.), Theoretical Aspects of Computer Software. Proceedings, 1994. XI, 887 pages. 1994.

Vol. 790: J. van Leeuwen (Ed.), Graph-Theoretic Concepts in Computer Science. Proceedings, 1993. IX, 431 pages. 1994.

Vol. 791: R. Guerraoui, O. Nierstrasz, M. Riveill (Eds.), Object-Based Distributed Programming. Proceedings, 1993. VII, 262 pages. 1994.

Vol. 792: N. D. Jones, M. Hagiya, M. Sato (Eds.), Logic, Language and Computation. XII, 269 pages. 1994.

Vol. 793: T. A. Gulliver, N. P. Secord (Eds.), Information Theory and Applications. Proceedings, 1993. XI, 394 pages. 1994.

Vol. 794: G. Haring, G. Kotsis (Eds.), Computer Performance Evaluation. Proceedings, 1994. X, 464 pages. 1994.

Vol. 795: W. A. Hunt, Jr., FM8501: A Verified Microprocessor. XIII, 333 pages. 1994.

Vol. 796: W. Gentzsch, U. Harms (Eds.), High-Performance Computing and Networking. Proceedings, 1994, Vol. I. XXI, 453 pages. 1994.

Vol. 797: W. Gentzsch, U. Harms (Eds.), High-Performance Computing and Networking. Proceedings, 1994, Vol. II. XXII, 519 pages. 1994.

Vol. 798: R. Dyckhoff (Ed.), Extensions of Logic Programming. Proceedings, 1993. VIII, 362 pages. 1994.

Vol. 799: M. P. Singh, Multiagent Systems. XXIII, 168 pages. 1994. (Subseries LNAI).

Vol. 800: J.-O. Eklundh (Ed.), Computer Vision – ECCV '94. Proceedings 1994, Vol. I. XVIII, 603 pages. 1994.

Vol. 801: J.-O. Eklundh (Ed.), Computer Vision – ECCV '94. Proceedings 1994, Vol. II. XV, 485 pages. 1994.

Vol. 802: S. Brookes, M. Main, A. Melton, M. Mislove, D. Schmidt (Eds.), Mathematical Foundations of Programming Semantics. Proceedings, 1993. IX, 647 pages. 1994.

Vol. 803: J. W. de Bakker, W.-P. de Roever, G. Rozenberg (Eds.), A Decade of Concurrency. Proceedings, 1993. VII, 683 pages. 1994.

Vol. 804: D. Hernández, Qualitative Representation of Spatial Knowledge. IX, 202 pages. 1994. (Subseries LNAI).

Vol. 805: M. Cosnard, A. Ferreira, J. Peters (Eds.), Parallel and Distributed Computing. Proceedings, 1994. X, 280 pages. 1994.

Vol. 806: H. Barendregt, T. Nipkow (Eds.), Types for Proofs and Programs. VIII, 383 pages. 1994.

Vol. 807: M. Crochemore, D. Gusfield (Eds.), Combinatorial Pattern Matching. Proceedings, 1994. VIII, 326 pages. 1994.

Vol. 808: M. Masuch, L. Pólos (Eds.), Knowledge Representation and Reasoning Under Uncertainty. VII, 237 pages. 1994. (Subseries LNAI).

Vol. 809: R. Anderson (Ed.), Fast Software Encryption. Proceedings, 1993. IX, 223 pages. 1994.

Vol. 810: G. Lakemeyer, B. Nebel (Eds.), Foundations of Knowledge Representation and Reasoning. VIII, 355 pages. 1994. (Subseries LNAI).

Vol. 811: G. Wijers, S. Brinkkemper, T. Wasserman (Eds.), Advanced Information Systems Engineering. Proceedings, 1994. XI, 420 pages. 1994.

Vol. 812: J. Karhumäki, H. Maurer, G. Rozenberg (Eds.), Results and Trends in Theoretical Computer Science. Proceedings, 1994. X, 445 pages. 1994.

Vol. 813: A. Nerode, Yu. N. Matiyasevich (Eds.), Logical Foundations of Computer Science. Proceedings, 1994. IX, 392 pages. 1994.

Vol. 814: A. Bundy (Ed.), Automated Deduction—CADE-12. Proceedings, 1994. XVI, 848 pages. 1994. (Subseries LNAI).

Vol. 815: R. Valette (Ed.), Application and Theory of Petri Nets 1994. Proceedings. IX, 587 pages. 1994.

Vol. 817: C. Halatsis, D. Maritsas, S. Theodoridis, G. Philokyprou (eds.), PARLE '94. Parallel Architectures and Languages Europe. Proceedings, 1994. XV, 837 pages. 1994.

Vol. 818: D. L. Dill (Ed.), Computer Aided Verification. Proceedings, 1994. IX, 480 pages. 1994.

Vol. 819: W. Litwin, T. Risch (Eds.), Applications of Databases. Proceedings, 1994. XII, 471 pages. 1994.

Vol. 820: S. Abiteboul, E. Shamir (Eds.), Automata, Languages and Programming. Proceedings, 1994. XIII, 644pages. 1994.

Vol. 821: M. Tokoro, R. Pareschi (Eds.), Object-Oriented Programming. Proceedings, 1994. XI, 535 pages. 1994.